THE PAPERS OF THOMAS A. EDISON

FINANCIAL CONTRIBUTORS

We thankfully acknowledge the vision and support of Rutgers University and the Thomas A. Edison Papers Board of Oversight.

This edition was made possible by grant funds provided from the New Jersey Historical Commission, National Historical Publications and Records Commission, and the National Endowment for the Humanities. Major underwriting of the present volume has been provided by the Cinco Hermanos Fund. We also recognize contributions from Aribex Inc., E. Thomas Arnn, Eugene Barretta, Thomas Berkenkamp, Kevin & Patricia Burke, Louis Carlat, Cathy Carpenter, Tara & James Condon, Paul Spehr & Susan Dalton, Peter DeSorcy, Helen Endick, Karen Ferrante, Bernard S. Finn, Dr. Michael Geselowitz, Paul & Lucy Israel, Thomas Jeffrey, David Katz, Gregory & Karen Kushla, David J. Michnal, Abhisek Panda, Crystal Paris, Alexandra Rimer, the Rutgers Alumni Association, David Sloane, Wayne Trester, Lawrence S. Waldman, Jane & Bernard Wallerstein, Daniel Weeks, Jackie Weeks, and Rachel Weissenburger. For the generosity of these organizations and individuals, the editors are most grateful.

THE PAPERS OF THOMAS A. EDISON

Volume 8

Plaster bust of Thomas Edison, made in Italy by American-born sculptor Longworth Powers in 1886.

Volume 8

The Papers of Thomas A. Edison

NEW BEGINNINGS

January 1885–December 1887

VOLUME EDITORS

Paul B. Israel

Louis Carlat

Theresa M. Collins

Alexandra R. Rimer

Daniel J. Weeks

GRADUATE ASSISTANTS

Scott Bruton

Dennis Halpin

Patrick McGrath

Ben Resnick-Day

Kristopher Shields

FOUNDERS

Rutgers, The State University of New Jersey

National Park Service, Edison National Historical Park

New Jersey Historical Commission

Smithsonian Institution

JOHNS HOPKINS UNIVERSITY PRESS

BALTIMORE

© 2015 Rutgers, The State University
All rights reserved. Published 2015
Printed in the United States of America
9 8 7 6 5 4 3 2 1

Johns Hopkins University Press
2715 North Charles Street
Baltimore, Maryland 21218-4363
www.press.jhu.edu

The paper used in this book meets the minimum requirements of the American National Standard for Information Sciences—Permanence of Paper for Printed Library Materials, ANSI Z 39.48-1984.

Library of Congress Cataloging-in-Publication Data
(Revised for volume 3)

Edison, Thomas A. (Thomas Alva), 1847–1931
    The Papers of Thomas A. Edison

    Includes bibliographical references and index.
    Contents: v. 1. The making of an inventor, February 1847–June 1873—v. 2. From workshop to laboratory, June 1873–March 1876—v. 3. Menlo Park. The early years, April 1876–December 1877.
    1. Edison, Thomas A. (Thomas Alva), 1847–1931. 2. Edison, Thomas A. (Thomas Alva), 1847–1931—Archives. 3. Inventors—United States—Biography. I. Jenkins, Reese.
TK140.E3A2   1989   600   88-9017
ISBN 0-8018-3100-8 (v. 1. : alk. paper)
ISBN 0-8018-3101-6 (v. 2. : alk. paper)
ISBN 0-8018-3102-4 (v. 3. : alk. paper)
ISBN 0-8018-5819-4 (v. 4. : alk. paper)
ISBN 0-8018-3104-0 (v. 5. : alk. paper)
ISBN 978-0-8018-8640-9 (v. 6. : alk. paper)
ISBN 978-1-4214-0090-7 (v. 7. : alk. paper)
ISBN 978-1-4214-1749-3 (v. 8. : alk. paper)

A catalog record for this book is available from the British Library.

Edison signature on case used with permission of the McGraw-Edison Company.

TO THE MEMORY OF
DAVID W. HUTCHINGS,
GREGORY JANKUNIS,
and SUSAN SCHREPFER

Contents

Contents   xii

Calendar of Documents

List of Editorial Headnotes

List of Maps

Preface

The years 1885 to 1887 were transformative in Edison's life and work, as his career branched out into diversified fields and he married for a second time. On the inventive side, for the first time since 1878, he no longer focused almost wholly upon electric lighting. He began 1885 continuing his research on long-distance telephony for the American Bell Telephone Company. After the firm decided not to employ him as a contract inventor, he turned instead to new telecommunications projects that included a precursor of radio technology (using "wireless" or induction telegraph systems to provide mobile communication to railways or ships). In another project, he combined induction telegraphy with telephone technology to develop the phonoplex, the first system capable of sending two or more signals between intermediate stations over a single wire. Edison's telecommunications research led him to undertake more basic scientific research as he became engaged with scientific debates over electromagnetic theory. In late 1885 and then in the spring of 1886, he filled several notebooks with speculations and plans for discovering an unknown physical force he called XYZ, an idea that he had investigated in 1874 and 1875. This effort was part of a larger research program on the conversion of forces or energy, including the relationship between gravity and electromagnetism. His interest in these subjects would lead Edison to a new approach for converting coal directly into electricity in the form of his "pyromagnetic generator," which he intended to produce electric energy directly from thermal energy.

Edison's work for American Bell had been undertaken at the behest of his old friend Ezra Gilliland, who headed the company's experimental department in Boston. Edison's renewed friendship with Gilliland set in motion events that would influence his work and fundamentally reshape his life. In February 1885, accompanied by his daughter and Gilliland's wife, Lillian, they set out on a circuitous journey to Gilliland's Michigan hometown, then to Chicago, New Orleans (where an industrial exposition was underway), and Jacksonville, Florida, before turning west again across central Florida to the Gulf Coast. They decided to buy thirteen acres of land near Fort Myers, an isolated but relatively affluent cattle entrepôt whose climate, natural surroundings, and opportunities for hunting and fishing all appealed to Edison. Each man planned to build his own house (and Edison a laboratory) to use as winter residences.

After returning north, Edison periodically visited the Gillilands in Boston, where they introduced him to nineteen-year-old Mina Miller, the daughter of Ohio inventor-industrialist Lewis Miller (a cofounder of the Chautauqua Sunday School Assembly), who was studying in the city. Edison was smitten with the young woman. Although Mina was not present in mid-July while he stayed with the Gillilands at their summer cottage in Winthrop, on the shore of Boston Harbor, she was present in the diary he kept during that visit. This diary, the only known place in which he deliberately wrote thoughts of a personal nature, was clearly intended to be read aloud for the amusement of the other guests, who were also keeping diaries. Edison used this parlor game to signal to his hosts and other guests his romantic interest in Mina. The diary also contains his observations on art, literature, and religion, along with comments about his dreams, health, and his feelings about his daughter. Edison spent the rest of the summer courting Mina. After visiting the Miller family at the Chautauqua Institution in western New York State, he traveled with Marion and Mina (and presumably her family) on an eastward journey through Niagara Falls and Montreal to the White Mountains of New Hampshire. In the foothills of Mount Washington, he proposed marriage (according to reminiscences by Mina and Marion) by tapping the question in Morse code.

The couple married in February 1886 and honeymooned at Edison's new winter home in Fort Myers, Florida. During their month-long stay, Edison spent part of nearly every day filling several notebooks with a remarkable range of ideas for experiments. Just as remarkably, Mina participated by witnessing each of these entries and occasionally writing or recopying some of them. After returning north they set up domestic life at "Glenmont," their new home in Llewellyn Park (located in West Orange, New Jersey), one of the first planned suburbs in the United States. Letters from the courtship, Edison's diary, and rare personal entries from his notebooks offer uncommon insight into Edison's relationship with his new wife. Later correspondence between Mina and her family also reveals an unsuccessful pregnancy in 1887, as well as Edison's typically difficult balance of devotions between family and work, and the challenge of blending the children of his first marriage into the changed family structure.

Soon after his return, Edison was faced with a strike at the Edison Machine Works over hours, overtime pay, and shop rules. When the men walked out, Edison and his partners in the business promptly reassessed their standing intention to expand the shop in the New York area. They decided instead to relocate to Schenectady, some one hundred seventy miles up the Hudson River valley. Edison arranged a personal loan of $45,000 to purchase the property, and the Machine Works moved north by the end of year, rupturing his long practice of manufacturing in the New York–Newark region under his close supervision.

Complaints about the premature failure of his incandescent lamps led Edison to move his laboratory to the lamp factory in June 1886. Here he again turned his energies to improving the lamps and their manufacturing process, including the development of a more efficient high-resistance lamp. After spending much of the summer and early fall working on electric lamps, Edison was forced to deal with another key element of his light and power system. His direct current system was expensive to build because of the limited distances over which it could operate economically. Extending the reach of the distribution network could be done by increasing the voltage at which current was distributed so as to diminish electrical losses. Edison had recognized this fact and made it the basis of his three-wire system of direct current (DC) distribution, but other inventors were prepared to take the idea much farther by using alternating current (AC), which could be transformed easily from high to low voltage before reaching customers. By the fall of 1886, George Westinghouse, who was building a rival electrical company based on alternating current distribution, was proving a direct threat to the Edison lighting interests. Between October and the end of 1886, Edison drafted nearly a dozen patent applications related to high-voltage conversion or distribution designed to make his direct-current system more effective for delivering electricity over longer distances. And in mid-November, he wrote a lengthy and uncompromising memorandum laying out a comprehensive case against AC on grounds of economics, engineering efficiency, public safety, and an unstated but unmistakable pride in his own system as a stand-alone entity.

By late 1886, Edison also faced the prospect of competition in an area in which he took a deeply personal interest—the phonograph. The graphophone, a recording and playback instrument devised at Alexander Graham Bell's Volta Laboratory in Washington, D.C., was receiving favorable publicity and the support of a nascent company. Edison and Gilliland (who had joined the laboratory staff in 1885) took notice and began to sketch ideas for a new standard phonograph in October, although sustained work on the improved phonograph did not begin until the following May. During the fall of 1887, Edison filed the basic patent on his new machine and prepared to begin manufacturing it at a factory in Bloomfield, New Jersey, under the supervision of Gilliland, who also became the sales agent for the newly organized Edison Phonograph Company.

By this time, the laboratory that Edison had planned near his home in West Orange was approaching completion. Edison intended it to be "the best equipped & largest Laboratory extant, and the facilities incomparably superior to any other for rapid & cheap development of an invention, & working it up into Commercial shape with models patterns & special machinery . . . there is no similar institution in Existence." His determination to build the world's leading private laboratory may have been prompted in part by the opening of Edward Weston's personal laboratory in Newark in 1886, which was described by *Engineering* as "probably the most complete in the world." Altogether Edison spent over $130,000 on the land, buildings, and equipment, struggling at times to find the funds. Nonetheless, Edison had ambitious plans for this facility, and during November he filled most of a notebook with experimental projects in preparation for the commencement of experimental work at the beginning of the new year.

The progress of the Thomas A. Edison Papers depends on the support of many individuals and organizations, financial contributors, academic scholars, Edison specialists, librarians, archivists, curators, and students. Representatives of the four

Overseeing Institutions (Rutgers, the National Park Service, the New Jersey Historical Commission, and the Smithsonian Institution) have assisted with this volume and the editors thank them for their continuing concern and attention. The strong support of public and private foundations and of their program officers has sustained the project and helped it remain editorially productive.

Preparation of this volume was made possible in part by grants from the Division of Research Programs (Scholarly Editions) of the National Endowment for the Humanities, an independent federal agency; the National Historical Publications and Records Commission; the New Jersey Historical Commission; the Charles Edison Fund; the Cinco Hermanos Fund; as well as through the support of Rutgers, The State University of New Jersey, and the National Park Service (Thomas Edison National Historical Park). The editors appreciate the interest and support of the many program officers and trustees, especially Elizabeth Arndt, Jason Boffetti, Lucy Barber, Timothy Connolly, Sara Cureton, Niquole Primiani, and Skylar Harris. Any opinions, findings, conclusions, or recommendations expressed in this publication are solely those of the editors and do not necessarily reflect the views of any of the above federal foundations or agencies, the United States Government, or any other financial contributor.

The Edison Papers project is indebted to the National Park Service. We are especially grateful for the support of the staff at the Thomas Edison National Historical Park in West Orange, New Jersey, especially Tom Ross, Jill Hawk, Theresa Jung, Michelle Ortwein, Leonard DeGraaf, Gerald Fabris, Edward Wirth, Beth Miller, Karen Sloat-Olsen, and Sheila Hamilton. The editors also want to thank the staff of the Northeast Region led by Regional Director Mike Caldwell

Many associates at Rutgers University have contributed significantly to the Edison Papers. The editors are grateful to Presidents Robert Barchi and Richard L. McCormick; Executive Vice Presidents for Academic Affairs Richard L. Edwards and Philip Furmanski; and Richard Falk and Douglas Greenberg, Deans of the School of Arts and Sciences, along with their dedicated staff, especially James Masschaele, Executive Vice Dean, and James Swenson, Dean of Humanities; Barbara Lemanski, Associate Dean for Policy and Personnel, and her staff; Jason Diapaolo, Director of Business Affairs, and Business Managers Heather DeMeo and Kevin Foran; and Chris Scherer, Director for New Program Initiatives & Digital Learning. In addition, we appreciate the support provided by Thomas Vosseler, David Motovidlak, Wade Olsson, and the staff of the School of Arts and Sciences IT Services. We also want to thank Monika Incze, of the Office of Research and Sponsored Programs; Frank Cotchan of the Division of Grant and Contract Accounting. A special thanks is due to Peter Shergalis, Supervisor Material Services, Susan O'Brien, Material Services Administrator, and Jeanne Schaab, Foreperson Custodial Services.

The editors value the support of colleagues and staff in the History Department, especially Michael Adas, Rudy Bell, Paul Clemens, Belinda Davis, James Delbourgo, Ann Fabian, Ann Gordon, David Greenberg, Jennifer Jones, Toby Jones, Samantha Kelly, Norman Markowitz, James Reed, Virginia Yans, Dawn Ruskai, Johanna Schoen, Candace Walcott-Shepherd, and Mary DeMeo. We also want to thank Michael Geselowitz, Rob Colburn, Sheldon Hochheiser, and Alex Magoun of the IEEE History Center. We are also deeply appreciative of the efforts of the Digital Humanities Steering Committee: Brittney Cooper, Ann Fabian, Andrew Goldstone, Meredith McGill, Andrew Parker, Jamie Pietruska, and Andrew Urban. Additional thanks go to members of the Rutgers University Libraries, notably Marianne Gaunt, Grace Agnew, Ron Becker, Thomas Frusciano, Francesca Giannetti, Tom Glynn, Linda Langschied, Jim Niessen, Rhonda Marker, and the Interlibrary Loan Office. A special thanks is due Michael Siegel, staff cartographer in the Department of Geography, who prepared the maps in this volume, and our web designer Bonnie Wasielewski.

Many scholars have shared their insights and assisted the editors in a variety of ways. For this volume, notable help came from Michele Wehrwein Albion, Christopher Beauchamp, Barbara J. Becker, Brian Bowers, David Heitz, Charles Hummel, Bruce J. Hunt, Gregory Jankunis, Sharon Kingsland, Robert Rosenberg, Tom Smoot, and Harold Wallace.

Institutions and their staff have provided documents, photographs, photocopies, and research assistance. The editors gratefully acknowledge Sarah Alexander at the Wells Library, Indiana University; Virginia Dunn at the Library of Virginia; Pamela N. Gibson of the Manatee County (Fla.) Public Library; Kirk Morrison and particularly LaSandra Adams, both at the Morgan Library of Suffolk (Va.); Bruce Saunders of the Nansemond Suffolk (Va.) Historical Society; Colleen Seale at the George A. Smathers Libraries, University of Florida;

and Sherri Xie of the Guggenheim Memorial Library, Monmouth University. They also thank staff and librarians of the AT&T Archives and History Center (Warren, N.J.); the Edison-Ford Winter Estates (Fort Myers, Fla.); the Port Huron (Mich.) Museum; the Charles Edison Fund Collection (Newark, N.J.); the Henry Ford Museum and Greenfield Village Research Center (Dearborn, Mich.); the Historical Society of Pennsylvania (Philadelphia); and the Smithsonian Institution's National Museum of American History Archives Center (Washington, D.C.). The editors also gratefully acknowledge assistance from the Milton S. Eisenhower Library of Johns Hopkins University; the Metropolitan Opera; and the Wisconsin Historical Society.

The editors are deeply grateful to Jon Schmitz and the staff of the Oliver Archives at the Chautauqua Institution for their hospitality and extensive research assistance; to Ted and Nancy Arnn (Pittsburgh, Pa.) for generous access to Miller family correspondence before it was accessioned by the Chautauqua Institution; to David E. E. Sloane (New Haven, Conn.) for generous access to other family correspondence and documents; and to Thomas Whitney (Boise, Idaho) for travel-related correspondence.

Staff members and students not mentioned on the title page but who have contributed to this volume include Thomas E. Jeffrey, Rachel M. Weissenburger, Eric Barry, Christina Chiknas, Kenny Moss, Jeremy Sam, and Randy Sparks.

As always, the project has had the benefit of the superb staff of Johns Hopkins University Press. For this volume, the editors are indebted to Robert J. Brugger, Mary Lou Kenney, and Julie McCarthy.

Chronology of
Thomas A. Edison

January 1885–December 1887

1885

c. 2–6 January	Makes brief trip to Boston, likely to continue negotiations regarding his work on American Bell Co.'s long distance lines and the Western Union's rights to his telephone research; probably revisits the Electrical Exhibition.
c. 23–26 January	Visits Boston.
20 February	Leaves New York with Ezra Gilliland on extended trip.
c. 23 February	Agrees with Ezra Gilliland to transfer joint patent rights for railway telegraphy to the Railway Telegraph and Telephone Co., a company incorporated in New York during this month.
c. 28 February	Visits the World Industrial and Cotton Centennial Exposition in New Orleans.
c. 5–21 March	Vacations in Florida with his daughter and Ezra and Lillian Gilliland; contracts to buy land in Fort Myers.
Winter–Spring	Conducts railway telegraph experiments.
3 June	Dispatches Francis Upton to Europe as his representative for electric light business.
2–c. 9 June	Visits Ezra Gilliland in Boston vicinity.
27 June–c. 8 July	Visit to Gilliland overlaps with visit of Mina Miller.
14 July	Arrives for extended stay at Gilliland's vacation cottage at Winthrop, Mass.
c. 26 July–5 August	Sails with Gilliland and others.
10–18 August	Visits the Miller family at the Chautauqua Institution.
18–c. 31 August	Travels with Marion, Mina, and others to Niagara Falls, Montreal, and New Hampshire; proposes marriage to Mina.
4 September	Authorizes settlement of major outstanding debts due him for construction of central station generating plants.

c. 19 September	Dispatches Alfred Tate to set up commercial trial of phonoplex system in Canada.
21–30 September	Visits the Gillilands in Boston.
28 September	Edison household moves to Normandie Hotel in New York.
30 September	Asks Mina Miller's father for permission to marry.
c. 8 November	With others, incorporates International Railway Telegraph and Telephone Co.
October–November	Designs improved phonoplex telegraph receiver. Executes five patent applications covering phonoplex and multiple telegraph inventions.
November	Joined in New York by Ezra Gilliland as a business and inventive partner.
8–10 December	Designs apparatus for experiments to discover an unknown force he calls XYZ.
13 December	Leaves to visit the Miller family in Akron, Ohio.
1886	
20 January	Purchases the former country home of Henry Pedder in Llewellyn Park, N.J.
30 January	Merges Electric Tube Co. and the Edison Shafting Co. into Edison Machine Works.
January	Agrees to pay judgment and court costs in long-running Seyfert lawsuit.
1 February	Participates in public demonstration of railway telegraph system on Staten Island.
16 February	Incorporates Sims-Edison Electric Torpedo Co.
24 February	Marries Mina Miller in Akron, Ohio.
c. 27 February	Arrives in Florida on honeymoon.
17 March	Arrives in Fort Myers.
18 March	Begins prolific series of notebook entries witnessed by Mina.
March–April	Plans extensive planting and landscaping of Fort Myers estate.
c. 29 April	Arrives in Akron with Mina.
c. 4 May	Takes up residence at Glenmont.
17 May	Strike at Edison Machine Works in New York City.
11 June	Begins lamp experiments at Edison Lamp Co. factory in East Newark and moves his laboratory there.
23 June	Announces decision to move Edison Machine Works to Schenectady, N.Y.
2 July	Incorporates the Edison United Manufacturing Co., which takes over sales and installation work of the Edison Co. for Isolated Lighting.
11–12 August	Attends meeting of Association of Edison Illuminating Cos. at Long Beach, Long Island.

5 October	Commences work on new "standard" phonograph.
25 October	Drafts first of several patent applications for transmission and distribution of high-voltage currents.
October	Reaches agreement in principle with Charles Porter for the development, manufacture, and sale of advanced steam engines.
	Drafts an informal critique of alternating current on engineering, economic, and public safety grounds.
c. 10 November	Drafts comprehensive memorandum to Edward Johnson against high-voltage alternating current.
26 November	Edison Electric Light Co. licenses Zipernowsky alternating current transformer patents in United States.
c. 18 December	Edison Machine Works moves to Schenectady.
23 December	Edison Electric Light Co. initiates infringement suit against Westinghouse Electric and others, including its representatives (the "Trenton Feeder Case").
c. 30 December	Becomes very ill with pleurisy.
31 December	Edison Company for Isolated Lighting is absorbed by the Edison Electric Light Company.
November–December	Completes at least ten patent applications related to high-voltage electric transmission and distribution.
1887	
January	Confined to Llewellyn Park home with pleurisy.
8 February	Decides to sell $80,000 of Edison Illuminating stock and loan the money to the Edison Machine Works.
	Resigns as president and director of Edison Electric Light Co. of Europe, Ltd.
9 February	Departs for Florida in care of a nurse.
16 February	Appointed to committee on uniform installation standards at convention of National Electric Light Association (Philadelphia).
21 February	Edison Electric Light Co. absorbs Western Edison Co.
1 March	Drafts first patent application and begins experiments on pyromagnetic motor.
8–21 March	Visited at Fort Myers by Charles Batchelor, who assists with pyromagnetic motor experiments.
24 March	Treated surgically for facial abcess.
March	Alfred Tate negotiates license of Edison's stencil patents allowing A. B. Dick Co. to market mimeograph machine.
6 April	Instructs Batchelor to begin planning new laboratory at Orange, N.J.
c. 8 April	Suffers recurrence of facial abcess.
30 April	Returns from the South in improved health.

1–2 May	Drafts notes for lamp experiments for laboratory staff.
3 May	Hires Henry Hudson Holly as the architect for new laboratory.
7 May	Begins sustained experiments on phonographs and recording cylinders.
18 May	Drafts first patent application for pyromagnetic generator.
19 May	Approves architectural plans for new laboratory. Goes to Schenectady.
21 May	Drafts patent application for improved phonograph cylinder and electrostatic playback mechanism.
5 June	Has William K. L. Dickson begin sustained magnetic ore separator experiments.
8 June	Drafts patent application for improved municipal lamp "cutout."
c. 5 July	Shops for laboratory equipment in Philadelphia; probably travels to Baltimore and Washington.
30 July	Dismisses architect Henry Hudson Holly and replaces him with Joseph Taft.
July	Begins planning staff of new laboratory. Decides to add four small laboratory buildings.
15 August	Has papers on pyromagnetic dynamo and magnetic bridge presented at meeting of the American Association for the Advancement of Science.
August	Solicits investment in prospective company for manufacturing inventions created at new laboratory. Devises five-wire electrical distribution system.
September	Designates George Gouraud his agent for the phonograph in Great Britain.
1 October	Contracts for the manufacture of phonographic dolls and toys.
8 October	Incorporates the Edison Phonograph Co.
25 October	Drafts patent application on battery device to convert alternating to direct current for motors.
27 October	Holds board meeting at Glenmont to reorganize the Edison United Manufacturing Co.
28 October	Assigns all U.S. rights for the phonograph to the Edison Phonograph Co., which appoints Ezra Gilliland its sole general agent.
11 November	Hosts visitors at laboratory from National Academy of Sciences meeting.
October–mid-November	Gives newspaper interviews about the forthcoming phonograph and new laboratory at Orange.
15–25 November	Makes extensive notes on experiments to be conducted at his new laboratory in Orange.

22 November	Executes patent application for new phonograph.
19 December	Replies to official inquiry regarding the use of electricity for capital punishment.
23 December	Lights Glenmont by electricity from new Orange laboratory.

Editorial Policy and User's Guide

The editorial policy for the book edition of Thomas Edison's papers remains essentially as stated in Volume 1, with the modifications described in Volumes 2–7. The additions that follow stem from new editorial situations presented by documents in Volume 8. A comprehensive statement of the editorial policy will be published later on the Edison Papers website (http://edison.rutgers.edu).

Selection

The fifteen-volume book edition of Thomas Edison's papers will include nearly 6,500 documents selected from an estimated 5 million pages of extant Edison-related materials. For the period covered in Volume 8 (January 1885–December 1887), the editors have selected 358 documents from approximately 7,500 available Edison-related documents. Most of those available from this period detail Edison's inventive work, scientific speculations, and business relations, but some directly concern his family life. While still small, the subset of family or personal documents is notably larger than in previous volumes (the first six, especially). This relative change is largely a consequence of Edison's courtship and marriage to Mina Miller and the couple's ongoing relations with her large family, but it is also a product of his continuing celebrity and his children having both the ability and desire to correspond.

The editors have sought to select documents that illuminate the full range of Edison's thought and activities and events in his life during this period. Those published here are primarily by or to Edison, his surrogates Samuel Insull and Alfred Tate, or others working in concert with him or on his behalf. Some third-party correspondence has been selected to highlight key events, to illustrate the context in which Edison worked, or (in some instances) to bridge gaps in Edison's own documentary record. He seems to have read most, if not all, of his voluminous correspondence, and his marginal notes on incoming letters frequently served as guides for the official responses that secretaries Samuel Insull and Alfred Tate prepared in his name. Where feasible, the editors have given priority to such incoming letters with Edison's comments.

For two distinctive groups of documents written by Edison in this period, the editors have altered their normal highly selective policy. One is the diary he kept for the amusement of friends while vacationing in July 1885. It is the only known volume that Edison kept specifically to record thoughts and feelings of a personal nature, and the editors have selected all ten entries in their entirety.

The other group is a set of notebooks that Edison used during and immediately after his Florida honeymoon with his bride, Mina Miller Edison, in March and April 1886. They are distinctive not only for the prolific nature of the writing and drawings across many hundreds of pages, but for the wide-ranging and speculative nature of many of his entries. These truly were idea books for Edison, and they are unusually rich in drawings. Some of the entries suggested experiments on practical objects such as lamp filaments that he might readily try on his return to the laboratory, but many others moved organically from one topic to another across several broad themes, loosely following the organization of Michael Faraday's third volume of *Experimental Researches*. Notable themes included the structure of matter, the nature of energy, possible as-yet undiscovered forms of energy, and the practical conversion of light and heat into electricity and magnetism. Because of the impracticality of applying normal selection criteria to these distinctive books, the editors have selected in their entirety most of Edison's original entries (many were also copied into other books by himself or Mina).

Note on Digital Sources

The number and scope of historical sources available electronically has continued to increase dramatically during the preparation of this volume. The editors have done their best to present state-of-the-art research as of the completion of this manuscript in mid-2014, but the tasks of culling available documents and revising annotations with new references are potentially endless. For this volume, the editors have made abundant use of proprietary collections of digitized newspapers, principally ProQuest databases, and NewspaperArchive .com. They have also relied on Ancestry.com (www.ancestry .com) to view census records and other genealogical and biographical information, and this service is acknowledged within specific bibliographic records at the end of this volume. Many of the contemporary sources listed in the bibliography, as well as many journal articles cited in annotation, have been viewed electronically through proprietary services like JSTOR and the HathiTrust Digital Library (hathitrust.org), and Google Books and similar free collections.

Transcription

The editors have again largely followed the transcription policies used in preceding volumes. Three changes adopted for Volume 7 are particularly relevant now: obvious stenographic errors in typed documents are silently corrected; Edison's idiosyncratic punctuation is standardized to a modest degree to make it reducible to type and intelligible to modern readers; and expressions of time (a subset of punctuation oddities) are transcribed in a standard form. The editors have standardized the highly irregular punctuation of Edison's associates, for the same reasons and to about the same extent.

Transcription practices have been modified to adapt to the particularities of several groups of documents in this volume. In his 1885 diary, Edison seldom divided his entries into paragraphs. In the early ones, he often left gaps in the text when changing subjects, and these spaces are transcribed as paragraph breaks. In the final pages, however, Edison largely stopped using the large gaps. Nevertheless, in the interest of enhancing the readability of those entries, the editors have broken the text into paragraphs where the subject obviously changes.

Other changes apply to Edison's 1886 Florida honeymoon notebooks. Because he used these as idea books rather than as records of laboratory research, he often jumped quite rapidly through a range of topics on a given day. To prevent confusion, the editors have transcribed all of the headings that Edison wrote in these books, even when they are repetitive. (The ordinary practice of transcribing each heading only where it first appears remains in place elsewhere.) Also, to better represent the character of these pages chock-a-block with Edison's ideas, the headings are transcribed on the same line as the text that follows, instead of each on its own line.

In technical drawings, where Edison does not refer in his text to the letters used to label elements of a drawing, the editors have generally not transcribed those letters into the endnotes as figure labels.

Abbreviations used repeatedly in document text are listed in a special section below, following the list of editorial abbreviations used to describe documents.

Annotation

In the endnotes following each document, citations are generally given in the order in which the material is discussed (as in previous volumes). However, when there are several pieces of correspondence from the same person or a run of notebook entries, these are often listed together rather than in the order they are discussed to simplify the reference.

Because of Edison's frequent travel in the years of this volume, the editors have supplied or conjectured a place on the dateline of an unusually large number of documents. The justification for each place is provided in a special note or in the document notes at large. A further comment regarding a few locations of particular importance: Edison moved in 1886 to a home named Glenmont in the community of Llewellyn Park in the township of West Orange, New Jersey, where he built a laboratory complex the next year. Although West Orange appeared on contemporary maps as a distinct entity, Edison and his correspondents typically referred to it simply as "Orange." They were probably following the practice of the Post Office, whose Orange office served the greater area. To avoid confusion, the editors have adopted the same convention in their conjectured datelines (and editorial notes). In cases where they conjecture that a document was written at Edison's home, the editors specify "Glenmont," though the name "Llewellyn Park" is used for the community at large. Similar overlaps in usage occurred in East Newark, New Jersey, also known contemporaneously as Harrison, where Edison had a laboratory in the factory of the Edison Lamp Company for more than a year. Following the company's adoption of one place or the other on its letterhead, the editors generally use "East Newark" until about May 1887 and "Harrison" thereafter.

Updated Biographical Information

The proliferation of digitized primary sources has afforded relatively easy access to a wealth of new biographical information in recent years. For this reason, the editors provide updated or expanded biographical entries for a number of individuals who were identified in previous volumes.

References to the Digital Edition

The editors have not provided a comprehensive calendar of Edison documents because the vastness of the archive makes preparation of such an aid impractical. Their annotations include, however, references to relevant documents in the Edison Papers image editions (digital and microfilm); the volume may therefore serve as an entree into those publications.

The Edison Papers website (http://edison.rutgers.edu) contains approximately 161,000 images from the first three parts of the microfilm edition of documents at the Thomas Edison National Historical Park. There are also nearly 35,000 additional images not found on the microfilm that come from outside repositories. Citations to images in the digital edition are indicated by the acronym *TAED*. The citations are in an alphanumeric code unique to each document (e.g., *TAED* D8314N). In this volume, for the first time, this rule of thumb generally applies to documents found in bound volumes such as notebooks. There are cases, however, such as account books, in which unique identifiers can be assigned only with great difficulty to particular entries in a book. In those few instances, the citation gives both the general identifier for the entire book and a specific image number or numbers (e.g., AB004 [image 60]). Image numbers are also used on occasion to direct the reader to a particular point in an unusually lengthy document (e.g., W100DEC002 [image 7]). All of these images can be seen by going to the Edison Papers homepage and clicking on the link for "Single Document or Folder" under "Search Methods." This will take the user to http://edison.rutgers.edu/singldoc .htm, where the images can be seen by putting the appropriate alpha-numeric code in one of the two boxes to retrieve either a document or a folder/volume. If retrieving a folder/volume, the user should click on "List Documents" and then "Show Documents" in the introductory "target" for that folder/volume. Then click on any of the links to specific documents in the folder/volume and put the appropriate image number in the box under the "Go to Image" link. Entering an image number in that box when viewing any document will take the reader to the specific image number. Similarly, citations are sometimes made to an entire folder of documents (e.g., *TAED* D8416). Pasting the alpha–numeric folder code into the "Folder/Volume ID" box on the search page, as above, will give the reader the option to display the contents of the entire folder.

The digital edition contains a number of other features not available in the book or microfilm editions, including lists of all of Edison's U.S. patents by execution date, issue date, and subject; links to pdf files; and a comprehensive chronology and bibliography. Other materials, such as chapter introductions from the book edition and biographical sketches, will eventually be added. Material from outside repositories, including items cited in this volume, will continue to be added. Under arrangements being made at press time, this volume and its predecessors will eventually be published in a digital format on the web.

The general availability of the Edison Papers digital edition through any standard web browser has led to the redefinition of the microfilm edition as a medium for archival preservation instead of a principal research tool. For this reason, and to simplify references in the endnotes, this volume does not give document-level citations to the Edison Papers microfilm edition. The exception to this practice is for documents not scheduled to be added to the digital edition until well after press time, for which references to the reel and frame are given following the *TAEM* acronym, as in previous volumes (e.g., *TAEM* 72:878).

Headnotes

Volumes of the Edison Papers typically include more introductory headnotes than is common in historical documentary editions, and the present one is no exception. Each chapter begins with a brief introduction outlining Edison's personal, technical, and business activities during that period. Within chapters, occasional headnotes appear before particular documents (see List of Editorial Headnotes) for several purposes. Some present specific technical issues (e.g., "Phonoplex Telegraph") or describe the characteristics and context of a set of documents (e.g., "Edison's Fort Myers Notebooks"). A headnote may also provide a coherent narrative of related events too intricate to follow only in scattered documents and endnote references (e.g., "Edison and High-Voltage Electrical Distribution"). In addition, headnotes present original research on major events in Edison's life that were under-represented in his personal papers (e.g., "Edison's Florida Vacation").

Just as chapter introductions and headnotes serve as guides for the general reader, discursive endnotes often contain annotation of interest to the general reader. Some of these discussions are broadly contextual. Other information of a biographical character suggests the tight and overlapping character of the business, technical, and social circles in which Edison's name circulated. The endnotes also include business or technical details likely to be of more concern to the specialized reader. In general, the editors provide more detailed information for technical issues that have received little scholarly attention than for topics that are already well treated in the secondary literature.

Citations

The citation practices used in previous volumes are carried forward to this one, including several conventions introduced in the preceding two volumes.

Appendixes

As in Volumes 1–7, we include relevant selections from the autobiographical notes that Edison prepared in 1908 and 1909 for Frank Dyer and Thomas Martin's biography of Edison (see App. 1). Appendix 2 identifies Edison's U.S. patent applications (successful and not) according to the case number assigned by his patent attorneys. A comprehensive list of Edison's U.S. patents is also available on the Edison Papers website (http://edison.rutgers.edu/patents.htm). A separate list identifies the U.S. patents secured by a number of Edison's colleagues and associates in this period.

Errata

Errata for previous volumes can be found on the Edison Papers website at http://edison.rutgers.edu/berrata.htm.

Editorial Symbols

~~Newark~~   Overstruck letters
   Legible manuscript cancellations; crossed-out or overwritten letters are placed before corrections
[Newark]   Text in brackets
   Material supplied by editors
[Newark?]   Text with a question mark in brackets
   Conjecture
[Newark?]ᵃ   Text with a question mark in brackets followed by a superscript letter to reference a textnote
   Conjecture of illegible text
⟨Newark⟩   Text in angle brackets
   Marginalia; in Edison's hand unless otherwise noted
[   ]   Empty brackets
   Text missing from damaged manuscript
[---]   One or more hyphens in brackets
   Conjecture of number of characters in illegible material

Superscript numbers in editors' headnotes and in the documents refer to endnotes, which are grouped at the end of each headnote and after the textnote of each document.

Superscript lowercase letters in the documents refer to textnotes, which appear collectively at the end of each document.

List of Abbreviations

ABBREVIATIONS USED TO DESCRIBE DOCUMENTS

The following abbreviations describe the basic nature of the documents included in the eighth volume of *The Papers of Thomas A. Edison*:

AD	Autograph Document
ADf	Autograph Draft
ADfS	Autograph Draft Signed
ADS	Autograph Document Signed
AL	Autograph Letter
ALS	Autograph Letter Signed
D	Document
Df	Draft
DS	Document Signed
L	Letter
LS	Letter Signed
M	Model
PD	Printed Document
PL	Printed Letter
TD	Typed Document
TL	Typed Letter
TLS	Typed Letter Signed
X	Experimental Note

In these descriptions the following meanings are assumed:

Document   Accounts, agreements and contracts, bills and receipts, legal documents, memoranda, patent applications, and published material, but excluding letters, models, and experimental notes

Draft   A preliminary or unfinished version of a document or letter

Experimental Note   Technical notes or drawings not included in letters, legal documents, and the like

Letter   Correspondence, including telegrams

Model   An artifact, whether a patent model, production model, structure, or other

The symbols may be followed in parentheses by one of these descriptive terms:

abstract   A condensation of a document

carbon copy   A mechanical copy of a document made by the author or stenographer using an intervening carbonized sheet at the time of the creation of the document

copy   A version of a document made by the author or other associated party at the time of the creation of the document

fragment   Incomplete document, the missing part of which has not been found by the editors

historic drawing   A drawing of an artifact no longer extant or no longer in its original form

letterpress copy   A transfer copy made by pressing the original under a sheet of damp tissue paper

photographic transcript   A transcript of a document made photographically

telegram   A telegraph message

transcript   A version of a document made at a substantially later date than that of the original, by someone not directly associated with the creation of the document

EDISON'S STANDARD ABBREVIATIONS IN DOCUMENTS

The following is a list of abbreviations that Edison used repeatedly, especially in technical notes and notebook entries. The editors have followed (to the extent possible) his inconsistent practices of punctuating and capitalizing these abbreviations. He sometimes used periods or occasionally hyphens or dashes between letters but he frequently did not include any punctuation. Edison typically used either all capital or all lowercase letters in these situations, but he occasionally capitalized just the first letter.

condsr	condenser
CP	candlepower

E.H.J.	Edward Hibberd Johnson
EL	electric light[ing]
EMF	electromotive force
EMG	electromotograph
EMW	Edison Machine Works
fil	filament
gal (or galv)	galvanometer
HP	horsepower
N.G.	no good
rcvr (or recvr)	receiver
res	resistance

STANDARD REFERENCES AND JOURNALS

Standard References

ACAB	*Appleton's Cyclopaedia of American Biography*
Am. Cycl.	*American Cyclopaedia: A Popular Dictionary of General Knowledge*
ANB	*American National Biography*
BDABL	*Biographical Dictionary of American Business Leaders*
BDUSC	*Biographical Directory of the United States Congress* [online edition]
CCB	*A Cyclopaedia of Canadian Biography*
Chambers's Ency.	*Chambers's Encyclopædia: A Dictionary of Universal Knowledge for the People*
Complete DSB	*Complete Dictionary of Scientific Biography*
DAS	*Dictionary of American Slang* (2nd. supp. ed.)
DNB	*Dictionary of National Biography*
DSB	*Dictionary of Scientific Biography*
DWB	*Dictionary of Wisconsin Biography*
Ency. ACIH	*Encyclopedia of American Cultural & Intellectual History*
Ency. Brit. 9	*Encyclopaedia Britannica*, 9th edition
Ency. Brit. 11	*Encyclopaedia Britannica*, 11th edition
Ency. Brit. 14	*Encyclopaedia Britannica*, 14th edition
Ency. Chgo.	*Encyclopedia of Chicago* (http://www.encyclopedia.chicagohistory.org)
Ency. Judaica	*Encyclopaedia Judaica*

Ency. Louisville	*Encyclopedia of Louisville*
Ency. NJ	*Encyclopedia of New Jersey*
Ency. NYC	*Encyclopedia of New York City*
ET	*Encyclopedia of Time: Science, Philosophy, Theology, & Culture*
GMO	*Grove Music Online* (http://www.oxford musiconline.com)
IDCH	*International Directory of Company Histories*
KNMD	*Knight's New Mechanical Dictionary*
LIMC	*Lexicon Iconographicum Mythologiae Classicae*
MEAR	*Melton's Encyclopedia of American Religions*, 8th edition
New Oxford Comp.	*New Oxford Companion to Literature in French*
NCFD	*New Cassell's French Dictionary: French– English, English–French*
NDWB	*Northeastern Dictionary of Women's Biography*
OCD	*Oxford Classical Dictionary*
ODR	*Oxford Dictionary of the Renaissance*
OED	*Oxford English Dictionary*
Oxford DNB	*Oxford Dictionary of National Biography*
TAEB	*The Papers of Thomas A. Edison* (book edition)
TAED	*The Papers of Thomas A. Edison* (digital edition, http://edision.rutgers.edu)
TAEM	*The Papers of Thomas A. Edison: A Selective Microfilm Edition*
WBD	*Webster's Biographical Dictionary*
WGD	*Webster's Geographical Dictionary*
WWW-1	*Who Was Who in America, Vol. 1*

Journals

Am. Jrn. Sci.	*American Journal of Science*
CPOR	*Canadian Patent Office Record*
Elec. and Elec. Eng.	*Electrician and Electrical Engineer*
NYT	*New York Times*
Proc. IEE	*Proceedings of the Institution of Electrical Engineers*
Sci. Am.	*Scientific American*
Sci. Am. Supp.	*Scientific American Supplement*

Teleg. J. and	*Telegraphic Journal and Electrical Review*
Elec. Rev.	
Trans AIEE	*Transactions of the American Institute of Electrical Engineers*

ARCHIVES AND REPOSITORIES

In general, repositories are identified according to the Library of Congress MARC code list for organizations (http://www.loc.gov/marc/organizations/). Parenthetical letters added to Library of Congress abbreviations were supplied by the editors. Abbreviations contained entirely within parentheses were created by the editors and appear without parentheses in citations.

DSI-AC	Archives Center, National Museum of American History, Smithsonian Institution, Washington, D.C.
(FFmEFW)	Edison & Ford Winter Estates, Fort Myers, Fla.
ICL	Cudahy Memorial Library, Loyola University, Chicago, Ill.
IMolD	Deere & Co. Archives, Moline, Ill.
(LMA)	London Metropolitan Archives, London
MdCpNA	National Archives and Records Administration, College Park, Md.
MdCpUHL	Library of American Broadcasting, Hornbake Library, University of Maryland, College Park, Md.
MiDbEI	Library and Archives, Henry Ford Museum & Greenfield Village, Dearborn, Mich.
MiPhM	Port Huron Museum, Port Huron, Mich.
N(ChaCI)	Oliver Archives Center, Chautauqua Institution, Chautauqua, N.Y.
NhD	Rauner Special Collections Library, Dartmouth College, Hanover, N.H.
NjHi	New Jersey Historical Society, Newark, N.J.
NjNC	Robert W. Van Houten Library, New Jersey Institute of Technology, Newark, N.J.
(NjWAT)	AT&T Archives and History Center, Warren, N.J.

NjWOE	Thomas Edison National Historical Park, West Orange, N.J.
NN	New York Public Library, Manuscripts and Archives Division.
NNNCC-Ar	Division of Old Records, New York County Clerk, New York City Archives, New York
(NScIS)	Museum of Innovation and Science (formerly Schenectady Museum), Schenectady, N.Y.
PHi	Historical Society of Pennsylvania, Philadelphia, Pa.
TxDaM-P	Bridwell Library, Perkins School of Theology, Southern Methodist University, Dallas, Tex.
(UkLIEE)	Institution of Electrical Engineers Archives, London, UK
Vi	Virginia State Library, Richmond, Va.

MANUSCRIPT COLLECTIONS AND COURT CASES

Accts.	Accounts, NjWOE
Bio. Coll.	Biographical Collection, NjWOE
BMI	BMI Collection, MdCpUHL
CEF	Charles Edison Fund Collection
Clippings	Unbound Clippings Series, NjWOE
CR	Company Records, NjWOE
DF	Document File, NjWOE
Diary	Thomas A. Edison Diary, NjWOE
ECB	Essex County Board of Chosen Freeholders Committee on Buildings, NjHi
Edison and Gilliland v. Phelps	*Edison and Gilliland v. Phelps*, Patent Interference File 8028, RG-241, MdCpNA
Edison Bio. Coll.	Edison Biographical Collection, NjWOE
Edison EL Coll.	Edison Collection, NScIS
Edison v. Thomson	*Edison v. Thomson*, Patent Interference File 12,332, RG-241, MdCpNA
EMFP	Edison-Miller Family Papers, FFmEFW
EP&RI	Edison Papers & Related Items, MiDbEI
FR	Family Records Series, NjWOE

GMHHP	Grace (Miller) and Halbert Hitchcock Papers, Miller Family Papers, N(ChaCI).
Heitz	David Heitz, New Hope, Pa.
Hummel	Charles Hummel, Wayne, N.J.
JHV	John Heyl Vincent Collection, TxDaM-P
Kilby	Kilby Family Papers, Vi.
Kruesi	John Kruesi Collection, Special Collections Series, NjWOE
Lab.	Laboratory notebooks and scrapbooks, NjWOE
LM	Miscellaneous Letterbooks, NjWOE
	5 Ore Milling Co. Letterbook (1881–1887)
	12 Phonoplex Letterbook (1886)
	19 Construction Dept. Letterbook (1884)
	22 Gilliland Letterbook (1888–1889)
MacKaye	Percy MacKaye Papers, MacKaye Family Papers, ML-5, NhD
Meadowcroft	William H. Meadowcroft Collection, Special Collections Series, NjWOE
MFP	Miller Family Papers, CEF
Miller	Harry F. Miller File, Legal Series, Nj-WOE
Misc. Legal	Miscellaneous Legal File, Legal Series, NjWOE
Misc. Notebooks	Miscellaneous Notebooks, Notebook Series, NjWOE
MMC	Mina Miller Collection, N(ChaCI)
MME-CD	Mina Miller Edison Correspondence and Documents, Edison–Miller Family Papers, Sloane
News Clippings	Newspaper Clippings, FFmEFW
NMC	Nancy Miller Collection, Miller Family Papers, N(ChaCI)
Pat. App.	Patent Application Files, RG-241, MdCpNA
PPC	Primary Printed Collection, CR, NjWOE
PS	Patent Series, NjWOE
Pub. Works	Published Works and Other Writings, NjWOE
Rice	E. W. Rice, Jr., Papers, NScIS
Scraps.	Scrapbooks, NjWOE
Sloane	David E. E. Sloane, New Haven, Conn.

Sprague	Frank J. Sprague Papers, NN
Swann	Swann Galleries, Inc., New York, N.Y.
TI	Telephone Interferences (Vols. 1–5), NjWOE; a printed and bound subset of the full Telephone Interferences
UHP	Uriah Hunt Painter Papers (Unbound Documents), Coll. 1669, PHi
Vail	Vail Papers, Special Collections Series, NjWOE
Vouchers (Household)	Mrs. Thomas A. Edison Vouchers [series 6], NjWOE (voucher number, box, and year are given in citations)
Vouchers (Laboratory)	Laboratory Vouchers [series 1], NjWOE (voucher number and year are given in citations)
Weston	Edward Weston Papers, NjNC
Weston v. Latimer v. Edison	*Weston v. Latimer v. Edison*, Patent Interference File, RG-241, MdCpNA
Whitney	Thomas Whitney Collection, Boise, Idaho
WHP	William H. Preece Papers, UkLIEE
WJH	William J. Hammer Collection, DSI-AC
WOL	West Orange Laboratory Records
Zipernowski v. Edison	*Zipernowski v. Edison*, Exhibits for Edison, Patent Interference File 13075, RG-241, MdCpNA

NEW BEGINNINGS
JANUARY 1885–DECEMBER 1887

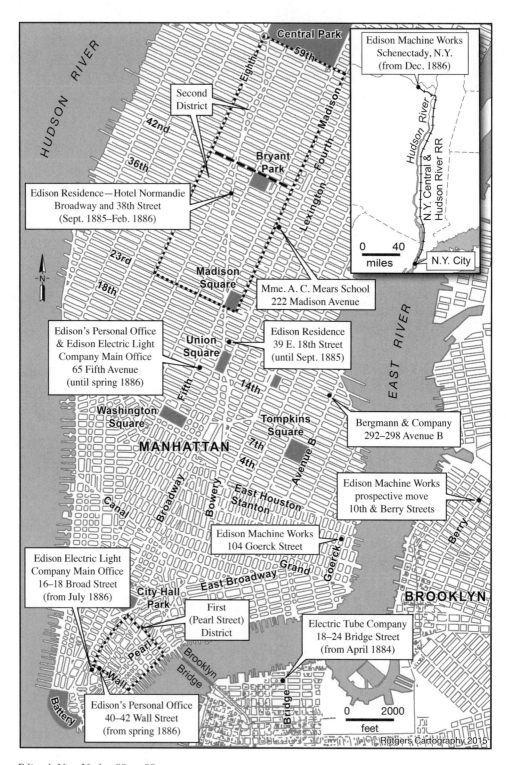

Edison's New York, 1885–1887.

–1– January–June 1885

Edison entered the new year with his inventive career in something of a lull. He had returned in the fall to work on telephone technologies, following his decision to leave the electric lighting business the previous summer. However, he found the American Bell Telephone Company, which he had hoped would sponsor his experiments, reluctant to make a contractual commitment, and in April the company ordered him to suspend all work on its behalf.[1] At mid-life, Edison had secured his inventive reputation (to the extent that a newspaper reviewer commented: "Mark Twain may be called the Edison of our literature") and was seen as a public authority on all things electrical, but he lacked either an ongoing project on which to focus his creative energy or the resources to launch such a project on his own.[2] He reportedly spent nearly every day at his personal laboratory atop the Bergmann & Company building in New York, where he was working on long-distance telephony and an improved stock printer, among other things, but there is a striking absence of technical notes and drawings in his hand from the winter months.[3] He let his trusted secretary and business adviser, Samuel Insull, handle his administrative and financial matters; these included, in addition to electric lighting business, negotiations for rights to Edison's ore separator.[4] He also dispatched Francis Upton, a German-trained physicist and mathematician who now managed the Edison lamp factory, to Europe to negotiate several business arrangements on behalf of his electric lighting enterprises, notably terms for the sale of American-made Edison lamps in Europe.

Edison's personal life had been unsettled as well, since the unexpected death of his wife in August 1884. He and his three children enjoyed relative financial security, with good prospects for reimbursement of many tens of thousands of dollars that he had expended for electric lighting ventures.[5] But he was living in rented rooms a few blocks from his New York offices while keeping an attachment to Menlo Park, New Jersey, where his mother-in-law set up residence and likely took care of his children at times.

Edison made at least two trips from New York to Boston in January.[6] These travels were partly to visit Ezra Gilliland, with whom he had renewed a friendship and started a collaboration in October. He was also seeing to business with American Bell, for whom he and Gilliland (who directed Bell's experimental department) were working on long-distance lines and transmitters, and attending an electrical exhibition in the city.[7]

Edison's association with Gilliland and his wife Lillian at this time set in motion several chains of events that would influence his work and fundamentally reshape his life. The most immediate result was a "shooting trip" that he and Gilliland planned to take to Florida.[8] Accompanied by Edison's daughter and Gilliland's wife, they set out in mid-February on a circuitous journey to Gilliland's Michigan hometown, then to Chicago, New Orleans (where an industrial exposition was underway), and Jacksonville, before turning west again across central Florida to the Gulf Coast. Once there, they rather impulsively agreed to buy thirteen acres of land near Fort Myers, an isolated but relatively affluent cattle entrepôt whose climate, natural surroundings, and opportunities for hunting and fishing all appealed to Edison. The two northern friends each planned to build his own house (and Edison a laboratory) to use in winters to come.[9]

Shortly before embarking to Florida, Edison and Gilliland discussed another inventor's patent (which Gilliland co-owned) for telephoning without wires from a train. One or both men saw in it the germ of an idea for a system to enable telegraphing without wires, a capability of great potential value for improving the safety and efficiency of railroad operation. They hastily sketched the fundamentals of a wireless telegraph system, and Edison mailed instructions from Michigan for a patent to be filed in Gilliland's name. From New Orleans a week later, he directed his laboratory assistant John Ott to make some basic experiments. This letter was one of several on various topics that he addressed to Ott during the trip.

Upon returning to New York in late March, Marion presumably went back to school,[10] and Edison returned to his lab with a new focus on communications technologies. He and Gilliland immediately signed two patent applications for the wireless telegraph (nicknamed the "grasshopper"). Soon after, Edison's brief experience with that system led him to conceive of applying one of its distinctive features—a series of rapid electrical pulses far briefer than the Morse dot and dash signals—to an entirely different telegraph problem. The result was what he would later call the "phonoplex"—essentially a set of different channels, each with its own frequency—that allowed the transmission of multiple signals among any stations (not just the end points) on a single wire. The phonoplex held out the possibility of greatly increasing the utility of wires already owned by telegraph companies (and railroads, especially), and Edison returned to its development in several creative bursts throughout the year.[11] By late springtime, he would again be working on telephone transmitters, this time despite, rather than on behalf of, American Bell.[12]

During the spring, the wireless and phonoplex telegraph systems further suggested to Edison parallels with contemporary questions about the transmission of energy through fields of force. These fundamental physical problems were discussed in technical journals like the *Electrician* that Edison read, and they clearly engaged with his own notions of fields of force. His conceptions had been shaped by early readings of Michael Faraday's lucid *Experimental Researches*, and they had been evident in his prior interest in discovering as-yet unknown physical forces. Sometime not long after his return from Florida, Edison tried to set out in a systematic way his understanding of forms of energy and the ways in which one (such as electricity) was convertible to another (magnetism, for example). His effort now was a brief one but it gave rise, at the end of 1885, to a more sustained project that he would again take up in Florida in 1886. In the short term, it also helped him to imagine (and patent) a system for telegraphing wirelessly over long distances between ships at sea.[13]

Edison visited Gilliland at least twice between April and June. On one of the later trips, Gilliland or Lillian introduced him to Mina Miller, the daughter of an Ohio manufacturer, who was studying in Boston. Possibly the two had met briefly at the Gillilands' home in January but that meeting, if it happened, left no lasting impression. At the end of June, however,

Edison was smitten with the young woman who would become his second wife.[14]

1. See Docs. 2770 and 2798.

2. "Literature. Mark Twain's Readable New Story," *San Francisco Chronicle*, 15 Mar. 1885, 6. Regarding his status as an authority see, for example, Doc. 2811 and "Electricity Man's Slave," *Electrical Review*, 24 Jan. 1885 (6): 8–9, Pub. Works (*TAED* PA013), originally published in the *New York Tribune*, 18 Jan. 1885.

3. His inventive work did produce a handful of patent applications in January; see App. 2.A. Charles Batchelor, his closest experimental assistant, was pursuing other projects, notably a motor for an electric-powered torpedo, at the Edison Machine Works; see Doc. 2779.

4. See Doc. 2784.

5. See Docs. 2771 and 2795. William Croffut, a newspaper writer friendly with Edison (and a past recipient of his generosity; see Doc. 1668), addressed published reports of Edison's alleged financial straits and a more general sense of the inventor's life in the public eye. Writing in the *New York World* on 14 June, Croffut asked, "How much is Edison worth? I do not know. He certainly does not own himself, for he has been public property, now, lo! these many years." On the basis of unspecified information, Croffut estimated Edison's annual income (mostly from royalties) to lie between $75,000 and $100,000. Cat. 1140:84a, Scraps. (*TAED* SB017084a).

6. Insull to John Tomlinson, 3 Jan. 1885, DF (*TAED* D8541A); TAE to Roscoe Conkling, 26 Jan. 1885, Lbk. 20:51B (*TAED* LB020051B).

7. According to a report in the *Electrician* about the Boston Electrical Exhibition, "Thos. A. Edison has been a frequent visitor. As he is a prominent exhibitor, his anticipated presence has always been announced, and he has in fact been made an attraction. To his and Mr. Ezra T. Gilliland's efforts what success the exhibition has had is largely due." "Correspondence. Boston," *Electrician* 3 (Feb. 1885): 75.

8. The descriptive phrase is Insull's. Insull to Charles Fitzgerald, 22 Mar. 1885, Lbk. 20:119 (*TAED* LB020199C).

9. See Doc. 2776 (headnote).

10. Edison had already paid a full semester's tuition. The whereabouts of Marion's younger siblings, Tom, Jr., and William Leslie, is not known; perhaps they remained in the care of Mrs. Stilwell at Menlo Park. Voucher (Laboratory) no. 49 (1885) for Mme. Mears.

11. See Doc. 2800 (headnote).

12. See Doc. 2813.

13. See Docs. 2804 (headnote) and 2807.

14. See Doc. 2824 (headnote).

–2770–

To John Tomlinson

Boston   Saturday Eve [January 3, 1885]

Tomlinson=[1]

These are two new contracts. They are apparently ok. but I send them on to you as I want to be <u>perfectly satisfied</u> that the WU[2] must continue the payment of 6000 yearly[3]   you will see that the Bell Co[4] take over the agreements in the 78 contract which relate to <u>future inventions</u> but leave the past as it was the WU continuing to pay the $6000. pls get the WU Contracts & see if this is so. Bell Co lawyer[5] explained to me that the particularly ~~state~~ drew it to leave no question about it= please write fully & return early monday[a] as am anxious to have matter closed so they will pay me past due money— I will wait for your points & the return of these Copies[6]— Morally Yours

Edison.

ALS, NjWOE, Miller (*TAED* HM850231D). Memorandum form of American Bell Telephone Co. [a]Followed by "over" as page turn.

1. John Canfield Tomlinson (1856–1927), a partner in the New York law office of Ecclesine & Tomlinson, represented Edison for several years in a variety of business and personal matters beginning about 1883. He was also the patent attorney of the Edison Electric Light Co. by about this time. Doc. 2552 n. 1; "Edison's Large Claims," *Philadelphia Times*, 3 May 1885, Cat. 1140:4a, Scraps. (*TAED* SB017004a).

2. Western Union Telegraph Co., owner of the nation's dominant commercial telegraph network, had once been a corporate pioneer of telephone development and service.

3. According to Tomlinson's acknowledgment of this letter, one of the draft contracts was between Edison and the American Bell Telephone Co.; the other was between Edison, American Bell, and Western Union (Tomlinson to TAE, 5 Jan. 1885, DF [*TAED* D8547C]). These were probably the proposed contracts (or more recent versions of them) discussed in Doc. 2745 nn. 2–3. Under consideration for well over a year, the agreements concerned the terms of Edison's work for American Bell and the company's assumption of Western Union's obligations and rights under an 1878 contract with Edison regarding his telephone research. Those obligations included a $6,000 annual salary and reimbursement of Edison's experimental and patent costs (TAE agreement with Western Union, 31 May 1878, Miller [*TAED* HM780045]; see Doc. 1317).

4. The American Bell Telephone Co., located in Boston, owned the principal telephone patents in the United States. Originally incorporated in early 1879 as the National Bell Telephone Co. to exploit the inventions of Alexander Bell, it achieved an effective monopoly under a November 1879 settlement with Western Union that gave it control of the telephone patents of Edison and Elisha Gray (Garnet 1985, 44–54; Bruce 1973, 260–71; Doc. 1822 n. 13). Edison began working on telephone technology for American Bell in October 1884 and frequently traveled to Boston in connection with his experiments (see Docs. 2743 [headnote] and 2767 n. 1).

5. Most likely Chauncey Smith.

6. Tomlinson replied promptly that "as they stand I think every thing is all right," though he noted suggestions for subsequent revisions. Tomlinson to TAE, 5 Jan. 1885, DF (*TAED* D8547C).

To Eugene Crowell[1]

Dear Sir:—

Referring to our several interviews, with reference to the various claims I have against your Co.,[2] and your request that I should make a statement of same to you in writing, I beg to draw your attention to the fact that under my contract with your Co.[3] I am entitled to receive from them, under certain conditions, $100,000. in cash. As I consider that this money is now due, and in view of the fact that there is a large balance due by me, under my subscription to the increase in the capital stock of the Co., I shall be glad to have an adjustment of my claim of $100,000., the same being set off against the amount which I owe on the stock.[4] In making a settlement of this matter, I desire Mr. Batchelor's stock to be considered as part of my own.[5]

With relation to the various disbursements made by me, on behalf of your Co., I beg to hand you herewith statement showing an expenditure of $38,261.04 in connection with experimental work, with the view to the perfection of a system of electric railroads. I would remind you that under my contract with your Co. the interest in electric railroad patents is divided as follows:— fifty two per cent to your Co., twenty four per cent to myself and twenty four per cent to raise the necessary capital for further work. In view of the changed circumstances with relation to the disposition of the Railroad interests I agreed to accept one third and to your Co. taking two thirds of any proceeds that may be received from the railroad patents. I would like to have this arrangement formally ratified by contract,[6] and I would impress upon you that care should be taken by your Co., so that my personal interests may be protected, with a view to my being reimbursed the amount of the account above referred to. Inasmuch as your Co. only spent one half the amount which I have spent, I think that they should agree to an arrangement by which they and myself shall be reimbursed out of pocket expenses, in connection with this business, in exact proportion to the amounts each has expended. That is, if $30,000. is received on account of this joint expense, $20,000. should come to myself and $10,000. to the Co. In addition to the amount of $38,261.04, I am entitled to receive from your Co. an amount equal to that paid me for my first experiments on electric railroading, which amounted to about $16,000.[7]

I also beg to hand you herewith account of the Construction Dept.[8] for $15,345.58, being for work done in connection with the engineering portion of your Co's. business.[9] At the time of the formation of the Construction Dept. arrangements were made between Major Eaton[10] and myself by which the Engineering Dept. of your Co. was practically abolished. I took over most of the employees of your Co's. Engineering Dept., and arranged to make canvasses of cities and do all the necessary engineering work in connection with the making of estimates for central station plants. If I obtained a contract for the installation of a plant I was to bear the engineering and canvassing expenses applying to such plant, but in the event of a failure to obtain a contract this expense was to be borne by your Co.[11] The above referred to account is for such expenses. I may mention that, under the instructions of the Auditing Committee appointed by the last Board of Directors, your Treasurer, Mr. Hastings,[12] made a careful audit of this account, for which purpose I placed at his disposal all my vouchers and books of account. He can bear witness to the fact that not a cent of profit was made on this account, but that it represents cash actually paid out by me.[13]

When installing the first village plant I found that many improvements of a minor character were necessary, in order to adapt our system to the requirements of town and village lighting. I accordingly engaged the necessary experts to improve the then existing instruments, and to produce others which were requisite. Mr. F. J. Sprague[14] was engaged entirely on this class of work, and in working up a system for the more rapid determination of conductors. I hand you herewith an account of $2,748.04, being the amount I paid him.

In consequence of defective instruments, and the nonexistence of others, I was compelled to keep my experts at stations for a period far beyond what would have been necessary had the system been complete, so far as it applied to village work. In this connection I spent $2,212.18, for which I also beg to hand you a statement.

I also hand you a statement of extraordinary lamp breakage, $194.10, alterations on pressure indicators,[15] $202.50 and an account for experimental work $169.04

The results of these various expenditures is that your Co. has to-day a system which they can readily put up in small towns and villages, without going to any further expense in the way of experimental work for this purpose, and although the several accounts above quoted by no means represents my losses in connection with the Construction Dept., owing to defective instruments, they form what I believe to be a fair claim against your Co. from which they have and will derive benefit. I may mention that the cost of the present pressure regulator, used by your Co., and so absolutely necessary for the proper working of its system, is in the above accounts. This instrument was invented by Mr. F. J. Sprague, whilst working for the Construction Dept., and will be assigned to your Co. on the settlement of the above accounts.[16]

It being necessary to obtain certain legislation in New Jersey, to enable your Co. to enter into the illuminating business in that State, Major Eaton, at my request, on behalf of the Co., agreed to bear the expenses in connection with getting the necessary bill passed through the N.J. Legislature.[17] The amount expended by me in effecting this was $1,753. Major Eaton gave a check for $1,000. on this account, and stated that this was all he was authorized to pay, as he had assured his Board that I had stated that the expenses would not be more than this amount. All I can say to this is that Major Eaton must have misunderstood me, as at the time I spoke to him about the matter I knew the expenses would have to be paid, and was in fact paying them, in addition to having promised $1,000., which was payable after the passing of the Act required. I have therefore paid $753. more than I received from Major Eaton, and hand you herewith an account for same.

The amount of $2,175.25 is for services rendered by Mr. Jas. A. Russell,[18] in connection your with your Co's. patent business. Mr. Russell was employed by me to do detective work in hunting up evidence with relation to our interference cases.

The account $1,389.42, for experimental work, is for a portion of my ordinary experimental expenses which your Co. has usually paid from week to week, but for some reason, which I do not understand, the amounts in the above referred to account were not paid to me.[19]

I disbursed the sum of $388. in working up an agitation in the daily press, having in view the injury of the gas interests, and more especially those of the water gas companies, for which amount I hand statement herewith.[20]

The account of $2,075., paid by me to Mr. John C. Tomlinson, is for services rendered in connection with your Co's. business, mainly in connection with interference cases, and the working up of evidence with a view to taking action against infringers of the lamp patents. I engaged Mr. Tomlinson because I was very much dissatisfied with the conduct of the legal portion of your Co's. patent affairs, and considered that my patents were not receiving the attention which they should have received, and the only means by which I could see my way clear to getting an improvement was by engaging counsel at my own expense.

I may mention that, in addition to the above, I have spent at the Edison Machine Works[21] upwards of $30,000. in improving the various types of dynamos now used by your Co. This experimental account has been written off by the Machine Works to profit and loss, and it is not my intention to make any claim for same. I have also expended ~~upwards of~~[b] $6,0200.[c] in my efforts to place the Central Station of the First District on a paying basis. I paid this sum to Mr. C. E. Chinnock,[22] in consideration of the extraordinary efforts he made to bring the station from a state of absolute failure to one of perfect success.

I am very anxious to have the various claims, set forth above, recognized at the earliest possible moment, and I desire to have a cash settlement for the various accounts which represent money paid out by me. Yours very truly[d]

TL (carbon copy), NjWOE, DF (*TAED* D8526B). [a]Date handwritten by Samuel Insull. [b]"~~upwards of~~" interlined above. [c]Obscured overwritten text; "$6200" also written in left margin. [d]"Yours very truly" handwritten by Insull.

1. Eugene Crowell (1817–1894) was elected president of the Edison Electric Light Co. in October 1884 in the management shakeup engineered by Edison. He was a stockholder in that and other Edison electric lighting companies, using the considerable wealth he had accumulated in the wholesale drug business in California. See Doc. 2753 esp. n. 7; *Appletons' Annual Cyclopaedia and Register of Important Events* 19 (1894): 572.

2. Edison had recently asked to meet with Crowell regarding the Electric Tube Co.'s claim against the Edison Electric Light Co. Doc. 2795 outlines a proposed settlement of this and other claims discussed in this document. TAE to Crowell, 3 Jan. 1885, Lbk. 20:16A (*TAED* LB020016A).

3. Edison meant the Edison Electric Light Co. Incorporated in October 1878, the firm owned and licensed Edison's U.S. patents for electric light, power, and heat. Doc. 1494 n. 4.

4. The sixth clause of Edison's 1878 contract with the Edison Electric Light Co. (Doc. 1576) provided that the company pay him $100,000 "out of its first net earnings." This phrase gave rise to competing interpretations; see Doc. 2795.

The editors have not more specifically identified the statements and enclosures mentioned in this letter. Edison's account records, though detailed, are incomplete, but entries in his ledger books for the Edison Electric Light Co. and related electric light enterprises provide a general picture of his transactions with these firms (Ledgers 5 and 8, both Accts. [*TAED* AB003, AB004]). The Edison Electric Light Co. raised its capital stock from \$480,000 to \$720,000 in January 1882 and to \$1,080,000 in September 1883. These increases were attained by calling on existing shareholders for cash, but in neither instance was the entire amount paid immediately (Edison Electric Light Co. annual reports, 24 Oct. 1882 and 23 Oct. 1883, in Edison Electric Light Co. Bulletins 15:37, 20:48, both CR [*TAED* CB015235, CB020442]). Although Edison had given or sold portions of his own holdings to family, friends, and associates over the years, he was the largest single shareholder in the company in September 1883 and almost certainly remained so at this time (list of stock transfers, n.d. [1884?], DF [*TAED* D8023ZAD1]; Edison Electric Light Co. list of shareholders, 29 Sept. 1883, Miller [*TAED* HM830194]).

5. Charles Batchelor (1845–1910) was Edison's longtime principal experimental assistant (Doc. 264 n. 9). Edison had given him 130 shares of Edison Electric Light Co. stock in 1879, as well as a 10 percent share of royalties paid by the company (Cat. 1318:60, Batchelor [*TAED* MBA001, image 26]; Docs. 1668, 1834 n. 3, 1866 n. 4). Batchelor was also a minority partner in Edison's manufacturing shops and, since June 1884, general manager of the Edison Machine Works (Docs. 2343 [headnote], 2725 n. 6; *TAEB* 7, chap. 5 introduction).

6. The 52-24-24 percent proportion was specified in an 1881 agreement with the Edison Electric Light Co. governing the ownership and commercial use of Edison's patents applicable to electric railways, which had not been specifically mentioned in his original 1878 contract with the company (TAE agreement with Edison Electric Light Co., 12 Jan. 1881, Miller [*TAED* HM810140]). With the phrase "changed circumstances," Edison probably referred to the joining of the Edison interests with those of capitalist Simeon Reed and inventor Stephen Field in April 1883 (see Doc. 2431). His suggestion here of a different division of any profits was embodied in a new contract executed on 4 February 1885, one in a series of agreements since January 1881 to reallocate obligations and potential profits arising from Edison's electric railroad patents (TAE agreement with Edison Electric Light Co., 4 Feb. 1885, Miller [*TAED* HM850237]).

7. The new contract with the Edison Electric Light Co. that Edison signed in February 1885 (see note 6) also dealt with his claim of \$38,261.04 for experimental expenses (this figure was very close to the amount that he had claimed in December 1883; see Doc. 2569). The February agreement stipulated that this sum was Edison's portion of a total reimbursement of \$60,497.87 payable jointly to him and the Electric Light Co. by the Electric Railway Co. of the United States under terms of Doc. 2431 and subsequent agreements (see Doc. 2431 nn. 5, 7). Although the numbers vary slightly from the stated proportion of $^{19}/_{30}$ths coming to Edison, all other reimbursements by the Railway Co. were to be divided in the same ratio. In the new contract, Edison agreed to waive a separate reimbursement claim of \$1,000 against the Edison Electric Light Co. Financier Henry Villard had loaned Edison \$12,000 for experimental expenses, but Edison repaid that amount with interest in 1882 (Doc. 2195 n. 1).

8. The Thomas A. Edison Construction Dept. was the name under which Edison did business as a contractor for building and equipping

central station electrical systems in the United States from May 1883 through August 1884, when its work was assumed by the Edison Co. for Isolated Lighting. Doc. 2437 (headnote).

9. The editors have not attempted to substantiate this figure from the tangled and incomplete financial records of the Edison Construction Dept., but the amount given here is roughly $1,500 more than claims that Edison had pending against the Edison Electric Light Co. since the spring of 1884 (Doc. 2704 n. 1; also cf. Samuel Insull to Frank Hastings, 14 Apr. 1884, DF [*TAED* D8416BGS]). Edison again pressed the matter at the end of January, when he requested from Crowell

immediate action with relation to that part of my accounts which cover work done by the Construction Department aggregating a total of $20 871 being cash spent by me. The bills for the greater part of this amount were rendered by me to your Predecessor long since and I think as a matter of simple justice a settlement should be made . . . I do not ask your Company to reimburse me my total outlay [greater than $50,000], although the benefit is derived by them alone, but only that portion (Engineering Expenses) which the then President, on behalf of the Company, agreed to pay, and also for some small amounts which would have been directly chargable to your company had I been conducting the experiments under my general arrangement with them as to experimental work . . . The delay is to me a very serious matte[r] indeed and it is causing me a great deal of worry and anxiety in connection with the finances of my business as from the shrinkage of my Capital owing to the losses above instanced and to other causes my resources are taxed to their utmost. [TAE to Crowell, 29 Jan. 1885, Lbk. 20:59 (*TAED* LB020059)]

10. Sherburne Blake Eaton (1840–1914), an attorney by trade, was president and de facto manager of the Edison Electric Light Co. from October 1882 until October 1884 (and an officer of other Edison lighting companies). Edison, who was dissatisfied with the company's policies, arranged to have Eaton removed at the conclusion of his annual term. Eaton remained connected with the company as its legal counsel. See Docs. 2120 n. 7, 2748 n. 4, 2753.

11. Docs. 2424 and 2437 (headnotes) provide overviews of the Construction Dept.'s purview and practices. Lists of completed and projected central stations are in *TAEB* 7 App. 2.

12. Frank Seymour Hastings (1853–1924) was treasurer of the Edison Electric Light Co. He also held that office in the Edison Co. for Isolated Lighting and the Edison Electric Illuminating Co. of New York. Doc. 2420 n. 24.

13. The audit took place in or after May 1884, following the Edison Electric Light Co.'s request for more detailed documentation of the expenses it was being asked to reimburse to Edison (Doc. 2704 n. 1). It was about that time that Edison had been planning to discharge his electrical staff; Sherburne Eaton arranged for the Edison Electric Light Co. to pay some of their salaries on a temporary basis (see Docs. 2655, 2672, 2677 esp. n. 3).

14. Electrical engineer Frank Julian Sprague (1857–1934) entered

Edison's employ about 1 June 1883. He left at the end of April 1884 to launch an independent—and ultimately very successful—career as an inventor and entrepreneur in electric traction (see Docs. 2433 n. 2, 2456 esp. n. 1, 2656–57). Although nominally employed to oversee the start of new central stations, Sprague made two distinctive contributions to the Construction Dept.'s work. He diagnosed design flaws in Edison's voltage indicator, a critical part of the system, and he helped devise a mathematical model for determining the conductors in any distribution network, doing away with the need for tedious physical models. See Docs. 2538 (headnote) and 2575.

15. The Edison Construction Dept. would have been obliged to replace customers' lamps during the break-in and training period, typically thirty days, that it had charge of a new central station before turning it over to the local illuminating company. High rates of lamp failure resulted directly from voltage regulation problems, typically caused by inaccurate instruments or inexpert operation. See Doc. 2505 (headnote).

16. Edison likely referred to Sprague's U.S. Patent 314,891, issued on 31 March 1885, for an "indicator of variations in electric current" (see Doc. 2575 esp. n. 8; *TAEB* 7 App. 4.C). The editors have not found evidence of its assignment to the Edison Electric Light Co.

17. See Doc. 2687 esp. nn. 3–5.

18. James A. Russell (b. 1837?–1897) had a variegated work history with Edison, including stints as a canvasser in New York, bodyguard, and detective, and his name appears repeatedly in Edison's voucher system for payments. In 1883, Russell had gathered affidavits denying allegations that Edison (or his companies) had instigated lawsuits against a rival firm, and he had been involved in a search for political malfeasance in lighting the new Brooklyn Bridge. He had also been deputed to obtain a legal release from a discharged Edison employee who was threatening to sue. Prior to his association with Edison, he had been a "confidential assistant" and bodyguard for Jay Gould. According to an obituary in the *Western Electrician,* Russell had served in the Army of the Potomac, reaching the rank of colonel, and died in New York of pneumonia after a long bout of invalidism caused by an illness he contracted through exposure during his wartime service. Doc. 2321 n. 6; Vouchers (Laboratory) no. 674 (1884); Russell to TAE, 28 Jan. 1884; Grosvenor Lowrey to TAE, 2 Jan. 1884; Ecclesine & Tomlinson to Russell, 27 June 1883; all DF (*TAED* D8403N, D8427A, D8327ZAJ); Samuel Insull to John Tomlinson, 27 Aug. 1883, Lbk. 13:23 (*TAED* LB013023); Tate 1938, 118–20; Obituary, *Western Electrician,* 27 Mar. 1897, 20:183.

19. Beginning in late 1878, Edison's office assistants itemized weekly experimental expenses to be reimbursed by the Edison Electric Light Co. Preparation of these lists, of which Doc. 1562 is an early example, kept up at least through late 1882; the practice of weekly payments evidently continued after that time.

20. This project appeared in Edison's financial records under the heading "Gas Statistics." Edison's effort apparently began in November 1883 and continued sporadically through the end of 1884, when the account was balanced by an internal transfer of $388.00. James Russell's name also appears in connection with this account (Ledger #5:560, Accts. [*TAED* AB003, image 277]; Vouchers [Laboratory] nos.

284 [1883], 674 [1884]). How many articles Edison tried to plant is not known, but proof of his success showed up in the Edison Electric Light Co.'s Bulletins for its agents, which published nearly a score of articles critical of gas between April 1883 and April 1884. Among them were several titled "Deaths from Gas" or "Danger from Gas" and a few on the hazards of carbonic acid given off by water (as opposed to coal) gas (Edison Electric Light Co. Bulletins 17–20, 22; 16 Apr. 1883, 31 May 1883, 15 Aug. 1883, 31 Oct. 1883, 9 Apr. 1884; CR [*TAED* CB017–20, CB022]).

21. The Edison Machine Works manufactured Edison dynamos and motors at 104 Goerck St., near the East River waterfront. It also had facilities for designing and testing electrical equipment and for training workers in the Edison lighting system. Established as a partnership in 1881, it was incorporated in February 1884. Doc. 2343 (headnote); Edison Machine Co. minutes, n.d. [Feb. 1884], DF (*TAED* D8431D).

22. When Charles Edward Chinnock (1845–1915) became superintendent of the First District central station (on Pearl St. in New York) in May 1883, Edison promised him $5,000 cash or seventeen stock shares in the New York illuminating company if the station returned a 5 percent annual profit. Edison paid him $5,000 at the end of 1884. In this document, Edison may have been looking ahead to the transfer of ten stock shares, worth $1,000 at par, recorded on 5 February 1885. Doc. 2456 n. 6; Vouchers (Laboratory) nos. 44–45 (1885).

–2772–

To Samuel Edison

[New York,] 13th Jan [188]5

My Dear Father,[1]

You really must stop the demand made on the Grand Trunk RR. for salary on my account.[2] The letters presumably signed by you but which look to me as if Symington[3] has written them have been sent to me. The whole thing makes me appear ridiculous & I must insist on nothing further being done[4]   Yours

Thos. A Edison   I[nsull].[5]

L (letterpress copy), NjWOE, Lbk. 20:28 (*TAED* LB020028). Written by Samuel Insull.

1. Samuel Insull addressed this letter to Edison's father, Samuel Ogden Edison (1804–1896), at his home in Fort Gratiot, Mich. *TAEB* 1 chap. 1 introduction, esp. n. 2; Obituary, *NYT,* 27 Feb. 1896, 4.

2. An officer of the Grand Trunk Railway had recently contacted Edison regarding a letter to the company purportedly written by Edison's father. The letter demanded payment of $28.60 in wages from 1863, when Edison briefly served as a telegraph operator at Stratford Junction, Ont. Edison had left the job abruptly after his negligence nearly caused a train collision. Samuel Edison to Grand Trunk Rwy., 10 Sept. 1884, enclosed with John Burton to TAE, 2 Jan. 1885; both DF (*TAED* D8514222, D8514111); *TAEB* 1, chap. 1 introduction (p. 15).

3. James Symington (b. 1818), a Scottish immigrant in 1836, was a

longtime friend of Samuel Edison in Port Huron. Edison undoubtedly recognized Symington's distinctive handwriting in the letter sent to the Grand Trunk Railway (see note 2; Doc. 1330 n. 1; U.S. Census Bureau 1982? [1900], roll T623_742, p. 9A [Port Huron, Saint Clair, Mich.]). He is not James Sanderson Symington (1845–1900) of Sarnia, Ont., an erstwhile Edison business associate in the Sarnia Street Railway Co. (Doc. 1686 n. 2; Beers 1906, 48–49; *Registrations of Deaths, 1869–1936,* MS935_97).

4. The matter did not end there. In May, Edison wrote again (through Samuel Insull) to his father about a letter (not found) forwarded from the railroad's general manager about the claim. Edison again "insist[ed] that [no] further letters of this character be sent as they make me appear very ridiculous." At the same time, he urged the railroad's general manager to "consign any further communications of this character to the waste paper basket." TAE to Samuel Edison, 18 May 1885; TAE to Grand Trunk Rwy., 18 May 1885; Lbk. 20:300, 300A (*TAED* LB020300, LB020300A).

5. Samuel Insull (1859–1938) became Edison's personal secretary in 1881. He was by this time a deeply trusted associate and Edison's de facto business manager. Doc. 1947 n. 2.

–2773–

From Charles Brooke

Toledo, Iowa, Jan 13th 1885.[a]

Dear Sir:—

The wave theory of sound is being questioned in our college. Some objections are appearing to be well taken as the subject is discussed.[1]

If you will please write <u>your opinion</u>, we as students of natural science will consider it an lasting favor.

If your time is too much employed to give your views in full please state whether you accept the common theory as presented in our text books or not?

If you reject the wave theory and will so state; it will be of great value to us as students. If you will in addition to the above give your views, if only in a few sentences, it will be a great favor to us. Very truly yours

C. M. Brooke.[2]

N.B. I inclose envelope ready for reply[b]

⟨I think the wave theory of sound correct but the explanation in text books & the results obtained from apparatus shewn in such books is wrong in nearly every particular[3]   Edison⟩

ALS, NjWOE, DF (*TAED* D8505A). Letterhead of Philophronean Literary Society, Western College. [a]"Toledo, Iowa," and "188" preprinted. [b]Postscript enclosed by braces.

1. The theory that sound is transmitted by waves moving through a medium, such as air, originated with the Greeks and had a dominant

place in the development of natural philosophy and physics in the West, including contemporary textbooks (see, e.g., Atkinson 1883, 182–84; Trowbridge 1884, chap. 13). Distinguished physicists such as John Tyndall and John William Strutt (Lord Rayleigh) took it for granted in their own research and elucidated it in cogent and—to most minds—convincing fashion (Deschanel 1873; Strutt 1945 esp. xvii–xxii; and Tyndall 1867).

Alexander Wilford Hall, apparently a New York publisher (*Trow's* 1885, 702), presented a combative critique of the wave theory, especially the case of transmission through solid materials, in his self-published 1877 book, *The Problem of Human Life: Embracing the "Evolution of Sound" and "Evolution Evolved."* The two chapters devoted to sound were central to Hall's ambitions for the book, in which he hoped, by attacking such a well-regarded physical explanation, to subvert scientific authority in general and the theory of human evolution in particular. He dismissed the wave theory as a sophistic conflation of cause and effect, and he put forward in its place a philosophical system in which sound could be explained by the motion of solid particles emanating in straight lines (a hypothesis somewhat reminiscent of that advanced by Pierre Gassendi in the seventeenth century; see Strutt 1945, xvii). This system, by which he attempted to explain all physical phenomena as the effects of "Corpuscular Emanations," came to be known as "substantialism" and Hall as its founder (Hall 1880 [1877], v, chaps. 5–6, esp. p. 231). Hall's arguments seem to have been ignored by leading scientific figures, but they found a receptive, if small, audience in other quarters. Henry Augustus Mott, Jr., who earned a Ph.D. from the Columbia School of Mines in 1875 or 1876 (*ACAB* 4:444; Columbia College 1888, 251), attracted some newspaper attention with a lecture to the New York Academy of Sciences in December 1884 and a small book on *The Fallacy of the Present Theory of Sound* published soon afterward ("The Theory Will Still Stand," *NYT*, 9 Dec. 1884, 4; Mott 1885). By 1884, Hall himself was putting out *The Microcosm*, a journal in which some contributors fused the substantialist philosophy with Christian doctrines (Swander 1885). With one of his followers, Hall later published *A Text-Book on Sound: The Substantial Theory of Acoustics* (Swander and Hall, 1887). His adherents seem to have been almost entirely American, at least until 1890, when accounts of his beliefs appeared in the British technical press ("American Opinion on the Authors of 'The New Theory of Sound,'" *English Mechanic and World of Science* 51 [11 Apr. 1890]: 138–39).

2. Charles Morgan Brooke (b. 1858?) wrote from Western College, a combined preparatory school and college affiliated with the United Brethren in Christ; the school moved to Toledo, Iowa, in 1881 (Parker 1893, 167). He graduated in 1886 with a degree in liberal arts. Brooke became president and professor of philosophy of Lane University in Kansas in 1891 and remained there for ten years. By 1920, he reportedly had taken up ranching in Oklahoma. U.S. Census Bureau 1970 (1880), roll T9_239, p. 11B, image 0270 (Eagle Point, Ogle, Ill.); ibid. 1992 (1920), roll T625_1482, p. 5A, image 637 (Clayton, Payne, Okla.); Ward 1911, 363; Berger 1910, 504–6; Connelley 1919, 2169–70.

3. In late 1875, Edison had begun to receive instruction in acoustic science from Robert Spice, a professor of chemistry and natural phi-

losophy at the Brooklyn High School. Spice remained a part of Edison's laboratory staff for several months. *TAEB* 2 chap. 10 introduction; Doc. 487 n. 1.

–2774–

To Henrietta Reiss[1]

[New York,] 14th Jan [188]5

Madam,

Referring to your favor of 12th[2] I think you will find that you have a difficult matter to defend your side of the debate you speak of.

I am strongly of the impression that steam has been, up to this time by far the most beneficial to the world   If it were a case of blotting out one or the other I think we could the better dispense with electricity. Without steam it would be next to impossible to cross the American continent in less than seven days or the Atlantic Ocean in about the same time   But for steam the field open to the uses of electricity would be but small as the steam engine is today essential as a motive power for the production of electricity for such purposes as Electric Lighting Electric motors for Rail Road Industrial & other purpos[a]

Of course I am speaking of what has taken place up to the present time   The future may, and I believe will, give us discoveries and inventions which will place electricity as a motive power in advance & independent of Steam[3]   Yours respectfully

Thos A Edison   I[nsull]

L (letterpress copy), Lbk. 20:41 (*TAED* LB020041). Written by Samuel Insull. [a]Copied off edge of page.

1. Not identified. Her letter to Edison (see note 2) gave a return address in care of a third party at Twentieth St. and Sixth Ave. in New York, where Insull addressed the reply.

2. Reiss wrote as an organizer of a New York "Debating Society" preparing to take up the question of "Whether Electricity or Steam is most beneficial to the country." She asked for information that would help her make the case for electricity. Reiss to TAE, 12 June 1885, DF (*TAED* D85050A).

3. Discussing the possibility of converting coal directly into electricity (see Doc. 2520 [headnote]), Edison was recently quoted as saying that such a process

would put an end to boilers and steam engines; it would make power about one-tenth the cost that it is now; it would enable a steamship to cross the Atlantic at a nominal cost; it would enable every poor man to run his own carriage; it would revolutionize the industrial world. The electric motor is the ideal motor for all kinds of work.

What we want is some means of producing the current cheaply.
["Mr. Edison to Devote Five Years to the Elimination of the Steam
Engine," *Electrical Review* 5 (11 Oct. 1884): 5]

–2775–

To Chester Lyman[1]

[New York,] Jan. 18, 1885

My Dear Sir:—

Referring to your favor of the 13th inst.[2] I would state that I
am now confining my efforts entirely to my Laboratory work,
and have comparatively little to do with the active management of our Electric Light business. I have no vacancies in my
Laboratory at the present time, and, even if I had, the class of
work that your son[3] would have to perform in it would by no
means prepare him for the business of Electrical Engineering.
In fact, I think that laboratory work has exactly the opposite
effect upon a young man, and is rather a drawback to him.

With relation to the probabilities of there being vacancies
on the Engineering Staff of the Edison Electric Light Companies, all I can say is that, owing to the general dull state of
trade, the electric light business is naturally very dull also, so
that our Companies have as much as they can do to take care
of the old members of their staff. If any considerable improvement should take place in business, I imagine there would
probably be openings for young men such as you say your
young son is, and if, then he desires a position, there should be
a vacancy, it will afford me very great pleasure in doing what I
can [by?][a] getting [it?][a] for him.

I may mention that at first he would not receive any [sal-
ary?][a], as of course his services would be of no great value until
he was familiar with our business, but, as his knowledge increased, and his services to the Company became valuable, he
would of course be paid accordingly. Yours very truly,

Thos. A Edison   I[nsull]

TL (carbon copy), NjWOE, Lbk. 20:49 (*TAED* LB020049). Signed for
Edison by Samuel Insull. [a]Faint copy.

1. Chester Smith Lyman (1814–1890) was the professor of astronomy
in the Sheffield Scientific School of Yale University, where he had been
appointed in 1859 as the professor of industrial mechanics and physics.
Lyman had himself graduated Yale in 1837. He became a pastor, then
traveled to Hawaii and California as a missionary, educator, and sur-
veyor. He wrote one of the earliest accounts of the California Gold Rush
and contributed to the revision of scientific terms for *Webster's Dictionary*. At the Sheffield School, Lyman invented several instruments, including one for determining latitude, and a wave motion detector. *DAB*,
s.v. "Lyman, Chester Smith;" *ACAB*, s.v. "Lyman, Chester Smith."

2. Lyman inquired whether his son, disappointed by job prospects in his chosen field of naval architecture, "could find a position in your establishment in which he might make a start as an electrical engineer." Lyman to TAE, 13 Jan. 1885, DF (*TAED* D85130A).

3. Chester Wolcott Lyman (1861–1926) graduated from Yale in 1882. After completing a special course at the Sheffield Scientific School in 1884, he entered the paper business in 1885 and eventually became a vice president of the International Paper Co. Chamberlain 1900, 5:330; Obituary, *NYT*, 8 Apr. 1926, 25; "Daughter Contests Will of C. W. Lyman," ibid., 5 Aug. 1926, 23.

EDISON'S FLORIDA VACATION   Doc. 2776

By mid-February, Edison was planning to spend part of the winter in Florida (see Doc. 2776), as he had in 1882 and 1884.[1] He would be accompanied by his daughter Marion, his friend Ezra Gilliland, and Gilliland's wife, Lillian, on what Samuel Insull described as a "shooting trip."[2] Edison and Gilliland also hoped to use their stay to find a suitable location for winter vacation homes. The trip was Edison's first extended period away from work since the death of his wife in August 1884. Mary had accompanied him on both previous visits to Florida, but he did not linger in places they had visited together in the eastern part of the state. He opted instead to explore the Gulf Coast.

Edison reached Florida only at the end of a circuitous journey. It included brief stops in Adrian, Michigan (Ezra Gilliland's hometown, where Edison and perhaps also Gilliland had been operators in 1863–1864), Chicago, Cincinnati (where Edison and Gilliland had worked together), and New Orleans.[3] They left New York on 20 February but a powerful snowstorm delayed their trip on the New York Central to Adrian, which they reached by the twenty-second.[4] After spending the day with Gilliland's father, the party traveled to Chicago by the twenty-third.[5] Edison had dinner that night with George Bliss, superintendent of the Western Edison Light Company, and left the next day for Cincinnati. Bliss regretted that Edison's short visit "gave very little opportunity to talk over matters of great importance to the business." Edison also did not stay for the opening of a convention of representatives of various electric lighting interests on the twenty-fifth.[6]

En route to Cincinnati, the Edison party made a brief stop in Indianapolis (another former residence), where Lillian Gilliland seems to have visited family and friends. Edison planned to stop in Cincinnati for just one day, the twenty-fifth, during which he saw friends and likely stayed at the St. Nicholas Hotel. It was probably at this time that he met Mrs. Gilliland's friend Lillie Fox, to whom he later sent a photographic portrait of himself.[7]

Edison reached New Orleans by the twenty-eighth, when he wired Samuel Insull to send his spring overcoat ahead to Florida (see Doc. 2785). He, Marion, and the Gillilands attended the World Industrial and Cotton Centennial Exposition, where the Edison Company for Isolated Lighting had built a power plant and provided some of the lighting. Edison's telephone inventions were also displayed by the Bell Telephone Company. The following year, the *New Orleans Picayune* erroneously claimed that Edison met his second wife Mina Miller at this exhibition.[8]

Edison left New Orleans and was en route to St. Augustine by 2 March.[9] Express rail service linked New Orleans to Jacksonville (about 550 miles) at the time, but there was no direct line from there to St. Augustine. From Jacksonville, Edison and his entourage would have had to take a ferry across the St. Johns River to South Jacksonville, where they could board a train on the newly built Jacksonville, St. Augustine, and Halifax River Railway for the 36-mile trip to St. Augustine. Alternatively, they could have taken a boat down the St. Johns River and made a short trolley trip due east to St. Augustine. Edison was in Jacksonville by 5 March and reached St. Augustine later the same day, where he was chagrined to learn that the rifles he had asked Insull to send had not yet arrived. The unseasonably cold weather also dimmed prospects for a pleasant hunt in the vicinity of St. Augustine.[10]

After depositing Mrs. Gilliland and Marion at the Hotel San Marco, the two men headed back to Jacksonville, where they could pick up the train for the short ride to Baldwin, nineteen miles west. Here they could transfer to a train headed southwest to Cedar Key on the Gulf Coast. The 107-mile trip west was fraught with difficulty, owing, as Edison later remembered, to the "deplorable condition" of the railroad. The train derailed three times, once occasioning a delay of a day and a half. Another time, there was no telegraph at the nearest station from which to wire for help. "I happened to have with me a pocket telegraph instrument," Edison recalled; "I cut into this station wire and got connection with Jacksonville. They sent on a little train to help us on our way."[11]

Edison and Gilliland spent one night in Cedar Key, likely at the Suwanee Hotel. From there on 14 March, Edison complained to Theodore Whitney about the delay in having the hunting guns forwarded from St. Augustine. The next day, after investigating the operations of the sawmill and lumberyard of the Faber Pencil Co., where guards stood watch over the lumber with Winchester rifles, Edison and Gilliland hired a fishing sloop, the *Jeannette,* to carry them south along the coast. They may have visited Sarasota Bay and almost certainly stopped in Tampa before they finally put in at Punta Rassa, a small port that served the region's cattle trade. Here, cowboys (known as "crackers" for the sound their whips made) herded long-horns into pens to await shipment to Cuba and elsewhere.[12]

Edison and Gilliland stayed at the Shultz Hotel, one of just two such accommodations in town. The Shultz was an abandoned army barracks from the Civil War. The proprietor, George R. Shultz, was an old telegrapher from Newark who came to Florida in 1866 as the chief of the International Ocean Telegraph Company crew charged with constructing the telegraph line from Jacksonville to Punta Rassa (from where it continued undersea to Key West and Havana). Shultz took over the yellow pine barracks and converted it to a rough-hewn hotel for cowboys and sportsmen. He initially did little more than chase out the snakes and rats and plug the bullet holes in the walls to keep out the mosquitos; when Edison visited, the Shultz was still a rickety, unpainted bunkhouse with no indoor plumbing, perched precariously on 14-foot pilings. Its proprietor spent a good deal of time on the veranda smoking cigars and bantering with his guests, providing just the sort of masculine camaraderie that appealed to Edison.[13]

The Shultz Hotel had gained greater notoriety and some better-heeled clientele in 1881 after the New York cartoonist and journalist Walt McDougall spread the word about the good fishing off Punta Rassa and highlighted the out-of-the-way appeal of the hotel itself. But it was only after 1885 that the hotel began to attract rich and famous sportsmen, including President Cleveland. On 19 March of that year, when Edison was still in Punta Rassa, the prominent New York sportsman W. H. Wood, who was lodging at the Shultz, hooked a five-foot, nine-inch, 98-pound tarpon just off the coast. The account of his exploit in *Forest and Stream* set off "tarpon fever" among sports fishermen and made the Shultz Hotel a fashionable destination. Shultz took full advantage, making some improvements in the amenities and even renaming his place "Tarpon House."[14]

It was Shultz who told Edison and Gilliland about Fort Myers, another little cattle town about twelve miles further up the south bank of the Caloosahatchee River. Intrigued perhaps by reports of bamboo growing abundantly at Fort Myers, Edison, Gilliland, and L. A. Smith, an electrician from New York who had joined their party, sailed upriver on the *Jeannette* to investigate. In 1885, Fort Myers had a population of 349. It boasted fifty houses, three general stores, and three lodging houses, including the Keystone Hotel, where Edison stayed. As he later recalled, rooms were 50 cents a day, and meals could be had for 25 cents each. The town also had its own newspaper—*The Fort Myers Press*—founded the year before.[15]

The *Jeannette* arrived about noon on 20 March, and in short order Edison must have inquired about the bamboo and the local fishing at Roan's General Store, where Edward L. Evans was the manager. Evans obligingly took Edison down to Billy's Creek to see the bamboo, and after their return to town introduced him to Clemens L. Huelsenkamp, a local real estate agent. Edison and Gilliland quickly decided that Fort Myers was a suitable place to build their winter homes. "We had everything in our favor," Edison later recalled, "a wonderful climate, beautiful surroundings, plenty of fish and game." In spite of its isolation, Fort Myers, as the home of some of the wealthiest cattlemen in Southwest Florida, also provided most of the modern amenities unavailable elsewhere on the southern Gulf Coast. The next day, Huelsenkamp took Edison to see the thirteen-acre Summerlin tract about a mile down river. Edison liked what he saw and later that day signed a contract to buy the property for $3,000, leaving a deposit of $100.[16]

That same day, Edison and Gilliland left on the *Jeannette* to make the return trip north along the Gulf Coast.[17] It is unclear how they got back to St. Augustine, but they could have disembarked at Tampa and then taken the Plant Investment Company Railroad to Sanford. From there they could have taken a steamboat up the St. Johns River to Palatka, where they could catch the trolley to St. Augustine. The editors have found no information about Edison's stay in St. Augustine or about his trip back to New York, which he reached by 26 March.[18]

1. See Docs. 2234 and 2607 (headnote).

2. Edison initially expected Oscar Madden of the Bell Telephone Co. to be one of his party; see Doc. 2776. Samuel Insull to Charles Fitzgerald, 22 Mar. 1885, Lbk. 20:119C (*TAED* LB020199C).

3. Edison had also paid a brief visit to New Orleans in 1866, between stints in Cincinnati. See *TAEB* 1 chap. 1 introduction.

4. In his testimony in a patent interference case involving the railway telegraph, Ezra Gilliland specifically recalled leaving New York for Adrian on the afternoon of 20 February and being delayed by heavy snow. That memory may not have been accurate, however, for he sent a letter from Boston to Edison on that date. Gilliland's testimony, *Edison and Gilliland v. Phelps*, p. 8 (*TAED* W100DKB); Gilliland to TAE, 20 Feb. 1885, DF (*TAED* D8503P).

5. Years later, Marion recalled that Edison fell ill in Chicago and that she was summoned to join him there. Aside from her description of

Southern itinerary, February–March 1885.

snow piled high in the city, her recollection cannot be reconciled with the contemporaneous record in which there is no mention of an illness. Far from being confined in bed for several days, Edison spent only about a day and a half in Chicago, during which he had dinner with Bliss, wrote a memorandum to John Ott (Doc. 2781), and dispatched a telegram to Insull. Oeser 1956, 9.

6. The convention was called in hopes of organizing a permanent national association of electric lighting interests, but the Edison and Brush electric light representatives reportedly did not participate. Bliss to TAE, 28 Feb. 1885, DF (*TAED* D8536D); "Electric Questions," *Chicago Daily Tribune*, 26 Feb. 1885, 8; "The Electric Light Convention," *Elec. and Elec. Eng.* 4 (Apr. 1885): 155; "Notes," *Teleg. J. and Elec. Rev.* 16 (28 Feb. 1885): 195.

7. "Personal and Social," *Indianapolis Evening Minute*, 26 Feb. 1885, 4; Louise Igoe to Mina Miller, 19 Apr. 1885, EMFP (*TAED* X104OE); TAE to Insull, 2 June 1885, DF (*TAED* D8503ZAT).

8. TAE to Samuel Insull, 28 Feb. 1885, DF (*TAED* D8503T); Israel 1998, 237; *New Orleans Picayune*, 2 Feb. 1886, Cat. 1140:124d, Scraps. (*TAED* SB017124d). The circumstances of Edison's acquaintance with Mina Miller are discussed in Doc. 2824 (headnote).

9. Insull to William Bell & Co., 2 Mar. 1885, LM 5:322 (*TAED* LM005322).

10. TAE to Insull, 28 Feb. and 5 Mar. 1885, all DF (*TAED* D8503T, D8503V, D8502G); Turner 2003, 69; Martin 2010, 103–104; "Florida: The State of Orange-Groves," *Blackwood's Edinburgh Magazine* 138 (Sept. 1885): 325; Grismer 1949, 114.

11. Lillian Gilliland to TAE, 12 Feb. 1929, NjWOE (typed transcript on file at Thomas Edison Papers); "Inventor Now Enjoying His 45th Season Here," *Fort Myers Tropical News*, 25 Apr. 1928, 1.

12. See Doc. 2789; "Inventor Now Enjoying His 45th Season Here," *Fort Myers Tropical News*, 25 Apr. 1928, 1; "Florida: The State of Orange-Groves," *Blackwood's Edinburgh Magazine* 138 (Sept. 1885): 325; Whitney to TAE, 6 Apr. 1885, DF (*TAED* D8503ZAC); Barbour 1884, 150; Board 2006, 34.

13. "Florida: The State of Orange-Groves," *Blackwood's Edinburgh Magazine* 138 (Sept. 1885): 325; "'Twas a Newark Pioneer Discovered Punta Rassa," *Fort Myers Press*, 25 Mar. 1909; McIver 1994, 236; Grismer 1949, 114, 136; Brown 1989, 148–49, 151.

14. McIver 1994, 235–39; Grismer 1949, 136–38.

15. *Blackwood's Magazine* reported that Edison was sitting on the hotel veranda smoking a cigar when he saw a sloop head upriver and inquired where it was going. This apparently elicited the information about Fort Myers from Shultz. "Florida: The State of Orange-Groves," *Blackwood's Edinburgh Magazine* 138 (Sept. 1885): 325; Fritz 1949, 6–7; "Distinguished Arrivals," *Fort Myers Press*, 28 Mar. 1885, News Clippings (*TAED* X104S001A); "Inventor Now Enjoying His 45th Season Here," *Fort Myers Tropical News*, 28 Apr. 1928, 1; Smoot 2004, 17; Grismer 1949, 113, 116, 276.

16. See Doc. 2790; Smoot 2004, 20–21; "Distinguished Arrivals," *Fort Myers Press*, 28 Mar. 1885, News Clippings (*TAED* X104S001A); Henshall 1884, 200.

17. L. A. Smith stayed behind and later reported that he shot a twelve-foot alligator. Smith to TAE, 12 May 1885, DF (*TAED* D8503ZAQ).

18. "Distinguished Arrivals," *Fort Myers Press*, 28 Mar. 1885, News Clippings (*TAED* X104S001A); Smoot 2004, 21–22; "Southern Florida," *Express Gazette* 9 (15 Feb. 1884): 35.

–2776–

To Samuel Insull

New York, [February 11,] 188[5][1a]

Insull—

Do you remember if my gun was returned by Express from St Augustine[2] If not telegh as follows   Theodore Whitney[3] St Augustine   Have you my ~~rifle~~ winchester rifle there if so keep until I arrive   answer paid—[4] also I want to Borrow of you and Johnson[5] your rifles—one is for Gill[6] the other for Madden[7] Secy Bell Co   Will You have them sent to office and also: 40 boxes of Cartridges to fit.[8] we are going to get some fishing rods etc & will have them packed & sent by Express to St Augustine to be called for— Answer if you can fix it

Edison

⟨Theo Whitney has one at St Augustine in his care—⟩[b]

ALS, NjWOE, DF (*TAED* D8503J1). Letterhead of Thomas A. Edison, Central Station, Construction Dept. [a]"New York," and "188" preprinted. [b]Marginalia probably written by John Randolph.

1. This date is conjectured on the basis of Edison's 11 February telegram to Whitney (see note 4).

2. Edison and his wife Mary spent about three weeks in St. Augustine at the end of their Florida vacation in March 1884. Doc. 2607 (headnote).

3. Theodore H. Whitney (b. 1861?), a self-described electrician who had worked at the Menlo Park laboratory, encountered Edison in St. Augustine in March 1884. He and his brother accepted the inventor's commission at that time to search for possible lamp fiber materials along the St. Johns River, borrowing Edison's rifle for their travels. Whitney had recently sent Edison a letter (possibly not yet received) seeking advice about setting up burglar alarms and a telephone exchange to serve St. Augustine and Jacksonville. Doc. 2607 (headnote, esp. n. 21); Whitney to TAE, 9 Feb. 1885, DF (*TAED* D8503J); unidentified obituary in possession of Thomas A. Whitney.

4. Edison's draft was the basis of the telegram sent in his name to Whitney. TAE to Whitney, 11 Feb. 1885, Lbk. 20:89 (*TAED* LB020089).

5. Edward Hibberd Johnson (1846–1917) was Edison's longtime friend, promoter, and trusted business associate. By Edison's arrangement, Johnson took de facto control of the management of the Edison Electric Light Co. in October 1884 and of the Edison Co. for Isolated Lighting a month later. *TAEB* 1–6 passim; see Docs. 272 n. 13, 2690, 2753.

6. Edison meant Ezra Torrance Gilliland (1848–1903), a former telegraph associate and business partner. Gilliland had been a major supplier (through Western Electric) and then a supervisor for the American Bell Telephone Co. He moved to Boston for the Bell Co. in 1883 and set up an experimental shop, soon reorganized as the Mechanical Dept., which he supervised until about this time. In October 1884, he and Edison renewed their friendship at the Franklin Institute's International Electrical Exhibition in Philadelphia. Gilliland encouraged him to work on telephone problems for American Bell, and Edison visited him in Boston in late 1884 and early 1885. At this time Edison and Gilliland were also collaborating on experiments in railway telegraphy. Adams and Butler 1999, 42–43, 48–52; Hoddeson 1981, 521; see Docs. 543 n. 8, 2730 n. 6, 2743 (headnote), 2745 n. 2, 2751, 2765, and 2780.

7. Oscar E. Madden (b. 1846) had worked in the sewing machine business and as a canvassing agent for the Bell Telephone interests in Buffalo. He was working in American Bell's Boston office as supervisor of the company's agents by 1882 and two years later was assistant general manager to Theodore Vail. U.S. Census Bureau 1970 (1880), roll T9_558, p. 238D, image 0785 (Boston, Suffolk, Mass.); U.S. Dept. of State n.d., roll 343, issued 10 Feb. 1890; Tosiello 1979 [1971], 360; Garnet 1985, 27; Madden to Richard Dyer, 20 Feb. 1882; Madden to TAE, 31 July 1884; both DF (*TAED* D8253H, D8403ZFF).

8. Insull shipped two rifles to Edison in care of Whitney at St. Augustine and consigned 500 boxes of cartridges purchased from a New York dealer. The guns reached Edison only after some delay. Insull to Whitney, 27 Feb. 1885, Lbk. 20:141A (*TAED* LB020141A); Vouchers (Laboratory) no. 72 (1885) for Hartley & Graham; see Doc. 2789.

–2777–

To William Lloyd Garrison, Jr.[1]

[New York,] 12th Feby [188]5

Dear Sir,

I did not reply to your letter of 3rd inst[2] because I did not wish to complicate matters as Mr Hastings had undertaken to deal with the matter[3] & further as the only channel of reply would have been through my legal adviser considering the menacing tone of your letter.

I now have your letter of 10th inst[4] & in as much as the matters in dispute have been settled I think it as well to say that I am not by any means satisfied with the course pursued by the Brockton Company & my reason for complying with their demands, as made by you, was not because I was uncertain as to my legal rights but simply because I was indisposed to spend the time necessary to successfully carry on suit.

So far as my personal feelings in the matter are concerned I consider it quite unnecessary to refer to them as the whole dispute is a mere matter of business. Yours truly

Thomas A Edison

LS (letterpress copy), NjWOE, Lbk. 20:96 (*TAED* LB020096). Written by Samuel Insull.

1. William Lloyd Garrison, Jr. (1838–1909) was the founding treasurer of the Edison Electric Illuminating Cos. in both Brockton and Boston, Mass. Garrison was also a director of the Edison Electric Light Co. and a New England agent of that company. Docs. 2436 n. 1 and 2752 n. 4; letterhead of Garrison to TAE, 3 Feb. 1885, DF (*TAED* D8523I).

2. Garrison's letter initiated a new episode in a long-running dispute between Edison and the local Brockton illuminating company over the sufficiency of its generating plant and the company's obligation to pay in full for its construction. The matter had seemingly been settled in October 1883, when the company gave Edison promissory notes to cover most of the debt. Now, however, Garrison stated that Edward Johnson, president of the Edison Electric Light Co., had reneged on promises he allegedly made to pay for upgrading the station's dynamos. Garrison, pointing out that the original construction contract had been made with Edison personally, threatened to dishonor the notes, due in late February, and begin legal action against him if the Brockton company did not receive satisfaction. The notes in question had since been given on Edison's behalf to one of his principal creditors. See Doc. 2734; Garrison to TAE, 3 Feb. 1885; Samuel Insull to Ansonia Brass and Copper Co., 5 Feb. 1885; both DF (*TAED* D8523I, D8523K).

Edison had entered the central station business in 1883 when he formed the Thomas A. Edison Construction Dept. That entity had no legal standing, however, and functioned merely as a contractual surrogate for Edison, who was personally liable for its obligations (see Doc. 2437 [headnote]). Brockton was one of several locales in which Edison either had Construction Dept. debts in dispute or trouble collecting from cash-poor local illuminating companies (see Electric Light—Edison Electric Illuminating Cos.—General, DF [*TAED* D8523], notably correspondence from Archibald Stuart and the Edison Electric illuminating Co. of Hazleton [Pa.]).

3. On behalf of the Edison Co. for Isolated Lighting, which had assumed the business of the Edison Construction Dept. in September 1884 (see Doc. 2725), Frank Hastings telegraphically negotiated a settlement under which the Brockton company would honor its notes and withdraw its threat of litigation, while Edison would exchange the station's two dynamos for larger ones, also paying freight and labor. Edison specifically retained his full legal rights regarding his original contract with the Brockton firm. He later expressed some confusion about the terms to which Hastings had agreed in his name, but eventually a contract was drawn up, presumably along these lines. Although the work was completed in late April, an accord about the final bills was not reached for another month, after a testy exchange, and the bills were not settled for more than a year. Hastings to TAE (with TAE marginalia to Samuel Insull), 10 Feb. 1885; Hastings to Charles Batchelor, 5 Mar. 1885; Garrison to TAE, 22 Apr., 1 and 18 May 1885 (with TAE marginalia on the latter); Garrison to Hastings, 7 May 1886; Garrison to Insull, 18 June 1886; all DF (*TAED* D8523Z, D8523Y, D8523ZAQ, D8523ZAU, D8523ZAX, D8622J, D8633N); Insull to Garrison, 22 June 1886, Lbk. 22:212A (*TAED* LB022212A).

4. Garrison expressed regret for any "personal annoyance & possibly

hard feelings" caused by his demands on the Brockton firm's behalf. However, he defended the company as "long suffering & patient," adding that having encouraged friends to invest in it, he now felt bound by "good faith" to defend their interests. Garrison to TAE, 10 Feb. 1885, DF (*TAED* D8523M).

–2778–

Agreement with Edison Electric Light Co.

[New York, February 13, 1885][a]

THIS INDENTURE, made this 13th[b] day of Feby[b] One thousand eight hundred and eighty-five, between THOMAS A. EDISON of the City of New York, party of the one part, and THE EDISON ELECTRIC LIGHT COMPANY, of the City of New York, of the other part,

WITNESSETH:

WHEREAS, the said Edison Electric Light Company has heretofore granted to said Edison certain right of exploiting the business of electric lighting in South America, Central America and Mexico, under certain conditions and provisions, and for considerations to be observed, performed and paid by said Edison;[1] and

WHEREAS, the parties hereto mutually desire to release each other from the agreements, conditions, provisions and considerations heretofore entered into and agreed upon between them;

NOW THEREFORE, in consideration of the premises and of the sum of One dollar, paid by each of the parties to the other, respectively, the receipt of which is hereby acknowledged, each of them, the said Thomas A. Edison and the Edison Electric Light Company, does hereby, for himself and itself, and for his and its legal representatives and successors, release and absolutely and forever discharge the other of and from all ~~agreements~~ claims[c] and demands, actions, causes of action of every name and nature, so that neither of them shall have any claim on the other directly or indirectly on any contract or thing undertaken, done or omitted to be done, in regard to the matters set forth in the first recital of these presents.

IN WITNESS WHEREOF, the said party of the first part has caused these presents to be executed by its President, and its corporate seal to be hereto affixed, and the said party of the second part has set his hand and seal hereto the day and year first above written.

Thomas A Edison

The Edison Electric Light Co. by
Eugene Crowell   President
Attest   F. S. Hastings   Secty

TDS (carbon copy), NjWOE, Miller (*TAED* HM850240). [a]Date from document; form altered. [b]Handwritten. [c]Interlined above.

1. Edison had become more actively involved in the Mexican and South American lighting business in 1883, while efforts to establish central stations in Europe proceeded slowly. His increased attention to the Americas during the winter of 1883–1884 coincided with the decline of Fabbri & Chauncey, the shipping and commission house that controlled his patents in much of South America. By springtime 1884, the Edison Electric Light Co. had formalized Edison's control of the rights in Mexico and Central and South America, effectively granting him license to make arrangements on a royalty basis. Edison now had promoters working on his behalf in Chile and Argentina. *TAEB* 7, chaps. 1 and 4 introductions; Docs. 2467 n. 32, 2602, 2677, 2688.

–2779–

*Charles Batchelor
Notebook Entry:
Torpedo*

New York, Feb 13 1885

32[1] Proposed <u>Electro Motor</u> for Sims'[2] Fish Torpedo.[3]
Conditions:—

It must not exceed 700 lb weight, nor turn faster than 700 revolutions to give 10 HP. at the propellor[4]

The electrical conditions (which are constant) of the circuit on which the motor must work are:—

Resistance of dynamo $\frac{1}{2}$ ohm
Volts at the dynamo   600
Resistance of the line   9 ohms

Armature:

Diameter of iron core   12.2"
    "   "   "   "   end   12"
    "   "   outside   13"
length of iron core   19" or 480 m.m.
#15 B[ritish]. W[ire]. G[auge] 8 times around
92 Commutator blocks
Total length of wire   3900 feet
    "   resistance   1.93 ohms

Field:[5]

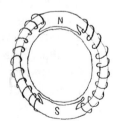

Shape of iron like sketch[6] and 19" long.

Width of pole piece

Iron ⅞ thick 19" long and 17" of room on inner side for
wire from pole piece to pole piece.

#10 734[7] B.W.G. 224 turns on each side.

2300 feet=1.32 ohms. 368 spires[8] on the Armature

448 spires on the magnet.

M.I.[9] should be at 700 turns[10]   .23

X, NjWOE, Batchelor, Cat. 1235:269 (*TAED* MBN012269). Written
by Charles Batchelor.

1. "32" is the entry number. Batchelor consecutively numbered the
entries in this notebook.

2. Newark inventor Winfield Scott Sims (1844–1918), a graduate of
Newark High School and a Union Army veteran, had devised a small
electric motor that he first adapted to operating the screw propeller of
a small boat in 1875, and two years later, he used this same apparatus to
propel a torpedo. Sims's experiments on the electrically propelled torpedo were aided by Moses Farmer, the electrician at the United States
Torpedo Station in Newport, R.I., who supplied a dynamo and gave
Sims advice on the arrangement of the cable connecting the dynamo
to the torpedo motor. Sims officially demonstrated the self-propelled
weapon for the Army Corps of Engineers in 1880, by which time he
was using a motor and dynamo developed by Edward Weston. The tests
were sufficiently promising by the end of 1882 that the Army agreed to
pay for the torpedoes and to request additional funding. The same year,
Sims filed patent applications for the electrically propelled torpedo and
for a device for paying out cables from a coil; he subsequently received
other patents for artillery and ordnance. He also formed the Sims Electrical Fish-Torpedo Co. of New York. *DAB*, s.v. "Sims, Winfield Scott";
Whitehead 1901, 2:177–80; U.S. Pats. 319,633 and 374,209; Bradford
1882, 85–92.

Congress agreed to appropriate $100,000 for torpedo defense in
1884, most of which was designated for the construction of Sims torpedoes to be tested at Willetts Point, N.Y., under the supervision of
Gen. Henry L. Abbot of the Army Corps of Engineers, who had overseen the previous tests. At the same time, Congress requested the Navy
to undertake competitive tests of torpedoes, and notices were sent to
torpedo inventors in the United States and Europe. With his funding
tied up in making torpedoes for the Army tests at Willetts Point, Sims
requested that the results of those trials be accepted by the Navy in lieu
of its own tests in Norfolk, Va. By this time, Sims's torpedo boat came
in both one-mile and two-mile versions that could travel at 10 miles
per hour, though this speed was still considered inadequate. The torpedo consisted of a copper cylinder 23 feet long and 18 inches in diameter constructed in four sections with a capacity for 200 pounds of
explosives. Brass braces secured the torpedo to a copper float that supported it under water. The torpedo was controlled by an operator with
a keyboard who was able to follow its path by watching staffs (lighted at
night) that projected above the water. U.S. Army 1886, 1:50–51; U.S.

Navy 1884, 1:155–52; Hughes 1887, 429; "The Sims Torpedo," *Engineering* 41 (5 Feb. 1886): 139.

Unfortunately for Sims, both the Navy and the Army found his torpedo inadequate. The Navy rejected it due to the insufficient means for keeping the cable clear of the propeller and the difficulty in putting together and handling the torpedo on a boat. The Weston motor also proved unsatisfactory, and Sims apparently acquired a motor from Siemens & Co. in London to use instead. The Army Board of Engineers, as a result of its trials, declined to expend the remaining $50,000 for additional Sims torpedoes. However, Sims was able to negotiate a new agreement with the Board in which he offered to assume all financial risk in exchange for this payment if he delivered five torpedo boats meeting all specifications by 1 July 1886; only three were finished by mid-summer. This new agreement became the basis for Sims's arrangement with the Edison Machine Works. *Report of the Chief of Engineers* 1885, 50; U.S. Navy 1884, 1:532–37, 549–53 (see also 1:237–44, 512–53); Maguire 1884, 187; Sims-Edison Electric Torpedo Co. circular, [1886], PP (*TAED* CA017B); Charles Batchelor to John Tomlinson, 7 Apr. 1886, enclosing draft agreement between Edison Machine Works and Sims [Apr. 1885]; Memoranda on Torpedo-Boats, 26 Aug. 1885; all DF (*TAED* D8627J, D8627K, D8531P); see Doc. 2897; "Highly Destructive Torpedoes," *New York Tribune*, 31 July 1886; "Torpedoes for Uncle Sam," *New York Sun*, 8 Aug. 1886; both Cat. 1057:46b, 46d (*TAED* SM057046b, SM057046d).

3. "Fish torpedo" was a generic term for a crewless self-propelled maritime weapon. An 1885 French report reprinted by the U.S. Navy identified this class of vessel as the equal of armor for importance in naval warfare and the one most in need of improvement (Normand 1885, 157). Two years later, in an article on the state of the art, Lieutenant W. S. Hughes of the U.S. Navy pointed to their role in coastal defense during the Civil War as a primary spur to research in Europe as well as the United States (Hughes 1887). According to Hughes, in Europe the primary focus was on "projectile" types, notably the Whitehead torpedo, that operated under water with no control by the operator. In contrast, efforts in the United States focused on "controllable" types, like the Sims, with guide-rods projecting above the water to enable an operator to maintain visual contact and control the vessel by means of a flexible cable and a steering mechanism. The general subject of torpedoes is well represented by clippings in a Menlo Park laboratory scrapbook (Cat. 1056, Scraps. [*TAED* SM056]; see also "A New Submarine Torpedo," *Mining & Scientific Press*, 11 Sept. 1880, Cat. 1057:26, Scraps. [*TAED* SM057026a]).

4. Batchelor recorded test results of a torpedo motor on 16 May, when it ran at or just under 700 rpm in a circuit of about 600 volts at the dynamo (only 350–390 volts at the motor brushes). The 700-pound motor was tested again a week later, when he reported that under similar electrical conditions it ran at 732 rpms, which "would have been less but sparking necessitated keeping the [commutator] brushes" in the same position. Batchelor noted that "We have decided to alter it to give more commutator blocks and do away with spark" (Cat. 1235:272, 275–76, Batchelor [*TAED* MBN012272, MBN012275]). Batchelor made brief

undated notes and a sketch of armature windings for the torpedo motor in Cat. 1305:5, Batchelor (*TAED* MBN013001A).

5. A circular field magnet would have made for a more compact machine than most motors, including Edison's. (Even the so-called "short core" dynamo design, intended for ships, used magnets about two feet high; see Doc. 2419 [headnote]). A circular design was unusual but not unprecedented; it may have required alternating current (see, e.g., U.S. Pat. 243,264 [Hussey]; "New Electric Motor," *Electrician* 5 [15 Jan. 1881]: 105; and Sprague 1884, 515–16).

6. Figure labels are "N" and "S."

7. Batchelor's notation is unclear but could refer to the number of feet of wire.

8. That is, the turns.

9. Batchelor's notation is unclear.

10. From other entries in this notebook, the editors infer that Batchelor probably meant revolutions.

WIRELESS TELEGRAPHY   Doc. 2780

Doc. 2780 is the first detailed contemporaneous evidence of Edison's interest in "railway signalling"—a system for sending telegraphic messages through the air, without connective wires, to or from a moving train. Such a system would have safety and efficiency benefits for controlling the movements of trains.[1]

Edison evidently became interested in the idea through his friend Ezra Gilliland, but there is considerable uncertainty about how he did so and what he initially contributed to their collaboration. Edison's directive in Doc. 2780 to attorney Richard Dyer to "take patent out for Gilliland" suggests that his friend could rightfully take credit for whatever they had done to that point. For some time, Gilliland had shared with inventor William Smith an ownership stake in an 1881 patent on a system for telephoning without wires from a train (U.S. Pat. 247,127). Gilliland testified in an 1886 patent interference proceeding that he had recommended railway telegraphy as a subject of interest to Edison in September 1884, while they were together at the Philadelphia Electrical Exhibition, but that Edison chose instead to take up long-distance telephony (see Doc. 2743 [headnote]). Gilliland then planned to develop and patent a wireless telegraph on his own, with Edison acting as an adviser, and he claimed to have had financial backing from his friend Frank Toppan. Both Edison and Gilliland testified that they discussed at length the possibility of railway telegraphy based on the Smith patent in December 1884 in Boston and had worked out what Edison called an essentially "complete" system by the end of that month. Their first two patent applications were drawn in Gilliland's name alone but, reportedly at the suggestion of legal counsel, Edison joined the applications before they were filed in March 1885.[2]

Edison offered to the public a more dramatic story in a *North American Review* article published in March 1886. There, he claimed to have identified the principle underlying the new telegraph while investigating new forces with an ordinary induction coil, when he noticed that he could "throw a very strong electric current *fifty feet* through the air, from one conductor to another, by means of a simple primary coil which gave no spark in the air." The *Scientific American* told a similar tale of this "new discovery in physics," but neither published version offered a sense of chronology or indicated whether Edison made these experiments specifically for railway telegraphy. These accounts seem more in accord with Edison's subsequent research interests than with the meager evidence of his work on the topic before February 1885.[3] In any case, Edison and Gilliland were not alone in working on schemes for wireless signaling. In addition to William Smith, inventors Lucius Phelps and Granville Woods and physics professor Amos Dolbear were each separately devising practical apparatus in the United States, while Willoughby Smith, president of the Society of Telegraph-Engineers and Electricians, worked along similar lines in Great Britain.[4]

The Phelps system worked by electromagnetic induction, a process familiar to all telegraphers. A large coil, composed of 1½ miles of wire wound around a long slender tube, was placed in the circuit of telegraph instruments in a rail car and suspended beneath the car. Working the key, the operator would send current pulsing through the coil. Those signals would induce corresponding electrical pulses in a straight conducting wire laid between the rails, and this line would carry the signals to stations along the way. The process could also work in the reverse direction. The Phelps system was demonstrated for officials of the Baltimore & Ohio in April, and it proved to be commercially feasible.[5]

The work of Phelps may have goaded Edison and Gilliland to action. The Phelps Induction Telegraph Company was incorporated on 2 February 1885 to commercialize the inventor's patents. Within days, Gilliland contacted William Smith, his partner in the 1881 telephone patent, and obtained consent to develop his idea with Edison.[6] As Doc. 2780 shows, Edison then moved swiftly to secure patent protection for Gilliland and himself. The two friends also wasted no time forming their own company to develop any system they might devise, agreeing by 7 February to have incorporation papers prepared.[7]

William Smith's system of inductive telephony used a sheet of metal covering the roof of a rail car as one conductor and a dedicated wire strung along telegraph poles as the other. In their adaptation of this system, Edison and Gilliland substituted a telegraph key for the telephone in the transmitter circuit and used ordinary uninsulated telegraph lines (from which strips of metal could be hung to increase their surface area) instead of a dedicated trackside wire (see Doc. 2833). Their first patents also identified two new components that would be crucial to the system. One was an induction coil. Its primary winding was in the telegraph circuit and the high-voltage secondary winding was connected to the metal on the car roof. The other element was a "musical vibrating reed," a common component of acoustic telegraphy that could open and close a switch hundreds of times per second, fast enough to produce an audible hum.[8] Edison and Gilliland placed this switch in circuit with the induction coil, where its action would produce high-voltage pulses at the same frequency in the secondary winding.[9]

It is not clear how Edison and Gilliland initially understood the operation of the system, but the vibrating reed ultimately provided the key for Edison to explain how electric signals could be transmitted over an astonishing distance through the air. Rejecting ordinary electromagnetic induction as the cause, he likened the entire system to a condenser, a device familiar to telegraph engineers at the time. In a condenser, two conducting surfaces are separated by a nonconducting (dielectric) medium. Applying a charge to one will, by electrostatic induction, create an opposite charge in the other (grounded) conductor. The resulting potential difference can be discharged suddenly by connecting the two conductors, typically through a spark gap, producing a brief current at high voltage. By 1886, Edison came to surmise that the railway telegraph system acted like a big condenser: the metal coverings on the cars and the bare telegraph wires served as conductors separated by a dielectric (the air). In the *North American Review* article, Edison explained that the system worked when charged by rapid pulses, such as the current passing through the switch of the vibrating reed:

the air seems to conduct electricity; but if the current were allowed to remain any longer the air would enter into such a state as to oppose any further transmission. If now an interval of time is allowed to elapse the air regains its normal condition and another wave can be transmitted. In sending a single Morse letter, for instance the letter E, which is a single dot, over fifty separate waves with waits between them have to be transmitted, at the rate of six hundreds per second. [Edison 1886, 287][10]

Having made invisible bursts of static electricity "jump," as he put it, hundreds of feet through the air, Edison nicknamed the system the "grasshopper" telegraph. He directed experiments on it through the spring (see Docs. 2786 and 2803) and hoped to apply it over many miles for maritime signaling (see Doc. 2807) but seems to have fallen far short of that goal due to the limits of electrostatic induction.[11] Although Edison was able to explain the railway telegraph in the familiar terms of a condenser, it likely piqued his long-standing curiosity about as-yet undiscovered forms of energy that could be transmitted through the air. He took up related questions about new forms of energy—and the convertibility of old ones—at intervals throughout the year and into 1886.[12]

1. Edison also suggested in 1886 that the system would be eagerly used by reporters, police officers, and business travelers of all kinds (Edison 1886, 290–91; see also Fahie 1971 [1901], 108–9). At some point in the early or mid 1870s, Edison gave at least cursory thought to a system for "The transmission of telegrams through railroad tracks" (Cat. 1214:2, Accts. [*TAED* A244]).

2. John Tomlinson, Edison's primary attorney, corroborated the basic elements of this story in his sworn deposition in the case. Testimony of TAE (pp. 35–36), Gilliland (pp. 3–6, 20–22), Toppan (pp. 25–26), and Tomlinson (pp. 30–32) on Behalf of Edison, *Edison and Gilliland v. Phelps*, MdCpNA (*TAED* W100DKF, W100DKB, W100DKC, W100DKE); Pat. Apps. 350,234 and 486,634.

3. Edison 1886, 285–86; "The Edison System of Railway Telegraphy," *Sci. Am.* 54 (20 Feb. 1886): 119; cf. Docs. 2804 and 2912 (headnotes). Undated drawings from a short series of "Induction Exper[iments]" may be evidence of the antecedent work on new forces that Edison mentioned. Sketched in an unidentified hand, the three drawings show condensers separated by sixty feet, through which Edison apparently hoped to send a charge. The circuit details vary, but what appears to be the transmitting condenser is in circuit with an induction coil in each case; in one, a motor was also indicated, perhaps to run a make-and-break mechanism. Telephone receivers appear to have been used as detectors on the other side. Unbound Notes and Drawings (1886), Lab. (*TAED* NS86ADK images 41–42).

4. Woods had not yet not publicly demonstrated the telegraph system he devised in 1881 and 1882 (Fouché 2003, 33–48). Dolbear, a professor at Tufts College, came up with a wireless telephone in 1882 that he thought worked by discharging earth-currents through the ground. By the time he showed it at the 1884 Philadelphia Exhibition, he was using a telephone to generate pulses in the secondary winding of an induction coil and discharging them through a point into the air (Hawks 1974 [1927], 129–38; Fahie 1971 [1901], 94–103; see also Moyer 1983 [43–45]). John Trowbridge was also conducting similar research at Harvard (Fahie 1971 [1901], 80–91); on W. Smith see Fahie 1900, 162–66 and Abernethy 1887, 388.

5. "Recent Progress in Electricity.—The Phelps System of Telegraphing from a Railway Train While in Motion," *Sci. Am.* 52 (21 Feb. 1885): 118–19; "The Phelps Induction Telegraph," ibid. 52 (2 May 1885): 275.

6. "To Telegraph to Moving Cars," *NYT,* 3 Feb. 1885, 8; William Smith to Gilliland, 7 Feb. 1885, Exhibit for Edison (p. 63), *Edison and Gilliland v. Phelps* (*TAED* W100DKP); Smith to TAE, 18 Feb. 1885, DF (*TAED* D8546D).

7. See Doc. 2782; Testimony of Tomlinson (p. 32) on Behalf of Edison, *Edison and Gilliland v. Phelps,* MdCpNA (*TAED* W100DKF, W100DKB, W100DKE).

8. In both the subsequent maritime version of this system and a later patent for the railway telegraph, Edison substituted a rapidly rotating circuit breaker for the vibrating reed. In the latter part of 1886, Gilliland tried instead a singing telephone transmitter. U.S. Pat. 350,235; see Docs. 2807 and 2989.

9. U.S. Pat. 486,634. Fahie 1900 (108–12) provides a substantial abstract of the corresponding British patent, which was evidently a longer specification with more figures. The *Electrician* published one of the earliest descriptions of the Edison system in its 12 June 1885 issue ("Edison's Latest Invention," *Electrician* [15]: 75–76). Rudd 1887, written by one of Edison's experimental assistants, and Maver 1904 (9–11) briefly outline both the Phelps and Edison systems.

10. Edison seems to have adopted static charge as the explanation for the system as early as June 1885 ("Edison's Latest Invention," *Electrician* 15 [12 June 1885]: 75–76). An explanation of this "new discovery in physics," similar to that in the *North American Review,* appeared in "The Edison System of Railway Telegraphy," *Sci. Am.* 54 (20 Feb. 1886): 119.

Edison described the system in greater or lesser detail in a number of newspaper interviews over the course of the ensuing year. Some of these were collected, along with other news accounts of the invention, by a clipping service and pasted into a scrapbook (Cat. 1140, Scraps. [*TAED* SB017]). One article notable for its clarity and comprehensiveness is "Telegraphing Through Air: How Edison Sent Dispatches From a Moving Railroad Train," *Washington Star,* 6 Feb. 1886, Cat. 1140:116b, Scraps. (*TAED* SB017116b).

11. See Doc. 2807 n. 2. Edison had not yet adopted the "grasshopper" nickname by late April. "The Wizard Edison," *Chicago Daily Tribune,* 30 Apr. 1885, 5.

12. See Docs. 2804 and 2912 (headnotes).

Memorandum to Richard Dyer[1]

ADRIAN, Mich. . . . . [c. February 22,] 188[5][2]

Dyer:[3]

Please take patent out for Gilliland on railway signalling.[4] He is to use a vibrating reed with self make and break. This is worked by a local battery, and runs through the primary of an induction coil. The self make and break is shunted by a condenser to stop sparks and sharpen waves in secondary.[5] The secondary is very high resistance, and gives ordinarily 1 or 2 inch spark. This is connected to strip of tin or galvanized iron on side or top of car. These strips are about foot wide, and run full length of the car. There may be one at top and bottom of side of car as well as one on top. They are thoroughly insulated from the car by glass insulation, the tin strip only touching car where it is on glass insulators.

These strips are all connected together by wires and couplings, which wires pass to the officer in the baggage car. Both sides of car are equipped, so they can go on both tracks.

Thus one or more strips can be used. The strips pass through the secondary, thence through a receiving telephone, which may be a magnet or motogph. chalk telephone,[6] thence to the wheels, where it makes earth with the track. A key with a back point short ckts. the secondary, and no waves go out. When the operator presses key down it makes dot or dash, according to length of depression—a telephone is interpolated. This telephone is of very high resistance and insulation.

The same device is used at the stations, the trains and stations being, so to speak, worked by the condensers in multiple arc. The line wire may be brought by looping right into the station, and there connected to a condenser instead of using a strip on top of station.

Another thing is, I think possible, and that is to use the regular telegraph wires, all the R.R. wires being connected to earth through condensers, and the regular Morse keys shunted by condensers. This will not interfere with the regular Morse working, while it will give increased induction surface on both sides of the track.

You might mention that if a separate special wire is to be used that it can be placed near the track made thus:

a separate flat strip being suspended from the regular wire so as to give increased induction surface, and thus render it unnecessary to place the wire between the track.

Claim the insulated strips on cars, several cars with strips with couplings, also controllable by wires.

The high tension coil, the musical vibrator and local battery. Transmitting by dots and dashes formed of musical waves and receiving on telephone; also specifically by short circuiting the coil.

Several stations and trains with condensers in multiple arc.

Claim use of several teleghic. wires bunched together by means of condensers, and condenser around keys and using whole for train signalling.

Also the special wires with increased surface, as shown.[7]

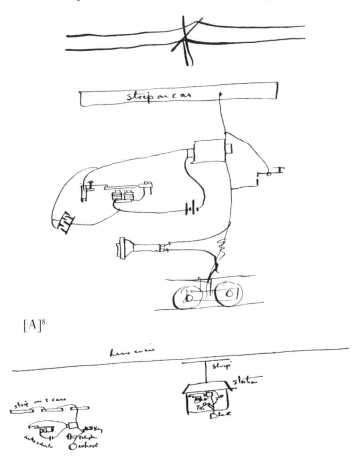

[A][8]

PD (transcript), MdCpNA, RG-241, *Edison and Gilliland v. Phelps* (*TAED* W100DKM). Letterhead of R. Gilliland, manufacturer of telegraph insulator pins and brackets.

1. See headnote above.

2. Edison and Ezra Gilliland were in Adrian to visit Gilliland's father, Robert, who had moved there from Indiana in 1882 (see Doc. 2776 [headnote]; Lindquist 2004, 80). According to Gilliland's later testimony regarding this document, they "were in Adrian only one day and the letter was written and mailed from there on that day." Edison only remembered that it was "about the 22d of February" (Gilliland's and Edison's testimony, *Edison and Gilliland v. Phelps*, pp. 8, 36–37 [*TAED* W100DKB, W100DKF]).

3. Richard Nott Dyer (1858–1914), an 1879 law graduate of Georgetown University, became Edison's principal patent attorney in 1882. He practiced with Henry W. Seely. Docs. 2440 n. 6, 2429 n. 3.

4. Edison's instructions became the basis for two joint patent applications that he and Ezra Gilliland executed on 27 March, following their return from Florida. Both applications were initially prepared in Gilliland's name alone and altered before completion to include Edison. One issued in October 1886 as U.S. Pat. 350,234 and the other in November 1892 as U.S. Pat. 486,634. See headnote above and Doc. 2782 n. 4; Pat. Apps. 350,234 and 486,634.

5. Mathematical physicists at this time, notably G. F. FitzGerald, explicitly modeled ideas of electromagnetic radiation on acoustic waves, but here Edison used the word "waves" to mean pulses of electric current through a circuit. Since the mid-1870s, having conceptualized battery current as a series of rapid pulses, he sometimes used the term in that manner. Hunt 1991, 41–44; cf., e.g., Docs. 409 n. 2, 754, 765, 768–69.

6. In 1874, Edison identified as a "New Force" the variability in friction between various materials under the action of an electric current. He devised a highly sensitive telephone receiver on this electrochemical principle in 1877 and later developed the instrument for commercial use in Great Britain. See Docs. 419, 463, 873, 888–89, 908, 1497, 1681 (headnote), 1784.

7. Figure label is "strip on car." This drawing became the first figure in U.S. Patent 486,634. The major elements are, from the top: metal strips on the car roof; induction coil and transmitter key to its right; vibrating reed; telephone receiver; and car wheels, grounding the system to the rails.

8. Figure labels are "Line wire," "strip on 3 cars," "[Auto vibrate?]," "wheel," "teleph," "Key," "Earth," "Tele," "station," and "strip."

–2781–

To John Ott

Chicago. Feby 23/85[a]

John Ott[1]—

Please tell Martin[2] to take those 30 fluff buttons[3] he made to Boston and go to Gillilands place & put them in telephones. also take some Lampblack there & cloth and show Gillilands man how to make them= ~~Tell Joh~~ when you[b] go over to Lamp factory[4] see Lawson "Basic"[5] & tell him I want to get some telephone Lampblack. Martin knows how to burn it & he can show Basic how to do it= He can start with ½ doz

Lamps= I will pay at the rate of [8?]^c 4 cents per regular Edison button or equivalent weight—^{b} I want [-]^c enough to make 500 buttons (regular Edison) right away that is to say the same weight of loose Lampblack

Edison

ALS, Heitz (*TAED* X225AD). Letterhead of Western Electric Co. ^{a}"Chicago." preprinted. ^{b}"when you" interlined above. ^{c}Canceled.

1. John F. Ott (1850–1931) was an expert machinist employed by Edison from 1870 to 1920. During much of that time he was Edison's principal experimental instrument maker and trusted laboratory assistant. Doc. 623 n. 1; "Ott, John," Pioneers Bio.

2. Martin Force (b. 1848?) began working for Edison in 1876, first as a carpenter and then as an experimental assistant. He helped set up Edison's exhibit at the International Exposition in Paris in 1881 and remained in Europe on behalf of Edison's electric lighting interests. After returning to the U.S. in 1882, he resumed work as an experimental assistant, at least part of the time at Menlo Park. Doc. 2425 n. 3.

3. Edison first devised a so-called fluff button as the key variable-resistance element for his carbon telephone transmitter in 1877. It consisted of a fibrous material (such as silk) saturated with plumbago, a form of graphite, and twisted or packed into a mass. (Lampblack—soot—became his preferred form of carbon.) The electrical resistance of the button varied according to pressure applied on it through the telephone diaphragm. The buttons produced at Menlo Park (from the soot of oil lamps) for commercial telephones were made entirely of packed carbon, without the fabric binding. See *TAEB* chap. 3 introduction, Docs. 997 esp. n. 3, 1016, 1226, and 1652 n. 20.

Edison had recently returned to using fabric saturated with loose carbon, perhaps in an effort to adapt the transmitter for long-distance use. In a patent application executed on 12 January (but not filed until October), he described the button's construction and operation. It

consists of a base of textile or woven fabric . . . whose meshes are filled or impregnated and whose surfaces are covered with lamp-black, plumbago, or carbon in any other suitable form. . . . I prefer to use veiling or other cloth of a similar texture. A flat piece of the material . . . is laid upon a quantity of the powdered carbon, and more of the carbon is then placed upon the flat piece. Pressure is then applied in any suitable manner to the carbon and fabric, and the carbon is thus forced into the meshes or interstices of the fabric and fills the same, so that the fabric is thoroughly impregnated with the carbon and is covered on each side with a layer of carbon. . . . I have found that carbon buttons of this character are more effective in use than those composed wholly of carbon, and I think the reason for this is that at those parts of the button where the carbon lies upon the threads or wires of the fabric the surface is higher than at those parts where it is forced into the meshes between the threads, and therefore the surface of the button is provided with a great number of minute raised contact parts . . . [U.S. Pat. 348,114]

4. Edison meant the plant in East Newark (Harrison), N.J., occupied by the Edison Lamp Co. since May 1882. See Docs. 2089 and 2260 n. 2.

5. Edison gave the nickname "Basic" to John Lawson (1857–1924), a self-taught chemist hired as a laboratory assistant in 1879, after Lawson settled an argument about the basic (non-acidic) properties of a particular oxide. Lawson was detailed to the lamp factory at East Newark in 1881, where he now superintended the carbonizing and electroplating departments. Docs. 1754 n. 1, 1879 n. 1, 2081 n. 4.

6. Edison sold carbon buttons at eight cents apiece in batches of 200 or 250 to the American Bell Telephone Co. during the first half of 1885. He also supplied several smaller lots to Bergmann & Co., to whom he had been selling intermittently for some time. Voucher (Laboratory) nos. 76, 86–87, 149, 204, 312, 325, 367 (1885); Edison Ledger 5:560, 573, Accts. (*TAED* AB003 [images 277, 283]).

Edison's proposed terms suggest that he may have wanted to make Lawson an inside contractor, an arrangement in which an employee had independent authority to hire and manage others for specific production tasks. Inside contracting was a common practice in American manufacturing at this time and one which Edison may have used on occasion at the lamp factory, including a possible prior instance with Lawson. Buttrick 1952; Hounshell 1984, 49–50; Docs. 2057 n. 3, 2081.

–2782–

Agreement with Ezra Gilliland and the Railway Telegraph and Telephone Co.

[New York,] February 23, 1885[1a]

Agreement[b] made this twenty third day of February, Eighteen hundred and eighty five, Between[b] The Railway Telegraph and Telephone Company,[2] a Corporation organized and existing under the laws of the State of New York, party of the first part, and Thomas A. Edison, of the City of New York, and Ezra T. Gilliland, of the City of Boston in the State of Massachusetts, parties of the second part.[c]

Whereas[b] the party of the first part has been organized with a view of acquiring, and exploiting when acquired, a complete system of telegraphing and telephoning between railway trains in motion and between moving trains and places and stations along their route, and is the owner of Letters Patent of the United States No 247 127, dated September 13th, 1881, and issued to William Wiley Smith,[3] of Indianapolis, in the State of Indiana, and Ezra T. Gilliland, then of the same place for an "Improvement in Car Telegraphs," and[c]

Whereas,[b] the parties of the second part have jointly made two certain inventions for which they have filed applications for Letters Patent of the United States, said applications being known respectively as Application No. 161 438, filed April 7, 1885, for an Improvement in systems of Railway Signaling, and Application No 161 437, filed April 7, 1885,[4] also for an Improvement in Railway Signaling, which said Inventions and the Letters Patent to be granted thereon, the said Company are desirous of acquiring, together with the right to use for the purposes aforesaid, any Improvements thereon or new Inventions which the parties of the second part, or either of them, may jointly or severally make within one year from the date hereof.[c]

Now it is agreed[b] as follows:—[c]

First:[b] The parties of the second part for and in consideration of the sum of One dollar to them in hand paid and of the issuance to them of Five thousand shares of the fully paid capital stock of the party of the first part to which they have subscribed at par and in full satisfaction thereof hereby agree to transfer, assign and set over, and by these presents do transfer, assign and set over to the said The Railway Telegraph and Telephone Company its successors and assigns, all their right, title, and interest in and to the application for Letters Patent aforesaid, Numbered respectively 161 438, and 161 437 and the Inventions thereon described and the Letters Patent to be granted thereon.[c]

Provided,[b] however that the said Company shall, and it hereby agrees, whenver[d] requested by the parties of the second part so to do, to grant to them, their heirs, administrators and assigns an exclusive license during the life of the said Letters Patent, and each of them, or any re-issues and extensions thereof, to manufacture, own or use, or license others to manufacture, own or use, the Inventions in said Applications described for all purposes other than Telegraphing or Telephoning between moving trains or between such trains and stations.[c]

Second:[b] The said parties of the second part, in consideration of the aforesaid, jointly and severally agree to and with the said Company that for One year from the date hereof they will devote a reasonable portion of their time, reference being had to their other occupations, to the conduct of investigations and the making of experiments looking toward the Invention and Perfection of a complete system of Telegraphy and Telephony between moving trains and between such trains and stations, and as inventions are made by them, or either of them, relating in any way thereto, they will promptly advise the Company thereof and will promptly prepare or cause to be prepared, applications for Letters Patent thereon, doing each and every act and thing necessary to be done to secure the issuance of such Letters Patent; all costs, charges and expenses incurred in obtaining the same to be borne by the said Company.[c]

Third:[b] In case any Invention as aforesaid, shall be applicable solely to the purposes for which said Company is formed, the said parties of the second part will assign the Invention and the Letters Patent granted thereon to the said Company, and in case any such Invention shall be applicable to purposes other than Telegraphing and Telephoning between moving trains and between trains in motion and stations and places along their route, then and in that case, the parties of the second part shall grant and convey to the said Company, its successors, administrators and assigns an exclusive license, during the term or terms of any such patent or patents and any extension or re-issue thereof, to manufacture, own and use, and to license others to manufacture, own and use, the Invention or Inventions described or claimed in such patent or patents so far as the same may be applicable to the purposes aforesaid; such percentage of the costs and expenses of obtaining such patents to be borne by the Company as the value of the license bears to the value of the patent. It being expressly understood that any invention made subsequent to one year from the date hereof shall be and remain the exclusive property of the parties of the first and second parts, as the case may be and shall be in no way effected hereby.[c]

Fourth: Whereas,[b] the parties of the second part are inventors by profession and engaged upon investigations and experiments relating to various kinds of instruments, apparatus, devices and methods used and employed, and to be used and employed in Telegraphy and Telephony generally and which, as improvements upon instruments, apparatus, devices and methods now used and employed therein, might and probably would be useful and valuable in a system of telegraphy and Telephony between moving trains and between such trains and stations.[c]

Now,[b] in order that no misunderstanding may arise as to the Inventions, Letters Patent, or Licenses thereunder assignable or to be granted to the party of the first part under this agreement, It is expressly undersigned, Provided and agreed that Inventions which are useful and valuable in Telegraphing and Telephoning between trains and stations as aforesaid only in so far as they may be Improvements upon instruments, apparatus, devices and methods now in use, shall be in no way affected by this Agreement, but shall be and remain the sole and exclusive property of the said parties of the second part, the only Inventions, Patents and Licenses coming under this Agreement being such as relate specifically to the solution of the problem and the accomplishment, practically, of Telegraphing and Telephoning between trains and stations as aforesaid as distinguished from Inventions chiefly applicable to Telegraphy and Telephony generally.

Fifth:[b] The parties of the second part expressly reserve to themselves the ownership of and the right to obtain Letters Patent in Foreign Countries on all Inventions covered by this Agreement; but the parties of the second part Hereby agree that they will deal in no way with such Foreign Patents as to shorten the life of any domestic patent.[c]

In Witness Whereof,[b] the party of the first part, by its officers, duly authorized therefor, has set its hand and corporate seal and the parties of the second part have set their hands and seals the day and year first above mentioned.[c]

Thomas A Edison[f]	Railway Telegraph and Telephone Company
Ezra T. Gilliland[f]	by John C. Tomlinson President
Witness to signatures of	
Edison & Gilliland	Attest Saml Insull Secretary
John F. Randolph[5]	

DS, NjWOE, Miller (*TAED* HM850243). [a]Date from document; form altered. [b]Multiply underlined. [c]Followed by dividing mark to right margin. [d]"n" interlined above. [e]Followed by dividing marks. [f]Followed by seal.

1. The contract was not signed on this date. Edison and Ezra Gilliland were in Adrian, Mich., and Chicago (see Doc. 2776 [headnote]). The agreement also refers specifically to the filing on 7 April of two patent applications that were completed in late March. It likely was signed in April, after Edison's return, and backdated close to either an informal agreement in principle or the day when it likely was drafted by John Tomlinson.

2. Papers to incorporate the Railway Telegraph and Telephone Co. were filed on 18 February on behalf of Edison, Sigmund Bergmann, and Charles Batchelor. Organized for the general purpose of developing inventions for electrical communication between moving trains and between trains and stations, its capital stock was listed as $1,000,000. Its original trustees consisted of the three incorporators, Ezra Gilliland, John Tomlinson, Samuel Insull, and William Wiley Smith ("Two New Corporations," *NYT*, 19 Feb. 1885, 8). An undated draft of its bylaws, evidently based on those of the Edison Machine Works, is in DF (*TAED* D8129ZAL). Attorney John Tomlinson later testified that between December 1884 and February, he had often discussed with Edison and Gilliland the idea of forming a company but had urged them to wait until they were confident of the success of their inventions. He claimed that the decision to incorporate was made prior to 7 Febru-

ary (Tomlinson's testimony, *Edison and Gilliland v. Phelps*, p. 32 [*TAED* W100DKE]). In November, Edison incorporated the International Railway Telegraph and Telephone Co. to work outside the United States and Canada. The other incorporators reportedly were Gilliland, Insull, Tomlinson, Charles Batchelor, John Kruesi, and John Randolph ("New Telegraphic Enterprise," *New York Advertiser*, 7 Nov. 1885, Cat. 1140:13b, Scraps. [*TAED* SB017013b]).

3. William Wiley Smith (1837–1894) worked at this time in the management of the United Telephone Co. in Kansas City, where he later became secretary and manager of the Missouri and Kansas Telephone Co. He had recently been in the telephone business in Indianapolis. Earlier, he had worked as a telegraph operator for Western Union and as a railroad telegraph superintendent. Obituary, *Electrical Engineer* 19 (9 Jan. 1895): 39; "Would Like Competition," *Kansas City Times*, 9 Jan. 1895, 5; "Funerals," ibid., 18 Apr. 1895, 3; online interment record #57804 of Spring Grove Cemetery (Cincinnati), http://www.springgrove.org/sg/genealogy/sg_genealogy_home.shtm, accessed 15 Dec. 2010; Smith to TAE, 18 Feb. 1885, DF (*TAED* D8546D); "Correspondence. Chicago," *Electrical Engineer* 7 (Apr. 1888): 186; "Important to Telegraph Managers," *The Telegrapher* 7 (26 Nov. 1870): 107; Taltavall 1893, 284–85.

In 1881, Smith obtained a patent for using induction between a metallic covering on a railroad car roof and a dedicated overhead wire to transmit spoken telephonic communication between trains or between a train and a station. (He reportedly came up with the idea after noticing a similar effect during a musical telephone demonstration in Indianapolis in 1880.) Smith immediately assigned a one-half interest in the patent to Gilliland. On 18 February 1885, he assigned his remaining half to Edison in a deal worked out by Gilliland. Smith's specification was brief but broad, and Edison and subsequent commentators identified it as the first for a workable system of induction telegraphy. "The Induction Telegraph," *Electrical World* 7 (3 Apr. 1886): 152–53; U.S. Pat. 247,127; Smith to TAE, 18 Feb. 1885, DF (*TAED* D8546D); Edison 1886, 285; "The Edison System of Railway Telegraphy," *Sci. Am.* 54 (20 Feb. 1886): 119; "Induction Telegraphy," *English Mechanic* 46 (4 Nov. 1887): 221.

4. Edison and Gilliland signed the applications on 27 March. The one identified by Patent Office serial number 161,437 was considerably broader than the other. It covered a complete system for sending and receiving telegraphic signals through the air on the principle of static induction, using trackside wires to collect and transmit signals to Morse receiving instruments at a station. This application was substantially revised at the Patent Office, in part because of four separate interference proceedings with applications of Lucius Phelps, and it did not issue until 1892 as U.S. Patent 486,634. The other case (serial number 161,438) concerned specific wiring arrangements, notably placing condensers in circuit with Morse telegraphic senders or with the telephones used as receivers. Noting that "The numerous wires running along trunk railway lines would give a large inductive surface, which is a point of great advantage," it also described the connection of multiple wires, each through its own condenser, to a single receiving instrument at each station. That application was initially rejected on the basis of prior patents to Phelps and Smith; after revisions, it was placed in interference with

an application from William King. It issued in October 1886 as U.S. Patent 350,234. Pat. Apps. 350,234 and 486,634.

5. John Randolph (1863–1908) began working for Edison as an office assistant in 1878. He was now keeping Edison's financial records and assisting with correspondence. Doc. 2647 n. 3.

–2783–

To Samuel Insull[1]

Chicago   feb 24 1885[a]

Samuel Insull
    Will be at St Nicholas Hotel[2] Cincinnati until Wednesday evening

Edison

L (telegram), NjWOE, DF (*TAED* D8503R). Message form of Western Union Telegraph Co. [a]"1885" preprinted.

1. See Doc. 2776 (headnote).
2. A small Cincinnati hotel in operation since about 1861 (and renovated in the 1870s), the St. Nicholas offered only twenty rooms but was distinguished by its restaurant, praised as the "Delmonico of the West." Kenny 1879, 59.

–2784–

From Samuel Insull

New York, February 26th. 1885.

My Dear Edison:—[1]
    After you left town I was very sick indeed and was absolutely unable to attend to any business until the day before yesterday. My Doctor was very much afraid that I would have pneumonia, but I have got through the most serious part of my severe cold and shall doubtless be all right in a few days, if I take decent care of myself.
    STOCK PRINTER.[2] As soon as I was able to get down town I called on Mr. David Seligman,[3] and he told me that owing to the holidays intervening practically nothing had been done with relation to the stock printer since you left. He expected, however, to get matters in shape in the course of a few days, and I do not doubt but what business will result. A few days before you left you told me there was a great deal I might do in connection with the stock printer in the way of preliminary work, so as to be able to push the matter actively immediately you returned. Will you please send me a memorandum of the various matters that I can look up! Do you think it advisable for Kenny[4] to go ahead and experiment with a seven dot letter? If so, will you please let me have your ideas on the subject, so that I can Kenny, or do you want any experiments conducted

with a view to improving the shape of the letters and figures of the fine line printer?

I am, through Russell, operating on the Police and Fire Departments, with a view to getting them to consider a bid for putting their wires underground.[5] I expect to go over to Philadelphia to see Mr. Bentley[6] early next week, with a view to getting him to consider the matter of putting his wires underground. Kruesi[7] is of course unable to do anything in the way of laying at the present, as we are having pretty severe weather here. In fact the earth at the moment is covered with quite a large amount of snow, and the frost is entirely too severe to allow of his starting his work. His tubes, however, are practically finished. To-day he got about three parts of the copper for his large Milan order.[8]

AUTOMATIC. I understand that Bonanza Mackey[9] leaves Paris for New York about the last day of this month. Can we not bring to his notice the Automatic, and with this end in view have you any suggestions to make as to Kenny conducting some experiments, with the object of getting things in good shape. I understand that Mackey will favorably entertain some deal about the stock printer should the Seligmans retire, and I think it would be a really good thing if, at the same time, we could make some arrangement with him, with a view of putting your own letter Automatic on the lines of the Postal Telegraph Co.[10]

RAILROAD TELEGRAPHS. I enclose you herewith copy of a report which I have got from Bradstreet's[11] on the "Phelps Induction Telegraph Co.,"[12] of which Cheever[13] is the main promoter. From this you will see that Cheever's Company has not raised any large amount of money so far. In fact they have simply put a stock in the hands of Trustees to sell, in order to provide them with capital, and you may depend upon it that their present exhibitions are made with a view to "boosting" up their stock and so enable them to dispose of it. I received a letter a few days ago, addressed to you from Cheever, drawing your attention to the fact that "The Railroad Telegraph Co." (which I imagine is one and the same thing as the Phelps Induction Telegraph Co.)[14] was incorporated in August, 1884, and Cheever asks that you will have the name of our Company, which is "The Railroad Telegraph & Telephone Co.," altered, as the two names are so much alike. Tomlinson says that, as a matter of law, there is no reason why we should alter our name, but as a matter of courtesy he presumes we should do so. Will you please let me have your ideas on this subject.

Will you also let me have some suggestions as to the methods I had better pursue, with a view to drawing public attention to our inventions. If you were here I should propose doing this by means of an interview with yourself, but your being absent makes the matter rather difficult to deal with. I presume our object should be to so draw attention to our Company, as to prevent Cheever from selling the stock of his Company, and thus compel him to make a deal with us, but I am in somewhat of a quandary to know what course I should pursue with a view to arriving at this desirable result.

ORE MILLING. You may remember that some time back the people backing Connolly[15] in his experiments opened some negotiations for a deal, by which they could get control of the ore separator, so far as the separating of iron sand is concerned. They have come to me within the last day or two, and the enclosed letter from them (William Bell & Co.)[16] is the result of my conversations with them. You will see that this proposition guarantees a minium income to the Ore Milling Co[17] of $1,000. for 1885, $2,500. for 1886, $3,500. for 1887 and $5,000. for 1888. You must remember that this income is only minimum, and that Bell & Co. anticipate separating quite a good deal of sand, and in fact pushing the business for all it is worth. They start on 25 cents per ton royalty and end up with 10 cents per ton, so that in order for us to get above $5,000. in the fourth year they have got to turn out upwards of 50,000 tons of pure iron sand. I think that the price which they propose depositing for each machine[18] ($400.) is amply sufficient. In your first negotiations with them you mentioned $500. I must have your opinion on this matter immediately. I want you to read Bell & Co.'s letter very carefully, and to write me fully on the subject, as should you be agreeable to their proposition I am going to the Cuttings'[19] and get them to agree and close up a deal immediately with these people. They mean business. They think they have a splendid thing in Connolly's process for making the sand into blocks, and they are going to push the matter for all it is worth. This matter not only means business to the Ore Milling Co., and a certain income to cover their current expenses, pending your producing results in connection with the working of precious ores, but it also means work for the Machine Works, as the Ore Milling Co. have not only to get ore separators from the Machine Works, but Bell & Co. propose buying our dynamos. I want to know what size dynamo will charge two separators, and also what [---][a] size will be used for charging three, four and six separators respectively. I have got to quote prices for dynamos to Bell & Co. to use in connection with these separators, providing a deal is made with them. Should shop prices be quoted? I notice in your previous negotiations with them that no profit beyond the Machine Works profit was put on the dynamos. One thing I like about this [---------]ᵃ proposed contract is that unless we get our money each year the contract is void. If we believe in the business of working the iron sands, providing the royalty is sufficient, I do not think that we can very well object to making the contract, as Bell & Co. will push the business, in connection with Connolly's new patent[20] (which I understand is a very good one), and if they make money out of it we are bound to. They propose first of all working on the Rhode Island Coast[21b] with a view to getting all their apparatus in perfect order, but immediately they close an arrangement with us they propose getting the control of all the Canadian deposits of sand, and later on they will get control of that on the Pacific Coast.

You will notice that Bell & Co. only propose to deal with the United States and Canada, but in talking to me they stated that should their home venture prove successful they propose going further afield and working the deposits in New Zealand and Norway and Sweden. If you decide to make the arrangement with them, which they propose and which is decidedly better, so far as policy is concerned, than the first proposition they made me, would it not be well to see if we cannot manage to get some kind of a patent in New Zealand and Norway and Sweden, with a view to being in a good position should Bell & Co. eventually decide to work in these countries. I cannot impress upon you too strongly that this matter requires immediate action, and I trust you will answer the various points in my letter fully, and immediately you get this.

LIGHT BUSINESS. The Isolated business is just as dull as when you left. It shows no improvement at all. The test took place at the "Chelsea" Flats[22] on the United States plant,[23] and they were not able to get more than from five to six lamps per horse power, notwithstanding that the U.S. people had guaranteed ten. What will be done in the matter is not known yet. I understand that exactly the same results were obtained from a plant in Philadelphia installed by the U.S. Co.

Batchelor is still negotiating with Coster[24] as to your various claims, but as Batchelor will write you about this himself, I do not consider it necessary to go into the matter.

I was round at 18th. Street yesterday,[25] and I understand

that Tommy[26] is very much better, and that he will be home in about a week.[27]

Hoping you are having a good time and benefiting from the vacation, I remain, Very sincerely yours,

Samuel Insull

TLS (carbon copy), NjWOE, DF (*TAED* D8503S). [a]Canceled. [b]Obscured overwritten text.

1. Insull evidently sent this letter to New Orleans. Realizing that he had missed Edison there, he sent a copy ahead to St. Augustine. Insull to William Bell & Co., 2 Mar. 1885, LM 5:322 (*TAED* LM005322).

2. Edison had worked extensively in the early 1870s on elements of automatic telegraphy, among them a roman-letter printer that used an array of special iron pens to reproduce alphabetic characters on chemically treated paper. It did so in response to signals produced by feeding paper strips with perforated outlines of the letters through a transmitting instrument (see, e.g., Docs. 183–84, 349 [headnote], 373–74, 394, 397, 482, 566, and 571). The speed and immediate legibility of such a process had obvious appeal to brokers and traders. Edison and Patrick Kenny had been working on what they called a chemical stock quotation instrument since 1882 and filed a patent application in March 1884 for an automatic system, including a five-pen printer, that would "supplant the present stock-printing telegraphs . . . [and] operate correctly and accurately, with much greater rapidity . . . and will have advantages of greater simplicity, durability, ease of repair, and a complete independence of the receivers" (U.S. Pat. 314,115; see also Doc. 2331). After November, when the patent was allowed by the Patent Office, but before March 1885, when it issued, Edison and Kenny drafted an agreement governing their future efforts to license the patent. They also drafted a power of attorney for Samuel Insull to negotiate and make arrangements on their behalf (draft agreement with Kenny, n.d. [Feb. 1885?]; TAE and Kenny draft power of attorney to Insull, n.d. [Feb. 1885?]; both DF [*TAED* D8546H, HM850248]).

3. David Joseph Seligman (1850–1897) was a partner of J. & W. Seligman & Co., the international investment bank founded in 1862 by his father, Joseph S. Seligman (1819–1880), and located in the Mills Building at Broad St. and Exchange Pl. in Manhattan ("Obituary," *NYT,* 28 Aug. 1897, 1; Muir and White 1964, 79–80, 88; *ANB,* s.v. "Seligman, Joseph"). The firm had recently been granted a sixty-day exclusive option for the commercial and corporate development of Edison's improved chemical telegraph, a device for stock quotations and financial news. Insull expected to receive a more definite sense of Seligman's plans in a few days (TAE agreement with Seligman & Co., 20 Feb. 1885; Insull to Tomlinson, 3 Mar. 1885; both DF (*TAED* D8546I, D8546E).

4. Patrick Kenny (b. 1830?) was a former superintendent of the Gold and Stock Telegraph Co.'s manufacturing shops. He lived in New York and likely was a machinist by trade. Kenny had worked intermittently with Edison on telegraph instruments since 1878. See Docs. 338 n. 21, 1328, 1388 n. 6, and 1638; U.S. Census Bureau 1970 (1880), roll T9_880, p. 103D, image 0148 (New York City, New York, N.Y.).

5. In April, the New York Police Dept. accepted an invitation to observe underground wires being buried, under Edison's auspices, for the Metropolitan Telephone and Telegraph Co. Insull had estimates prepared for both the police and fire departments the following month. TAE to New York City Police Commissioners, n.d. [Apr. 1885?]; New York City Police Dept. to TAE, 24 Apr. 1885; John Langston to Insull, 18 May 1885; all DF (*TAED* D8546N, D8546M, D8533M).

6. Henry Bentley (1831–1895) was the founding president of the Philadelphia Local Telegraph Co., established about 1873. A friendly colleague since at least 1878, Bentley had responded favorably to Edison's 1884 request to set up a dynamo to run telegraph instruments. Insull had recently written to him in Edison's name offering to arrange a deal for the Electric Tube Co. to put the Philadelphia company's wires underground, as it reportedly was doing in New York for the Metropolitan Telephone and Telegraph Co. See Doc. 2651 esp. n. 1.

7. A highly skilled machinist, John Kruesi (1843–1899) had worked for Edison for many years and supervised the Menlo Park laboratory's machine shop. Kruesi had been secretary and manager of the Electric Tube Co. since its formation in 1881, and he had taken general charge of installing underground conductors for the Edison Construction Dept.'s central stations in 1883–84. Docs. 659 n. 6, 2125 n. 3, and 2543 n. 3.

8. John Kruesi had expected to receive about 7,000 pounds of copper the previous day, about half the amount ordered from the Ansonia Brass & Copper Co. for underground conductors to extend the service area of the Edison central station in Milan, Italy. Powered by the first permanent Edison central station in Europe (opened in June 1883), the Milan district had some 5,000 lamps in service. In 1885, it was expanded along the Corso Vittorio Emanuele and the number of incandescent lamps increased to 7,921, plus 101 arc lights. Kruesi to Insull, 25 Feb. 1885, DF (*TAED* D8533E); Insull to Alfred Cowles, 9 Mar. 1885, Lbk. 20:151C (*TAED* LB020151C); see Doc. 2481; "Usine Centrale pour l'Éclairage Électrique à Milan," *La Lumière Électrique* 19 (6 Feb. 1886): 242–45.

9. John William Mackay (1831–1902) made a fortune in 1873 in the Consolidated Virginia mine, the so-called "Big Bonanza" of Nevada's Comstock Lode. (Edison consulted with the mine's operators in 1881; see Doc. 2199.) With publisher James Gordon Bennett, Mackay started the Commercial Cable Co. in 1881 to build a transatlantic cable to rival the American Union Telegraph Co.'s line (leased to Western Union); the Commercial cable entered service in December 1884. Mackay also owned a controlling interest in the Postal Telegraph Co., which in 1884 formed a working alliance with two other firms in opposition to Western Union for domestic traffic. Mackay and his wife lived largely in Paris and London. Although they did return to the United States with some frequency, Insull may have been mistaken about a trip at the end of January because Mackay's daughter was married in Paris in mid-February. *DAB*, s.v. "Mackay, John William"; Coe 2003 [1993], 90–92; "Triple Telegraphic Alliance," *Washington Post*, 18 July 1884, 2; "Messages over the New Cable," *NYT*, 13 Dec. 1884, 8; "Miss Mackay Married," ibid., 13 Feb. 1885, 1.

10. The Postal Telegraph Co. was organized in 1881 to apply Elisha Gray's patents in harmonic telegraphy. John Mackay used his personal fortune to acquire a controlling interest in 1883 so that the Postal Co.,

rather than Western Union, could provide domestic service in the United States in conjunction with his Atlantic cable venture. Coe 2003 [1993], 91.

11. Bradstreet & Sons, one of the principal credit-reporting services, had been headquartered in New York since 1885, when founder John M. Bradstreet moved the business from Cincinnati. The agency published reports on credit worthiness for its subscribers in several forms, including reference books, pocket books, weekly circulars, and detailed compilations on specific businesses. Edison subscribed to the service until July 1885 (Olegario 2006, 65, 157–64; TAE to Bradstreet Co., 11 Mar. 1885, Lbk. 20:159 [*TAED* LB020159]). The copy to which Insull referred in this document is probably of a 21 February report on the organization and leadership of the Phelps Co. (DF [*TAED* D8546F]).

12. The Phelps Induction Telegraph Co. was incorporated about 2 February 1885 to control the patents of electrical inventor Lucius J. Phelps. Phelps had devised a system of railroad telegraphy in 1883 and filed patent applications for it by this date (Richard Dyer to Insull, 9 Feb. 1885; Bradstreet[?] report on Phelps Induction Telegraph Co., 21 Feb. 1885; both DF [*TAED* D8546C, D8546F]; "Railroad Telegraphing," *Electrician* [N.Y.] 2 (Oct. 1883): 304; *Decisions of the Commissioner* 1887; U.S. Pat. 312,506). The Phelps system, described and illustrated in the 21 February *Scientific American,* was currently being tested on a twelve-mile branch of the New York, New Haven, and Hartford Railroad. The Phelps company merged with the Railway Telegraph and Telephone Co. in April 1887 to form the Consolidated Railway Telegraph Co. (Phelps Induction Telegraph Co. circular, n.d. [Feb. 1885]; Consolidated Railway Telegraph Co. to TAE, 28 Apr. 1887; both DF [*TAED* D8546G, D8752AAG]; "Recent Progress in Electricity: The Phelps System of Telegraphing from a Railway Train While in Motion," *Sci. Am.* 52 [21 Feb. 1885]: 118–19).

13. Charles A. Cheever (b. 1854?), a self-described "promoter of enterprises," came from a prominent New York family. His father, John H. Cheever, was associated with the New York Belting and Packing Co. (among other businesses), whose financial difficulties in 1878 had strained the family's resources. Charles, who had connections with the Bell telephone interests, organized the Telephone Co. of New York with Hilborne Roosevelt in 1877 and was one of the founders of the Edison Speaking Phonograph Co. in 1878. U.S. Census Bureau 1970 (1880), roll T9_875, p. 259A, image 0162 (New York City, New York, N.Y.); Cheever's statement in [Bradstreet?] report on Phelps Induction Telegraph Co., 21 Feb. 1885, DF (*TAED* D8546F); Crowell 1918, 14–15; see Docs. 1173 n. 11 and 1190.

Cheever had recently invited Edison to witness a demonstration of the Phelps system. Edison did not attend but used the opportunity to advise Cheever of his own interest in induction telegraphy. TAE to Cheever, 14 Feb. 1885, Lbk. 20:107 (*TAED* LB020107).

14. The editors have not found Cheever's letter, but Insull acknowledged receipt of one dated 21 February from Cheever to Edison (Insull to Cheever, 25 Feb. 1885, Lbk. 20:137A [*TAED* LB020137A]). In 1887, the Railway Telegraph Co. merged with the International Railway Telegraph Co. to become the Consolidated International Railway Telegraph Co. (Railway Telegraph Co. agreement with International

Railway Telegraph Co., 3 Mar. 1887; separate certificates of proceedings, both 7 Apr. 1887; all NNNCC-Ar [*TAED* X119F2A, X119F2, X119F2C]).

15. Michael R. Conley (b. 1843?) had been superintendent of the Edison Ore Milling Co.'s operations at Quonocontaug Beach, R.I., since November 1881. He later identified himself for census records as an electrical engineer, but had in the meantime obtained several patents on furnaces and processes for iron ore. Doc. 2591 n. 6.

16. William Bell and Co., a commission and export house at 57 Broadway, was a partnership of William Bell and C. Seton Lindsay. William Bell was also president of the Ocean Magnetic Iron Co. and an associate of Michael Conley (see Doc. 2591 n. 6). The firm's proposed terms, summarized in this document, were laid out in a 24 February letter to Edison. Edison gave his consent in March, as did the Edison Ore Milling Co. Edison delayed signing the contract, however, and did so only in December, after making a minor modification. *Anuario* 1885, 457; *Wilson's New York City Co-Partnership Directory* 1879, 27:37; Bell & Co. to TAE, 24 Feb. 1885, DF (*TAED* D8543A); Insull to Bell & Co., 16 Mar. and 10 Dec. 1885, LM 5:323, 330 (*TAED* LM005323, LM005330); Insull to Bell & Co., 5 Nov. 1885, Lbk. 21:78 (*TAED* LB021078).

17. The Edison Ore Milling Co. was organized in December 1879 to exploit Edison's inventions for concentrating ferrous and precious metal ores by electromechanical processes. Its business had come to a standstill, and Insull was trying to exempt it from taxes on the grounds that it had no assets, cash, or tradable stock but was simply "an experimental Co." Sherburne Eaton was its general manager. Docs. 1844 n. 5, 2393, and 2591; Edison Ore Milling Co. association papers, 9 Dec. 1879; Sherburne Eaton report to Edison Ore Milling Co. stockholders, 20 Jan. 1885; both CR (*TAED* CG001AAD, CG001AAI5); TAE to John Tomlinson, 25 Feb. 1885, DF (*TAED* D8543B).

18. Edison's ore separator worked on the general principle of using powerful electromagnets to deflect the trajectory of ferrous particles in a stream of falling sand or other fine material. A commercial-scale prototype was working on Long Island and another in Rhode Island by mid-1881. An improved version was processing some sixty tons of Rhode Island sand each day by early 1882. See Docs. 1921, 1950 n. 7, 2093, 2175 n. 3, and 2246.

19. Robert Livingston Cutting, Jr. (1836–1894), part of a socially prominent New York family, bore the same name as his son (1868–1910) and his father (1812–1887). Cutting was treasurer and a major shareholder in the Edison Ore Milling Co. and was involved with the Edison Electric Light Co. and other Edison companies. The two younger Cuttings were in the brokerage business of R. L. Cutting, Jr. & Co. at 19 William St. in New York, and they shared a residence at 141 Fifth Ave. Doc. 2724 n. 1; Edison Ore Milling Co. list of stockholders, 23 Oct. 1883; Cutting, Jr., to TAE, 4 June 1886; both DF (*TAED* D8368ZAA1, D8603ZBP); *Trow's* 1885, 377.

20. Insull probably referred to U.S. Patent 314,113, for which Conley applied in February 1884; it issued on 17 March 1885. It covered processes for aggregating fine ores by mixing them with molten pitch or tar and then molding them into bricks or blocks.

21. William Bell & Co. likely planned to process iron-bearing sand at

Quonocontaug Beach, R.I., where the Edison Ore Milling Co. had set up an ore separator in July 1881. See Docs. 2175 and 2591.

22. The eleven-story Chelsea building on Twenty-third St. between Seventh and Eighth Aves. likely was, in the view of one architectural historian, New York's first high-rise apartment tower when completed in 1883. A monolithic structure, the only natural light for its ninety cooperative apartments came from the exterior facades. Philip Hubert designed it with modern conveniences, however, including elevator service and a novel ventilation system. Plunz 1990, 71; "City and Suburban News," *NYT,* 17 Jan. 1883, 8.

23. The United States Electric Lighting Co., formed by the New York Equitable Life Assurance Co. in 1878, controlled the lighting patents of several inventors, including William Sawyer and Albon Man, Hiram Maxim, and Edward Weston (presently the superintendent of its manufacturing shops in Newark). One of the chief rivals to the Edison Electric Light Co., the firm gained publicity for its arc lamps when its operating affiliate, the United States Illuminating Co., won the contract for the new Brooklyn Bridge in 1883. According to a recent credit report, the United States Electric Lighting Co. was "prosperous" and paid a 15 percent annual dividend (its first) in 1884. Docs. 1617 n. 4, 2455 nn. 5, 8; advertisement, Berly 1883, 584; Pope 1889, 77; R. G. Dun & Co. report, 28 Mar. 1885, DF (*TAED* D8526K1).

24. Charles Henry Coster (1853?–1900) became a partner in Drexel, Morgan & Co. at the start of 1884, soon after having been named a trustee of both the Edison Electric Light Co. and the Edison Co. for Isolated Lighting. Having engineered a reorganization of the Edison lighting interests in 1884, Coster was still negotiating a comprehensive settlement of the financial claims of its various constituents. See Docs. 2690 n. 7, 2725, and 2795.

25. Edison and his children moved to the third floor of a rented house at 39 E. 18th St., New York, in September 1884. Doc. 2721 n. 2.

26. Thomas Alva Edison, Jr. (1876–1935) was Edison's second child with his late wife, Mary.

27. The editors have no information about Tom, Jr.'s, possible sickness, nor of his whereabouts. Dr. Edwin Chadbourne, one of the family's physicians, billed Edison $94 in May for medical services. The doctor did not itemize this substantial charge or indicate the dates covered by his bill (Voucher [Laboratory] no. 268 [1885]; see Doc. 2712 [headnote n. 12] on Chadbourne's service to the family). Young Tom had suffered a serious but unidentified illness in early 1879, followed by a lengthy convalescence in Florida, perhaps with his mother (see Doc. 1674; "Down in Dixie," *Atlanta Daily Constitution,* 5 Mar. 1879, 1).

–2785–

To Samuel Insull

Exposition Grounds New Orleans. Feb 28 1885[1a]

Saml Insull

Forgot Spring Over=Coat   Send by Express quick San marco hotel Augustine   duplicate your letter[2] there also

Edison

L (telegram), NjWOE, DF (*TAED* D8503T). Message form of Western Union Telegraph Co. ᵃ"1885" preprinted.

1. See Doc. 2776 (headnote).
2. Possibly Doc. 2784.

NEW ORLEANS, March 1, 1885.

John Ott:

Get three lengths of foot pine board each about 15 feet long and paste tin foil over one surface the total length of the three boards, so the whole surface is electrically connected. Nail these boards up on supports in center of laboratory, then run an iron wire, No. 10 or 12 or 6, about twenty feet away.[1]

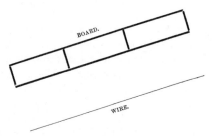

Get a ¼ inch spark coil[2] and get one of the short pronged auto-tuning forks that gives a high note & has magnet,[3] has one set spring to work local magnet so tuning fork go all time use the other spring to open and close the primary of the induction coil, then connect up thus:[4]

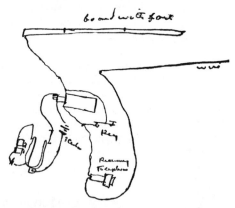

When key is opened the induced waves start out. What I want to find out is how far you can hear dots and dashes in telephone, that is to say, how far apart can you place the iron wire from the tin foil surface of the boards and still read dots and dashes, you better start about foot away and then keep going further and further until you can't hear. Also you might shunt the points that break the primary on the fork by a condenser.[5] The object I have is to telegraph to a train in motion, using an insulated strip about a foot wide along the side of each passenger car, connecting this with the coil and fork telephone ground through the wheels and track, while a telegraph wire running near the track forms the other part of the condenser. The question is, how far apart can this wire be placed from the strip on the car?

Let me know the results at San Marco Hotel,[6] St. Augustine, Florida.

T.A.E.

PL (transcript), MdCpNA, RG-241, *Edison and Gilliland v. Phelps* (*TAED* W100DKN). Document multiply dated; transcribed from letterhead of Hotel Royal.

1. Figure labels are "BOARD." and "WIRE."

2. That is, a coil able to generate a spark across an air gap of this length.

3. Edison had extensive experience using tuning forks, especially in his work on acoustic telegraphy in the 1870s. Each fork would vibrate at a characteristic frequency, typically several hundred times per second. In the arrangement he sketched for Ott, as in some of his earlier telegraph instruments, Edison would sustain the fork's motion by the attraction of an electromagnet. As one prong approached the coils, it would move a switch to break the circuit of the magnet, thereby allowing the fork to oscillate. (For a few examples, see Docs. 652, figs. 8 and 9, and 659 esp. n. 4.) The fork's other prong would operate a switch in the main circuit by which the battery charged the foil surface, like a condenser plate. Edison explained that the purpose was to create a "wave of electricity" lasting just a fraction of a second, during which the air was conductive (see Doc. 2808 [headnote]).

Edison showed Ott's experimental apparatus (or one very much like it) to a reporter sometime before the end of May. He explained that by using static electricity in this way, he could "make electricity jump 35 ft., and carry a message. This is something quite new, no induction has ever been known that extended over three or four or five feet." "Edison's Latest Invention," *Electrician* 15 (12 June 1885): 75.

4. Figure labels are "board with foil," "wire," "Key," "3 carbon," and "Receiving Telephone."

5. The condenser was to "stop sparks and sharpen waves in secondary." See Doc. 2780 and U.S. Pat. 486,634.

6. The San Marco Hotel, located near the Spanish fort of that name and the City Gate, opened in late 1884 but was probably not quite complete when Edison arrived there on 5 March. Advertised as a significant improvement in St. Augustine's lodgings, it had five main floors and three towers and could accommodate six hundred guests. It was the property of Isaac Cruft, who also owned the Magnolia Hotel in Mag-

nolia, Fla., where Edison and his wife Mary had stayed in 1884; both establishments were operated by Osborne Seavey. The prominent real estate developer Henry Flagler reached the San Marco about a week before Edison. TAE to Insull, both 5 Mar. 1885, DF (*TAED* D8502G, D8503V); "Pleasure Travel and Resorts," *Outing* 5 (Jan. 1885): 314; Graham 2003, 12–13; *Appleton's Handbook* 1884, n.p. [156 and advertisement p. 168]; Martin 2010, 103–104; Lomax 2003, 93; Braden 2002, 143–54; Doc. 2607 (headnote).

–2787–

To John Ott

St. Augustine, Fla., Mch 6 *1885*[a]

John—

You must make that little Siemens Magneto[1] work on the spark coil. [p-][b] you can try Larger[c] & also smaller wire on armature as the present wire may not fit the wire of the induction coil— also you better see if Bergmanns[2] man cant cast Some field magnets of steel the same as Bergmann used in his magneto calls or if he cannot you can make some of steel= I think you better fool with armature until you get best results from that then make field magnets ½ again as strong—then twice as strong & then 3 times as strong until you get about ⅛ to 3/16 spark— I want you to press up about 100 chalks[3] so you can have them ready by the time I return= you might also make some experiments with chalks by pressing ½ dozen of each Kind commencing with moderate pressure & increase until four of you on the press could put no more pressure on about ½ doz at say pressure of 1 man on short lever ½ doz pressure 1 man on long lever ½ doz pressure [-----][b] 2 men & ½ doz with all the pressure you can get on=

Let me know result of the Experiment on Repeater with the relays=[4] Dont forget to try a plate Copper over the Ends of the soft carbon buttons for Mendenhall—[5]

wish you would start some one working the individual calls Keeping a record until about 5000 calls are made. Keeping record of every <u>miss</u> and in case it gets out of order the boy to call you and you record exactly what trouble was.[6]

address all Letters to me at San Marco Hotel St Augustine florida

AL, (MiPhM), (*TAED* X059AA). Letterhead of Hotel San Marco. [a]"*St. Augustine, Fla.*," and "*188*" preprinted. [b]Canceled. [c]Obscured overwritten text.

1. The distinctive drum armature invented by Werner von Siemens in 1857 had been used in magnetos for ringing alarms, signals, and telephone call bells. Without a commutator, the Siemens magneto pro-

duced alternating current, which would make regular pulses through an induction coil. It also had "a very short magnetic axis, and so changes its magnetic condition rapidly" so that it would generate the sharp and rapid pulses Edison would have wanted for wireless telegraph experiments. Schellen 1884, 110–14.

2. A skilled machinist and longtime associate of Edison, Sigmund Bergmann (1851–1927) was the principal in Bergmann & Co., a New York manufacturer of electrical equipment and fixtures in which Edison was a silent partner. Since 1882, Edison had his laboratory in rented space on the top floor of Bergmann's factory. See Docs. 313 n. 1, 2091, 2343 (headnote), and 2356.

3. Edison referred to the small cylinders of a pressed phosphate compound that had been manufactured at the Menlo Park laboratory for his electromotograph telephone receiver. When moistened, the "chalk" exhibited what Edison termed the electromotograph principle, by which an electric current varied the friction of a stylus moving over the chalk's surface. Before applying it to the telephone, Edison had used the phenomenon in a variety of other instruments, including a sensitive relay. See, e.g., Docs. 463, 881, 1440 n. 7, 1681 (headnote), and 1693; see also Doc. 2799.

4. Edison may have had in mind experiments related to a patent for a telephone repeater, a subject on which he had returned to work in December 1884 (Doc. 2759 nn. 2–3). That month, he filed a patent application for a design incorporating both the electromotograph receiver and relays (U.S. Pat. 422,579). The device used an acoustic (non-electrical) coupling between an electromotograph receiver and a carbon transmitter to convey the signal from one circuit to the next, with the direction of transmission controlled by switches (relays). He signed a related application on 31 December for an arrangement that would, by additional switches, allow multiple repeaters to work through a series of circuits (U.S. Pat. 478,743). He did not file that application until 14 October 1885, however, an unusually long interval in which he presumably made more experiments.

5. Physicist and mathematician Thomas Corwin Mendenhall (1841–1924) had recently joined the U.S. Army Signal Corps as professor of electrical sciences and head of its Instrumentation Division (Doc. 2755 n. 1). He had corresponded intermittently since 1884 about acquiring telephone transmitter carbon buttons plated with copper, which he hoped to use to advance a definitive explanation for the effect of pressure on the conductivity of carbon (Doc. 2755 n. 3). Samuel Insull, responding to Mendenhall's inquiry, apologized for Edison's failure to provide the buttons before leaving for Florida, and he likely brought the matter to Edison's attention (Mendenhall to TAE, 14 Feb. 1885, DF [*TAED* D8518F]; Insull to Mendenhall, 2 Mar. 1885, Lbk. 20:142A [*TAED* LB020142A]). In early April, in response to a 30 March letter from Edison (not found), Mendenhall conceded that the copper plating "must be given up." At his request, Edison later instructed Ott to send a dozen unplated buttons, which Mendenhall seems to have used in making conclusive experiments, the results of which he published in 1886 (Mendenhall to TAE, 7 Apr. 1885; TAE to Ott, 9 Apr. 1885; both DF [*TAED* D8503ZAD, D8503ZAF]; Mendenhall 1886).

6. Edison referred to tests of a selective signaling device—that is, for ringing only the call bell of the intended recipient instead of every party on a line—a subject to which he had turned in late 1884. See Doc. 2763.

–2788–

Memorandum to
John Ott

[St. Augustine, Fla.] March 10 1885

John Ott=

I suppose you have the auto regulator for Duplex ready   I now want you to set up a Quadruplex thus and see if an automatic will work on that—[1] all the Extra apparatus necessary is a polarized relay with Centered tongue so when no current passes theree tongue[a] will be between the 2 points and touch neither   use differential relays.[2]

The polarized relay that works the ratchet motor resistance is in the bridge.[3] when the distant man sends on the common relay the armature opens the circuit of the polarized relay in the bridge wire, as it is not in a bridge in relation to the incoming Current but only to the outgoing— thus Every time there is no dot or dash on the common relay from the distant station the relay is connected & in the bridge, hence ~~should the~~ ~~lConstant battery on the line or the~~ every time the home station closes & the line is out of balance the polarized relay will work until the resistance is adjusted so practically not enough Current passes through the bridge wire to work the polarized relay[4]

Have both the Duplex & Quad set up so I can see them working when I return— there is no need to put relays at the other End—[b]

Get out that old type writer and fix it up so all the Keys and mechanisms will work then have 6 or 7 letters cut by Bergmanns type wheel[a] man[5] formed of shallow points just sufficient to strike through $3/10000$ French Electric pen stencil paper= put a stripper on paper just around where[a] type strikes   cover the Cylinder with Lead— Rubber, Raw Hide or other material that will work well, then see if you can pound out a stencil & print it on a regular pen press.[6] if you get it to work perfectly satisfactory get a good type writer from Insull & change all of the letters to perforating ones & make a good type writer— you better take the letter N say also n and Cut them in various ways to see which is the best, as the perforation need be only $3/1000$ thick you will see that the points can be very shallow[7c]

Have some resonators[8] of thick [---][d] $1/16$ brass tubing ~~thus~~[9]

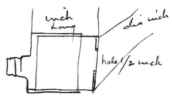

1   1 same[a] dia ½ long
1   "   ¼[b]

   also one[a] each kind ½ inch dia   inch ½ inch ¼ inch long—
hole in Sliding tube ¼ inch[b]

T A Edison

ADS, NjWOE, Lab., Unbound Notes and Drawings (1885) (*TAED* NS85ADB01). Document multiply signed and dated. [a]Obscured over-written text. [b]Followed by dividing mark. [c]Obscured overwritten text; followed by dividing mark. [d]Canceled.

   1. Edison refined a system of duplex telegraphy (for sending two sig-nals in opposite directions on a wire) in 1872–1873; in 1874, he devised the quadruplex, the first practical system for transmitting four signals (two each way) over a single line (see *TAEB* 1 chap. 11 passim; Docs. 275 (headnote), 348, 446, 449 [headnote]). Duplex and quadruplex in-struments required an operator to regulate or adjust an internal resis-tance in response to changing line conditions; see notes 2 and 4 below.

   2. This point occurs at the end of the page numbered "1" by Edi-son; the text continues at the top of the page numbered "3." The edi-tors have not determined if this apparent discrepancy is due to Edison's mistake or to a missing page. The polarized relay, invented by Werner von Siemens and readily adopted by Edison in his early telegraph work, was essentially a switch that moved one way or the other depending on the direction of current flowing through it. The differential relay discriminated between two currents of different strength flowing in the same direction. Its magnet cores were wound in opposite directions with an equal number of turns of wire of the same resistance so that equal currents would produce evenly counterbalanced magnetic forces, while those of different strength would allow one magnet to dominate the other in acting on the armature. Such relays were fundamental to differential systems of multiple telegraphy such as the Stearns duplex and Edison's quadruplex, but they required careful adjustment to maintain the correct relationship with the resistance of the outside line. See Docs. 7 (headnote "Multiple Telegraphy"), 12 n. 1, and 446; Sloane 1892, s.vv. "Relay, Differential," "Relay, Polarized"; Maver and Davis 1890, 7–53.

   3. That is, a Wheatstone bridge, named after British telegraph inven-tor Charles Wheatstone. When the currents in two branches of a circuit are balanced, no current will flow in a "bridge" connection between the branches. Many duplex telegraph systems employed this principle, put-ting a relay in the bridge to eliminate the effect of outgoing signals while remaining responsive to incoming signals. See Docs. 285 n. 17 and 392 fig. 14.

   4. This description appears to fit a drawing that Edison made and labeled "auto balancing for Duplex" on 9 February. The ratchet motor

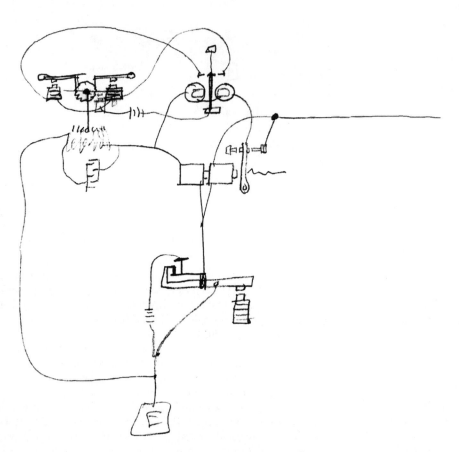

Edison's 9 February "auto balancing" sketch.

at left controls a variable resistance used to keep the resistance of the differential relay properly calibrated with respect to the line. The motor would be actuated by the polarized relay shown to its right. He made a related but less detailed sketch of an "Auto adjust Duplex" arrangement on 13 April. Unbound Notes and Drawings (1885), Lab. (*TAED* NS85ADA1, NS85ADB03).

5. Not identified but presumably a machinist making print wheels for printing telegraph instruments.

6. That is, a press for making copies from a stencil created by the electric pen (see Doc. 721).

7. It is not clear how the device Edison envisioned would differ from the perforating typewriter developed for him by James MacKenzie, on which Edison applied for a patent in 1878. That instrument created a stencil using chisel point type "in a type-writer worked with the fingers like a pianoforte." The stencil could be placed in an electric pen press or similar apparatus to create as many copies as desired (Doc. 1629 nn. 2–3). The idea was organically related to Edison's extensive work in the early 1870s on perforators for automatic telegraphy and his more sporadic efforts to improve the typewriter.

8. An instrument devised by Hermann von Helmholtz, the resonator is a spherical or cylindrical vessel with a small opening (or sometimes

two) used to detect or amplify faint sounds of a particular tone. Edison was familiar with it through working on acoustic telegraphy and his electromotograph telephone receiver. Doc. 708 n. 5; Turner 1983, 142–43; Sen 1990, 121–22.

9. Figure labels are "inch Long," "dia inch," and "hole ½." On 6 February, Edison sketched what appears to be a similar resonator. He gave instructions to have twenty-four of them made in various dimensions of "the thickest tin that can be got in the market." Unbound Notes and Drawings (1885), Lab. (*TAED* NS85ACZ).

-2789-

To Theodore Whitney

Cedar Key, Fla. Mar 14 1885.[1a]

Theodore

I find from the Way bill dated 12th that the guns did not leave there until next day after we left, although man promised they would go same day=[2] There is no mortal Excuse why they could not have been sent. There was another delay at Jacksonville—guns arrived tonight= Im going to have some explanation from the head office of the Co[3]   Yours

T A Edison

We leave Early tomorrow morning   address Tampa

ALS (photocopy), Whitney (*TAED* X512). Letterhead of the Suwanee hotel. [a]"Cedar Key, Fla." and "188" preprinted.

1. The Suwanee's accolade on its letterhead as "the largest & best appointed hotel on the Gulf Coast" may have been somewhat misleading. Some nine years later, a traveler described the Suwanee as "a big whitewashed house set squarely by what looked like a lot of sewage ooze." In 1885, one writer found Cedar Key to be "an old and filthy town almost surrounded by water." "On the Gulf of Mexico," *All the Year Round: A Weekly Journal,* 3rd ser., 12 (24 Nov. 1894): 487; Stevens 1885, 106.

2. Whitney later explained that the express agent in St. Augustine had simply forgotten to send the guns by the first available train. Whitney to TAE, 6 April 1885, DF (*TAED* D8503ZAC); see also Doc. 2776 (headnote).

3. Edison probably meant the Southern Express Co., in care of whose office in St. Augustine he addressed this letter to Whitney. Southern Express was created from the extensive southern lines of the Adams Express Co. at the start of the Civil War in 1861. Its headquarters were in Augusta, Ga. Stimson 1881, 127, 160–61.

-2790-

Agreement with Huelsenkamp & Cranford[1]

[Fort Myers, Fla.,] March 21, 1885[a]

Received of Thomas A Edison the sum of one hundred dollars as part payment on the property known as the Summerlin tract,[2] consisting of 416 feet of river front containing thirteen acres in all. The balance twenty nine hundred dollars ($2900) to be paid by the said Edison within 90[b] days from date. But if the said Edison fails to pay the said twenty nine hundred dollars within such period The one hundred dollars ($100) shall be forfeited[3]   Huelsenkamp & Cranford[4] agree to deliver a full warrantee deed for said property clear of all encumberances whatsoever upon payment of the balance twenty nine hundred dollars ($2900)

Dated this day of March 21 1885.

Thomas A Edison   Heulsenkamp & Cranford

ADS, NjWOE, DF (*TAED* D8539A). [a]Date from document, form altered. [b]Circled.

1. See Doc. 2776 (headnote).

2. The Summerlin tract fronted on the Caloosahatchee River in Monroe County, on Florida's southwest Gulf Coast. The land was acquired in 1879 for five hundred dollars by Samuel Summerlin, son of cattle baron Jacob Summerlin, who used nearby Punta Rassa as a transhipment point for trade with Key West and Cuba. The previous owner was Francisco Abril of Havana, Cuba. Abril to Samuel Summerlin, 3 June 1879; Monroe County (Fla.) court clerk's abstract of title, 25 July 1885; both DF (*TAED* D8539F, D8539E); Smoot 2004, 9–11, 19–23.

3. Huelsenkamp & Cranford prepared a purchase contract by the end of March for the reduced price of $2,750 (without explaining the change). However, because Edison did not complete the transaction by 19 June, he subsequently forfeited the $100 deposit. Edison paid three-quarters of the total; Ezra Gilliland paid the remainder. Huelsenkamp & Cranford to TAE, 31 Mar. 1885, DF (*TAED* D8539B); Florida land account summary, [n.d.], Cat. 1165:11–12, Accts., NjWOE; see Doc. 2843 n. 9.

4. Huelsenkamp & Cranford, land agents in Fort Myers, promoted the region as "The Italy of America" and "The Only True Sanitarium of the Occidental Hemisphere." One of the firm's principals seems to have been Clemens (variously Clement) J. Huelsenkamp (1854?–1932), a Missouri native who became a founding director of the Key of the Gulf Railroad Co. in 1887 and a Florida state representative to the 1893 Columbian Exposition in Chicago. Smoot 2004, 20–21; *Acts and Resolutions* 1887, 231–32; Handy 1893, 76; U.S. Census Bureau 2002 (1930), roll 309, p. 21B, image 586.0 (Miami, Dade, Fla.); Death record for Clemens J. Huelsenkamp, *Florida Death Index, 1877–1998*, online database accessed through Ancestry.com, 6 May 2013.

–2791–

From John Trowbridge

Cambridge March 30 1885

My dear Sir

I have already become so much your debtor by the receipt of the lamp[1] that I hesitate to ask another favor so soon. I do not know however where I can get the following apparatus constructed except through you—

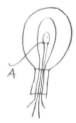

The thermo electric arrangement you placed inside one of your lamps for me is not sensitive enough and I wish to ask if you could have a very fine strip of platinum or steel placed at the centre A of one of your 16 candle power lamps[2]   My idea is to get the temperature inside the lamp by the principle of change of resistance of this this platinum wire or strip abc

I wish to have in this way a check upon a standard of light. Hoping to be able to determine the temperature at the moment the candle power of the lamp is measured. The platinum wire should be very fine between a and c

Hoping that I am not asking too great a favor— I remain Very Sincerely yours

John Trowbridge[3]

⟨John Ott or Martin has it[4]   make it [¼?][a] ½ or ¾ inch long Make 3 Lamps   Send to Trowbridge⟩

ALS, NjWOE, DF (*TAED* D8518G). [a]Canceled.

1. John Trowbridge had received from Edison in February a lamp with an "enclosed thermo electric junction," as described below, along with two alternative "thermo electric junctions" (Trowbridge to TAE,

7 Feb. 1885, DF [*TAED* D8518E]). He used it in experiments to determine a standard of light that would be at once more accurate for practical purposes and commensurate with the scientific definitions of physical and electrical units. The problem of obtaining uniform photometric measurements arose with the introduction of commercial incandescent lights and was taken up at the first meeting of the Congress of Electricians in Paris in 1881. Within three years, the Congress decided to adopt as a standard the light emitted by a square centimeter of platinum at its point of solidification. While this remained the official benchmark until 1937, its limitations were recognized from the start. The National Conference of Electricians, meeting concurrently with the International Electrical Exhibition in Philadelphia in 1884, adopted it but at the same time established a Committee on the Standard of Light to seek a better definition. That committee consisted of Edison, Trowbridge, William Preece from the British Post Office, George Barker of the University of Pennsylvania, Edward Pickering from Harvard, Major David P. Heap of the Corps of Engineers, and Charles Cross of the Massachusetts Institute of Technology (Quinn 2012, 129–30, 224; "Scientific Intelligence," *Am. Jrn. Sci.*, 3rd ser., 28 [1884]: 389).

Trowbridge reported his work with the modified Edison lamp to the American Academy of Arts and Sciences in May 1885. Following unsatisfactory experiments to relate the strength of current through an incandescent platinum strip to the light emitted, he

> next endeavored to ascertain if a thermal junction enclosed in an Edison incandescent lamp, at the centre of the carbon loop, would be sensitive to changes in the heat radiation of the lamp. It is evident that, if this were the case, the carbon loop might be raised to the same point of incandescence in successive times, assuming that the thermal junction at this point of incandescence receives the same amount of radiant energy. Mr. Edison kindly provided me with a lamp in which one thermal junction of an alloy of iridium platinum and platinum was inserted at the centre of carbon loops. The other junction was placed in ice and water. The thermo-electric force of this combination, however, was extremely feeble. The difficulty of inserting wires of other metals into glass prevented me from carrying this idea further. [Trowbridge 1885b, 496]

2. Edison evidently obliged with a lamp like the one sketched in this document. In the May 1885 account of his experiments to the American Academy, Trowbridge explained that

> Instead of the thermal junction, a small loop of extremely fine platinum wire was placed at the centre of a carbon loop in an Edison lamp. This fine wire constituted a bolometer strip and made one branch of a Wheatstone's bridge, it being my intention to place a similar strip in another branch of the bridge, thus making a bolometer. The lamp was placed in a photometer box, and its light was compared with that of a candle as it was raised from a red glow to a light of fifteen-candle power. [Trowbridge 1885b, 496]

These experiments led Trowbridge to conclude favorably that "For a practical standard, a carbon loop in an exhausted vessel raised to such a point of incandescence that it will radiate a definite amount of energy,—this energy being measured by a bolometer strip or the thermopile at a definite distance from the carbon loop . . . would have a greater range than an incandescent strip of platinum placed in free air" (Trowbridge 1885b, 499).

Edison's prior work with platinum in incandescent lamps also figured in the present search for a standard of light. Committee member Charles Cross attempted to determine if the melting point of platinum would provide a more useful reference point than its solidification. Cross found, however, in working with commercially prepared platinum, that "successive heating and cooling, or continued heating of the wire even under ordinary atmospheric pressure, tend slightly to raise its point of fusion, an action which Edison has shown to be carried to an extreme degree when continued heating in vacuo is employed" (see Doc. 1796). This observation led him to conclude that "any photometric standard based on the luminosity of *melting* platinum will present no advantage over those standards now in ordinary use, unless specially prepared platinum freed from gases and consolidated by Edison's process of heating in vacuo is found to be available." Cross 1886, 225–26.

3. John Trowbridge (1843–1923) was professor of physics at Harvard University. A graduate of the Lawrence Scientific School at Harvard, he served for two years as an assistant professor of physics at the Massachusetts Institute of Technology before taking a similar position at Harvard in 1870. He received his doctorate from Harvard three years later and became a full professor of experimental physics. In that position, he was responsible for founding the Jefferson Physical Laboratory and introducing laboratory experiments into the undergraduate curriculum. Much of Trowbridge's own research concerned electricity and magnetism, and Edison drew on his designs in constructing his own dynamo and dynamometer in 1879. Lyman 1925; Hall 1931; "Sketch of Professor John Trowbridge," *Popular Science Monthly* 26 (Apr. 1885): 836–39; Docs. 1770 n. 5, 1851 n. 3.

4. Figure label is "fine platinum wire $^1/_{1000}$—."

–2792–

Notebook Entry:
Miscellaneous

[New York,] April 1 1885

Have Lamp Co[1] carbonize some plates of ½ & ½ mixture oxide manganeese <u>precip</u> & hard Carbon with Tar [~~or s?~~][a] as binder plates for Storage—[2b]

Also some porous plates Carbon made with Tar, hard Carbon & baking soda—[b]

See if pure black Tin Monoxide[3] is conductor[b]

Try precip Mono Sulphide Tin—

hard rubber cylinder[4]

Copper wire coiled around it like Siemens[5]

Beam charges surface rubber statically[6]   Copper banded
wires takes it up give Reverse Currents   use telephone

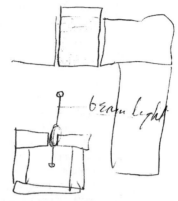

Rubber suspension vibrate coil by Expansion[7]
[A][8]

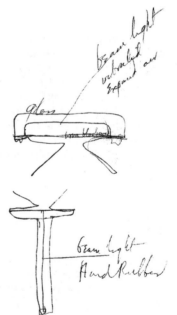

[B][9]

TAE

X, NjWOE, Lab., N-82-05-26:238 (*TAED* N204238). Document multiply signed and dated. ªCanceled. ᵇFollowed by dividing mark.

    1. The Edison Lamp Co. manufactured incandescent lamps in East
Newark (Harrison), N.J., under license from the Edison Electric Light

Co. Formed in 1880 under a slightly different name, the firm was originally a partnership among Edison and several close associates but was incorporated in January 1884. Docs. 2018, 2343 (headnote), 2536 n. 2; *TAEB* 7 chap. 4 introduction.

2. Edison probably meant storage batteries.

3. Stannous oxide.

4. Figure label is "Beam heat."

5. Edison referred to the general pattern of winding wire loops longitudinally around a cylindrical dynamo armature, devised by Werner von Siemens.

6. See Doc. 2804 (headnote) regarding Edison's interest in energy conversion. In the case of this particular proposed experiment, Edison may have been thinking about fundamental issues of photosensitivity raised by the photophone invented by Alexander Graham Bell and Charles Sumner Tainter in 1879 and 1880. Their instrument used an intermittent beam of light to translate a sound wave pattern to a distant receiver, where it fell on a photosensitive medium (initially selenium) that reproduced the original sound wave pattern in an electrical circuit. Bell and Tainter soon discovered that a modulated light beam acting on a thin diaphragm made from any number of materials, including hard rubber, lampblack, and wood, created an audible sound directly, without the aid of an electrical circuit. Bell attributed this phenomenon to thermal expansion and contraction of the diaphragm (and the air around it) under the influence of a pulsating light beam. Other experts, including the acoustical authority John Tyndall, ascribed no special role to light and suggested that any focused heat rays would cause the same effect. Bruce 1973, 335–43; Prescott 1972 [1884]: 313–25; Bell 1881; "Radiophony," *Electrician* 6 (26 Feb. 1881): 180–81.

7. Figure label above is "beam light." This proposed experiment drew on Edison's experience with the thermal sensitivity of hard rubber, a phenomenon he had exploited in designing (in 1878) the tasimeter, an extremely sensitive heat-measuring device. See Doc. 1316.

8. Figure labels are "glass," "beam light vibrated   Expand air," and "iron blackened."

9. Figure label above is "beam light   Hard Rubber." This sketch resembles the hard rubber case used for an experimental Edison telephone transmitter in 1877. The instrument worked poorly because of the rubber's great sensitivity to temperature changes. See Doc. 1125.

–2793–

Notebook Entry:
Miscellaneous

[A][1]

[New York,] April 2nd 1885

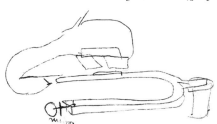

[B][2]

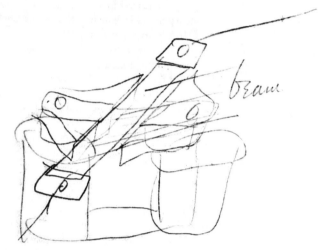

Cut right angles—
    [C][3]

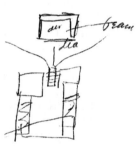

Have Center wire put in 1 candle Lamp—[4]

try effect beam arc light   also magnet   also everything to see if can diminish Carrying Current

    get some Luminous paint. put inside globe exhaust high vac see if sensitiveness greater   also if electrification make it Luminous   if so make for watch chain see time day nights

    Look up all the Colored Salts that are in flat scales like plumbago & try Compression & mixture with clay like reg led pencils—Red & blue needed   also try Bronze powder pencils—[a]

[D][5]

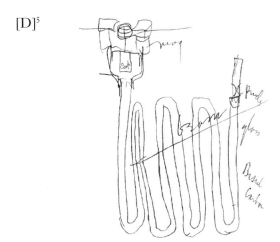

or liquid greatest expansion

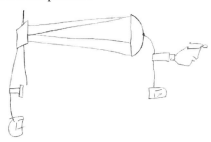

focus inductive Current see if it will go as a beam

Quad Cable

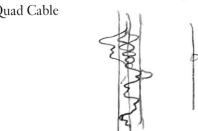

unpolarized Syphon & polzd

try self dischg for mag instead of induction Coil—

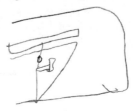

X, NjWOE, Lab., N-82-05-26:242 (*TAED* N204242). Document multiply signed and dated. [a]Followed by dividing line.

    1. Figure label is "mirror."

    2. Figure label is "beam."

    3. Figure labels are "[dev?]," "dia[phragm]" and "beam."

    4. Edison had inserted a third wire into one of his lamps in February 1880 in an experimental effort to stop what he called "carbon carrying," or the transfer of carbon from the filament to the glass bulb. He later observed that an electrical potential could be induced in the wire by bringing the filament to incandescence, and he designed a voltage regulator on this principle in October 1883. In one of the earliest papers on the subject, read to the 1884 meeting of the American Institute of Electrical Engineers, which Edison attended in Philadelphia, Edwin Houston provisionally but suggestively referred to the Edison Effect as a case of "electrical convection" and a "new source of electrical excitement." The British electrician William Preece, describing the phenomenon in March 1885, termed it the Edison Effect. Docs. 1898 (headnote) and 2538 (headnote); Houston 1884, 2, 4; Preece 1885, 229; Hong 2001, 119–27.

    5. Figure labels are "mag[net]," "cork," "beam," [powder?], "glass," and "Bisul Carbon."

–2794–

From Henry Craven

Tarrytown, N.Y., April 4th 1885[a]

My dear Sir

I am convinced that in our subterranean work, and in all underground surveys in general, a small incandescent lamp to be worn in the hat, with a battery carried somewhere about the person, would be a very great convenience; and I take the liberty of writing to ask you for information in the matter. If such lamps are manufactured, could you tell me where to find them; and if they are not made, could not something of the kind be originated?[1]

Might not, also, some contrivance be made by means of which, a plumb-line would[b] be rendered incandescent; for instance, by the insertion of a small piece of platinum wire (or some less costly metal, if possible), in the line?

Trusting that I am not trespassing too much upon your

valuable time, and that you will kindly give me the information I seek, believe me   Very Respectfully

H. S. Craven   Engr of Construction[2]

⟨Dear Sir= We manufacture one candle Lamps[3] at our factory & I am at the present moment Expmtg to get a suitable battery to be carried on the person— I intended it for coalminers
but it will undoubtedly suit your purpose   I expect to have it
within a month & will let you know when ready—[4]

The [main?][c] ~~trouble in putting small Lamp~~ [------][c]   A[b]
plumb line of fine platina wire could be rendered visible if
you had a Dynamo or Electric Light plant ~~in a~~ at top of shaft
& wires run in tunnell   It takes a great deal of power   perhaps by using large cord painted with Luminous paint & ig-
niting piece of magnesium wire for instant would give light
Enough   the light would last ½ to 1 hour & if[b] tunnell very
dark would shine[b] out plain & distinct⟩[5]

ALS, NjWOE, DF (*TAED* D8503ZAB1). Letterhead of Office of Chief
Engineer, Aqueduct Commissioners. [a]"Tarrytown, N.Y.," and "188"
preprinted. [b]Obscured overwritten text. [c]Canceled.

    1. The main problem in the development of this sort of portable light
was the size and weight of the battery (see Doc. 2796). The editors have
found nothing to suggest what, if anything, Edison knew about the work
of French electrical engineer Gustave Trouvé and his so-called "frontal electric photophore," a headlamp for health practitioners that used
Trouvé's bichromate of potash battery to power a small Swan lamp.
After an 1883 demonstration at the French Academy of Sciences, the
device was described in electrical and medical journals as well as in the
New York Times ("Hélot and Trouvé's Frontal Electric Photophore,"
Electrician 2 [6 June 1883]: 179; "Notes," *Teleg. J. and Elec. Rev.* 12
[5 May 1883]: 379; "News Items," *Medical News* 42 [19 May 1883]: 575;
"Scientific Gossip," *NYT,* 13 May 1883, 12). Underground gasworks
technicians in Paris had another version of Trouvé's portable headlamp
in use by late 1884 ("Applications of Trouvé's Lamps," *Electrician* 14
[27 Dec. 1884]: 129; "Recent Advances in Electrical Science," *Science* 5
[16 Jan. 1885]: 45). Joseph Swan developed a portable lamp for mines,
though it was not suitable to be worn. In 1887, the Edison & Swan
United Co. introduced a portable safety lamp weighing seven pounds
(Alglave and Boulard 1884, 176–77; Hospitalier 1883, 1:398–402;
"Notes," *Electrician* 19 [24 June 1887]: 134–35; "Electricity in Mines,"
ibid. [16 Sept. 1887]: 396–97; "Safety Lamps," *Tel. J. and Elec. Rev.* 23
[19 October 1888]: 439).

    2. Henry Smith Craven (1845–1889), a graduate of Hobart Col-
lege, became the construction engineer of the New York City Aqueduct Commission in January 1884 while on leave from the U.S. Navy.
His credits as an inventor already included patents for an automatic trip
for mining buckets and a tunneling machine. Craven was a son of Rear
Admiral Thomas Tingey Craven and a nephew of civil engineer Alfred Wingate Craven, an earlier engineering commissioner of the Cro-

ton Aqueduct Board. Obituary, *NYT,* 8 Dec. 1889, 2; Obituary, *New York Tribune,* 8 Dec. 1889, 5; *NCAB* 12:371; Johnson and Brown 1904, s.vv. "Craven, Henry Smith," "Craven, Alfred Wingate," and "Craven, Thomas Tingey."

3. The Edison Lamp Co. produced miniature lamps on at least a custom, if not routine, basis (see Docs. 2502 and 2664; also Edison Lamp Co. Catalogue of Lamps, n.d. [c. 1887], pp. 22–23, PPC [*TAED* CA041J]). Stout, Meadowcroft & Co. marketed Edison lamps from one-half to six candlepower at this time. They also manufactured and sold various batteries, including the Meadowcroft & Guyon Pocket Battery (a modified Bunsen cell), to provide portable power for illuminated scarf or tie-pins ("The Electric Light as a Scarf Pin," *Sci. Am.* 52 [7 Feb. 1885]: 81; Advertisement, ibid. 52 [28 Feb. 1885]: 140; Advertisement, ibid. 52 [28 Mar. 1885]: 204; "The Electric Scarf-Pin," *Electrical World* 5 [21 Feb. 1885]: 72). Edison one-candlepower lamps were favorably noted in a contemporary review of devices for medical and microscopic illumination ("Electrical Illumination,"*American Monthly Microscopical Journal* 6 [5 May 1885]: 96–97).

4. The first clear evidence of Edison's work on a battery for a miner's lamp came more than a week later, in Doc. 2796. An undated measured drawing shows a mining lamp in the form of a hand lantern (Oversize Notes and Drawings: 1879–1886, Lab. [*TAED* NS7986BAV, image 49]). Edison had also previously been involved in costume lighting in 1883 that included illuminated helmets, wired to a dynamo, for parades (Doc. 2502 n. 2, *TAEB* 7, chap. 7 introduction).

5. Edison's marginal notes formed the basis for his reply on 9 April. Lbk. 20:234 (*TAED* LB020234).

−2795−

*Charles Batchelor
and Charles Coster
Report to Edison
Electric Light Co.*

Exhibit A.[1]

New York City, April 8th 1885.

To the Board of Trustees of the Edison Electric Light Co., New York.

Your Committee appointed to investigate and recommend a plan of settlement of certain pending questions and claims between Mr. Edison, the T. A. Edison Construction Department, the Edison Lamp Company, the Edison Machine Works and the Electric Tube Company[2] severally on the one side and this Company, the Edison Company for Isolated Lighting[3] or the Edison Electric Illuminating Co. of New York[4] severally on the other side, respectfully report as follows:—

The Claims presented by Mr. Edison and his shops against the three Companies named, amount to $136,340.68, and are made up as detailed below.

1st $1,077.94[a] claimed by the Edison Lamp Co., from the Edison Electric Light Co. for altering sockets

2nd $5,689.05[a]   claimed by Edison Machine Works from the Edison Electric Illuminating Co., for balance due for labor and material on 24 dynamos ordered by the Ill'g Co.[5]

3rd $629.09[a]   claimed by the Electric Tube Company from the Edison Electric Light Co., for experimenting on compound used in tubes.

4th $1,020.11[a]   claimed by the Edison Machine Works.

$172.18[a]   from the Edison Electric Light Co. for testing dynamos, &c.

6th $100.00[a]   claimed by Edison Machine Works from Edison Electric Light Co., for experimental work on bar and disk armature.[6]

The foregoing items excepting that of the dynamos ordered by the Ill'g Co., seem all to be directly or indirectly for experimental work, or for changes and improvements of devices, and while it is doubtful whether they were explicitly authorized, there does not appear any reasonable doubt that they were undertaken in good faith to promote the interests of the companies. The most questionable charge, is that of the Tube Co., for $629.09 for experimenting on Tube compounds.

7th $17,106.74[a]   claimed by the T. A. Edison Construction Dept. from the Edison Electric Light Co., mainly for canvassing and similar work done for the purpose of organizing local Illuminating companies.[7]

While this work was certainly very crude and imperfect in its character, there is no doubt that it brought about the formation of most of our local illuminating companies which exist to-day. It seems to have been done in pursuance of requisitions made on the T. A. Edison Construction Dept., by the former Executive of the Edison Electric Light Co.,[8] and the Edison Co. for Isolated Lighting, and the figures have been reported upon by your Secretary as substantially correct.[9]

In the opinion of your Committee they should be apportioned as follows:

$16,668.97[a]   to the Edison Elec. Light Co.

437.77[a]   "   "   "   Co. for Isolated Ltg.

8th   Embraces claims of Mr. Edison's against the Edison Electric Light Co., as follows:

Legal fees to Mr. Tomlinson[a] $2,075.00

9th   payments to Mr. Sprague, while experimenting, & c. and from which your Committee understand valuable inventions were secured.[a]   2,748.04

10th   Expert work at various village plant stations, correcting errors of Engineering & c.[a]   2,212.18

11th   Lamp breakage from errors of engineering[a]   194.10

12th   Defective work, procuring information for patent suits   2,175.25[10]

13th   J. F. Fischer   (this is claimed by Mr. Edison's Construction Department).   753.00[11]

14th   Russell & c.   for gas statistics.[12]   388.00

The first two items in the foregoing list seem to have been incurred in good faith for the benefit of your Company. The justice of the others is in the main dependent on whether the distinction which Mr. Edison makes as to work done by him individually and work done by him as Engineer of the Light Co., has any foundation in fact, or otherwise.

15th   T. A. Edison against the Edison Electric Light Company $100,000.

This claim is based upon Sec. VI. of the original contract between the Light Co. and Mr. Edison, Nov. 18, 1878,[13] which reads "The Company further agree out of its first net earnings remaining after reserving $50,000, or so much thereof as may be required to repay the sums paid in cash by the subscribers to its capital stock, to pay to the said Edison, the further sum of $100,000 in cash."

The supplemental contract between the Light Co. and Mr. Edison, Jan. 12, 1881,[14] by which time the capital of the Company had been increased to $480,000, says in Sec VI, "After the Company shall have repaid to the subscribers to its capital stock the sums paid in by them in cash, and shall have paid the $100,000 provided by said first agreement to said Edison."

From these extracts it appears

1. That there are clearly $50,000 ahead of Edison's claim.

2. That the wording of the supplemental contract suggests the intent of increasing such prior claim to the amount of capital that had then been paid in in cash, which is stated by the Treasurer as about $177,200.

3. That the payment of Mr. Edison's claim is dependent on the Company having net earnings, and on those net earnings having enabled the Company to pay the prior claim of $50,000 or $177,200.

The language of the contract is "first net earnings" and your Committee have sought to ascertain what was contemplated at the time the expression was used. They do not propose to discuss the matter here; suffice to say that while they do not believe Mr. Edison's claim to be well founded, inasmuch as in the ordinary sense of the term this Company has never had any net earnings, the claim has some foundation in the fact that the Company nevertheless earned money which as Mr. Edison says, and as the expression might possibly suggest, was all that was contemplated when the contract was made.

The claims presented by the several companies against Mr. Edison, and his associates Mr. Batchelor and Mr. Insull, are as follows:—

16th[a]   Edison Co. for Isolated Ltg against Mr. Edison for supplies &c., $1,727.35, which is admitted by Mr. Edison.

17th[a]   same against T. A. Edison Construction Department for $2905.64, of which Mr. Edison admits $2,830.12.

18th[a]   Edison Electric Light Co., against Mr. Edison for supplies, cash advanced, rent, & c., $8,126.39 of which Mr. Edison admits 2073.63 of the balance $2,700 charged him as cash paid has been credited by him in another account. A good part of the remainder is for excessive rent, &c.

19th   Edison Electric Light Co., against Chas. and Rosanna Batchelor[15] for balance due on stock subscription on increase of capital from $480,000 at $720,000— $3810. ⟨C.B.⟩[b]

20th   Edison Electric Light Co., against Chas. Batchelor, Rosanna Batchelor, T. A. Edison and Samuel Insull for stock subscribed to on last increase, 70% called, nothing paid, viz:—

684   shares Edison
$117^{7}/_{10}$ "   Batchelor
$6^{42}/_{143}$ "   Mrs. Batchelor.
$7^{5}/_{10}$ "   Insull.

21st   Edison Electric Light Co., against Mr. Edison for stock issued to him, full paid, and held in escrow, as per resolution of Board of Trustees, dated $470^{1}/_{2}$ shares $47,050.—

22d   Edison Electric Illuminating Co., against T. A. Edison for balance due on stock subscriptions $3,932.54

23d   Edison Co. for Isolated Ltg against T. A. Edison for balance due on 50% called on increased stock $8,161.34 ⟨[Dr?][16c] E[dison] M[achine] W[or]ks⟩[d]

Detailed statements of these 23 claims as furnished by the parties in interest are attached hereto, together with various explanatory and other letters regarding same, and are all made part of this report.

Your Committee understand the wishes of the Board to be that they submit some plan of settlement of these pending claims which shall be based upon fairness to all concerned, so as to avoid any recourse to litigation or other extreme measures, and which shall also have in view the importance to this Company at present of not parting with any ready cash or assuming any early maturing obligation.

They accordingly recommend the following, to which Mr. Edison assents as a full settlement.

1. Cancel the subscriptions of Mr. Edison, Mr & Mrs. Batchelor and Mr. Insull to the last increase of Light Co's stock and restore this stock to the Treasury.

2. Issue to Mr. Edison the 470½ full paid shares of the Light Co's stock, and to Mr. & Mrs. Batchelor the stock on which $3810 are due charging Mr. Edison for same with

	$50,860.00
Issue to Mr. Edison his stock with the Illuminating Co. charging him amount due	3,932.54
Edison Co. for Isol. Ltg. amt. due 50 per cent paid	8,161.34
Total	$62,953.88

Charge him also with claims of

Isolated Co., as admitted[a]	$1,727.35
&[a]	$2,830.12
& of Light Co., as admitted	$2,073.63
say	$6,631.10
Making total to be charged him by the Light Co., of	$69,584.98
Credit him per contra with his claims against the three companies	$36,340.68
Leaving	$33,244.30

to be charged Mr. Edison on account of his $100,000. claim, and agreement to be made at the same time, by Mr. Edison deferring the balance of said $100,000 claim which the Light Co. shall not be obliged to pay at present, but which it will pay in full before it pays any dividends on its stock.

Owing, as your Committee is informed, to an arrangement of long standing between Mr. Edison and Mr. Francis R. Upton,[17] the latter has a certain interest in Mr. Edison's claim for $100,000[18] and while of course such interest is entirely a matter of arrangement between the two gentlemen named with which your committee has nothing to do, Mr. Upton and Mr. Edison both desire that, as the plan of the present report if accepted, is a virtual payment of $33,244.30 on account of said $100,000 without benefit to Mr. Upton, an allowance equal to the latter's interest in said virtual payment, viz: an allowance of $1,650*[19] shall be provided for to him conditionally and credited on account of the uncalled instalments on subscriptions (70% paid, 30% uncalled) to Stock of this Company held by him proportionally whenever and as calls are made on such subscriptions and your Committee recommends that this be done.

The Light Co., to adjust with the Illuminating Co., and the Isolated Co., all matters growing out of this settlement, and Mr. Edison to make like adjustment with the T. A. Edison Construction Department, the Edison Lamp Co., the Edison Machine Works, the Electric Tube Co., Mr. & Mrs. Batchelor and Mr. Insull and Mr. Upton. Respectfully submitted

Charles H. Coster[e]
Charles Batchelor[e]   Committee of Edison Electric Light
Company.

ADDENDUM[f]

New York City, April 8th 1885.
Entries.
To be made in books of the several companies if the foregoing plan is adopted.
In E.E.L. Co's Books.

Patents, U.S. & Canada	Dr.	$33,244.30
Experimental Expenses	"	15,347.40
Legal Expenses	"	4,250.25
Canvassing & Estimates	"	16,668.97
Edison Elec. Illg. Co.	"	1,756.51

To Edison Co. for Isol. Ltg.	$12,281.04
" Capital Stock	50,860.00
" T. A. Edison	8,126.39

In E. E. I. Co's Books.

| Experimental work 1st Dist | Dr. | $5,689.05 |

| To Capital Stock | $3,932.54 |
| " E. E. Light Co. | 1,756.51 |

In Isolated Co's Books

| Edison Elec. Lt. Co. | Dr. | $12,281.04 |
| Profit & Loss | | 513.29 |

To Capital Stock	$8,161.34
" T. A. Edison	1,727.35
" " " " Construction Dept.	2,905.64

D (copy), NjWOE, Miller (*TAED* HM850252A). Copy written by William Meadowcroft. [a]Followed by checkmark written in blank space. [b]Charles Batchelor enclosed this paragraph in a brace and initialed it in the left margin. [c]Illegible. [d]Marginalia written in an unknown hand. [e]Signatures enclosed by brace at right. [f]Addendum is a D.

1. This report was appended as "Exhibit A" to the 23 April 1885 contract by which all the parties named below assented to its proposed terms. Edison's records reflected a credit in May of $33,244.30, the full amount due him from the Edison Electric Light Co. (see below), though a full cash settlement of the account (on slightly different terms) was not accomplished before November 1887. TAE agreement with Edison Electric Light Co. and others, 23 Apr. 1885; Frank Hastings to Coster, 10 Dec. 1887; both Miller (*TAED* HM850252, HM850253); Voucher [Laboratory] no. 194 (1885) for Edison Electric Light Co.

2. The Electric Tube Co. manufactured the underground electrical conductors used in the Edison lighting system. It was incorporated in early 1881 with Edison as president and John Kruesi as superintendent. Kruesi moved the manufacturing works from New York to Brooklyn in April 1884. See Docs. 2343 (headnote) and 2624.

3. The Edison Co. for Isolated Lighting was formed in late 1881 to install individual lighting plants (such as for residences, factories, and mills). In September 1884, it assumed the former Edison Construction Dept.'s business of building small central station plants. It was a stock company in which the Edison Electric Light Co. held a controlling interest. See Docs. 2189 n. 2, 2420 n. 20, and 2725.

4. The Edison Electric Illuminating Co. of New York was incorporated in December 1880 to provide electric light and power service in Manhattan. It operated the Pearl St. central station and distribution system. Its officers and directors closely overlapped those of the Edison Electric Light Co., with the notable difference that Spencer Trask now served as its president. See Docs. 2037 and 2243 (headnote); TAE agreement with Edison Electric Illuminating Co. and others, 23 Apr. 1885, Miller (*TAED* HM850252).

5. The basis for this charge is not clear. The Pearl St. station in New York, the sole plant built or operated by the Edison Electric Illuminating Co., used only six dynamos of the large "Jumbo" design. One possible explanation is that the Edison Machine Works began fabricating equipment for a central station to be constructed along the lines that Edison had envisioned in 1880, with dozens or scores of small dynamos driven by belts from a few steam engines. Edison's preliminary plans for a 10,000-lamp central station had called for 180 such dynamos (see, e.g., Docs. 1890, 1897 n. 3), but his development of the large direct-connected steam dynamo completely altered this conception. The new type of machine was successfully demonstrated in February 1880, just a few months after the incorporation of the Illuminating Co. (see Doc. 2057). However, had the Machine Works already started work, the parent Edison Electric Light Co. could presumably have transferred the liability to the new firm.

6. In 1881, Edison had designed a disk dynamo with radial armature bars in the expectation that it would run more efficiently than a drum-type armature. He had planned to make a large prototype for the Expo-

sition Internationale de l'Électricité in Paris that year but instead completed a smaller version in 1882. See Docs. 2082 n. 2, 2150, and 2228.

7. Cf. Docs. 2704 and 2771.

8. As Edison Electric Light Co. president until November 1884, Sherburne Eaton had actively managed the company's affairs.

9. As both secretary and treasurer of the Edison Electric Light Co., Frank Hastings had reviewed Edison's claims against the company for expenses related to central station construction. See Docs. 2704 and 2736.

10. The word "Defective" was probably miscopied from "Detective." The amount of this item appears in Doc. 2771 in connection with James Russell's detective work in patent interference cases, but the editors have not determined particulars of this charge. There are a few acknowledged instances of Edison or the Edison companies having surreptitiously gathered information on rival inventors or competing firms. See *TAEB* 6 App. 1.B.20; Docs. 2278 esp. n. 10, and 2455 n. 9.

11. Joseph F. Fisher (1835?–1885) was a disgraced former public official and banker hired in 1884 to secure passage of New Jersey laws favorable to the Edison lighting interests. Edison characterized the amount noted here as Fisher's "personal expenses," unrelated to a $150 charge, likely for a bribe, that Edison personally reimbursed to Fisher after the Edison Electric Light Co. refused to do so. See Doc. 2687 esp. nn. 3–4.

12. It is not clear whether this line refers to James Russell's work as a canvasser of gas usage in New York City in 1880 (see Doc. 1995 esp. n. 2) or a possible role in compiling an undated book of "Gas Statistics." The book included information about the ownership, capitalization, and fees of scores of gas companies in a number of states. Gas Statistics Book No. 1 [n.d.], Electric Light Companies—Domestic: Edison Construction Department, CR, NjWOE.

13. Doc. 1576, dated 15 November 1878.

14. This contract concerned Edison's future inventions in electric light and power, as well as the control of patents related to electric railroads, a subject not specifically covered in the 1878 agreement. Miller (*TAED* HM810140).

15. Rosanna Batchelor (1849–1942), née Donohue, a Newark native, was the wife of Charles Batchelor since 1871 and the mother of their two daughters. Birth record for Rosanna Donohue, *New Jersey, Births and Christenings Index, 1660–1931*, online database accessed through Ancestry.com, 13 Jan. 2014; see Docs. 619 n. 8 and 870 n. 1; "Mrs. Charles Batchelor," *NYT*, 20 Mar. 1942, 19; U.S. Census Bureau 1982? (1900), roll T623_1111, p. 17A, enumeration district 682 (Manhattan, New York, N.Y.).

16. "Dr" was a standard bookkeeping notation for "debtor" and often used as a heading for the left-hand or debit column of an account. *OED*, s.vv. "dr," "debtor."

17. Francis Robbins Upton (1853–1921), a mathematician and physicist, joined Edison's laboratory staff in 1878 and made essential contributions to the development of the incandescent lamp and dynamo. He was presently general superintendent and treasurer of the Edison Lamp Co. Docs. 1568 n. 1 and 2260 esp. n. 2.

18. Edison sometimes promised his closest collaborators a percentage of receipts from specific inventions. Upton made a verbal agree-

ment in June 1879 to forgo his salary in return for a share in profits from the electric light. Upton looked forward to 5 percent of the money due Edison under the sixth clause of the contract with the Edison Electric Light Co. (Doc. 1576), as well as fifty paid shares in the company. He also expected his interest in the proceeds to be recorded by the company. See Docs. 1709, 1762, and 1775.

19. No explanation of the "*" symbol appeared in this copy of the report. A 5 percent share in the "virtual payment" to Edison would have been $1,622.20.

–2796–

Notebook Entry: Electric Lighting

[New York,] April 13 1885

Experiments on storage batteries for Miners Lamps=[1]

Lead Electrodes inch wide. 23 inches long & ¼ inch apart ¹⁄₁₆ thick

No 1   Lead & Lead— Ordinary Red fluid Bichrom—
   "   3:35 PM. April 13. deflctn on strap 50—[2] By mistake 3 cells ch[ar]g[in]g was used for 2 mins   now changed— defln 27 strap at   At 3:45 disconnted & put through 2 ohm coil gal— gave 50 then down to 10 in 10 seconds   20 seconds gave 4— N-G—

No 2   Lead & Lead— Chromic Acid & Sul Acid— 22 on strap 3:48 pm— pol[ari]z[ed] in 10 seconds   15=chargg with 2 Bichom Cells— 30 seconds goes to 11 on strap= stays at 11— Disconnected at 3:58— 15 on strap2 ohm coil—[a] in 3 seconds   10 in 5 seconds 8. 5 in 20 seconds— N.G.[b]

No 3— Lead & Lead— Strong Solution Bisulphate Potash= at 4:02 pm—strap gives 8 deflection. stays there same defln at 4:05. disconnected at 4:12 pm— in in 2 seconds 10 deg on 2 ohm coil. 8[c] seconds 5 deg   20 seconds 3 deg= Reversed & Rechg 4:14½ PM. 8 deg on strap— at 4:16—10 on strap   at 4:20=2 seconds 12— 10 seconds 5 on 2 ohm coil[b]

No 4   Phenomenon   Lead & Lead   Hyposulphite Potash on[c] at 4:33—pm—def on strap 6. Solution almost instantly turns black at top. in one or 2 seconds shaking this black clouds so thick that by shaking bottle its almost like ink, pbly go[od] fluid for chemical telegraph.[3] one Lead pole coated heavily with white substance, viscous & dont rub off.= perhaps sul-phur. deflectin at 4:41 2 on strap= this coating evidently insu-lates. off at 4:42= in 2 seconds 8 on 2 ohm coil. in 30 seconds 5 on 2 ohm coil   1 minute to 3 on 2 ohm coil= [---------][d]

 5= Lead & Lead— Cyanide Potash— On[c] at 4:24. gives 12 on strap   off at 4:33— 10 in 2 seconds. 3 in 5 seconds   1 in 8 seconds[b]

 6   Sulphate Manganeese   Lead & Lead—on at 4:45— strap 4 deg—4 deg at 4:55—off— 30 deg 2 secs— 15 at 5 secs— 30 seconds 10 deg— 60 secs—5 deg on 2 ohm coil— 3 minutes 3 deg on 2 ohm coil[b]

In discharging found that bright plate got blackened ie[e] aparantly the discharging re-charges it thus giving C[ounter] EMF & poor results   the peroxide black after dischag ap-pears as black & thick as it was be4 discharging   I now re-verse direction at 5 oclock   something peels off in flakes & drops to bottom— deftn on strap 4 deg—off at 5:05.

2 seconds 30   5 seconds 20   10 seconds 15— 20 seconds 11 25 seconds 10— 30 sec 9— 3.5 sec 8— 40 seconds 60— 50 seconds 5   1 min 5—added strong Sul Acid— Rechgd— 5:15 pm—Rechg— The Sul A cleans plates & disolves the flakes at bottom= def on strap 12—[f] at 5:20—strap gives 20 deftn— off at 5:20— 2 sec 20   5 seconds 15   10 sec 10— 15 seconds 7   20 seconds 6   25 seconds 4— 30 seconds 3—[b]

 No 7— Ferrocyanide of Potassium   Lead & Lead[g] acidulated slightly with SO_4 strap—5—at 5:24 pm— at 5:28—7. def—off at 5:30   Defln "2" on 2 ohm coil?? N.G.[b]

 No 8— Double Sol Sul Zinc & Manganeese   Lead & Lead[g]   5:34 pm—strap dcf. 10 dcg— 5:44 off   5 secons 70— on 2 ohm coil   10 secons   60— 15 secs 50. 20 seconds. 45— 25 second 35   ~~5060~~ se 1 min 32   1½ min   30—2 mins. 23— 32½ minutes. 19— 3 minutes. 15   3½ minutes. 14— 4 minutes. 12   4½ minutes 11— 5 minutes. 10   5½ minutes. 9   6 minutes 8   6½ minutes. 8   7 minutes. 7. 8 minutes 6— 8½ minutes. 6 deg[4]

 No 9   Lead & Lead. Hypophosphate soda   6 pm. 4 on strap. one plate turns blacks almost instantly. off at 6:09— Runs down instantly to 2 on 2 ohm coil & zero in 10 sec-onds—[b]

TAE   M N Force

X, NjWOE, Lab. N-80-04-17:101, 103–15 (*TAED* N050101). Document multiply signed and dated; some expressions of time standardized. [a]"ohm coil—" interlined above. [b]Followed by dividing mark. [c]Obscured overwritten text. [d]Canceled. [e]Circled. [f]Followed by "over" as page turn. [g]"Lead & Lead" interlined above.

1. Edison wrote this entry on pages following and overlapping with John Ott's undated list of nearly sixty "Solutions used in Storage battery experiment for Miner's Lamp." The start of Ott's list corresponds roughly with the nine solutions Edison tried on this date and with others tried on 14 and 15 April (see note 2). N-80-04-17:99, 100, 102, Lab. (*TAED* N050099).

2. This measurement and subsequent similar ones refer to the degrees of deflection of the needle in a tangent galvanometer. The needle lay in the horizontal plane within a circular strap (or hoop) of conducting metal (often copper) placed in the vertical plane, through which the current to be measured would pass. *KNMD*, s.v. "Tangent galvanometer"; Urquhart 1881, 92–93.

3. That is, for use as a recording solution.

4. Edison continued to record results of a few dozen similar experiments during the next several days. He became excited over a solution of glacial phosphoric acid on 14 April, noting: "Every bit off the peroxide reduced   plates perfectly clean after discharge   this seems to be <u>big advance over SO_4 . . . This should be investigated by constructing large battery.</u>" He judged the current from at least two batteries on 17 April by putting a one-half candlepower lamp in the circuit (N-80-04-17:115–65, Lab. [(*TAED* N050115, N050133, N050152, N050159, N050163]). Martin Force continued to make trials in the manner of Edison into mid-May, using many of the solutions listed by John Ott (see note 1; N-80-04-17:168–281, Lab. [(*TAED* N050168–274]). Force and Montgomery Waddell worked on storage battery experiments for the miner's lamp at least through late May (N-85-05-28:1, N-82-06-21:156, all Lab. [*TAED* N30900A, N238156]).

–2797–

To David Day[1]

[New York,] April 20th. 1885

Dear Sir:—

Referring to your favor of 11th. inst.[2] I will reply to your questions <u>seriatum</u>.

FIRST. In 1879 had expert travelling through the South & west, gathering ore in situ and placer sands, which were sent to Menlo Park for analysis, circulars were sent to all parts of the U.S., thousands of samples of ore & placer sands were received.[3] In no instance was a trace of platinum or its allied metals found from ores or placer sand from the South, none east of the Sierra Nevadas. But on the Pacific Coast platinum exists in nearly all placer sands north of the Central Pacific Railroad and as far up as British territory. The further north the greater the amount of platinum. Large quantities of platinum

are in the placers around Oroville, Butte Co., for every 9 parts
of gold 1 part of platinum and its allied metals are found, on
the Oregon coast placers, one part of platinum metals to 5 of
gold and in some localities as much platinum metals as gold
found—at one time Miners saved the platinum and sold it at
San Francisco but of late years no attempt whatever has been
made to save it. I have never yet found platinum in ores in situ.
Platinum is found in the placers on the Chaudiere River, Canada, but in small quantity.

SECOND. I do not think an ounce of platinum has been
mined in the U.S. in year 1883 or 1884.[4]

THIRD. Platinum is worked by a firm at Newark, N.J. I do
not remember their name, they are the largest workers in this
Country.[5]

FOURTH. I am quite sure no crude platinum ore is imported
into this Country, only pure platinum. The parties who work
platinum here, use scrap only. About 80 per cent of platinum
ore comes from the Ural Mountain placers in Russia, 15 per
cent from the U.S. of Colombo. It is found on the head waters
of the Atral River and is bought by traders in the inland towns
and sent to Bueneventura the seaport town on the Pacific all of
this platinum goes to Paris. 5 per cent of the ores come from
Borneo.

Johnson, Mathay & Co.[6] of Covent Garden, London, England, practically control the platinum of the world and do
probably 80 percent of the Refining.

I would like to have a copy of the book.[7]

Should you desire any further information at any time I
shall be glad to furnish it. Yours truly,

T. A. EDISON, per Mc.G[owan].[8]

TL (carbon copy), NjWOE, Lbk. 20:245B (*TAED* LB020245B).

1. David Talbot Day (1857?–1925) received his Ph.D. in chemistry
from Johns Hopkins University in 1884. Presently a demonstrator of
chemistry at the University of Maryland and a continuing graduate
student at Johns Hopkins, he was also, since 1883, a special agent of
the Division of Mining and Mineral Resources of the U.S. Geological
Survey. Day became chief of the division in 1886 and held that position
for twenty years, during which he became distinguished for his exper-
tise in both mining and petroleum. Obituary, *NYT,* 17 Apr. 1925, 21;
Obituary, *Washington Post,* 18 Apr. 1925, 8; U.S. Census Bureau 1982?
(1900), roll T623_158, p. 8A, enumeration district 12 (Washington,
D.C.); Brown 1926, 85.

2. Writing from Johns Hopkins University on letterhead of the
U.S. Geological Survey, Day requested information for an article on
platinum he was preparing for a new edition of the Survey's *Mineral*

Statistics of the United States (Day to TAE, 11 Apr. 1885, DF [*TAED* D8543C]). The unsigned article that resulted (Williams 1885) incorporated, without attribution, much of the information Edison provided.

3. The "expert" was probably Frank McLaughlin, who prospected on Edison's behalf for platinum, mainly in California, starting in September 1879. He had also investigated the region of the Chaudière and the Rivière du Loup in Quebec earlier that year (see Docs. 1756, 1776 n. 6, 1938 n. 1). Edison already had some knowledge of platinum sources at that time, when, anticipating a great need of it for his electric light, he embarked on an all-out search that included personal inquiries and a circular letter (Doc. 1734) distributed widely throughout the American West. These efforts yielded hundreds—if not thousands—of samples that Edison and his staff painstakingly assayed (see Doc. 1941 esp. n. 2).

4. The published report stated that 200 troy ounces of platinum were recovered from placer deposits in the United States in 1883, and 150 ounces in 1884, though Edison may not have construed that process as mining. Williams 1885, 578.

5. Edison probably meant Baker & Co., identified in Day's report as importers of a "trifling amount" of platinum. The firm, listed in an electrical directory under the heading of platinum manufacturers, importers, and refiners, was located in Newark at 408 New Jersey Rail Road Ave. Williams 1885, 578; Berly 1884, 56, 194.

6. Johnson Matthey & Co., of London, was a leading refiner of precious metals and a manufacturer of platinum devices, with which Edison had dealt since 1878. Doc. 2149 n. 18.

7. Williams 1885.

8. Frank McGowan (1849?–1890?) entered Edison's business world about 1882 as a stenographer for Sherburne Eaton, then president of the Edison Electric Light Co. McGowan remained connected with the Edison interests for several years and later undertook an expedition to Brazil, but at this time he seems to have been assisting Samuel Insull with correspondence and other office duties. His timesheets from late spring and early summer 1885 indicate he was acting then as Edison's "librarian." Doc. 2261 n. 2; timesheets, Vouchers (Laboratory; May–July 1885).

–2798–

From Chauncey Smith

[Boston,] April 24th 85

My Dear Sir

You remember I presume that I gave you about three months ago an order for a [break?][a] wheel which would say "Hello"   Can you send it to me immediately?[1] I have been to busy with other matters to give much attention to the theorys of the telephone but must do so soon, and I wish to get your instrument for experiment.[2] Yours truly

Chauncey Smith[3]

⟨Gill— I wrote him that Vail[4] had ordered all Expmtg stopped= You know I had that Lead covered phonograph made to prepare wheel but Im dam'd if Im going ahead at my Expense[5]   Edison⟩

ALS, NjWAT, Box 1221 (*TAED* X012F2H). Letterhead of Chauncey Smith. [a]Illegible.

1. Chauncey Smith evidently made this request of Edison in person (Smith to TAE, 5 May 1885, Box 1221, NjWAT [*TAED* X012F2J]). As part of his research on speaking telegraphs in 1877, Edison proposed using a similar sort of break wheel to reproduce speech by electrical means on a "Keyboard Talking Telgh" (see Docs. 921, 964). At least one historian contends that Edison's proposed device was a conceptual antecedent of the phonograph, and he points out that Edison was drawing on a tradition of experimental tonewheels—rotating devices whose protrusions could mimic specific sounds by plucking at stiff paper—dating to the seventeenth century (Feaster 2007, 19–24).

2. A primary architect of the Bell Co.'s successful legal strategy of developing a broad interpretation of Alexander Graham Bell's initial telephone patent, Chauncey Smith was intimately involved with the company's ongoing patent litigation. The company had beaten back the major challenges, though appeals were being consolidated at this time in federal courts into what became known simply as "the telephone cases" (decided by the U.S. Supreme Court in 1888). Beauchamp 2010, 858–75.

One possible reason for Smith's wish to "experiment" with the break wheel at this time is the interest in the make-and-break telephonic device invented by Phillip Reiss in the 1860s, with some authorities arguing that it could produce speech (Beauchamp 2010, 864–65). Smith may also have anticipated a legal matter in the offing but not, as yet, in the public record. It concerned a pair of patents, still pending in the Patent Office, to Chicago inventor James Freeman, a business associate of former Patent Commissioner Edgar Marble. Freeman's patents were for a make-and-break telephone transmitter; after they issued on 19 May, his invention received considerable publicity as a possible alternative to the Bell company's instrument (U.S. Pats. 318,423; 318,424; "The Freeman Telephone Patents," *Electrician* [New York] 4 [July 1885]: 257–58; "A Rival of Bell's Telephone," *Washington Post*, 24 May 1885, 5).

3. Prominent Boston attorney Chauncey Smith (1819–1895) was a principal advisor to the American Bell Telephone Co. (and its successors). After attending the University of Vermont and then studying law with a Burlington attorney, he was admitted to the bar in 1848 (*DAB*, s.v. "Smith, Chauncey"; see also note 2). His office at 5 Pemberton Sq. was a short distance from the Bell Co.'s headquarters.

4. Theodore Newton Vail (1845–1920), a cousin of telegraph pioneer Alfred Vail, was superintendent of the U.S. Railway Mail Service before becoming general manager of the American Bell Telephone Co. in 1878. He served in the latter capacity until 1885 (not 1887, as stated in *TAEB* 7), when he became the first president of American Telephone & Telegraph, the Bell system's new long-distance provider. Doc. 2745 n. 7; *ANB*, s.v. "Vail, Theodore Newton."

5. Edison sent a handwritten reply to Smith on 1 May, saying only: "Mr Vail has ordered that all experiments in New York shall cease un-

til further orders are received from Boston," a directive recently received through Ezra Gilliland (TAE to Smith, 1 May 1885, Box 1221, NjWAT [*TAED* X012F2I]; Gilliland to TAE, 28 Apr. 1885, DF [*TAED* D8547ZAA]). Vail addressed a memorandum to the American Telephone Co.'s Executive Committee soon afterward, conveying Smith's opinion that "no advantage would be derived by this Co. retaining, or making any arrangement to retain, the services of Mr Edison, either as Expert, Adviser, or Experimenter" (9 May 1885, Box 1221, NjWAT [*TAED* X012F2K]). Smith, contending that Vail's order did not apply to his personal requisitions, reiterated his request that Edison supply a wheel for twenty to thirty dollars. Edison asked Gilliland to consult with Vail about the matter, noting that "it would take a big lot of my time to get a perfect record to make a <u>wheel</u> by. Smith must have highly developed ideas of the ease & quickness with which this can be done   I wouldnt do it except for the Co for $50" (Smith to TAE, with TAE marginalia, 5 May 1885, Box 1221, NjWAT [*TAED* X012F2J]).

Edison had been designing other telephone devices in recent weeks. Among these were an "Individual Call" signaling apparatus and transmitter with a spun metal diaphragm and the new "cloth button" mentioned in Doc. 2781. Unbound Notes and Drawings (1885); Oversize Notes and Drawings (1879–1886); all Lab. (*TAED* NS85ADB05, NS85ADB06, NS85ADB10, NS7986BAO).

–2799–

Memorandum to John Ott

NEW YORK, 25 april 1885[a]

John

Have been Experimenting on recording on Motograph[1] fine weight of iron wire $^{10}/_{1000}$ resting on regular paper[2] gives splendid mark. please have little mercury Cup made of iron into which a globule mercury is placed   have a pin about $^1/_{32}$ dia put in middle extending above mercury then make iron marking wire bend End around so it fits the pin rather loose & lays down on the paper on drum of its own weight= On the End of the long lever put Couple pins with the smallest lost motion between them to move the wire back & forward on the paper   You will understand by the drawing— get this ready by tomorrow night sure=[b]

Also get the Motograph that works a Sounder ready using[c] ⟨Finished⟩[d]

AD (fragment), NjWOE, Unbound Notes and Drawings (1885) (*TAED* NS85ADC). Letterhead of Thomas A. Edison. [a]"NEW YORK," and "188" preprinted. [b]Diagonal line drawn through entire paragraph. [c]Line drawn through paragraph. [d]Marginalia written on reverse by John Ott.

1. Edison seems to be describing a method for using his electromotograph to register the fluctuations of very small currents. His descrip-

tion to Ott suggests that he wished to couple the electromotograph, an instrument highly sensitive to rapid electrical oscillations, with an automatic telegraph recorder so as to make a visible record of wave signals. The editors have not determined his intent for doing so, but possibly he wished to study the wave nature of sound, about which there was some controversy at this time (see Docs 2773 and 2833). Starting in about 1874, Edison had used a chemical recorder—an automatic telegraph receiving instrument—to investigate static discharges from telegraph lines. His electrochemical experiments with the recorder, in which an electric current discolored chemically treated paper passing under a conducting stylus, led him to the electromotograph principle, in which a current varies the friction between a specific material and a stylus rubbing over it. Edison ultimately exploited this phenomenon in a variety of ways, notably in telegraph relays and telephone receivers (Telegraph Notes and Essays, n.p., Unbound Notes and Drawings [NS-74-002], Lab. [*TAED* NS7402AD, image 35]; see, e.g., Docs. 342, 873, 888–89). He appears to have drawn undulatory waves in an unclear sketch at this time possibly related to wireless or railway telegraphy, and he represented the electromotograph relay in another sketch of a telegraphic circuit on 22 April (N-80-04-17:167; Unbound Notes and Drawings [1885]; both Lab. [*TAED* N050163, NS85ADB12]).

2. That is, Edison's standard chemical recording paper.

PHONOPLEX TELEGRAPH   Doc. 2800

With the wireless railway telegraph fresh in his mind, Edison returned in April to an old problem. His goal was to create a system of multiple telegraphy (for sending more than one message at a time on a single wire) that could be used by every station on a line, not just the two at either end. Because existing multiple telegraph systems (duplex for two messages; quadruplex for four) depended on the careful adjustment of each instrument to the line conditions, including the device at the other end, they were restricted to working only between a matched pair of instruments. These systems, while advantageous on commercial trunk lines, offered less benefit where the intermediate (or "way") stations generated a lot of traffic themselves, such as for controlling the movements of railroad trains. Edison had tried to overcome that limitation as early as 1874, and he gave thought to a "Way duplex" system in 1878 and again while vacationing in Florida in 1884. Some of these previous attempts foreshadowed the approach he took now, which blended his prior knowledge of acoustic telegraphy with recent experiments on the wireless (induction) telegraph.[1]

Edison recognized that the principle of the induction telegraph could work through wire as well as air. That is, the relatively high frequency discharges that jumped to nearby telegraph wires without disturbing the transmission of make-and-break Morse pulses could just as well travel through a wired connection to the receiver circuit.[2] In either case, the induction discharges would activate a receiver that was essentially a telephone. Such receivers could be independent of each other without the fussy tuning required of duplex or quadruplex sounders, opening the possibility of using several of them on the same line. The other key components were the induction coil to generate the signals and, notably, the condenser. The condenser would act as a filter, blocking steady or slowly pulsing Morse signals but allowing higher frequency currents to pass.[3] A well-known element, the condenser already played a similar role in the wireless telegraph. Edison had clearly been thinking about its operation, using it as the basis for an analogy by which he understood the wireless telegraph to work by a similar process of electrostatic induction.[4]

Doc. 2800 is the first coherent outline of such a multiple telegraph system, one that Edison would modify and refine throughout the year. The system consisted of two different types of matched transmitter and receiver pairs. One set of instruments was a standard Morse transmitter and sounder working in the normal way. The other transmitter was a key in circuit with a local battery and primary winding of an induction coil. As the key interrupted the current, rapid high-voltage pulses would be induced in the secondary winding and discharged onto the line. The distinguishing feature at the receiving end was a telephone receiver that would respond audibly to those induced currents much the way an ordinary sounder responded to Morse signals. The sharp induction impulses would act on that instrument, sensitive as it was to rapid variations of current, but pass too quickly to overcome inertia in the ordinary Morse sounder. Edison initially used his motograph receiver as the telephone. He covered these elements in two patent applications developed from the sketches and notes in Doc. 2800. Edison executed the first and more general application on 27 April; three days later, he signed the other, pertaining to a transmitter with a rapidly rotating circuit interrupter like that used in the wireless railway telegraph.[5] Both were filed on 8 May and passed quickly through the Patent Office.[6]

Edison initially conceived the system as a stand-alone

method for duplexing way lines. Just as he was finishing with the two patent applications, he thought of an approach for applying quadruplex technology to way lines (see Doc. 2801). It was midsummer, however, before he began to make notes on using the phonoplex in tandem with existing duplex or quadruplex systems, which led to two more patent applications in October (see Doc. 2849). The new circuit designs effectively added independent channels to existing wires so that, as Edison explained in one of the patents, "duplex induction transmission can be carried on over the line alone simultaneously with the ordinary duplex or quadruplex transmission, producing a new system for duplex, quadruplex, or sextuplex transmissions."[7] This flexibility, he hoped, would allow telegraph engineers to divide long lines into segments and manage their traffic in creative ways, increasing the utility of each wire beyond the simple addition of channels (see Docs. 2879 and 2902).[8]

Edison decided by late summer to make a practical trial of his system. He designated Alfred Tate, who had just finished a stint drumming up electric lighting business in Canada, to set it up on a line of the Great North Western Telegraph Company from Toronto to Hamilton, Ontario. It is not clear why he chose this venue. His friend Erastus Wiman, with whom he had been collaborating on the railway telegraph, was president of the Great North Western, but it is obvious that they had their eyes on a larger prize: a contract with Western Union (see Docs. 2840 and 2850). It was also around this time that Edison formally gave the name "phonoplex" to the new system, which by now was potentially more versatile than a mere "way duplex."[9] In any event, the Hamilton wire proved too short for a good test, so Tate began using a much longer line to Ottawa along the tracks of the Grand Trunk Railroad.[10]

Edison was uncharacteristically absent from this first field trial of a major invention.[11] Not only was it taking place hundreds of miles from his base in New York, but he was still shuttling to Boston in September and October. Tate, a former telegrapher, was a good choice for the task but he ran into problems, as might be expected, and did not at first appreciate the importance of frequent and full communication. Samuel Insull impressed on him that Edison "wants you to go into every detail and on every little point. . . . [Y]ou should remember that it is absolutely necessary for Mr. Edison to be posted on every possible detail. You may not think a thing important whereas it may give him the clue to the resolution of one of your greatest difficulties."[12] One of Tate's difficulties was the interference of induction from Morse signals on adjacent lines. Edison advised him on the strategic placement of condensers to block the unwanted currents (see Doc. 2863). By mid–November, Edison was sufficiently satisfied with the system to quote royalties for its use by the Baltimore & Ohio Railroad (see Doc. 2866).

Edison also modified the receiver, substituting a magnetic telephone for his motograph and placing a weight on its vibrating diaphragm loosely enough to hop up and down noisily. This instrument discriminated better against induction and produced distinctive timbres on the upstroke and downstroke, like the Morse receivers to which all operators were accustomed.[13] After making tests over a telephone line to Boston in mid–November, Edison refined the transmitting apparatus to produce sharper electrical impulses. He embodied both sets of changes in a patent application executed on 12 November. A second application twelve days later covered a phonoplex repeater, while a third at the end of the month protected his arrangement for dividing a long Morse line for phonoplex "way" use (see Docs. 2867 and 2879). All three applications encountered delays in the Patent Office, in part because the examiner objected to the word "phonoplex" as "new in the art" and consequently "vague and indefinite." Edison complied by replacing every occurrence with "induction."[14] The change represented a subtle semantic shift in emphasis from the receiver (and the importance of its sound) to the transmitter (and its distinctive impulses), but in Edison's mind and business plans the system remained the phonoplex.[15]

Because Edison based his phonoplex on new arrangements of familiar components like condensers, induction coils, and telephone receivers, it is hardly surprising that similar systems emerged about the same time (as was also the case with wireless telegraphy, where his first step was an adaptation of another inventor's patent).[16] François Van Rysselberghe invented a method of simultaneous telephony and telegraphy that was used commercially in his native Belgium by about 1885, and British inventor Charles Langdon-Davies patented a similar system (named the "phonophore") in 1884, though it did not see practical service for several years.[17] Nor is it surprising that rival claimants came forward, including several with inside connections to the Great North Western Telegraph Company. Tate had reason for concern about testing the phonoplex under their eyes while Edison's Canadian patents were not yet complete, at least with respect to the latest improvements (see Doc. 2874 esp. n. 5).

Edison swore in May 1886 to "bring the invention to perfection."[18] When it became an ongoing business concern shortly afterward, Samuel Insull took the role of manager, and Tate acted as electrician. During that summer, Insull solicited interest and answered numerous inquiries, mostly from railroad companies. At least six railroads, the Great North Western Telegraph Company, and the United Lines Telegraph Company reportedly had the system in use on their lines by October, on royalty terms favorable to Edison. By the end of 1886, Edison had spent some $15,634 developing and testing the system.[19] The phonoplex remained in service for years and ranks as one of Edison's more successful and broadly used inventions.

1. On Edison's earliest efforts toward a way duplex, see, for example, Doc. 392. His 1877 work on a sextuplex system suggestively combined elements of quadruplex and acoustic telegraphy (see Docs.754 [headnote] and 808), and his 1878 idea for a "Way duplex" incorporated a telephone receiver, a crucial element of what would become the phonoplex (see Doc. 1415 esp. n. 2). See Doc. 2632 regarding Edison's renewed interest in the way duplex in 1884.

2. In fact, Edison's new way-station telegraph system was similar enough to the wireless telegraph that he had to consider the possibility that one might interfere with the other when used on nearby lines. See Doc. 2857.

3. As in the case of the wireless telegraph, Edison's work with these components took him suggestively near questions that were vexing the best scientific minds about the nature of electrical charges and their transmission through conductors and seemingly empty space. His continued involvement with the phonoplex throughout 1885 likely kept such questions in mind and perhaps contributed to his renewed interest in the physics of energy at the end of 1885 (Docs. 2780 and 2804 [headnotes]).

4. See Doc. 2780 (headnote). Dr. Abner Rosebrugh, a Canadian inventor and potential rival to Edison, explicated the critical role of condensers in both the wireless and phonoplex systems in the course of staking his own priority claim before the Canadian Institute (Toronto) in 1886. Rosebrugh 1886; Doc. 2874 n. 5.

5. Two figures from Edison's fundamental patent show the basic arrangement of any number of stations. The first drawing shows three stations (**A**, **B**, and **C**) for transmitting and receiving phonoplex signals. As shown in more detail in the second figure, each would have the ordinary Morse instruments (relay **a** and key **b**) shunted by condenser (**G**); a telephone receiver diaphragm (**H**); and induction coil (**D**) and key (**F**) for transmitting. The resistance provided by the electromagnet (**K**) would help route induction currents through the telephone and Morse pulses

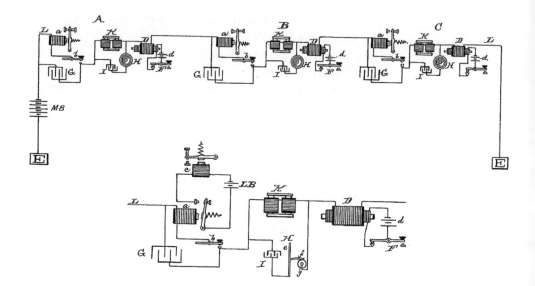

to the sounder. The second drawing includes a local battery (**LB**), used with a back-point sounder to generate a distinctive sound on the key's backstroke. U.S. Pat. 333,289.

6. U.S. Pats. 333,289 and 333,290. A number of contemporary (or nearly so) publications offered lucid descriptions and illustrations of the mature phonoplex system, including the refinements discussed below. Among them are an undated promotional pamphlet published in the names of Samuel Insull and Alfred Tate. "Edison Phonoplex System of Telegraphy," n.d. (1886?), PPC (*TAED* CA012A); "The Edison 'Phonoplex' or 'Way-Duplex,'" *Electrical World* 7 (17 Apr. 1886): 177; "The Edison 'Phonoplex' or 'Way-Duplex,'" *Electrician* 16 (7 May 1886): 516–17; "The Edison Phonoplex," *Engineering* 42 (22 Oct. 1886): 411–13; Hammer 1889, Maver 1899, 353–55; and International Textbook Co. 1901, 2:60–67.

7. U.S. Pat. 437,422; see also Doc. 2857.

8. The Patent Office saw not the flexibility but a duplication of established multiplex technologies. It initially refused to grant Edison's claims because "No patentable combination is believed to exist between the two sets of telegraphic instruments." Edison's attorneys successfully countered that the invention "makes it possible to utilize duplex and quadruplex lines for 'way' business, which is an end that has been unsuccessfully sought for years. Edison was the first to reach the goal. He was obliged to invent the phonoplex or induction instruments, and then to provide special means for making them operative in connection with duplex or quadruplex apparatus. An invention of the highest character has been produced." As was the case with three later applications (discussed below), the Patent Office also objected to the term "phonoplex." Charles Kintner to TAE, 3 May 1886; Dyer & Seely to Commissioner of Patents, 31 Dec. 1886; both Pat. App. 422,072.

9. See Docs. 2849 and 2850.

10. Tate to TAE, 3 Oct. 1885, DF (*TAED* D8546ZAV); Israel 1998, 241–43.

11. Edison's remove from the commercial installation of a new technical system was not unprecedented. In 1883, after having set up village plant electrical systems in Sunbury and Shamokin, Pa., and Fall River and Brockton, Mass., he largely left their break-in periods to his deputies, whom he also authorized to construct additional plants. The vexations and inefficiencies of those experiences, however, might reasonably have led Edison to foreswear repeating them. See, e.g., Docs. 2424 (headnote), 2490, 2563, 2615, 2709.

12. Doc. 2862.

13. See Docs. 2859 esp. n. 10, 2869, and 2870; the instrument is presented as Doc. 2868. Tate reported that the new instrument eliminated the "frying pan" sounds and "morse hash" of induction (see Doc. 2870 esp. n. 5). Among the advantages later claimed for the phonoplex was the ability to work unimpaired in wet weather that nearly disabled Morse apparatus ("Edison Phonoplex System of Telegraphy," n.d. [1886?], PPC [*TAED* CA012A]).

14. The Patent Office's linguistic objections—and Edison's revisions—in these cases were similar to those in the two October applications. Charles Kintner to TAE, 9 Apr. and 7 May 1886, both Pat. App. 422,074.

15. The term "phonoplex" carried explanatory connotations to contemporary ears. As one descriptive article noted, the name "suggests immediately an application of the telephone, and such is indeed the case." "The Edison 'Phonoplex' or 'Way-Duplex,'" *Electrical World* 7 (17 April 1886): 177.

16. See Doc. 2780 (headnote). After the speedy issuance of Edison's first two U.S. phonoplex patents, it took him and his attorneys considerable time to persuade the Patent Office of the novelty of improvements in his next five applications, all of which issued in 1890. U.S. Pats. 422,072; 437,422; 422,073; 422,074; 435,689.

17. Tucker 1978; Tucker 1974. Edison somehow acquired copies of the results of tests on the phonophore made by telegraph expert Josiah Latimer Clark in the latter part of 1886. Clark to J. H. Duncan, 30 Oct. 1886; Clark to Langdon-Davies, 30 Nov. 1886; Clark report, Dec. 1886; all Miller (*TAED* HM860292, HM860295, HM860296).

18. Doc. 2948.

19. "The Edison Phonoplex," *Engineering* 42 (22 Oct. 1886): 412; TAE agreement with United Lines Telegraph, 23 Mar. 1886, Miller (*TAED* HM860283); also cf. Doc. 2866 nn. 2–3; Ledger #5:388–90, 469–73, Accts. (*TAED* AB003, images 203–4, 231–33). Regarding royalties, see Insull's correspondence from August and September 1886 copied into Phonoplex Letterbook LM 12 (*TAED* LM012); cf. Doc. 2866. Records of phonoplex contracts with roughly two dozen railroads were kept in a large bound volume. The records include the number and length of circuits, number of offices, and royalty rates; many also have schematic drawings of the lines showing relative locations of offices and batteries. Most of the contracts date from the late 1880s to mid-1890s, with some from the early twentieth century. "Cases Vol 1," Edison Phonoplex System, CR (see *TAED* CK509).

Technical Note:
Phonoplex¹

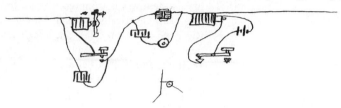

Dyer— I find that with Regular Key arranged this way it is as good & much simpler than with sounder closing on both back & front points³   You Know that when you open a primary of a coil the wave is twice as strong as on closing as the primary closed ckt absorbs the induction— hence on depressing key primary opens loud sound is heard in telephone representing down stroke   on opening key primary closed a weaker sound is heard in telephone representing up stroke— This does away with necessity of dash pot & sounder⁴

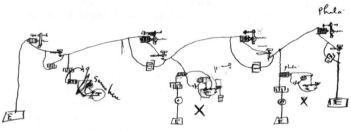

One X at each station.⁵

at terminals only condenser round Key   at all intermediate stations Condensers around Key & Relay—
   Type No 1⁶

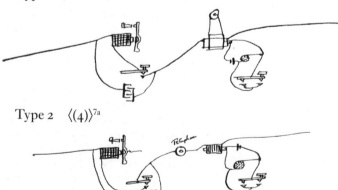

Type 2   ⟨(4)⟩⁷ᵃ

Condenser may be around key only— other way preferred—
Type 3 ⟨(5)⟩[8a]

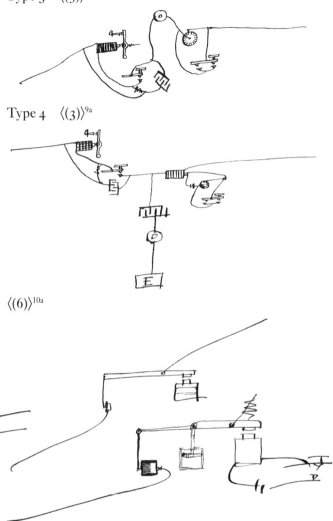

Type 4 ⟨(3)⟩[9a]

⟨(6)⟩[10a]

may have res[is]t[ance]. instead of condenser. not so good. Prefers electro-Motograph—Diaphragm sounders Condenser works opposite to Magnet—no resistance at first[11]

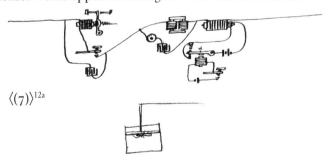

⟨(7)⟩[12a]

⟨(8)⟩[13a]

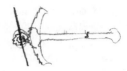

ADDENDUM[b]

[New York, c. April 28, 1885[14]]

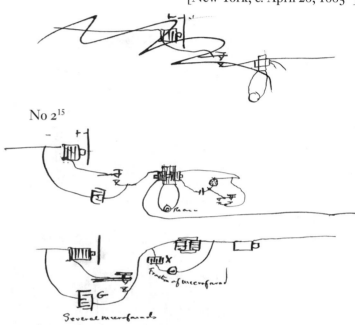

No 2[15]

Dyer= Here is important fact to base claim on—[16] The sound on telephone is just as loud with $^{10}/_{1000}$ of microfarad in X as with several microfarads, while the morse writing heard on the telephone increases in loudness & [distrubing?][c] faster as we increase the microfarad capacity of X   Hence claim such a capacity at X as will diminish the ~~mors~~ regular morse below the [distrubing?][c] point— on the other hand the disturbing power of regular morse is diminished if capacity of condenser G is increased—

Edison

X, NjWOE, Lab., Cat. 1151 (*TAED* NM020AAE). Document multiply dated. [a]Marginalia likely written by Richard Dyer. [b]Addendum is an X. [c]Illegible.

1. The editors have assembled this document from ten loose-leaf pages outlining a multiple telegraph system that would become known as Edison's "phonoplex" (see headnote above). The drawings have been grouped first by their association with two related patent applications that Edison completed about this time and then arranged according to the

numbers provided by Edison (where available) and someone else (probably patent attorney Richard Dyer); the final section concerns the proposed claim, which Edison usually wrote at the end of a patent draft. Edison executed one application for a fundamental patent (U.S. Pat. 333,289) on 27 April and another for a supplemental patent three days later (U.S. Pat. 333,290); both issued on 29 December 1885. The latter was mistakenly published under the title "Duplex Telegraphy" instead of simply "Telegraphy," an error corrected by the Patent Office in 1886. Preserved with the pages of this document was another sheet with three numbered drawings very similar to those published in the first specification; these probably were made by draftsman Edward Rowland (image 20).

2. Draftsman Edward Rowland dated several of these pages 27 April, likely at some stage in the preparation of the resulting patents.

3. This drawing is similar to the one immediately above but also includes a local battery and back-point key, as shown in the fundamental specification's first figure.

4. Edison nonetheless included the "dash pot & sounder" in his supplemental application (see below).

5. Edison indicated earth as "E" in the drawing above; figure labels are "same here," "X," "Phila," "X," and "Phila." Edison appears to have added the mechanism shown in the drawing below at "X" in the center of the drawing above. Also on the drawing above where the figure label "same here" appears, Edison crossed out the figure and wrote "N[o] G[ood]." The drawing above corresponds generally to figure 1 of the supplementary patent (U.S. Pat. 333,290). That specification covered a specialized transmitter "constructed to make and break circuit with great rapidity, so as to send for each signal a large number of such momentary and sharply-defined waves," further differentiating signals to the telephone receiver from those intended for the Morse sounder. Edison's preferred mechanism was a rapidly revolving break wheel (represented by the small circle near each "X"), akin to that used in the railway telegraph for a similar purpose. The sketch below appears to show small components incompletely erased above. The sounder, "with both its front and back points connected in circuit and operated by a key and local battery," was described in the supplemental application, where the components appeared in figure 1.

6. The arrangement shown in this drawing is a variation of the supplemental patent's figure 2, depicting Edison's preferred placement of the "circuit controller" (represented by the small circle at right). The unusual arrangement of the induction coil may have confused Edison's patent attorneys, and he redrew it more clearly on a separate sheet (see addendum below).

7. Figure label is "Telephone." This drawing shows the circuit controller in a shunt circuit, which, depending on the action of the key, could send rapid impulses through the induction coil. It became figure 4 of the supplemental patent.

8. This drawing was the basis for figure 5 of the supplemental specification. It shows another variation of the circuit controller, shunt circuit, and induction coil.

9. This drawing, another variant of those above, became figure 3 of the supplemental patent.

10. This drawing, corresponding to figure 6 of the supplemental patent, shows an alternative to the rotating make-and-break circuit con-

troller mechanism. Here a reciprocating mechanism, modulated by a dashpot, would slide up and down a contact block.

11. The drawing (which Edison made on the same sheet as the two sketches immediately following) incorporates a variation represented in figure 4 of the fundamental specification. It shows a back-point sounder (in lieu of a regular key) and local battery in the induction coil circuit. By working the sounder, "signals are thrown upon the line in the form of momentary and sharply-defined waves, which are responded to only by the [distant] diaphragm-sounders, the regular Morse relays not acting quick enough to respond to these waves."

12. The device shown in this sketch appeared in the supplemental patent as figure 7. The dashpot would retard the movement of the contact block mechanism shown previously.

13. This device, shown in figure 8 of the supplemental patent, was a ratchet-driven retarding fan intended for the same purpose as the dashpot. Draftsman Edward Rowland signed and dated a larger sketch of the same mechanism on 27 April. Cat. 1151, Lab. (*TAED* NM020AAE [image 21]).

14. The editors conjecture that Edison made these drawings (on a single sheet) to provide more detail to his patent attorneys about the unusual induction coil arrangement shown above as "Type No 1." The arrangement appeared in the supplementary specification, which Edison signed on 30 April.

15. Figure label is "[Recver?]." The completed drawing, a completed version of the crossed-out one immediately preceding it, became the basis for figure 2 of Edison's supplemental patent. Its distinctive feature is an induction coil with two separate primary windings. Edison's application did not describe this coil or its connections, which are unlike those in the other drawings, and the Patent Office apparently raised no questions about it. Edison did, however, illustrate and describe such a coil in an application that he executed in October 1885 for a sextuplex telegraph system. Pat. App. 333,290; U.S. Pat. 437,422; see Doc. 2857 n. 3.

16. Figure labels above are "Several microfarads" and "Fraction of microfarad." The drawing represents a partial diagram of a phonoplex office. The first figure in Edison's fundamental patent (U.S. Pat. 333,289) showed three such offices along a telegraph line. (The second patent drawing showed in detail a variant arrangement of a single office.) The specification explained that the effect of regular Morse signals on the telephone instrument was proportional to the capacity of the condenser in the telephone circuit but inversely proportional to that of the condenser in the key circuit. These relationships formed the basis of one of the patent's twelve claims.

–2801–

Memorandum to Richard Dyer: Multiple Telegraphy

New York, April 29 1885[a]

Dick=

I send you diagram for another patent its on the lines of the Quadruplex   one message is sent by reversing the Current the other by increasing and decreasing Current. To use

this principle on way wires I struck a cute thing= Say I have
no reverse men are working & keys X X all along the line are
open   I have no battery from their apparatus= on the line
there is constantly connected say 16 cells M. ½ of these may
be at one terminal & ½ at other or all together anywhere in
line. This Current is used to signal on common relay by increase & decrease by means of Key & Resistance[1]

The battery S S'; is exactly twice as many cells as M but
always thrown to line in the opposite direction to M   hence
of the 32 cells 16 neutralize the 16 of the main battery   put
16 on the line but in opposite direction thus giving a reversal
& working polarized relay but not effecting the Common relay as there is always the same [-----][b] amount of Current=
You know that in the Quad as in all cases the Common relay
at the moment of reversal loses its magnetism for an instant
and this causes the relay lever to leave its front point for an instant   thus mutulating[2] signals [----][b] it is very quick but still
it muttilates   to obviate this in the Quaudruplex The back
point is used to close the Sounder hence at the moment of
reversal the lever jumps back nearly striking the back point &
does frequently[c] strike & give a false sound.[3] To reduce this
defect to the minimum I have struck very good thing   I place
a large magnet n in connection with the regular magnet &
shunt the latter with it pBut placing a condenser of large Capacity in circuit with it   The Resistance of this magnet is 2 or
3000 ohms or more while the regular magnet is only ordinary
Relay resistance 150 ohms   at the moment of reversal a powerful waves due to the discharge & recharge of the condenser
takes place and this wave has its greatest power at the Exact
moment the ordinary magnet has no magnetizm   This wave
charges the big mag n & holds it for an instant= This wave
is momentary, thus you see at the moment when the regular
magnet has no magnetism n has which ceases when the common magnet has—

Claim this=

also number of stations on one line, provided with common
& polarized—

Combin throwing in out Res to work common

The manner of reversing the direction of the Current on
line at a number of stations

This method of reversing for use in teleghy.

This way station system is entirely original & you o[ugh]t
get some very broad claims=

The only defect is that while the Common relay op[erato]rs

can interrupt the polarized men must only close or interrupt while the sender is open   this is no objection as they can hesistate every 10 words or so—[4]

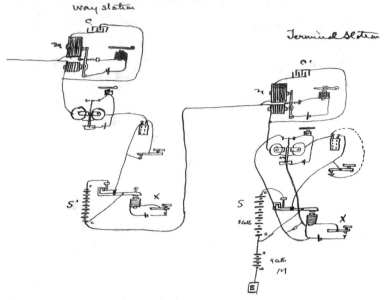

Edison

ADS, NjWOE, Lab., Cat. 1151 (*TAED* NM020AAF). Letterhead of Bergmann & Co. Electrical Works. [a]"New York," and "188" preprinted. [b]Canceled. [c]Obscured overwritten text.

1. This document outlines a means for extending quadruplex telegraph service to intermediate ("way") stations. Richard Dyer incorporated most of its elements (except as noted below) into a patent application that Edison signed on 6 May. The application contained seven claims, two of which were deleted during the examination process. It issued otherwise essentially unchanged in December 1885 as U.S. Patent 333,291 for a "Way-Station Quadruplex Telegraph." Pat. App. 333,291.

2. Possibly an inadvertent misspelling, but this is also an archaic spelling of "mutilating." *OED*, s.v. "mutilate."

3. This unwanted movement of the relay armature was a longstanding problem with the quadruplex. Edison had tried addressing it in a number of ways, including the use of a condenser in the circuit (see, e.g., Doc. 449 [headnote] and Maver and Davis 1890, 49). His proposed solution here was not included in the May patent application that Dyer prepared (Pat. App. 333,291).

4. Figure labels are "way station" and "Terminal station."

–2802–

From Richard Dyer

My dear Mr Edison—

I leave second case on telegraphs—[1]

How about the Ry stock? We haven't seen anything of it yet

Dyer

⟨Insull   give Dyer 75 shares of treasury stock   Edison⟩[2]

ALS, NjWOE, DF (*TAED* D8544L). Letterhead of Thomas A. Edison. [a]"New York," and "188" preprinted.

1. Dyer referred to the patent application for the phonoplex transmitter (later issued as U.S. Patent 333,290) discussed in the notes accompanying Doc. 2800. Edison signed it the same day.

2. Edison sometimes presented stock shares to friends and associates as gifts, and to employees as either gifts or compensation (see, e.g., Doc. 1668). Dyer held forty-eight shares of the Railway Telegraph & Telephone Co. as of November 1885. Edison directed Samuel Insull to give journalist William Croffut fifty shares of the company in the hope "that it may be worth something some day." Other stockholders, apart from the principals and financiers, included Dyer's partner Henry Seely (15 shares), Charles Batchelor (163), Samuel Insull (116), John Ott (20), Martin Force (20), and office assistant John Randolph (10). TAE to Insull, undated 1885; Railway Telegraph & Telephone Co. list of stockholders, 21 Nov. 1885; both DF (D8503ZEX, D8546ZCK).

–2803–

Technical Note:
Wireless Telegraphy

Menlo Park— April 30th to May 6th [1885][1]

100 percent is based on 140 × 1 feet of sheet iron 20 feet from telegraph line (of 7 wires bunched) and 15 feet from ground

Sheet iron		back	high	wires	opposite	percent
Sheet iron		20 feet	15 feet	4 wires	opposite	60 to 70 percent
" "		20 "	" 15 "	" 1 "	"	15 to 20 " "
" "		20 "	" 12 "	" 7 "	"	95 to 100 " "
" "		20 "	" 10 "	" 7 "	"	90 to 95 " "
" "		20 "	" 5 "	" 7 "	"	75 to 80 " "
" "	flat	20 "	" 5 "	" 7 "	"	75 to 80 " "
" "	" "	20 "	" 2inches"	7 "	"	15 to 20 " "
" "	upright 30[a] "		" 15 feet "	7 "	"	70 to 75 " "
" "		30 "	" 15 "	" 7 "	" 98feet of sheet iron[b]	50 to 55 " "
" "		30 "	" 15 "	" 7 "	" 70feet " " "	30 to 35 " "
" "		30 "	" 15 "	" 7 "	" 42 " " " "	25 " "
" "		30 "	" 15 "	" 7 "	" 28 " " " "	15 " "
" "		60 "	" 15 "	" 7 "	" 140 " " " "	50 to 60 " "
" "		60 "	" 2 "	" 7 "	" 140 " " " "	25 to 30 " "
" "		180 "	" 15 "	" 7 "	" 140 " " " "	15 to 20 " "

on last experiment tried low resistance Coil and found it increased volume of sound to 25 to 30 per c

Double Jump= one end 70 feet of sheet iron 20 feet from
line & 15 feet from ground and other end 205 feet from line
and 5 feet from ground with fine rain falling 15 to 20 percent
by doubling up condenser boxes on telegraph line.

W. T. King[2]

X, NjWOE, Lab., Unbound Notes and Drawings (1885) (*TAED*
NS85ADC01). Letterhead of Thomas A. Edison, Central Station,
Construction Dep't.; written by William King. [a]Obscured overwritten
text. [b]"98 feet of sheet iron" interlined above.

1. Edison later testified that the principle underlying his and Ezra
Gilliland's railway telegraph patents were tested at Menlo Park a few
weeks after his return to New York in late March 1885. Edison's
testimony, *Edison and Gilliland v. Phelps*, p. 38, MdCpNA (*TAED*
W100DKF).

These tests were conducted outdoors (probably on Western Union
wires), where a storm on 1 May "completely wrecked" the experimental
apparatus. William King rebuilt the equipment with new lumber but
then had to fix an erratic motor. He finally had "everything in good
shape" by the end of 4 May when, too tired to see Edison personally in
New York, he prepared a short written report of his activities. King to
TAE, 4 May 1885, Unbound Notes and Drawings (1885), Lab. (*TAED*
NS85ADC03).

A few days later, Edison sketched an experimental setup apparently
intended to focus the discharge from the induction coil into a narrow beam. The coil would discharge through a tube "2 foot long   8
inch dia" towards a receiving plate with a galvanometer in the circuit
(Unbound Notes and Drawings [1885], Lab. [*TAED* NS85ADC04]).
Edison made at least one other undated sketch likely connected with
these experiments (Unbound Notes and Drawings [1885], Lab. [*TAED*
NS85ADC01 image 57]).

2. The editors have not positively identified King apart from his
role in the railway telegraph experiments. According to Ezra Gilliland,
Edison hired King in mid-April, after the latter's discharge from "the
Telephone Company," presumably American Bell. Possibly he was the

*Edison's 6 May drawing
of a spark coil experiment,
apparently for discharging
it in a focused beam.*

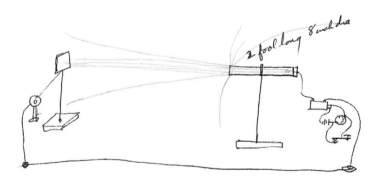

same William T. King listed in New York directories as a draftsman living in Brooklyn and working at 55 Broadway in Manhattan, home office for the Petroleum Exchange and an electrical promotion and brokerage firm, among other concerns. *Trow's* 1884, 905; ibid. 1885, 936; Gilliland's testimony on behalf of Edison, *Edison and Gilliland v. Phelps,* pp. 17, 20, MdCpNA (*TAED* W100DKB); Berly 1884, 84, 102, 120, 142; "A Broker in the Tombs," *NYT,* 16 Mar. 1885, 8.

ENERGY RESEARCH   Doc. 2804

Doc. 2804 is a tantalizing assemblage of Edison's undated writing on three large sheets of drafting paper. Edison used five of their six sides, completely filling three sides with notes arranged in two- or three-column lists. The sheets were each folded once from left to right, possibly one into another, creating eight more or less distinct written pages.[1] It is not easy to discern the order in which Edison produced them, and the absence of any date (contrary to his usual meticulous practice) raises the obvious possibility that he did so at different times.

Reasonable conjectures may nonetheless be made about the sequence, approximate date, and purpose of these notes. Treating each folded half-page as a unit, the editors have transcribed them in a rough progression that starts with Edison's comparative observations on the behavior of various electromagnetic appliances (including condensers, magnets, and conductors) and the effects of energy in them. The sequence ends with more general considerations of energy, including physical manifestations by which its passage might be detected (though the train of thought is not so neat as this dichotomy suggests). This progression also corresponds with a decrease in the neatness of Edison's handwriting, a decline evident in many documents where his fingers seemed barely able to keep pace with his accelerating thoughts.

Edison probably made all these notes at or nearly at the same time, most likely in April or early May 1885. Although they do not refer to specific objectives or experiments, nor to particular scientific publications, it is easy to see them as directly related to his work on railway telegraphy in the winter and the phonoplex system in the spring.

The notes are associated, at least in a general way, with the translation of energy from one form to another. From the start of April 1885, when he sketched schemes for deriving electricity from a beam of heat or light, this topic would recur in

Edison's speculations, notably in the notebooks from his honeymoon in 1886.[2]

It is also possible to view these writings as the foundation of a renewed search for unidentified forces or forms of energy. This idea had been largely dormant in Edison's mind since he started intensive work on electric lighting in 1878, but he clearly had such a project in view by May 1885 (the latest of the likely dates for these notes), when he undertook "Experiments to discover a new form of Energy" (Doc. 2805). He also told a writer about a new search for a force he labeled simply "xyz" (Doc. 2811). In December, he dedicated a notebook to the "discovery of a new mode of motion or energy and also to the conversion of heat directly into electricity" (Doc. 2872), though he did not systematically pursue this program until his 1886 honeymoon in Florida. He found then the leisure to let his mind play, as he had on previous visits to that state.[3] Some of the otherwise puzzling experiments he outlined at that time (such as the effects of passing beams of light or heat through liquids) can be linked to ideas in this document.

Edison seems to have used these notes as a springboard into what he did not know, either about particular devices or more abstract properties of energy. Although they contain statements that he would have accepted without question, such as the increase of heat in a resistor with the square of the current (Joule's law), many are more provisional, written as conditional statements or as suppositions drawn (as by analogy) from accepted facts.[4] Reading across from one column to another in this document, one can often see him using knowledge of one type of energy, such as an electric current in a circuit, to reach plausible conclusions about a related form.

One noteworthy example is the juxtaposition of remarks on condensers and magnets, devices that exemplify seemingly unrelated phenomena. By late April, however, Edison was using the two types of devices as complementary filters or switches for line signals in his experimental phonoplex telegraph. His explanation to patent attorney Richard Dyer that "Condenser works opposite to magnet" in the system evokes the reciprocal behaviors described in these notes (see Doc. 2800).

Edison's work on the telegraph systems in early 1885 came in quick bursts, giving him both the stimulus and the time to take up open-ended questions. He had enjoyed a similar period in 1874–1875 before becoming absorbed in sustained telephone research. He explained the electromotograph principle as a "new force" in 1874, and his search for similar effects culminated in his announcement of an "etheric force" the next year. That "force" was widely (though probably incorrectly) dismissed as an uninteresting effect of familiar electromagnetic induction, and Edison dropped the claim in embarrassment.[5] During that period, he also dabbled with at least one of the phenomena cited in this document.[6]

Edison was predisposed to speculate broadly about energy and its transmission through space. Even in 1878, the etheric force sprang readily to his mind after a three-year hiatus. According to a newspaper interview at the end of that year, recent electric light experiments showed "some more indications of the presence of some subtle, evasive force that I could not call electricity or anything else with which we are acquainted." Asked if it seemed to be like the mysterious etheric force, Edison answered, "Yes, it is the same—if that was anything, and I rather think it is something—a new radiant force, lying somewhere between light and heat on one hand and magnetism and electricity on the other." Edison concluded with the wish that "some ambitious students would take hold of it and solve the problem."[7]

The bent of Edison's thinking toward as-yet undiscovered natural forces seems to have been shaped by the constructive interaction of technical experience with intuition (no doubt nurtured by his reading of Michael Faraday).[8] His imagination was stirred by the intertwined notions, prevalent though not fully explained, that energy is indestructible and exists in forms that are convertible from one to another. He also apparently embraced the less conventional position, synthesized some twenty years earlier by John Tyndall, that energy is essentially kinetic in nature and intrinsically associated with the production of heat.[9] Nearing the end of his fourth decade, Edison was primed by years of inventive work and a lifetime of wide reading to recognize unexpected concordances between practical devices and theories about the physical world.

One topic that recurs—implicitly or explicitly—across Edison's columns is the passage of electrical energy through a dielectric medium—that is, an insulator. Edison's most immediate point of contact with this general subject was his considerable experience with condensers, the devices, familiar to telegraphers but still unstandardized and incompletely understood, for storing electrical charge. The condenser, with its conducting plates and intervening dielectric layers, furnished

Edison with a mental model to explain the operation of the railway telegraph (the so-called grasshopper system) he developed in January and February (see Doc. 2780 [headnote]). In that system, bursts of electrical charge jumped from a railroad car to trackside wires, in defiance of the obvious insulating properties of the air, a phenomenon which Edison explained as a condenser operating on a grand scale. The storage and release of electrical charge through the condenser could also serve as an example of the more abstract case of energy translating through space or the unseen electromagnetic field. Edison, slipping in these notes from the particulars of the condenser to the general properties of the dielectric, may have been doing much the same thing in a more oblique way.

Questions about the generation and transmission of energy through space were at the forefront of physics at this time. Edison was familiar with Faraday's *Experimental Researches*, and his own subsequent investigations suggest an acquaintance with Maxwell's *Treatise on Electricity and Magnetism*, the touchstone work for a generation of physicists. Most of the recently published papers on the subject were highly mathematical and beyond Edison's reach, but two particular sources may have prompted him to think about it in a general way. One was Harvard professor John Trowbridge, who, in a vice presidential address to the American Association for the Advancement of Science in Philadelphia in 1884 on the question "What is Electricity?" speculated provocatively on the connections among electricity, magnetism, light, heat, gravity, and other forces of attraction.[10] Edison had been in Philadelphia for the International Exhibition and would at least have heard informal discussion or comment on the presentation.[11]

A second source was Oliver Heaviside, a reclusive Englishman who defied the boundaries between mathematics, engineering, and physics. A former telegrapher, Heaviside was among a small coterie of "Maxwellians" intent on understanding the late Scottish professor's work on the propagation of energy. Where Maxwell had provided a rigorous mathematical language for Faraday's vision of the lines of force crisscrossing space, it might be said that Heaviside's goal at this time was to elaborate the physical implications of those mathematical abstractions for the actual world in which energy and matter exist. His work appeared regularly in the *Electrician,* a respected weekly trade journal to which Edison subscribed.[12] In early January 1885, the *Electrician* began publishing a long series of papers (ultimately twenty-four) on

"Electromagnetic Induction and Its Propagation." The first four installments constituted what one historian has aptly termed an "introduction to field thinking for the intelligent non-mathematical electrician."[13] Heaviside himself called the first, third, and fourth parts a "rough sketch of Maxwell's theory," and these were technical enough. The second part, however, framed as a discursive digression, was qualitatively different. Heaviside's engaging thought experiments and witty prose made this article a model of lucid scientific exposition, one plainly accessible to Edison's understanding. Its subtitle, "On the Transmission of Energy through Wires by the Electric Current," was misleading because Heaviside set out to demonstrate by reasoned argument the conductive properties of the space around a wire—the electromagnetic field—through which the energy would actually pass.[14] It appeared in the 10 January 1885 issue of the *Electrician* and would have reached Edison's library about the middle of the month, just before he began working on railway (induction) telegraphy.[15] Although Edison does not seem to have used Heaviside's name in connection with his own speculations nor made any acknowledgment of the author, the article is highly suggestive of his own line of thought and could well have catalyzed his thinking.[16] Even if it did not do so, the article (its unusual rhetorical style excepted) typified the fundamental questions about electric charges and currents percolating through scientific and technical publications in these years, questions that, from whatever sources, clearly engaged Edison's imagination.[17]

1. Of the two remaining page equivalents (each half of a large sheet), one is blank. Edison filled the other with calculations and columns of numbers, seemingly data points. Unable to learn anything about their meaning, the editors have not transcribed them.

2. See Docs. 2792, 2793, and 2912 (headnote).

3. See Docs. 2912 and 2609 (headnote).

4. The notes to Doc. 2804 identify sources for some of Edison's factual statements, but the editors have not attempted to identify them all.

5. Historian Ian Wills provides a useful narrative of Edison's etheric force investigations in the course of arguing that they represented a particular way of doing scientific research. See Docs. 419, 425, 462–63, 499, 542 (headnote), 570, 581, 665–66, 678–70, 673, 678–80, 685, 690, 693, 701, 718, and 726; Wills 2007, Wills 2009.

Inventor David Hughes had an unhappy experience in 1879–1880 in Britain with an "Etheric Force" that registered audible clicks in a telephone not connected to any wire. Like Edison's so-called "etheric force" in 1875, Hughes's discovery was dismissed as an inductive effect, and he dropped the subject (Hughes and Evans 2011, chap. 9

esp. pp. 202–205; Hawks 1974 [1927], 168–74). Brown Ayers, a student at the Stevens Institute, independently made observations similar to Hughes's in 1878 (Doc. 1242 n. 4).

6. Namely, the effects of magnetism and electricity on the physical dimensions of iron wire; see Doc. 597. Sometime in or about 1874, prompted by a lack of published data on phenomena noticed by various other investigators, Edison undertook a series of experiments on the basis of which he drafted two pages of an essay on "laws" correlating the charge and discharge of an iron magnet core with the charge and discharge of a long insulated conductor. Notes and Drawings: Telegraph—Notes and Essays, n.d. [1874?], Lab. (*TAED* NS7402BB).

7. "Two Hours at Menlo Park," *New York Daily Graphic*, 28 Dec. 1878, Cat. 1241:1091, Batchelor (*TAED* MBSB21091); see Doc. 1651 n. 12.

8. The distinction in physics between the terms "force" and "energy" was the product of a relatively recent—and ongoing—conceptual transformation. Edison's frequent use of "force" for what would now be understood as "energy" suggests both his almost visceral understanding of the relationship between electricity and magnetism and his indebtedness to the intellectual framework provided by Faraday. Nahin 2002, 115–16; Coopersmith 2010, chap. 15; Smith 1998, 1–2.

9. Tyndall 1863; cf. Yavetz 1995, 144–45, 270.

10. Trowbridge titled his talk "What Is Electricity?" (Trowbridge 1885a). In 1896, he expanded his ideas into a book of the same name (Trowbridge 1896). He directly addressed the abstract question of his title only at the end of the book, after twenty-one chapters dealing with magnetism, other forms of energy, batteries, condensers, and Maxwell's theory of electromagnetism, among other things. In October 1884, soon after the Philadelphia meetings, William Thomson delivered his much-anticipated "Baltimore Lectures" at Johns Hopkins on "Molecular Dynamics and the Wave Theory of Light" (Moyer 1983, 48–49, 75–79).

11. Edison and Trowbridge were both appointed in 1884 to a committee charged with establishing a uniform standard of light. Though Edison does not appear to have actively participated in the committee's activities, he did send Trowbridge some specially designed lamps and the two men communicated in early 1885. "Scientific Intelligence. Physics," *American Journal of Science*, 3rd ser., 28 (no. 167): 389; see Doc. 2791.

12. Yavetz 1995, 174–76; Nahin 2002, chap. 7 esp. pp. 100–106; Mahon 2009, chaps. 5–6; Hunt 1991, 33–47; Smith 1998, 290–99.

13. Yavetz 1995, 77–87.

14. Heaviside began puckishly with the premise of a thought experiment:

> Consider the electric current, how it flows. From London to Manchester, Edinburgh, Glasgow, and hundreds of other places, day and night, are sent with great velocity, in rapid succession, backwards and forwards, electric currents, to effect mechanical motions at a distance, and thus serve the material interests of man.
> By the way, is there such a thing as an electric current? Not that it is intended to cast any doubt upon the existence of a phenomenon so called; but is it a current—that is, something moving through a wire? [Heaviside 1894c, 434]

Then he was off, tracing the movement of energy through the hypothetical telegraph circuit in a manner by turns autobiographically reflective and resolutely analytical. Heaviside reached this provocative conclusion, divorcing the particular identity of electrical current from a broader notion of energy flow:

> Had we not better give up the idea that energy is transmitted through the wire altogether? That is the plain course. The energy from the battery neither goes through the wire one way nor the other. Nor is it standing still. The transmission takes place entirely through the dielectric. What, then, is the wire? It is the sink into which the energy is poured from the dielectric and there wasted, passing from the electrical system altogether. [Heaviside 1894c, 437]

15. Heaviside 1894b.
16. In 1878, Edison referred specifically to an article in which Heaviside attributed the so-called etheric force to effects of electromagnetic induction. See Doc. 1651 esp. n. 33.
17. J. J. Thomson prepared an expansive review of theorizing about the nature of electricity for the September 1885 meeting of the British Association for the Advancement of Science. He gave particular attention to Maxwell's theories about the dielectric, their relation to Faraday's work, and their subsequent interpretation by others (though not Heaviside). Although published too late to have a bearing on Edison's thought at this time, Thomson's article remains a useful synthesis of the state of the art at mid-decade. Thomson 1886, esp. pp. 123–42.

–2804–

Technical Note:
Energy Research[1]

[New York, April 1885?]

Condenser— charges best open— Iron best closed.[2]
    "   discharges quickest shortest circuit— Iron open ck
    "   charge current same direction— Fe opposite direction—
    "   Discharge Current opposite direction— Fe same direction—
    "   better Insulation longer hold charge & ~~open~~
                 Fe better Condr longer hold charge   See if so
Open holds charge   Fe closed holds charge
charge heats dielectric— Fe charge heats metal.
Discharges slower through high Res ck   Fe—low res ckt.
Charges quicker through ~~high~~Low Res ckt— Fe chgs quicker thro high   See
Capacity ½ by doubling weight— Fe Capacity doubled by double weight

Discharge time square of Res—4 times slower double Res—
Fe 4 times slower ½ Res— with given Res twice Capacity
Res gives 4 times time.

with high Res—long time charge   Fe with high Res short time
to charge or Vice Versa   with low Res long time charge[a]

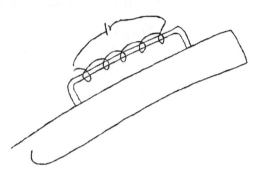

<table>
<tr><td>

Condr
L̶Small Low[b] Res—quick charge—

High Res   Slow discharge

Twice w̶e̶i̶g̶h̶t̶ the [length?][c] dielectric [¼?][d] charge 4 times quicker charge.[e]

High Res same initial volts slow fall=[g]

Two condensers Multiple arc twice quantity. same Emf—4 times slower charge—

two in series twice Emf ½ quantity twice as quick discharge—

charging time independent of strength battery o̶n̶l̶y̶ ̶o̶n̶ ̶R̶e̶s̶ ̶ ̶ ̶S̶l̶o̶w̶ ̶a̶t̶ ̶h̶i̶g̶h̶ ̶R̶e̶s̶ depends on capacity[j]

</td><td>

Mag
High Res—quick charge

Low Res   slow discharge

Twice w̶e̶i̶g̶h̶t̶ length[b] Fe t̶w̶i̶c̶e̶ ̶t̶h̶e̶ c̶h̶a̶r̶g̶e̶4 times slower[f]

Low Res. Same[h] initial volts slow fall.[g]

two mgts series twice quantity ½ E̶m̶ s̶a̶m̶e̶ 4 times slower discharge[h]

Two mag multiple arc t̶w̶i̶c̶e̶ ½[b] the quantity ½ ½ twice the Emf

same[g]   only on Magnetic Res [quic?][i]

</td></tr>
</table>

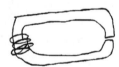   

closed ck quickest[k] discharge

If perfectly insulated open will hold charged

The greater the surface Exposed & less thickness greater the charge—

charge Increases[k] with close promity— discharge on low Res

open circuit quickest discharge

If perfect Conductor closed will hold charge

The less the surface Exposed & the greater the thickness the greater the charge

charge increase by greater distance discharge on low res circle of iron

[Circ---?]^[i] of condenser Rapid bigger the [Circ---?]^[i] lower must be Res

[Thermo?]^[d] Rapid the higher res must be

D̶i̶Time of discharge at the square of the length of the Magnetic [circuit?]^[i] with a given Res to discharge—

T̶discharge time^[b] also at the square of the Res ½ the Res around core four times longer to discharge—^[1]

Then making fine bundles of dialectric should lengthen the discharge in a condenser; non insulated from each other

If making magnetic circuit of very fine wires to quicken the discharge—insulated from each other

Notes— If field mag of Dynamo b̶e̶ c̶l̶o̶s̶e̶d̶ with no back be enveloped field pieces & all with solid Copper say 4 inches thick it will be charged up to saturation gradually by the armature & remain that way without any Exterior Energy & the Volts will fall with the load   it can be regulated by opening the Copper or Counter winding probably require in hour charging it—

If we double the length of the core we h̶a̶v̶e̶ 4 times the air space a̶f̶t̶e̶r̶ ̶t̶h̶e̶ ̶R̶e̶ w̶h̶i̶c̶h̶ whose Res should Equal the magnet.

The small the Source of E Res & greater the outer Res greater the Energy available

The Small^[k] the source of M & the lesser the outer or air Res the greater the available L[ines] of force—^[m]

Electric Ckt

A great deal of energy can be transferred with little heat in the Conductors

Heat is as the square of the Current^[3]

when the internal Res of source is equal to External maximum amount available outside.

Magnetic Ckt

A great deal of energy can be stored up with little loss in the iron

when the res of Magnetic circuit is equal to air space maximum magnetic^[k] arc or greatest length of air space permissible—

A static device may be charged from Ends of circuit—ie[n] condenser

If such device have plates seperated it will hold chg indefinitely—

A static permanent magnet may be charged from Magnetic arc

If such per mag have poles closed it will hold charge indefinitely—[4]

If poles closed charge goes of as heat & matter is free

If poles open doesnt go off but probably would if whole of steel was glass hard & pole piece off—the softened portion probably acts as a keeper by induction through air space.

A sound is heard when charging & discharging the Condenser

i cell gives one pressure

2 cells two pressure

2 cells will overcome twice the space in liquid

Charging matter between sides circuit quantity or[k] work stored is in proportion to surface & thinness of matter

Rapidity of discharge at the square of the thiness—

E passing through heated wire heats it more[o]

a sound is heard when charging & discharging a Magnet

i magnet gives i magnetic pressure

2   "   "   2   "   "

2 magnets in series will overcome twice the air space—

charging iron by wire work stored up proportion to smallness of surface & thickness of matter.

Rapidity of discharge at the square of the thickness—

Magnetism passing through heated[k] magnetic[h] ckt cools it more—[o]

Ckt

Length of arc proportionate to Emf—

Conductivity directly proportion to metal

Doubling current 4 times the heat.

Magnetic Ckt

Ditto[k] magnetic arc

Doubling Magnetism 4 times the cold while being magnetized—

Doubling current twice magnetizing power

Electricity produces heat in circuit continuously—[p]

Closed circuit no diference Emf between any one foot & another except slight amount due to Res that foot

passage E in circuit charges air between the outgoing & ingoing

Electricity passes through a liquid Compound if interpolated in circuit

nothing being static[g]

no dif magnetism[g]

same

Passage of magnetism will discharge it

Magnetism passes through both liquid & solids—

Res of liquid same section[q] ~~1.0~~Million[k] times greater than the metal—amount of current depends on pressure

Res of Liquid Coductor[k] diminished by heat.

Two Ends dipping in the ocean very close Res proportion to distance But[k] at great distance Res independent of distance

heating metal of circuit diminishes amount current by raising resistance

The two sides of the circuit repel each other.

~~Several~~ two separate circuits with battery 1 volt[r] each of ~~1 ohm~~ 1 unit Res give each 44 foot lbs seperately put in series— give on the 2 ohms 88— ~~or same total. on double or 4~~ if in Multiple arc on $\frac{1}{2}$ ohm 88. but if on 2 ohms 22—ie[n] $\frac{1}{4}$ the projectile force.[50]

~~depending~~ not so high depends on pressure—

Res of same for Mag increased by heat

Res of air space very close directly as distance   at great distance independent of distance.

Resistance to magnetism increased by heat

Two sides of circuit attract[o]

Electric Ckt

Electricity passing through iron wire increases it diameter & shortens its length

The charge matter[k] will take varies with each insulator

The amount of E storable is ~~independent of the~~ [Res?][d]— is directly as to the surface of unchargeable[k] & thinness of chargeable matter[μ]

The dielectric of a Condenser ought to have a saturation point—& vary with each kind.[t]

~~An electric circuit around iron magnetizes it~~

The charging & dischrg time of a long Electric Ckt is infinitely quick if away from chargeable nonconducting[u] matter[7]

Pressure increases resistance to E of chargeable matter—[p]

Magnetic ckt

Magnetism passing through an iron Wire[k] diminishes[k] its diameter & increases its length[6]

The magnetism[k] ~~iron w~~ metals will take varies with each metal—

The amount of E or M storable is directly as to the smallness of the unchargeable matter & thickness of the chargeable matter ie[n] greatest amount of matter surrounded by the smallest amount of wire.[s]

Iron soon becomes saturated & Cobalt Nickel etc vary in this point

[can Iron?][d]

The charging time of a magnetic circuit of any length should be instantaneous if away from unchargeable Conducting matter.

Pressure decreases resistance to magnetizable matter—[p]

But with given size wire in circuit The charging & discharging time in close proximity to chargeable non condctg matter varies at the square of the distance

~~Conduction in chargeable matter causes charge to be loss & quickly~~

The smaller the amount of chargeable nonconducting[b] matter between the parallel wires of the circuit & the greater the surface of the wire The greater the charge storable[8]

The Capacity of chargeable matter should increase by heat & be lost at intense cold[t]

Conduction of E through compound liquids should ~~diminish~~ increase[b] by heat, in proportion to the res it bears to the whole circuit.

Electricity

chargeable matter sets itself axially between Ends of cicruit—& magnetizable matter equatorially

anything attractable[k] by poles of E circuit

an Electric Condenser has superimposed charges—

An Electric circuit should be solid conductor If Electric condr broken up & made of numerous wire insulated with chargeable matter charging & discg times lengthened[v]

~~An~~[b] Electricity can be insulated by a perfect non conductor.

Doubling Current on Electric circuit gives 4 times heat, or loss of power—[9u]

Conducting matter cannot be charged by E

The best chargeable compound for E—

But with a given size iron core the charging & discharging times when in close proximity to unchargeable Conducting matter varies at the square of the[k] ~~length~~ distance—[b]

The greater the unchargable conducting[b] matter between the two parallel sides of the magnetic circuit & the smaller the surface the greater the charge storeable

The Capacity to Store[k] in the iron should diminish by heat & increase by intense cold

Conduction of mag through a compound solid or liquid should diminish by heat, in proportion to the Res it bears to the whole circuit[o]

Magnetism

~~Magnetizable matter~~ Electro chargeable matter should set itself equatorially between magnetic poles & megnetizable[k] matter axially & Electric unchargeable matter indiferent.

should be repelled by magnetic circuit & vice versa—

magnetism should have same

magnets should be broken up in fine wires— If magnetic conductor broken up in wires & surrounded in the unchargeable conducting matter charging & dischg time diminished, but if solid & surrounded with unchable condctg matter time lengthened[v]

Magnetism can be insulated by a perfect conductor.

Conducting matter can be charge by M—

Should be the least of M

Condenser   The better the insulation of Condenser the longer the sucking in charge & residual charge[w]

The amount of E stored in chargeable matter depends on the Emf of E independent of quantity—

Energy in any Electric device depends on Emf independent of Current—[w]

If there is no saturation for a diaelectric= There probably is[k] but this can be made greater by doubling the amount but ~~but~~

Electricity can only pass through ~~liqu~~ Some insulators when they are liquid

Electricity Causes decomposition of compounds through which it passes

If passage of E through a liquid produces heat[p]

If heat going from a hot to a cold Junction produces E[p]

The amount of storeable E or M depends on the Current & is independent of Emf—

Energy Mag depends on Current & independent of Emf

There should be none for [~~closed magnets?~~][d] iron— There is for iron but you can double the amount.

Magnetism that passes through solids cannot pass through when they are[x] liquid—

Hence magnetism should produce combination in liquids or solids which it can pass.

passage of magnetism through a liquid should produce cold[g]

The magnetism going from a cold to hot junction should produce E[y]

Heat[k]	~~M~~Electricity	Magnetism
Expands matter through which it passes—	Electricity should ~~contract~~expand lengthens & contract in diamter	ditto
Heat stays longest in large wire & longer as it is poor condr & greater capacity for heat	instantly disappears as from conducting matter	Magnetism stays longer in large wire as it is a better conductor & greater capacity for magnetism

X, NjWOE, Lab., Oversize Notes and Drawings (1879–1886, Undated) (*TAED* NS7986BAV1). Miscellaneous sketches not reproduced; columns defined by vertical lines and column heads set off by horizontal lines. [a]Drawing appears at end of first page. [b]Interlined above. [c]"the [length?]" illegibly interlined above. [d]Canceled. [e]"4 times quicker charge" interlined above. [f]"4 times slower" overwritten on "twice the charge" and into space above line. [g]Connected by line to corresponding section in facing column. [h]"same 4 times slower discharge" interlined above. [i]Illegible. [j]"depends on capacity" interlined above. [k]Obscured overwritten text. [l]Paragraph written at end of two-column section. [m]Drawing appears at end of two-column section. [n]Circled. [o]End of two-column section. [p]Paragraph enclosed by brace and connected by line to corresponding paragraph in facing column. [q]"same section" interlined above. ʳ"1 volt" interlined above. ˢParagraph preceded and followed by horizontal lines connecting it to corresponding paragraph in facing column. ᵗParagraph marked by large question mark in left margin. ᵘ"chargeable nonconducting" interlined above. ᵛParagraph interlined and connected by line to facing column. ʷParagraph interlined above. ˣ"they are" interlined above. ʸEnd of two column section; paragraph enclosed by brace indicating connection to facing column.

1. See headnote above. With a few exceptions in the notes below, the editors have not attempted to identify the basis for Edison's statements in this document.

2. Edison wrote this list on a piece of paper with several incomplete graphs, none of which seems related to this text. There is also a rough sketch of a circuit (accompanied by calculations) that may have some bearing on another section of text (see note 5 below).

3. Joule's first law.

4. Figure labels are "N," "S," "iron," and "iron."

5. On the edge of the first page transcribed here, Edison made a rough sketch and some calculations seemingly related to these statements.

6. Doc. 597 n. 1.

7. An obvious example of the opposite case was the Atlantic cable, in which "chargeable nonconducting matter"—insulation—surrounding an exceptionally long conductor grievously slowed the transmission of signals from one end to the other. The counterintuitive ideas of an insulating material being "chargeable" and a conductor being "unchargeable" were inextricably related to field theory.

8. This statement perhaps reflected both Edison's direct experience with condensers and his awareness of recent theory (again, propounded by Oliver Heaviside) on the importance of the surface of a conducting wire for the flow of energy. Yavetz 1995, 167–69.

9. Joule's first law.

Edison's circuit sketch was accompanied by rough calculations of repulsive force.

–2805–

Notebook Entry: Energy Research[1]

[New York,] May 3 1885

Experiments to discover a new form of Energy using the Motograph Mirror[2] as indicator   following Electrodes to be used each pair immersed in each of the substances

Electrodes.

Sulphur—	Zinc
"	Lead
"	Copper
Copper[a]	Zinc
"[a]	Carbon
"[a]	Iron
"[a]	Tin
Lead[a]	Carbon
Lead[a]	Zinc

"a	Antimony
"a	Bismuth
Antimony[a]	Bismuth
Iron[a]	Carbon
Tin[a]	Iron
Zinc[a]	Carbon

Liquids—

Carbon DiSulphide[a]
Turpentine[a]
Alcohol[a]
   "   & chloral Hydrate[a]
Wood Naptha
Benzine[a]
Benzol[a]
Nitro-Benzol
Kerosene[a]
Boiled Linseed[a]
Oil Anise[a]
glacial Acetic Acid[a]
Alcohol & Salycililc acid[a]
Ether
Chloroform
Formic acid[a]
Ammonia
Sulphuric anhydride[a]
glycerine pure[a]
Sugar strong sol in alcohol[a]
Carbolic acid
Creosote[a]
Fusel oil.[b]

Lead peroxidized & Lead in Solution of Linseed oil— also Solution Sugar in alcohol— also protoxide iron sulphate in alcohol— ditto same in glycerine—

Tin & hard rubber Electrodes in Carbon Disulphide— also same Carbon substituted for tin—

J. F. Ott

X, NjWOE, Lab. N-82-06-21:143–45, 153 (*TAED* N238143). Document multiply signed and dated. [a]Entry preceded by check mark in left margin. [b]Followed by "Continued page 153"; two intervening pages of miscellaneous calculations and one of text related to submarine telegraph cables have not been transcribed.

1. See Doc. 2804 (headnote).
2. Edison referred to a galvanometer he devised in 1876 that used

the motograph principle of variable friction to deflect the mirror. See Docs. 774–76.

–2806–

To Compagnie Continentale Edison[1]

[New York,] May 5th. 1885.

Dear Sirs:—

As I informed you some time since, I referred the letter of the German Edison Co.,[2] copy of which you sent me on the 26th. February,[3] to my Geneva friends.[4] I have a communication from them, dated the 31st. March,[5] in which they repudiate absolutely the charge made against them by the German Co. At the same time they make a counter charge to the effect that efforts are constantly made by Edison Companies in Europe to enter into competition with them in Switzerland.[6]

Under these circumstances, I would suggest that you enter into correspondence direct with the Societe D'Appareillage Electrique, 17 Place Cornavin, Geneva, and I feel assured that you will be able to arrive at some friendly understanding with them. Yours very truly,

T.AE   I[nsull].

TL (carbon copy), NjWOE, Lbk. 20:257J (*TAED* LB020257J).

1. The Compagnie Continentale Edison was the holding company that licensed Edison's electric light patents in France and the French Colonies, Belgium, Denmark, Germany, Austria-Hungary, Russia, Italy, and Spain (Samuel Insull to Francis Upton, 23 Mar. 1885, Lbk. 20:199D [*TAED* LB020199D]). It was incorporated in Paris in February 1882 concurrently with the Société Industrielle et Commerciale Edison, which pursued the isolated lighting business and made trial installations, and the Société Électrique Edison, which manufactured lamps and other equipment. The three entities were referred to collectively as the "French Companies" or the "Paris Companies." Described as "apparently distinct, but really very closely united by common interest," the companies had been the subjects of negotiations since at least 1883 toward a "fusion" or merger that would permit their recapitalization and reorganization (Docs. 2182 n. 2, 2574 n. 2, 2593 n. 6; "Our Paris Correspondence," *Electrician* 2 [Aug. 1883]: 246– 48).

2. The Deutsche Edison Gesellschaft für angewandte Elektricität (DEG, or German Edison Co. for Applied Electricity) was organized in Berlin in early 1883 by Emil Rathenau and Oskar von Miller. It controlled Edison's lighting patents in the German Empire and shared, through interlocking agreements with Siemens & Halske, among others, the German manufacturing rights for Edison lamps and electric lighting equipment. Docs. 2448 n. 1, 2480 n. 5, 2555 n. 1; Hughes 1983, 68; "The Edison Light in Europe," *Electrical World* 3 (21 June 1884): 202.

3. Complaints of intrusions into various exclusive European markets had been festering for more than a year (see Doc. 2642). In this latest flare-up, the German Edison Co. alleged in February that its rights were

infringed by Arthur Achard and the Swiss Société d'Appareillage Électrique, which it accused of selling Edison lamps at discounted prices in Germany. The Compagnie Continentale forwarded the matter to the Edison Electric Light Co. of Europe, Ltd. (Louis Rau to Edison Electric Light Co. of Europe, Ltd., 26 Feb. 1885 CR [*TAED* CE85007]; Deutsche Edison Gesellschaft to Compagnie Continentale Edison, 24 Feb. 1885, DF [*TAED* D8535J]). Samuel Insull had addressed the Geneva company in March on behalf of Edison (who was in Florida) and so advised the Compagnie Continentale (TAE to Compagnie Continentale Edison, 9 Mar. 1885, Lbk. 20:158C [*TAED* LB020158C]).

4. The Société d'Appareillage Électrique was incorporated in Geneva in May 1883. It became the exclusive agent for Edison's lighting system in Switzerland by an arrangement with Edison superseding his earlier contract with Ernst Biedermann and others. Doc. 2642 n. 1.

5. Arthur Achard, president of the Société d'Appareillage Électrique, also suggested that the German firm might use this unfounded allegation to further its own goal of expanding into Switzerland. When the Swiss firm later complained of French marketing incursions into Switzerland, Louis Rau questioned the exclusivity of its marketing rights. Achard to TAE, 31 Mar. 1885; Upton to TAE, 15 Sept. 1885; Rau to TAE, 3 Oct. 1885; all DF (*TAED* D8535T, D8535ZBQ, D8535ZBT); TAE to Upton, 21 Sept. 1885, Lbk. 20:499C (*TAED* LB020499C).

6. Writing to the Swiss firm the same day, Edison specifically declined to endorse DEG's charges against it and made a general promise, in case of unauthorized incursions into its sales territory, to "take the same measures to protect you that I would in the case of any other Company operating my inventions." Edison did intercede on its behalf with the Compagnie Continentale in September. TAE to Société d'Appareillage Électrique, 5 May and 21 Sept. 1885, Lbk. 20:257I, 20:499B (*TAED* LB020257I, LB020499B).

–2807–

Memorandum to Richard Dyer: Wireless Telegraphy

New York, May 6 1885[1a]

Dyer—

Take out patent for new method telegraphing without wires especially available for communicating between ships at see across Rivers from Island to Island.[2]

[Fig.] 1   <u>Broad claims</u>   30 to 50 miles—as long as Curvature of Earth is taken in consideration[3]

big drum coated with metal raised & lowered by rope & block[4]

You can telegh from ships 24 miles apart when metallized strip is 100 feet from sea lever[5] & then be 50 feet over the line of Curvature of the Earth[6]

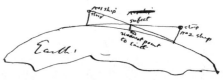

Spose two ships were 24 miles apart they could telegh & If there was another ship 24 miles further the messages could be repeated from the 1st to last ship 48 miles apart & so on if in the lines between Liverpool & NYork Communications cld be exchanged & repeated from ship to ship ½ way across ocean— at sea its very quiet & one only has to contend with the absorpbtion due to the sea, & not to trees houses, hills etc.

E[dison]

ADDENDUM[b]

[New York, c. May 10, 1885][7]

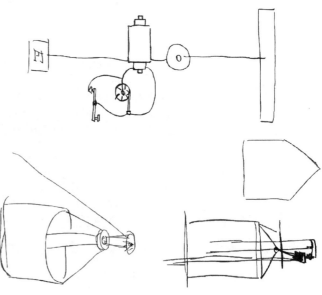

ADS, NjWOE, Lab., Cat. 1151 (*TAED* NM020AAG). Letterhead of Bergmann & Co. Electrical Works. ᵃ"New York" preprinted. ᵇAddendum is an X.

1. Edison dated and initialed this document probably sometime after he finished writing it.

2. Edison executed a patent application prepared from these instructions on 14 May. Covering improvements in the "Art of Electrical Telegraphing and Signalling," it had three broad claims related to the "art of communicating electrically between distant points without the medium of connecting conducting wires," based on the transmission of "electro-static impulses" and their reception at an "elevated condensing surface." The application was rejected and substantially amended at least three times. The Patent Office referred to earlier patents, including those of Wiley Smith and Lucius Phelps, that could pertain under these broad headings. Attorneys for Edison subsequently argued that he had "invented—or perhaps discovered—something beyond these. He is the first to transmit signals to a distance by induction without the use of conducting wires." In the end, the claims were narrowed from a general process to a specific apparatus, and the patent finally issued in 1891 under the title "Means for Transmitting Signals Electrically." The patent was acquired by the Marconi Wireless Telegraph Co. in 1903. Dyer & Seely to U.S. Patent Office, 16 July 1889, Pat. App. 465,971; Dyer and Martin 1910, 830; Guarini 1903.

The final specification (U.S. Pat. 465,971) included figures similar to the first two drawings below and another based loosely on the third drawing. It also illustrated a foil-covered captive balloon for transmitting and receiving messages. Edison's explanation of the wireless system was very similar to that of the railway telegraph, with electrostatic impulses transmitted from one "condensing" surface to another, as in an ordinary condenser, through the dielectric medium of the air (see Doc. 2780 [headnote]). In a popular account of what he called the "air-telegraph" (Edison 1886, 288–89), Edison explicitly equated the two systems.

Unsurprisingly, given its similarity to the railway telegraph, Edison evidently had been talking about this subject with Ezra Gilliland. Following one such conversation, Gilliland sent him a formula and table for determining the curvature of the earth. Earlier, Gilliland had sent a piece of the "tin conductor cable about which we were talking the other day." He added that he did not believe "the tin-foil is of any great consequence. I am not sure, but think they have a patent on the use of it for this purpose." According to account records, experiments seem to have begun in mid-June and continued to early August; Martin Force's experimental notes date from July. Ledger #5:579, Accts. (*TAED* AB003 [image 286]); Gilliland to TAE, 27 and 18 Apr. 1885, both DF (*TAED* D8503ZAL, D8547Y); see Doc. 2836.

3. Figure labels are "Canvass with metallic coating," "motograph," and "Earth plate."

4. Figure labels are "Earth," "Island," "water," and "Islan[d]."

5. Edison probably meant "sea level."

6. Figure labels are "Earth," "No 1 ship," "strip," "~~No 1 ship~~," "50 feet," "nearest point to Earth," "strip," and "No 2 ship."

7. These drawings were likely added in the offices of Dyer & Seely.

The first drawing became figure 3 of the patent and the lower set of drawings were transformed into figure 5.

–2808–

To J. W. Butler
Paper Co.[1]

[New York,] May 7th [188]5

Dear Sirs

I notice in the Chicago Press Dispatches of the 5th inst.[2] reference to a Patent Case, in which you were Defendants, with relation to the manufacture of Paraffine Paper by the Bancroft process,[3] and that you were defeated. I have a paper which is not only a good substitute for, but in many respects vastly superior to Paraffine paper. The Paper I speak of is water and air proof will receive and <u>hold</u> any colors. It can also be perfumed and will hold its boquet for several months.

If you are disposed to arrange with me for the right to manufacture it I shall be glad to negotiate with you on some basis of royalty.[4]

I may mention that I was the first man to design and build a machine for the manufacture of Paraffine paper which I used to use in Condensers but I now use the paper above spoken of[5]   Yours truly

Thos. A. Edison   I[nsull].

L (letterpress copy), NjWOE, Lbk. 20:267 (*TAED* LB020267). Written by John Randolph; signed for Edison by Samuel Insull.

1. Oliver Butler and B. T. Hunt, operators of a paper mill in St. Charles, Ill., opened Chicago's first paper store, Butler & Hunt, in 1844. Butler's brother, Julius Wales Butler, succeeded Hunt as partner in 1856, when the company took its present name. After another intervening name change, the firm was incorporated in 1876. "'Indefatigable.'—Fiftieth Anniversary of the J. W. Butler Paper Company," *The Inland Printer,* 12 (Oct. 1893–Mar. 1894): 321; Butler 1901, 42–43.

2. The *New York Times* published a report from Chicago on 5 May about a patent infringement case, *Hammerschlag Manufacturing Co. v. J. W. Butler Paper Co. and George W. Bancroft,* before the federal court there. The Hammerschlag Co. sued the Butler Co., claiming that George Bancroft's 1882 patent for a machine for waxing paper, used by the Butler Co., infringed Siegfried Hammerschlag's 1877 patent for an improvement in waxing paper. Hammerschlag's patent, reissued in 1878 and 1879, had been upheld in three previous cases, as it eventually was in this matter. "Roscoe Conkling in Chicago," *NYT,* 5 May 1885, 5; "The Patent Holds Good," ibid., 18 Sept. 1885, 3; "Ex-Senator Conkling Argues an Interesting Patent Case Here," *Chicago Daily Tribune,* 5 May 1885, 10; U.S. Pat. 193,867; "Hammerschlag Manuf'g Co. v. Spalding *et al.,*" *Federal Reporter* 35 (July–Oct. 1888): 67.

3. George W. Bancroft of Lynn, Mass., filed a patent application in 1879 for a machine to produce wax paper. Bancroft claimed that his

machine improved on existing processes by producing a smoother and more thoroughly saturated finished product. He received a patent in March 1882. U.S. Pat. 255,129.

4. Julius Wales Butler responded with interest to Edison's offer. Noting that a temporary injunction prohibited his firm from using the Bancroft machine to supply its customers, he requested a sample and more information about Edison's process, its probable costs, and whether it might run afoul of other patents. Butler to TAE, 9 May 1885, DF (*TAED* D8503ZAP).

5. Edison and Charles Batchelor designed and used, perhaps as early as January 1875, a machine for applying a paraffin coating to paper to be used in condensers (see Doc. 698). In *Hammerschlag Manufacturing Co. v. Wood*, heard by a federal court in Massachusetts, the defendants argued that Edison had anticipated Hammerschlag's process for manufacturing waxed paper. Rejecting that claim, the court ruled in Hammerschlag's favor in 1883. The question of Edison's priority arose in another case in 1886 in which Hammerschlag's patent was again upheld. In the meantime, Edison's account records show that he made some experiments in late May 1885 on "Caromel Paper," which could be related to the paraffin paper. Decisions of the Commissioner 1884, 442–43; "Hammerschlag Manuf'g Co. v. Spalding *et al.*," *Federal Reporter* 35 (July–Oct. 1888): 66–68; Ledger #5:579, Accts. (*TAED* AB003 [image 286]).

-2809-

To Edison & Swan United Electric Light Co., Ltd.[1]

[New York,] May 14th [188]5

Dear Sirs:

I beg to advise you that I have sent you by this mail[2] under separate cover copies of the Complaints in various suits which the Edison Electric Light Company has recently brought against the following Infringers[3]

1. United States Electric Lighting Company.
2. The "   "   Illuminating[4]   ".
3. The Consolidated Electric Light[5]   ".
4. Swan Incandescent[6]   "   "   ".
5. William H. Jackson. Ebenezer C. Jackson. and John H. Hankinson. Copartners composing [firm][a] of Wm. H. Jackson and Co.[7] New York. Customer of 3.[b]
6. Michael Costello.[8] New York. Customer of 4.[c]
7. Van. Derveer and Holmes Biscuit Company.[9] Customer of Remington Co[10d]
8. First National Bank[11] and National Bank of the Republic.[12] New York. Customer of 3.[e]
9. Wm G. Mortimer and Richard Mortimer as Trustees under the last Will and Testament of Richard Mortimer Deceased[13] New York.

10. The National Park Bank.[14] New York. Customers of 1 and 2[f]   Yours Very truly,

Thos. A Edison   I[nsull]

L (letterpress copy), NjWOE, Lbk. 20:282 (*TAED* LB020282). Written by John Randolph; signed for Edison by Samuel Insull. [a]Text from alternate versions. [b]"Customer of 3" faintly written in left margin. [c]"Customer of 4." faintly written in left margin. [d]"Customer of Remington Co" faintly written in left margin. [e]"Customer of 3." faintly written in left margin. [f]"Customers of 1 and 2" faintly written in left margin with brace enclosing items 9 and 10.

1. The Edison & Swan United Electric Light Co. was formed in London in October 1883 by a consolidation that allowed the Edison Electric Light Co., Ltd., and the competing Swan United Electric Light Co., Ltd., to avert costly infringement battles over the incandescent lamp patents of Edison and British inventor Joseph Swan. The company reported a profit of £12,354 for the current year ending June 30. See Docs. 2483 and 2514; Hughes 1962, 33; Carosso and Carosso 1987, 271; Edison & Swan United Electric Light Co., Ltd., annual report, 21 July 1885, DF (*TAED* D8534M).

2. Edison sent similar letters the same day to at least four other foreign companies that used his patents. Samuel Flood Page, secretary of Edison & Swan United, acknowledged receipt of his copy and the enclosed documents (not found) about two weeks later. TAE to Société d'Appareillage Électrique; TAE to Deutsche Edison Gesellschaft; TAE to Comitato per le Applicazioni dell'Elettricita; TAE to Compagnie Continentale Edison; all 14 May 1885; Lbk. 20:280–81, 283–84 (*TAED* LB020280, LB020281, LB020283, LB020284); Flood Page to TAE, 27 May 1885, DF (*TAED* D8534J).

3. The Edison Electric Light Co. initiated a series of patent infringement suits in early May. "I have waited for years," Edison reportedly told the *New York World*, "for these imitators and pirates to die out and withdraw, but they managed to get money by hook or by crook and keep in the field. I'll see now if I can't put them where they will stay put." The litigation strategy targeted competitors and their customers in the manufacture, sale, and use of incandescent lamps, fixtures, and other lighting components, which the company's annual report stated were causing "Great injury" to its business. The United States Electric Lighting Co. initiated similar legal action against the Edison firm at nearly the same time, alleging infringement of its own patents. The Edison Electric Light Co. eventually chose, from among its several defendants (not all named in this document), to pursue only the U.S. Electric Lighting Co. Edison Electric Light Co. circular letter, 23 May 1885, UHP (*TAED* X154A4DQ); "The Edison Claims Summarized," *Elec. and Elec. Eng.* 4 (June 1885): 205–207; Edison Electric Light Co. annual report (p. 11), 27 Oct. 1885, PPC (*TAED* CA001A); "Mr. Edison's New Move," *NYT*, 3 May 1885, 2; "Edison on the Warpath," *New York Herald*, 3 May 1885, 25; "Electric Light Patents," *NYT*, 3 May 1885, 2; *New York World*, 3 May 1885, Cat. 1140:4b Scraps. (*TAED* SB017004b); "Notes," *Electrician* 15 (5 June 1885): 55; Bright 1972 [1949], 87–88.

4. The United States Illuminating Co., the operating licensee of the U.S. Electric Lighting Co., had three central stations in Manhattan by the end of the year from which it supplied power for arc and incandescent lighting. It also operated a number of isolated plants, including one at the U.S. Post Office in New York. Doc. 2455 n. 8; "The United States Illuminating Company," *NYT,* 1 Jan. 1886, 6; R.G. Dun & Co. report, 28 Mar. 1885, DF (*TAED* D8526K2).

5. The Consolidated Electric Light Co. owned the electric lighting patents of William Sawyer and Albon Man. Organized in Brooklyn in 1882, it initiated infringement suits against the Edison companies and its customers the following year. In June 1885, it sought an injunction to prevent the Edison companies from using carbonized fibrous filaments. Berly 1884, 68, 235; "Electricians at Law," *Chicago Daily Tribune,* 18 June 1883, 9; "Events in the Metropolis," *NYT,* 14 June 1883, 8; "Electric Light Litigation," ibid., 23 June 1885, 1; "Electric Light Companies at Odds," ibid., 30 Oct. 1885, 8.

6. Established in New York in 1882, the Swan Incandescent Electric Light Co. controlled the Swan electric lighting patents in the United States. Berly 1884, 235, 251.

7. The brothers William H. (1829–1908) and Ebenezer C. (1835–1904) Jackson were partners with John H. Hankinson (1847?–1900) in William H. Jackson & Co., the successor to W. & N. Jackson, a firm of fireplace equipment craftsmen, manufacturers, and merchants in New York since 1827. The company's showrooms were located near Union Square, and its business was growing to include architectural metal work. Of late, some of the firm's capital was apparently being used for private real estate investments by Hankinson and the elder Jackson. Obituary [Wm. Jackson], *New York Tribune,* 26 Nov. 1908, 7; Obituary [Wm. Jackson], *Mantel, Tile and Grate Monthly,* 3 (July 1908): 25; Obituary [E. Conover Jackson], *New York Tribune,* 19 Feb. 1904, 9; Cole 1908, 60; Obituary [Hankinson], *New York Tribune,* 2 Apr. 1900, 5; Obituary [Hankinson], *NYT,* 5 Apr. 1900, 7; Putnam 1886; "Advertisement," *Architect, Builder and Woodworker* 19 (12 Dec. 1883): i.

8. Michael Costello of 929 Broadway. "Electric Light Patents," *New York Tribune,* 3 May 1885, 2.

9. The Van Derveer & Holmes Biscuit Co. was organized in 1876 and operated its factory and salesrooms at 54–58 Vesey St. in lower Manhattan. *New York's Great Industries* 1885, 155; "Electric Light Patents," *New York Tribune,* 3 May 1885, 2.

10. The Remington Electric Light Co. was identified as a defendant in one of the Edison Light Co.'s suits. The firm was part of the manufacturing constellation of E. Remington & Sons, the famous makers of firearms in Ilion, N.Y., which had diversified into sewing machines, typewriters, and agricultural equipment after the Civil War. It had offices at 283 Broadway in New York City. Remington did not appear in an 1884 directory of electrical manufacturers and suppliers but did advertise the "Remington (low tension) system of electric lighting" by 1886, as the Remington organization was sinking into bankruptcy. "Electric Light Patents," *New York Tribune,* 3 May 1885, 2; *Trow's* 1885, 1450; *American Electrical Directory* 1886, 145, 200, 326; Doukas 2003, 82–89; cf. Berly 1884, 230–36; Hoke 1990, 146–48; Hatch 1956, 167–83.

11. One of the leading American banks of this time, the First National

Bank of New York was formed in 1863 with the participation of George Baker, who served as its president from 1873 to 1909. Its building at 94 Broadway was near the National Bank of the Republic (see below). *ANB*, s.v. "Baker, George Fisher"; "The Profits of Baker's Bank," *Moody's Magazine* 15 (Apr. 1913): 107–108; *Appleton's Dictionary* 1889, 15.

12. The National Bank of the Republic was located at 2 Wall St., near the corner of Broadway. Founded with a state charter in about 1850, it became a national bank in 1865; its current president was John Jay Knox, former U.S. Comptroller of the Currency. *Appleton's Dictionary* 1889, 15; "Proposed New Bank Building," *NYT*, 13 Feb. 1880, 3; *Finance and Industry* 1886, 103; cf. King 1893, 724.

13. Richard Mortimer (1791–1882), a native of England, made his fortune in New York real estate and erected several of the city's business buildings, including the Mortimer Building at 11 Wall St. (Hall 1895, 456; Obituary, *New York Tribune*, 31 May 1882, 5). His son, William Yates Mortimer (1823–1891), and grandson, Richard Mortimer (1852–1918), were heirs and executors of his estate. William had recently razed the Mortimer Building and replaced it with a new eight-story structure equipped with electric lighting (Obituary [Wm. Mortimer], *New York Tribune*, 5 Dec. 1891, 5; "Electric Light Patents," ibid., 3 May 1885, 2; "Richard Mortimer's Millions," *NYT*, 6 June 1882, 8; "The Mortimer Building," ibid., 15 Feb. 1885, 7; U.S. Dept. of State n.d., roll 253, R. Mortimer passport issued 9 Feb. 1883; Obituary [R. Mortimer], *NYT*, 16 May 1918, 13).

14. National Park Bank was founded with a state charter in 1856 and became a national bank in 1866. It was located at 214–16 Broadway in a building designed by architect Griffith Thomas. Landau and Condit 1966, 57–58; *Appleton's Dictionary*, 1889, 15.

–2810–

Draft to Erastus Wiman

NEW YORK, May 14 1885[a]

Friend Wiman[1]

I am ~~prepard~~ ready to take any way wire through or local or Railroad with 2 or 20 offices and make two absolutely independent wires of them telegraphically. The second circuit being ~~so~~ as[b] independent of the regular circuit ~~that it~~ as if it never Exitsted.[c] There is no drawbacks as in Quadruplex & duplex where it is necessary to adjust the balance from time to time. If have offered it to the W[estern] U[nion] at an anual royalty of $10. per year per station. But in Canada I think ~~five six~~ six $8[c] per year would be a fair figure, ~~whic This would probably give me an~~ [---][d] ~~income of 25 $2500 a year 300 offices~~ The cost of fitting each station is about $50.[2]

ADf, NjWOE, DF (*TAED* D8546P). Letterhead of Thomas A. Edison. [a]"NEW YORK," and "188" preprinted. [b]Interlined above. [c]"six" interlined above, with "six" and "$8" written successively above it. [d]Canceled.

1. Canadian-born financier and land developer Erastus Wiman (1834–1904) attained prominence and wealth in New York (where he immigrated about 1866 or 1867) as a partner and general manager of the R. G. Dun & Co. credit reporting firm. Wiman held considerable power in the telegraph business as the president and cofounder of the Great North Western Telegraph Co. of Canada and a director of Western Union, and he was also a director of the Edison Electric Light Co. and related Edison companies. Wiman exercised great influence on the development of Staten Island. He gained control of the Staten Island Railway Co. about 1883 and by this time also operated the ferry service to Manhattan. Docs. 2692 n. 1, 2752 n. 4; "Death of Erastus Wiman," *NYT,* 10 Feb. 1904, 7; "A New Route to Coney Island," ibid., 9 July 1885, 2; *American Electrical Directory* 1886, 322–23; Croffut 1886, 175.

2. No reply from Wiman has been found, but see Docs. 2840 and 2866.

–2811–

George Parsons Lathrop Article in the New York Union and Advertiser

NEW YORK, May 22 [1885]
NEW YORK LETTER[a]
An interview With the Wizard of Menlo Park.[a]
Thomas A. Edison, the Remarkable Scientific
Investigator— A Plan to Telegraph
from One Ship to Another at Sea— Other Wonders.[a]
Special Correspondence of the Union and Advertiser

Last week I talked with a Brahmin. This week I have conversed with Thomas A. Edison in his laboratory. When the greater part of the city is asleep; when graveyards yawn— though why they should be addicted to yawning I cannot see since they have no harder work than sustaining epitaphs to fatigue them; when respectable citizens have mostly gone to bed, and gamblers and criminals, together with people of extreme fashion who attend late parties, begin the sweetest part of their life; then it is that the most remarkable of recent scientific investigators goes to work.

We make our way down to a certain corner of Avenue B,[1] where a huge brick building stands unresponsive in the midst of the nocturnal blackness. It would appear dead but for a gleam or two of light in the highest story. According to a preconcerted plan, we knock at the big door and are admitted by a sleepy youth who carries a lantern. "You'll find a light on the next floor," he says, reassuringly, and forthwith lets us in through the trap or weir of the woodwork compartment which forms the office. We clamber up two half-flights of rough wooden steps; find only a single gas jet in the midst of a wilderness of machinery; and proceed to light matches. By their fitful illumination we are able to achieve flight after flight until we reach the fifth story. Still, darkness everywhere, except for glimpses of wide spaces in which machines, belting and posts stand out suddenly in the match glare. At last, stepping upon the final landing, we walk securely under the rays of a few gas jets; for Edison, curiously enough uses gas instead of the electric lamp, in his own workshop. Here, in a small room partitioned off from the rest of the area, we meet the presiding genius of the whole place. The room is perfectly plain. At one end is the new "ticker"[2] from whom the occupant expects to draw a small fortune. Along one side are shelves partly filled with books on chemistry and physics, opposite a complication of wires and apparatus which are explained as an arrangement for transferring messages from one telephone line to another.[3] At the nearer end is a big table, at which Edison is sitting. The first impression is that of a medium-sized, quiet man, with a very large head and a pale, smooth-shaven face. Black hair interspersed with white, surmounts the high forehead in a careless, short trimmed mass oblivious of a parting. One is inclined to say "Here is an actor, a musician, an artist of some sort." The features are strong and clear cut, with that air of mobility under self-control which is so frequent in the physiognomies of actors. The lips are firmly shut when in repose; the chin is decisively molded; yet the general effect is an open and genial one. The moment Mr. Edison begins to speak his face becomes as frank and eager as a boy's but at the same time his eyes—especially when he is describing some scheme for an invention—grow bright with an intense intellectual gaze. Concentration seems to be the great secret of his strength, concentration in a marvelous degree. At one moment he was talking placidly; at the next he was up and active, wholly absorbed in the working of the motograph, which he had had rigged up for the occasion. This is the contrivance for magnifying sound from a telephone, which Edison sold to the Western Union for $100,000, only to find it locked up and withheld from use by that company because it interfered with another patent.[4] He is just as much interested in it as if this had not happened. He quitted his study and hurried to and fro from one end of the building to another, trying experiments. He rushed out to the organ[5] in the main room and played various melodies on it to test the power of the receiver. Some gentlemen came in on business, but he gave no heed to them until he had ended the experiment. All at once he shut himself up in his office and became immersed in business. That done, he once more emerged, and entered into a mock fencing bout with one of the party in waiting; listened to and told stories; danced gayly about among the benches; then resumed sober discourse. All this without the slightest affectation. The man throws himself absolutely into the mood of the moment. He is as ready for play as for work; but the instant he fixes his mind on one thing he is oblivious of all others.

It was rather a weird experience, meeting him there in the great gloomy building, where there are but two men besides himself, at night. He was chiefly engaged with his new idea of telegraphing from railroad trains in motion. This is not to be done by a cable laid along the track, on the Phelps plan, but by throwing the electric current, by induction, to one of the wires alongside the railroad. His experiments have already shown that the spark can be thrown 180 feet. The regular Morse instrument, with certain appliances, will be used. The battery is to be grounded in the wheels of the car, and on the top of the car there will be condensers of tin foil spread upon long strips of wood. Arrangements are also progressing for an experiment in telegraphing by the same method from one ship to another at sea.

"But is that possible?" I asked. "How far do you think you can throw the current over the water?"

"I am afraid to say how far," was the answer. "From the data already obtained, the theoretical conclusion is that we can throw it twenty-four miles. Possibly we can throw it more than that."

Then Edison rapidly sketched on paper a map of the two continents and the Atlantic, and illustrated his plan of telegraphing from ship to ship so as to establish certain communication between the shore and any part of the frequented seas. Not content with this projected miracle, which seems to be near its fulfilment, he is also busy upon improvements in submarine telegraphy. The method now generally in vogue of reckoning words sent through cable by the flicker of a flame thrown upon a mirror is amazingly insufficient, as is shown on a diagram which Edison displayed. The number of dots indicating letters often has to be judged by operators from the length of time that the flame hesitates. Even the siphon receiver, invented by Sir William Thompson and used by one of two of the new cables, is not quite satisfactory, although it marks the dots pretty nearly.[6] Edison is trying to devise some means of attaining a higher or better regulated rate of speed so that the record may be made clearer. But "it's a tough job" he says.

Perhaps the most interesting thing he had to say was respecting his exploration for a "new force." At present he calls it simply xyz.[7] He does not pretend to know what it is. But he says that there are many phenomena which are not explained by any force yet recognized, and it is these which he is going to investigate. Vibrations of matter at the rate of 30,000 a second produce the highest sound we can hear. Between those and the vibrations which, at the rate of trillions per second, cause the sensation of heat there is a large gap; and between these and the vibrations that give sensations of color there is another gap.[8] These gaps, Edison believes, are filled by vibrations as yet unmeasured, which constitute the new or unnamed force he is in search of. He brought out from a drawer sundry loose sheets on which he had sketched a number of machines he had projected, which respond to some influence still undefined.[9] "I jot these down as they occur to me," he said, "and when I get enough of them together I shall have the machines made and try to generalize my observations."

Think of it! A man in this skeptical century who dares believe in a discovery beyond all discoveries! Here is a student of nature who is not afraid to have the spirit of a Galileo or a Keppler or an Isaac Newton. Perhaps we shall learn from him that in returning to faith and insight, aided by bold and patient experiment, we may go forward by going backward.

"What do you think as to the nature of matter?" I asked, unscrupulously. The answer was prompt: "I do not believe that matter is inert, acted upon by an outside force. To me it seems that every atom is possessed of a certain amount of primitive intelligence. Look at the thousand ways in which atoms of hydrogen combine with those of other elements, forming the most diverse substances. Do you means to say that they do this without intelligence? When they get together in certain forms they make animals of the lower orders. Finally they combine in man, who represents the total intelligence of all the atoms."

"But where does this intelligence come from originally?"

"From some power greater than ourselves."

"Do you then believe in an intelligent Creator, a personal God?" was the next question.

"Certainly," said Mr. Edison. "The existence of such a God, in my mind, can almost be proved from chemistry."

Lucretius[10] thought that all atoms were moved by feelings of love or hate—what we call attraction or repulsion. Edison's idea is far more subtle, since he allows the atoms only a germ of intelligence. It also seems to be quite in keeping with the doctrine of evolution, while it contains nothing that is not in harmony with the idealism of the Platonists. And so we discover down on Avenue B, in the prosaic city of New York, a philosopher who believes in a personal God, and is at the same time the foremost exponent of applied science. Curious that he should be at work here, night after night, in the midst of a million people, only a few hundreds of whom know how he is employed during the nocturnal hours! As a usual thing he works until 5 or 6 o'clock in the morning, his supper basket remaining untouched beside him; and sometimes it is 9 o'clock of the next day before he leaves the bench or the laboratory. "I can't think out anything," he says, "except when I'm experimenting. I have a library of 5,000 scientific works, but somehow I can't find what I want in books. How do I make calculations? Well, I don't know exactly. I can't do it on paper. I have to be moving around."

So there he goes, moving around, thinking and working with his hands, in the big sombre building, while the city is asleep. He is the controlling power of several large factories, a millionaire, a man of business, a marvelous inventor; yet he is as simple and happy as a child, when wrapped in an old seersucker dressing gown, he can manipulate as at will and without interruption the mysterious forces and properties of nature. In meeting him I though of him more as a poet or a musician than as a machinist and electrician. Like the Brahmin I saw last week, he deals with occult powers, in quite a different way, but perhaps to the same end, of perfecting man's control over the elements that shape life. It was significant that we climbed a dark stairway to reach his top most place of light and intelligence. Americans are practical and skeptical. It ought to amuse them greatly to learn that the champion of their inventive genius is largely a believer in things unseen and unknowable.

George Parsons Lathrop.[11]

PD, NjWOE, Scraps., Vol. 45, Cat. 1062 (*TAED* SM062045). ^aFollowed by dividing mark.

1. The factory of Bergmann & Co. at Avenue B and Seventeenth St., where Edison had his laboratory on the top floor.

2. Probably the automatic printer referred to in Doc. 2784 esp. n. 2.

3. Probably the repeater discussed in Doc. 2787.

4. Edison sold the electomotograph relay to Western Union, for use in telegraphy, in 1880. He originally intended the device as an alternative to a patent of Charles Page which, at least in principle, covered

the electromechanical relays so common in telegraph systems (see Doc. 1913). Edison's relay was a distinctive device, and it is not clear what other patents it could be said to infringe. Perhaps Lathrop was confused by the Page patent (also controlled by Western Union), but his statement that the company "locked up" the motograph, unused, was essentially correct. Charles Batchelor repeated that claim in a reminiscence years later, and it made its way into at least one Edison biography (see *TAEB* 2 App. 2.5; Jones 1907, 77).

5. This was probably the same pipe organ that Edison had used for telephone and phonograph experiments, as well as amusement, at his Menlo Park laboratory. It was a gift from Hilborne Roosevelt, a New York builder of the instruments, in 1878. Doc. 1190 n. 3; Jehl 1937–40, 137.

6. Lathrop referred first to the mirror (or reflecting) galvanometer of William Thomson and second to Thomson's siphon recorder. Both instruments used a coil suspended between the poles of an electromagnet so as to twist in response to faint incoming signals. In the first instance, the coil transferred its motion to a mirror, delicately suspended from it, that reflected a beam of light onto a screen. The siphon receiver, patented by Thomson in 1867 (though not practically adapted until 1870), ingeniously overcame the obstacle friction posed to making a written record of the receiver's slight motions. It used, in place of the mirror, a capillary tube from which a thin jet of electrostatically charged ink discharged onto a moving strip of paper. Thompson 1910, 1:572–75.

7. See Doc. 2804 (headnote).

8. This belief was roughly consistent with contemporary understanding of the electromagnetic spectrum. The familiar case of sound moving through air was sometimes introduced as an analogy for electromagnetic waves traveling through the ether, while visible light and radiant heat were seen as similar phenomena occurring at different frequencies (see, e.g., Schellen 1872, 55–61; also Doc. 2804 [headnote]). Among the interpreters of James Clerk Maxwell's theorizing about energy and fields of force, George Fitzgerald was making direct use of the electromagnetic-acoustic analogy by this time. It also figured implicitly in the work of Oliver Heaviside and the published lectures of John Tyndall (Hunt 1991, 41–43; Smith and Wise 1989, chaps. 12–13; Heaviside 1894a, 298; Tyndall 1977 [1895]). William Thomson, in an 1880 article published in the *Encyclopaedia Britannica* (9th ed.), argued that

> This process of transference from one body to another body at a distance through an intervening medium is called radiation of heat. The condition of the intervening matter in virtue of which heat is thus transferred is called light, and radiant heat is light if we could but see it with the eye, and not merely discern with the mind, as we do, that it is perfectly continuous in quality with the species of radiant heat which we see with the material eye through its affecting the retina with the sense of light. . . . Thus radiant heat is brought under the undulatory theory of light, which in its turn becomes annexed to heat as a magnificent outlying province of the kinetic theory of heat. [*Ency. Brit.* 9, s.v. "Heat."]

9. The editors have not identified the "loose sheets," but see Docs. 2792, 2793, and 2805.

10. The Roman philospher Lucretius Carus (c. 99–55 B.C.E.) advanced an atomistic explanation of the world in an epic poem, *On the Nature of Things. ET,* s.v. "Lucretius."

11. George Parsons Lathrop (1851–1898), journalist, novelist, and poet, was a former associate editor of the *Atlantic Monthly.* He published several studies of Nathaniel Hawthorne, whose daughter, Rose, he married in 1871. His *Complete Works of Nathaniel Hawthorne* (1883) remained the definitive collection for decades. Lathrop remained associated with Edison for years as an interviewer, prospective business partner, and would-be coauthor of a proposed novel. *ANB,* s.v. "Lathrop, George Parsons"; Israel 1998, 288, 365–69.

–2812–

Notebook Entry: Miscellaneous

[New York,] May 22 1885

I propose to form some new nitrogen Compounds by Electrolysing different Electrodes Coated with various substances capable of being nitrogenized—using Ammonia as the electrolyte   This giving off [am?][a] Nacent Nitrogen by Secondary Action—[1] possibly I can thus produce artificial silk—[2]

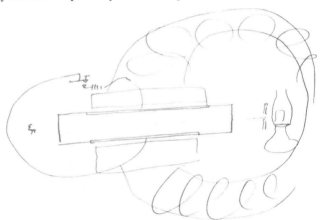

Produces a pressure on the Liquid a la ~~m~~Hg-amperemeter[3] & alters the ray of light   perhaps should be polarized
Process for preparing Oxygen[4]

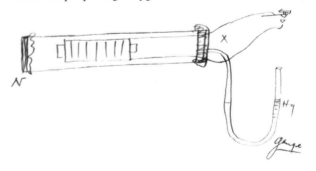

glass tube magnet inside closed at both ends   End N has porous material say bladder—stucco etc fitted air tight= Oxygen being magnetic is[b] drawn through the pores while nitrogen is neither Magnetic[a] or Diamagnetic[b] being[b] absolute Zero=

Red hot platinum for porous filter also cocanut charcoal carbonized paper charcoal—etc

This magnet may condense the gas already in and produce a movement of the diaphragm— idea for Telephone or teleghic instrument[c]

Standard Emf=

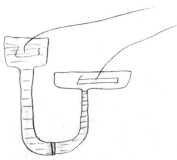

Water (distilled) passing through meerschaum plug[5] by gravity give Emf—plat Electrodes=

Thermo[6]

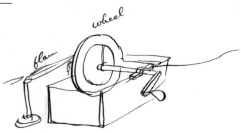

See Preeces Noad p 406.[7]

[A][8]

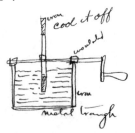

fused salt Borax say

[B][9]

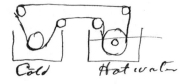

X, NjWOE, Lab., N-85-05-22:3 (*TAED* N308003). Document multiply dated. ªCanceled. ᵇObscured overwritten text. ꞓFollowed by dividing mark.

1. A fairly chemically inactive gas, nitrogen forms compounds more readily in its nascent state; that is, at the moment of its release from another compound, such as by electrolysis. Watts 1873, s.v. "Nitrogen" (esp. p. 63); *OED*, s.v. "nascent" 2.c.

2. Figure label is "Eye." A process for producing artificial silk from nitrated cellulose fibers (commonly used to make highly explosive guncotton) was patented in France in 1884 by Comte Hilaire de Chardonnet. British chemist Joseph Swan, whose wife was crocheting with a silk-like fiber of his creation about 1883, developed a similar process simultaneously, and his announcement of an "artificial silk" in December 1884 is cited by one authority as the first published use of that term. Like Swan, Edison was also interested in creating such fibers to use as lamp filaments, and he had some experience with the process of nitration (as did his rival Edward Weston, toward the same end), which was used commercially in the manufacture of celluloid. *Ency. Brit.* 14, s.v. "Silk, artificial"; see Docs. 583, 586, 645, 655, 2609, and 2627 esp. n. 2; Mossman 1997, 40–43; Meikle 1995, 18–20.

3. This instrument, apparently devised in 1884 by Gabriel Lippmann, professor of mathematical physics at the Faculty of Sciences in Paris, operated on the principle that a current passing through a conductor in a magnetic field will tend to move the conductor parallel to itself and perpendicular to the lines of force. Instead of a solid conductor, as in a motor armature, Lippmann used mercury in a U-shaped tube. Current passing through the bottom raised the liquid on one side and lowered it on the other. After Lippmann published descriptions of this instrument, at least one other person claimed to have made a similar experimental device earlier (in 1881). *DSB*, s.v. "Lippmann, Gabriel Jonas"; "Lippmann's Apparatus for Electric Measurements," *Sci. Am.* 53 (25 July 1885): 50; Hering 1893, 214; "Abstracts. G. Lippmann—A Mercury Galvanometer," *Journal of the Society of Telegraph-Engineers and Electricians* 13 (1884): 618; "Abstracts. An Experimental Mercury Galvanometer," ibid. 13 (1884): 618–19.

4. Figure labels are "N," "X," "Hg," and "gauge." Two common methods of producing oxygen at this time were by the electrolysis of water and the combination of chlorine with steam. It was also evolved by heating or burning various oxides. *Ency. Brit.* 9, s.v. "Chemistry. Oxygen"; Watts 1873, s.v. "Oxygen."

5. That is, a disk or wafer made of the mineral compound meerschaum.

6. Figure labels are "flame" and "wheel." Cf. Doc. 2875.

7. Edison referred to the 1879 edition of Henry Noad's 1867 *Student Text-Book of Electricity*, which William Preece finished revising and brought to press after Noad's death. The page cited was the last in the chapter on thermo-electricity, containing a paragraph on "Thermo Currents from Fused Salts" pertinent to Edison's electrochemical experiments:

> an electrical current is always produced when a fused salt capable of conducting electricity is brought into contact with two metals

at different temperatures, and . . . powerful affinities can be overcome by this current, quite independently of chemical action. The
direction of the current is not influenced by the nature of the salt
or metal, being always from the hotter metal through the fused salt
to the colder. Its intensity is greatly superior to that of the common
thermo-currents, and is capable of decomposing with facility water
and other electrolytes. [Noad and Preece 1879, 406]

See Docs. 1724 and 2271 regarding Edison's prior interest in thermo-
electric generators.
    8. Figure labels are "iron," "cool it off," "insulated," "iron," "metal
trough."
    9. Figure labels are "Cold" and "Hot water."

–2813–

To Edward Johnson

[Boston, c. June 5, 1885][1]

EHJ—

    Took your transmitter[2] amalgamated the two platina Elec-
trodes (tough job)—put it in with the Coil Bell Receiver.
Talked loud but somewhat <u>uneven</u>   put tissue paper over
mouth piece this didnt perciptablely diminish loudness but
stopped air rushes from bouncing points—Evened the talk-
ing= found diaphagm [g̶r̶n̶?][a] of Receiver Vibrated very
strongly & there was little harshness   Put 2 microfarads
Condenser around receiver   this put <u>Velvet</u> sound on talk-
ing= Man talked for one hour & twenty minutes 10 mins of
which was with two Carbon Cells[3] 10 minutes with 1 Carbon
Cell bal[ance] time with One Bergman[4] at the End of the time
talking was just as good & loud as when starting— now from
your experience does this show anything if not what am I to
do to develope the <u>bug</u> = platina cannot be roughened in the
presence of mercury—

                           Edison

    ⟨I should let it stand over until tomorrow & then talk on it
again for a half hour   If that does not settle its hash—then
you have done it. I think you have any way as it is a more severe test than I have ever put them to successfully   EHJ⟩[b]

ALS, NjWOE, DF (*TAED* D8547ZAG). [a]Canceled. [b]Marginalia written by Edward Johnson.

    1. This approximate date is based on a notation written on the letter
in an unknown hand: "June—before 9th, 1885." Edison spent most (if
not all) of the first week of June in the Boston area with Ezra Gilliland.
    2. Johnson filed two patent applications in November and Decem-
ber 1883 for a make-and-break telephone transmitter with "contacts or
electrodes of platinum or other metals, which instrument will maintain

its adjustment and can be readily duplicated." The patents specifically covered the diaphragm, to which one electrode was attached, and means for "locking the metal contact-electrodes together positively." Both patents issued on 25 July 1887 (U.S. Pats. 356,688 [inadvertently omitted from *TAEB* 7 App. 4.C] and 356,689), and both were related to applications filed in November 1883 jointly by Edison and Sigmund Bergmann and in December 1883 by Bergmann individually (see Doc. 2839 esp. n. 3). Edison and Ezra Gilliland viewed all these devices as potential legal weapons against the Bell patents (see Docs. 2833 and 2839). Johnson had also experimented earlier with a thermo-electric telephone. Although this device likely produced undulatory currents, possibly it offered another way around the Bell patents, and a patent application was prepared for him in early 1883 (Johnson draft application, 24 Jan. 1883, DF [*TAED* D8370Q]).

3. This was a common name for a battery developed by German chemist Robert Wilhelm Bunsen, who substituted carbon for platinum in the cathode of the Grove battery. Like the Grove cell, it used a zinc anode in a porous cup filled with a dilute solution of sulfuric acid. Silliman 1871, 579.

4. The Bergmann & Haid battery, placed on the market in 1883 or early 1884, was advertised for telephone circuits as "the greatest open circuit battery in the world." Similar to the Leclanche cell, it consisted of a glass jar with a porous cup containing manganese and the carbon (negative) pole. Surrounding the cup in the jar was a solution of chloride of ammonium and the zinc (positive) pole. Doc. 2743 n. 5; "New Instruments," *Medical Record* 25 (22 Mar. 1884): 334.

–2814–

To Ezra Gilliland

[New York,] June 10th [188]5

E. T. Gilliland
    Am on witness stand   dont see how I can come.[1]

                                 Edison

L (telegram, letterpress copy), NjWOE, Lbk. 20:348 (*TAED* LB020348). Probably written by John Randolph.

1. Gilliland evidently invited Edison to join him and his wife in Boston or at their summer cottage in nearby Winthrop, Mass. (see Doc. 2824 [headnote]). Gilliland answered that they were "greatly disappointed" but hoped Edison could get there on Saturday, 20 June (Gilliland to TAE, 18 June 1885, DF [*TAED* D8503ZAZ]). Edison started giving a deposition in a patent interference case on 9 June. He resumed testifying the next day but did not finish, and his deposition carried into the following week. The case concerned a method of joining carbon filaments to the conducting wires in incandescent lamps (testimony of TAE, 9 June 1885, *Weston v. Latimer v. Edison*, pp. 1–39, MdCpNA [*TAED* W100DFA001]).

*Letter of Introduction
for Francis Upton*

[New York,] June 10 [188]5

To My European Friends[1]

My friend (Mr Francis R Upton) the bearer of this letter has been closely associated with me in my laboratory work ever since my early experiments in Electric Lighting

Mr Upton is visiting Europe in connection with the business of the Edison Lamp Co of which he is General Manager

If any of my friends to whom he may present this letter can in any way further his wishes I shall esteem their efforts the same as if the services were rendered to myself.

Thomas A. Edison.

ALS (letterpress copy), NjWOE, Lbk. 20:342 (*TAED* LB020342).

1. Due to Edison's absence from New York City, letters of introduction were not ready before Upton departed for Liverpool aboard the steamship *City of Rome* on 3 June. Samuel Insull sent at least four letters on 9 and 10 June, along with instructions and Edison's power of attorney for Upton's work in Switzerland ("Outgoing Steamships," *NYT,* 2 June 1885, 8; Insull to Upton, 9 and 10 June 1885; Ernst Biedermann to Upton, 5 June 1885; all Miller [*TAED* HM850262, HM850264, HM850261]; TAE to Guiseppe Columbo, 9 June 1885; TAE to Samuel Flood Page, 9 June 1885; TAE to Werner von Siemens, 10 June 1885; TAE to William Preece, 10 June 1885; Lbk. 20:336–37, 340–41 [*TAED* LB020336, LB020337, LB020340, LB020341]; the original letter to Preece is in WHP [*TAED* Z005BL]; a photocopy of the original to Siemens is in Upton [*TAED* MU082]). Upton reached London by 16 June, from which, five days later, he confirmed receipt of Edison's instructions and letters of introduction (J. S. Morgan & Co. to Upton, 16 June 1885, Miller [*TAED* HM850266]; Upton to Insull, 21 June 1885, DF [*TAED* D8535ZAN]).

Upton had Edison's power of attorney in Switzerland for this trip, permitting him to negotiate with Ernst Biedermann and to examine the accounts of the Société d'Appareillage Électrique in Geneva, which in 1883 had become the exclusive agent for Edison's lighting system in Switzerland. Edison signed the power on 13 June; a notarized copy was presented at the Swiss consulate in New York two days later (TAE to Upton, 9 June 1885; TAE power of attorney, 13 June 1885; both Miller [*TAED* HM850263, HM850265]; Doc. 2642 n. 1). Upton also carried introductory letters from Charles Batchelor (undated to Siemens) and several each from Edward Johnson and New York businessman Robert Hewitt (see correspondence in Upton [Unbound Documents—1885] from 29 May to 1 June 1885 [*TAED* MU08]).

To Franklin Pope

[New York,] 19th June [188]5

My Dear Pope,[1]

I have your letter and should be very glad to write the endorseme[nt][a] you ask for.[2] The difficulty is however that I am largely interested & practically "father" the underground system of the Electric Tube Co just put down for the Met Telephone Co[3] and if you are appointed commissioner you will probably have to pass on the System.[4] Cannot you fix the matter some other way say for insta[nce][a] use my name as a reference. If however you think there is no objecti[on][a] under the circumstan[ce][a] to my writing the letter I shall be delighted to do s[o][a]   Yours Truly

Thos A Edison

LS (letterpress copy), NjWOE, Lbk. 20:364A (*TAED* LB020364A). Written by Samuel Insull. [a]Copied off edge of page.

1. Franklin Leonard Pope (1840–1895) had been in a short-lived partnership with Edison that ended in 1870 with lingering acrimony. Pope went on to a successful career as an inventor and head of Western Union's patent department before going into patent practice himself. He became managing editor of the *Electrical Engineer* in 1884. *DAB,* s.v. "Pope, Franklin Leonard"; "Franklin Leonard Pope," *Trans. AIEE* 7 (1896): 680–82; "Franklin Leonard Pope," *Electric Power* 8 (Dec. 1895): 503–504; Doc. 59 n. 1; *TAEB* 1, chap. 7 introduction.

2. Pope was seeking an appointment to Brooklyn's Electrical Subway Commission, where, according to him, "merit counts and not politics." He asked Edison to endorse him to mayor Seth Low "as strong as you consistently can and oblige an old friend & well wisher." Pope's application to Low was evidently unsuccessful. Pope to TAE, n.d. [c. 18 June 1885], DF (*TAED* D8503ZEY); "The Brooklyn Underground Problem," *Elec. and Elec. Eng.* 4 (Oct. 1885): 380–82; *Minutes and Reports* 1896.

3. The Metropolitan Telephone and Telegraph Co., formed by consolidating Bell and Western Union telephone interests in 1880, was the American Bell Co.'s licensee in New York. The company contracted with Edison in October 1884 to place some of its lines underground in the vicinity of Broadway and Twenty-first St., where it had a switching office. While the work was being carried out in April 1885, Edison invited the New York Board of Police Commissioners, then considering whether to place police wires underground, to come and observe. "City and Suburban News," *NYT,* 16 May 1880, 5; Garnet 1985, 78, "Putting Wires Underground," *NYT,* 21 Apr. 1885, 8; John Cahill to Electric Tube Co., 21 Dec. 1884; TAE to N.Y. Board of Police Commissioners, n.d. [Apr. 1885]; both DF (*TAED* D8533ZAA, D8546N).

The dense tangle of overhead electric lighting, telephone, and telegraph wires in New York City (reportedly some 5,000 miles of them) was a perennial subject of contention over their safety, appearance, and intrusion on public and private spaces. Indeed, the Metropolitan Co. had been sued successfully by the state in 1882 over its lines along Twenty-first St. The jury's verdict against the company was modified and only partially sustained on appeal in early 1884. "The Attorney General Must Appear," *NYT,* 13 Oct. 1882, 8; "The Telegraph Pole Nuisance," ibid., 19 Oct. 1882, 2; "The Wires on the Roofs," ibid., 19 Dec. 1883, 3; *Reports of Cases* 38 (1884): 596–605; see also Doc. 2568 n. 1.

4. The general question of overhead electric lines came before the New York legislature on several occasions. In 1884, it passed a law requiring all telegraph and electric lighting wires to be placed underground by the first of November 1885. Then on 13 June 1885, legislators approved a related bill compelling civic authorities (in cities with at least 500,000 residents) to appoint a supervisory board of commissioners within twenty days of that date. Each board was to consist of three "disinterested persons" and be responsible for implementing and enforcing the underground requirement. Thompson 1891, 71–72; *Minutes and Reports* 1896, 3–10; Sullivan 1995, 133–34.

–2817–

To Samuel Insull

[Boston,][1] June 27 [18]85

Insull—

Tell Hamilton[2] at Laboratory he may have the Vacation he asks— I think Stewart[3] of Chili is honest, and if Johnson can fix up an arrangement between him & W R Grace & Co[4] to work the West Coast of S.A. he better do so by all means getting money in NYork on all goods but make the price reasonable i.e.[a] less than Isolated[5] prices   Isolated say at least 25 pct less—Were not those Lamps sent   Upton promised to ask[b] Dyer[6]= Keep[b] me posted about orders that Lamp factory— Isolated & Batch gets.= Could you come over here to spend 4th at[b] Gills—there is lots pretty girls—[7]

Edison

ALS, NjWOE, DF (*TAED* D8503ZBD). [a]Circled. [b]Obscured overwritten text.

1. The editors have not determined if Edison wrote this in Boston proper or the adjacent town of Winthrop, where his hosts, Ezra and Lillian Gilliland, rented a summer cottage. Edison was reported to be in Winthrop by 28 June. "Table Gossip," *Boston Daily Globe,* 28 June 1885, 13; see also Doc. 2824 (headnote).

2. Edison hired H[ugh?]. de Coursey Hamilton, a native of Suffolk, England, as a laboratory assistant in late 1882. Hamilton had worked on a range of experimental projects. He spent parts of 1886 and 1887 traveling for Edison (probably in South America) seeking plant fibers to use as lamp filaments. Doc. 2425 n. 9; Edison Lamp Co. expense report, 1 Aug. 1888, DF (*TAED* D8828ACQ2).

3. Willis Stewart (b. 1854), trained in dynamo installation and repair at the Edison Machine Works, went to Chile in 1881 on behalf of Edison lighting interests and worked there until 1889. Stewart formed the Compañía Eléctrica de Edison during 1885 as a general agency for Chile, Peru, and Bolivia, but his ambition to control the central station in Santiago was often frustrated by investors in La Compañía de Luz Eléctrica Edison-Santiago, which had been founded in 1882 by Edward Kendall & Co., a former business correspondent of Fabbri & Chauncey. Stewart had recently arrived in New York and met with Samuel Insull,

who summarized their meeting and the proposed arrangements in a 25 June letter to Edison:

> He has got in with W. R. Grace & Cos. Valpariso House which is run by W R Grace's brother. They are willing to go in with him either to back him up with money (receiving his goods & paying for them here) or else they will (Grace & Co) make a contract with the Edison Co. & then make a partnership contract by which Stewart will get a half interest in their contract with the Edison Co. Stewart believes in the South American business as much as ever & he wants to go ahead. Will you please let me have your ideas on the subject so that I can talk to Johnson. [Insull to TAE, 25 June 1885, Lbk. 20:380 (*TAED* LB020380)]

In mid-1885, Stewart secured cooperation for his Chilean endeavors from W. R. Grace & Co. in New York, and he often worked through its branch in Valparaíso, Grace & Co. (Docs. 2602 nn. 1–2, 6, and 2711 n. 3).

4. Building on his early experience as a dealer of supplies and equipment for vessels in Peruvian ports, William Russell Grace (1832–1904) began in the mid-1860s to operate as W. R. Grace & Co., a shipping and merchant agency in New York. By the mid–1880s, the firm and its numerous branches had an extensive trading, investment, and industrial portfolio. It was also the dominant owner-operator of shipping lines between the United States and Chile and Peru, particularly after its Merchant Line absorbed Fabbri & Co.'s business on the west coast of South America. In addition to William (now the mayor of New York), other partners at this time included his brother, Michael Paul Grace (1842–1920), who had authority over the branch in Valparaíso. Charles Ranlett Flint, another partner, was a graduate of the Polytechnic Institute in Brooklyn with long-standing interests in the electric lighting industry (and had one of the first New York homes so lighted). Flint was elected president of the United States Electric Lighting Co. in 1880, and two years later, he unsuccessfully sought a merger of the Edison, Weston, Thomson and Houston interests. Flint also had a knowledge of torpedo guns and boats, developed during his role with Grace as an arms and war matériel supplier to the Peruvian government during the War of the Pacific (1879–1883). *ANB*, s.v. "Grace, William Russell"; Wilkins 1970, 175–76; de Secada 1985, 598–603, 618–621; Doc. 2602 n. 5; *Who's Who in American Finance* (1922), s.v. "Flint, Charles Ranlett"; McPartland 2006, 98; Carlson 1991a, 185–87.

5. That is, the Edison Co. for Isolated Lighting, the principal sales agent for the Edison Electric Light Co. Articles for export to the Americas would presumably have to be licensed by the latter firm. See Doc. 2725 esp. n. 6.

6. Philip Sidney Dyer (1857–1919), bookkeeper for the Edison Lamp Co. since January 1881. Doc. 2435 n. 4.

7. See Doc. 2824 (headnote).

[Boston,][1] June 28, 1885—

Sammy

Tell the World man to he may count my subscn in the general fund—[2a]

I ~~could~~ wouldnt like to answer the telegram reporter[3] until I had more[b] <u>data</u>[a]

E[dison]

ALS, NjWOE, DF (*TAED* D8503ZBE). [a]Followed by dividing mark. [b]Obscured overwritten text.

1. Edison could have written this note, like Doc. 2817, from Boston or nearby Winthrop, Mass.

2. John R. Reavis (1848–1914), an editor for the *New York World*, managed the paper's campaign to raise money for the multi-storied pedestal upon which the Statue of Liberty would stand in New York Harbor. Reavis invited Edison to subscribe to the fund on 9 June and, on 15 June, also asked him to suggest the names of additional donors. In marginalia on the second letter, Edison addressed an unidentified reader: "[Dr?]— if you don't feel too poor here is a chance for a little patriotism." Edison added $250 to the effort that ultimately raised $100,000, mostly in small contributions from readers of the one-cent *World* (Reavis to TAE, 9 and 15 June 1885; Reavis to Insull, 25 July 1885; all DF [*TAED* D8503ZAW, D8503ZAX, D8503ZBQ]; "Decay of Circuses," *Boston Daily Globe*, 14 June 1885, 12; Juergens 1966, 193–94; *ANB*, s.v. "Pulitzer, Joseph"). Reavis had been a journalist for the *St. Louis Post-Dispatch* and founded the *Spectator*, another St. Louis newspaper, in 1880 (Scharf 1883, 944; Obituary, *Hayti [Mo.] Herald*, 15 Jan. 1914, 6; Smythe 2003, 114; Swanberg 1967, 104; Rammelkamp 1967, 67).

3. Edison possibly referred to the *New York Telegram*.

Hotel Chatham[1]   Paris June 28, 1885.

Dear Edison:

I have seen Mr. Rau[2] twice and have found him willing to tell me regarding the European Co.[3] They have lost money heavily during the year 1884 as they were compelled to mark down their[a] assets. The total loss was 500,000 francs.[4] They have no intention to give the Lamp Co. any chance, if it is in their[a] power to prevent our selling lamps.[5] I am now driving at them for information and avoiding the subject of lamps. I represent here merely the financial side of the Lamp Co. and have to refer them to Holzer[6] for technical details.

Biedermann[7] has evidently through his family connections made a good stroke in Vienna. He has the Gas Cos.[8] behind him to put in the installation for the Opera House Theatre[9] and ~~Co~~ Palace[10] in Vienna. This much seems to be a fact. The

Gas Cos. are waking up here to a realization that there is a public demand for the electric light from incandescent lamps. If you had told me to trust Biederman I could tell you a very long story as to possible results and matters already accomplished. Under my instructions though, I am simply putting two and two together and trying to get from outside sources proof of Biederman's assertions.

The power of attorney for Geneva is here.[11] I shall stop there and examine matters.

I do not think there has been much money made as I find they have done very little business.

I shall try to get a concession for Austria, so that the Lamp Co. can supply the lamps for the large installation now going on there. Biederman says there is an order for 20,000 lamps ready to be placed as soon as ~~the~~ we can show our right to sell in Austria.[12]

Unless he is playing a regular Bunco game[13] on me I think he tells the truth. I know from outside sources that a very large sum of money is being put into the plant at Vienna. There is a chance for the Tube Co. as the mains go underground.[14] Yours Truly

Francis R. Upton

ALS, NjWOE, DF (*TAED* D8535ZAW). [a]Obscured overwritten text.

1. International travelers sought out the Hotel Chatham for its location (at 17–19 rue Daunou) near the Paris Opera, theaters, and fashionable shopping districts, as well as for its luxurious appointments and quietude. Constructed in the late 1860 for H. J. Holzschuch, at this time the Chatham featured Otis elevators and electric lights throughout. Phillips 1891, 126–27; Advertisement, *Old World and European Guide* 6 (Spring 1886): 85.

2. Louis Rau (1841–1923) played a major role in the commercial introduction of Edison's electric light in France and Germany. A naturalized Frenchman, he was managing director of the Compagnie Continentale Edison and the Société Électrique Edison. His brother-in-law in New York, Henry ("Harry") Wallerstein, was also the agent for various financial and legal affairs of these companies. Docs. 2249 n. 5, 2581 n. 3, 2593 n. 2.

3. Upton referred to the Compagnie Continentale Edison. The terms "European Company" or "European Edison" were more customarily applied to the Edison Electric Light Co. of Europe, Ltd., in New York City.

4. The Compagnie Continentale Edison's losses were due, in part, to its efforts to build a Parisian central station during an era of credit contraction. Doc. 2667 n. 10; Lanthier 1988, 2:314–15; "The Edison Light in Europe," *Electrical World* 3 (21 June 1884): 202.

5. While the Edison Lamp Company was generally seeking to improve its sales abroad, Upton negotiated with the French companies

to revise and clarify its rights to sell American-produced Edison lamps in countries (other than France and its colonies) where the Compagnie Continentale Edison controlled lamps made under Edison's patents, including those manufactured by the Société Industrielle et Commerciale at Ivry-sur-Seine, near Paris. French patent laws prohibited the resale of imported Edison lamps (Doc. 2806 n. 1; Insull to Upton, 30 Nov. 1885, DF [*TAED* D8316BMC]; Insull to Edison Lamp Co., 23 Mar. 1885; TAE to Compagnie Continentale Edison, 9 June 1885; TAE to Clemens Herschel, 2 July 1885; Lbk. 20:199D, 190, 389C [*TAED* LB020199D, LB020345, LB020389C]; *Ency. Brit.* 9, s.v. "Patents"). The parties reached a provisional agreement in early July (Upton to TAE, 4 and 8 July 1885, both DF [*TAED* D8535ZBA, D8535ZBD]; see Doc. 2822). Similar situations regarding the European rights to sell Edison lamps arose with other companies. In addition to the Deutsche Edison Gesellschaft, these included the Société d'Appareillage Électrique and the Swan United Electric Light Company, Ltd. The latter reached an agreement with the French firms in 1888 (see Doc. 2806; "Companies Meetings. Swan United Electric Company," *Electrical Engineer* 1 [n.s.; 8 June 1888]: 549–50]).

6. Glassblower and machinist William Holzer (1844–1910) was the manufacturing superintendent of the Edison Lamp Co. since March 1882. He was also Edison's brother-in-law. Doc. 2518 n. 2.

7. The editors have not established the place of Ernst Biedermann in the extensive Biedermann family tree, whose wealthy Pressburg (Bratislava)/Vienna branch included founders of M. L. Biedermann & Co., an old Viennese investment bank (*Ency. Judaica*, s.v. "Biedermann, Michael Lazar"; *Banker's Almanac* 1881, 303; McCagg 1992, 64, 67, 83). Biedermann had promoted Edison's electric light and power interests in Switzerland prior to the formation of Société d'Appareillage Électrique and was now negotiating toward the sale of that company to the Imperial Continental Gas Association (see Docs. 1878 n. 2, 1962, 2642 n. 1; Société d'Appareillage Électrique to TAE, 16 May 1885, Miller [*TAED* HM850263A]). He had initially wanted to organize a syndicate for the exploitation of Edison's light in multiple European nations (TAE to James Banker, 16 Nov. 1880, MiDbEI [*TAED* X001A1BM]; Biedermann to TAE, 9 June 1891, DF [*TAED* D9102ACC]). In 1882, Biedermann secured (jointly with Charles Havemeyer) the Swiss rights for Edison's electric railway (TAE agreement with Biedermann, 19 June 1882, Cat. 2174, Misc. Scraps. [*TAED* SB012ACG]).

8. Upton had recently reported that Biedermann had "got at the inner side of the great Imperial Continental Gas Association and they have taken the Vienna job in their hands." The British-based Association, the dominant gas utility company in Vienna, held the municipal concession for electric lighting and wished to use American-made Edison lamps there. Upton to TAE, 22 June 1885, DF (*TAED* D8535ZAR); Weaver 1884, 20–21; Csendes 1998, 75–76; Jussen 1886, 479.

9. The International Electric Co. (formerly the Anglo-Austrian Brush Electric Light Co.) had previously expected to secure this business, but by May 1885 the contract for lighting the Vienna State Opera (Hofopern) and the new Burg Theatre was awarded to the Imperial Continental Gas Co., with the British electrical engineering firm of R. E. B. Crompton & Co. as a third-party signatory. The Schenkenstrasse cen-

tral station, planned for a capacity of 12,000 lamps, was being designed in collaboration with Crompton, British engine builder Peter William Willans, and electrical engineering consultant Demeter Monnier (from the École Centrale in Paris). Intended to operate at a high voltage (700 to 1,200 volts), it required secondary batteries to reduce the voltage at the point of use; the batteries could also supply a small amount of light during the daytime. The theater and its 3,000 lamps were a substantial distance (about three-quarters of a mile) from the station ("International Electric Company, Limited," *Electrician* 12 [29 Dec. 1883]: 168; "Notes: Electric Lighting in Vienna," *Tel. J. and Elec. Rev.* 16 [2 May 1885]: 401; "Notes: The Electric Light at the Vienna Opera," ibid., 21 [26 Aug. 1887]: 217; Jussen 1886, 479; "Lighting of the Vienna Opera," *Times* [London], 1 Sept. 1887, 3; Hedges 1892, 160–62; Crompton 1928, chap. 7; Crompton 1922, 394; "Obituary" [Monnier], *Proc. IEE* 55 [July 1917], 545; Monnier to TAE, 2 Jan. 1886; Upton to TAE, 7 July 1885; both DF [*TAED* D8630A, D8535ZBC]). Biederman's portion of the work apparently involved his rights to the so-called Turrettini high-voltage distribution system and to the secondary batteries of Farbaky and Schenck; by the end of 1887 his role in the project was the subject of public criticism, and he declared bankruptcy in early 1888 (Upton to TAE, 21 and 22 June 1885, both DF [*TAED* D8535ZAO, D8535ZAQ]; Waltenhofen 1886, 600–601; "Failure of a Contractor," *Teleg. J. and Elec. Rev.* 22 [24 Feb. 1888]: 201; "Rundschau," *Centralblatt für Elektrotechnik*, 9 [1887]: 729–33).

10. Portions of the Imperial Palace (Hofburg) were to be lighted with 5,000 lamps. "Electric Light and Power Foreign," *Electrician and Electrical Engineer* 4 (May 1885): 200.

11. The power of attorney permitted Upton to examine the financial accounts of the Société d'Appareillage Électrique regarding royalties due Edison and to deal with the proposed sale of the company to the Imperial Continental Gas Association. TAE to Upton, 9 June 1885; TAE power of attorney to Upton, 13 June 1885; Société d'Appareillage Électrique to TAE, 16 May 1885; all Miller (*TAED* HM850263, HM850265, HM850263A); TAE to Upton, 9 June 1885, Lbk. 20:343 (*TAED* LB020343).

12. The rights for the commercial introduction of Edison's electric lighting technologies in Austria were presently a matter of negotiation between the Edison Lamp Co. and the Compagnie Continentale Edison (Upton to TAE, 22 June and 7 July 1885, both DF [*TAED* D8535ZAR, D8535ZBC]). The Austrian-Hungarian rights had been a recent subject of negotiations between the Edison Electric Light Co. of Europe and the Deutsche Edison Gesellschaft, which wanted to establish agencies beyond Germany, but as recently as January 1885, the European company had refused to allow DEG into these markets (Doc. 2724 nn. 2 and 4). The Swan company had also initiated an infringement suit against the owners of Edison's patents in Austria in late 1883 (Doc. 2598 n. 5).

13. Bunco (or bunko) is American slang for swindles or confidence tricks. *OED*, s.v. "bunco."

14. Upton cabled on 7 July for estimates on three miles of underground copper conductors insulated for 1,200 volts. Upton to TAE, 7 July 1885, DF (*TAED* D8535ZBC).

July–December 1885

The months of July and August were unusual ones in Edison's adult life. Leisure trumped work, although he pursued it with a similar intensity, and even an incomplete account of his activities reads like an itinerary. Accompanied by his daughter Marion, Edison returned to the Massachusetts hospitality of Ezra and Lillian Gilliland about the middle of July, having been there as recently as the first week of the month. This time he visited at their rented summer cottage in Winthrop, on the shore of Boston Harbor a short distance from the city. And this time, Lillian Gilliland, reportedly acting on Edison's expressed intention to seek a wife, arranged for several eligible young women to stay with them, too.

Edison kept a diary during this vacation, as did at least some of the other visitors. Although it is the only known place in which he deliberately wrote thoughts of a personal nature, it was clearly intended to be read aloud for the amusement of others. Edison used this parlor game to signal to his hosts and other guests his romantic interest in Mina Miller, who was not present but with whom he had overlapped at Winthrop a few weeks earlier.[1]

Edison came back to New York about 21 or 22 July, but only for a few days. Then he set off again on a sailing trip with Gilliland and John Randolph (his bookkeeper and office assistant), among others.[2] He briefly returned to his home and office, but by 10 August he and Marion were at the Chautauqua Institution in western New York State. There he visited the family of Mina Miller, whose father, Lewis Miller, was a cofounder of the Institution. The meeting must have gone well; after little more than a week, he left with Marion and

Mina (and presumably her family) on an eastward journey through Niagara Falls and Montreal to the White Mountains of New Hampshire. In the foothills of Mount Washington, he proposed marriage (according to reminiscences by Mina and Marion) by tapping the question in Morse code.[3] Back in New York in early September, he bought sets of six tickets to two operatic or theatrical performances, perhaps to entertain Mina's family.[4] (Others less favored took a back seat. Edward Johnson, a long-time intimate, complained that Edison would not commit to a social engagement "for the simple reason that he is in love and dont want to make any appointments in advance that might possibly conflict with Cupid's demands upon him.")[5] Edison waited until the end of the month to formally ask Lewis Miller to ratify his engagement with Mina.[6]

Amid these comings and goings, it fell to Samuel Insull to oversee the move of the Edison household to the Normandie Hotel (at Broadway and Thirty-eighth Street) in late September.[7] At some time, most likely during his summer traveling and socializing, Edison created an enigmatic set of tables in which he listed various personal and physical traits of friends and associates and speculated on ways that some of the characteristics might combine.[8]

The extended period of leisure, unusual as it was for Edison, was not devoid of business. For example, he wrote seven pages of notes in Boston for Charles Batchelor about armature design and asked about the performance of Edison dynamos in competitive trials at the Franklin Institute in Philadelphia; entertained a proposal to use the phonoplex on Western Union commercial lines; and corresponded with Samuel Insull about tests of the wireless railway (grasshopper) telegraph on Staten Island. He also plotted with Johnson and Gilliland (who left the American Bell Telephone about this time) to devise an alternative form of transmitter that would "bust the undulatory theory" of sound transmission on which the Bell company's legal defense of its patents rested. By 20 July, he wrote Insull from Winthrop, he had "Committed myself too deep to withdraw" from the scheme.[9] And although the timing may have been coincidental, promptly after returning from the trip on which he and Mina became engaged, he approved financial settlements with local electric companies whose central stations he had built at his own expense; the settlements freed him to repay some of his own debts related to the construction.[10] Resuming something more like his normal pace of work over the course of September, he dispatched Alfred Tate, a secretary with telegraphic experience, to Canada to set up practical trials of the phonoplex on the Great North Western Telegraph Company's commercial lines, where success might give him an advantage in negotiations with Western Union.[11]

When Edison resumed his normal schedule in October, it was the phonoplex that occupied much of his time. He urged Tate to write in detail about his experiences and troubles installing the system on lines from Toronto. Inductive interference with regular Morse lines proved to be a major problem. To overcome it, Edison redesigned the receiving instrument, enhancing its sensitivity to the high-frequency phonoplex signals and making it more impervious to spurious impulses. These changes necessitated recalibrating the entire phonoplex circuit, where condensers played crucial roles as electrical filters. By mid-November, the new receiver—and redesigned system generally—were largely complete in the form in which they entered commercial service in 1886, though Edison continued to communicate refinements to Tate through the end of the year.[12] In December, with no immediate technological problems to solve, Edison turned his attention back to general physical questions of force and energy and the search for an elusive "XYZ" force. He dedicated a fresh notebook to speculative "ideas as to the discovery of a new mode of motion or energy and also to the conversion of heat directly into electricity."[13]

In early November, Edison finally completed the purchase of the Fort Myers, Florida, property where he and Gilliland intended to build winter homes and a laboratory. Having previously engaged an architect, now the men contracted for precut materials from which the buildings would be assembled on-site. They also indulged in buying sprees (mostly by Gilliland) for furnishings, supplies of all kinds, and entertainment items, including a self-playing Gally organ. At least six schooners embarked for Florida with their goods over the winter, one of which was struck by lightning and its cargo lost.[14]

Late in the year, three events consummated important relationships that Edison had developed. For one, Ezra Gilliland moved to New York in November with the intention of becoming his partner in business and inventing. Second, Lewis Miller, his future father-in-law, seems to have spent a good deal of time in November and early December in the New York area, where the two men repeatedly went out to dinner and bought beer, sarsaparilla, and cigars. Lastly, Edison returned the favor by visiting Mina's family in Akron, Ohio, the week before Christmas.[15]

1. See Doc. 2824 (headnote).
2. See Doc. 2838 n. 2.
3. See Docs. 2843 and 2844.
4. Voucher (Laboratory) no. 442 (1885) to Brentano's.
5. Johnson to Uriah Painter, 12 Oct. 1885, UHP (*TAED* X154A4FA).
6. Edison's letter to Lewis Miller is Doc. 2853. He sent it immediately upon his return from a week in Boston, where he probably overlapped at the Gillilands' home with Mina, who was expected there on 23 September. See Doc. 2844; TAE to Insull, 25 and 29 Sept. 1885; both DF (*TAED* D8541ZAH3, D8503ZCY); Insull to John Weir, 21 Sept. 1885, Lbk. 21:3 (*TAED* LB021003).
7. See Doc. 2854.
8. The tables are Doc. 2852.
9. See Docs. 2820, 2823, and 2833.
10. See Doc. 2847.
11. See Doc. 2851.
12. See Doc. 2868 (headnote).
13. Doc. 2872.
14. Cat. 1165:11–12, Accts., NjWOE; see Docs. 2860 esp. n. 3 and 2907.
15. Gilliland's testimony, p. 6, *Edison and Gilliland v. Phelps*, MdCpNA (*TAED* W100DKB); Cat. 1165:11–12, Accts., NjWOE; Insull to Erastus Wiman, 14 Dec. 1885, Lbk. 21:170 (*TAED* LB021170).

–2820–

To Charles Batchelor

Boston= July 1 85[1]

It seems to me that if bare cylinder shows rise of 15° covering it with a thick[a] coating of wire through which conduction of heat must take place not [endwise?][a] through the metal wire but cotton must necessarily raise this 15° by accumulation until the external surface of the armature or wire has reached such a temperature as the rapidity of radiation equals the generation. This would account for the Extra 20 degs when the wire was on which is due to the original 15° in the iron & not to any in the wire itself—[2] I dont believe much in whirlpools in large wire=[3] You could prove the amount of heat in wire by using a wooden armature & winding it in the regular way with Commutator etc & see the heat= Even if you did get heat, this may not be due to whirlpools but to an entirely different cause to wit— If You know the wire on an armature when brushes are off is a <u>closed Loop</u>—if by unsymetrical winding <u>on ends</u> etc one side of the loop is longer than the other[4] there will be a diference of electromotive force & this flowing around the loop generate heat= You could prove this by putting two rings on a block of wood over the commutator & breaking this loop where it is soldered to a Commutator block, where both ends of the loop meet, [one?][b] instead of having them together, separate & Connect one end to one ring & the other end to other ring   with two brushes on rings you could run machine not using the regular brushes & thus get <u>reversals</u> & ascertain just how much heat was thus generated= again another way is to run an armature without brushes—get temperature—then disconnect The two ~~branches~~ wires from one of the Commutater blocks & leave them apart ie[c] insulated from each other thus breaking the[a] loop—then start with the temperature of air & make a heat test   if the 2nd test shows less heat than there is a circulation of current within the loop due to want of <u>symmatry at the ends</u>—or pole pieces or somewhere=

Regarding the primary heating of the iron cylinder alone This is the true bug I think[5]   ~~You~~ How would it do to put ~~tape the shaft~~ in a simple shaft & tape it & run it in regular field— If it is the bolts the amount of heat could perhaps be reduced by using brass with phosphor [& to?][b] or arsenic or even plain brass instead of iron— With[a] iron the lines of force are conducted right through it & of course cut the metal—[a] with brass the iron of the plates would act as a shield I think & if you could get the conducting power of the brass as poor as iron this would be an improvement if its the cause=

=Cant you make a cylinder in the regular way with alternate plates & tissue & with flat tools take a <u>cut</u> over [-][b] surface of iron so the tissue will show & with two wires & sounder you cannot get a circuit on surface of cylinder= The [burr?][b] Burr thrown by the tool connects the plates together—[d]

I think that clutch a good one= ~~Why not go ahead with~~ [this?]—[b] I spose you have got the plans for the new motor for the other truck [ne?][b] nearly ready=[6] If I remember right the <u>shape of your field on</u> the truck motor is of such a character as not to bring the neutral point[7] very close to the theoretical   its thus—

shouldnt it be like this

glad you got money from W.U. are they using the machines yet=[8]

Did we beat Weston[9] at Phila[10] & will our bad luck with machines go against us as related to Weston

Should find out who is responsible for Des Moins & charge Walters[11] time & fare to them—

If you could get Sims work it might turn out big thing some day—[12] do you mean to build the whole or only motor= How did other one come out   Have they tested it— If you have struck a non-commutator machine[13] you have the biggest thing for the century but Im afraid there is a bug

Edison

ALS, NjWOE, Batchelor (*TAED* MB170). [a]Obscured overwritten text. [b]Canceled. [c]Circled. [d]Followed by dividing mark.

1. Edison wrote this date with a different ink, possibly at a different time.

2. Whatever report Batchelor sent to Edison has not been found but during the latter part of June, while Edison was in Massachusetts, Batchelor made a series of experiments to determine the relative significance of several factors in the heating of dynamo armatures. These investigations were likely prompted by the 18 June test of a 1,000 ampere low-voltage dynamo that "was found to heat up terribly." Two days later, Batchelor began subjecting small armatures to a series of tests designed to isolate the causes, carefully describing each new set of conditions and the corresponding temperature changes. After testing an armature core bare of windings, he concluded that the "Foucault [eddy] currents are very small. The first 20° are probably due to the changes of magnetism," and he attributed the next twenty degree increase to "the difference of potential between the different coils of the same section and eddy currents." A similar test later gave comparable results. Cat. 1235:272, 276, 278; Batchelor (*TAED* MBN012272, MBN012276, MBN012278, MBN012278a).

3. Edison referred to eddy currents, local circulations of electrical charge typically dissipated as heat.

4. The unevenness may be a result of Edison's 1881 decision to design armatures for relatively small isolated plant dynamos with an even number of commutator bars, which required an asymmetrical pattern of connecting the armature coils at the commutator. He found that this arrangement reduced sparking at the commutator. See Docs. 2124 esp. n. 1 and 2126 (headnote).

5. Edison had proved that eddy currents in the rotating mass of an armature core were largely responsible for its heating, and he and his associates paid great attention to the issue (as did other early dynamo designers) in constructing the basic Edison dynamo. See Docs. 1899 esp. n. 1, 2419 (headnote); Thompson 1886, 82–83, 146.

6. The context of this paragraph suggests that Edison was referring to an electric railroad motor. According to a March 1885 report to the board of the Electric Railway Co. of the United States by its president, Edward Johnson, a committee of Johnson, Edison, Batchelor, and Stephen Field decided to adopt a system of car-mounted motors and friction clutches. Batchelor participated on behalf of the Edison Machine

Works, which planned to build the motors; one was installed on a car and nearly ready for testing by the first of October. Johnson report to Electric Railway Co., 27 Mar. 1885, UHP (*TAED* X154A4DE); "Edison's Electric Motor," *New York Telegraph*, 1 Oct. 1885, Cat. 1138:16a, Misc. Scraps. (*TAED* SB016016a).

7. That is, the point in the rotation of an armature coil where it lies between the opposing fields of the magnets and induces little or no current.

8. The editors have not positively identified this subject but conjecture that Edison was inquiring about Western Union using dynamos on telegraph circuits. He had promoted this idea in 1884, and Edison's assistants worked on a dynamo for this purpose in late 1885 and early 1886. Western Union did widely adopt Edison dynamos sometime before 1892. See Docs. 2651 and 2676 n. 5; Cat. 1305:13; Henry Walter test report, 24 Mar. 1886, Unbound Documents; both Batchelor (*TAED* MBN013013, MB200).

9. Electrical inventor and manufacturer Edward Weston (1850–1936) of Newark was an engineer and factory superintendent for the United States Electric Lighting Co., which had absorbed the Weston Electric Light Co. Though entering the lighting field as a dynamo designer and builder, Weston had also made significant contributions to both arc and incandescent lighting, including processes for coating filaments with hydrocarbons and for producing filaments from squirted cellulose. Over several years of commercial rivalry, Edison had come to view Weston with particular hostility. Docs. 2218 n. 2, 2479 n. 2.

10. The Franklin Institute had planned a competitive trial of dynamo efficiency in conjunction with the International Electrical Exhibition in Philadelphia in September and October 1884 but was obliged to postpone the tests until 1885. The Edison Electric Light Co. submitted four machines (80, 100, 200, and 400 amperes) and the U.S. Electric Lighting Co. three of Edward Weston's models. The first to be tested was an Edison 100-light model, but its "commutator went all to pieces," in the phrase of George Prescott, who was attending on Weston's behalf and sent him frequent reports. (John Howell looked after Edison's interests.) After repairs it carried a full load for ten hours before "the machine gave out" just as measurements were being made. Problems with electrical instruments caused more delays, and the event continued into the last week of June. The final report, published late in the year, showed mixed results, with neither Edison's nor Weston's machines securing a decisive advantage. In addition, enough doubts were raised about the accuracy of the measurements to ensure that questions would linger. Gibson 1980, 172–73; Gibson 1984, 124–26, 137–42; Prescott to Weston, 1 June 1885, Weston (*TAED* X250ACX); "The Dynamo Tests at Philadelphia," *Elec. and Elec. Eng.* 5 (Jan. 1886): 2; Franklin Institute 1885b, 58; John Vail to TAE, 17 June 1885, DF (*TAED* D8522H); more generally, see Weston's correspondence, notably with Prescott, in Weston Papers (*TAED* X250).

The dynamo tests followed a similar competition among incandescent lamps, which had also been postponed from the fall. That event stirred controversy from the start when, after having secured only lamps from Edison and the U.S. Electric Lighting Co. (whose factory Weston supervised) for the contest, the organizers surreptitiously entered other lamps without their makers' consent. Trials of both efficiency and durability were scheduled. Before the latter could begin, Weston asked to withdraw his lamps on the grounds that they were defective. Francis Upton, who looked after Edison's interests, objected, and weeks of wrangling and ill-will followed before the supervising committee compelled Weston to go forward in mid-April. The tests continued until late May. Results were released on 8 July and published in the fall by the Franklin Institute as a 127-page supplement to its *Journal*. Gibson 1980, 172–73; Gibson 1984, 124–26, 127–36; Woodbury 1949, 133–37; particulars may be found in Weston's correspondence with William Marks during this period (Weston Papers [*TAED* X250]); Edison Co. for Isolated Lighting reports, 21 and 25 May 1885, both DF (*TAED* D8522E1, D8522E2); Edison Co. for Isolated Lighting Bulletin, 8 July 1885, 10:10–26, CR (*TAED* CC010010); Franklin Institute 1885a.

11. Edison likely meant Henry Walter, an electrician and assistant superintendent at the Edison Machine Works. At the end of 1884, the Machine Works evidently shipped dynamos to Des Moines, Iowa, where a small Edison station was being constructed after more than a year of planning and delays, but they did not fit the plans prepared earlier by the Edison Construction Dept. "Edison's Electric Motor," *New York Telegraph*, 1 Oct. 1885, Cat. 1138:16a, Scraps. (*TAED* SB016016a); Jehl 1937–41, 1024–25; Edison Co. for Isolated Lighting Bulletin 7:4, 19 Aug. 1885, CR (*TAED* CC007); F. E. Cruttenden to Western Edison Light Co., 13 Feb. 1885, ser. 5, EP&RI (*TAED* X001M2AC); TAE to Western Edison Light Co., 29 Dec. 1884, Lbk. 20:10 (*TAED* LB020010); *TAEB* 7 App. 2.B.

12. Edison referred to Winfield Scott Sims, who about this time secured a contract to produce a number of his "fish torpedoes" for government tests. *DAB*, s.v. "Sims, Winfield Scott."

13. A non-commutator or unipolar motor (or dynamo) could use (or produce) direct current without sending it through a commutator and brushes. Batchelor presumably sought to do away with brushes, which seem to have been particularly troublesome on the motors he was testing for the Sims torpedo (see Doc. 2779). Although Edison had experimented with little success on unipolar machines as generators in 1878 and 1879 (see Doc. 1683 esp. n. 1), Batchelor and Henry Walter executed a patent application on 23 June for a unipolar machine with a radial armature. They filed the application on 6 July and it issued in April 1886 (U.S. Pat. 339,839).

–2821–

From Ichisuke Fujioka

Tokei, Japan, July 1st *1885*[a]

Dear sir,

We have duly received with much pleasure a pair of your telephone apparatus and three dozen of your incandescent lamps,[1] which you were so very kind as to present to our College.[2]

Sometime ago I shewed them to some Senate Members & other high authorities of this country (including of course the

Minster of Public Works Department)[3] songs being transmitted by the telephone and the room lighted with some 8 c.p. incandescent lamps. This shew was successful to such a degree that the gentlemen & ladies present on this occasion were highly interested and nearly all daily news of this City described this aspect with full explanations of the telephones & incandescent lamps of your invention, and your esteemed name was thus at once made familiar among our country people. It is nearly every day since then that some people come to our College to see these marvelous things.

I may be permitted to tell you that to shew you our utmost gratitude, the Minister of Public Works Department (to which our College belongs) is now preparing to present to you some Japanese Articles which may interest you.[4] Yours respectfully

I. Fujioka.[5]

ALS, NjWOE, DF (*TAED* D8503ZBH). Letterhead of Imperial College of Engineering. a"*Tokei, Japan*," and "*188*" preprinted.

1. These items were donated to the Imperial University during Ichisuke Fujioka's 1884 visit to the United States (see note 5). Subsequent correspondence from W.W. Hastings, secretary to the Japanese Consul in New York (Shinkichi Takahashi) suggests that Bergmann & Co. handled the shipment. Doc. 2678 n. 6; Segawa 1998 [1933]; Hastings to Bergmann & Co., 17 Nov. 1884, DF (*TAED* D8424ZBK).

2. The Imperial College of Engineering (Kobu-daigakko) was founded in 1873 by the Ministry of Public Works (Kōbushō), an agency of the Japanese government. It quickly became a center of technological and engineering education, particularly electrical engineering, and of Western influence generally (Takahasi 1990, 199–200). At the end of 1885, when the Ministry of Public Works was abolished, the Ministry of Education took charge of the College, and in 1886 it was merged with the University of Tokyo as one of the world's first engineering colleges (Nelson 1993, 93; Wada 2007, 27–29, 33, 51–52).

3. The Minister of Public Works at this time was Sasaki Takayuki (1830–1910). Nishikawa 2002, 234.

4. The editors have not identified these items. In 1886 the Department of Education sent Edison a pair of copper vases. Segawa 1998 [1933]; Jiro Yoshida to TAE, 16 June 1886, DF (*TAED* D8603ZBV).

5. Ichisuke Fujioka (1857–1918) graduated from the Course in Telegraphy at the Imperial College in 1881 after completing a thesis on galvanometers. He immediately joined the College faculty and worked to develop a more diversified electrical engineering curriculum. The Ministry of Public Works sent him to the Philadelphia Electrical Exposition in 1884, and during this trip he seems to have toured the Edison Illuminating Co. of New York and perhaps also the Edison Lamp Co.; he may have met Edison and visited his laboratory as well. Fujioka resigned his faculty position in 1886 and became chief engineer for the Tokyo Electric Light Co. (Tokyo Dento), which introduced the Edison system of electric lighting into Japan and imported machines for the

manufacture of incandescent lamps. He also arranged for the purchase of a 200-lamp municipal system, possibly on behalf of the Imperial College. In 1890 Fujioka cofounded Hakunetsusha, a lamp and light electrical equipment company that was a forerunner of Tokyo Shibaura Electric (Toshiba). Takahashi 1990, 200; Obituary, *Proc. IEE* 57 (1919): 617; Chokki 1988, 27; Segawa 1998 [1933]; Okamura 1994, 70; Jenks 1887, 18; Odagiri and Goto 1996, 156–60.

–2822–

From Francis Upton

Paris July 9, 1885.

Dear Edison:

I have received a signed memo. from Mr. Rau as follows.

"Memorandum of agreement between Mr. Upton, on behalf of the Edison Lamp Company in Newark (U.S.) of the one part and Mr. Louis Rau on behalf of the Comp. Continentale Edison of 8 Rue Canmartin Paris acting for the other Paris Edison Cos.[1] of the other part

It has been agreed between the parties hereto.

1— Mr. Upton as well as Mr. Rau enter into this agreement subject to the eventual ratifications of their respective boards, which they agree to notify to each other, within six weeks of these presents.

2   The Company continentale Edison agrees to offer indiscriminately American or other lamps[2] to all their customer outside of France, quoting the same price for American or other lamps per unit of candle power volt and ampere.[3]

3— The American Lamp Company engages to sell exclusively to the Company Continentale Edison and with its authority on the continent of Europe except in Sweden, Norway, Switzerland Portugal[4] and to invoice such lamps to the Company Continentale at the conditions granted to their most favored customer or purchaser whoever he may be.

4   The Company Continentale agrees to buy from the Edison Lamp Company all lamps ordered from said Company Continentale for other countries in Europe except Germany, Italy, Austria, Denmark, Russia, Spain and Belgium.[5]

The Lamp Co agrees to bind parties to whom they sell in Sweden, Norway, Switzerland and Portugal not to sell outside of these countries and will furnish the Comp. Continentale with a list of quantities and the prices of all lamps shipped to these places, the Lamp Company agrees to cease supplying any parties who infringe the thus made stipulations and reserve the right of legal action to the Comp. Continentale.

5. This agreement is made for five years and afterwards un-

til either party has notified the other with six months notice of its termination.

Paris July 9 1885. Lue et apprové[6]   signed Louis Rau"

I think very well of this agreement as it prevents them from keeping us out of the European market. We really give up very little as the German and Italian Co. can obtain lamps of the Paris Co. at the same price we sell at. We cannot hope to come into France for five years as it would [be?][a] a clear case if we attempted to introduce the Edison Lamp unmodified.

They want the figures of our sales to enable them to collect the royalties properly.

If you approve this I wish you would cable me "Edison Paris Upton Agree." E̶ If you do not like it cable "~~disagree~~" "disfavor" and then if your points of disagreement are short state them.[7]

I consider[b] it a great concession to have our Lamp Factory put on an equality with the Edison factory in Paris.[8]

If we can make better lamps the "unit candle power volt ampere"[9] will trouble them greatly.[c]

I cable you this evening[10]

Edison New York   Send me or Siegel[11] powers ratify German contract[12] increase his power French fusion covering small changes, then apparently honest royalty certain   Upton.

If Mr. Rau is reasonable honest the European Co. of New York[13] are going to have an honest royalty.[14] Mr. Siegel is thoroughly outside of any ring here and is only interested to have all he can for the New York Co. He can be trusted fully.

There is the strongest desire to ~~hold~~ fuse as[b] there are 2,000,000 francs promised them on fusion.

We never did a better stroke of business than when we held strictly to the letter of our proposals. Now to finally conclude it is needful to let up a little on the strict powers given Mr. Siegel as he cannot now[d] do anything more than sign his name, and there are a few little points that should from mutaul interest be modified.

I leave for Holland tonight and go to Berlin next week. Yours truly

Francis R. Upton

⟨Sammy   Better telegh Upton "Edison Paris Upton agree" — Would also give Siegel full powers   Edison⟩[15]

ALS, NjWOE, DF (*TAED* D8535ZBE). [a]Word evidently omitted at end of page. [b]Obscured overwritten text. [c]Followed by dividing mark. [d]Interlined above.

1. The Société Électrique Edison and the Société Industrielle et Commerciale Edison.

2. The editors have not determined whether this clause relates to use of non-Edison lamps. A change to the Continentale Co.'s bylaws proposed in 1883 would have permitted it to use patents other than Edison's. Doc. 2574 n. 2.

3. Edison and Upton had wished for some time to put American-made lamps on a competitive basis in Europe (Docs. 2553 n. 2 and 2819 n. 5). The relative quality of goods from the U.S. and French Edison factories apparently had an uneven history, but by this time Upton was confident that the quality of American lamps would provide a market advantage (see Docs. 2540 and 2554 n. 2; Upton to TAE, 4 July 1885, DF [*TAED* D8535ZBA]). In an 1886 business forecast for the Edison Lamp Co., Upton predicted that Siemens & Halske would be the most important foreign competitor (Upton to Edison Lamp Co., 16 Jan. 1886, Upton ([*TAED* MU089]). By the end of 1885, Siemens & Halske claimed gains in efficiency and durability for its lamps by coating filaments with a hydrocarbon (using a modification of processes patented by Hiram Maxim, Edward Weston, and William Sawyer and Albon Man, already adapted by the Swan interests in Great Britain and by the United States Electric Lighting Co.). Edison experimented off and on for several years with this practice but never adopted it for commercial use (Siemens 1885; Docs. 2479 n. 2 [Weston], 2017 n. 1, 2021 [Maxim], 1907, 1961, 2005 n. 1, 2033, 2101 n. 6, 2307, 2425 n. 4 [TAE]).

4. Switzerland was presently controlled by Société d'Appareillage Électrique, which purchased lamps directly through Edison. Sweden, Norway, and Portugal (among others) were controlled by Drexel, Morgan and Co. Docs. 1920, 2161 n. 4; Insull to Upton, 23 Mar. 1885, Lbk. 20:199D (*TAED* LB020199D).

5. The Edison-affiliated companies in Germany and Italy were contractually free to buy from either the American or French lamp companies. Samuel Insull to Upton, 23 Mar. 1885, Lbk. 20:199D (*TAED* LB020199D).

6. A standard French phrase meaning "read and approved." *NCFD*, s.v. "approver."

7. The editors have not found the reply sent by Edison but see his marginal note below. The board of the Edison Electric Light Co. of Europe agreed on 17 July to give Upton wider authority on its behalf to "obtain the best terms for the fusion of the three Paris Cos." Samuel Insull advised Edison of this fact and forwarded Upton's correspondence to him in Massachusetts. Edison conveyed, through Insull, his instructions to Upton: "in all these matters such as the business in England & on the Continent & a contract with the Cie. Cont Edison of Paris—you must be guided by your own judgement." Insull promised to see that "the Paris people in the future have no reason to complain." Neither the French fusion nor the German contract progressed much further during the summer. Pressure to conclude these matters continued through the rest of the year, but they dragged into 1886. Insull to TAE, 17 July 1885, DF (*TAED* D85270); Insull to Upton, 22 July 1885, Lbk. 20:427 (*TAED* LB020427); see also Doc. 2835.

8. The Société Industrielle et Commerciale Edison was the first for-

eign manufacturer of Edison lamps, having begun production in 1882 in the Paris suburb of Ivry-sur-Seine. The decision to start a factory there reflected both the manufacturing requirements of French patent law and Edison's hope of supplying the projected Paris central station (*Ency. Brit.* 9, s.v. "Patents"; see Docs. 2153 and 2463 n. 1). When Charles Batchelor, the original manager, returned to the U.S. in mid-1884, the plant was turning out 800 lamps per day ("The Edison Light in Europe," *Electrical World* 3 [21 June 1884]: 202).

According to Batchelor's analysis of production costs in January 1884, the Ivry factory spent about 40 cents to make each standard lamp. The Edison lamp factory in East Newark offered lamps to the German company at 40 cents in June 1884; that price dropped to 33 cents in February 1886. Cat. 1235:64, Batchelor (*TAED* MBN012064); Emil Rathenau to TAE, 23 June, 1884, DF (*TAED* D8436ZCW); TAE to Oscar Siegel, 2 Feb. 1886, Lbk. 21:214 (*TAED* LB021241).

9. This phrase referred to a lamp's efficiency at converting electrical power into visible light. The volt-ampere became, by definition, equivalent to the watt as the standard unit of electrical power. William Siemens proposed the watt about 1882; the name was championed by William Preece but not yet universally used. Edison was named in 1884 to the National Conference of Electricians' Committee on the Standards of Light that considered the question of light and power units, as did a separate Committee on a Unit of Power. "National Conference of Electricians," *Electrical World* 4 (27 Sept. 1884): 106; "National Conference of Electricians," ibid. 4 (8 Nov. 1884): 181; "Philadelphia Electrical Congress," *Electrician* 13 (4 Oct. 1884): 477; Preece 1884; United States Electrical Commission 1886, 44–45.

10. The text of Upton's 9 July cable follows. The message received and transcribed in New York is in CR (*TAED* CE85019).

11. Oscar Othon Siegel (1848–1917) was an associate of Drexel, Harjes in Paris since approximately 1871 and an original subscriber in the Paris companies' syndicate (Archives électorales [Seine]: Electeurs, from France Electoral Rolls [1891], online database accessed through Ancestry.com, 23 Jan. 2012; Obituary, *NYT*, Dec. 21, 1917, 11; Banque Centrale du Commerce to TAE, 9 Jan. 1882, DF [*TAED* D8238J]; Carosso and Carosso 1987, 442). By May 1885, he had been assigned a special power of attorney to represent the Edison Electric Light Co. of Europe in negotiating the fusion of the French companies (Rau to TAE, 8 and 9 May 1885, Siegel to TAE, 19 Oct. 1885; all DF [*TAED* D8535Z, D8535ZAB, D8535ZBX]). The European company gave a general power of attorney to John Harjes in January 1885, reflecting Edison's preference for him over Joshua Bailey, under terms that permitted Harjes to appoint an associate attorney (Edison Electric Light Co. of Europe, Ltd., Minutes, 17 Jan. 1885, DF [*TAED* D8527F]).

12. Changes proposed to the Edison Electric Light Co. of Europe's contract with the Deutsche Edison Gesellschaft (see Doc. 2724 esp. nn. 2–4) were bound up with the prospective reorganization of the French companies. The new German terms would yield the European company an immediate cash infusion of 25,000 marks but the issue was not settled by October 1885, when Rau complained about indecision on Edison's part and warned that the Germans had made their own ratification contingent on the outcome of the long-awaited French fusion.

Frustrated, Rau threatened to resign from the Compagnie Continentale board if both matters were not settled soon (Compagnie Continentale Edison to Edison Electric Light Co. of Europe, Ltd., 20 June 1885, CR [*TAED* CE85017]; Rau to Batchelor, 14 Oct. 1885, Batchelor [*TAED* MB173]; Rau to Siegel, 3 Mar. 1886, DF [*TAED* D8625666]). About the same time, Edison, acting for himself and as president of the European company, instructed Siegel to "ratify Germany contract, also Bailey fusion agreement on basis 14 000 instead of 8,000 founders [shares], providing our percentage shares same in either case. Lamp royalty unchanged. You must secure payment 25 000 marks to us from payment Germany contract." Siegel declined to sign the German agreement on the strength of the special power of attorney given him for French matters. By year's end, the French interests proposed yet another revised contract. Siegel refused to sign but later sent it to New York for review, along with a letter summarizing its numerous new provisions. Neither the contract nor Siegel's letter has been found, but Batchelor referred to both in an undated four-page critique of the proposed terms that he prepared for the board of the Edison Electric Light Co. of Europe (Siegel to TAE, 19 Oct. 1885; Batchelor to Edison Electric Light Co. of Europe, Ltd., n.d. [Dec. 1885?]; both DF [*TAED* D8535ZBX, D8527Q]).

13. The Edison Electric Light Co. of Europe, Ltd., was formed in 1880 to exploit Edison's electric lighting patents in France and a handful of other European companies. Based in New York, it was often referred to as the "European Company." It had assigned its patent rights to the Compagnie Continentale Edison in 1882, retaining the right to veto all contracts. It endorsed a plan to reorganize the French companies in March 1884. See Docs. 1736 esp. n. 6, 2182, and 2667.

14. One item in the long-running negotiations for a fusion of the French companies was the conversion of Edison's ownership interest to a per-lamp royalty basis (see, e.g., Docs. 2574 esp. n. 3, and 2593 esp. n. 1; Batchelor to Charles Coster, 6 Apr. 1885, Batchelor [*TAED* MBLB2073]). These royalties also were at issue in discussions about lamp prices and market rights (Insull to Upton, 30 Nov. 1883, DF [*TAED* D8316BMC]; Joshua Bailey to Sherburne Eaton, 29 Dec. 1883, article 44, Batchelor [*TAED* MB103]).

15. The editors have found neither this cable to Upton nor correspondence that might have increased Siegel's authority.

–2823–

From George Hamilton

New York July 10th, *1885.*[a]

My dear Edison,

Mr. Van Horne[1] has handed me a paper containing a proposition from you to equip <u>any</u> way morse line so as to make two independent circuits, and asks what I think of it.

As I would, naturally, prefer to know what the thing is before committing myself I would be glad to have such details from you as you may think proper to communicate. Truly Yours,

George A. Hamilton.[2]

⟨Its not patented yet ~~hence~~ But if WU want to accept my proposition in which they take no risk, will of course show it in actual operation[3] Edison⟩

ALS, NjWOE, DF (*TAED* D8546X). Letterhead of Western Union Telegraph Co., Executive Office. [a]"*New York*" and "*188*" preprinted.

1. John Van Horne, born in New Jersey in 1827, became a Western Union vice president in 1878. His purview included the electrical department, statistics, and the company's contracts. Doc. 786 n. 3; Reid 1886, 218–19, 668–69.

2. George A. Hamilton (b. 1843) was chief electrician for Western Union in New York, having joined the company in 1875. He left it for Western Electric in 1889. Doc. 2326 n. 3; Reid 1886, 674–76; "Okonite Conductors," *Western Electrician* 3 (28 July 1888): 48; "New York Notes," *Electrical Engineer* 4 (29 Nov. 1889): 433.

3. The editors have found neither the terms of Edison's offer nor what, if anything, came of it. His two pending applications for a phonplex telegraph system issued in December. U.S. Pats. 333,289; 333,290; see Doc. 2800 (headnote).

EDISON'S DIARY AND COURTSHIP   Docs. 2824, 2825, 2826, 2827, 2828, 2829, 2830, 2832, 2834, and 2837

Edison enjoyed part of the summer of 1885 in Boston and environs, in the company of long-time friend Ezra Gilliland and his wife Lillian. On his third trip to the area, he spent the week of 14–21 July with his daughter Marion and other guests at Woodside Villa, the Gillilands' rented cottage in Winthrop, on the shore of Boston Harbor.[1] Edison and others in the party each kept a diary of the holiday to be shared for their mutual entertainment.

Edison's diary is the only known volume that he kept specifically to record thoughts and feelings of a personal nature. It begins with events on 12 July, when he was still at Menlo Park, and continues with an account of his travels on 13 July (through New York City and Providence, Rhode Island). Edison composed these early entries after arriving in Massachusetts.[2] Thereafter, the journal continues with accounts of the week's recreations, which included sailing, listening to music, and reading aloud. Edison incorporated observations on literature, art, and religion, along with comments about his dreams, his health, and the weather.

Edison did not take the diary-keeping enterprise wholly seriously, noting on 16 July that he had "Diaried [a] lot of nonsense" (Doc. 2828). He seemingly was good-natured about it, despite any disorientation he may have felt during an unaccustomed period of enforced leisure in mixed company, far from his familiar milieu. One may see Edison's legendary storytelling skills in the patent exaggerations and jocular retelling of events already known to his audience. His lively accounts of temperatures "hot enough to test Safes" (Doc. 2834), for example, brought a splash of humor to days when the mercury reached ninety degrees in central Boston.[3] Modern readers may wonder whether he was expressing social discomfort or satiric egotism when, in the company of people with finer educations, he described himself as "a literary barbarian . . . not yet educated up to the point of appreciating fine writing" (Doc. 2824).

The diary may also be read as a courtship document, and in this respect it is profoundly serious. Edison's first wife, Mary, had died in August 1884. His daughter Marion later recalled that it was while traveling with the Gillilands to New Orleans and Florida in February and March 1885 that her father concluded "he wanted a home, a wife and a mother for his three children and asked Mrs. Gilliland, who lived in Boston, to introduce him to some suitable girls."[4] The Gillilands seem to have arranged Edison's stay at Winthrop in order to introduce him to marriageable women of appropriate background. Mrs. Gilliland invited two—Grace "Daisy" Gaston and Louise M. Igoe—from her hometown of Indianapolis and another—Mina Miller, of more recent acquaintance—from Akron, Ohio.[5] Edison used the diary to signal—clearly but gently—to the Gillilands and the other young women his preference for Miss Miller, who would become his second wife in February 1886.[6]

Mina Miller, the seventh child of Lewis and Mary Valinda (née Alexander) Miller, was born in Akron, Ohio, on 6 July 1865. Her father was the inventor of the Buckeye mower and reaper and the cofounder and manager of Aultman, Miller & Co., a manufacturer of farm equipment. He was also a cofounder of the prestigious Chautauqua Institution, a summer learning community in upstate New York.[7] Mina graduated from Akron High School in 1883 and enrolled at Miss Abby H. Johnson's Home and Day School for Young Ladies at 18 Newbury Street, Boston, beginning with the fall semester that year. Miss Johnson, formerly the preceptress of Bradford Academy, was widely recognized as one of the top educators of young women in the eastern United States. Her school seems to have catered to the daughters of wealthy industrialists—young women such as Mina and her best friend there, Margaret Hall, the daughter of a proprietor of the Jamestown Worsted Mills of Jamestown, New York.[8] Miss Johnson's school attracted top teachers and lecturers, including the renowned historian, philosopher, and Harvard professor John Fiske, who was under contract to lecture there during Mina's attendance.[9] Mina also studied music in Boston with Stephen Emery, a distinguished professor of harmony at the New England Conservatory of Music since 1867.[10]

As part of her education, Mina also made a truncated tour of Europe, sailing from New York in late June 1884 with her older brother Lewis and sisters Jane, Grace, and Mary. During their summer abroad, the Millers toured Ireland, Scotland, the English Lake District, and Wales. They also visited Edinburgh, London, and Paris before sailing home at the end of September without, apparently, having gone to Italy as they intended. Although Mina expressed some reluctance to return to school at the end of her trip, in October she was back at Miss Johnson's for the remainder of the fall semester.[11]

In the spring of 1885, Mina was romantically involved with George E. Vincent, the son of Chautauqua Institution cofounder and Methodist bishop John Heyl Vincent. The couple was informally engaged by April (though this fact was concealed from Mina's younger sisters), and there seems to have been no change in this status when she attended his graduation from Yale on 25 June.[12] Certainly, marriage between Mina and the younger Vincent had been a topic of discussion within the family since summer 1883, but Mina seems to have harbored some doubt about the relationship almost from the start. As late as May 1885, Mary Valinda advised that Mina and George "studey each other" in order to make "a lasting and true decision" before marrying.[13]

Precisely when Edison and Mina were introduced has been a matter of dispute. Newspapers at the time of their marriage made conflicting claims, which historians and biographers have sometimes used to reconstruct the story of their courtship. Two of the more persistent stories are that they met either at the World's Industrial and Cotton Exhibition at New Orleans in late February or early March 1885, or at the Miller family's summer house in Chautauqua, New York, during the ensuing summer.[14] These reports are erroneous, however. Mina's account book for 1885 shows that she did not travel to New Orleans in the late winter but remained in Boston, and

Edison does not appear to have been in Chautauqua until August, well after he wrote of her in his diary.[15]

Mina herself, in a 1947 interview with Associated Press reporter Milton Marmor, said that she first met Edison at a "friend's" Boston apartment during a dinner party. "I was taking music at the time," Mina noted. "I never played for anybody but they asked me to play, and so I said, 'Oh, well, I'll never see these people again.' That was the first time we knew each other." She found Edison to be, on the basis of this first encounter, "a genial, lovely man." Marmor then asked, "Do you remember after you first met Mr. Edison at the dinner in Boston whether anything developed that night? Did the romance develop then? To this, Mina replied somewhat ambiguously:

> No, I thought I would never see him again. Oh yes, then Mrs. Gilliland (that was just the close of the year of the term and they were going up to the beach home) and they wanted me to go for a week and so I wrote and asked father if I could go for a week and while I was there Mr. Edison came up.

"I don't think that was an accident," Marmor observed. To which Mina responded, "I guess it was. Anyway, we enjoyed it there. I was there a week and he went there once in awhile."[16]

It is clear from Mina's account, which must be reckoned the most definitive, that she and Edison were introduced at the Gillilands' Boston apartment in the Hotel Huntington, 101 Milk Street, though the time of the meeting remains uncertain.[17] They could have met one Sunday in late January when they both were in Boston, though it is also possible that they met some other time in the winter or spring. Edison visited the Gillilands in Boston or Winthrop from the end of May through the first week of June and could have encountered her in that interval. In any case, Edison and Mina had occasion to deepen their acquaintance at the end of June or beginning of July. He returned to the Gillilands' hospitality about 20 June and overlapped with Mina as their guest into early July.[18] The *Boston Daily Globe* noted on 28 June that Edison and Daisy Gaston were staying with the Gillilands at Winthrop, and his correspondence shows that he did not return to New York until 7 or 8 July.[19] For her part, Mina attended the Yale commencement on 25 June and the Yale-Harvard Boat Race in New London, Connecticut, the next day but was in Winthrop before 1 July.[20] She seems to have stayed with the Gillilands

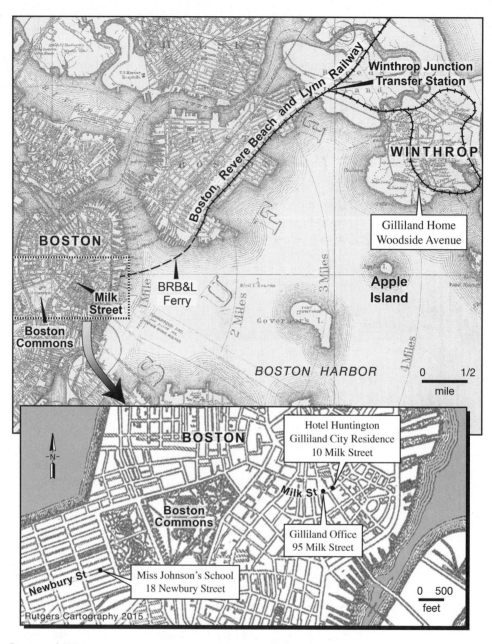

Boston and vicinity,
summer 1885.

after Edison departed for New York. However, she left Winthrop before Edison returned on the fourteenth. What Mina's
interest in Edison might have been as she returned home
from Massachusetts is not known. Edison, however, was already making plans to court her. Although his acquaintance
with Mina at Winthrop was brief, it was long enough for her
to have made an indelible impression on the young widower,

who soon after publicly signaled his affection toward her as "the Maid of Chataqua" (Doc. 2829).

1. Edison's first visit to the Gillilands stretched from perhaps the very end of May through the first week in June; he had returned to New York by 9 June. He set out once again for Winthrop via Boston on 18 June, was back in New York by 8 July, and seems to have returned to Menlo Park by 11 July. Two days later, Edison started again on an overnight journey, this time with his daughter Marion, to Boston and ultimately Winthrop. Samuel Insull to Francis Upton, 9 June 1885, Miller (*TAED* HM850262); Insull to Eugene Kenebel, 18 June 1885; TAE to Marion Edison, 8 July 1885; both Lbk. 20:362A, 396 (*TAED* LB020362A, LB020396); see Docs. 2824–2826.

2. Edison purchased the diary books and pencils in Boston on 15 July (see Doc. 2827).

3. According to reports in the *Boston Daily Globe*, the city temperature touched this threshold on 16–18 July but dropped below seventy degrees each night. Remarking more matter-of-factly on the weather in Doc. 2833, Edison unsurprisingly noted that it was cooler in Winthrop. "Almanac," *Boston Daily Globe*, 17 July 1885, 10; "Almanac," ibid., 18 July 1885, 10; "Almanac," ibid., 19 July 1885, 2.

4. Oeser [1956], 9–10.

5. Mina apparently was invited to Winthrop sometime after Louise Igoe and Daisy Gaston. Those two longtime Indianapolis friends of the Gillilands knew their intentions by mid-April; Mina, six or seven years younger and a more recent acquaintance, made her plans a month later. Igoe to Mina Miller, 19 Apr. 1885; Igoe to Mina Miller, 18 May 1885; both EMFP (*TAED* X104OE, X104OE1); Oeser [1956], 9–10.

6. At Winthrop, Marion apparently developed a preference for Louise Igoe, "more because she was a blonde like my Mother than for any other reason. I had the impression, however, that my Father was in love with the Ohio girl, Mina Miller, whom he had previously met." Oeser [1956], 10.

7. Edison-Miller family *Holy Bible*, Cat. 104948, NjWOE (Glenmont); Hendrick 1925, 67, 72, 82–87.

8. Mina Miller Graduation Announcement, 22 June 1883, Miller Family, Box 2, Mina Miller-Edison, 1883–1947 and Mina Miller to Grace Miller, 24 Jan. 1885, RG III-A9-1 (*TAED* X465F), GMHHP; both N(ChaCI); Hill 1903, 352–53; Hazeltine 1887, 436; U.S. Census Bureau 1970 (1880), roll T9_815, p. 338D, image 0678 (Jamestown, Chautauqua, N.Y.).

9. Fiske 1940, 510.

10. Stephen Albert Emery (1845–1891) numbered among his students three members of the influential "Second New England School" of composition: George W. Chadwick, Horatio W. Parker, and Arthur Foote. The composer Ethelbert Nevin was another prominent student. Emery to Mina Miller, 19 Dec. 1885 and 16 Feb. 1886; both MME-CD (*TAED* X401BA, X401BB); *GMO*, s.v. "Emery, Stephen Albert."

11. The Millers sailed 25 June for Liverpool aboard the *Oregon*, which the Cunard line had recently purchased. "Marine Intelligence," *NYT*, 26 June 1884, p. 8; "Oregon to Disappear," *NYT*, 22 June 1884; Jane Eliza Miller to Mina Miller, 16 June 1884, FR, NjWOE (*TAED*

FM001AAC); Mina Miller to Lewis and Mary Valinda Miller, 25 June and 2 July 1884; Mina Miller to Mary Valinda Miller, 20 July, 1, 5, 12, 20, and 31 Aug., 14 Sept. 1884; all CEF (*TAED* X018A700A–X018A700I); *Passenger Lists* 1962, microfilm, M237_481, line 18, list number 1307.

12. George Vincent seems to have continued to pursue Mina even after she became engaged to Edison. George Vincent to Elizabeth Vincent, 28 Aug. and 31 Aug., 10 Sept. 1885, JHV; see also Hendrick 1925, 184; Mary Valinda Miller to Mina Miller, 16 Apr. 1885, FR (*TAED* FI001AAA18); "The Yale Seniors," *NYT,* 31 May 1885, 7.

13. Louise Igoe to Mina Miller, 4 Apr. 1884, EMFP (*TAED* X104OC); Mary Valinda Miller to Mina Miller, 8 May 1885, CEF (*TAED* X018D1AB).

14. *New Orleans Picayune,* 8 Feb. 1886, Scraps., NjWOE (*TAED* SB017124d); see, e.g., Conot 1979, 230–33; Baldwin 1995, 147; Israel 1998, 244–45; for details about Mina's education and wedding, see Jeffrey 2008, 162; Advertisement, *Literary World* [Boston], 5 Sept. 1885, 312; "Marriage in High Life," *Republic County Pilot* [Cuba, Kansas], 11 Mar. 1886; and "Thomas A. Edison's Bride," *Boston Daily Globe,* 25 Feb. 1886.

15. Mina Miller Edison account book (1885–1886), Accts., NjWOE (*TAED* AB011).

16. Mina Edison interview for the Associated Press, 10 Jan. 1947, typescript in Edison Bio. Coll. Mina's interview was reprinted in newspapers throughout the country. Mina's account, as it appeared in the newspapers, is slightly altered from the typescript interview (see, for instance, "Wizard of Menlo Park First A Family Man, Widow Writes," *La Crosse [Wisc.] Tribune,* 6 Feb. 1947). An article from the Winthrop, Mass., paper published at the time of Edison's death, however, claims that the meeting took place in Winthrop ("Edison First Met Wife in Winthrop," *Winthrop Review,* 23 Oct. 1931, 1; cf. Albion 2008, 12–13).

17. *The Boston Directory* 1885, 450. In her 1934 biography, *Thomas A. Edison: A Modern Olympian,* Mary Childs Nerney, who was head of the Historical Research Department at the Orange Laboratory in the last years of Edison's life, also says the first meeting with Mina took place in Boston. Nerney knew the Edisons personally, so her account, though undocumented, carries considerable weight. She says Edison called at the Gillilands' home in Boston at a time when Mina was also paying them a visit. "After getting her consent," Nerney wrote, "[Ezra] Gilliland came into the room where the inventor was talking with some gentlemen and said to him: 'Mina Miller is here and she is going to play and sing for you'." In this telling, Edison was impressed with Mina's beauty and composure, though he did not think she played the piano particularly well. Nerney's account does not specify when the incident occurred beyond saying that it took place "one Sunday." In all other respects, her story is entirely reconcilable with Mina's remembrance (Nerney 1934, 272–73). Mina had met the Gillilands in Boston in the spring of 1884, so it is conceivable that the Gillilands introduced her to Edison in the winter of 1885. Edison was in Boston from 3–6 January to attend the 1885 Electrical Exhibition, but Mina did not purchase a train ticket to return to Boston from Akron until 5 January, so that a meeting on that occasion is unlikely (Mina Miller Edison account book [1885–1886], Accts., NjWOE [*TAED* AB011]). Edison did make

a trip to Boston between 20 and 26 January, at a time when Mina, who was also in Boston, contemplated paying a visit to the Gillilands. She had not done so by 24 January, leaving the possibility that both she and Edison called on the Gillilands the next day, a Sunday (TAE to Roscoe Conkling, 26 Jan. 1885, Lbk. 20:51B [*TAED* LB020051B]; Mina Miller to Grace Miller, 24 Jan. 1885, MMC [*TAED* X465F]).

Biographers' interpretations that Edison and Mina met at the Gilliland's Boston home about February 1885 are not necessarily inconsistent with Mina's own ambiguous chronology (see, e.g., Israel 1998, 244–45; Baldwin 1995, 147–48; Conot 1979, 230–31; and Josephson 1959, 304–305). Francis Jehl repeated the February date in his *Menlo Park Reminiscences*. His account seems to have been taken in part from newspaper reports of Edison's engagement. According to one story filed from Akron, "Miss Miller first met the Menlo-park wizard last February, while visiting at the home of Mrs. Edward [sic] Gilliland" (Jehl 1937–41, 1006; "Edison's Love-Story," *Richmond Dispatch,* 3 Feb. 1886, 3).

18. Edison was with Ezra Gilliland in Boston early in June, perhaps for as long as a week, before he was back in New York on 9 June, and he could easily have gone to Winthrop. TAE to Insull, 2 and 3 June 1884; Gilliland to TAE, 18 June 1885; all DF (*TAED* D8503ZAT, D8533O, D8503ZAZ); Insull to Francis Upton, 9 June 1885, Miller (*TAED* HM850262).

19. "Table Gossip," *Boston Daily Globe,* 28 June 1885, 13; Gilliland to TAE, 18 June 1885; Insull to Alfred Tate, 2 July 1885; both DF (*TAED* D8503ZAZ, D8541S); TAE to Marion Edison, 8 July 1885, Lbk. 20:396 (*TAED* LB020396).

20. Mary Valinda Miller to Mina Miller, 1 July 1885, CEF (*TAED* X018D1AC2); "Harvard Freshmen Win," *NYT,* 26 June 1885, 1.

–2824–

Diary Entry

Menlo Park N.J. Sunday july 12 [15?] 1885[1]

Awakened at 5:15 AM. My eyes were embarassed by the sunbeams—turned my back to them and tried to take another dip into oblivion—succeeded—awakened at 7 A.M. thought of Mina,[2] Daisy,[3] and Mamma G__[4]   put all 3 in my mental kaledescope to obtain a new combination a la Galton.[5] took Mina as a basis, tried to improve her beauty by discarding and adding certain features borrowed from Daisy and Mamma G. a sort of Raphaelized[6] beauty, got into it too deep, mind flew away and I went to sleep again. Awakened at 8:15 AM. Poweful itching of my head, lots of white dry dandruff— what is this d__mnable material. Perhaps it is the dust from the dry literary matter I've crowded into my noddle lately   Its nomadic. gets all over my coat, must read about it in the Encyclopedia.[7] Smoking too much makes me nervous—must lasso my natural tendency to acquire such habits—holding heavy cigar constantly in my mouth has deformed my upper lip, it has sort of a Havanna curl. Arose at 9 oclock came down stairs expecting twas too late for breakfast—twas'nt. could'nt eat much, nerves of stomach too nicotinny. The roots of tobacco plants must go clear through to hell. Satans principal agent Dyspepsia must have charge of this branch of the vegitable kingdom.— It has just occured to me that the brain may digest certain portions of the food, say the etherial part, as well as the stomach—perhaps dandruff is the excreta of the mind— the quantity of this material being directly proportional to the amount of reading one indulges in. A book on German metaphysics would thus easily ruin a dress suit. After breakfast start reading Hawthorne's English Note Book[8] dont think much of it.— perhaps Im a literary barbarian and am not eyet educated up to the point of appreciating fine writing— 90 per cent of his book is descriptive of old churches and graveyards and coronors— He and Geo Selwyn[9] ought to have been appointed perpetual coroners of London.

Two fine things in the book were these.

Hawthorne shewing to little Rose Hawthorne[10] a big live lobster told her it was a very ugly thing and would bite everybody, whereupon she asked "if the first one God made, bit him"— again "Ghostland is beyond the jurisdiction of veracity"— I think freckles on the skin are due to some salt of Iron, sunlight brings them out by reducing them from high to low state of oxidation—perhaps with a powerful magnet applied for some time, and then with proper chemicals, these mud holes of beauty might be removed

Dot[11] is very busy cleaning the abode of our deaf and dumb parrot—she has fed it tons of edibles, and never got a sound out of it. This bird has the taciturnity of a statue, and the dirt producing capacity of a drove of buffalo.

This is by far the nicest day of this season, neither too hot or too cold.—it blooms on the apex of perfection—an Edenday Good day for an angels pic nic. They could lunch on the smell of flowers and new mown hay, drink the moisture of the air, and dance to the hum of bees. Fancy the Soul of Plato astride of a butterfly, riding around Menlo Park with a lunch basket

Nature is bound to smile somehow. Holzer has a little dog which just came on the veranda. The face of this dog was a dismal as a bust of Dante,[12] but the dog wagged its tail continuously— This is evidently the way a dog laughs— I wonder if dogs ever go up to flowers and smell them—I think not— flowers were never intended for dogs, and perhaps only incidentally for man, evidently Darwin has it right   They make themselves pretty to attract the insect world who are the transportation agents of their pollen, pollen freight via Bee line   There is a bumblebees nest somewhere near this veranda, several times one came near me—some little information (acquired experimentally) I obtained when a small boy causes me to lose all delight in watching the navigation of this armed flower burglar.

Had dinner at 3 P.M. ruins of a chicken, rice pudding— I eat too quick— at 4 oclock Dot came around with her horse "Colonel" and took me out riding—beautiful roads—saw 10 acre lot full cultivated red raspberries, "A burying ground" so to speak.—got this execrable pun off on Dot   Dot says she is going to write a novel, already started on—she has the judgement of a girl of 16 although only 12   We passed through the town of Metuchen, this town was named after an Indian chief, they called him Metuchen the chief of the rolling lands,[13] the country being undulating. Dot laughed heartily when I told her about a church being a heavenly fire-escape.

Returned from drive at 5 PM commenced read short sketches of life's Macauley Sidney Smith,[14] Dickens[15] & Charlotte Bronte.[16] Macauley when only 4 years ago omnivorous reader, used book language in his childish conversation. when 5 years old, lady spilled some hot coffee on his legs, after a while she asked him if he was better—he replied—"Madam the agony has abated"   Macauleys mother must have built his mind several years before his body. Sidney Smiths flashes of wit are[a] perfect to call them chesnuts would be literary blasphemy— They are wandering jewlets to wander forever in the printers world— Dont like Dickens—dont know why— I'l stock my literary cellar with his works later.

Charlotte Bronte was like DeQuincy,[17] what a nice married couple they would have been   I must read Jane Eyre.—[18] played a little on the piano[19]—its badly out of tune—two keys have lost their voice.

Dot has just read to me outlines of her proposed novel, the basis seems to be a marriage under duress—[b] I told her that in case of a marriage to put in bucketfulls of misery. This would make it realistic. speaking of realism in painting etc   Steele Mackaye[20] at a dinner given to H H Porter,[21] Wm Winter[22] and myself told us[b] of[b] a difinition of modern realism given by some frenchman whose name I have forgotten. "Realism, a dirty long haired painter sitting on the head of a bust of Shakespeare painting a pair of old [bo-][c] boots covered with dung"

The bell rings for supper   Igoe[23]   Sardines the principal attraction—on seeing them was attacked by a stroke of vivid memory of some sardines I eat last winter that caused a rebellion in the labyrinth of my stomach—could scarcely swallow them today   They nearly did the "return ball" act.

After supper Dot pitched a ball to me several dozen times—first I ever tried to catch. It was a hard as Nero's[24] heart—nearly broke my baby finger—gave it up— learned Dot and Maggie[25] how to play "Duck on the rock"[26]   They both thought it great fun, and this is sunday— My conscience seems to be oblivious of sunday—it must be incrusted with a sort of irreligious tartar. If I was not so deaf I might go to church and get it taken off or at least loosened—eccavi[27]   I will read the new version of the bible

Holzer is going to use the old laboratory for the purpose of hatching chickens artificially by an electric incubator. He is very enthusiastic—gave me full details—he is a very patient and careful experimenter—think he will succeed—everything succeeded in that old laboratory—

Just think electricity employed to cheat a poor hen out of the pleasures of maternity— Machine born chickens— What is home without a mother

I suggested to H that he vaccinate his hens with chicken pox virus, then the eggs would have their embryo heriditarily innoculated[28] & none of the chickens would have the disease, for economys sake he could start with one hen and rooster. He being a scientific man with no farm experience I explained the necessity of having a rooster, he saw the force of this suggestion at once. The sun has left us on time, am going to read from the encyclopedia Brittanica to steady my nerves and go to bed early. I will shut my eyes and imagine a terraced abyss, each terrace[d] occupied by a beautiful maiden   to the first I will deliver my mind and they will pass it down down to the uttermost depths of silence and oblivion. Went to bed worked my imagination for a supply of maidens, only saw Mina Daisy & Mamma   Scheme busted—sleep.

Woodside Villa[29]   Boston Harbor

AD, NjWOE, TAE Diary, Cat. 117 (*TAED* MA001). [a]"re" interlined above. [b]Obscured overwritten text. [c]Canceled. [d]Interlined above.

1. Edison refers in this document to Woodside Villa, Ezra and Lillian Gilliland's summer home in Massachusetts. He purchased diary books and pencils in Boston on 15 July, and that evening he recorded that "Everyone [the Gillilands and other guests] started their Diary" (see

Doc. 2827). The editors infer that he retrospectively wrote this entry (and the two following) on or soon after that date; see also Doc. 2828.

2. Mina Miller (1865–1947); see headnote above.

3. Grace "Daisy" Gaston (1859–1940), daughter of John M. Gaston, a prominent Indianapolis physician, was Ezra and Lillian Gilliland's guest at their rented summer home during Edison's earlier visit from late June to early July. She was still there, perhaps according to a plan to introduce Edison to eligible young women, when he returned on July 14. In a 1970 interview with historian Kathleen McGuirk, Lillian Gilliland's niece Lillian Warren identified Gaston as one of Ezra's "girls" (Warren interview in "Gilliland, Ezra T.," Bio. Coll.). In April 1887, Daisy was among the guests vacationing with the Edisons and Gillilands in Florida. The next year, she and her younger sister accompanied the Gillilands to Europe. She married Indianapolis attorney John B. Sherwood in 1894. Her brother, George B. Gaston, worked with Edison for a time on the promotion of the phonograph. Doc. 2852 includes Edison's assessment of Daisy Gaston's physical appearance and character. U.S. Census Bureau 1967 (1860), roll M653_279, p. 316, image 321 (Indianapolis Ward 1, Marion, Ind.); ibid. 1970 (1880), roll T9_295, p. 276B, image 0254 (Indianapolis, Marion, Ind.); Crown Hill Cemetery Lot Interment Order for Grace Gaston Sherwood, 27 Mar. 1940, Crown Hill Cemetery, Indianapolis, Ind.; "Table Gossip," *Boston Daily Globe*, 28 June 1885, 13; "George B. Gaston," *NYT*, 29 Nov. 1942, 64; Oeser [1956], 9–10.

4. Lillian Gilliland (b. 1858?), the daughter of Sidney Johnson, a carpenter and machinist of Madison, Ind., became Ezra Gilliland's second wife in 1880. According to Edison's daughter Marion, she played a central role in introducing Edison to eligible young women (see headnote above). Two of them were Daisy Gaston and Louise Igoe, both of Indianapolis. U.S. Bureau of the Census 1967? (1860), roll M653_270, p. 649, image 353 (Madison, Jefferson, Ind.); ibid. 1970 (1880), roll T9_295, p. 296B, image 0294 (Indianapolis, Marion, Ind.); Indiana Marriage Collection, 1800–1941, online database accessed through Ancestry .com, 27 Apr. 2011; "Married," *Telegrapher* 1 (1 Nov. 1865): 185; Albion 2008, 13–14; "Obituary Notes," [Ezra Gilliland], *Electrical World and Engineer*, 41 (23 May 1903): 912.

5. The English polymath Francis Galton (1822–1911; knighted 1909), a cousin of Charles Darwin, was an early contributor to the study of human genetics and a promoter of eugenics. Galton also pioneered composite photography, the technique of superimposing portrait photographs of individuals in order to draw out common physical traits (cf. Doc. 2852). Galton thought these common features might be markers of hereditary psychological traits or susceptibility to certain diseases. He therefore began making composite portraits of criminals and later made them of people suffering from various physical and mental illnesses. Galton published works on the making and interpretation of these portraits in 1878 and 1879. *Oxford DNB*, s.v. "Galton, Sir Francis"; Brookes 2004, xiii, xv; Gillham 2001, 215–19.

6. Raphael or Raffaello Sanzio da Urbino (1483–1520), painter and architect of the High Italian Renaissance, is credited with deeply influencing the western ideal of feminine beauty with his "Madonna del

Granduca" and "Sistine Madonna." This ideal is characterized by regularity, symmetry, simplicity of hairstyle, innocence of expression, and youthfulness. Oberhuber 2001, 39.

7. Edison likely meant the *Encyclopaedia Britannica,* to which he refers at the end of this entry.

8. *Passages from the English Note-books,* detailing the sojourn of American writer Nathaniel Hawthorne (1804–1864) in England from 1853 to 1857, had appeared in several editions since 1870. *ANB,* s.v. "Hawthorne, Nathaniel"; Stewart 1935, 3.

9. George Augustus Selwyn (1719–1791), member of Parliament from 1747 to 1780 and a noted wit, was reputed to have been obsessed with corpses and executions, though his personal friends denied the imputation. *DNB,* s.v. "Selwyn, George Augustus."

10. Rose Hawthorne Lathrop (1851–1926), daughter of Nathaniel Hawthorne, was a minor poet. In 1871 she married George Parsons Lathrop, who became associate editor of the *Atlantic Monthly* in 1875. He edited *The Works of Nathaniel Hawthorne,* published in fifteen volumes between 1882 and 1884, and Edison could have been reading Hawthorne's *English Notebooks* from that collection. Both Lathrops converted to Catholicism in the 1890s. Rose founded a religious sisterhood, the Servants of Relief for Incurable Cancer, in Hawthorne, N.Y., which she served under the name Sister Alphonsa for the remainder of her life. *DAB,* s.vv. "Alphonsa, Mother," "Lathrop, George Parsons."

11. In honor of the Morse telegraphic alphabet, Edison gave this nickname to Marion Estelle Edison (1873–1965), the eldest child and only daughter by his first wife, Mary.

12. Edison almost certainly referred to Thomas Parsons's well-known poem, "On a Bust of Dante." The poem appeared as a preface to Parsons's translation of the first ten cantos of the "Inferno," which he published anonymously in 1843 along with an engraving of a bust of Dante. In the poem, Parsons remarks, "How stern of lineament, how grim / The father was of Tuscan song" and describes the bust in these melancholy terms: "The cheeks with fast and sorrow thin,/The rigid front, almost morose." The poem had been recently republished in an anthology of English and American works. Parsons 1843; Head 1884, 446–48.

13. The village of Metuchen was named by seventeenth-century European settlers for Chief Matochshoning. Part of Raritan Township and a commercial and transportation hub for the rural area around it, Metuchen was linked by rail to New Brunswick and nearby Menlo Park. *Ency. NJ,* s.v. "Metuchen"; Spies 2000, 7–8.

14. English author and wit Sydney Smith (1771–1845) was a co-founder of the *Edinburgh Review. Oxford DNB,* s.v. "Smith, Sydney."

15. Charles Dickens (1812–1870), English novelist. *Oxford DNB,* s.v. "Dickens, Charles John Huffam."

16. Edison was likely reading *Personal Traits of British Authors,* which includes sketches of Thomas Babington Macauley, Charles Dickens, Sidney Smith, and Charlotte Brontë. The story of the scalding is recounted, in slightly different words in the sketch of Macauley. Mason 1884, iii, 40.

17. Thomas De Quincey (1789–1859), English author, is chiefly

known for his *Confessions of an English Opium Eater,* first published in book form in 1822. *Oxford DNB,* s.v. "De Quincey, Thomas Penson De."

18. Charlotte Brontë (1816–1855), English novelist, authored the classic *Jane Eyre* (1847). *Oxford DNB,* s.v. "Brontë [*married name* Nichols], Charlotte [*pseud.* Currer Bell]."

19. Edison had some piano instruction as a child and, during his married life with Mary Stilwell Edison, kept an instrument in their homes. Daughter Marion Edison remembered him playing self-composed tunes during her childhood. *TAEB* 1 chap. 1 introduction, 3 chap. 1 introduction n. 2, 7 chap. 3 introduction; Docs. 218 n. 2, 2671, and 2683 n. 4.

20. Actor, playwright, and theater manager James Morrison Steele MacKaye (1842–1894) was a social companion of Edison. He had also shown strong interest in electric lighting for dramatic purposes. See Doc. 2762 esp. n. 2.

21. Henry Hobart Porter (1829–1904) was the commissioner of charities and corrections for New York City from 1881 until 1896. "City and Suburban News," *NYT,* 15 Dec. 1881, 8; "Henry Hobart Porter," ibid., 16 Apr. 1904, 9.

22. The drama editor for the *New York Tribune* since 1865, William Winter (1836–1917) was by this time considered the most powerful drama critic in the United States. Years later, Edison recalled meeting both Winter and Steele MacKaye on several occasions, including one in November 1884. *ANB,* s.v. "Winter, William"; Doc. 2762 n. 1; TAE questionnaire responses, n.d. [Sept. 1925], MacKaye (*TAED* X009AF).

23. A punning reference to Louise Igoe.

24. Nero Claudius Caesar Drusus Germanicus (37–68), Roman emperor (54–68) notorious for wanton cruelty. Among the acts attributed to him, directly or indirectly, were the murders of his mother, two wives, and a sister-in-law, and starting a conflagration that consumed much of Rome. *WBD,* s.v. "Nero."

25. Margaret Stilwell (b. 1871?), the younger sister of Edison's first wife. Doc. 2646 n. 1.

26. "Duck on a rock" is a children's game that combines stone throwing and tag. Each participant chooses a large stone suitable for throwing. The player who is "it," also known as the keeper, sets his or her stone, called a "duck," on a flat surface such as a rock or tree stump. The remaining participants stand behind a line some twelve feet away and pitch their stones at the duck to knock it down. If all the players miss, they run to retrieve their stones while avoiding the tag of the keeper. If a player knocks the duck off the rock, all players can rush from the line to retrieve their stones. The keeper once again tries to tag one of the others, but only after having replaced the duck. Newell 1883, 189.

27. Edison likely meant "peccavi," Latin for "I have sinned." The Latin word, which appears in the vulgate edition of the Bible, came into popular use after 1844, when the British humor magazine *Punch* published a notice suggesting that the British general Sir Charles Napier had telegraphed the one-word message "peccavi" to his superior after having conquered the Indian province of Sindh without having had permission to do so. Although apocryphal, the punning allusion to both "I have sinned" and "I have Sindh" became a well-known example of succinctness and wit. Farwell 1985, 30; Poulos 2004, 271.

28. Edison was likely referring to sore-head disease in chickens, which though called "chicken pox" is unrelated to the varicella virus that causes chicken pox in humans. Sore-head chicken pox or contagious epithelioma killed up to ninety percent of the chicks in an infected flock, and it had reached epizootic proportions in the United States by the late nineteenth century. The typical veterinary response at this time was to treat the sores locally with iodine or silver nitrate, which proved ineffective. Edison's hope for a vaccine that would protect yet-unborn generations is consistent with a tradition of scientific belief in the inheritance of acquired characteristics, a belief exemplified in the early nineteenth-century theorizing of Jean Baptiste Lamarck, the French botanist and zoologist. It was only in 1910 that German bacteriologist Paul Manteufel developed a vaccine for sore-head disease. Bealey 1884, 1:123; Nörgaard 1916, 155; Hadley and Beach 1913, 704–8; *DSB*, s.v. "Lamarck, Jean Baptiste Pierre Antoine de Monet de."

29. The Gillilands' rented home in Winthrop, Mass.; see headnote above.

–2825–

Diary Entry

Menlo Park NJ July 13 [15?] 1885[1]

Woke (is there such a word) at 6 oclock—slipped down the declivity of unconcienceness again until 7. arose and tried to shave with a razor so dull that every time I scraped my face it looked like I was in the throes of cholera morbus.[2] by shaving often I too a certain extent circumvent the diabolical malignity of these razors—if I could get my mind down to details perhaps could learn to sharpen it, but on the other hand I might cut myself— As I had to catch the 7:30 am train for New York I hurried breakfast, crowded meat potatoes, eggs, coffee, tandem down into the chemical room of my body. Ive now got dyspepsia in that diabolical thing that Carlyle calls the stomach,[3] rushed and caught train— Bought a New York World[4] at Elizabeth[5] for my mental breakfast— Among the million of perfected Mortals on Manhattan island two of them took it into their heads to cut their naval chord from mother earth and be born into a new world, while two other less developed citizens stopped two of the neighbors from living— The details of these two little incidents conveyed to my mind what beautiful creatures we [liv?][a] live among, and how with the aid of the police, civilization so rapidly advances— Went to New York via Desbrosses Street ferry[6]—tooks cars[7] across town—saw a women get into car that was so tall and frightfully thin as well as dried up that my mechanical mind at once conceived the idea that it would be the proper thing to run a lancet into her arm and knew joints and insert automatic self feeding oil cups to diminish the creaking when she walked— Got off at Broadway—tried experiment of walking two miles to our office at 65 5th Ave[8] with idea it would alleviate my dyspeptic pains— It didnt— Went into Scribner & Sons[9] on way up saw about a thousand books I wanted right off   Mind No 1 said why not buy a box full and send to Boston now— Mind No 2 (acquired and worldly mind) gave a most withering mental glance at mind No 1 and said You fool, buy only two books, these you can carry without trouble and will last until you get to Boston. Buying books in NYork to send to Boston is like "carrying coals to Newcastle" of course I took the advice of this earthly adviser— Bought Aldrich's Story of a bad boy[10] which is a sponge cake kind of literature, very witty and charming—and a work on Goethe & Schiller by Boynsen[11] which is soggy literature, a little wit & anacdote in this style of literature would have the same effect as baking soda on bread, give pleasing results.

Waited one hour for the appearance of a lawyer who is to cross-examine me on events that occurred 11 years ago[12]— went on stand at 11:30— He handed me a piece of paper with some figures on it, not another mark, asked in a childlike voice if these were my figures, what they were about and what day 11 years ago I made them— This implied compliment to the splendor of my memory was at first so pleasing to my vanity that I tried every means to trap my memory into stating just what he wanted—but then I thought what good is a compliment from a 10 cent lawyer, and I waived back my recollection. A lawsuit is the suicide of Time.— Got through at 3:30 PM— waded through a lot of accumulated correspondence mostly relating to other peoples business— Insull saw Wiman about getting car for Railroad Telegh experiment—will get costs in day or so.— Tomlinson made Sammy mad by saying he Insull was Valet to my intellect= Got $100 met Dot and skipped for the Argosy of the Puritan Sea: ic[b] Sound Steamboat.—[13] Dot is reading a novel—rather trashy, Love hash.— I completed reading Aldrich's Bad Boy and advanced 50 pages in Goethe then retired to a "Sound" sleep.

AD, NjWOE, TAE Diary, Cat. 117 (*TAED* MA007). [a]Canceled. [b]Circled.

1. Edison presumably wrote this entry, like the preceding and following ones (Docs. 2824 and 2826) at the Gillilands' Woodside Villa home on or after 15 July; see Doc. 2824 n. 1.

2. Contemporary medical authorities distinguished between Asiatic cholera, an epidemic and frequently fatal communicable disease, and cholera morbus, an episodic condition resulting from acute gastro-

intestinal irritation of various causes. Outward symptoms were often similar, however, and included "a cadaverous hollowness of the cheeks and eyes, a livid color of the face, hands, and feet, . . . [and] a loss of the elasticity of the skin." The condition was sometimes referred to as simple cholera; use of the word "morbus," Latin for disease, was declining generally at this time. Stillé 1885, 14–17, 106–9; Quain 1883, s.vv. "choleraic diarrhoea," "morbus."

3. The Scottish historian and essayist Thomas Carlyle (1795–1881) famously suffered from recurrent dyspepsia. *Oxford DNB*, s.v. "Carlyle, Thomas."

4. Joseph Pulitzer purchased the *New York World* in 1883 and quickly remade it into one of New York's most popular and distinctive newspapers, blending sensational human interest and crime stories with high-minded public spiritedness and a one-cent price. *ANB*, s.v. "Pulitzer, Joseph"; Juergens 1966, chap. 1.

5. The manufacturing and shipping city of Elizabeth, N.J., was the location of a busy junction between main lines of the Pennsylvania Railroad and the Central Railroad of New Jersey. *Ency. NJ*, s.v. "Elizabeth"; Sipes 1875, 48–49.

6. The ferry terminal at Desbrosses St., directly across the Hudson River from the Pennsylvania Railroad's Jersey City depot, was one of the major landings for Manhattan-bound passengers. Cudahy 1990, 120–21; *Ency. NYC*, s.v. "Ferries."

7. Edison meant one of New York's many horse-drawn street railroad lines.

8. Edison kept an office in this building, which he had shared with the Edison Electric Light Co. since 1881. See *TAEB* 5, chap. 9 introduction and Doc. 2699.

9. Charles Scribner's Sons, the renowned publishing house and bookseller, had its offices at 743–45 Broadway, near Eighth St. *Ency. NYC*, s.vv. "book publishing," "booksellers"; *ANB*, s.v. "Scribner, Charles."

10. Thomas Bailey Aldrich (1836–1907), American poet and novelist, published his semi-autobiographical *The Story of a Bad Boy* in 1869 (in serial form) and 1870 (as a book). The novel was an early representative of what has been called a "post–Civil War boom in 'bad boy' books" and, according to later critics, an influence on Mark Twain's *The Adventures of Huckleberry Finn*. *ANB*, s.v. "Aldrich, Thomas Bailey"; Obenzinger 2005, 402.

11. Norwegian-born poet and author Hjalmar Hjorth Boyesen (1848–1895) was professor of Germanic languages and literature at Columbia University. His pathbreaking study, *Goethe and Schiller: Their Lives and Works*, was published in 1879. *ANB*, s.v. "Boyesen, Hjalmar Hjorth."

12. The patent interference case of *Edison v. Latimer v. Weston* officially resumed at 10 a.m. on 13 July. Edison continued his testimony, adjourned since June, under cross-examination by Edwin Brown, an attorney for Edward Weston, about his early incandescent lamp experiments. Edison's attorneys stated for the record that the case (concerning the attachment of carbon lamp filaments to lead-in wires) was one of some seventy-five to one hundred such proceedings or lawsuits presently involving Edison or his associates about electric lighting. TAE's

testimony, 16 June and 13 July 1885, pp. 16–38, 73, 77, *Weston v. Lat-imer v. Edison*, MdCpNA (*TAED* W100DFA).

13. Several steam ferry lines operated overnight service from New York, through Long Island Sound, to the cities of southern New England. According to Doc. 2826, Edison landed in Providence, R.I. Dunbaugh 1992, esp. map following acknowledgments.

–2826–

Diary Entry

Woodside Villa July 14 [15?] 1885.[1]

Dot introduced me to a new day at 5:30 am. Arose—toileted quickly—breakfasted—then went from boat to street car—asked colored gentleman, how long before car left—worked his articulating apparatus so weakly I didnt hear word he said.— its nice to be a little deaf when travelling   you can ask everbody directions then pump your imagination for the answer, it strengthens this faculty.— Took train leaving at 7 from Providence for the metropolis of culture—arrived there 9 AM   "Coupaid"[2] it to Damons[3] office[4]—waited ¾ hour for his arrival. Then left for the Chateau-sur-le-Mer— If I stay there much longer Mrs. G__ will think me a bore —perhaps she thinks I make only two visits each year in one place each of 6 months—[5] Noticed there was no stewardess on the ferryboat,[a] strange omisssion considering the length of the voyage and the swell made by the tri-monthly boat to Nantasket.—[6] Man with a dusty railroad Co Expression let down a sort of portcullis and the passengers poured themselves out— Arrived Winthrop Junction[7] found Patrick[8] there according to telephonic instructions,[a] another evidence that the telephone works sometimes. Patrick had the Americanized Dog cart and incidentally a horse, suppose Patrick would forgetten[b] the horse, because last week he went into Boston to Damons city residence[9] and turned on the gas & started up the meter from a state of innocence to the wildest pevarication,—& forgot to turn gas off[c]—arrived at Woodside Villa and was greeted by Mamma with a smile as sweet as a cherub that buzzed around the bedside[a] of Raphael— A fresh invoice of innocence and beauty had arrived in my absence in the persons of Miss Louise Igoe[10] and her aunt[11]   Miss Igoe like Miss Daisy is from Indianpolis, that producer of hoosier venus's   Miss Igoe is a pronounced blonde, blue eyes, with a complexion as clear as the concience of a baby angel, with hair like Andromache[12] Miss Igoes aunt is a bright elderly lady who beat me so bad at checkers that my bump of "Stragetic combination"[13] has sunk in about two inches— for fear that Mrs G__ might think

I had an inexhaustable supply of dirty shirts, I put on one of those starched horrors procured for me by Tomlinson— put my spongy mind at work on life Goethe— Chewed some Tulu gum[14] presented me by Mrs G__ conceived the idea that the mastication of this chunk of illimitable plasticity—a dentiferous tread-mill so to speak, would act on the salivial glands to produce an excess of this necessary ingredient of the digestive fluid and thus a self acting home made remedy for dyspepsia would be obtained— believe[a] there is something in this as my dyspeptic pains are receeding from ~~from~~ recognition

—Dot is learning to play Lange's "Blumenlied"[15] on the piano—

Mis Igoe I learn from a desultory conversation is involved in a correspondence with a brother of Miss Mina[16] who resides at Canton Ohio being connected with the Mower & reaper firm of Aultman & Miller[17]   The letter received today being about as long as the bills at the Grand Hotel at Paris are I surmise of rather a serious character, cupid-ly speaking   The frequency of their reception will confirm or disaffirm my conjectures as to the proximity of a serious catastrophe— A post office courtship is a novelty to me, so I have resolved to follow up this matter for the experience which I will obtain— This may come handy should "My head ever become the dupe of my heart" as Papa Rochefacauld[18] puts it.— In evening went out on sea wall—noticed a strange phosphorecent light in the west, probably caused by a baby moon just going down Chinaward   thought at first the Aurora Boreallis had moved out west— Went to bed early dreamed of [~~deamon~~?][d] Demon with eyes four hundred feet apart.

AD, NjWOE, TAE Diary, Cat. 117 (*TAED* MA011). [a]Obscured overwritten text. [b]"ten" interlined below. [c]"& forgot . . . off" interlined above. [d]Canceled.

1. Edison presumably wrote this entry, like the preceding ones (Docs. 2824 and 2825) at the Gillilands' Woodside Villa home on or after 15 July; see Doc. 2824 n. 1.

2. Edison was probably coining a verb from the anglicized noun "coupé," a small four-wheeled carriage seating one or two passengers. Fennell 1892, s.v. "coupé"; *OED*, s.v. "coupé," n. 3.

3. Damon was Edison's literary nickname for Ezra Gilliland, a reference to the Greek mythological figure who, with Pythias, personified unstintingly loyal friendship. Notable recent literary treatments of these stock characters included Louisa May Alcott's 1871 novel *Little Men*, which included a chapter titled for them. Another was an irreverent 1884 essay by popular humorist Bill Nye (Nye 1884), "The True Story of Damon and Pythias," satirizing the capriciousness of a modern corporation (with allusions to a powerful communications company like

American Bell, Gilliland's current employer, or Western Union) toward its employees.

4. That is, Gilliland's office at the American Bell Telephone Co.'s headquarters at 95 Milk St. in Boston. *Boston Directory* 1885, 1365.

5. See Doc. 2824 (headnote).

6. Nantasket Beach, a summer resort on a peninsula in Massachusetts Bay, about ten miles southeast of Boston. Sweetser 1888, 59–80; *WGD*, s.v. "Nantasket Beach."

7. Winthrop Junction was the connecting point between the Boston, Revere Beach and Lynn Railroad line to East Boston (with ferry service to Boston proper) and the Boston, Winthrop and Shore Railroad, which made multiple stops on the peninsula. Nason 1878, s.v. "Winthrop"; Bacon 1886, s.v. "Boston, Revere Beach, and Lynn Railroad."

8. Not identified; possibly a household servant.

9. The Gillilands resided in the Hotel Huntington (room 20) at 101 Milk St. in Boston. *Boston Directory* 1885, 450.

10. Louise M. Igoe (1858?–1936) of Indianapolis was visiting Boston as a guest of Lillian Gilliland. A friend of Mina Miller's older sister Jane, Louise introduced Mina to the Gillilands. She later married Mina's older brother Robert Anderson Miller. Her brother Philip married Lillie Fox, Lillian Gilliland's friend, in 1886. "Mrs. Robert A. Miller," *NYT,* 7 June 1936, 46; "Obituary," *N.A.R.D. Journal* 29 (12 Feb. 1920): 864.

11. Rachel Kinder Clark (1828?–1890) was Louise Igoe's maternal aunt. Family Bible, acc. no. 104948, NjWOE; U.S. Census Bureau 1963? (1850), roll M432_159; p. 226A, image 156 (Indianapolis, Marion, Ind.); Burial record at Crown Hill Cemetery (Indianapolis), Find A Grave memorial no. 45905391, accessed through www.findagrave.com on 24 Jan. 2013); Louise Igoe to Mina Miller, 18 May 1885, EMFP (*TAED* X104O E1).

12. Andromache, wife of the Trojan warrior Hector, appeared in ancient Greek and Roman depictions of scenes from Homer's *Iliad*. She was typically represented with short hair pulled back from the front, a depiction that carried through to classical English scholarly commentary on the *Iliad,* such as that by Alexander Pope. Andromache famously pulled out her hair in grief at her husband's death. *OCD*, s.v. "Andromache"; *LIMC* 1/1:767, 774; Homer 1796, 6:59 (Book 12, l. 600).

13. Edison likely referred to a presumed phrenological feature of his skull.

14. As early as 1873, Louisville pharmacist John Colgan developed what he called Taffy Tolu, reputedly the first sweetened chewing gum, by combining balsam tolu extract (a cough syrup flavoring) with chicle. Colgan formed a partnership with James McAfee about 1880 and together they manufactured and successfully commercialized Taffy Tolu to a growing consumer market. *Ency. Louisville,* s.v. "Colgan, John."

15. German composer Gustav Lange (1830–1889) wrote several hundred minor compositions. His "Blumenlied" ("Flower Song") for solo fortepiano achieved wide popularity. Cooke 1910, 216.

16. Robert Anderson Miller (1861–1911), born the fifth child of Lewis and Mary Valinda Miller, worked at the farm equipment manufacturing plant of C. Aultman & Co. as an assistant to the president, Jacob Miller (his uncle). He married Louise Igoe in 1887. In 1900, Miller

was appointed postmaster of Ponce, Puerto Rico, a post he kept until 1911. Hendrick 1925, 70–73, 80–81, 85–86, 111.

17. Aultman, Miller & Co. was organized in 1863 to expand the capacity of C. Aultman & Co., of Canton, Ohio, to manufacture the Buckeye mower and reaper. Its new factory in Akron, superintended by Lewis Miller, was close to the older firm's plant in Canton. Its principal, the late Cornelius Aultman, was Miller's stepbrother. Hendrick 1925, 69–73; *NCAB* 31:136.

18. François de la Rochefoucauld (1613–1680), French nobleman and memoirist, published *Réflexions ou Sentances et maximes morale,* a collection of taut epigrams, in various versions between 1665 and the issuance of a definitive edition in 1678. Widely read and translated, the work was often referred to in English as simply the *Maxims.* Edison quoted maxim 102: "The mind is always the dupe of the heart." *New Oxford Comp.,* s.v. "La Rochefoucauld, François, duc de"; La Rochefoucauld 2001, vii–xvi, 22.

-2827-

Diary Entry

Woodside Villa july 15 1885

Slept well—breakfasted clear up to my adams apple—took shawl strap[1] and went to Boston with Damon with following memorandum of things to get.

Lavatar on the human face[2]—Miss Clevelands book[3]—Helosie by Rosseau[4]—short neckties—Wilhelm Meister[5]—Basket fruit—Sorrows of Werther[6]—Madam Recamiers works[7]—Diary[a] books[8]—pencils   Telephone documents   Mark Twains gummed Potentiality of Literature ie[b] scrap book.[9]—also book called "How success is won"[10] containing life of Dr Vincent[11] & something in about Minas father and your humble servitor.

Found that only copy of Lavatar which I saw the other day had been sold to some one who was on the same lay as myself   Bought Disraeli's Curriosities of Literature[12] instead— Got Miss Cs book—Twains scrap book—Diary books, How Success is Won also fruit among which are some peaches which the vendor said came from California—think of a lie 3000 miles long—There seemed to be a South Carolina accent in their taste— Started back to office with fruit, apparantly by the same route I came, brought[a] up in a strange street saw landmark and got on right course again   Boston ought to be buoyed and charts furnished strangers— Damon suggests American District Messenger buoys with uniform—[13] Saw a lady who looked like Mina—got thinking about Mina and came near being run over by a street car— If Mina interferes much more I will have to take out an accident policy— Went to dinner at a sort of No-bread-with-one-fishball restaurant[14] then came up towards Damons office, met Damon Madden and Ex Gov Howard of Rhode Island.[15] The Governor whom I know and who is very deaf greeted me with a boiler yard voice.[16] He has to raise his voice so he can hear himself to enable him to check off the acuracy of his pronounciation The Governor never has much time, always in a hurry—full of business, inebriated with industry— If he should be on his death bed I believe he would call in a short hand clerk to dictate directions for his funeral, short sketch of his life, taking a press copy of the same in case of litigation

Madden looks well in the face but I am told its an Undertakers blush— Went to Damons office he was telling me about a man who had a genius for stupidity when Vail came in dressed like Beau Brummel,[17] both went into another room to try some experiments on Damons Phonometer[18]

—Saw Hovey[19] a very very bright newspaper man told me a story related to him by a man who I never would have imagined could or would have told such stories I refer to a gentleman in the employment of the Telephone Co who Tomlinson nicknamed "Prepositum" because he got off that word in a business conversation, his eminent respectability so impressed Tomlinson that when he came out of his office asked me to take him quickly somewhere disreputable so he could recover. This story would have Embarassed Satan— I shall not relate it, but I have called it "Prepositums Turkish Compromise" Hovey told me a lot about a 6th sense, mind reading etc made some suggestions about Railroad Telegraph

—Came home with Damon at 5 oclock— Damon has an ulcerated molar— Just before supper Mrs Roberts[20] and another lady came in to visit[a] Mrs G. Mrs R is a charming woman—Condensed [shi?][c] sunshine—beautiful—plays piano like a long haired professor—played several of Lange's pieces first time seeing them. This seems so incomprehensible to me as a man reciting the Lords prayer in four languages simultaneously— Mrs R promised to come tomorrow evening and bring with her a lady who sings beautifully and a boy dripping with music

—Everyone after supper started their Diary Mrs G Igoe —Daisy and Dot went to bed at 11:30. Forgot two nights running to ask Damon for night shirt— That part of my memory which has charge of the night shirt department is evidently out of order.

AD, NjWOE, TAE Diary, Cat. 117 (*TAED* MA015). [a]Obscured overwritten text. [b]Circled. [c]Canceled.

1. That is, a pair of straps (usually leather) attached to a transverse handle, for carrying objects such as a shawl or books. *OED*, s.v. "shawl-strap."

2. Johann Kaspar Lavater (1741–1801), a Swiss Protestant minister, published *Physiognomische Fragmente, zur Beförderung der Menschenkenntniss und Menschenliche* in the 1780s. The work appeared in English translation from 1788 to 1799 as *Essays on Physiognomy, Designed to Promote the Knowledge and the Love of Mankind* and subsequently in many English editions during the nineteenth century. Lavater contended that a scientific examination of facial features could reveal a person's character. An English-language pamphlet, *How to Read the Face; or, Physiognomy Explained According to the Philosophy of Lavater*, was published in New York in 1883. Stemmler 1993, 151; Graham 1961, 297.

Edison sometimes drew on phrenology to judge the worthiness of job applicants (see Docs. 2475 esp. n. 3 and 2485 n. 1), and he may have used Lavater's ideas in compiling a chart of the physical and character traits of his associates and acquaintances (Doc. 2852). Late in life, however, Edison dismissed the usefulness of phrenology for judging a person's acumen or morality. "Men cannot be judged rightly by looks," he noted: "I have tried it and been fooled many times." Israel 1998, 445; Edison and Runes 1948, 132.

3. Rose Elizabeth Cleveland (1846–1918), the sister of President Grover Cleveland, was a graduate of Houghton Academy and a former headmistress. She published *George Eliot's Poetry, and Other Studies* in 1885 (Cleveland 1885). At this time, she was serving as the president's official White House hostess. Caroli 1987, 102–3.

4. *Julie, ou La Nouvelle Heloise* (1761), Jean-Jacques Rousseau's epistolary novel, proved to be one of the most important best sellers of the eighteenth century in both France and Britain. It helped to turn literary taste away from adventure stories toward works that focused on personal intimacy, the life of the emotions, and the relationship between the individual and nature. *La Nouvelle Heloise* is also generally credited with being one of the seminal works of the Romantic Movement. *New Oxford Comp.*, s.v. "Nouvelle Héloïse, Julie ou la"; Cranston 1991, 247; Rousseau 1968, 1–2, 10.

5. *Wilhelm Meisters Lahrjahre* was first published by Johann Wolfgang von Goethe in 1795 and 1796. It became known to the English-speaking world through Thomas Carlyle's 1824 translation as *Wilhelm Meister's Apprenticeship* and was subsequently republished many times. A new translation by R. Dillon Boylan that appeared in 1855 was itself often reissued. Bohm 2004, 40, 42; Goethe 1855.

6. Goethe's *Die Leiden des jungen Werthers*, published in 1774, was first translated into English from the French as *The Sorrows of Young Werther* in 1779 and, in 1786, directly from the original German. Bohm 2004, 35, 40.

7. Jeanne-Françoise Julie Adélaïde Bernard Récamier (1777–1849) hosted a famous Parisian salon. After a decade of exile (1805–1815) imposed by Napoleon for her royalist sympathies, Récamier returned to Paris and resumed her salon, by then largely centered on Chateaubriand. She published nothing during her lifetime but her niece Amélie Lenormant brought out *Souvenirs et Correspondance tires des papiers de Madame Récamier* in 1859, consisting mainly of letters to Madame Ré-

camier. Twelve years later, Lenormant brought out a second volume, *Madame Récamier et les amis de sa Jeunesse, et sa correspondance intime,* which included more than forty of Récamier's own letters. These works were translated into English and published in Boston in 1867 as *Memoirs and Correspondence of Madame Récamier* (Récamier 1867) and *Madame Récamier and Her Friends* (Lenormant 1867), respectively. Both had been republished by this time. *NDWB*, s.v. "Récamier [née Bernard] (Jeanne-Françoise-Julie-Adélaide)"; *New Oxford Comp.*, s.v. "Récamier, Jeanne-Françoise"; Récamier 1867, v.

8. Presumably the books Edison used for his diary entries.

9. Samuel Clemens patented a self-pasting scrapbook in 1873. The pages were coated with mucilage or other adhesive requiring only moisture for items to be pasted on them. Marketed as "Mark Twain's Patented Self-Pasting Scrapbooks," they sold well (peaking in 1878) and proved to be the author's only profitable invention. U.S. Pat. 140,245; Rasmussen 2007, 874; Krass 2007, 93–95.

10. *How Success Is Won*, an 1885 collection by noted temperance advocate and educator Sarah Bolton (1841–1916), consisted of short biographies of twelve contemporary (or nearly so) men, including Edison (whom she had asked for biographical information in 1884). Her essay was not up-to-date, however, because it named only Edison's two older children, neglecting William Leslie (b. 1878). It included a version of the apocryphal story that had rankled Edison's wife, to the effect that he had forgotten to leave his workshop on their wedding night. Bolton 1885, 174–94; *ANB*, s.v. "Bolton, Sarah Knowles"; Bolton to TAE, 16 Apr. 1884, DF (*TAED* D8406L); see Doc. 2683 (and headnote).

11. John Heyl Vincent (1832–1920), an elder of the Methodist Episcopal Church and the editor of influential religious publications, was widely recognized as a leader of the American Sunday School movement in Protestant religious education. Seeking a wider scope for his educational vision, Vincent cofounded (with Lewis Miller) the Chautauqua Assembly in 1874 and the Chautauqua Literary and Scientific Circle in 1878. Vincent visited Menlo Park that same year and corresponded with Edison for some months afterward, arranging (or so he thought) for the inventor to lecture at Chautauqua, though Edison was returning from a trip through the West when the appointed day came and went. Vincent moved to New York City in 1866 and was elected a bishop in 1888. He and his wife, Elizabeth Dusenbury, had one child, George Edgar Vincent. His profile in *How Success Is Won* included several references to Lewis Miller. *ANB*, s.v. "Vincent, John Heyl"; Doc. 1303 n. 2; *TAEB* 4 App. 1.G.28; Vincent to TAE, 12 and 25 Apr., 12 and 13 Aug. 1878; Stockton Griffin to Vincent, 13 Aug. 1878; all DF (*TAED* D7802ZHM, D7802ZIT, D7802ZVF, D7802ZVH, D7802ZVG); Bolton 1885 (221–45).

12. The English writer (and father of a future prime minister) Isaac D'Israeli (1766–1848) published *The Curiosities of Literature*, a six-volume collection of anecdotes about literary figures, between 1791 and 1834. It was republished multiple times within his lifetime. *DNB*, s.v. "D'Israeli, Isaac"; *Oxford DNB*, s.v., "D'Israeli, Isaac."

13. The American District Telegraph Co. was formed in New York in 1872 to provide private customers with intracity messenger and security services in Manhattan and Brooklyn. The firm greatly expanded

its geographic scope in 1874 by contracting with Western Union, which specialized in intercity telegraphic service, to hand deliver telegrams and packages in towns and cities across the nation. The company's messenger boys wore distinctive military-style blue uniforms with shiny metal buttons, a peaked cap with a badge, and a metal shield pinned to the chest in the manner of the police. "The American District Telegraph Company.—A Novel Enterprise," *Telegrapher* 8 (13 Apr. 1872): 268; Downey 2002, 40–42, 64.

14. Edison alluded to a ballad, "The Lay of the One Fish Ball," written by George Martin Lane (a Harvard professor of Latin) that became popular after it appeared in *Harper's Magazine* in 1855 and was frequently sung by Union troops. The epithet "No bread with one fish ball," taken from the song, became synonymous with a cheap, stingy eatery. "Editor's Drawer," *Harper's New Monthly Magazine*, 11 (July 1855): 281; "Music: 100-Year-Old Hit," *Time*, 9 Apr. 1945, 58; Goodwin 1900, 2, 11.

15. Henry Howard (1826–1905) served two terms as Republican governor of Rhode Island from 1874 to 1876. He was the founding president of the Armington & Sims Engine Co., one of Edison's major suppliers, among other business interests. Doc. 2430 n. 3; *NCAB* 9:404.

16. The work in a typical boiler yard in the 1880s, which included hammering together and riveting iron plates, was notably noisy, and boilermakers often became deaf at an early age. Campin 1883, 119; Oliver 1902, 752.

17. George Bryan "Beau" Brummell (1778–1840) was a socialite and arbiter of dress in Regency England from about 1800 until he retired to France (in financial distress) in 1816. His manner of dress epitomized the fastidious elegance of contemporary "dandyism." *Oxford DNB*, s.v. "Brummell, George Bryan."

18. The word "phonometer" referred to at least two distinct kinds of devices. One was an instrument to measure the intensity of sound. The other was a combination of compass and timing clock used to provide visual indications of the number and length of whistle-blasts needed for signaling the direction of a steamship's travel, according to maritime convention. The editors have not identified Gilliland's apparatus, but he applied for patent applications in October 1885 on two devices, either of which Edison may plausibly have construed as a phonometer in this latter sense. One was for generating telephone call signals of uniform duration; the other was a clock mechanism to synchronize signaling devices in multiple telephone offices. *OED*, s.v. "phonometer"; *KNMD*, s.v. "phonometer; U.S. Pats. 335,693 and 336,562.

19. Formerly the editor of the *Boston Evening Transcript*, William Alfred Hovey (1841–1906) was involved in electric lighting enterprises by 1883 and at this time was an editor on *The Electrical Review*. Earlier in 1885, Hovey published *Mind-Reading and Beyond* (Hovey 1885), reporting on the investigations of paranormal phenomena by the Society for Psychical Research, in London. Daniel Hovey Association 1913, 342–43; Chamberlin 1969 [1930], 151; "For the Busy Man," *New York Tribune*, 25 Feb. 1906, 7; Norton 1883, 16; "Electrical News and Notes," *Elec. and Elec. Eng.* 3 (May 1884): 113; *Book Notes* [Providence, R.I.] 2 (28 Mar. 1885): 111.

20. This was very likely Hinda Roberts, née Barnes (1839–1909),

whose husband, George Litch Roberts, was a patent attorney with offices at 95 Milk St., Boston, in the same building as Ezra Gilliland and the American Bell Telephone Co., for which he did extensive defense work. Hinda had attended Wesleyan Academy in Wilbraham, Mass., and Mount Holyoke (apparently without graduating), and she and her husband were known to be accomplished musicians. Hinda Roberts was the sister of the feminist writer Zadel Barnes Gustafson. Barbour, White, and Crossley 2000, 40; "Mrs. Hinda B. Roberts Dead," *Boston Daily Globe,* 29 May 1909, 9; "George Litch Roberts, Xi '59," *The Diamond of Psi Upsilon* 15 (June 1929): 295; "George L. Roberts, Noted Lawyer, Dies," *NYT,* 1 May 1929, 25; *Alumni Record of Wesleyan University* 1883, 166; Fiske 1940, 198; Herring 1995, 2–3, 314 n. 15.

–2828–

Diary Entry

Woodside Villa July 16 1885

I find on waking up this morning that I went to bed last night with the curtains up in my room=glad the family next door retire early— I blushed retroactively to think of it— Slept well—weather clear—warm. Thermometer prolongatively progressive—day so fine that barometer anaethized— breakfasted— Diaried al lot of nonsense—[1] Read some of Longfellows Hyperon,[2] read to where he tells about a statue of a saint that was attacked with somnambulism and went around nights with a lantern repairing roofs, especially that of a widow woman who neglected her family to pray all the day in the church

Read account of two murders in morning Herald to keep up my interest in human affairs— Built an air castle or two— Took my new shoes[3] out on trial trip— Read some of Miss Clevelands book where she goes for George Eliot for not having an heavenly amen streak of imaginative twaddle in her poetry— The girls assisted by myself trimmed the Elizabeth collars on twelve dasies, inked eyes nose & mouth on the yellow part which gave them a quaint human look, paper dresses were put on them and all were mounted on the side of a paper box labelled "The Twelve Daughters of Venus"   I hope no College[a] bred dude[4] will come down here and throw out insinuations that Venus was never married, and never had any children anyway—

Girls went in bathing. Me and Damon went out in the steam Yacht sailed around over the lobster nursury[5] for an hour or so— In the evening Damon started a diary—very witty— Miss Igoe told Damon she couldnt express her admiration, whereupon he told her to send it by Express freight.

Lunched our souls on a Strauss[6] waltz played by Miss

Daisy, then we all set around the table to write up our diaries. I learned the girls how to make shadow pictures by use of crumpled paper= we tried some experiments in mind reading which were not very successful. Think mind reading contrary to common sense, wise provision of the Bon Dieu that we cannot read each others minds twould stop civilization and everybody would take to the woods= in fifty or hundred thousand centuries when mankind have become perfect by evolution then perhaps this sense could[a] be developed with safety to the state. Damon and I went into a minute expense account of our proposed earthly paradise in the land of Flowers,[7] also a duplicate north and we concluded to take short views of life and go ahead with the scheme. It will make a savage onslaught on our bank account. Damon remarked that now all the wind work is done there only remains some little details to attend to such a "raising the money" etc.

Mrs Roberts hurt her Sophrano arm[8] and could not come over an play for us as promised and thus we lost her perfumed conversation lovely music and serephic smile— La femme qui-rit—[9]

Since Miss Igoe has been reading Miss Clevelands book her language has become ~~disyllibic~~ dissyllabic,[b] [and?][c] ponderous, stiff and formal, each observation seems laundried—

If this weather gets much hotter, Hell will get up a reputation as a [---][c] summer resort. Dot asked how books went in the mail. Damon said as second class ~~matte~~ mail matter. I said Me and Damon would go at this rating—suggested that Mina would have to pay full postage. Damon thought she should be registered— This reminds me that I read the other day of a man who applied for a situation as sexton in the Dead letter office.— Daisies sisters[10] photograph rests on the mantel, shews very beautiful girl   every fly that has attempted to light on it has slipped and fallen. going to put piece of chalk near it so they can chalk their feet. this will permit with safety the insectiverous branch of nature to gaze upon a picture of what they will attained after ages of evolution. Ladies went to bed, this removed the articulating upholstery, then we went to bed.

AD, NjWOE, TAE Diary, Cat. 117 (*TAED* MA019). [a]Obscured overwritten text. [b]Interlined above. [c]Canceled.

1. Edison may have been referring to Docs. 2824–26, each of which he apparently wrote retrospectively.

2. American poet Henry Wadsworth Longfellow (1807–1882) published *Hyperion* in 1839. The loosely autobiographical novel traces the German sojourn of a young American, Henry Flemming. An early inci-

dent in the novel concerns the tale of an impoverished old woman who could not afford to fix the broken tiles of her roof. One night, as rain poured in, she heard a hammering on the roof and later saw a figure with a ladder and lantern. The mysterious figure turned out to be the statue of Christ from the town chapel; as the story goes, the statue would come alive at night and wander through the town aiding the poor. *ANB*, s.v. "Longfellow, Henry Wadsworth"; Longfellow 1848, 13–18.

3. On June 18, Edison ordered a fourteen dollar pair of shoes from Adam Young, a "French Boot Maker" at 1144 Broadway in New York. Voucher (Laboratory) no. 323 (1885).

4. "Dude" was a recent American slang term for "a man affecting an exaggerated fastidiousness in dress, speech, and deportment," synonymous with a dandy. Its usage had become common in New York to describe "swells" who paraded themselves in fashionable attire along the avenues and at summer resorts. The activities of these "dudes" were often covered in the press, so that they became arbiters of fashion. In 1885, E. Berry Wall was identified in New York as "The King of the Dudes." *OED*, s.v. "dude"; Lerer 2003, 482; Hill 1994, 321–23; "The King of the Dudes," *NYT,* 5 Sept. 1885, 1.

5. Edison and Gilliland most likely visited a natural nursery in Boston Harbor, as the U.S. Commission of Fish and Fisheries was just in the process of establishing a managed lobster hatchery at Woods Hole, Martha's Vineyard, which in any case was too distant for their day's expedition. Lobster larvae float with the tide until reaching the post-larval stage of development, when they settle to the bottom in nurseries. Boston Harbor was notable for its natural nurseries, though these were becoming depleted. Rathbun 1886, 17–18, 24–25; Woodard 2004, 249; Cowan 1999, 738.

6. It is unclear to which member of the famed Strauss family Edison referred. Johann Strauss (1804–1849) and his three sons all composed waltzes in the Viennese style, but the most popular were those by Johann Strauss II (1825–1899). *GMO*, s.v. "Strauss."

7. Edison presumably referred to the homes and laboratory that he and Gilliland proposed to build in Florida.

8. That is, the right arm for playing the higher notes on the piano.

9. Meaning literally "the woman who laughs," Edison's phrase is similar to the beginning of a French proverb that seems to have had some currency in English at the time: "Femme qui rit quand elle peut, et pleure quand elle veut" ("Woman laughs when she can, and cries when she wishes"). Fenton 1873, 436; Houstoun 1872, 2:221.

10. Daisy Gaston had two sisters—Olive and Amelia Love—who respectively would have been about fifteen and thirteen years old. U.S. Census Bureau 1970 (1880), roll 295, p. 276B, image 0254 (Indianapolis, Marion, Ind.).

–2829–

Diary Entry

Woodside Villa July 17 1885 ~~July 18 1885~~[1]

Slept so sound that even Mina didn't bother me as[a] It would stagger the mind of Raphael in a dream to imagine a being comparable to the Maid of Chataqua[2] so I must have slept very sound— As usual I was the last one up. This is because Im so deaf—found everybody smiling and happy— Read more of Miss Clevelands book, think she is a smart woman—relatively— Damons diary progressing finely— Patricks went to city get tickets for Opera[b] of Polly,[3] we can comparrot with Sullivans=[4] We are going out with the ladies in Yachts to sail perchance to fish. The lines will be bated at both ends.

Constantly talking about Mina who me an Damon use as a sort of yardstick for measuring perfection makes Dot jealous. She threatens to become an incipient Lucretia Borgia.[5] Hottest day of season—must Hell must have sprung a leak. at two oclock went out on yacht—cooler on the water. Sailed out to the Rock-buoy.[6] This is the point when Damon goes to change his mind, he circles the boat around several times, like a carrier pigeon before starting out on a journey, then we start right dropped an anchor in shady part of the open bay— I acted as Master of the fish lines, delivered them bated to all. The clam bouquets were thrown to the piscatorial[c] actors— Miss Daisy caught the first   he came up smilingly to seize the boquet when she jerked him into the dress circle, genus unknown— I caught the next—genus uncertain. The next was not caught.

Fish seem rather to be conservative around this bay, one seldom catches enough to form the fundamental basis for a lie— Dante left out one of the Torments of hades— I could imagine a doomed mortal made[b] to untangle wet fish lines forever— Everybody lost patientce at the stupidity of the fish in not coming forward promptly to be murdered— We hauled up anchor, and Damon stearing by the compass, (he being by it) made for the vicinity of Apple island—[7] While approaching it we saw a race between two little model vessels full rigged and about 2 feet long= Two yawl boats filled apparently with US naval officers and men were following them. Are these effeminate pursuits a precursor of the decline and fall of the country as history tells us.—[8] Landed at dock 4:30. Came into Villa and commenced reading Lavatar on Facial Philosophy— Dot saw a jelly fish and vehemently called our attention to this translucent chestnut.— Barge called to take us to theatre via Winthrop Junction and Railroad. when we arrived at Junction found we should have to wait some time, so we took an open street car for City— while passing along saw man on Bicycle, asked Damon if he ever rode one— He said he did, once practiced riding in large freight shed where floor was even with the door of cars and three feet from the ground, one day from reason he never could explain he went right through one of doors to the ground. I remarked that I supposed he kept right on riding   No said Damon I jumped back?

Arriving at Ferry boat I asked Damon if it was if it was further across River[9] at[a] high tide[a] said he thought it was a he noticed the piles in the slip were at a slight angle— Arriving on the other side took street Gondola, arrived near top of Hanover Street   when horses were unable to pull cars to the top of the hill, car slipped back. The executive department of my body was about to issue a writ of ejectment when some of the passengers jumped out and stopped car. one passenger hallooed out to let her go they would get more ride— Arriving a little too early for theatre, went to an Ice cream bazar, frigidified ourselves. Then went to Theatre, where we found it very hot. Solomon[10] the Composer came from the cellar of fairies and sprung a chestnut overture on the few mortals in the audience chamber. Then the Curtain arose shewing the usual number of servant girls in tights— The raising of the panapoly of fairyland let some more heat in—a rushing sound was heard and[b] Damon said they were turning on[d] the steam [--]—[e] The fairies mopped their foreheads—perspiration dripped down on stage from the painted cherubs over the arch—after numerous military evolutions by the chorus Miss Lillian Russell[11] made her apperance— Beautiful woman, sweet voice. Wore a fur lined cloak which I thought about as appropriate in this weather as to clothe the firemen on the Red Sea Steamers in sealskin overcoats— noticed one or two original strains the balance of the music seemed to be Bagpipean Improvastarationes— Didnt hear anything that was spoken except once when I thought I heard one of the actors say that his mother ~~sang~~ sung in the Chinese ballet— Our seats were in the baldhead section.[12] After theatre walked to ferry boat— Saw a steamer passing brilliantly lighted   Mrs G asked what could be nicer that a lighted steamer on the waters at night= somebody suggested two steamers— arrived at sister ferry, took RR train. Saw Miss Russell with he last husband Mr Solomon get on train, they stop I believe at the sea shore near us— Home— Bed— Sleep—

AD, NjWOE, TAE Diary, Cat. 117 (*TAED* MA023). [a]Interlined above. [b]Obscured overwritten text. [c]"s" interlined above. [d]Written in left margin. [e]Canceled.

1. Edison wrote and canceled this later date on the fourth page of the diary entry.

2. Lewis Miller (Mina Miller's father) and John Heyl Vincent, both

deeply involved with Methodist Sunday school education, founded the Chautauqua Lake Sunday School Assembly at Fair Point, N.Y., a rural camp-meeting site in the western part of the state, in 1874. The founders broadened their considerable ambitions, quickly turning the Chautauqua Assembly into a large experiment in out-of-school learning, self-improvement, and uplift for those members of the middle class with the means and inclination for a summer vacation. By 1885, the campus occupied at least 130 acres and hosted hundreds of attendees at the summertime assembly programs. It was also the seat of popular non-resident programs such as the Chautauqua Literary and Scientific Circle (a nationwide program of guided reading) and the Chautauqua College of Liberal Arts. Morrison 1974, 31–54; *Ency. ACIH*, 3:355; Rieser 2003, chap. 1; Gould 1961, 3–11; Vincent 1886, 37–40, 44–53, 283.

3. *Polly, the Pet of the Regiment*, an 1882 operetta in two acts by composer Edward Solomon and librettist James Mortimer, opened in New York in April with Lillian Russell taking the title role in the Gaiety Comic Opera Co.'s production. One review dismissed it as a "faint shadow" of popular works by the famous duo Gilbert and Sullivan. The Gaiety Co.'s production opened in late June at the Boston Museum, again with Russell in the lead. *GMO*, s.v. "Solomon, Edward"; "Record of Amusements," *NYT*, 28 Apr. 1885, 5; Advertisement, ibid., 3 May 1885, 15; "The Summer Stage," *Boston Daily Globe*, 28 June 1885, 10.

4. English composer Sir Arthur Seymour Sullivan (1842–1900) was renowned for his light operas written in collaboration with librettist William Gilbert, and his scores have been viewed as models for those of Edward Solomon, a younger contemporary. The works of Gilbert and Sullivan were widely produced in the United States. In 1879, Edison had attended *HMS Pinafore* in New York, where the *Pirates of Penzance* premiered at the end of that year. *Oxford DNB*, s.v. "Sullivan, Sir Arthur Seymour"; *GMO*, s.v. "Solomon, Edward"; *Annual Register* 1885, 88; see Doc. 1711.

5. Lucrezia Borgia (1480–1519), duchess of Ferrara, acted as an unofficial regent for her father, Pope Alexander VI, when he left Rome. Her father arranged for her a series of four marriages of increasing political value to him and the family, with the result that she acquired a lasting reputation (likely the result of propaganda against the family's Spanish origins) for political conniving, including assassination and incest. During the final seventeen years of her life, she emerged as a patron of the arts. *ODR*, s.v. "Borgia, Lucrezia."

6. Possibly the warning buoy at Nash's Rock, a well-known marker in the entrance to Boston Harbor. Pratt 1879, 44.

7. A local landmark with prominent trees, Apple Island occupied nine or ten acres in Boston Harbor near Winthrop. Frequently used at this time as a staging area for scavenging shipwrecks, it was covered in 1946 by Boston's Logan International Airport. Sweetser 1888, 177–78; Shurtleff 1871, 456; Kales 2007, 91.

8. Some enlightenment-era historians, most famously Edward Gibbon in *The History of the Decline and Fall of the Roman Empire* (which Edison had read as a boy [Israel 1998, 8]), ascribed the fall of Rome (and other empires) in part to "effeminate" traits of their rulers or military leaders. For Gibbon, the word was synonymous with the "luxury of Oriental despotism." *OED*, s.v. "effeminate" 1.b; Gibbon 1831, 1:83

(chap. vi), 171 (chap. xi), 183 (chap. xii); see also, e.g., Whelpley 1808, 1:41, 162.

9. That is, the mouth of the Charles River, crossed by a ferry route from East Boston to Boston. Sweetser 1888, frontispiece.

10. Edward Solomon (1855–1895), born in London to Jewish parents, was a promising but erratic composer of operettas in the vein of his renowned older contemporary, Arthur Sullivan. Romantically linked for some time with the actress Lillian Russell, with whom he had a daughter, Solomon composed *Polly* for her, and he brought the show from London to New York in search of financial rewards. He married Russell in May 1885 but abandoned her within sixteen months and was arrested in Britain in 1886 on bigamy charges (later dropped). *Oxford DNB*, s.v. "Solomon, Edward"; "Lillian Russell Married," *NYT*, 11 May 1885, 2.

11. Lillian Russell (1861–1922), born in Iowa as Helen Louise Leonard, was an actress, singer, and stage beauty. Having spent the last two years making a name for herself in England, including the title role in *Polly*, Russell had recently returned to New York with composer Edward Solomon, who became her second husband in May. *ANB*, s.v. "Russell, Lillian"; "Lillian Russell Married," *NYT*, 11 May 1885, 2.

12. A slang expression for the front rows of seats in a theater (particularly a burlesque house), typically occupied by older, wealthy men. *DAS*, s.v. "bald-headed row."

–2830–

Diary Entry

Woodside July 18 1885

Last night room was very close, single sheet over me seemed inch thick— Bug proof windows seems to repel obtrusiveness on the part of any prowling Zephyr that might want to come in and lunch on perspiration. Rolled like a ship in a Typhoon, if this weather keeps on I'll wear holes in the bed clothes. Arose early   Weather blasphemingly hot— went out in sun, came back dripping with water, tried to get into the umbrella rack to drain off, took off two courses of clothes   This would be a good day to adopt Sidney Smiths plan of taking off your flesh and sitting down in your bones.[1] Mem—go to a print cloth mill and have yourself run through the calico printing machine.[2] This would be the Ultama Thule[3] of thin clothing. Read some in Lavatar, Mm Recamier, Rosseaus Emilé.[4] Laid down on sofa—fell asleep— Dreamed that Damon had the sunstroke and was laid on the floor of his office, where he swelled up so that he broke the floor above and two Editors of a baseball journal fell through and were killed. Thought the chief of the fire department came in and ordered holes to be bored in him   Then something changed the dream saw a lot of animals which such marvellous characteristics as would be sufficient to bust up the whole science of paleontology—

Cuvier,[5] Buffon[6] & Darwin[7] never could have started their theories had a few samples of these animals ever browsed around on[a] this little mud ball of ours— After a survey of this vast imaginative Menagerie I woke up by nearly falling off the sofa. Found the heat had reached the apex of its malignity— Went out yachting=all the ladies in attendance— I was delightfully unhot. Ladies played a game called memory— Scheme[b]   No 1 calls out name of prominent author No 2 Repeats this name and adds another & son on, soon one has to remember a dozen names all of which must be repeated in the order given— result Miss Daisy had the best & I the poorest memory— We played another game called "pon honor," resultant of which is that if you are caught you must truthfully answer a question put by each player. These questions generally relate to the amours of the players— Arrived home at 7:30   Yacht brought in too far and left stranded by the receding of the tide. suppered, went out and [s--][c] saw some fire works set off by an unknown sojourner in these ozonic parts. afterwards went over to Cottage Park[8] at the kind invitation of the Charming Mrs Roberts to hear the band play pro bono publico and her boy exclusivemento. Boy is quite a progedy on the piano, plays with great rapidity, his hand and fingers went like a buz saw, played a solemn piece which I imagined might be God Kill the Queen. In walking back Miss Igoe got several boulders in her shoes. Miss Daisy smiled so sweetly all the evening that I imagined a ray of [--][c] sunshine tried to pass her and got stuck. Mrs Roberts caught cold in her arm its cough is better. home—bed—oblivion—

AD, NjWOE, TAE Diary, Cat. 117 (*TAED* MA028). [a]Interlined above. [b]Obscured overwritten text. [c]Canceled.

1. When Sydney Smith sat for a portrait on a hot day, he reputedly said he would "prefer to take off his flesh and sit in his bones." The anecdote was reprinted or referred to countless times. *Oxford DNB*, s.v. "Smith, Sydney"; see, e.g., Duyckinck 1856, 75.

2. Edison referred to the method of printing cloth between rotating cylinders, invented by Thomas Bell about 1783. It was first applied— and still widely used—in the calico trade. *Oxford DNB*, s.v. "Bell, Thomas."

3. The Latin phrase "ultima Thule," referring to land beyond Thule, which the ancients supposed to lie well north of Britain, means the limit of attainment. *OED*, s.v. "Thule" b.

4. *Émile, ou De l'éducation*, a 1762 novel by Jean-Jacques Rousseau describing the education of a solitary boy, was a vehicle for Rousseau's philosophy of education as a means for developing inherent abilities rather than fitting the individual into society. *New Oxford Comp.*, s.v. "*Émile, ou De l'éducation*."

5. French zoologist Georges Cuvier (1769–1832), professor of comparative anatomy and comparative physiology at the Muséum d'Histoire Naturelle, was instrumental in establishing comparative anatomy as a scientific discipline. Not sympathetic to evolutionary ideas, Cuvier strove to define apparent variations among individuals in a way that would preserve species as fixed biological categories. *DSB*, s.v. "Cuvier, Georges"; Coleman 1964, chap. 6; Ruse 1979, 13–15; Appel 1987, 238–40.

6. As a way of accounting for the overwhelming diversity he observed in nature, Georges-Louis (comte de) Buffon (1707–1788), a mathematically inclined French zoologist and naturalist, advocated the idea of the mutability of species. Acquired characteristics could be inherited, he thought, and species could adapt to changes in their environment, such as those encountered during migration. Buffon's works exercised substantial influence on naturalists before Darwin. *DSB*, s.v. "Buffon, George-Louis Leclerc, comte de"; Roger 1997, chap. 18.

7. English naturalist and geologist Charles Darwin (1809–1882) proposed a theory of biological evolution animated by the principle of natural selection in *On the Origin of Species* (1859). *Oxford DNB*, s.v. "Darwin, Charles Robert."

8. Cottage Park was a Winthrop summertime cottage community facing Boston Harbor and Apple Island. Sweetser 1888, 132.

–2831–

From Francis Upton

GÖTEBORG[1] [Sweden,] July 19, 1885.[a]

Dear Edison:

I find your picture in many of the shop-windows here. It looks very strange to see you and our lamps so far from home.

You should see the new stations in Berlin.[2] They are built, each of the two, for 6000 16 c.p.[b] lamps lighted at the same time. The work is splendidly done. We have nothing in America to compare with them in solidity and perfection of steam fitting. The German Co. is in the hands of very strong practical men, and are carrying on the business in a way that is doing your name a great deal of good. They say that they do no poor work, and will not take contracts except at full prices.

Siemens[3] makes a splendid machine for them. His cable[4] appears to be very practical and he guarantees it for five years, which seems to prove that Siemens considers it good.

I shall try to keep the German Co. friendly with the Lamp Co. as they are evidently going to [do?][c] do the Central Station business on a large scale in the next few years. Yours Truly

Francis R. Upton

ALS, NjWOE, DF (*TAED* D8535ZBF). Letterhead of Hotel Christiania. [a]"GÖTEBORG" preprinted. [b]"16 c.p." interlined above. [c]Canceled.

1. A Swedish commercial center with a relatively ice-free harbor on the Göta Älv, Göteborg (Gothenburg) had 80,000 residents in 1885. The local electrical firm of Wockatz & Co. did a large business in Edison products. Karl Baedeker 1882, 281; "Notes. Cheap Telephony in Sweden," *Teleg. J. and Elec. Rev.* 17 (11 July 1885): 31; "Manufacturing and Trade Notes," *Elec. and Elec. Eng.* 6 (Feb. 1887): 80.

2. Upton likely referred to the central stations at 44 Markgrafenstrasse (8,000 lamp capacity) and 80 Mauerstrasse (6,000 lamps), which became operational in 1885 and 1886, respectively. They were built by the Städtische Elektrizitätswerke Berlin, a stock company founded by Deutsche Edison Gesellschaft. Two older stations continued to operate at 9 Schadowstrasse and 85 Friedrichstrasse. Hughes 1983, 72–73; Schilling 1886, 20–21, 33; "Éclairage électrique," *La Lumière Électrique* 16 (Apr. 1885): 250; Zacharias 1887, 161–62; cf. Doc. 2487.

3. Werner von Siemens (1816–1892), preeminent German electrical engineer, inventor, and entrepreneur, was a cofounder and principal of Siemens & Halske, a major manufacturer and developer of electrical equipment, most famously of submarine telegraphic cables. Docs. 1851 n. 1, 2173 n. 19, 2448 n. 3.

4. Upton may have meant the electrical cable, manufactured by Siemens & Halske, that was insulated with jute and sheathed in lead. Black 1983, 61; Verity 1889, 355.

–2832–

Diary Entry

~~I've dated wrong~~ Woodside Villa July 19th 1885
Slept as sound as a bug in a barrell of morphine.[1] Donned a boiled and starched emblem of respectability=[2] Eat food for breakfast, Weather delightful— Canary seed orchestra started up with same old tune, ancestors of this bird sang the self same tune 6000 years ago to Adam down on the Euphrates, way back when Abel got a situation as the first angel— Read Sunday Herald, learned of John Roache's[a] failure[3]—am sorry—he has been pursued with great malignity by newspapers and others, from ignorance I think— Americans ought to be proud of Roach who started life as a day laborer and became ~~the~~ giant of industry and the greatest shipbuilder in the United States employing thousands of men and feeding innumerable families— what has he now for this 40 years of incessant work and worry   People who hound such men as these I would invent a special Had~~a~~es. I would stricken them with the chronic sciatic neuralagia and cause them to wander forever stark naked within the artic circle— Saw in same paper account of base ball match, this struck me as something unusual—[4] Read more about that immeasurable immensity of tact and beauty Madame Recamier. I would like to see such a woman= ~~n~~Nature seems to be running her factory on another style of goods nowdays and wont switch back until long

after Im baldheaded— Damon went out to assist the tide in— Daisy told me something about a man who kept livery stable in Venice.

In afternoon went out in yacht, on first trip all our folks and lot of smaller people sailed around for an hour returned and landed the abreviated people— Started for Cottage Park where we took on board the Charming Mrs Roberts brévet Recamier, and a large lady friend whose name has twice got up and jumped out of my mind. Then sailed away for Rock buoy and for some ocult reason Damon didnt stop and change his mind but headed for Liverpool went out two miles in ocean, undulations threatened to disturb the stability of the dinner of divers persons, returned at 7 pm. Then Damon took out a boat load of Slaves of the Kitchin— Damon and I after his return study plans for our Floridian bower in the lowlands of the peninsular Eden, within that charmed zone of beauty, where wafted from the table lands of the Oronoco[5] and the dark Carib sea, perfumed zephyrs forever kiss the gorgeous flora, Rats!— Damon took the plans to Boston to place them into the hands of an archetectualist to be reduced to a paper reality—[6] Damon promised to ascertain probable cost chartering schooner to plough the Spanish main loaded with our hen coops— Dot came in and gave us a lot of girlish philosophy which amused us greatly— Oh dear this celestial mud ball has made another revolution and no photograph yet received from the Chataquain Parragon of Perfection. How much longer will Hope dance on my intellect

Miss Igoe told me of a picture she had taken on a rock at Panama NY.[7] There were several others in the group, interpolated so as to dilute the effect of Mina's beauty, as she stated the picture was taken on a rock I immediately brought my scientific imagination to work to ascertain how the artist could have flowed collodion over a rock and put so many people inside his camera. Miss Igoe kindly corrected her explanation by stating that a picture was taken by a camera of a group on a rock. Thus my mind was brought back from a suspicion of her verbal integrity to a belief in the honesty of her narrative

After supper Mrs. G, Daisy and Louise with myself as an incidental appendage walked over to the town of Ocean Spray,[8] went into a drug store and bought some alleged candy, asked the gilded youth with the usual vacuous expression, if he had any nitric peroxide, he gave a wild stare of incomphrensibility   Then I simplified the name to nitric acid,[9] which I hoped was within the scope of his understanding   a faint gleam of intelligence crept over his face whereupon he went into another room from which he returned with the remark that he didnt keep nitric acid— Fancy a drug store without nitric acid. A drug store nowdays seems to consist of a frontage of Red blue and green demijohns a soda fountain, Case with candy and toothbrushes, a lot of almost empty bottles[b] with death and stomachatic destruction written in latin on them, all in charge of a young man with a hatch shaped head, hair ~~parted~~ laid out by a civil engineer, and a blank stare of mediocrity on his face, that by comparison would cause a gum indian in the Eden Musée[10] look intellectual— On our return I carried the terrealbian gum drops,—moon was shining brightly—girls called my attention several times to beauty of the light from said moon shining upon the waters, couldnt appreciate it, was so busy taking a mental triangulation of the moon the[c] two sides of said triangle meeting the base line of the earth at Woodside and Akron, Ohio.[11]

Miss Igoe told us about her love of ancient literature, how she loved to read latin, but couldn't. I told her I was so fond of Greek that I always rushed for the comedies of Aristophanes to read whenever I had the jumping toothache. Bed—Mina, Morning.

AD, NjWOE, TAE Diary, Cat. 117 (*TAED* MA031). [a]"c" interlined above. [b]Interlined above. [c]Obscured overwritten text.

1. Like many contemporaries, Edison had an everyday familiarity with morphine in his household, and it may have contributed to the death of his wife in 1884. He also stocked the compound among his laboratory chemicals and used it in his commercial analgesic preparation. See Docs. 2712 (headnote) and 1287 esp. n. 3.

2. Edison presumably meant a freshly laundered shirt.

3. Irish immigrant John Roach (1813–1887) was the preeminent builder of iron ships in the United States, and the principal works of John Roach & Sons, in Chester, Penn., employed well over a thousand men. Roach placed his firm in receivership on 18 July after the U.S. Navy, following months of dispute, refused to take delivery of a completed cruiser and appeared ready to repudiate the balance of its contract. Some newspapers viewed the Navy's actions as a combination of personal animus and partisan politics by the new Democratic administration of Grover Cleveland, while others, including the *Boston Herald,* saw Roach's enterprise as emblematic of Republican corruption. The Edison Machine Works had occupied a Roach industrial property in New York since 1881. *DAB,* s.v. "Roach, John"; Swann 1980 [1965], 55, 202–29; "Failure of John Roach," *Boston Herald,* 19 July 1885, 1; see Doc. 2060; Samuel Insull to James Potter, 7 Mar. 1881, Lbk. 8:14 (*TAED* LB008014).

4. Edison may have had in mind the *Boston Herald's* rather mirthful account of a game notable not only for poor fielding by the Chi-

cago team but for its players' protests against an umpire's decisions; five players surrounded the official at one point, to the delight of spectators. "Boston Yields to Chicago," *Boston Herald,* 19 July 1885, 2.

5. Edison meant the Orinoco River, which drains from much of Colombia and Venezuela into the Atlantic. *WGD,* s.v. "Orinoco."

6. Ezra Gilliland hired Alden Frink (b. 1833), an independent architect in Boston since 1860 who designed residential and commercial structures, as well as a number of stations for the Boston & Maine Railroad. Frink worked at 28 State St., a few blocks from Gilliland's office. Frink charged Edison and Gilliland $200 for drawing the laboratory and two houses. Gilliland later gave the plans to the Kennebec Framing Co. Albion 2008, 14–15; Roberts 1901, 4:115; *Boston Directory* 1885, 427; Alden & Frink invoice, 23 Sept. 1885, Cat. 1165 (Fort Myers accounts): 11–12, Accts., NjWOE; see Doc. 2848.

7. A village in Chautauqua County.

8. The small resort village of Ocean Spray, constructed in 1875 on the ocean side of the peninsula, was considered one of Winthrop's most attractive communities. It was said to be a particular favorite among stage actors. Sweetser 1888, map frontispiece, 134–35.

9. A highly reactive compound, nitric acid was commonly used to treat liver problems and digestive difficulties more generally, either by taking internally as a tonic or applying externally in conjunction with hydrochloric acid. In its strong form, nitric acid was also used externally to remove warts and similar skin growths. Nothnagel and Rossbach 1884, 2:335–36; *Chambers's Ency.* 1876, s.v. "nictric acid."

10. The Eden Musée, a dime museum opened in 1884 on Twenty-third St. (between Fifth and Sixth Aves.) in New York, featured a large collection of wax figures. The Eden Musée quickly proved itself a commercial success and became a model for similar museums in other cities. Dennett 1997, xii, 44–46.

11. The family home of Mina Miller, named Oak Place, was in Akron, Ohio. Hendrick 1925, 103–4.

–2833–

To Samuel Insull

[Woodside Villa,] July 20 [1885]

Sammy

I send you letter received from John Ott[1]   you better make it your business to get these two fellows wkg in harmony and have an understanding just what is to be done to cars ground wires etc— [2] Sooner this is done the better—[a]

Regarding Uptons Letter—did you add enough to prices of tubes to give a Comsn[3]   I've nothing to say about Lamp Co affairs except I think he is on right track—[a]

E[dison]

Its not very hot down[b] here but it is in Boston— Tell E[ward] H J[ohnson] Im pretty certain to[b] sell that telephone but will have to put in 4 or 5 months on it myself to get it stable and Commercial & a good receiver— tell him do nothing as I have Committed myself too[c] deep to withdraw and am going to bust the undulatory theory[4]   E

ENCLOSURE[d]

NEW YORK, July 20 1885[e]

Dear Sir

I have not bin able to do enything on insulated car roof As Mr Insull has not received word.

Last Thursday I went to Totenville[5] to see about 4 cars that are in construction two of wich have roofs on. I asked the supt of Staten Island railroad[6] how soon they would be finn–ished   He said they wood be finished in a few days but if I wanted him to hold back with the roof he would do so.

Whereupon King broke in and said he did not see whare this was going to do a dam bit of good, judgeing from his ex–periments   I told him they were the orders of Mr Edison and we will have to exicute them.

Where upon he said that he was as much to blame as the next one as he had sugested the idea, and the old man did not know any thing about this. didnt he and Gill come down the other day and tryed to make some sugestions, and they could not make a dam one. He not only made this sugestion to one person but Informes everybody of such

I think he is a fit subject for some instution   Yours truly

J. F. Ott

This is no free lecture on general usefullness on Laboratory boys

ALS, NjWOE, DF (*TAED* D8503ZBP). [a]Followed by dividing mark. [b]Obscured overwritten text. [c]Interlined above. [d]Enclosure is an ALS on letterhead of Thomas A. Edison. [e]"NEW YORK," and "188" preprinted.

1. See enclosure below regarding Ott's railway telegraph experiments.

2. Payroll and expense records from summertime work on the railway telegraph, including that done on Staten Island, are in Edison's Bill Book Vol. 2 (Apr. 1884–Dec. 1888), Accts., NjWOE.

3. Edison likely referred to Upton's 7 July letter about price estimates for underground conductors for Vienna and about efforts to negotiate rights to sell American-made lamps in Europe. Upton to TAE, 7 July 1885, DF (*TAED* D8535ZBC).

4. Johnson's telephone transmitter and a similar instrument that was the subject of a pending patent application by Edison and Sigmund Bergmann (see Doc. 2813 n. 2) were make-and-break devices. Unlike the carbon transmitter, in which pressure altered the resistance of carbon granules to create an electrical signal of correspondingly undulating intensity, the metallic electrodes of the Johnson and Edison-Bergmann instruments cleanly opened the circuit at a certain pressure threshold, creating a signal of on/off pulsations. Such binary action was charac-

teristic of some early telephones, notably that of Phillip Reiss, whose antecedence of Alexander Graham Bell's instrument had been put forward in a spate of attempts to break the Bell patents in court. American Bell built its defense largely on the claim that only undulating currents, not make-and-break pulses, could transmit speech telephonically. That distinction had prevailed in every major case, and the counter-example of a practical make-and-break instrument could have been a serious blow to the Bell interests (Beauchamp 2010, 858–65; Doc. 2798 n. 2). Puzzled by the resemblance, however superficial, to the Reiss transmitter, the *Electrical World* conjectured that Bergmann saw his new device as advantageous because it could produce larger currents that would activate a receiver more effectively than a conventional carbon transmitter (*Electrical World* 7 [20 Mar. 1886], 96).

5. A village on the southwest tip of Staten Island, opposite Perth Amboy, N.J., and the terminus of the Staten Island Railroad. New York State Senate 1885, 2:311–12.

6. John W. Wilbur superintended the railroad after having been a contractor for its construction. Opened about 1860, the line ran southwest from Vanderbilt Landing, opposite Manhattan. *Prominent Men* 1893, 168; Gold and Weintrob 2011, 21.

–2834–

Diary Entry

Woodside July 20 1885

Arose before anybody else—came down and went [----][a] out[b] to look at Mamma Earth and her green clothes— Breakfasted— Read aloud from Madame Recamiers memoirs for the ladies— Kept this up for an hour got as [hoare?][a] hoarse as a fog horn. Think the ladies got jealous of Madame Recamier—

its so hot—I put everything off— Hot weather is the mother of procrastination— my energy is at ebb tide—Im getting Caloricly stupid— Tried to read some of the involved sentences in Miss Clevelands book, mind stumbled on a ponderous perioration and fell in between two paragraphs and lay unconscious for ten minutes— Smoked a cigar under the alaias alias[b] of Reina Victoria[1]   think it must have been have been made seasoned in a sewer—

Mrs Clark[2] told me a story about Louise's mother[3] singing in a company a song called I have no home, I have no home,[4] somebody halloed out that he would provide her with a good home if she would stop. I understood Mrs. Clark to say that this gentlemen was a bookeeper in a small pox hospital—

Mrs. G. Has placed fly paper all over the house. These cunning Engines of insectiverous destruction are doing a big business— One of the first things I do when I reach heaven is to ascertain what flies are made for—this[c] done Ill be ready for business. perhaps I am too sanguine and may bring up at the other terminal and one of my punishments will be a general ukase from Satan to keep mum when Edison tries to get any entomological information— Satan is the scarecrow in the religious cornfield=

Towards sundown went with the ladies on yacht— Talked about love, cupid, Appollo, Adonis, ideal persons   One of the ladies said she had never come across her ideal— I suggested she wait until the second Advent.— Damon steered the galleon. Damons heart is so big it inclines him to embonpoint— On shore it was hot enough to test Safes but on the water twas cool as a cucumber in an artic cache—Mrs G has promised for three consecutive days to have some clams a la Taft.[5] She has perspired her memory all away—

Been hunting around for some ants nests, so I can have a good watch of them laying on the grass— Don't seem to be any around—don't think an ant could make a decent living in a land where a yankee has to emigrate from to survive.—

For[c] the first time in my life I have bought a pair of premeditately tight shoes— These shoes are small and look nice   My No 2 mind, (acquired mind) has succeeded in convincing my No 1 mind, (primal mind or heart) that it is pure vanity, conceit and folly to suffer bodily pains that ones person may have graces the outcome of secret agony— Read the funny column in the Traveller[6] and went to bed.

AD, NjWOE, TAE Diary, Cat. 117 (*TAED* MA036). [a]Canceled. [b]Interlined above. [c]Obscured overwritten text.

1. "Reina Victoria" was a term applied to a variation on the perfecto, a generously sized cigar tapered at each end. "Cigar Facts for Retail Druggests," *Pharmaceutical Era* 49 (Oct. 1916): 392–93.

2. Rachel Kinder Clark.

3. Nancy Kinder Igoe (1829?–1891), a life-long resident of Indiana. U.S. Census Bureau 1965 (1870), roll M593_340, p. 21A, image: 45 (Indianapolis, Marion, Ind.); online burial record of Crown Hill Cemetery (Indianapolis), www.crownhill.org/locate/index.html (accessed 30 Jan. 2013).

4. "I Have No Home," an orphan's lament in three stanzas, was written by William Shakespeare Hays and published in New York in 1873. Hays 1895, 76–77; Library of Congress (American Memory) Music for the Nation online database (http://memory.loc.gov/ammem /smhtml), accessed 24 May, 2011.

5. The Taft (or Taft's) Hotel, also called the Point Shirley House, was located at the southernmost tip of Winthrop. Famous for its game and fish, it was a regional favorite said to rival Delmonico's. Sweetser 1888, 136–37, 140–41.

6. The *Boston Traveller*, established in 1845, published morning and multiple evening editions. Mott 1941, 560; Lee 1973, 278.

[Woodside Villa, c. July 21, 1885][1]

Think you better sound Adams[2] & get best price you can from London[3] would take 50c per share if couldnt do better= How about Colonial E Light—[4] Whats chances—

Can you sell any Edison Field Electric RR[5a]

Tell Upton I cannot very well advise him in Lamp matters on the Continent Im too much involved but for him to look out for the best interest of the Stockholders of the Lamp factory & I will be satisfied[6]

E[dison]

ALS, NjWOE, DF (*TAED* D8503ZBT). [a]Followed by dividing mark.

1. During his vacation in Winthrop, Mass., Edison made excursions into Boston and may have written from there. Insull addressed identical telegrams to him at Winthrop and Boston on 20 July (Insull to TAE, 20 July, Lbk. 20:416, 416A [*TAED* LB020416, LB020416A]). Edison probably wrote this note after his 20 July letter to Insull (Doc. 2833) but certainly on or before 22 July, when Insull sent letters based on it (Insull to Edward Adams, 22 July 1885; Insull to Francis Upton, 22 July; Lbk. 20:424, 427 [*TAED* LB020424, LB020427]).

2. A financier with an engineering background, Edward Dean Adams (1846–1931) was currently associated with the New York banking firm of Winslow, Lanier & Co., as well as a trustee of the Edison Electric Light Co. and a director of the Edison Electric Illuminating Co. Doc. 2690 n. 2.

3. Exercising powers of attorney for Edison, Charles Batchelor, and Edward Johnson, Adams had transferred their holdings of deferred ("B") shares of the Oriental Telephone Co. to London in December 1884, where they were used in a successful proxy fight for control of the company. The bulk of the 15,635 shares belonged to Edison (Doc. 2757 n. 2; Insull to Adams, 25 Nov. 1884, Lbk. 19:405 [*TAED* LB019405]; Adams to Insull, 3 Dec. 1884, DF [*TAED* D8472ZBU]). On 16 July 1885, when Adams inquired whether he should have the stock certificates returned to New York, he noted that his estimation of their value as "entirely nominal—no transactions" accorded with Insull's own view (Adams to Insull, 16 July 1885, DF [*TAED* D8547ZAH]).

Insull subsequently conveyed Edison's instructions to Adams, telling him to seek a buyer in London. When Adams offered to take all the shares himself for $3,000, Insull insisted that they "would be a bargain at $9,000" and again asked him to test the market in London (Insull to Adams, 22 and 28 July 1884, Lbk. 20:424, 433 [*TAED* LB020424, LB020433]; Adams to Insull, 24 July 1885, DF [*TAED* D8547ZAI]). Adams did so without effect, and the owners decided to hold the shares in hopes that their value would rise (Adams to Insull, 31 July 1885,

DF [*TAED* D8547ZAK]; Insull to Adams, 3 Aug. 1885, Lbk. 20:434A [*TAED* LB020434A]).

4. Edison referred to Edison's Indian and Colonial Electric Co., a London concern organized in 1882 to control Edison's patents in British colonies around the Indian Ocean basin. The firm was moribund; having recently reduced its capital, its future was in limbo while the directors sought to merge with another company working in the region. Doc. 2624 n. 1; Edison's Indian and Colonial Electric Co., Ltd., report, 2 June 1886, DF (*TAED* D8630ZBA).

5. Edison meant the Electric Railway Co. of the United States, formed in April or May 1883 through a combination of patents owned by Edison and inventor Stephen Dudley Field, among others (see Doc. 2431 esp. n. 6). As of May 1885, Edison owned at least 1,333 shares in the company; in June he sought to sell 100 shares at twenty-five dollars per share (Meadowcroft to TAE, 22 May 1885; TAE to Insull, 22 June 1885; both DF [*TAED* D8545I, D8503ZBA]; TAE to Coster, 19 June 1885, Lbk. 20:364 [*TAED* LB020364]).

6. Insull relayed Edison's directive to Upton. Insull to Upton, 22 July 1885, Lbk. 20:427 (*TAED* LB020427); see also Doc. 2822 n. 7.

–2836–

Martin Force to
Samuel Insull

Menlo Park, July 21st 1885[a]

My Dear Sir

I cannot report as great results as ~~I~~ would like to my experiments so far is rather up hill work but still I keep pegging away at it in hopes of getting some data to start from.[1] Two and a half miles is quite a gap to Jump but I have not the least doubt, ~~of it,~~ but[b] what it can be done it is only a question of experimenting. a while. I am now making a portable condenser to see Just how far I can work so I can move it to any place I wish. Will write you again in a day or two. Have written Mr Edison each days experiments.[2] Your Resptf

Martin N Force

ALS, NjWOE, DF (*TAED* D8546Y). Letterhead of Thomas A. Edison. [a]"188" preprinted. [b]Obscured overwritten text.

1. Force was experimenting on the ship telegraph system described in Doc. 2807. It took him at least several days to set up at Menlo Park, and on 9 July he reported to Edison his trouble raising the elevated "banners," presumably constructed of sheet metal or foil. Force to Insull, 6 and 8 July 1885; Force to TAE, 9 July 1885; all DF (*TAED* D8546U, D8546V, D8546W).

2. Other than the 9 July letter quoted in note 1, the editors have not found Force's reports to Edison on this subject. On 1 August, after a delay caused by high winds, he told Insull that the best results "so far is at a distance of 200 feet and that is only one way. At first after making my portable banner I placed it at a distance of 940 feet from large

banner and at that point I did not succeed in getting anything either way consequently I had to come nearer." Force was still working on the project through August and into mid-September, when a timesheet indicated that he spent "6 days flying Kites for Ship Tel." Force bill to TAE, 3 Sept. 1885, DF (*TAED* D8546ZAI1); Force timesheets, 10 and 17 Sept. 1885, Voucher (Laboratory) nos. 450, 454 (1885).

—2837—

Diary Entry

Woodside July 21 1885

Slept splendidly—evidently I was innoculated with isomnic bactilli when a baby—arose early went out to flirt with the flowers.=[a] I wonder if there are not microscopic orchids growing on the motes of the air— Saw big field of squashes throwing out their leafy tenticles to the wind preparing to catch the little fleeting atom for assimutation[1] into its progeny the squash gourd— A spider[b] weaves its net to catch an organized whole, how like this is the ~~vegitable~~ living plant, the leaves and stalk catch the primal free[c] atom all ~~an~~ are[c] then arranged in an organized whole.[2]

Heard a call from the house that sounded like the shreick ⟨How is this spelled⟩ of a lost angel, it was a female voice three sizes too small for the distance and was a call for breakfast— After breakfast laid down on a sofa, fell into light draught[b] sleep dreamed that in the depth of space, on a bleak and gigantic planet the solitary soul of the great Napoleon was the sole inhabitant. I saw him as in the pictures, in contemplative aspect with his blue eagle eye, amid the howl of the tempest and the lashing of gigantic waves high up on a jutting promontory gazing out among the worlds & stars that stud the depths of infinity   Miles above him circled and swept the sky with ponderous wing the imperial condor bearing in his talons a message. Then the scene busted— This comes from reading about Napoleon in Madame Recamiers memoirs.[3] Then my dream changed— Thought I was looking out upon the sea, suddenly the air was filled with millions of little cherubs as one sees in Raphaels pictures each I thought was about the size of[b] a fly. They were perfectly formed & seemed semi-transparent, each swept down to the surface of the sea, reached out both their tiny hands and grabbed very small drop of water, and flew upwards where they assembled and appeared to form a cloud. This method of forming clouds was so different from the method described in Ganots Physic's[4] that I congratulated myself on having learned the ~~truth~~ true method and was thinking how I would gloat over the chagrin of those cold blooded Savans[5] who would disect an angel or boil a live baby to study the perturbations of the human larynx. Then this scheme was wrecked by my awakening—

The weather being cool went out on Veranda to exercise my appreciation of Nature. Saw bugs, butterflies as varied as Prangs Chromos,[6] Birds innumerable, flowers with[b] as great a variety of color as Calico for the African market. Then to spoil the whole two poor miserable mortals came, who probably carry the idea that this world was created for them exclusively and that a large portion of the Creators time was specially devoted to hearing requests, criticisms and complaints about the imperfection of the natural laws which regulate this mud ball— What a wonderfully small idea mankind has of the Almighty. My impression is that he has made unchangeable laws to govern this and billions of other worlds and that he has forgotten even the [exh?][d] Existence of this little mote of ours ages ago. Why cant man follow up and practice the teachings of his own conscience, mind his business, and not obtrude his purposely created finite mind in affairs that will be attended to without any volunteed advice.— Came into the house at the request of the ladies and read aloud for two hours from the Memoirs Recamier— Then talked on the subject of the tender passion,[7] the ladies never seem to tire of this subject— then supper some

Some[e] Trovotores du Pave[8] made their appearance and commenced to play— I requested the distinguished honor of their presence on the Veranda— After supper weather being cool but rather windy, took our trovotores on the yacht and all hands sailed out on the bay— Had to go around an arm of the bay to get coal— water splashed so I got dashed wet. Three several times the water broke loose from the iron grasp of gravitation and jumped on my 65 dollar coat   But when one of the ladies got a small fragment of a drop on her dress [of course?][d] orders were issued to make for port—landed and took our Trovotores to house   several ladies hiring houses for the summer brought their husbands with them and helped sop up the music— afterwards Mrs & Mr G hospitablized[b] by firing off several champaign bottles and some of those delightful Cookies. I do believe I have a big bump[9] for Cookies. The first entry made by the recording angel on my behalf was for stealing my mothers cookies. 11 o'clock came and the pattering of many footsteps upon the stairs signalled the coming birth of silence only to be disturbed by the sonorous snore of the aimable Damon and the demonic laughter of the amatory family Cat.

AD, NjWOE, TAE Diary, Cat. 117 (*TAED* MA039). [a]Unknown mark over equal sign. [b]Obscured overwritten text. [c]Interlined above. [d]Canceled. [e]Added in pencil at an unknown later date.

1. Edison's unclear writing of this word leaves his meaning ambiguous. He may have misspelled "assimilation" or intended a neologism coined from "assimilation" and "mutation," a word used at this time particularly in discussions of evolutionary theory. *OED*, s.v. "mutation" 1.a; see, e.g., Krauth 1878, s.v. "species."

2. Cf. Doc. 2811.

3. Mme. Récamier, exiled for a decade by emperor Napoleon Bonaparte (1769–1821), mentions him repeatedly in her *Memoirs and Correspondence,* including an incident in which she found the eyes of the then-First Consul "fixed upon her with a persistency that finally ended in making her uncomfortable." Récamier 1867, 19–20; Englund 2004, 10, 115, 166–69, 229–30, 444–45, 455.

4. After Adolphe Ganot (1804–1887), a science teacher in France, published *Traité élémentaire de physique expérimentale et appliqué* in 1851, it quickly became the standard textbook for experimental and applied physics. Ganot sold his rights to Hachette and retired in 1881, long after the work had been translated multiple times. An English translation by Edmund Atkinson appeared by 1863 and went through numerous editions. The chapter on meteorology in Atkinson's 1883 edition contained two sections devoted to clouds and their formation. Simon and Llovera 2009, 536–37; Atkinson 1883, sec. 932–33.

5. Edison used a variant spelling of "savant." *OED*, s.v. "savant."

6. Louis Prang (1824–1909), American lithographer born in Silesia, became well known after the Civil War for what he called "chromolithographs" (or "chromos"), which were mass-produced color lithographic reproductions of famous works of art. *ANB*, s.v., "Prang, Louis."

7. Euphemism for love, used variously in the nineteenth century to stand for idealized notions of love, for true emotional attachment, and for sexual desire. For nineteenth-century literary usage, see Collins 1879, 83; Holmes, 1867, 320; for a full discussion of nineteenth-century custom and belief in reference to love and love-making, see Gay 1986.

8. Probably Edison's attempt at "street musicians."

9. That is, a phrenological feature.

–2838–

To Samuel Insull

[Woodside Villa?, c. July 22, 1885][1]

Sammy—

Tell Russell to [let?][a] up on the yacht= I have changed my mind the Expense is too great when you get the proper yacht—[2] You will have to write the German Co that Letter recd too late to get anything ready to reach there in time but if they in future will give me 3 or 4 months notice will be happy to send working models of any new inventions for Exhibition purpose—[3] give Little taffy=Etc   Hope you have Ott and King in harmony— You better come on as you say— Gil has written Tomlinson to come on on his affair[4]

Edison

ALS, NjWOE, DF (*TAED* D8503ZBS). [a]Obscured by ink blot.

1. Edison's comment below regarding the cooperation of John Ott and William King suggests that he wrote this document soon after his 20 July directive to Insull on that topic (Doc. 2833). At the time, Edison was largely at Woodside Villa in Winthrop, with occasional trips into Boston; he had not yet returned to New York as of 22 July. Insull to Francis Upton, 22 July, Lbk. 20:427 (*TAED* LB020427).

2. Edison's instructions notwithstanding, he (or someone on his behalf) apparently found a suitable craft about this time. On 25 July, David Kirby, a well-known yacht builder in Rye, N.Y., submitted a modest bill for paint, rope, and five days of work fitting up the *Triton* (Kirby to Insull, 25 July 1885, DF [*TAED* D8541Y1]; "Aquatic Notes," *Brentano's Monthly* 3 [Apr. 1880]: 148). This expenditure followed several weeks of searching for a vessel. On Edison's instruction, James Russell placed a notice in the 5 July *New York Herald* about chartering a 65-foot steam yacht with two staterooms and full crew. Without referring to Edison by name, the advertisement stated that he wished to "sail during the day and put up at hotels at night," and it invited interested parties to reply with details and monthly rates to "PLEASURE" in care of the *Herald*. Edison, visiting the Gillilands in Massachusetts, asked that all responses be sent to him there, and on 6 July Russell forwarded him the only one he had received: Manning's Yacht and General Shipping Agency offered a 92-foot steam vessel with a crew of five for at least three months at $1,000 per month. Edison seems to have also inquired about using the personal yacht of Stephen Wilcox (of Babcock & Wilcox Co., the New York boiler-making firm), but was told that it was not in condition to sail. Russell also made appointments to meet with several yacht owners but these inquiries and a separate one by John Randolph led nowhere (Russell to TAE, 5 and 6 July 1885; TAE to Insull, n.d. [c. 5 July 1885]; Thomas Manning to Russell, 6 July 1885; Wilcox to TAE, 11 July 1885; Fletcher Roberts to Randolph, 24 July 1885; all DF [*TAED* D8503ZBI, D8503ZBK, D8503ZBJ, D8503ZBJ1, D8503ZBN, D8503ZBP2]).

Absent direct proof of Edison's whereabouts for the next week, circumstantial evidence suggests that he sailed through Long Island Sound. On 27 July, John Randolph wired Insull from City Island, just offshore of the Bronx: "Laid here all night. Go up the Sound tomorrow. All well, having a splendid time." The next day he telegraphed from New London, Conn., about plans to start for Bridgeport, further up the coast, on 29 July. The journey likely continued to Boston, from where Ezra Gilliland sent word to Insull on 3 August (Monday) that "We will all be in New York Wednesday morning" (Randolph to Insull, 27 and 28 July 1885; Gilliland to Insull, 3 Aug. 1885; all DF [*TAED* D8541Y2, D8541Z1, D8503ZBW]). Hiring a vessel for such an excursion would have been in character for Edison. He and Gilliland had chartered a boat in Florida in March for a trip to Fort Myers. There was also a short

Florida fishing expedition in 1884, a much longer trip on the Hudson and St. Lawrence Rivers in 1882, and closer to home, a three-day fishing outing in Raritan Bay in 1877 ("Distinguished Arrivals," *Fort Myers Press*, 28 Mar. 1885, News Clippings [*TAED* X104S001A]; see Docs. 2607 [headnote], 2311, and *TAEB* 3, chap. 6 introduction).

3. Writing from Berlin on 29 June, Emil Rathenau of the Deutsche Edison Gesellschaft invited Edison to send material for the Electrotechnic Society's exhibit samples of telegraph and electric lighting equipment at the International Conference of Telegraph Engineers, an affair scheduled for 10 August to mid-September. He advised that Siemens & Halske was the only potential exhibitor aware of the event. Rathenau to TAE, 29 June 1885, DF (*TAED* D8535ZAX).

4. The editors have not determined the nature of Gilliland's "affair." Possibly it was related to his own telephone inventions, for which he filed separate patent applications in September on a transmitter and a receiver; both patents were assigned to the American Bell Telephone Co. U.S. Pats. 384,201 and 343,449; see Doc. 2848 n. 7, also cf. Doc. 2839 n. 4.

–2839–

Memorandum:
Telephone Patents

[Woodside Villa, July 1885?][1]

1st— Edison as agent to sell to Simmonds[2] the patents & application of Bergman Bergmann & Edison & Johnson as per Copies furnished—[3]

2nd   that Edison will place at the disposal of Simmons agent[4] his laboratory whereby Said agent is to perfect the telephone transmitter working by make & break with metallic points.[5] Making the same Commercially available for competition with the present Bell Co   no charge for use of Laboratory $5000. cash is to be allowed by Simmons to be expended by Simmons agent for this purpose under direction of Edison—

3rd   Edison[a] only agrees personally to experiment & devise apparatus for Litigation which apparatus shall prove conclusively to the mind of the said Simmons that the present theory of the transmission of articulate speech by Electrical undulation over wires is incorrect ~~& that all the necessary~~ & that the transmitter when finished does not work ~~ont~~ the undulatory theory but by make & break

4. That when the said agent of S has perfected the transmitter & Edison has completed his apparatus then S[a] is to come at once to Inspect & investigate the same and is then to decide if he is perfectly satisfied   if so he is to pay in cash $100,000. whereupon E as agent will cause title to all the inventions to be transferred to Simmons—[6]

Simmons is to pay all necessary Expenses of obtaining the patents & the money is not to be dependent on the ~~pallow-~~

ance of the patents Simmons must satisfy himself that ~~the~~ we are entitled by law to the patents and there is every reason to suppose they will ultimately be allowed, but as to this he must attend.

Edison does not agree personally to attend any suit & explain apparatus because this would hamper him if he wanted to travel but agrees to furnish an expert to so explain & work apparatus at Expense of S.

Edison does not agree to assign his ½ interest in the joint patent but will assign any interest which he may have in it in view of his contract with W.U.[7]

Edison cannot bind Bergman or Johnson to any future invention or improvement.

In case of failure the 5000 is to be lien on said patents repayable when they are sold

In addition to the 100 000 we are to get the ¼ cap Stock etc as he has it—[8]

AD, NjWOE, DF (*TAED* D8547ZBR). [a]Obscured overwritten text.

1. During their holiday in Winthrop, Mass., Edison and Ezra Gilliland likely talked over a proposed joint business arrangement with Zalmon Simmons for the metallic telephone transmitters discussed in Doc. 2813 (see esp. n. 2), and Edison referred implicitly to such a commitment in his 20 July letter to Samuel Insull (Doc. 2833). Gilliland's plan to confer with attorney John Tomlinson toward the end of July (see Doc. 2838) could plausibly have concerned his involvement in the project. Tomlinson drafted in his own hand at least two versions of a contract based on Edison's memorandum (one of which appears to be the basis for a typed version) before James Howe, a Simmons associate, reportedly produced his own finished agreement (not found) about the first of September (TAE draft agreements with Simmons, Gilliland, Edward Johnson, and Bergmann, all n.d. [Aug.?] 1885; Howe to TAE, 22 Sept. 1885; all DF [*TAED* D8547ZBS, D8547ZBT, D8547ZBU, D8547ZAS]).

2. Zalmon Gilbert Simmons (1828–1910) of Kenosha, Wis., headed the Simmons Manufacturing Co., a highly successful maker of wire mattresses that dominated the business into the next century. While cultivating that enterprise in the 1870s, Simmons also transformed a small stake in the Wisconsin State Telegraph Co. into the much larger Northwestern Telegraph Co., which he leased to Western Union in 1881. He became a director of Western Union in 1884. Simmons was elected to the state legislature in 1865 and served as Kenosha's mayor in 1884 and 1885. With prior experience in railroading, he later organized a company that built and successfully operated a cog railway line up Pike's Peak. *DWB*, s.v. "Simmons, Zalmon Gilbert"; *NCAB* 15:136–37.

3. Edison, Edward Johnson, and Sigmund Bergmann had among them at least four pending patent applications related to make-and-break telephone transmitters with metallic (instead of carbon) elec-

trodes. In November 1883, Edison and Sigmund Bergmann jointly filed an application on such a similar transmitter having a small fluid reservoir to dampen the motion of the diaphragm and electrodes. The following month, Bergmann filed a closely related application for dampening reservoirs on both sides of a diaphragm. Both patents issued on 2 March 1886 (U.S. Pats. 337,254; 337,231). Edward Johnson also had two similar applications from the same time at the Patent Office; they are discussed in Doc. 2813 n. 2.

4. The agent was to be Ezra Gilliland, who was named in this capacity in the draft agreements. In the typed version, Edison promised to

> extend the facilities of his laboratory in the City of New York to the said Gilliland, or such other person or persons, for the conduct of experiments with the view of perfecting the same as aforesaid. The said Gilliland hereby agreeing to forthwith enter the laboratory of the said Edison, and to devote his time and best efforts to the perfection of said inventions, and to the invention of a complete system of telephony . . . [TAE draft agreement with Simmons, Gilliland, Johnson, and Bergmann, n.d. [Aug.?] 1885 (Art. 1), DF (*TAED* D8547ZBU)]

A rough draft in Tomlinson's hand granted the privileges of Edison's laboratory to Johnson and Bergmann as well as to Gilliland (TAE draft agreement with Simmons, Gilliland, Johnson, and Bergmann, n.d. [Aug.?] 1885, DF [*TAED* D8547ZBS image 6]).

5. Such a transmitter could undermine American Bell Telephone Co.'s legal defense of its basic patents. See Docs. 2813 and 2833.

6. Emendations to a subsequent typed draft indicated that Bergmann and Johnson were to receive $36,000, Edison $24,000, and Gilliland $40,000 if Simmons accepted their work. They were also to receive stock shares in a new company that Simmons pledged to form within sixty days of his acceptance. TAE draft agreement with Simmons, Gilliland, Johnson, and Bergmann, n.d. [Aug.?] 1885 (Art. 4), DF (*TAED* D8547ZBU).

Simmons evidently planned on additional backing from industrialist and railroad financier Henry H. Porter, whose connections included the Pullman Palace Car Co. as well as timber and steel interests. Howe described him as "one of the richest men in Chicago and [he] is as enterprising as he is wealthy. We could not have a better man." Howe to TAE, 24 Aug. 1885, DF (*TAED* D8547ZAN); Pierce 1957, 157 esp. n. 36; *WWW-1*, s.v. "Porter, Henry H."

7. This clause alluded to the 1878 contract by which Edison sold to Western Union his telephone patents and any that he might obtain within seventeen years, an agreement that Edison believed entitled American Bell to his future inventions by virtue of its 1879 settlement with Western Union (TAE agreement with Western Union, 31 May 1878, Miller [*TAED* HM780045]; see Docs. 1317, 1822 n. 13). The ninth article anticipated that Edison might devise a telephone based on an entirely new principle or different mechanical construction that could be used as an alternative to his carbon transmitter. Such new inventions were to fall within the scope of the contract and to be assigned to the company. This provision contained a qualifying term, however,

that seemed to invite conflicting interpretations. Namely, it would apply only in cases where the new device "shall have no such practical advantage over the [prior] invention covered by such patent as to make the new invention of any greater value than the invention already covered by such patent." Subsequent drafts of the Simmons agreement (see note 1) stipulated that the Western Union contract had no bearing on the patents of Bergmann or Johnson or those they obtained together with Edison, but the joint Edison-Bergmann patent was evidently a cause for concern to James Howe. Contemporary law required all parties to a joint patent to have contributed to the invention, and Howe cautioned that "the Bell people may rely upon their ability to prove that this is all your invention & the joint patent would be void, and that they are entitled to the whole, under your contract." In preparation, Howe asked Edison and Bergmann to give statements to an attorney, which they did. After researching the question, John Tomlinson expressed his opinion that the joint patent was legally sound (Howe to TAE, 7 and 8 Sept. 1885, n.d. [Sept.?] 1885; statements of TAE and Bergmann, n.d. [Sept.?] 1885; Tomlinson to TAE, n. d. [Sept.?] 1885; all DF [*TAED* D8547ZAP, D8546ZAJ, D8547ZBP, D8547ZBQ, D8547ZBO]).

8. Simmons decided in October not to sign the final version of the agreement sent him by Tomlinson because of two (unspecified) objectionable provisions. Howe to TAE, 16 Oct. 1885, DF (*TAED* D8547ZAV).

–2840–

From Erastus Wiman

New York, August 7th 1885.

My Dear Mr. Edison;—

Yours of August 6th is received,[1] in which you say that if I can "put the Western Union way wire affair through this summer you will assign to me one quarter of the royalties."[2a] This is indeed a very liberal offer on your part, and I accept it gratefully, and I promise to use my best and most active influence in this direction.[3] I have been quietly working at it for some time,[b] and will do the best I can. Nothing would be so influential in accomplishing the purpose we both have in view,[b] as a practical demonstration of the efficiency of your invention. There is no place in the world where it could be done better than in Canada. Mr Dwight,[4] the General Manager, is, as you know, in thorough sympathy with you, has the best possible facilities, and would give a practical and anxious oversight to any practical demonstration of your views. Would it not be possible, for instance, to put your device on wires between Toronto and Hamilton, a distance of 40 miles, and illustrate by practical daily work the advantages which your invention promises?[b] It is true it might be unwise to apply for a patent in Canada prior to your application here, as that might injure it owing to the rule in regard to foreign patents.[5] We have, however, lines in the United States belonging to the Great North Western Co.,[6] for instance in northern New York, and we could use the device between Whitehall and Watertown.[7] Anything you suggest in this regard I will be prompt to carry out, only, let us get to work at it and get it demonstrated. Why should you not take a run up to Canada and have a conference with Mr Dwight about it?[b] I saw him last week for three days at the Thousand Islands,[8c] and he was very much in earnest over the whole matter. He asked me many questions about it which I was unable to explain. He has some first rate electricians in his employ.

Did you not reside long enough in Canada to be a member of the Canadian Club?[9] The enclosed speech may have some interest for you.[10]

I had great pleasure in entertaining Johnson and our friend Coster at my house the other night.[11] Johnson thinks there is no difficulty in illuminating base ball grounds at night by the incandescent light, so that the game could be played. If this could be done we have a bonanza in the grounds at Staten Island[12] and transportation to and from them beyond the dreams of avarice. Let the subject of illuminating grounds for athletic sports have some thought from you now and again.

I trust you will have a pleasant vacation, which I am sure you greatly need. I am, Faithfully yours

Erastus Wiman

TLS, NjWOE, DF (*TAED* D8546ZAC). Letterhead of Great Northern Western Telegraph Co. of Canada; typed in uppercase. [a]Underline and quotation marks added by hand. [b]Punctuation added by hand. [c]"s" added by hand.

1. Not found.

2. Wiman referred to the phonoplex telegraph system. He subsequently offered to share the cost of the proposed demonstration (see Doc. 2854).

3. Edison evidently wished to use Wiman, who was a director of Western Union (as well as president of the Great North Western Telegraph Co.) as an intermediary for leasing or selling the phonoplex to the company. Wiman reported a few weeks later that after "several conferences with parties in the Western Union," he expected "no difficulty in coming to some arrangement, provided a satisfactory illustration of the device can be afforded and a thorough test. I have prepared the way for this, but nothing can be done under the existing vague understanding of the matter" (Wiman to TAE, 17 Sept. 1885, DF [*TAED* D8546ZAL]). Years later, Alfred Tate recalled that Edison did not want to deal directly with Western Union because of his long-standing enmity toward Thomas Eckert, the vice president and general manager since 1881. Al-

though the editors have not corroborated it, this suggestion is at least plausible (Tate 1938, 104–6; *ANB*, s.v. "Eckert, Thomas Thompson"; cf. Doc. 1713).

4. Harvey Prentice Dwight (1828–1912) became general manager of the Great North Western Telegraph Co. in 1881 and eventually president of the company (in 1892). A native of Belleville, N.Y., he learned telegraphy as an operator with the Montreal Telegraph Co., where he had a distinguished career, becoming manager of its Toronto office in 1850 and superintendent of its western division about 1865. Dwight was also a founding member (1874) of the American Electrical Society. *CCB*, s.v. "Dwight, Harvey Prentice"; Hedley 1905, 313; "Death Notice," *Toronto World*, 5 July 1912, 7; "Formation of an Electrical Society," *NYT*, 22 Oct. 1874, 4.

5. Statutory and case law provided generally that a U.S. patent obtained after one in a foreign country for the same invention would expire with the foreign specification. The maximum life of Canadian patents was fifteen years, though Edison followed what was apparently the common practice of taking out specifications for three years at a time and subsequently renewing them. Edison's two pending U.S. applications for the phonoplex were presumably immune to the limitations entailed by any foreign specifications he might submit at this time, but he may reasonably have anticipated filing other applications in the United States which, in fact, he did. Abbott 1886, 2:481–82; see Docs. 2800 (headnote) and 2859 n. 6.

6. The Great North Western Telegraph Co. of Canada was incorporated in 1880. Erastus Wiman acquired it in 1881 and soon thereafter the company took operational control of the Montreal Telegraph Co. and the Dominion Telegraph Co., ending a rate war between the former competitors and creating a unified system with an exclusive working agreement with Western Union. By the end of 1884, the Toronto-based Great North Western managed some 3,000 employees, nearly 2,000 offices, and about 44,000 miles of wire. It became a branch of the Canadian Telegraphs Co. in 1921. *Acts of the Parliament* 1880, 91; Reid 1886, 609, 611; Babe 1990, 49–50; Rens 2001, 1:22–23; Wiman 1893, 98, 101–3; U.S. Dept. of State 1887, 53, 59; "Great North Western Telegraph Company Becomes Canadian National Telegraphs," *Telegraph and Telephone Age* 39 (1 Feb. 1921): 51.

7. Whitehall, N.Y., near the eastern shore of Lake George in the north-central part of the state, was a contact point with Western Union on the old Montreal Telegraph Co. line. Watertown, N.Y., to the west near Lake Ontario and the St. Lawrence River, was home to a superintendent's office on the Great North Western's New York line. Reid 1886, 338; Berly 1883, 538.

8. The Thousand Island House in Alexandria Bay was one of the largest resort hotels on the St. Lawrence River. It was built during 1872–1873 by hotel magnate Orrin Staples, who sold the property in 1882 to Richard Southgate. (Southgate also owned the Long Beach Hotel on Long Island, N.Y., at which Edison's family stayed in August 1883; see Doc. 2489.) Recent renovations had added such conveniences as Otis elevators, Weston electric lights, and steam heating ("Thousand Islands House," *NYT*, 24 Mar. 1897, 7; "The Thousand Islands," *NYT*, 22 July 1883, 10; *Appleton's General Guide* 1882, [509, 524]; Emerson 1898,

439–40). Edison's traveling party stayed there toward the end of August 1885 (see Doc. 2844).

9. The Canadian Club was organized in April 1885 as a New York social organization, with Erastus Wiman as its first president. Its rooms opened on 1 July 1885 at 3 N. Washington Sq. but soon moved to 12 E. Twenty-ninth St. Membership was ordinarily open to natives or former residents of Canada, though its bylaws did not specify a minimum residency period. Edison and Samuel Insull were elected to membership in August 1885. Canadian Club 1885, 4; ibid. 1887, 2; Canadian Club to TAE, 19 Aug. 1885 and 2 June 1886; Canadian Club to Insull, 16 July and 7 Aug. 1885; Canadian Club circular, n.d. [1885]; all DF (*TAED* D8511G, D8611G, D8541V, D8541ZAB, D8541W).

10. Wiman likely referred to his speech on "The Canadian Club: Its Purpose and Policy" given to a Dominion Day dinner on 1 July (and later privately printed as Wiman [1885]).

11. Wiman owned several Staten Island residences. He often entertained at the mansion known as "Tantallon," where this dinner may have taken place. Coles 1964; "New Brighton in the Guilded [sic] Age: Social Life at 'Tantallon,'" 15.

12. The Staten Island Amusement Co., a syndicate headed by Wiman, purchased a harbor-side playing field near the St. George Ferry in 1884 and began planning its redevelopment as a summertime amusement destination. The lease of the Staten Island Cricket and Baseball Club, an important amateur association that had long held its events there, was terminated by October 1885 ("Staten Island Improvements," *NYT,* 31 Oct. 1884, 8; "The Staten Island Cricket Club," ibid., 24 Oct. 1885, 8; Bayles 1887, 669). Two months later, the Staten Island Amusement group acquired an American Association baseball team, the New York Metropolitan Club, and initiated plans to relocate it from Manhattan's Polo Grounds to the St. George Grounds, where they wished to install a football team as well. The group hoped to have night games played under electric lights and expected to spend as much as $50,000 for 12,000 incandescent lamps (20 candlepower each), including a crisscross arrangement to prevent shadows on the playing field. However, the company's general manager, George F. Williams, decided in March 1886 that games would be played during the day and lighting spectacles would be a night-time attraction for other resort entertainments. An illustration of the new ballfield published in May 1886, showing arc lights hung at regular intervals from the grandstand roof, nonetheless suggests the possibility of night games, and an *Electrical World* report the next month substantiated that intention ("Realms of Pastime, Baseball by Electric Light," *Rocky Mountain News,* 14 Dec. 1885, 2; "Hard By The Ferry. The New St. George's Baseball Grounds on Staten Island," *NYT,* 14 Mar. 1886, 14; "New Grounds of the Metropolitan Baseball Club on Staten Island," *Harper's Weekly* 30 [15 May 1886]: 309; "Baseball on Staten Island," ibid. 30 [15 May 1886]: 311).

Before Wiman sold the team at the end of the 1887 season, Edison incandescent lamps were put in the grandstand. The main lighting attraction, however, was an illuminated fountain that could propel water 160 feet high and, with its colored effects, would suggest a display of water fireworks when viewed from the grandstand and the harbor. Although Wiman had invited Edison in November 1885 to discuss illuminated fountains ("in which I am sure there is more money than in anything else just now"), Edison did not seem to take part in the project. The St. George Grounds installation ultimately used the patents of Sir Francis Bolton adapted by consulting engineer (and former Edison employee) Luther Stieringer; it combined the Edison incandescent and Brush arc lighting systems. The fountain was sold to Chicago traction magnate Charles Yerkes by the end of 1886 and relocated to Chicago. Months later, the Edison Electric Illuminating Co. of Staten Island acquired the Edison isolated generating plant. Wiman to TAE, 10 Nov. 1885, DF (*TAED* D8546ZCC); "The Great Illuminated Fountain on Staten Island," *Electrical World* 7 (16 June 1886): 292; "An American Illuminated Fountain," *Engineering* 42 (6 Aug. 1886): 142; "Base-Ball Notes," *Baltimore Sun*, 10 Oct. 1887, 6; "Coming Up The Bay," *Puck's Library* No. 12 (1888): 7; Stieringer 1901, 417; "Fairy Land on Staten Island," *Frank Leslie's Illustrated* 62 (17 July 1886): 348; Maltbie 1911, 167; Leng and Davis 1930, 1:319.

-2841-

To Archibald Stuart[1]

[New York,] August ~~7th.~~ 10[a] [188]5

Dear Sir:—

Your favor of the 3rd. inst. came duly to hand.[2] When I was in Cincinnati last I intended to go down to your office the day after I saw you, but we had a lot of women folks with our party,[3] I got ~~worried~~ flurried and forgot all about the appointment. I am expecting to be in Cincinnati before the summer is out and when I do go there, it is my intention to sit in your office and talk you to death. If the lightning arrester which we designed for the village plant business is used, it is impossible for any accident to occur.[4] Of course no instrument constructed will provide against the carelessness of man. Very truly yours,

T.AE   I[nsull].

TL (carbon copy), NjWOE, Lbk. 20:434C (*TAED* LB020434C). Initialed for Edison by Samuel Insull. [a]"10" interlined above by Insull.

1. Archibald Stuart (1846–1938) was secretary of the Ohio Edison Electric Installation Co., a Cincinnati firm that contracted with Edison in 1884 to build village system electric lighting plants. Docs. 2497 nn. 1, 4.

2. Stuart urged Edison to investigate lightning suppression in electric light systems. His concern was prompted by a lightning strike on the Edison system in Piqua, Ohio, and more famously, reports of one at the crowded Balmoral Hotel in Mount McGregor, N.Y., where the body of the late Ulysses S. Grant lay in repose in an adjacent cottage. That bolt traveled around the property on the lines of the hotel's Edison isolated lighting plant, causing several injuries both indoors and out. Stuart also chided Edison for failing to visit on his way through Cincinnati in February and reminded him of a subsequent promise to return to the city in the summer. Edison's colorful phrases in his marginal notes on the letter found their way into Samuel Insull's typed reply. Stuart

to TAE, 3 Aug. 1885, DF (*TAED* D8523ZBM); Pitkin 1973, 105–6; "Looking Upon the Body," *NYT*, 31 July 1885, 1; "The Great Soldier's Body," ibid., 1 Aug. 1885, 1; Edison Co. for Isolated Lighting Bulletin 6:27, 25 July 1885, CR (*TAED* CC006).

3. Edison presumably referred to his brief stay in Cincinnati in February 1885 (see Doc. 2776 [headnote]). The editors have not identified the women in his party, though Lillian Gilliland and Marion Edison were certainly on hand. Lillie Fox, a friend of Lillian's who resided in Cincinnati and was the daughter of a local lumber merchant and lawyer, could well have joined them. Edison approvingly appraised Lillie in Doc. 2852 and had a large photo of himself sent to her in June. "Obituary," *N.A.R.D. Journal* 29 (12 Feb. 1920): 864; Louise Igoe to Mina Miller, 8 Feb. 1885, EMFP (*TAED* X104OD); TAE to Samuel Insull, 2 June 1885, DF (*TAED* D8503ZAT).

4. See Doc. 2496 n. 9 regarding Edison's lightning safety system. The Mount McGregor incident was later attributed to a damaged lightning rod. Pitkin 1973, 105–6.

–2842–

To Samuel Insull

Chautauqua NY Aug 11 1885[a]

Samuel Insull
    Will be at Chautauqua[b] for four days[1]

Edison

L (telegram), NjWOE, DF (*TAED* D8503ZBZ). Message form of Western Union Telegraph Co. [a]"1885" preprinted. [b]Obscured overwritten text.

1. Edison went to Chautauqua to see Mina Miller and meet her family. According to Mina, after she had returned home to Akron, Ohio, from her visit to Winthrop, Edison asked her to join him, the Gillilands, and Marion on a trip to the White Mountains in New Hampshire. Her parents would not consent to this arrangement, which prompted Edison and his companions to stop in Chautauqua in an attempt to convince them. "They made the trip sound so attractive," Mina recalled, "that father at last consented to my going." Edison remained at Chautauqua until 18 August when he and his party, which now included Mina and Louise Igoe, left for Alexandria Bay, N.Y. (see map p. 218). Mina Edison article for Associated Press, 10 Jan. 1947, typescript in Edison Bio. Coll.; TAE to Insull, 18 Aug. 1885, DF (*TAED* D8503ZCG); Mary Valinda Miller to Mina Miller, 30 Aug. 1885, CEF (*TAED* FI001AAA20A); see Docs. 2843 and 2844.

–2843–

From Samuel Insull

New York, August 12th. 1885.[a]

My dear Edison:—
    I received the enclosed letter from King[1] yesterday, but was unable to send it to you until your telegram giving your

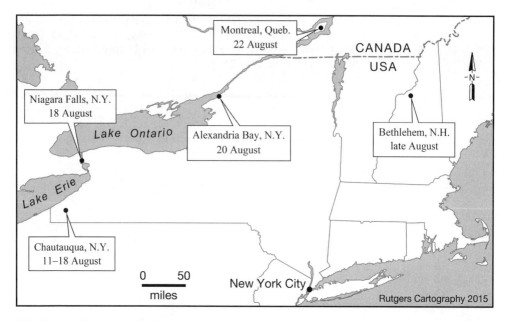

Edison's courtship trip with Mina Miller, August 1885.

address reached me.[2] I enclose letter from Prof. Barker[3] and would state that I have written him saying that I would forward his letter to you with the request that you reply direct. I also enclose letter from Mr. Dixon.[4] If you have anything to say to him please let me know. The gentleman who saw you with relation to one of your electric pen patents has been in here and desires something a little more formal,[5] so as to enable him to go ahead prior to your return: he was talking about royalty and asked me what I thought you would think of 10 per cent on the selling price of the apparatus. Please give me some kind of an authority in this matter so I that I can fix the thing up to the gentleman's satisfaction and then the formal papers can be executed on your return. I want you to understand that this matter is pressing. Keep me posted as to where you go so that I can wire you again. Yours[b]

Sammy

⟨I think 20 percent is about right as the profit is great on these goods=

Tell King I want to try the insulated car roof against the other roofs in very wet weather—

Tell Tomlinson I leave here[6] Monday the 17th for Niagara[7] etc will probably be at 1000 Island House Thursday next & he [----][c] or he [&?][c] Dyer might run up with the papers about Lamp Case—[8] will keep you posted where I will be & you can tell him— He can do what he wants in this manner cant he—

Have[d] Tomlinson look over the deed or search of the Florida property & see if its OK)[9]

TLS, NjWOE, DF (*TAED* D8503ZCA); carbon copy in Lbk. 20:442B (*TAED* LB020442B). Letterhead of Thomas A. Edison. [a]"New York," preprinted. [b]Handwritten. [c]Illegible. [d]Obscured overwritten text.

1. Insull probably meant William King, but the editors have not found the letter.

2. The telegram is Doc. 2842.

3. George Frederick Barker (1835–1910), professor of physics at the University of Pennsylvania, had a long (and generally friendly) association with Edison going back to 1874. He was partially responsible for renewing Edison's interest in electric lighting in 1878, though rivalries in its subsequent commercial development had caused some strains in their relationship. The editors have not found Barker's letter nor identified its contents. See Docs. 500 n. 8, 2022, and 2033.

4. The editors have not found the letter from Theron Solimon Eugene Dixon (d. 1891), a Chicago patent attorney. Dixon was of counsel in an 1886 appeal to the U.S. Supreme Court by the People's Telephone Co. and the Overland Telephone Co., defendants in major infringements suits brought by the Bell company. He had sent Edison unsolicited suggestions about electric lighting in 1880. "Obituary," *Chicago Daily Tribune,* 16 Sept. 1898, 5; *People's Telephone Co.* 1886, 235; Beauchamp 2010, 865–67; Dixon to TAE, 2 Feb. 1880, DF (*TAED* D8020ZBE).

5. Henry Herman Unz (1856–1905) was a native of Philadelphia and former manager for E. Remington & Sons in Chicago before transferring to the firm's headquarters in Ilion, N.Y. He was at work on a new typewriter and corresponded with Insull from New York and, in November, from Staten Island. By 1890, Unz was vice president and general manager of the National Typewriter Co. in Philadelphia and had several patents issued or pending. U.S. Dept. of State n.d., roll 377, Henry H. Unz passport issued 3 July 1891; *Philadelphia, Pennsylvania, Death Certificates Index, 1803–1915,* online database accessed through Ancestry.com, 13 Mar. 2012; Advertisement, *Chicago Daily Tribune,* 15 Nov. 1882, 5; Advertisement, ibid., 15 Dec. 1883, 5; Charles Clarke to Insull, 16 Sept. 1885; Unz to Insull, 22 Aug. and 7 Nov. 1885; all DF (*TAED* D8537F, D8537C, D8537H); Advertisement, *National Stenographer* 1 (Sept. 1890): 330; U.S. Pats. 409,340; 507,189; 613,178.

Unz sought to license Edison's U.S. Patent 224,665 on a method of cutting stencils, but he objected to Edison's suggestion in this document of a 20 percent royalty on sales. He proposed either a 10 percent or a flat seventy-five cent royalty on each device sold, with a promise to acknowledge the patent's validity (which he regarded as questionable). Edison accepted the flat payment offer and signed a three-year license agreement on 16 September. Unz apparently intended to form a company for the purpose but carried out little or no business before he and Edison annulled the contract in June 1887. Insull to TAE, 18 Aug. 1885, Lbk. 20:442F (*TAED* LB020442F); TAE agreement with Unz, 16 Sept. 1885, DF (*TAED* D8537G); see Doc. 3034.

6. Edison was in Chautauqua; see Doc. 2844 regarding his itinerary.

7. The city of Niagara Falls, N.Y., on the Niagara River at the famous falls, faced the smaller Ontario, Canada, town of the same name across the river just below the falls. It is also possible that Edison meant the town of Niagara, Ont. (also known as Niagara-on-the-Lake), fourteen or so miles downstream where the river empties into Lake Ontario. *WGD*, s.vv. "Niagara Falls," "Niagara-on-the-Lake."

8. Edison probably referred to either of two adjudication proceedings at the Patent Office concerning applications for incandescent lamps. One was the interference case *Weston v. Latimer v. Edison*, relating to the joining of the filament and lead-in wire, in which Edison, represented by John Tomlinson and Richard Dyer, had testified in June and again in mid-July. His latter appearance nearly concluded the evidence on his behalf, so it seems improbable that Dyer would have conferred with him about it now. Edison's testimony, 9 June and 13 July 1885, pp. 1–38, 73–81, *Weston v. Latimer v. Weston*, Lit. (*TAED* W100DFA001, W100DFA073).

The other, more likely, case involved Edison's application Case 187 for a carbon filament, filed in December 1879, that had been caught since 1880 in a tortuous Patent Office interference proceeding with a similar application by William Sawyer and Albon Man. Though Edison claimed in 1883 that his application no longer had any material value, he continued to contest it "on principle" until all avenues of appeal were exhausted (see Docs. 2508 n. 2 and 2555). On the basis of its ruling against Edison in the interference, the Patent Office rejected his application in February 1885 and, three months later, issued a patent in the name of Sawyer (now deceased) and Man (Patent Office to TAE, 19 Feb. 1885, Charles Kintner statement, 30 July 1885; *Edison Electric Light Co. v. U. S. Electric Lighting Co.*, defendant's depositions and exhibits [Vol. IV], pp. 2275, 2283; both Lit. [*TAED* QD012E2275, QD012E2283]; U.S. Pat. 317,676). In June, patent attorney Richard Dyer tried to revive Edison's application by submitting an amended version. He simultaneously pressed a lawsuit in equity, which prompted examiner Charles Kintner to refuse consideration of the revised application. An appeal of Kintner's decision was scheduled to be heard by the Commissioner of Patents on 31 July but Dyer requested a postponement because of his wife's illness, and the date was changed to 28 August. After numerous other delays, the commissioner ruled in February 1888 against Edison's request to have an interference declared with the Sawyer and Man patent. The editors have not determined the result of the lawsuit (Dyer to Patent Office, 19 June, 8 and 30 July 1885; Patent Office to TAE, 27 June, 14 and 30 July 1885, 27 Feb. 1888; Kintner statement, 30 July 1885; *Edison Electric Light Co. v. U. S. Electric Lighting Co.*, defendant's depositions and exhibits [Vol. IV], pp. 2276, 2281, 2286, 2279, 2282, 2287, 2315, 2283; all Lit. [*TAED* QD012E2276, QD012E2281, QD012E2286, QD012E2279, QD012E2282, QD012E2287, QD012E2315, QD012E2283]).

9. John Tomlinson declined to give an opinion based on the information he had, which included the deed and presumably the abstract of title, sent by the county clerk in July. In October, he instructed Insull not to complete the purchase until he could talk about it with Edison, who in the meantime had given the go-ahead to his agents. As a result

of Edison's delays, the seller asked to back out at the last minute but was denied; Edison did, however, forfeit his one hundred dollar deposit. The transaction was completed by 2 November, when Tomlinson sent the deed, signed over to Edison on 19 September, to be recorded by the county clerk. Insull to TAE, 18 Aug. and 20 Oct. 1885, Lbk. 20:442F, 21:117D (*TAED* LB020442F, LB021117D); Monroe County (Fla.) Court Clerk to TAE, 25 July 1885; Huelsenkamp & Cranford to TAE, 18 and 26 Sept. 1886; TAE to Insull, 23 Oct. 1885; Tomlinson to Monroe County (Fla.) Court Clerk, 2 Nov. 1885; all DF (*TAED* D8539D, D8539H, D8539I, D8539K, D8539L).

–2844–

To Samuel Insull

Alexandria Bay NY[1] Aug 20 1885[a]

Saml Insull

Express my spring over coat to maple wood Hotel[2] Boethh-lehem[b] white mountains N.H.[3]

Edison

L (telegram), NjWOE, DF (*TAED* D8503ZCH). Message form of Western Union Telegraph Co. [a]"1885" preprinted. [b]"eth" interlined above the "o."

1. After leaving Chautauqua on 18 August, Edison and his party stayed at the Thousand Island House in Alexandria Bay, a summer resort on the St. Lawrence River. The town was promoted both for its "romantic and highly picturesque" setting near the Thousand Islands section of the river (increasingly popular with American tourists) and for the excellent "boating and fishing" in the region, described as a "young Venice." TAE to Insull, 18 Aug. 1885, DF (*TAED* D8503ZCG); Insull to TAE, 21 Aug. 1885, Lbk. 20:442 (*TAED* LB020442); Pennsylvania Railroad 1881, [1]; *WGD*, s.v. "Alexandria Bay"; "The Thousand Islands," *NYT*, 22 July 1883, 10.

2. Insull replied the next day that he had done as requested (Insull to TAE, 21 Aug. 1885, Lbk. 20:444A [*TAED* LB020444A]). Edison and his party spent a night or two in Montreal at the Windsor Hotel before reaching the Maplewood Hotel, probably on 23 August. They stayed until the last day of the month, though Edison himself seems to have made a brief business trip to New York on the 24th ("From the Star Files Thirty Years Ago," *Montreal Star*, 23 Aug. 1915 [reprinted from 22 Aug. 1885], Scraps. [*TAEM* 287:180]; Voucher [Laboratory] no. 458 [1885] for Seavey; TAE to Insull, 18 Aug. 1885; Ezra Gilliland to Insull, 24 Aug. 1885; both DF [*TAED* D8503ZCH, D8503ZCJ]). While at the Maplewood, Edison famously proposed to Mina Miller in Morse code. As Mina later recalled in a newspaper interview printed under her byline: "One evening, after spending the day on top of Mount Washington, we were sitting around the hotel in the foothills. Mr. Edison wrote down for me the Morse code characters and by the next morning I had memorized them." A little while later Edison tapped out a "sacred" message said Mina, that was "one of the steps that led to our marriage."

Edison's daughter Marion later claimed to have witnessed Edison tapping the code in Mina's hand, after which Mina "blushed and nodded 'Yes.'" "Wizard of Menlo Park First A Family Man, Widow Writes," *Lacross (Wisc.) Tribune,* 6 Feb. 1947 (article likely based on Mina Miller interview by the Associated Press, 10 Jan. 1947, typescript in Edison Bio. Coll.); Oeser [1956], 10.

The Maplewood Hotel was part of a resort complex developed by the Boston merchant Isaac S. Cruft in the 1870s. Built in stages in an eclectic blend of Victorian architectural styles, by 1885 the four-story hotel could accommodate 500 guests. It still depended on gas lighting but offered several dining rooms, gaming rooms, postal and telegraph offices, and a large entertainment hall, among other amenities. One of its attractions was pure spring water drawn from the surrounding hills. The larger complex also included the Maplewood Cottage, which served as an annex to the hotel for 100–150 guests, and Maplewood Hall, which could accommodate 150 more (Doc. 2640 n. 1; Tolles 1998, 125–27; Holden 1883, 2:37–38; Sweetser 1882, 164). Cruft also built and owned both the Magnolia Hotel in Magnolia, Fla., where Edison and his wife had stayed in February 1884, and the San Marco Hotel in St. Augustine, Fla. (Doc. 2607 [headnote]; Braden 2002, 143–54).

3. Bethlehem, situated within sight of the Franconia and Presidential mountain ranges, began to be developed as a resort in the 1870s. Its year-round population of about 1,400 swelled to more than 4,000 in the summer months when tourists took up residence in more than 25 hotels and boardinghouses, of which the Maplewood was the largest. Tolles 1998, 124–25; Sweetser 1882, 164–65; Holden 1883, 37–38.

–2845–

*Samuel Insull to
William Pitt Edison*

[New York,] August 21st. 1885.

Mr dear Pitt:—[1]

T.A.E. has been away quite a great deal lately, but he has just returned your letter to me (which I had forwarded to him) with instructions to send you a check for $200.,[2] I shall be able to do this in a day or two when our finances are in a little better shape than they are now. Yours very truly,

Saml. Insull.

TLS (carbon copy), NjWOE, Lbk. 20:444C (*TAED* LB020444C).

1. Edison's brother, William Pitt Edison (1831–1891). *TAEB* 1, chap. 1 introduction, n. 4.

2. Pitt Edison had a long history of periodically seeking financial help for various business ventures. He did so again on 29 July, when he requested $230, apparently to meet a mortgage on either his home or his farm (see Docs. 2385 and 2503). The editors have not found Edison's reply to Insull. Pitt Edison to TAE, 29 July 1885, DF (*TAED* D8514J).

Ithaca, N.Y., September 3rd, *1885*[a]

My dear Mr. Edison:

The Trustees of Cornell University have asked me to come here and assist them in building up a great school of engineering, and promise to stand by me to the end. I am at work, therefore, fitting up lecture rooms, gathering collections in the museums, arranging workshops, and gradually getting ready for a good start. I think that, in a year or two, we shall have the best material facilities in the world for teaching the mechanic arts and the higher work of engineering.[1]

In looking about here, on my arrival, I noticed that all the dynamos were of other styles than yours, and that, so far as I have observed, there is not a single Edison machine or an illustration of any one of your inventions here. There may, however, be something in the Physical Department that I have not seen. Can you not see a way to let me have a 5-light[2] machine of your best pattern? and such other of your inventions as you may like to have represented in our museums? I am thinking of lighting up the shops with the incandescent lamp of high power, and that would be a good way of making the young men familiar with them. I can buy the machine from the company if you think it necessary; but would much rather have it as a contribution from you personally, if possible. The other companies seem to have monopolized this kind of contribution to our collections.

I hand you, herewith, a circular describing the course that I propose to attempt to teach.[3] I should very much like you to look it over, and give me any suggestions that may occur to you. You will observe that a number of gentlemen are promised to come here at their own convenience, in the course of the Winter, and give talks to the boys, at their weekly meetings for debate, or formal lectures, as they may find most convenient and agreeable. Could you not come in some day in the same way. I am sure that the young men would receive you very enthusiastically. You could make your talk what you might think on the whole best suited to bring you into good fellowship with your audience.

I would apologize for taking your time in this manner; but that I know your interest in all that is proposed to help the boys of the coming generations to get what you and I could not get, though we needed it so much, when we were youngsters.

I was at the meeting of the American Association, last week, and hoped that you might be there.[4] I hope that you will come here, by and by, and see what we are doing. The Sci. Am. will give some account with illustrations of our college that will interest you, in a few weeks.[5] Very truly yours

R. H. Thurston[6]

⟨Present the College through Thurston with a 25 Light machine 50 Lamps & sockets Res. box—cut outs[b]   say don't believe will have time to go up— E[7]⟩

TLS, NjWOE, DF (*TAED* D8503ZCK). Letterhead of Cornell University Schools of Mechanical Engineering, R. H. Thurston, Director. [a]"*Ithaca, N.Y.,*" and "*188*" preprinted. [b]"cut outs" interlined above.

1. Cornell University established the nation's second higher education program in electrical engineering in 1883. University president Andrew White made a forceful plea at that time for Edison to donate the equipment he had displayed at the 1881 Paris International Exposition. Edison was noncommittal, but in 1884 he referred a young advice-seeker to Cornell as a "good place [to] Learn" practical electrical engineering. Answering a similar inquiry in 1886, he recommended Cornell's Sibley College of Engineering as the best place for electrical training, placing it ahead of two peer programs and the alternative of gaining experience at a commercial workshop. Doc. 2434 n. 2; TAE marginalia on Watson Hurlburt to TAE, 21 Mar. 1884, DF (*TAED* D8403ZAQ); TAE to Reid Miller, 17 July 1886, Lbk. 22:245 (*TAED* LB022245); cf. Docs. 2427 n. 3 and 2643.

2. This phrase was probably a stenographic mis-transcription of "25-light"; see Edison's marginalia below.

3. Thurston likely referred to a pamphlet containing his nine-page printed description of the courses of study, requirements, and equipment of Sibley College. It did not mention training or instruments specific to electricity. An article in the trade press characterized the Cornell program as designed to give both practical and theoretical training, with particular attention to the use of instruments for measurement and testing. Sibley College pamphlet, 20 Aug. 1885, DF (*TAED* D8516A); "The Course in Electrical Engineering at the Cornell University," *Electrician* 2 (Nov. 1883): 336.

4. The American Association for the Advancement of Science held its annual meeting from 26 August to 1 September in Ann Arbor, Mich. Edison had attended its meetings in 1878 in St. Louis, where he was warmly received as a new member, and 1879 in Saratoga, N.Y. *AAAS Programme* 1885; see Docs. 1406 and 1793 n. 4.

5. *Scientific American* filled the cover of its 17 October issue with illustrations of the Sibley College facilities. The accompanying article showed the "Sibley College Dynamo and Electrical Room" used in both the mechanical engineering training and the "special course of 'electrical engineering.'" The unidentified dynamo "furnishing the electric lights for the grounds of the university is placed here, as will be the beautiful machinery lately presented Sibley College by Mr. Edison." According to the article, the entire room was soon to be refurbished and a new engine installed. "Sibley College, Cornell University," *Sci. Am.* 53 (17 Oct. 1885): 239, 247.

6. Robert Henry Thurston (1839–1903), a pioneering mechanical engineering educator, had recently been appointed both the professor of mechanical engineering and the founding director of the Sibley College of Engineering at Cornell. He previously was professor of mechanical engineering at the Stevens Institute of Technology from 1871. In 1881, while recovering from a breakdown, Thurston had offered his services as a consulting engineer to Edison, who was deeply involved in planning electrical central stations. *ANB*, s.v. "Thurston, Robert Henry"; Doc. 591 n. 2; Thurston to TAE, 14 Sept. 1881, DF (*TAED* D8104ZCZ).

7. Edison signed a brief reply (prepared by Samuel Insull) making this promise to Thurston. It said nothing about visiting Cornell (TAE to Thurston, 10 Sept. 1885, Lbk. 20:479 [*TAED* LB020479]). The Edison Co. for Isolated Lighting made an estimate to install a 25-light dynamo, associated equipment, and wiring at Sibley in October 1885. The machine was not shipped until the following May and the wiring not completed before October 1886; the editors have not determined who paid for the plant. In the course of related correspondence, Thurston again invited Edison to speak to students in late May 1886. Edison replied that he was too busy to leave his laboratory; in a later letter, he held out hope of making an informal visit during the winter (Edison Co. for Isolated Lighting estimate, 6 Oct. 1885; Thurston to TAE, 17 and 22 May 1886; [*TAED* D8522M, D8603ZBI, D8603ZBL]; TAE to Thurston, 21 and 26 May, 10 Aug. 1886, Lbk. 22:99, 22:111B, 22:307 [*TAED* LB022099, LB022111B, LB022307]). Edison had declined an 1883 invitation from physics professor Anthony White, who also taught in the new program, to visit Sibley College then. White also inquired at that time about purchasing two custom dynamos for experimental purposes (White to TAE, 10 May 1883 and 17 July 1884; Sherburne Eaton to Frank Hastings [with TAE marginalia], 22 July 1884; all DF [*TAED* D8303ZDD, D8420X1, D8420Z1]; TAE to White, 15 May 1883, Lbk. 16:328A [*TAED* LB016328A]).

–2847–

To Phillips Shaw[1]

[New York,] September 4th. [188]5

Dear Sir:—

You are hereby authorized to make settlements on my behalf with The Edison Electric Illuminating Co. of Shamokin; The Edison Electric Illuminating Co. of Sunbury; The Edison Electric Illuminating Co. of Mt. Carmel, and The Edison Electric Illuminating Co. of Bellefonte, Penn.[2] The settlements to be made in accordance with confidential memorandum of even date, which I enclose you herewith.[3] Yours truly

ENCLOSURE[a]

[New York, c. September 4, 1885]

Memorandum for Mr. P. B. Shaw with Relation to Mr. Thomas A. Edison's Accounts Against Various Local Companies in Pennsylvania.[b]

The Edison Electric Illuminating Co. of Bellefonte.

The amount due by this Company to Mr. Edison is $1256.42 as per statement enclosed.[4] The Bellefonte Co. do not dispute Mr. Edison's claim, having as recently as the 28th. of March 1885 made a cash payment on account.[5] The amount should be paid in full.[6c]

The Edison Electric Illuminating Co. of Sunbury.

The unsettled indebtedness of this Co. is $2416.68. Mr. Edison is prepared to take stock in the present Co. for this amount as agreed between himself and the Board of Directors nearly eighteen months ago.[7c]

The Edison Electric Illuminating Co. of Mt. Carmel.

The amount due by this Co. is $2813. Mr. Edison has made a number of allowances as claimed by Mr. Schwenk[8] on behalf of the Co., aggregating a total of $872.98, as will be seen by accompanying statement.[9] The contract between Mr. Edison and the Mt. Carmel Co. only called for a $8\frac{1}{2} \times 10$ engine. Mr. Edison installed without any extra expense to the Co. a $9\frac{1}{2} \times 12$ engine, thus donating the difference between the cost of the former and the latter to the Co. In view of the trouble that the Mt. Carmel Co. has had, and for the purpose of getting an immediate settlement Mr. Edison will accept the sum of $2500 in full settlement of the account.[10c]

The Edison Electric Illuminating Co. of Shamokin.

This Co. owes Mr. Edison the sum of $2238.45. On July 2nd. of this year the President of the Shamokin Co. wrote Mr. Edison claiming that Mr. Edison had not fulfilled his contract,[11] but Mr. Douty[12] seems to have overlooked the fact that on June 12th. 1884 this matter was settled at a meeting attended by Major Eaton, on behalf of the Edison Electric Light Co., Mr. Edison and Mr. Insull on behalf of the Construction Co., and by Mr. Douty and two other gentlemen on behalf of the Shamokin Co.[13] In consideration of the trouble the Shamokin Co. have been put to, and the possible breach of contract by Mr. Edison, the Edison Electric Light Co. returned to the Shamokin Co. their share of the bonds of the latter Co. This settlement was made on a memorandum of agreement signed by the Shamokin Co., The Light Co. and by Mr. Edison, and immediately it was concluded the Shamokin Co. handed Mr. Edison a check for $805.48, on account, and agreed to pay the balance due him immediately after certain alterations had been completed. These alterations involved the return to Mr. Edison of two engines and one dynamo, which he has credited to the Shamokin Co. at the price at which they were originally billed, as will be seen from accompanying accounts.[14] By the above referred to memorandum Mr. Edison undertook to supply certain other material including a $14\frac{1}{2} \times 13$ engine and 2 "S" dynamos at cost; this he did. Inasmuch as a basis of settlement has already been arrived at, and remembering the heavy sacrifice Mr. Edison made in taking back the dynamo and two engines which had been used for upwards of a year, Mr. Edison does not feel that he ought to make any further concessions on this account. If however it is possible to get a settlement immediately, he will accept a $2000 check in full payment of all his claims.[c]

General.[15d]

With relation to the method of settlement, of course it is desirable that cash should be obtained in each case if possible. In the case of the Bellefonte and Mt. Carmel Companies Mr. Edison will accept notes, running for as short a period as possible, but in any event not longer than six months.

In the case of the Shamokin Co. if they are not prepared to pay cash down, Mr. Edison will accept a note providing it is for the full amount of the account enclosed.[c]

TL (carbon copy), NjWOE, Lbk. 20:467C, 467D (*TAED* LB020467C, LB020467D). [a]Enclosure is a TD (carbon copy). [b]Heading typed in upper case; followed by dividing mark. [c]Followed by dividing mark. [d]Heading typed in upper case.

1. Phillips B. Shaw (c. 1848–1937), the Edison Electric Light Co.'s agent for Pennsylvania, had been instrumental in organizing local illuminating companies to purchase and operate Edison village plant central stations. Docs. 2350 n. 1 and 2424 n. 1 (and headnote).

2. The obligations of the four companies named above stemmed from their contracts with Edison to build central stations in 1883 and 1884, and they had endured Edison's efforts to collect them in cash or stock (see, e.g., Docs. 2705, 2708, 2737; *TAEB* 7 App. 2.A). Shaw's role in settling these accounts may have been discussed at a special directors' meeting of the Edison Electric Light Co. on this date, called in part to consider an unspecified "proposition from P. B. Shaw." When he acknowledged this letter on 7 September, Shaw promised to report to Edison after meeting with the board of the Sunbury company the next day (Frank Hastings to TAE, 2 Sept. 1885; Shaw to TAE, 7 Sept. 1885; both DF [*TAED* D8526ZAB, D8523ZBQ]).

In a separate letter on this date, Edison directed Shaw to accept stock in the Sunbury company, with the understanding that in a subsequent reorganization he would receive one share in a new company for every two shares held in the old one, provided both that other stockholders accepted the same terms and that the Edison Electric Light Co. surrendered its entire interest. He also promised Shaw a five percent commission on funds collected in cash or notes from the Shamokin, Mount Carmel, and Bellefonte companies. Edison specifically excluded the small debt of the Hazleton, Pa., company, which he planned to collect himself (TAE to Shaw, 4 Sept. 1885, Lbk. 20:467A [*TAED* LB020467A]).

In a June 1885 accounting of debts owed to Edison on behalf of the Edison Construction Dept., Samuel Insull listed nearly $13,000 due from nine illuminating companies. Insull identified Shamokin and Bellefonte as the "most aggravated cases" and suggested that Edison should engage a lawyer unless settlements were made quickly. Edison was also involved at this time in negotiations between the Sunbury firm and the Ansonia Brass & Copper Co. (one of his largest suppliers, to whom he owed around $9,200), to settle its account of about $260. Insull to Edward Johnson, 3 June 1885; Lbk. 20:315A (*TAED* LB020315A); Ansonia Brass & Copper to TAE, 16 Apr., 24 June, 31 Aug., and 14 Nov. 1885; all DF (*TAED* D8518H1, D8518K1, D8519A, D8518O1).

3. The enclosure follows. Complaints from local illuminating companies about the quality and performance of central stations had dogged the Edison Construction Dept. The unresolved disputes had complicated the transfer of the village plant business to the Edison Electric Light Co. and the Edison Co. for Isolated Lighting in 1884 and now hindered Edison's efforts to collect money due him (see Docs. 2424 [headnote], 2496 nn. 2–3, 2677, and 2709).

4. Not found.

5. James Harris to TAE, 28 Mar. 1885, DF (*TAED* D8523ZAC).

6. Edison apparently was willing to accept only $750, payable in notes of two, four, and six months. TAE to Shaw, 19 Sept. 1885, Lbk. 20:498A (*TAED* LB020498A); Shaw to TAE, 21 Sept. 1885, DF (*TAED* D8523ZBV).

7. Edison assented in 1884 to a deduction of $750 from the full amount due him, as an allowance toward construction of a new station (see Doc. 2737). In September 1885, after Shaw provided proof that the company's entire assets totaled only $1,745.78, Edison agreed to accept stock in settlement of his claim. Even so, the company borrowed $400 from Shaw in order to provide the shares. Shaw to TAE, 9 and 25 Sept. 1885; all DF (*TAED* D8523ZBR, D8523ZBS, D8523ZBW); Vouchers (Laboratory) no. 476 (1885) for Sunbury.

8. William Schwenk, a prominent coal mine operator, was the founding president (in November 1883) of the Edison Electric Illuminating Co. of Mt. Carmel. At the same time he became involved with that enterprise, Schwenk also helped to organize a water company in Mt. Carmel. Bell 1891, 665; Letterhead of Schwenk to TAE, 26 June and 10 Dec. 1884; both DF (*TAED* D8456ZBI, D8456ZBL).

9. Not found.

10. Edison accepted 48 stock shares, subsequently valued at $2,400. Shaw to TAE, 23 Oct. 1885, DF (*TAED* D8523ZBY); Vouchers (Laboratory) no. 101 (1886) for Mt. Carmel.

11. The editors have not found such a letter to Edison, but company president William Douty did write defiantly to Edward Johnson on that date that he was not "disposed in any way to pay Mr Edison one penny more than we have paid him." Douty to Johnson, 2 July 1885, DF (*TAED* D8523ZBG).

Among Edison's relations with the local illuminating companies, perhaps none was more tangled than that with the Shamokin firm, whose protestations were similar to those made from Brockton, Mass. (summarized in Doc. 2734 n. 2). The Shamokin dispute dated to December 1883 and initially concerned construction defects but later expanded to include the performance of engines and dynamos and a high rate of lamp failures. After Edison promised to replace some of the machinery, the parties agreed on settlement terms in June 1884. Problems with lamp breakage and the remaining original engine persisted (TAE to William Douty, 27 Dec. 1883; Douty to TAE, 29 Dec. 1883; Samuel Insull to Douty, 13 Feb. 1884; correspondence from Douty in Electric Light—TAE Construction Dept.—Stations—Pennsylvania—Shamokin; all DF [*TAED* D8316BWA, D8360ZDC, D8416ANA, D8457]; Memorandum of agreement, 11 June 1884, Miller [*TAED* HM840222]; William Brock to Douty, 20 June 1885, DF [*TAED* D8523ZBE]).

12. William H. Douty (b. 1837), owner of a Shamokin dry goods store and a local coal operator, was president of the Edison Electric Illuminating Co. of Shamokin. Doc. 2438 n. 4.

13. A memorandum of this meeting gives its date as 11 June, while subsequent correspondence puts it a day later. Accompanying Douty were his associate Andrew Robertson and company treasurer John Mullen. The settlement terms recorded in the memorandum are in essence those summarized below. Memorandum of agreement, 11 June 1884, Miller (*TAED* HM840222); cf. Insull to Douty, 13 June 1884, LM 19:427 (*TAED* LBCD6427); Doc. 2438 nn. 5–6.

14. Not found.

15. Shaw billed Edison $135.25 as his commission on the Shamokin and Bellefonte accounts. Edison promptly sent him signed releases for the companies in Sunbury, Bellefonte, and Mt. Carmel (pending fulfillment of specified terms). Shaw to TAE, 18 Sept. 1885, DF (*TAED* D8523ZBT); TAE to Shaw, 19 Sept. 1885, Lbk. 20:498A (*TAED* LB020498A); TAE agreement with Edison Electric Illuminating Co. of Sunbury, 22 Sept. 1885; TAE agreement with Edison Electric Illuminating Co. of Bellefonte, undated Sept. 1885; TAE agreement with Edison Electric Illuminating Co. of Mt. Carmel, 23 Oct. 1885; all Miller (*TAED* HM850268, HM850269, HM850270).

–2848–

From Ezra Gilliland

Boston. 1 pm   Sept 17 ⟨—85⟩[a]

My Dear Edison

Yours recvd, also the borrowed letter[1]   I cannot of course leave here much before the first of the month it was my intention to come on about 2 days before Mina is due here and then you can come back with me   I understand she is to leave there[2] next Tuesday that would probably bring her[b] here Wednesday 3 pm. Haskins[3] is in nyrk stopping at the Normandie[4]   he is going to be here Sunday or I would come over Saturday Evening probably Monday will do as well   I am having the New York wire that runs into the flat put in order and one of the new transmitters connected in and have written to Bergmann to put one at your End so that we talk over the wire during the week we will be at work in New York nights before I close up the house here[5]

Lockwood[6] has been after me to take out Foreign patents on new transmitter   I fear some of the other [Concerns?][c] will do so if we dont   Cant Dyer go ahead and do it   I am sure it will pay—[7]

Why dont Jim[8] answer my letter or telegrams— I enclose herewith a clipping out of a Chicago paper on the subject of Florida you will observe the writer goes for Meyers pretty rough—[9] It was my intention to put the drawings in the hands of the lumber Co[10] then[b] ask them to make a bid at once— Yours very truly

E. T. Gilliland

Saturday mrng[11]   It might be well to send these newspaper clippings[12] to Hulsenkamp and see what he has got to say. Yours very truly   E. T. Gilliland

ALS, NjWOE, DF (*TAED* D8503ZCQ). Letterhead of E. T. Gilliland. [a]"Boston." preprinted. [b]Obscured overwritten text. [c]Illegible.

1. The editors have not found Edison's letter nor identified the "borrowed" one.

2. It is not clear if Mina stayed with her family in Akron or with her sister Jane in New York before returning to Miss Johnson's school in Boston. For his part, Edison left New York for Boston on Monday, 21 September, and evidently did not come back until the last day of the month. Samuel Insull to John Weir, 21 Sept. 1885, Lbk. 21:003 (*TAED* LB021003); TAE to Insull, 29 Sept. 1885, DF (*TAED* D8503ZCY).

3. Telephone inventor and executive Charles H. Haskins (1830–1910) became the local representative of the National Bell Telephone Co. in Milwaukee in 1879, the same year he organized the Milwaukee Telephone Exchange. At this time, Haskins was president of that firm's successor, the Wisconsin Telephone Co. He corresponded occasionally with Edison. *DWB*, s.v. "Haskins, Charles H."; Barsantee 1926, 155.

4. New York hotelier Ferdinand Earle opened the Hotel Normandie, on the corner of Broadway and Thirty-eighth St., in October 1884. Promoted as fireproof and thoroughly modern throughout, the Normandie became one of the city's most fashionable hotels. "City and Suburban News," *NYT*, 5 Oct. 1884, 7; "Hotel Normandie Sold To Builder," ibid., 25 July 1925, 18.

5. Probably the summer home at Winthrop, outside Boston.

6. Thomas Dixon Lockwood (1848–1927) headed the Patent and Technical Information Bureau of the American Bell Telephone Co. in Boston from 1881 until 1911. Lockwood had emigrated from England with his family in 1865 and engaged in a variety of trades in Canada and the United States. He worked for the Automatic Signal Telegraph Co. and the Gold and Stock Telegraph Co. in New York before join-

ing the nascent National Bell (later American Bell) Telephone Co. in 1879. Lockwood published at least three books on practical electricity before 1885. "Electrical World Portraits—VIII," *Electrical World* 14 (6 July, 1889): 178; *NCAB* 22:439; "Obituary," *Boston Daily Globe*, 6 Apr. 1927, 1.

7. Gilliland likely referred to the carbon transmitter for which he executed a U.S. patent application on this day. The corresponding British preliminary specification was filed on 29 January 1886. The invention was an adjustable transmitter that kept the diaphragm in a horizontal position, which would help prevent the carbon granules from clumping or packing. The U.S. patent was assigned to the American Bell Telephone Co. and presumably was independent of Gilliland's ongoing collaboration with Edison to find ways around American Bell's fundamental patents on carbon transmitters. U.S. Pat. 384,201; Brit. Pat. 1,310 (1886).

8. James F. Gilliland (b. 1852?), Ezra's brother, was a machinist by trade and, until this month, managing director of the Gilliland Manufacturing Co. of Indianapolis. He planned to arrive in New York on 21 September and intended to meet with Edison. He seems to have spent the winter there and in March was arranging with Bergmann & Co. for the manufacture of telephone transmitters. U.S. Census Bureau 1970 (1880) roll T9_296, p. 639D, image 0194 (Indianapolis, Marion, Ind.); "Miscellaneous Notes," *Electrical World* 3 (16 Feb. 1884): 55; James Howe to TAE, 1 Sept. 1885; James Gilliland to TAE, 5, 10, and 13 Sept. 1885; all DF (*TAED* D8547ZAO, D8503ZCL, D8503ZCM, D8503ZCN).

9. The *Chicago Daily Tribune* recently reprinted a New York physician's lengthy appraisal of the health benefits and hazards of Florida's climate and geography. Fort Myers, the writer argued, deserved its beneficent reputation only in the winter and spring. The prevalence of "rotten limestone" universally contaminated wells and forced residents to rely on inadequate cistern supplies in the dry summer months. "Going To Florida," *Chicago Daily Tribune*, 14 Sept. 1885, 3.

10. Gilliland referred to the Kennebec Framing Co. of Fairfield, Maine, which he and Edison subsequently hired. The company specialized in producing building materials that were pre-cut and numbered to allow for rapid assembly at building sites. It did business in Boston at 7 Exchange Pl. and 172 Washington St., both about midway between Gilliland's office and that of architect Alden Frink. Kennebec billed $5,820.53 for the structures, a total that rose to $6,714.85 with the addition of freight charges. "Kennebec Framing Co.," National Publishing Co. 1883, 189; Albion 2008, 14–15; Roberts 1901, 4:115; *Boston Directory* 1885, 427; Mercantile Publishing 1889, 231–32; Rosenblum 2000, 1:4–5; Kennebec Framing Co. invoices, 25 and 27 Nov., and 3 Dec. 1885, Cat. 1165 (Fort Myers accounts): 5–6, Accts., NjWOE.

11. The nineteenth of September.

12. Not found but see note 9.

*Memorandum to
Richard Dyer:
Phonoplex*

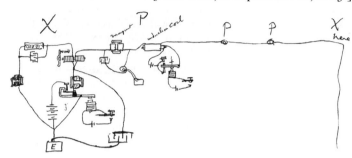

condenser[2] around key to prevent duplex
from bothering phonoplex

Dyer—

This is to take a Duplex wire and at intermediate & terminals add several phonoplex apparatus so we can make a way
wire out of a Duplex[3]   patent this in US & England[4]

E[dison]

⟨Germany France—⟩[5]

ADS, NjWOE, Lab., Cat. 1151 (*TAED* NM020AAJ). Letterhead of
Edison Central Station Construction Dept.

1. Draftsman Edward Rowland, who prepared the patent drawings,
initialed and dated the document on this day.

2. Figure labels above are "X," "magnet," "P," "induction coil," "P,"
"X," and "here."

3. Edison apparently had this general idea by 25 August, when he
sketched one such "Way Duplex" (Unbound Notes and Drawings
[1885], Lab. [*TAED* NS85ADC05]). Edison executed two patent applications on 7 October for using the phonoplex multiple telegraph system
on existing duplex or quadruplex lines in terminal and way stations.
The more general specification (U.S. Pat. 422,072) included a drawing based on the sketch in this document. In the text, Edison explained
that the system would transmit duplex or quadruplex signals "between
terminal stations at the same time that the line may be operated as a
way-line, the two sets of signals being transmitted and received simultaneously and without interference." He elaborated this objective in the
narrower specification (U.S. Pat. 437,422), stating that "duplex induction transmission can be carried on over the line alone simultaneously
with the ordinary duplex or quadruplex transmission, producing a new
system for duplex, quadruplex, or sextuplex transmissions." Together,
the applications covered the modifications to duplex or quadruplex apparatus necessary to prevent their reversing currents from affecting the
phonoplex receiver and also to preserve their delicate electrical balance despite the addition of phonoplex instruments to the circuit. The
Electrical World illustrated and described "The Edison 'Phonoplex' or
'Way-Duplex'" in its 17 April 1886 issue (7:177), and this article appeared under the same title in Great Britain's *Electrician* (16:516–17)
and *Telegraphic Journal and Electrical Review* (18:413–15) in their is-

sues of 7 May 1886; see also "The Edison Phonoplex," *Engineering* 42 (22 Oct. 1886): 411–13.

The new hybrid system would work because "The relays of the duplex or quadruplex apparatus prove too sluggish in their action to respond to the rapid vibrations produced by the [phonoplex] induction-transmitters, and hence the duplex or quadruplex transmission will not be interfered with." He found it necessary to "to prevent false sounds in the [phonoplex] diaphragm receivers, due to the responding of such receivers to the extra vibrations produced by the duplex or quadruplex transmitting instruments," particularly "the extra vibrations accompanying the reversals of current" by quaduplex transmitters. Accordingly, he decided to place an electromagnet, with a condenser shunted around it, in the transmitter circuit to damp out "the extra vibrations caused by the rebound of the contacts," as suggested in Doc. 2801. (He soon considered adapting this arrangement to telephone call bells; see Doc. 2856.) He also stipulated that each artificial line in the existing multiplex sets be equipped with an electromagnet of the same inductance as the "induction apparatus" in order "to preserve the balance necessary for the proper operation of the duplex or quadruplex apparatus" (U.S. Pat. 422,072). The sketch in this document (and one of the patents) represents both features. The second drawing in Edison's U.S. Patent 422,072 shows this arrangement on a quadruplex circuit.

Edison continued to play with these ideas through late October. He made an unclear sketch on 19 September, a pair of more complete drawings two days later showing arrangements for both the quadruplex (marked "England Germany France") and sextuplex (the latter incorporated into U.S. Patent 437,422), and related drawings dated 21 October (sextuplex), 22–23, and 27–29 October. Unbound Notes and Drawings (1885), Lab. (*TAED* NS85ADC06, NS85ADC07, NS85ADC08, NS85ADC09, NS85ADC10); Cat. 1151, Lab. (*TAED* NM020ABD, NM020AAM); N-85-10-03:47, Lab. (*TAED* N311047).

4. The editors have found no evidence that Edison (or his agent) filed a British patent specification for this arrangement. He had, however, submitted British provisional specifications for the basic phonoplex (7,583) and a way-wire quadruplex (7,584, which corresponded to U.S. Pat. 333,291), both on 22 June 1885.

5. The editors have not found evidence of specifications filed in these countries.

–2850–

To Erastus Wiman

[New York,] Sept. [----].ª 19thᵇ [188]5

My Dear Mr. Wiman:—

Yours of the 17th. came only to hand.[1]

You will have heard from Mr. Tate[2] what we are proposing to do with relation to the Way Duplex—the new name is the "Phonoplex," by which name I desire the invention to be known from my old duplex system.[3] I do not think that it will be possible for me to come and lunch with you on Monday, as I am going to Boston some time on that day and shall be there the rest of the week.[4] By the time I return, however, I hope to [~~report?~~][a] be able to report progress to you with relation to Tate's work in Canada. He will go right to Mr. Dwight and arrange for the placing of the apparatus immediately.[5] Yours very truly.

T.AE.   I[nsull]

TL (carbon copy), NjWOE, Lbk. 20:499A (*TAED* LB020499A). [a]Canceled. [b]"19th" interlined above by Samuel Insull.

1. After "several conferences with parties in the Western Union regarding your new invention, the way wire," Wiman expected "no difficulty in coming to some arrangement, provided a satisfactory illustration of the device can be afforded and a thorough test." He asked Edison to give him specific instructions soon for making such a demonstration. At the end of October, Wiman reiterated his belief that any agreement with Western Union would depend on a successful test in Canada, which Alfred Tate had not yet been able to provide. As of 1893, Western Union had evidently taken no steps to adopt or test the phonoplex, perhaps a result of the company, which enjoyed a strong monopoly, having embraced a deeply conservative policy towards technological innovations. By 1898, however, it was starting to use it on its lines. Wiman to TAE, 17 Sept. and 30 Oct. 1885; William Logue to TAE, 25 Sept. 1893; all DF (*TAED* D8546ZAL, D8546ZBR, D9347AAC); "Phonoplex Telegraph," *Electrical World* 31 (15 Jan. 1898): 103; Hochfelder 2012, 42.

2. Alfred Ord Tate (1863–1945), a Canadian-born former telegrapher and railroader, entered Edison's employ as an assistant to Samuel Insull in 1883 through the intercession of Charles G. Y. King, an engineering assistant to Edison. In addition to helping with correspondence and accounting, Tate had since become considerably involved in Edison's business affairs as an agent for the Edison Machine Works and a promoter of electric lighting, especially in Canada. Docs. 2456 n. 17, 2753 n. 1; Tate 1938, 43.

3. That is, the way wire duplex (see Doc. 1415).

4. Edison left for Boston on Monday, 21 September. He was still there eight days later but planned to return to New York on Wednesday, 30 September. Insull to John Weir, 21 Sept. 1885, Lbk. 21:3 (*TAED* LB021003); TAE to Samuel Insull, 29 Sept. 1885, DF (*TAED* D8503ZCY).

5. Tate had been in Canada since 1884 on behalf of the Edison Electric Light Co. but wrapped up that business in early September. He was clearly engaged on telegraph matters by 22 September and spent the rest of the month importing instruments and supplies. Doc. 2753 n. 1; Insull, 8, 22, and 29 Sept. 1885; Tate to TAE, 3 Oct. 1885; all DF (*TAED* D8546ZAK, D8546ZAN, D8546ZAT, D8546ZAV).

[New York,] Sept 21st [188]5

Miss Marion Edison

You must be at house by nine oclock Tuesday as men will be there to pack[1]

Edison

L (telegram, letterpress copy), NjWOE, Lbk. 21:001 (*TAED* LB021001). Written by John Randolph.

1. Edison addressed this telegram to his daughter at Menlo Park. On Tuesday, 22 September, workers from the Lincoln Safe Deposit Co. arrived to pack up the New York apartment at 39 East Eighteenth St. where the Edison family had lived for thirteen months. (Edison considered the flat "convenient" but, as he explained to a prospective new tenant, he wished to make "alterations in my domestic establishment"). The company was to place the furniture in storage while the Edisons occupied rooms at the Normandie Hotel at Broadway and Thirty-eighth St., across from the Metropolitan Opera House. Because the packing was incomplete at the end of the week, the move was postponed until the following Monday, 28 September. Samuel Insull, who arranged the details for Edison, was evidently on warmly familiar terms with John Van Wormer, the Lincoln company's secretary and general manager (Doc. 2721 n. 2; Van Wormer to Insull, 25 and 26 Sept. 1885, both DF [*TAED* D8503ZCU, D8503ZCV]; TAE to Silas Burt, 30 Sept. 1885, Lbk. 21:35C [*TAED* LB021035C]; Oeser [1956], 10). The fashionable Normandie operated on the so-called "European plan," in which guests paid for rooms only and were free to take their meals elsewhere. This was unlike the family's earlier residence at the Clarendon Hotel, which charged for both rooms and meals on the "American plan" (Kobbé 1891, 39–41).

CHART OF PHYSIOGNOMIC TRAITS   Doc. 2852

Doc. 2852 is one of Edison's more enigmatic documents. He seems to have created this chart over an indeterminate period, likely in several locations, using three unnumbered looseleaf pages that the editors have arranged based on their appearance and the internal coherence of groups of entries. Edison gave no direct indication of why he would characterize particular personal and physical traits of more than fifty friends, associates, and acquaintances. There are, however, clues that connect the chart with his activities in the summer of 1885. It is likely that Edison wished to test the claims of the eighteenth-century Swiss physiognomist Johann Kaspar Lavater, whose book he was reading during his vacation that summer (see Docs. 2827 and 2829). Lavater held that a scientific examination of facial features could reveal a person's character; as if to test that theory, Edison listed facial characteristics alongside personality traits as a basis for comparison.[1]

Edison likely started the chart sometime in July 1885, when he was reading Lavater's book.[2] During the summer, he came into contact with most of the people listed here, some for the first time. He clearly did not make the entire chart at a single sitting and may have added to it over a period of months. Differences in the writing implements he used, in the handwriting, and in the groupings of names indicate that Edison expanded the document as he encountered or thought about various people. For example, the upper third of the presumptive first page, extending from Charles Batchelor to Josephine Reimer, includes business associates and friends from New York and New Jersey. All of those names (and only those) are written in black ink. The next set—from Lillian Gilliland through Mina Miller—includes people with whom Edison vacationed in Winthrop, Massachusetts, during late June and early July 1885. All of these names are written in pencil with a dull point.

Members of other groupings are identified less certainly. The next three names—Lilly Fox, Nettie Johnson, and Miss Elder—are those of close associates of Lillian Gilliland. They are written with a sharply pointed pencil, which differentiates them from the others. The editors have not identified Miss Hughes or the Hamiltons, who immediately follow, but because they appear between Miss Elder and the Holsteins, known to be Indianapolis acquaintances of Louise Igoe, it may be that they are also people Edison came to know from that city. The grouping that extends from the Roberts through George Vincent may have been people Edison encountered in August at Chautauqua, where he definitely met Alice Miller and George Vincent. How he came to know the Roberts and Mrs. Comstock is not known, and the editors have not positively identified them. It could be that they were Indianapolis friends of Louise Igoe or Lillian Gilliland, rather than friends of the Millers.

Seven names are missing from the top of page two as a result of the corner of the paper having been torn off. This act seems to have been deliberate; the rectangle neatly torn from the chart excised only the names while leaving the characteristics intact. The eleven people listed immediately afterwards (beginning with Phillip Igoe and ending with Mollie Landers) have, with the exception of Jessie Condon, been identified as family and friends of Louise Igoe from Indianapolis, though it is not at all clear how or when Edison would have met them. The next five names after Landers are members of Mina Miller's family, whom Edison likely met during his August trip to Chautauqua. Edison listed his business associates Samuel Insull and Frank Toppan at the top of the presumptive third page, perhaps to correct an inadvertent omission from the group of colleagues at the beginning. The Winings and the Wellses of Indianapolis, who lived near each other on New Jersey Avenue a short distance from the Igoe home, might similarly have been added later.

Edison seems to have used the last five entries to make composite studies of himself and Louise Igoe and of himself and Daisy Gaston. Both women had been introduced to him by Lillian Gilliland as prospective brides, which might suggest that he was considering them as potential wives. Oddly, however, he also made composite studies of Mina Miller and Jim Gilliland, of Daisy Gaston and Jim Gilliland, and finally of Jim Gilliland and Lillian Gilliland. While it is perhaps tempting to see the composite studies as connected to his desire in 1885 to remarry, in all likelihood he made these notes after he had already decided to pursue Mina Miller. A more likely conjecture is that Edison made these studies in response to reading the work of Francis Galton in the summer of 1885 (see Doc. 2824). Galton, a eugenicist, had made composite photographic portraits with the aim of revealing common facial characteristics that he believed were indicators of psychological traits.

1. Edison had a long-standing interest in phrenology and sometimes evaluated prospective employees on that basis. See Doc. 2827 n. 2.

2. Edison was reading Lavater's book by 17 July, according to his diary entry that day (Doc. 2829). However, he seems to have had some familiarity with Lavater's ideas, which had recently been published in an English-language pamphlet (see Doc. 2827 n. 2), so it is possible that he began the chart sometime before that.

Chart of Physiognomic Traits[1]

	Light Eyes	Dark Eyes	Eyes Beautiful or Otherwise	Blonde	Brunette	Temper	Happily or Unhappy Maried	Nose
Batchelor		1	ord		1	mild	Un	ord
Mrs Batchelor[2]	1		ord	1		Bad	Un	Pug—
Tomlinson	1		ord		1	Mild	Un	good
Mrs Tomlinson[3]	1		Beau		1	mild	Un	fair
Chas F Clarke[4]	1		ord	1		Mild	Un	Large
Mrs C F Clarke[5]	1		ord		1	Bad	un	Pug
Mr E H Johnson	1		ord		1	Quick	Fair	ord
Mrs E H Johnson[6]			ord		1	Quick	"	Pug
Mr Reimer[7]		1	Bad		1	Bad	un	big
Mrs Reimer[8]	1		ord	1		mild	Un	good
Mrs Gilliland		1	Fair		1	Quick	Fair	ord
Mr Gilliland	1		Fair	1		Quick	Fair	good
Edison	1		ord		1	[Stubborn?][b]		Large
Daisy	1		ord	1		good		ord
Igoe	1		ord	1		Quick		ord
Mina		1	Beau		1	Mild		ord
Lilly Fox[9]		1	"		1	Mild		homly
Nettie Johnson[10]	1		ord	1		Quick		homly
Miss Elder[11]	1		ord		1	Quick		Large
Miss Hughes[12]		1	Beau		1	~~Mild~~Quick		fair
Mrs Hamilton[13]	1		ord	1		Mild	Fair	Med
Mr "[14]		1	good		1	Quick	"	Large
Mrs Holstein[15]		1	Beau		1	Mild	un	med
Mr Holstein[16]		1	good		1	Quick	un	Med[a]
Mr Roberts[17]		1	med		1	stuborn	fairly	Large
Mr Roberts[18]	1		ord	1		Quick	fairly	med
Mrs Comstock[19]	1		ord	1		Quick	[--][c]	Large
Mrs Jacbo Miller[20]	1		ord	1		Quick	Hap.	med
Geo Vincent[21]	1		ord	1		Quick		Large

Effectionate or not	Tall or Short	Edu-cated or Not	Mouth	Handsome or Homely	Lips Thin or thick	Ambitious		Remarks
yes	ord	ord[a]	good	Med	ord	yes		
no	ord	no	Not good	Homly	thin	no		
yes	ord	High	good	Med	thick	very		
no	ord	ord	fair	Hand	Med	none		unreasonable—
No	short	High	Bad[a]	Homly	thin	yes		
No	ord	Fair	Bad	Homly	thin	No		
Fairly	short	no	Bad	Homly	ord	Very		
no	short	ord	Bad	Pretty	thin	no		ambitious socially—
no	short	ord	Bad	Homly	thin	no		
yes	Tall	Fairly	good	Med	ord	yes		
yes	short	"	Fair	Med	thick	very		Unreasonable
not very	short	no	good	Hand	thick	very		
Effectionate	tall	no	good	Hand	Med	yes		
Effectionate	Tall	ord	good	Hand	Med	yes		somewhat unreasonable
Effectionate	tall	ord	good	ord	thick	Very		
Effectionate	ord	ord	ord	Hand	thick	yes[a]		
Effectionate	short	ord	Homly	ord	thick	Very		
Effectionate	Tall	ord	good	Hand	med	yes		
no	Tall	ord	good	med	med	Very[a]		
Effectionate	Tall	ord	good	Hand	med[a]	yes		
Effectionate	Med	ordFair	good	Hand	thick	ord		
Effectionate	Tall	ord	Large	med	med	no		
Effectionate	Tall	ord	Homly	Hand	thick	Very		
Med	Tall	High	good	med	thick	no		
med	Tall	ord	fair	Hand	thick	no		
Very Effect	Tall	ord	good	Hand	Med	Fairly		
no	Tall	ord	Fair	ord	thin	ord		
Affectionate[a]	Tall	ord	Fair	ord	thick	Very		
doubtful	Tall	High	Large Homly	Homly	thick	very		

	Light Eyes	Dark Eyes	Eyes Beau or otherwise	Blonde	Brunette	Temper	Hap or Unhap Mar	Nose
		1	Beau		1	Fearful		ord
	1		ord	1		Quick		ord
		1	good		1	QK		Large
		1	~~good~~Beau[d]		1	mild		Large
		1	Fair		1	mild		Small
	1		ord		1	Fearful		Large
[N--][e]	1		ord		1	Fearful		Large
Phillip Igoe[22]	1		ord	1		QK		Large
Ted[a] Elder[a23]	1		Beau		1	Mild		ord
Mr Geiger[24]	1		ord		1	QK	fair[a]	ord
Mrs Geiger[25]	1		Pretty	1		QK	fair.[a]	ord
Mr Ohr[26]	1		ord		1	QK	fairly	ord
Mrs Ohr[27]	1		ord	1		QK	Fairly.	ord
Miss Jessie[28] [Condon?][b]	1		ord		1	QK		ord
Annie Baggs[29]	1		ord		1	QK		ord
Mrs Beck[30]	1		Pretty	1		QK	Fairly	ord
Mr Beck[31]	1		ord	1		Mild	"	ord
Mollie Landers[32]		1	Pretty		1	QK		ord
Minas Father		1	Hand		1	Mild	Hap	Large
" Mother	1		ord		1	Lovly	"	Small
Minas Uncle		1	Hand		1	Mild	Hap	Large
Edward Miller		1	Hand		1	mild		ord
Lewis Miller	1		nice	1	1	QK		Large
	Light Eyes	Dark	Eyes Beau or Otherwise	Blonde	Brunette	Temper	Hap or Unhap Married	nose
Sammy		1	no		1	Quick		ord
Toppy	1		no		1	mild		ord
Mrs Winnings[33]	1		ord	1		Mild	Fairly	ord
Mr "[34]	1		Nice		1	Mild	"	ord
Mrs Wells[35]	1		ord	1		QK	Un	ord
Mr Wells[36]		1	Nice		1	QK	Un	Small
Edison Igo	1		Homly		1	Fearful		Big
Edison D[aisy?][b]	1		B[--][b]		1	Mild		Big
	1		ord		1	Mild		Big
Mina—Jim G		1	fair		1	None		Bad
Daisy "	1		Beau	1		Splendid		Lovely
Jim Gilliland Mrs ETG[f]	1		ord		1	Stubborn		Large

Aff[a] or not	Tall or short	Ed or not	Mouth	Hand or Homly	Lips thin or thick	Ambitious		Remarks
no	Med	ord	good	Pretty	Med	~~yes~~Very		
no	~~short~~ Med	med	Large	ord	med	no		
no	Tall	ord	Large	Med	thick	yes		
Affec	Short	ord	Large	Med	thick	Very		
Affec	Tall	ord	Large	ord	~~med~~thin	yes		
no	Tall	ord	Large	Hand	thick	Very		
no	Med	good	Large	Homly	thick	ord		
Affec	short	ord	Large	~~ord~~Homly[d]	thick	yes		
Affec	Tall	ord	Large	Hand	Med	yes		
no	Tall	ord	Large	ord	thick	med		
no	med	ord	Large	Hand	thick	no		
Affec	Tall	ord	Med	ord	thin	ord		
Affec	Tall	ord	Med	ord	Med	ord		
no	Tall	ord	Med	Med	thick	yes		
no	Tall	ord	Large	ord	Med	no		
Affec	Short	ord	Med	Pretty	Med	no		
Indif	Short	ord	Large	ord	thin	no		
Mild	Tall	ord	Large	ord	med	yes		
Affec	Short	ord	Large	Hand	thick	yes		
Affec	Short	ord	Small	ord	thin	no		
Very	Tall	ord	Large	ord	thick	yes		
Very	Short	ord	Med	Hand	thick	ord		
Very	Tall	ord	Large	Hand	~~Thick~~ Med	~~y~~ord		

Aff	Tall Short	Ed or not	Mouth	Handsome or not	Lips Thin	Ambitious	Conceited	
~~yes~~no	Short	ord	Bad	no	thin	yes		
no	ord	ord[a]	Bad	ord	thin	no		
no	Short	ord	ord	Pretty	ord	no		
no	Short	ord	Large	Hand	thick	no		
Affec	Tall	Yes[a]	Large	Homly	thin	Very		
no	Med	ord	ord	Nice	thin	no		
no	Tall	no	Large	Homly	thick	ord	fearful	
Very	Tall	yes	Large[a]	Hand	thick	yes	Large	
affec	Tall	Med	ord	ord	thick	Very	some	
no	Med	ord	Large	med	med	no	Very	
Affec	Tall	Ord[a]	Beau	Hand	thick	Very	no	
Affec	Tall	ord	Large	ord	thick	yes	Rather	

AD, Sloane (*TAED* X401AA). Portions of table written in pencil. [a]Obscured overwritten text. [b]Illegible. [c]Canceled. [d]Interlined above. [e]Paper torn. [f]"Mrs ETG" interlined above "Jim Gilliland."

1. See headnote above regarding the character and approximate date range of this document.

2. Rosanna Batchelor.

3. Frances Tomlinson (d. 1886), née Adams, the first wife of John Tomlinson. Obituary for Franny Adams Tomlinson, *New York, Death Newspaper Extracts, 1801–1890 (Barber Collection)*, online database accessed through Ancestry.com, 15 Jan. 2014.

4. Not identified; possibly Edison meant his former associate Charles Lorenzo Clarke.

5. Not identified, but if Edison meant the wife of Charles Lorenzo Clarke (see note 4), this would have been Clarke's first wife Helen, née Sparrow, whom he married in 1881. She was the daughter of the merchant John Sparrow of Portland, Me. Charles and Helen Clarke divorced in 1893. Marriage record for Charles L. Clarke and Helen E. Sparrow, *Maine Marriage Records, 1713–1937*, online database accessed through Ancestry.com, 15 Jan. 2014; *NCAB* 30:44.

6. Margaret Virginia Johnson (1856–1925), née Kenney, of Philadelphia married Edward Hibberd Johnson in 1873. Doc. 2258 n. 3; Cutter 1916, s.v. "Johnson, Edward Hibbard"; "Died," *NYT*, 27 Oct. 1925, 23.

7. Henry C. F. Reimer (1851–1893), a bookkeeper in Newark, N.J., who had worked for Edison in the 1870s, was a native of Bremen, Germany. Doc. 2608 n. 11; Birth record for Heinrich Carl Friedrich Reimer, *Germany, Select Births and Baptisms, 1558–1898*, online database accessed through Ancestry.com, 22 Jan. 2014; Death record for Henry C. F. Reimer, *New Hampshire Death and Burial Records Index, 1654–1949*, online database accessed through Ancestry.com, 23 Jan. 2014.

8. Josephine Lawall Reimer (1852–1915), née Stucky, was the wife of Henry Reimer and a friend of Edison's first wife. Doc. 2608 n. 11; Wells College 1868–1894, 109.

9. Lillie Fox (1859–1920) of Cincinnati, was a friend of Lillian Gilliland and later the wife of Phillip Igoe. Doc. 2841 n. 3; Certificate of Death for Lillie Fox Igoe, *Kentucky, Death Records, 1852–1953*, online database accessed through Ancestry.com, 27 Jan. 2014; U.S. Census Bureau 1965 (1870), roll M593_1208, p. 745B, image 761 (Symmes, Hamilton, Ohio); ibid. 1970 (1880), roll T9_1030, p. 524C, image 0418 (Cincinnati, Hamilton, Ohio); "Obituary," *N.A.R.D. Journal* 29 (12 Feb. 1920): 864.

10. Nettie Johnson (b. 1863) of Indianapolis, was Lillian Gilliland's sister. In 1887, she married Danforth Brown, an insurance agent in Indianapolis. Widowed in 1903, she afterward lived with her sister in Pelham Manor, N.Y. She died after 1940 when she was still living in Pelham Manor. U.S. Census Bureau 1982? (1900) roll 388, p. 5A (Indianapolis, Marion, Ind.); ibid. 1992 (1920), roll T625_1281, p. 11A, image 711 (Pelham Manor, Westchester, N.Y.); ibid. 2012 (1940), roll T627_2812, p. 6A (Pelham, Westchester, N.Y.); Marriage record for Danforth Brown, *Indiana, Select Marriages, 1780–1992*, online database accessed through Ancestry.com, 19 Feb. 2014; Polk 1887, 893.

11. Mary Elder (1858–1923) of Indianapolis, a friend of Louise Igoe,

was the daughter of John R. Elder, an Indianapolis Railroad executive. In 1889, she married Frank H. Blackledge, an Indianapolis attorney who had served from 1880 to 1884 as the private secretary of Indiana governor Albert Porter. John Elder Blackledge application to the Indiana Society Sons of the American Revolution, 29 Feb. 1940, *Sons of the American Revolution Membership Applications, 1889–1970,* online database accessed through Ancestry.com, 12 Nov. 2012; Dunn 1919, 5:2033–34; "Obituaries," *Indiana Law Journal,* 2 (1927): 440.

12. Jessie J. Hughes (b. 1858?) of Indianapolis was a friend and occasional guest of the Miller family. "Personal and Society," *Indianapolis Journal,* 16 Feb. 1889, 3; Polk 1887, 423; Jane Miller to Mina Edison, 28 Dec. 1887; Robert Miller to Mina Edison, 14 Feb. 1896; both FR (*TAED* FM001ABB2, FR001AAC); Mary Miller to Mina Edison, 4 July 1891, CEF (*TAED* X018C9AO).

13. Not identified.

14. Not identified.

15. Magdalena Holstein (1845–1916), née Nikum, of Indianapolis, was a next-door neighbor of the Igoe family. The Holsteins lived at 528 Lockerbie St. in a house Magdalena's father built in 1872. The poet James Whitcomb Riley began boarding with the family in 1893; he lived there for the next twenty-three years, and the house is now a National Landmark. U.S. Census Bureau 1970 (1880) roll T9_295, p. 328C, image 0357 (Indianapolis, Marion, Ind.); Louise Igoe to Mina Miller, 8 Feb. 1886, CEF (*TAED* X018C8A); Van Allen 1999, 247; National Parks Service, National Historic Landmarks Survey, online database accessed through http://www.nps.gov/history/nhl/designations/lists ofNHLs.htm, 27 Nov. 2012.

16. Charles Holstein (1843–1901), husband of Magdalena, was an Indianapolis attorney and poet. From 1859–1861, he attended the Kentucky Military Institute and afterward served in the Civil War as a sergeant-major in the Sixth Indiana Volunteers and later as a first lieutenant, captain, and major in the Twenty-second Indiana Volunteers. He was a graduate of Hanover College and earned an LL.B. degree at Harvard Law School. He was the assistant U.S. district attorney for Indiana from 1871 to 1880 and district attorney from 1880 to 1885, when he returned to private practice. U.S. Census Bureau 1970 (1880) roll T9_295, p. 328C, image 0357 (Indianapolis, Marion, Ind.); Louise Igoe to Mina Miller, 8 Feb. 1886, CEF (*TAED* X018C8A); "Charles Louis Holstein Jr.," *U.S. Civil War Soldier Records and Profiles,* online database accessed through Ancestry.com 19 Nov. 2012; "Charles L. Holstein," *The Green Bag,* 10 (Sept 1898): 366.

17. Not identified.

18. Not identified.

19. Not identified.

20. Alice N. Miller (1855–1930), née Newton, of Canton, Ohio, was the second wife of Jacob Miller, Lewis Miller's older brother. After Jacob's death in 1889, she moved to Washington, D.C., having inherited his estate, estimated at more than $1 million. She married businessman Norman H. Chance in 1892 and moved back to Canton in the early 1890s. By 1910, she had moved with her husband to Tacoma, Wash., where she died. Headstone photo for Alice N. Chance, Mountain View Memorial Park, Lakewood, Wash., Find A Grave memorial no.

107048442, online database accessed through Findagrave.com, 20 Feb. 2014; Death record for Alice N. Chance, *Washington Deaths, 1883–1960,* online database accessed through Ancestry.com, 20 Feb. 2014; "An Unwilling Indorser," *Washington Post,* 28 Dec. 1893, 1; U.S. Census Bureau 1982? (1900) roll 1321, p. 5A (Canton, Ward 1, Stark, Ohio); ibid. 1982 (1910) roll T624_1665, p. 8A (Tacoma Ward 7, Pierce, Wash.).

21. George Edgar Vincent (1864–1941) was the only child of Elizabeth Dusenbury and Bishop John Heyl Vincent. Formerly a suitor of Mina Miller, Vincent graduated from Yale in 1885. At this time, he was the manager of the Chautauqua Institution's press and also the co-editor of *Our Youth,* a Methodist weekly. Vincent served as president of the Chautauqua Institution from 1907 to 1915. Vincent earned a Ph.D. in sociology in 1896 from the University of Chicago, where he subsequently served as a dean of faculty. He was later president of the University of Minnesota and then of the Rockefeller Foundation. *ANB,* s.v. "Vincent, George Edgar."

22. Phillip (variously Philip) Francis Igoe (1856–1914), a bookkeeper at the First National Bank of Indianapolis, was the older brother of Louise Igoe. Certificate of Death for Phillip Francis Igoe, *Kentucky Death Records, 1852–1953,* online database accessed through Ancestry .com, 27 Jan. 2014; Polk 1885, 325; U.S. Census Bureau 1965 (1870), roll M593_340, p. 21A, image 45 (Indianapolis, Ward 1, Marion, Ind.).

23. Edward C. Elder (1863–1924), the brother of Mary Elder, was at this time a student at the Rose Polytechnic Institute in Terre Haute, Ind., from which he graduated in 1886. He later studied medicine at the Indianapolis Medical College and became a practicing physician in Indianapolis. U.S. Dept. of State n.d., roll M1372_398, Edward C. Elder passport issued 16 July 1892; Death record for Edward C. Elder, *Ohio Deaths, 1908–1932, 1938–1944, & 1958–2007,* online database accessed through Ancestry.com, 13 Nov. 2012; Rose Polytechnic Institute 1909, 122.

24. Possibly George W. Geiger (1833–1900) of Indianapolis, who lived just around the block from the uncle of Mollie Landers (see below). He was a wholesale dry goods salesman. Headstone photo for George W. Geiger, Crown Hill Cemetery, Indianapolis, Ind., Find A Grave memorial no. 45906654, online database accessed through Ancestry.com, 19 Feb. 2014; U.S. Census Bureau 1970 (1880), roll 294, p. 38A, image 0553 (Indianapolis, Marion, Ind.).

25. Possibly Kate F. Geiger (1830–1913), née Russell, the wife of George Geiger. "Obituary," *Fort Wayne Journal Gazette,* 19 July 1913.

26. Likely John H. Ohr (1827–1911), the brother of John R. Elder's first wife, Julia Ohr (see note 11). In the late 1840s, Ohr and John Elder had been partners in publishing the weekly Indianapolis newspaper *The Locomotive.* From 1852 until at least 1878, Ohr was an agent for the Adams Express Co. in Indiana. In the 1880s, he was an insurance agent. Headstone photo for John H. Ohr, Crown Hill Cemetery, Indianapolis, Ind., Find A Grave memorial no. 45984737, online database accessed through Findagrave.com, 6 Dec. 2012; Nowland 1870, 350; Indianapolis Board of Trade 1990 [1857], 42; "Adams Express Co.," *The Expressman's Monthly,* 1 (Oct. 1876): 318; Polk 1878, 404.

27. Likely Sarah I. Ohr (1832–1903), née Deavers, who married John Ohr in South Carolina in 1851. In 1884, she and Martha Landers,

the stepmother of Mollie Landers (see note 32) served together on the board of managers of the Indianapolis Home for Friendless Women. Headstone photo for Sarah I. Deaver Ohr, Crown Hill Cemetery, Indianapolis, Ind., Find A Grave memorial no. 45984737, online database, accessed through Findagrave.com, 6 Dec. 2012; Marriage record for Sarah I. Deavers, *South Carolina Marriages, 1641–1965*, online database accessed through Ancestry.com, 5 Dec. 2012.

28. Not identified.

29. Anna W. Baggs (b. 1862) of Indianapolis was the daughter of revenue collector Frederick Baggs. She married Louis Koehne in 1891. U.S. Census Bureau 1970 (1880) roll T9_295, p. 292B, image 0286 (Indianapolis, Marion, Ind.); ibid. 1982? (1900) roll T623_175, p. 5B (Gotha, Orange, Fla.); Marriage record for Anna W. Baggs, *Indiana Marriage Collection, 1800–1941*, online database accessed through Ancestry.com, 6 Dec. 2012.

30. Frances M. Beck (b. 1854?), sister of Anna Baggs and a friend of the Miller family. Marriage record for Frances M. Baggs, *Indiana, Marriage Collection, 1800–1941*, online database accessed through Ancestry.com, 27 Nov. 2012; U.S. Census 1970 (1880) roll T9_295, p. 292B, image 0286 (Indianapolis, Marion, Ind.); Jane Miller to Mary Valinda Miller, 8 Dec. 1887, FR (*TAED* FM001ABB1).

31. Joseph W. Beck (b. 1853?), an Indianapolis gunsmith and firearms dealer. U.S. Census Bureau 1970 (1880) T9_roll 295, p. 292B, image 0286 (Indianapolis, Marion, Ind.).

32. Mary "Mollie" Landers (1860?–1892) was a prize-winning artist and craftswoman. Her father, Franklin Landers, a prominent merchant and owner of a pork-packing business in Indianapolis, was a prominent Democrat who had served in the state senate and a single term in the U.S. House of Representatives. In December 1885, Mary Landers married John E. Beall, a real estate broker in Washington, D.C., and moved to that city. U.S Census Bureau 1970 (1880) roll T9_295, p. 184C, image 0069 (Indianapolis, Marion, Ind.); "Death of Mrs. John E. Beall," *Washington Post*, 26 Nov. 1892; *BDUSC*, s.v. "Landers, Franklin"; Indiana 1883, 24:126, 129; ibid. 1885, 26:127, 130–31; ibid. 1886, 27:94; "The City," *Indianapolis Sunday Critic*, 29 Nov. 1885, n.p.; "Personal and Social," *Indianapolis Evening Minute*, 2 Dec. 1885, p. 1.

33. Probably Nellie P. Winings (1859–1950), née Patterson, of Indianapolis. She married Daniel P. Winings in 1877. Headstone photo for Nellie P Winings, Crown Hill Cemetery, Indianapolis, Ind., Find A Grave memorial no. 46051635, online database accessed through findagrave.com, 17 Feb. 2014; Marriage record for Daniel P. Winnings, *Indiana, Select Marriages, 1780–1992*, online database accessed through Ancestry.com; U.S. Census Bureau 1970 (1880) roll T9_295, p. 319A, image 0339 (Indianapolis, Marion, Ind.).

34. Probably Daniel P. Winings (1849–1903), a commercial agent for A. B. Gates & Co., an Indianapolis dealer in coffee and spice mills. Headstone photo for Daniel P. Winings, Crown Hill Cemetery, Indianapolis, Ind., Find A Grave memorial no. 42311625, online database accessed through findagrave.com, 17 Feb. 2014; U.S. Census Bureau 1970 (1880) roll T9_295, p. 319A, image 0339 (Indianapolis, Marion, Ind.); Polk 1885, 636; *Manufacturing and Mercantile Resources of Indianapolis, Indiana* 1883, 430.

35. Probably Amelia H. Wells (1843–1910), née Smith, a Cincinnati native who married Graham Wells in 1864 as his second wife. U.S. Dept. of State n.d., roll M1372_641, Amelia H. Wells passport issued 29 Jan. 1904; Headstone photo for Amelia H. Wells, Spring Grove Cemetery, Cincinnati, Ohio, Find A Grave memorial no. 79068292, online database accessed through Findagrave.com, 17 Feb. 2014; "Dr. Graham A. Wells," *Dental Cosmos: A Monthly Record of Dental Science* 36 (1894): 665.

36. Probably Dr. Graham A. Wells (1827–1894), an Indianapolis dentist and near neighbor of the Winings. Headstone photo for Graham A. Wells, Spring Grove Cemetery, Cincinnati, Ohio, Find A Grave memorial no. 79068346, online database accessed through Findagrave .com, 17 Feb. 2014; Death record for Graham A. Wells, *Indiana Deaths, 1882–1920*, online database accessed through Ancestry.com; Polk 1885, 620; "Dr. Graham A. Wells," *Dental Cosmos: A Monthly Record of Dental Science* 36 (1894): 665.

–2853–

To Lewis Miller[1]

New York Sept 30 1885[2]

My Dear Sir

Some months since, as you are aware, I was introduced to your daughter, Miss Mina. The friendship which ensued became admiration as I began to appreciate her gentleness and grace of manner, and her beauty and strength of mind

That admiration has on my part ripend into love, and I have asked her to become my wife. She has referred me to you, and our engagement needs but for its confirmation your consent.

I trust you will not accuse me of egotism when I say that my life and history and standing are so well known as to call for no statement concerning myself. My reputation is so far made that I recognize I must be judged by it for good or ill.

I need only add in conclusion that the step I have taken in asking your daughter to intrust her happiness into my keeping has been the result of mature deliberation, and with the full appreciation of the responsibility I have assumed, and the duty I have undertaken to fulfil

I do not deny that your answer will seriously affect my happiness, and I trust my suit may meet with your approval. Very sincerely yours

Thomas A Edison

ALS, Swann (*TAED* B037AAA).

1. Lewis Miller (1829–1899), a manufacturer, inventor, educator, and philanthropist, was a partner with his stepbrother, Cornelius Aultman, in an agricultural equipment manufacturing firm in Canton, Ohio. After Miller devised an improved reaper, called the Buckeye mower,

the firm expanded into a second plant in nearby Akron in 1863 under the name of Aultman, Miller & Co., with Miller as superintendent. Miller also became president of the original firm, C. Aultman & Co., in 1882. Two years later, he became an incorporator and president of the American Locomotive Electric Headlight Co., of Akron. A devout Methodist, Miller was interested in both public and religious education. He developed standard teacher training and curricula for Sunday school instruction, as well as the so-called "Akron plan" of church architecture to provide dedicated spaces for religious teaching. In 1874, he joined with John Heyl Vincent to organize a Sunday school assembly at Fair Point, on the shore of Lake Chautauqua in upstate New York. Combining popular education with recreation, music, physical activity, and religious instruction, the assembly was an immediate popular success, and Miller and Vincent quickly put it on a permanent basis as the Chautauqua Institution. Married to the former Mary Valinda Alexander since 1852, the couple had eleven children, of whom Mina was the seventh. *ANB*, s.v. "Miller, Lewis"; *DAB*, s.v. "Miller, Lewis"; *NCAB* 31:136–37; "Electric Light and Power," *Elec. and Elec. Eng.* 3 (July 1884): 155; "News, Notes, and Comments," *American Engineer* 7 (2 May 1884): 179; Morrison 1974, 17–38.

2. Edison used the return address of Bergmann & Co., where his laboratory was located.

–2854–

To Erastus Wiman

[New York,] Oct. 1st. 1885.

My dear Wiman:—

I have your favor of the 28th.[a] of Sept.[1] and [~~would?~~][b] in reply would state that the expense of sending Mr. Tate to Canada will be so comparatively small that I do not think it worth while to trouble you to bear any part of it, I however very much appreciate your offer in this connection.

I am not very often at Goerck St., but I will see Mr. Batchelor about your son[2] and will arrange that every possible opportunity is given him to make himself thoroughly proficient in our business. The Staten Island work[3] has lagged somewhat owing to my continued absence from the city. I am now arranging however to have a public trial of the system made and will communicate later as to this. I may say that the experiments have been absolutely successful and that I see no reason why the system should not be a perfect success. With kind regards I remain, Very truly yours,

Thos. A. Edison[4]

TL (carbon copy), NjWOE, Lbk. 21:35D (*TAED* LB021035D). [a]Date corrected by hand. [b]Canceled.

1. Wiman offered to bear half the costs related to Alfred Tate's tests of the phonoplex in Canada because, he wrote Edison, "You very gener-

ously said that I should have an interest if I should succeed in getting Western Union to take up this matter. I shall succeed if the project is successful, but if it is not successful I ought to bear one half the expense." Wiman to TAE, 28 Sept. 1885, DF (*TAED* D8546ZAR).

2. Wiman told Edison in his 28 September letter (see note 1) that he would "be much indebted for any interest you are kind enough to take" in his son, who had recently begun working at the Edison Machine Works. William Dwight Wiman (1861–1914) was born in Canada and immigrated to New York as a young boy. He attended Lehigh University, where he was among the thirteen students enrolled in the special one-year sequence in electrical engineering established in 1884 by the physics department. After graduating in 1885, he reportedly spent two years in Edison's shops (presumably at the Edison Machine Works) and was helping to test the railway telegraph in 1886. Wiman and his father both became deeply involved in the electrification of Staten Island; by 1892, the son was general manager of the Richmond [County] Light, Heat and Power Co., Ltd., incorporated in 1887. He became enamored of electrical fountains and participated in work on them done by both his father and Charles Yerkes. Wiman married Anna Deere, an heiress to the agricultural equipment manufacturing company of that name, in 1890. U.S. Census Bureau 1982? (1900), roll T623_338, p. 10B (Moline Ward 2, Rock Island, Ill.); Family Genealogy, IMolD; Yates 1992, 86; William Wiman to TAE, 2 Feb. 1886, Cat. 1140:134, Scraps. (*TAED* SB017134); Maltbie 1911, 199–200; Stieringer 1901, 417–18; "William D. Wiman, Electrician," *Chicago Daily Tribune*, 6 Oct. 1890, 3.

3. In his 28 September letter (see note 1), Wiman also inquired about the Staten Island trials of the railway telegraph, noting: "I do not hear of it now a days."

4. This letter was signed with Edison's distinctive umbrella signature by an assistant.

–2855–

*Samuel Insull to
A. C. Mears*[1]

[New York,] Oct. 1st. 1885.

My dear Madam:—

Your favor of the 26th. ult.[2] was received during Mr. Edison's absence from the city, on his return to-day I showed it to him and he desires me to state that his daughter will return to school, as soon as he settles in the city again, which will be at the latest on the 10th. of the month. It is impossible for him to place her in your boarding-school as he has made arrangements for her to spend the winter with him. Your very truly

Saml Insull   Private Secy

TLS (carbon copy), NjWOE, Lbk. 21:35E (*TAED* LB021035E).

1. Mme. A. C. Mears operated the Mme. C. Mears boarding and day school for girls in New York, founded in 1840 by her mother-in-law. Mears, the former Louise See, seems to have used both her husband's first name, Albert, and her own given name. Marion Edison had at-

tended the school, at 222 Madison Ave., as a day student since 1883. Doc. 2535 n. 1; "Mme. C. Mears," *NYT,* 26 Nov. 1877, 5; "Obituary Notes," *New York Herald,* 15 Jan. 1892, 4; *Trow's* 1884, 1139; Mears to TAE, 7 May 1886, DF (*TAED* D8614C).

2. Mears inquired of Edison if Marion would be in school for the fall term, scheduled to begin on 1 October. Noting that Marion was old enough to move to a higher grade, Mears promised the "most vigilant endeavors to advance her rapidly this year," and she strongly encouraged Edison to board his daughter at school so as to avoid the distractions of "engagements at home." Mears to TAE, 26 Oct. 1885, DF (*TAED* D8514P).

–2856–   [New York,] Oct 1 1885

Notebook Entry:
Telephony

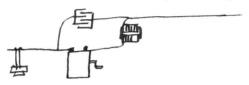

Device applied to Telephone ringer so it will dull waves and prevent them being heard in Telephone,[1] ⟨wks ok tried by Gilliland—⟩[a]

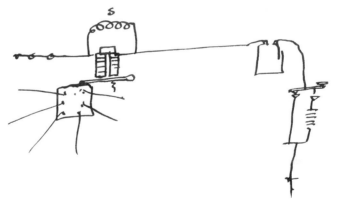

Village Telephone System[2]   Shunt around releasing magnet on commutator switch shunted to retard & prevent magneto call waves from acting on it   ⟨OK tried by Gilliland⟩[a]
TAE.   E[zra] T G[illiland]
J F Ott

X, NjWOE, Lab. N-85-10-01:1 (*TAED* N310001). Document multiply signed. [a]Followed by dividing mark.

1. Edison intended the condenser (*top*) and electromagnet (*right*) in the drawing above to damp out or cancel rapid signaling pulses to which a telephone receiver might respond falsely, the same purpose to which he had put those components a few weeks earlier in the phonoplex. In

one of the pertinent telegraph patents, he noted that such an arrangement "prevents inductive disturbances in telephones connected with adjoining lines, and the principle is generally applicable to telegraphing or other signaling-instruments for the purpose of preventing disturbances in telephones on the same or adjoining lines." U.S. Pat. 437,422.

2. Ezra Gilliland invented the telephone "village system," a method of connecting subscribers in lightly populated areas where the use of a conventional manual switchboard would be uneconomical. Consisting of several parallel circuits looped through each customer's home or office, its operation relied on what has been described as the first automatic switching system. Gilliland received the first of several patents (U.S. Pat. 306,238) in 1884 and he applied for another on an "automatic circuit-changer" in October 1885 (U.S. Pat. 334,014) for making connections between two village systems. That device did not incorporate Edison's suggestion for suppressing spurious signals, but it entered commercial service the same year between Worcester and Leicester, Mass., where the first village system opened in September. The village plan proved economical only for small numbers of subscribers over short distances; about fifty such installations were made. Fagen 1975, 546, Hill 1953, 23–24; "Correspondence. Boston," *Elec. and Elec. Eng.* 5 (June 1886): 234.

-2857-

Notebook Entry: Telephony and Phonoplex

Telephone[1]

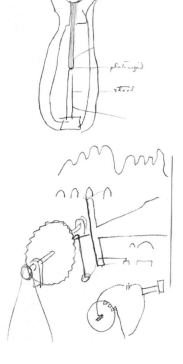

am trying today what effect musical sounds from RR telegh will have on the phonoplex & trying devices to modify its action so the 2 systems will not interfere with each other[2]

Sextuplex on phonoplex principle[3]

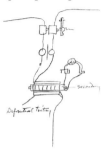

Sextuplex

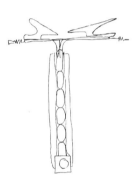

Edison's 4 October drawing of a transmitter akin to the one in this document.

Edison's drawing, one of several, of a transmitter with a spherical electrode behind the diaphragm.

draw core in & out to balance exact.[4]
TAE   J. F. Ott

X, NjWOE, Lab., N-85-10-03:35 (*TAED* N311035). Document multiply signed and dated.

1. Figure labels are "platinized" and "steel." This drawing of what appears to be a make-and-break transmitter is one of a number sketched in recent days by Edison and Ezra Gilliland. Among them is a transmitter (dated 4 October) with a column composed of short rods resting end-to-end behind the diaphragm. Edison and Gilliland also considered several iterations of a transmitter with a ball free to vibrate between the diaphragm and the electrode, a variant form of which was the subject of a patent application that they jointly executed in December. N-85-10-01:1, N-85-10-03:9–19, N-85-02-22:23, all Lab. (*TAED* vN310001, N311009, N311011, N308023); U.S. Pat. 438,306.

2. Edison may have had in mind something like the arrangements of condensers and relays that he and William King sketched on 3 October "To overcome multiple induction from wires above those grounded," although one of the wires is marked "R. R. Telephone line." Interference between the phonoplex and railway telegraph systems was a potentially serious problem, given that both systems used similar types of induction transmitters and telephone receivers. The railway telegraph could induce currents in any telegraph wires within range, including those that might be using the phonoplex. Difficulty may have been encountered in the other direction as well. A 1901 technical reference

noted that the phonoplex created such strong induction currents it could be used on only one line per set of poles, and its energy presumably could have interfered with receiving apparatus on passing trains. Cat. 1151 (1885–1886), Lab. (*TAED* NM020AAK); International Textbook Co. 1901, 2:67.

3. Figure labels are "secondary" and "Diferential tertiary." Sextuplex telegraphy—a system for simultaneously sending three messages in each direction over a single wire—was an idea that Edison tried to develop in 1877, after flirting with it during the previous two years. Among his attempts in 1877 was a combination of acoustic telegraph systems (which discriminated on the basis of signal frequency) and the quadruplex (based on the strength and polarity of the current). His present idea was to work the phonoplex (which, like acoustic systems, operated on the basis of frequency) in conjunction with the quadruplex. See Doc. 754 (headnote).

Edison signed a patent application covering such a hybrid phonoplex system on 7 October. Among the features it covered was an arrangement to prevent outgoing induction signals from affecting the receiver at the same station. The sketch below appears to show one such method that was illustrated (fig. 5) and described in the patent as a "differential diaphragm-sounder." The circuit contained differential and polarized relays as well as an induction coil with two primary windings, one connected to the main line and the other to the artificial line used to balance the circuit. Outgoing signals passed through the primary windings in

Edison's patent drawing showing differential and polarized relays (P and Q), induction coil (V), condenser (H), and receiver (B).

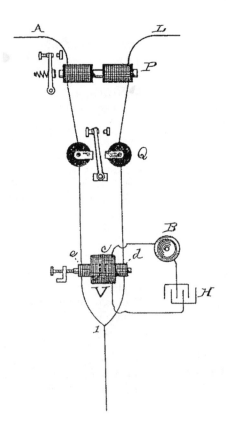

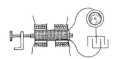

Patent drawing of a variant arrangement for equalizing the main and artificial lines and isolating the receiver from outgoing signals.

opposite directions and canceled each other (effectively giving the coil a "differential tertiary" winding) without energizing the receiver in the secondary winding. U.S. Pat. 437,422.

4. This text relates to the magnet cores near the bottom of the drawing. The receiver circuit appears to be a variation of one represented in the patent drawing showing "the adjustment of induction-coils of differential diaphragm-sounder to balance [main and artificial] lines inductively." The secondary windings were placed on a common adjustable shaft. U.S. Pat. 437,422.

–2858–

To Alfred Tate

[New York,] 12th Oct [188]5

Dear Sir,

Referring to your letter of 3rd to myself & subsequent letters to Insull I am glad to hear of the progress you are making & shall await impatiently a report as to the first trial of your Toronto-Ottawa line.[1]

With relation to Fuller Cells we should use them on the Motograph Motors[2] if we can get them— Please write often

Yours truly

Thos. A. Edison   I[nsull].

L (letterpress copy), NjWOE, Lbk. 21:21 (*TAED* LB021021). Written by Samuel Insull.

1. In letters sent on 3 October to Edison and on 6 and 8 October to Samuel Insull, Tate described his efforts to import necessary equipment into Canada and his progress setting up a trial phonoplex circuit. He and Harvey Dwight initially planned to test the phonoplex between Toronto and Hamilton, Ont., but found that wire too short; they instead began installing it in Toronto and Ottawa and in eight other stations along the 275-mile line. In his own reply to Tate's subsequent reports, Samuel Insull conveyed Edison's "opinion that you have chosen a very long line for the first experiments." As Tate later discovered, the line was even longer—333 miles—because it continued from Ottawa to Montreal, albeit without intermediate stations. Tate reported the equipment ready, save for connecting the last condensers, on 8 October. Tate to TAE, 3 and 8 Oct. 1885; Tate to Insull, 6 and 8 Oct. 1885; all DF (*TAED* D8546ZAV, D8546ZAZ, D8546ZAX, D8546ZAY, D8546ZBA); Insull to Tate, 12 Oct. 1885, Lbk. 21:23 (*TAED* LB021023).

2. Edison's notes on Tate's 3 October inquiry (see note 1) about using Fuller batteries on the electromotograph became the basis for this reply. A powerful bichromate of potassium cell devised by John and George Fuller, the Fuller battery held up well over periods of infrequent or intermittent use. The electromotograph telephone required power to rotate the chalk cylinder continually "so that the instrument will be ready at all times to respond to induction signals." Edison illustrated but did not otherwise describe this generic component in an October patent application on the integration of the phonoplex with other forms of multiple telegraphy. Doc. 1511 n. 4; U.S. Pat. 422,072.

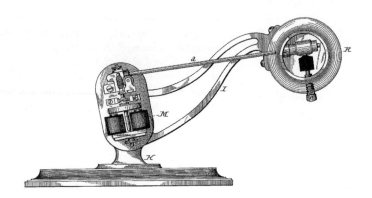

Patent illustration (U.S. Pat. 422,072) of the "diaphragm receiver" R, a motograph telephone in which the motor M rotated the chalk cylinder through shaft d.

–2859–

[New York,] Oct 22 1885[a]

Memorandum: Phonoplex[1]

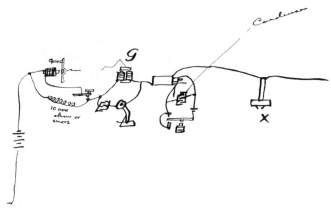

New patent on phonoplex[2]

The new features are a Condenser across the primary of the sending coil—[3]

2nd   using a Regular Magnetic telephone X right in circuit without the magnet G—[4]

3rd   a High resistance around the Relay and Key, arranged in the way shewn[5]

⟨Sound on opening only used   File Canadian applns[6]

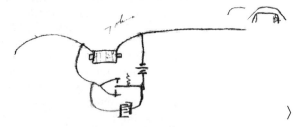

⟩

AD, NjWOE, Cat. 1151, Lab. (*TAED* NM020AAL). Letterhead of Thomas A. Edison, Central Station, Construction Dep't. [a]"188" pre-printed.

1. Edison presumably intended this document for Richard Dyer, his principal patent attorney. It precedes by one day a burst of related drawings of phonoplex permutations. Unbound Notes and Drawings (1885), Lab. (*TAED* NS85ADC08).

2. Figure labels above are "10 000 ohms or more," "G," "Condenser," and "X." The reverse of this page contains instructions, probably by Richard Dyer to himself, to research questions regarding the life of Edison's patents, a patent by Elisha Gray on a condenser around a relay, and Edison's Case No. 132 (an 1877 application on multiple telegraphy still pending at the Patent Office). Pat. App. 377,374.

3. Edison made a variant sketch of this arrangement the same day without the condenser shunt around the coil (Unbound Notes and Drawings [1885], Lab. [*TAED* NS85ADC07]; see also Doc. 2849 n. 3 regarding other phonoplex drawings from this period). He executed a patent application incorporating some elements of this document (and the major innovation shown in Doc. 2861) on 12 November, soon after making the tests described in Doc. 2867. Covering several alterations to simplify and improve the phonoplex, it was the first of three related applications that he signed in mid- and late November. After initially rejecting it, the Patent Office insisted on a number of modest revisions before issuing the patent in February 1890. Among the changes was the elimination of the word "phonoplex" from this and the two companion applications (Pat. App. 422,073; see Doc. 2800 [headnote]).

4. See Doc. 2868 (headnote).

5. The "high resistance" was a shunt circuit around the regular Morse key and relay (**A** and **B** in the patent drawing). In the specification, Edison described this circuit as having enough resistance to "keep the line constantly closed for the induction signals" (cf. Doc. 2863). It would also "prevent the relays from absorbing the induction impulses" without itself affecting the operation of the Morse instruments. Edison sketched several variations on 28 October. U.S. Pat. 422,073; Unbound Notes and Drawings (1885), Lab. (*TAED* NS85ADC09).

The patent application that Edison signed on 12 November included

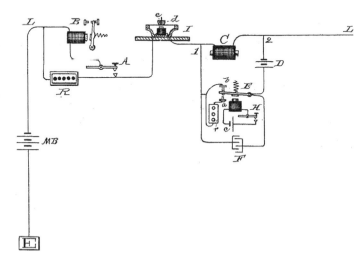

*Phonoplex circuit diagram from the specification that resulted from this document (U.S. Pat. 422,073). The Morse relay (**B**) and key (**A**) are at left, the receiver (**I**) in the center, and the electromagnet (**C**) and transmitter-controlling circuit below it at right.*

another significant use of a resistance circuit, not mentioned in this document. In the induction (phonoplex) transmitter, operated by a local battery (**D**), the resistance (**r**) was connected to the front point (**a**) of the key (**E**). The resistance would attenuate the signal from that point, enabling the receiver to produce different sounds for the upstroke and the downstroke (cf. Doc. 2870).

6. The figure label is "7 ohms," which refers to the induction coil. The sketch shows a simple form of the transmitter circuit shunt. As explained in the patent application that Edison signed on 12 November, the condenser "absorbs the spark at the points, and also sharpens the impulses, so as to materially improve the operation of the induction apparatus." Perhaps as a consequence of his extensive tests with induction coils on 10 November, the application stated Edison's preference for replacing the coil with an electromagnet. U.S. Pat. 422,073; see Doc. 2867 n. 2.

Edison received Canadian patent 24,084 on 18 May 1886, the claims of which were identical to those in the U.S. application he signed on 12 November (before the revisions mandated by the Patent Office). He also received at least three other Canadian patents on related subjects in January 1886, though the editors have found no information about the filing of those applications. Pat. App. 422,073; Canadian Pats. 24,084 [U.S. Pat. 422,073]; 23,119 [U.S. Pat. 333,289]; 23,120 [U.S. Pat. 333,290]; 23,121 [U.S. Pat. 333,291]; *CPOR* 13 (June and Feb. 1886): 271, 43.

–2860–

To Huelsenkamp & Cranford

New York, Oct. 23d 1885[1]

GENTS

please have prepared at once a map drawn to scale, of my lots, showing everything that it will be necessary to know in order to locate our buildings: also show the position of the old house and the avenue or street that passes through the property. We will erect two dwelling houses on the river front, and will place the laboratory building and dwelling for workman on the other side of the avenue. If practicable will move the old house[2] and make it over or repair it and use it for employees. Our steam launch draws about three feet and a half and we will build a pier our far enough to reach that depth, and would like to know about how far that will be, in order to provide necessary material.

Our buildings are being made in Maine and will be loaded on board a ship at Boston. The ship will touch at New York and take on board the engine, boilers, machinery and apparatus for laboratory and furniture, &c., for dwelling houses.[3] The ship will also bring a steam launch and a small lighter.

We propose to unload everything at Punta Rassa[4] and tow it up the river in the lighter. We were unable to procure vessels sufficiently light in draught to go up the river. We would like to have you ascertain what arrangements can be made for unloading the ship on the docks at Punta Rassa and necessary storage room for furniture and such material as cannot be exposed to the weather during the time the buildings are being erected. We will send three or four of our employees to superintend the work.[5] They will probably require assistance in handling the material as well as in construction of the buildings, &c. Will probably require four carpenters and four laborers can they be obtained at Fort Myers?

The ship will leave here in three weeks. We hope to have the buildings completed and ready for occupation in January.[6] An agent will leave here in about a week, who will come direct to Fort Myers and do what he can to prepare the grounds, build the pier &c.[7] Having explained to you fully what we propose to do and about how we propose to do it, I would ask that you furnish us with all the information you can to aid us in carrying out our plans and greatly oblige   Yours very resp'y,

THOS. A. EDISON, per Gilliland

PD, *Fort Myers News Press,* 7 Nov. 1885, FFmEFW, EMFP, News Clippings (*TAED* X104S003B).

1. This letter was published in the *Fort Myers News Press* under the heading "Letter from Thos. A. Edison." It was pasted into a scrapbook below a blurb marked 31 October 1885 from the same paper, reporting that Edison "is loading a vessel for the construction of his buildings and the erection of his works when he arrives in Fort Myers." News Clippings (*TAED* X104S003A).

2. This was the cottage the previous property owner Samuel Summerlin built sometime after 1879. In 1928, Edison expanded the cottage, adding a second floor and a garage. During his lifetime, it continued to be used as a home for the caretaker and chauffeur. The Caretaker's Cottage, located on the far northeast edge of the Edison property, is now used for special programs and luncheons at the Edison-Ford Winter Estates historic site. Rosenblum and Associates 2000, ii, I-1–I-2, I-68; http://www.edisonfordwinterestates.org, accessed 17 Jan. 2012.

3. At least six ships were eventually involved in transporting Edison's goods during the winter and early spring. In December, Edison insured the cargo aboard the schooner *Julia S. Bailey* for $8,000 (later increased to $9,000) and arranged payment of $1,150 to captain John Gould upon the safe arrival of his and Gilliland's goods at Punta Rassa. The *Julia S. Bailey* took on a cargo of building materials at Portland, Maine, including lumber and millwork from the Kennebec Framing Co., cement, 50,000 bricks, 66 bushels of hair for plastering, 95 casks of lime, and 100,000 pounds of Lehigh egg coal. The schooner made port in New York and took on additional cargo, including an engine, boilers,

other machinery and apparatus for the laboratory, and furniture for the houses. It also took on board the steam launch *Lillian,* which Gilliland had built by Ambrose Martin in Boston. The schooner reached Punta Rassa in late January. Rosenblum 2000, 1:6; Boston Marine Insurance Co. certificates, 3 and 23 Dec. 1885, DF (*TAED* D8509A, D8509B).

About a week after insuring the cargo on the *Bailey,* Edison and Gilliland gave instructions for the delivery of "our goods to the schooner Fannie A Milliken now loading at Pier #14 East River" in New York. That vessel did not reach Florida. Replying to a request for a phonograph in February, Edison told the seeker that "The last one I had was shipped with other effects to my Florida Laboratory but the schooner carrying the goods was unfortunately struck by lightning and abandoned at sea." According to a newspaper account, the *Milliken* left New York on 14 January for Key West, Tampa Bay, and Pensacola but was disabled a week later in a storm well off the North Carolina coast. A passing ship rescued the crew after four days adrift (TAE to Albert Keller, 29 Dec. 1885; TAE to P. Belford, 5 Jan. 1886; both DF [*TAED* D8539N, D8603B2]; TAE to John DeMott, 9 Feb. 1886, Lbk. 21:257 [*TAED* LB021257]; "Wrecked By A Bolt," *NYT,* 3 Feb. 1886, 1). In February, Edison reportedly telegraphed Gilliland about another shipment: "All our things have gone down. Schooner stove up this side of Hatteras. Captain and crew safe." This message, if it existed, was merely a prank. According to a newspaper, Edison, in high spirits about his upcoming wedding, explained to Gilliland that their belongings "had gone down—down South, I meant" ("Edison Alarms His Partner. The Great Inventor in a Playful Frame of Mind," *Brooklyn Daily Eagle,* 26 Feb. 1886, 6).

Edison and Gilliland also hired the schooner *Fostina,* which sailed from New York for Key West at the end of February. The *Fostina* likely brought materials to replace the laboratory supplies lost aboard the *Milliken.* The two men hired other schooners to ship goods in the ensuing months, including the *Lily White* in March and the *New Venice* and the *Miss Johnson* in April. Rosenblum 2000, 1:6.

4. Punta Rassa, at the southern extreme of Charlotte Harbor, provided access to Fort Myers via both the Caloosahatchee River and telegraph lines. Barbour 1884, 150; Brown 1989, 145, 147, 149.

5. Among this group were probably J. S. Knowles of Boston and L. S. Petris and Albert Keller, both of New York, who arrived the first week of January. "More Men for Edison," *Fort Myers News Press,* 9 Jan. 1886, News Clippings (*TAED* X104S004C).

6. Cf. Doc. 2907.

7. A local newspaper identified the agent as Eli Thompson. By December, Thompson was supervising a "force of men building a wharf and preparing the ground." *Fort Myers News Press,* 12 Dec. 1885; "Edison and Gilliland Form a Copartnership," ibid., 26 Dec. 1885, News Clippings (*TAED* X104S004A, X104S004B).

[New York, c. October 24, 1885[1]]

Dick—

In view of that patent

assigned[a] to WU change the diagram thus[2]

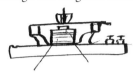

pin in centre diaph—[a] & Little weight[3]

Edison

ALS, NjWOE, Cat. 1151, Lab. (*TAED* NM020AAK1). Letterhead of Thomas A. Edison, [Central Station, Construction Dep't.], W. D. Rich Supt. of Construction. [a]Obscured overwritten text.

1. Edison likely wrote this directive within a few days of the instructions for a new patent application that he gave to Dyer on 22 October (Doc. 2859). He and Dyer presumably had talked or corresponded about the matter in the intervening day or two, but the editors have found no record of their exchange.

2. Edison referred to his 1878 patent for a telephone call signal that was assigned to Western Union. The device consisted of a magneto telephone on whose diaphragm rested the end of a metal lever. When a signal came through, the lever would be thrown "violently" from the surface and, "in returning, strikes the diaphragm a blow, and produces a sharp penetrating sound like that of a Morse sounder" (U.S. Pat. 203,017; see also Doc. 877 n. 5). The editors have not found either the drawing or other correspondence from this time about the 1878 patent; presumably Edison and Dyer discussed the matter on or soon after 22 October. The phonoplex patent application that Edison signed on 12 November included the form of receiver shown here (U.S. Pat. 422,073).

3. This "weight" was the loose metallic ring described in Doc. 2868 (headnote).

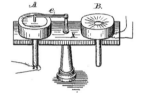

Portion of drawing from Edison's 1878 patent for a call signal device, showing the receiver (A) and swinging lever (e).

[New York,] October 26th. [1885]

My Dear Tate:—

I have your favor of October 23rd.[1] I have seen your various letters to Mr. Edison, have handed them right over to him, and in each case he has promised to send you detailed replies.[2] Whether he has done so, you know better than I do. He says that you do not give him all information. He wants you to go into every detail and on every little point. If you suffer from induction, he wants to know the nature of that induction and whether you get the   sounds on motograph at all. Is the induction from Morse wires, from through duplex wires, from quadruplex wires or from automatic wires. In all experimental work, you should remember that it is absolutely necessary for Mr. Edison to be posted on every possible detail. You may not think a thing important whereas it may give him the clue to the resolution of one of your greatest difficulties. I am simply writing all this on general principles.[3] I fancy that the Old Man has got an idea that you have bitten off more than you can chew.

Many thanks for your detailed information as to the "WIDOW."[4] It look very suspicious, your pinning the various pages of your epistle with a safety pin. WHERE HAVE YOU BEEN TO GET SUCH AN ARTICLE? Very truly yours,

TL (carbon copy), NjWOE, Lbk. 21:117C (*TAED* LB021117C).

1. Tate wrote three pages on mostly personal matters (see note 4). He also mentioned his letter of the same day to Edison concerning his suggested remedy for unwanted induction on phonoplex lines (see note 2).

2. Tate dispatched at least four letters and three telegrams to Edison in the preceding week about his phonoplex trials, especially the problem of induction from Morse telegraph lines. He described his efforts to isolate the cause and determine the greatest distance through which the phonoplex could operate. Doc. 2863 is Edison's reply to the last of these extant letters, from 23 October. Tate to TAE, 19, 20, 21, 22, and 23 Oct. 1885; all DF (*TAED* D8546ZBF, D8546ZBG, D8546ZBG1, D8546ZBG2, D8546ZBK, D8546ZBM, D8546ZBN).

Edison did not immediately make good on his pledge to keep Tate informed. On 19 November, after taking Edison's dictation of Doc. 2870, Insull told Tate that he had "been trying for weeks to get him to dictate letters to you so that you may get full instructions & he promises to do so in future." Insull to Tate, 19 Nov. 1885, Lbk. 21:121 (*TAED* LB021121).

3. Insull advised Tate at the end of September that Edison was complaining about the dearth of letters from him and urged him to "write fully and often." Despite the recent spate of letters received, Insull again admonished in early November: "You must write more frequently." Insull to Tate, 30 Sept. and 4 Nov. 1885, Lbk. 21:14, 68 (*TAED* LB021014, LB021068).

4. Tate devoted two substantial paragraphs of his letter to a person identified only as "my widow." He promised to enclose one of her recent letters "To show you she was not bogus" and advised Insull that she would seek him out on a forthcoming visit to New York. According to Tate, this woman "carries a six shooter and has informed me [that] she would bore a hole in a man she loved who did not reciprocate her affection!!" but he confided a few weeks later that he had not heard from her and feared his letters had not reached her. Tate to Insull, 23 Oct. and 17 Nov. 1885, both DF (*TAED* D8546ZBO, D8546ZCF).

[New York,] 27th Oct [188]5

Dear Sir,

Yours of 23rd received.[1] If you will recall your original instructions you will find that I told you that if a wire gave trouble from induction you should locate the troublesome wire & put a condenser round the key &[a] relay, & that this would stop your trouble. So much for your discovery!!! Yours truly

Thos. A. Edison   I[nsull].

L (letterpress copy), NjWOE, Lbk. 21:61 (*TAED* LB021061). Written by Samuel Insull. [a]Obscured overwritten text.

1. Tate excitedly described having "made a discovery tonight which promises to result beneficially" in better operation of the phonoplex system. He found that interference on the phonoplex came from Morse lines, not out on the poles, but within or near the telegraph office, where the battery current was strongest. Unsure "whether to congratulate myself upon having made this discovery tonight or to curse my stupidity for not having found it earlier," he apologized in the end for being a "brainless jackass." He planned to put condensers on the adjacent lines in the office to see if that would correct the problem. Tate to TAE, 23 Oct. 1885, DF (*TAED* D8546ZBN).

Fort Gratiot Oct 30th 1885

My Dear Boy

I Shal be ready and Start for N. York on Thursday Next and Bring James Symanten[1] with me Except I hear from you— if all Things is rite than offer Europ[2]   Ever yours

S. Edison

ALS, NjWOE, DF (*TAED* D8514U).

1. James Symington.
2. Edison had offered to pay for his father and Symington to visit Europe in 1878 but that trip never materialized. Samuel Edison had wanted in particular to visit Amsterdam to investigate his Dutch ancestry (see Doc. 1330; Symington to TAE, 14 May 1878, DF [*TAED* D7802ZLL]). The idea for the trip was revived in 1885 and to pay for it, Edison gave his father a $1,500 line of credit. Before leaving, the senior Edison was quoted in the *New York World*, perhaps with some embellishment, that "A young fellow of sixty is going along with me, and he and I are going to see old Europe for the first time. . . . Alva? No, he won't go with us. Pretends he is too busy. He is always fussing over some new patent jigger." During their three and a half months abroad, Samuel and Symington visited Scotland, England, Germany, Belgium, France, and Italy, as well as Amsterdam. "A Young Man at 90 Years," *New York Mail*, 5 Aug. 1893, Clippings (*TAED* SC93033a); "Important Trifles," *New York World*, 2 Nov. 1885, 2; Symington to TAE, 24 Nov.

and 21 Dec. 1885, 20 Jan. and 20 Nov. 1886, all DF (*TAED* D8514W, D8514ZAF, D8614A, D8614O).

-2865-

Samuel Insull to
Michael Moore

[New York,] Oct. 31st. 1885

My Dear Mr. Moore:—[1]

You may remember that last Saturday afternoon I suggested to you that you should ask Mr. Edison whether it would be advisable for you to do anything with what we call our "Grasshopper Telegraph."[2] The ground patent in this matter is not an invention of Mr. Edison's, but one which we purchased from an associate of his.[3] It is for the purpose of enabling the man in charge of a train to communicate whilst in motion with a stationary point or vice a versa, that is it places the conductor of a train and the train dispatcher in constant communication with one and other whatever may be the position of the train on a section of railroad or whatever speed it may be running at. Our experiments on a short railroad at Staten Island proved eminently successful and shows our system will be entirely practicable. Mr. Edison has added materially to the original invention and reduced it from what was a mere laboratory experiment to its present practical state. This progress has of necessity developed a number of points on which we are able to get very good patents. The first money for the experiments was supplied by Messrs Seligman & Co.[4] the bankers of this city, and about a month ago they visited Staten Island and saw the invention in practical operation. They took with them a Mr. Kinsley[5] who is one of the State Commissioners of Railroads for Massachussets. Mr. K. gave it as his opinion, as a railroad expert, that the invention would prove of very great value in connection with the modus operandi usually adopted by railroads in the dispatch and the regulation of the running of their trains. He said that it would prove especially valuable [in][a] the far West where there is a very heavy freight traffic where [a][a] single block on a railroad[6] often costs from 25 to 50,000 dollars and he gave it as his opinion that the invention was the most perfect of any railroad invention that had been brought to his notice. [The][a] objection has been raised to the system that it would be necessary to [-----][b] a trained telegraph operator on the train but this objection is dispelled by the fact that it is possible to arrange a code of signals so that any railroad conductor of ordinary intelligence would be able to communicate with his superior officers at the main station, or these superior officers be enabled to communicate with their subordinates on a train by the [use?]ᵃ of arbitrary systematic signals to read which would require no knowledge of the telegraphic alphabet. Both the Seligmans and [many?]ᵃ of our friends considered the invention of very great importance in this country and as a proof of this I may mention that [moneyed]ᵃ friends of Mr. Edison, who were more or less associated [with]ᵃ some of the other enterprises, have bought the stock of [the]ᵃ company at $25 a share placing the present value of the invention in its undeveloped condition at $250,000, there being [10,000]ᵃ shares of stock. Will you please make some inquiry among your English railroad friends and ask them whether it would be possible to market such an invention in England. In this country where we have everything in our own hands we propose to hold on to the invention and simply authorize ~~others~~ railroads to use it on a given royalty per mile. With our experience in connection with electric lighting and other enterprises, our inclination is to sell our foreign [~~pat-ent?~~]ᵇ interests in each different country for a given sum of money down. If you find amongst railroad men of your acquaintance that such an invention would be of value in connection with the working of the various English railroad systems do you think that you could form a syndicate to purchase from us the English patent rights? I may mention that the cost of construction in order to operate the system is practically nil. We use the ordinary telegraph lines running by the side of a railroad and our use of them for our induction telegraph does not at all interfere with their simultaneous use for the ordinary purposes of telegraphic business. In proof of this I may mention that all our experiments have been conducted on a set of wire lines which run parallel with the Staten Island railroad and which lines are owned by the Baltimore & Ohio Tel. Co. The telegraph company's business has not been interfered with, no one in fact but the local operators on the line of the road where we are working having any knowledge of the fact that we are using the lines for the purpose of our experiments, and their business continues just the same as if our experiments were not being tried. The only expense for apparatus in connection with the working of the induction telegraph is some small telegraph instruments placed on a train at a cost of at the very outside about $100 per train and similar apparatus placed in such stations as might be necessary at the same cost. You might make the inquiry suggested above and write me in relation to this matter.[7]

PHONOPLEX. Our Canadian reports from Mr. Tate to-day

are far more encouraging than we have heretofore received. It appears that owing to his want of acquaintance with the invention necessarily arising from the fact that this is the first time it has been attempted to be put in operation, that he has neglected to carry out some of the instructions given him in case he should meet with any trouble and his obtaining results has consequently been delayed. That is, if he had followed closely the instructions laid down by Mr. Edison before he left New York he would have been able to report long ere this very good results as to the working of the invention. This is the conclusion Mr. Edison has arrived at to-day from reading Mr. Tate's recent letters. I am, therefore, under the circumstances very sanguine of being able to send you at an early date reports of the successful working of the phonoplex system. I may mention that Mr. Edison has been trying in the laboratory to-day a devise for receiving messages on the phonoplex side of a wire which device will enable him to dispense with the chalk motograph.[8] If it is decided to use this device the apparatus will of necessity be very much simplified as of course the use of the motograph introduces a class of apparatus which it is desirable to be dispensed with if some other means of operating the system are found to be equally successful. The new receiving device has a sound more nearly resembling the ordinary Morse receivers by which telegraphic signals are ordinarly received. I am expecting in a few days to get Mr. Edison's description of the Phonoplex and it shall go forward to you immediately I can get him to give it to you. Mr. Edison has written to Hanford[9] authorizing him to give you any information you may desire as to the patents in this matter. He has also written to Preece[10] stating that you will possibly call on him in the near future in relation to a telegraphic invention which might prove of great benefit to the Postal Telegraph Dept.[11] and requesting of Mr. Preece the same consideration for you that he would extend to Mr. Edison himself.[12] Very truly yours,

TL (carbon copy), NjWOE, Lbk. 21:117B (*TAED* LB021117B). [a]Copied off edge of page. [b]Canceled.

1. Michael Miller Moore, an American doing business in London, was associated with the Mercantile Trust Co. of New York. Moore was a veteran of Edison's business enterprises in Great Britain, having invested in the original Edison telephone and electric light companies in London and been a director of the Glasgow telephone company. Letterhead of Moore to Insull, 21 Nov. 1885, DF (*TAED* D8546ZCM); Doc. 2467 n. 5.

2. That is, the railway (or wireless) telegraph.

3. See Doc. 2780 (headnote).

4. J. & W. Seligman & Co.

5. Edward Wilkinson Kinsley (1829–1891) had served as a member of the Massachusetts Board of Railroad Commissioners (with a brief interruption) since 1878. Raised by abolitionist parents, during the Civil War Kinsley was a personal representative of Massachusetts governor John Andrew. The published record of his Civil War military service is ambiguous, but he seems to have been a regimental lieutenant-colonel of the Massachusetts Volunteer Militia, an officer training corps. He became a notable defender of freedmen's rights during Reconstruction. "Joined the Majority," *Boston Daily Globe,* 27 Dec. 1891, 1; Board of Railroad Commissioners 1892, 123; Bates 1892, 446–48; Downs 2006, 97 n. 20.

6. That is, a section of track governed by a specific signal. *OED,* s.v. "block," n. 19.c.

7. Moore reported in November that he was soliciting railroad officials' opinion of the grasshopper telegraph but the editors have not learned what, if anything, else he did. Moore to Insull, 21 Nov. 1885, DF (*TAED* D8546ZCM).

8. Regarding the new diaphragm see Doc. 2868 (headnote).

9. Thomas John Handford (1842?–1890) was a London patent agent whom Edison retained in 1882. Edison wrote him about this matter on 2 November, in reply to which Handford promised to share with Moore details from two pending patent cases. Doc. 2738 n. 1; TAE to Handford, 2 Nov. 1885, Lbk. 21:117F (*TAED* LB021117F); Handford to TAE, 16 Nov. 1885, DF (*TAED* D8544V).

10. William Henry Preece (1834–1912) was electrician and assistant chief engineer for the British postal telegraph network, as well as a Fellow of the Royal Society. His long association with Edison was punctuated by a bitter public falling-out in 1878 but had recently been warmly collegial (Docs. 2600 n. 5, 2756 n. 3). On 2 November, Edison sent Preece a typed letter of introduction explaining Moore's representation of him and stating his wish to bring the phonoplex to the attention of the British Post Office. TAE to Preece, 2 Nov. 1885, Lbk. 21:117A (*TAED* LB021117A).

11. All domestic British telegraph systems came under the control of the Post Office's Postal Telegraph Dept. in 1870. Baker 1976, 88–94; *Hazell's Annual* 1886, s.v. "Postal Telegraph Department"; "Telegraph Street" 1871, 502.

12. On 27 October, Moore signed an agreement to become Edison's agent for negotiating the use of the phonoplex by the British government. He was to receive one-third of the total royalties, based on a minimum rate of one pound sterling per set. The contract was in effect until the first day of 1888. Edison did not sign the copy retained in his records but John Tomlinson's notarial oath indicates that he did so. On the same date, Edison executed a power of attorney granting Moore authority to act on his behalf (TAE agreement with Moore, 27 Oct. 1887; TAE power to Moore, 27 Oct. 1887; both Miller [*TAED* HM850271A, HM850271]). Nothing seems to have come of the matter before July 1886, when Insull prompted Moore to approach Preece; he soon after suggested terms on which the British Post Office might buy the pho-

noplex outright. Insull also corresponded with Moore during the fall about the less likely prospect that British railways would use it. In mid-1887, Moore reported that Preece was considering the phonoplex but that Edison's system faced competition from a similar British invention. In the end, Preece decided not to adopt it, and there the matter seems to have rested (Insull to Moore, 28 July, 23 Aug., and 21 Sept. 1886; LM 12:82, 201; 13:18 [*TAED* LM012082, LM012201, LM013018]; Insull to Moore, 18 Oct. 1886, Lbk. 23:24 [*TAED* LB023024]; Alfred Tate to Moore, 20 July 1887; Moore to Tate, 14 Nov. 1887; both DF [*TAED* D8717AAA, D8753AEW1]).

–2866–

Ezra Gilliland to
Charles Selden

New York Nov 10 1885[a]

"Copy"
My Dear Selden[1]
    Yours of the 7th to Edison duly received.[2] He desires me to reply to your inquiries also to give you a few additional facts concerning the Phonoplex—
    It is proposed to charge a royalty for use upon Railroad lines of ($20)—twenty dollars per year per complete set—[3] Is willing to put it on operation in short line at own expense, and turn it over if it works satisfactorily—
    The cost of fitting out a line with 6 offices and two terminals (8 stations) all equipped with Phono's Fig 1 the cost would be 33 per station or total of $264— If only 5 of the offices are to be worked Phono see Fig. 2[4] then the other offices will only cost $14— each or a Total of $207— Battery is additional, 4 cells to each Phono station— Fig 3 shows Phono worked from through stations only. Fig 4 a short section of a line worked Phono's— Fig 5 a regular duplexed line with all the way stations worked Phonoplex— If operatorr leaves his Morse key open he can be called on the Phonoplex—[5] It is excedingly simple, no adjustment, and when once set up requires no further attention   Don't confound it with the complicated and troublesome duplex apparatus— Edison does not want to make any arrangements at present for your commercial lines— If you should use it only from terminal points getting a through wire out of every way wire, it would cost you only $40—per circuit, you would probably not introduce it in more than 100 offices the royalty on which would amount to only $2000—per annum or about the salary of one good man— now this is the price which he wants it to net him, to this must be added whatever it is decided to give to those who assist us in getting the thing introduced. you will remember our conver-

sation, and will therefore know about how to be governed on this point—

I would be very glad to see you to discuss any points any points upon which you are not perfectly clear. It will be impossible to leave here for a week or ten days any further information you may desire will be furnished promptly   Yours very truly

(signed)   E. T. Gilliland

L (transcript), NjWOE, DF (*TAED* D8966ABH2). Letterhead of Baltimore & Ohio Co. Telegraph Dept., General Superintendent's office. ª"188" preprinted.

1. Charles L. Selden (1849–1930), a former Western Union district superintendent, took over management of the Baltimore & Ohio's railroad and commercial telegraph lines east of the Ohio River in 1884. As a young man, Selden had been an operator in Cincinnati, where he worked with Edison (and perhaps also with Gilliland). Doc. 2474 n. 8; "Elections and Appointments," *Elec. and Elec. Eng.* 3 (Mar. 1884): 69.

2. Selden made a "Confidential" inquiry about royalty rates for what he called the "Electro Phonoplex." He suggested that Edison set his figures low to attract the business of the B&O, which he anticipated would be "an entering wedge for other Railroads to follow." This prediction proved accurate, as Samuel Insull acknowledged the following August when he told Selden that "the early use of the System by the B. & O. people has materially helped us in pushing our business" on at least five other railroads. Selden to TAE, 7 Nov. 1885, DF (*TAED* D8966ABH1); Insull to Selden, 30 Aug. 1886, LM 12:224 (*TAED* LM012224).

3. Nine months later, when Samuel Insull answered a raft of inquiries about the phonoplex, he quoted substantially higher royalties: $100 annually for a simple Morse circuit, $150 for a duplex circuit, and $200 for each line using quadruplex instruments. Cf. Doc. 2923; see, e.g., Insull to W. K. Morley; Insull to Charles Jones; Insull to William Kline; Insull to J. W. Lattig; all 12 Aug. 1886, LM 12:136, 138, 141, 144 (*TAED* LM012136, LM012138, LM012141, LM012144).

4. Figures not found with the transcription.

5. Edison pointed out in one of his November patent applications that high resistance shunts would keep the line closed regardless of the key's position (see Doc. 2859 n. 5). An enthusiastic article in *Engineering* later emphasized the point: "Another great advantage lies in the fact that phonoplex circuits can never be left open, either through their own instruments or those of the Morse line." "The Edison Phonoplex," *Engineering* 42 (22 Oct. 1886): 412.

–2867–

Notebook Entry: Phonoplex[1]

[New York,] Nov 10 1885

Phonoplex Experiments on proportioning Coils[2] made on Boston Telephone Line grounded here & looped a[t] Providence—[3a]

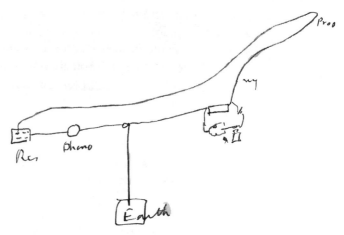

7 ohm Coil—

    1 cell battery no condenser—no res   nothing dont move reg weight.[4] 2½ m[icro]f[arad] condenser makes it growl—

    2 cells imperfect writing   70 sheets condsr writing good— no res just in margin

    3 cells. 200 ohms   wrtg good—no condsr   85 sheets gets it thro 900 ohms—

    4 cells 600 ohms no condr   1400 ohms with 85 sheets.

    5 cells 6400 ohms—1400 ohms with 85 sheets—[b]

17 ohm Coil—primary only—

    no res no condr nothing   85 sheets nothing

    2 cells growls no res no condr   Condsr weakens it so no growling

    3 cells—just get fair writing only no Res no condr   Condr weakens it

    4 cells 600 ohms Condr weakens

    5 cells—800 ohms Condr weakens

2–17 ohm Coils Multiple arcd—[5]

    1 cells ntg   growls with 85 sheets Condr

    2 cells—just get it no res   85 sheets 200 ohms.

    3 cells 400 ohms no condr   85 sheets 600[c] ohms

    4 cells 1000 ohms no condr—1400 ohms with 85 sheets

2–17 ohm Coils multiple arcd & batteries multiple arcd ie[d] cells ½ Res each—so as to run from 1 to 5 cells each of ½ the internal Res—

    1 big cells notg   get[c] it shaky with 85 sheets no res—

    2 big cells little shaky no res— 85 sheets OK thro 300 ohms—

    3 cells—500 ohms— 1000 ohms with 85 sheets

    4 cells—1200 ohms— 1400 ohms on 85 sheets—

2–7 ohm Coils multiple arcd   Double size battery
   4 cells OK no res   1000 ohms—85 sheets—
   3 cells—OK no res   800 on 85 sheets[c]
Single batteries—Different coils—longer than 7 ohm coils
about inch—iron twice dia[meter]— Res .48 Res. 4 layers—
   1 cell— Notg on no res   notg on 85 sheets—
   2 cells— Notg on no res   growls on 85[c] sheets—
   3 cells—notg on no res   growls better on 85 sheets
   4 cells—ntg on no res   ng on 85—
Big Coil. using primary only   Res 4.25 ohms marked No 1
   1200 ohms 4 cells No Res   72 sheets Condr 1800 ohms
   3 cells 300 ohms no condr   900 ohms—
   2 cells—little shaky in no res   400 ohms on 72 sheets—[c]

Another larger Coil—No[c] 2—using primary only   Res primary 6.05. shorter coil ¼ amount iron core—
   2 cells growls slightly   85 sheets writing only fair—
   3 cells no res just get writing   300 ohms with 85 sheets—
   4 cells 100 ohms no res   400 ohms with 85 sheets—
Another big Coil No 3   Res primary 4.25—¾ inch iron
core—
   4 cells 400 ohms no resis   1200 ohms with 85 sheets
   3 cells—200 ohms no res   600 ohms with 85 sheets=[c]
Big Coil No 4   primary only used Res. 1.23.
   3 cells weak on no res   300 ohms on 85 sheets=
   4 cells—weak on no res   300 on 85 sheets=[c]
Big Long Coil—broken wood Ends— Res primary 1.32
   4 cells 400 ohms no condr [--][f]   1400 ohms 85 sheets=
Same coil using Secondary 64 ohms in Line—
   4 cells 600 no res— 1200 with 85 sheets.[c]
Coil No 4 using Secondary to Line— Res secondary—77½
ohms—
   4 cells—growls at no res— 85 sheets 300 ohms—
No 1 coil using secondary in line   Res secondary [30?][f] 78
ohms
   4 cells—weak on no res   85 sheets 400 ohms[c]
No 3 Coil   Secondary to Line— Res Secondary—78—
   4 cells: weak no res— 85 sheets 300 ohms[c]
No 2   secondary to line   Res Secy 75 ohms
   growls no res   300 ohms 872 sheets=[c]
The two 7 ohm coils in series Reg way
   4 cells 200 ohms   40[c] sheets 600 ohms
   5 cells—800 ohms no condr   1100 with 32 sheets
   6 cells—1000—no condr— 1600 32 sheets.[c]

Long coil .41   No 7—

64 cells just growls with 72 sheets

Experiments on jumping weights   7 ohms Coil reg 4 on line— Condsr plug 1000. 72 sheets condr.[e]

Ring[c] split 700.

flat ring 600   [–][g]

 1000

800

1000

The small weight on weak current needs no condr only heavy weight[c]   on strong ckts improved by condr condrs—[e]

The Little Johnny Jump up[6] opens & closes the primary of a second on same ck & repeats[7]

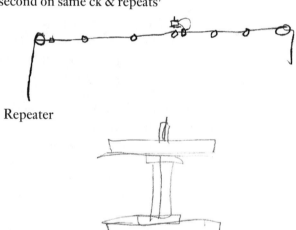

Repeater

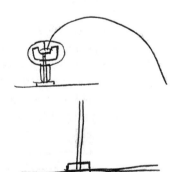

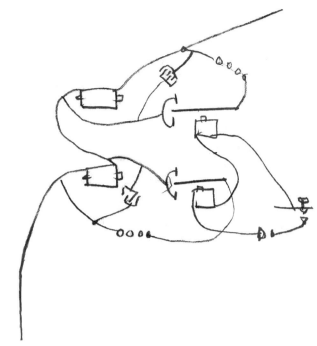

Dont improve but little—[c]

Tried 11 .48 primaries. 4 cells no better than the single 7 ohms   on weak ckt use light weight or no weight at all= if no weight at all use no condenser as it weakens it greatly=[c]

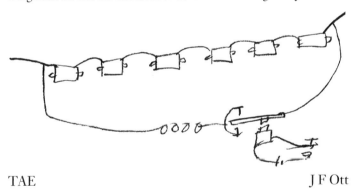

TAE   J F Ott

X, NjWOE, Lab., N-85-05-22:35 (*TAED* N308035). Document multiply signed and dated. [a]Drawing followed by "over" as page turn. [b]Followed by dividing mark and "over" as page turn. [c]Obscured overwritten text. [d]Circled. [e]Followed by dividing mark. [f]Canceled. [g]Illegible.

1. This document shows Edison attempting to optimize phonoplex performance by trying different sizes and combinations of induction coils. He evaluated each in combination with other variables, including battery strength, circuit resistance, and condenser capacity. He seems to have taken the clarity of his "writing"—the distinctive sounds produced in his receiver by his own messages passing to the outside line—

as the principal gauge of his adjustments, as he instructed Alfred Tate to do in a 19 November letter (Doc. 2870) following tests over a New York–Boston telephone line. See also Doc. 2869; Tate to TAE, 31 Oct. 1885, DF (*TAED* D8546ZBS).

2. Edison included improvements to the transmitter in a patent application that he started outlining in October (see Doc. 2859) but executed two days after making these tests. In the patent, he stated that although an induction coil in the main circuit would suffice, he would instead

> prefer to make this induction transmitter circuit simply a shunt around an electro-magnet in the line. This shunt includes a local battery and the transmitting-circuit controller [key], with points shunted by condenser, and by opening this shunt the magnet will discharge upon the line, producing a momentary and sharply-defined impulse, while the closing of the shunt will form a complete local circuit, including the battery and magnet, and the magnet will be energized, ready to be discharged at the next opening of the shunt. [U.S. Pat. 422,073]

3. Figure labels are "Res," "Phono—," "Earth," "NY," and "Prov." This test appears to have been made on a Bell Telephone Co. line (see Doc. 2870); in years past, Edison tested both telegraph and telephone instruments with the cooperation of Western Union over the company's telegraph lines. See, e.g., *TAEB* 3 chap. 1 introduction; Docs. 1194, 1204, 1223 n. 3, and 1247 n. 5.

4. To get the desired motion on the receiver weight or ring (see Doc. 2868 [headnote]), Edison tried to calibrate battery strength and the capacity of the condenser in the transmitter shunt circuit.

5. That is, two 17 ohm coils wired in parallel.

6. Edison probably whimsically adopted this name for the instrument because of a resemblance to pansies or violets commonly known as "johnny jump-ups," distinguished by large flowers on long, erect stems. *OED*, s.v. "Johnny-jump-up"; Wood and Bache 1885, s.v. "Viola tricolor."

7. The sequence of drawings immediately preceding and following this line of text illustrates Edison's ideas for a phonoplex repeater. He signed a patent application for such an instrument on 24 November, though it was not filed until 19 February 1886. Akin to the ordinary phonoplex receiver, the repeater consisted of a small loose weight resting on a metallic block on the horizontal diaphragm of a telephone receiver. Free to slide up and down on a spindle, it would bounce in response to the diaphragm's vibrations, alternately making and breaking the circuit of the repeating transmitter electrical connection through the metal block. Contact with the other side of the circuit came through a cup of mercury atop the spindle. A metal hoop extended from the top of the weight and dipped into the mercury, permitting the weight to move while maintaining its electrical connection. U.S. Pat. 422,074.

Edison described (but did not claim) how the instrument could alternatively be used to amplify induction signals in the same line, and his circuit sketch following this text became the basis for a patent drawing illustrating this point. The repeater or relay is at the center; the circles represent other induction transmitters, each with its own local battery.

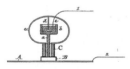

*Figure from Edison's U.S. Patent 422,074 for a phonoplex repeater. Shown in cross-section, the loose weight (**C**) connects through attached hoop (**c**) and wire (**d**) to the mercury cup (**b**). The weight rides on a spindle (**a**) and rests on a block (**B**) on the receiver diaphragm (**A**).*

With little fanfare or explanation, Edison made significant changes to the phonoplex receiver between the last week of October and about 12 November, when he signed a patent application related to it. He may have been motivated by complaints on 19 and 20 October from Alfred Tate, who had just set up the phonoplex for commercial trials on a line from Toronto, about a mechanical "grating sound" and, more urgently, the unintelligibility of messages amid heavy background induction noise during a rainstorm.[1] Edison seems to have had the ideas in mind by 22 October and, about 24 October, he sketched a new receiver and instructed his patent attorney to include it in a forthcoming application (see Doc. 2861). Edison tried the new design over a New York–Boston telephone line on 10 November (see Doc. 2867) and signed the related patent application two days later.[2] With the design largely settled, he sent a description of the receiver and its operation to Alfred Tate on 19 November (Doc. 2870).

Edison's innovation was an adaptation of an 1878 device for signaling the arrival of incoming telephone calls. It could be attached to the diaphragm of either an electromotograph or magnetic telephone in the phonoplex to produce a clear, loud sound. In practice, Edison used it only with the magnetic telephone (as shown in Doc. 2868), which was less sensitive than the electromotograph to outside induction.[3] The *Telegraphic Journal and Electrical Review* later provided an able description and illustration of the complete instrument:[4]

> Resting upon a wooden base is a hollow column of brass, within which is placed a magnet with the bobbins facing the diaphragm. At the lower end of the magnet there is a rack and pinion, by which it can be adjusted with reference to the diaphragm. To the centre of the latter there is attached a screw-threaded pin, with thumb nut and binder at the top. Encircling the pin loosely is a split-hardened steel ring which rests upon the diaphragm. When the latter is snapped by the attraction of the momentary current in the magnet, it throws the ring violently against the stop-nuts, and produces a sharp click, far louder than that from the ordinary sounder. The steel ring has a pin projecting from its side that passes between two prongs, which, while permitting free up and down motion, prevent the ring from turning and altering the sound. Over the top of the "phone" there is clamped a thin brass plate as a

protection for the projecting screw. ["The Edison 'Phono-plex' or 'Way-Duplex,'" *Teleg. J. and Elec. Rev.* 18 (7 May 1886): 414–15]

The telephone receiver was connected in the telegraph line. When Edison began to license the phonoplex system for commercial use in 1886, Bergmann & Company manufactured the components, presumably including the complete phonoplex receiver.[5]

Edison tried to calibrate phonoplex circuits to emphasize the difference in sound made by the metal ring (sometimes referred to as a weight) at the top and bottom of its travel. This difference corresponded to the upstroke and downstroke of the sending key and was an important distinction to operators, who were accustomed to it by conventional sounders.[6] He also tried to exploit the receiver's acoustic and mechanical properties by adapting it to a phonoplex repeater instrument and later to a phonoplex sextuplex system.[7]

Cutaway illustration in the Telegraphic Journal *of the phonoplex receiver, essentially a magnetic telephone with a loose weight resting on the diaphragm.*

1. Tate to TAE, 19–20 Oct. 1885, all DF (*TAED* D8546ZBF, D8546ZBG, D8546ZBG2).

2. U.S. Pat. 422,073.

3. See Doc. 2870. Samuel Insull apprised Alfred Tate on 10 November that Edison had "made extensive improvements in the apparatus which will enable him to dispense with the motograph altogether." Insull to Tate, 10 Nov. 1885, Lbk. 21:82 (*TAED* LB021082); U.S. Pat. 422,073.

4. An engraving of an instrument seemingly identical to the one in Doc. 2868 was also published, with a full description of the phonoplex system, in *Engineering* ("The Edison Phonoplex," 42 [22 Oct. 1886]: 411–13).

5. See, e.g., Samuel Insull to Bergmann & Co., 19 July, 3 and 23 Aug. 1886, LM 12:34, 92, 197 (*TAED* LM012034, LM012092, LM012197).

6. See Doc. 2870; cf. Doc. 2911.

7. See Docs. 2896 and 2902; U.S. Pat. 422,074.

–2868–

Production Model: Phonoplex Receiver[1]

[New York, c. 12 November 1885]

M (17.5 cm × 17.5 cm × 22.5 cm), Hummel.

1. See headnote above.

*Memorandum to
John Ott*

John—

Make an adjustment on one of the new instruments[1] so you can adjust the Cores to & from the diaphragm[2b]

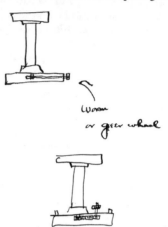

The steel diaphragm & ring is the best instrument. Mica is not good

Also instead of that thin spring in the repeater put a light lever—[3]

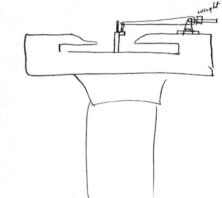

⟨This is finished   Nov 18, 85⟩[c]

J. F. Ott

AD, NjWOE, Lab., Unbound Notes and Drawings (1885) (*TAED* NS85ADC14). Letterhead of Thomas A. Edison; document multiply signed and dated. [a]"NEW YORK," and "188" preprinted; remainder of date written by John Ott. [b]Drawing followed by dividing mark. [c]Marginalia written by John Ott.

1. Edison referred to the new phonoplex receiver; see Doc. 2868 (headnote).

2. Figure label is "worm or geer wheel."

3. Figure label is "weight." Regarding the repeater see Doc. 2867 n. 7.

[New York,] 19th Nov [188]5

Dear Sir,

I find it impossible for me to leave New York just now as I am very busy in fact I hardly think I will be able to get to Toronto at all.[1]

I shall send you tomorrow three Phone Receivers. There is a pin in the centre of the diaphragm upon which upon which is a weight.[2] When a Strong move comes the weight jumps up and in coming back makes the down stroke: the next wave makes the up stroke. To make the up stroke weaker, I put in a small resistance on the down point so that when the lever leaves the down [stro?]ᵃ point to make the up stroke the battery is weaker owing to the four little spools [inserted?]ᵇ in the branch running to the down point. The amount of condenser is best ascertained when you get your own writing strongest.[3] The phone is put right in the [line?]ᵇ itself. There is an adjusting screw on the bottom whereby you can adjust the magnet to & from the diaphragm. This screw works left handed. Where the phone current is very weak a very light weight is used; where it is stronger heavier weight can be used. We send you some different weights for experimental purposes. I find this magneto phone while not so sensitive as the motograph works very satisfactory on a line One Hundred and twenty five miles long.ᶜ I have now tested it four nights between New York & Boston on a copper wire grounded at each end with induction so strong that they are unable to work it (the line) with a telephone when for talking when grounded but have to use metalic circuit[4] to enable them to talk. This circuit is 300 miles long but its equivalent in No 8 Iron wire is 120 miles   It works perfectly satisfactorily over this distance therefore do not attempt to phonoplex a line longer than from 100 to 125 miles with this new apparatus. You will find that it is not near so sensitive to outside induction as was the motograph.[5] Be careful when you work to have the magnets as close to the diaphragm as they will stand without sticking. This you can readily ascertain by adjusting while the distant man is writing. Use four cells of good carbon battery on the phone centre. Be careful that every relay in the line has got a Condenser on it: should there be a relay at some test office unbeknown to you this would almost wipe out the Phone on account of its self discharge.[6] Be sure that every magnet has a condenser round it on the line and be sure that they do not plug in an extra instrument at the terminal Station. If you desire to try this on a line from Toronto to Montreal and if they will work

Morse from Montreal to Toronto while you work phonoplex from Toronto one hundred miles out do as follows. Shunt all the Keys and relays of the Morse apparatus at <u>all</u> the stations from Toronto one hundred miles out but never mind shunting beyond that. Now at the last station at the end of the hundred miles put in an extra relay beyond the phonoplex & Morse apparatus. Then take two condensers: put the the end of one condenser on the side of the magnet pointing towards Toronto and the other end of that condenser to the earth: now at the other side of the magnet pointing toward Montreal put another condenser with the other end to the ground as shown in diagram No 2. Put 48 sheets in each of these condensers. You want to use the heaviest weight on the phone that you possibly can to get the writing because the heavier the weight which will still do the work the greater the Induction impulse must be to lift it. Thus you see you can work the phone on this Montreal wire without interferring with the through traffic— However you had better try it grounded first   Yours truly

Thos A Edison   I[nsull]

L (letterpress copy), NjWOE, Lbk. 21:122 (*TAED* LB021122). Written by Samuel Insull. ᵃCanceled. ᵇIllegible. ᶜ"One hundred . . . long" multiply underlined and written in large, emphatic script.

1. Samuel Insull had suggested to Tate on 10 November that Edison, after finishing improvements on phonoplex instruments, "will probably go to Toronto himself with Mr Gilliland. They will probably leave here in about a week." When Tate mentioned this possibility to a businessman friend, that person promptly began to organize a dinner in Edison's honor at the Toronto Club. Insull remained noncommital about Edison's plans. Insull to Tate, 10 and 17 Nov. 1885, Lbk. 21:82, 109 (*TAED* LB021082, LB021109); Tate to Insull, 17 Nov. 1885, DF (*TAED* D8546ZCF).

2. See Doc. 2868 (headnote).

3. Though the audibility of outgoing phonoplex messages was advantageous in this test setting, Edison noted in a patent application filed in July 1886 (about the time he was starting to sign commercial license agreements for the phonoplex) that he had "found by practice that the diaphragm receivers or sounders respond so loudly to the signals transmitted from their own sets that the noise is confusing to operators at adjoining tables, and hence some provision for stopping or reducing the noise becomes desirable." The application pertained to circuit arrangements to isolate the receiver; it was rejected by the Patent Office and eventually abandoned by Edison. Patent Application Case 674, PS (*TAED* PT016AAA).

4. In telegraph or telephone practice, a "metallic circuit" used a wire or other conductor for the return path. An "earth circuit," in contrast, was grounded at each end. *OED*, s.v. "metallic" 7.

5. Tate enthusiastically endorsed the new receiver, which "enables

one to at once locate the line which interferes because any wire which is loud enough to disturb the phone comes out <u>distinctly</u> in it so the writing can be <u>read</u>— There is no 'frying pan' induction or 'morse hash' to drown the writing of the phone." Tate to TAE, 17 Dec. 1885, DF (*TAED* D8546ZDG).

6. Edison wanted the condenser to damp out "the extra vibrations caused by the rebound of the contacts" in the relays, as explained in Doc. 2849 n. 3.

–2871–

Jane Miller to Mina Miller

N.Y. Nov. 21. 1885

My Darling.

Your note came this morning am glad you have decided to go home. It is still my opinion to wait a few months at least before marrying. I would not promise to be married in Feb. before going home any way. Seems to me Mr. Edison ought to wait for you a year or six months at least.[1] It will be ~~the~~ best for us to go to the hotel and I think we had better stop at the Victoria.[2] Let me know at what hour you will arrive in N.Y. so that I can meet you. You can find out at the station or Mr. Edison can tell you. You know my dear I love you truly. And I want you always to be happy. And I want your married life to be most happy. I wish too I were with you. Am going down street now.

Will send this note by Mr. E.

With a heart full of love for you— I am, Affectionately

Jennie[3]

ALS, NjWOE, FR (*TAED* FM001AAD).

1. Jane Miller returned to this thought in a December letter to Mina. After recounting a social meeting with George Vincent (to whom Mina may have been previously engaged) and his mother, Jennie counseled Mina not to

> worry about the Vincent family any more. . . . I think you have not been so unkind to any of them so as to make you feel unhappy. If you truly love Mr. Edison I cannot help but feel if you wait one year will know so much better just how you do love him— You know one so often makes a mistake and you are yet so young. You know I like him and I have no doubt but what you love him but whether it is because you love him for himself or because he loves you. [Jane Miller to Mina Miller, 17 Dec. 1885, FR (*TAED* FM001AAE)]

In this letter, Jennie also related a conversation in which Lillian Gilliland expressed misgivings that she and her husband had "influenced" Mina and Edison. Jane reportedly agreed that this was true but also "quite natural."

2. Occupying an entire block of Twenty-seventh St. between Broadway and Fifth Ave., the Victoria Hotel was built about 1868 as a family residential hotel, reportedly the first of its character in New York. A favorite with British and European travelers, after a management change in 1882 the Victoria also cultivated a reputation for hosting American politicians, notably Grover Cleveland. "Victoria Hotel To Be Torn Down," *NYT,* 3 June 1911, 8.

3. Jane Eliza Miller (1855–1898), known to her family as Jennie, was Mina's older sister by ten years and reportedly a maternal figure toward her younger siblings. Unmarried until 1892 (when she married Richard Pratt Marvin, Jr.), Jennie was well traveled and had spent about a month in and around New York in early 1884. She sent both this document and her 17 December letter (see note 1) from the same address at 137 E. Fifty-seventh St. Jeffrey 2008, 171; Hendrick 1925, 84–85; Jane Miller to Mina Miller, 1 Jan. and 6 Feb. 1884, both FR (*TAED* FM001AAA, FM001AAB).

–2872–

Notebook Entry:
Energy Research

[New York,] Dec 8 1885

This book is to contain ideas as to the discovery of a new mode of motion or energy and also to the conversion of heat directly into electricity[1]

Thos A Edison

X, NjWOE, Lab., N-85-12-08:1 (*TAED* N313001).

1. This document marks Edison's planned start of a systematic exploration of these subjects, the foundation for which he had begun preparing some seven or eight months earlier (see Doc. 2804 [headnote]); account records associated with "New Force Experiments" start in mid-January 1886 (Ledger #5:587, Accts. [*TAED* AB003 image 290]). In his 1878 interview with the *New York Daily Graphic* concerning prior experiments with etheric force, Edison listed reasons for thinking that the phenomenon was an unrecognized form of energy; some of those reasons could have served as rationales for the experiments he was now planning:

1. It does not respond to any of the physical tests of electricity, except the spark.
2. It produces no perceptible or demonstrable physiological effects like electricity, save on the frog.
3. It gives no evidence of polarity.
4. It passes through the air and other resistances by large surface at the terminals, even when the apparatus is not insulated.
5. When connected with the earth or walls of a room it can yet be drawn off from the conductor. ["Two Hours at Menlo Park," *New York Daily Graphic,* 28 Dec. 1878, Cat. 1241:1091, Batchelor (*TAED* MBSB21091)]

Notebook Entry:
Energy Research[1]

X all the different metals insulated in form of wire the ends connected to every form of XYZ inst made for detecting a new form of energy— S S′ Copper wire spools. ⟨Res of S each spool from 1 to 5 ohms   for each metal a separate board—⟩ ⟨12 Bases   5 × 4 in ½   52 Binding posts⟩[2a]

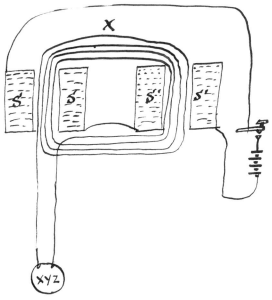

C is various metals   X Hard rubber & other bushings   ⟨1 Base 4 × 5 = ½   4 Binding posts⟩[3a]

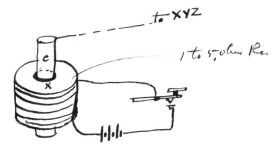

Use etheriscope[4] as an XYZ detector.

Roll Sulphur with every metal in every solution, every metal with another in every solution, both insulating and non-insulating fluids.

X a mirror C a pan for laying every kind of matter. Bifilar suspension— N coil of glass tubing filled with various solutions conducting and non-condctg   G G Electrodes of various materials condctg & non— ⟨Bifilar suspension as long as can get it.⟩[5]

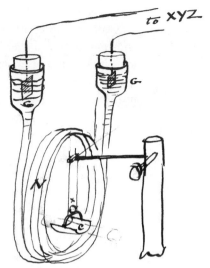

⟨Thermo pile[6]—german silver & Copper—⟩ Conduction of radiant heat through hard rubber   X is ½ stick h[ard] rubber one end a thermo[b] couple with mirror gal other end disc rubber. The Stick & Thermo covered with felt etc prevent conduction to air. ⟨Bases $16 \times 16 = \frac{5}{8}$⟩[7a]

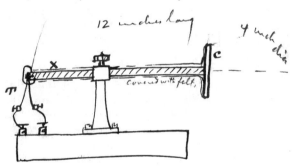

Lead and Lead, one plate peroxidized in every solution. Ditto Zinc & Carbon also peroxidized by Manganeese in every solution.[8]

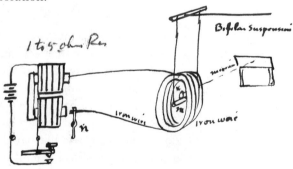

N for making and b[rea]k[in]g contact with iron core   M cores of all materials ⟨Base $10 \times 5 = \frac{5}{8}$⟩[c]

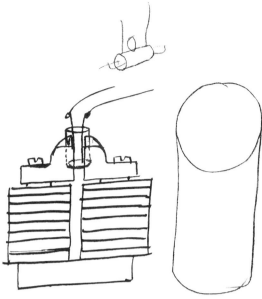

Try all the metals in and out of a magnetic field, also open and close the magnetic field to get an independent effect— ⟨Base $8 \times 10 = 1$ in⟩[a]

Passing a beam of heat or light through a liquid causes absorption of energy, hence devise conditions, that instead of heat, it goes to conductors.[9]

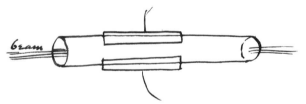

glass ends, Rubber tube, also jointed metallic tubes, also glass tubes wire sealed in side and tube blackened. Fill tubes various Liquids

Hard rubber disc $\frac{1}{16}$ narrow chamber filled with liquid, small specific heat, index tube, of glass   Heat thrown on disk moves column liquid— ⟨Base $5 \times 4 = \frac{1}{2}$⟩[a]

Expansion mirror indicator.[10]

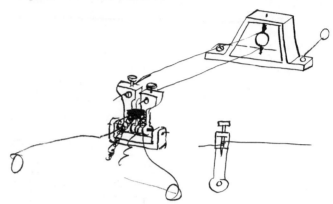

Solutions, ie[d] disolving substances Liberate or absorb heat, hence use all solutions with all metals and disolve various things in the solutions—

<u>XYZ</u>[11]   ⟨Base $6 \times 8 = \frac{5}{8}$⟩[a]

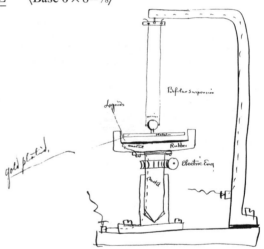

Different Liquids   Platina Electrodes [12]

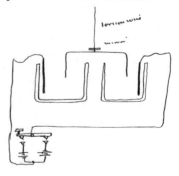

$\underline{XYZ}$[13e] $\langle 5 \times 6 = \frac{1}{2} \rangle$[a]

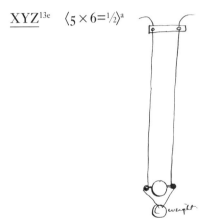

Cord moistened with various solutions, also animal strings—

Two different liquids which combine ie[d] acid & alkali; also other liquids with react on each other & exchange places.[14]

Act of re-combination gives energy hence try all solutions and all metals with every XYZ indicating instrument. ⟨Base 5 × 6½⟩[a]

XYZ Indicator. All connection wires same material as spool is wound with—[15]

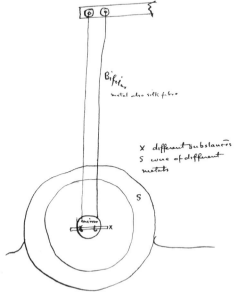

spools of every dif metal insulated.

Get Tyndals work on Thermics—[16] make Light and heat filters.[17]

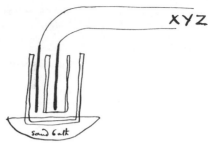

Carbon & Carbon—also various metals like and unlike—water in one & Ethylic Iodide ie[d] best & poorest conductor heat.

One part of cell water acidulated porous cell,[18] good conducting liquid having lowest specific heat. hence by heating the water will for long time, on account of its high specific heat be the coolest & we should get a big thermo current   Carbon electrodes to be used. Cooling gives also a Current

Best method of ascertaining new source of energy without use of unknown indicating apparatus is to take for instance plate of Zinc and a plate of every known other substance, and immerse these in every kind of Liquid, first carefully weighing each plate, then short circuit them and leave them in the liquid for one month or less. if there has been any action due to association the scales will show—a pair not connected ie[d] short cktd should be put in the same solution.

Suppose we take zinc as one plate and Hard rubber, Sulphur, Lead tin Zinc & all the metals, Carbon Sulphides, phosphides Selenides Arsenides, wood glass, glue, Dextrin, slate, Gutta Percha mica, peroxides, paper, moulded sticks of various material, parafine, Rosin, asphaltum, moulded oxides.

Place these in solutions of salts where galvanometer shows no electricity. Use Bisulphide Carbon, Benzine, Carbolic acid, & kinderd solutions.

Then substitute for the Zinc some other material & so on this will require several thousand Experiments.

Supposing the Conducting wire of a submarine cable was coated $1/32$ of an inch thick with a selenide phosphide or amorphous phos that was a true conductor but very high, so the restance between the wire for one mile of cable if the exterior was coated with copper over the selenide was 500 ohms would not the electrostatic capacity of the cable be enormously reduced when the dialectric was in contact with the selenide. It would have to charge the dialectric through a high resistance.[19]

It seems to me if an instrument such as an Electromotograph Mirror inst was used on the cable and a battery of iron & copper in water or even less electromotive force that the cable would not be charged on signalling as the diaelectric is already charged to the potential of Copper & iron. it is in fact a single cell of Copper and iron, the dialectric being charged accordingly   consequently if you signal with an emf equal to this it shouldnt charge the cable— I have constructed a condenser of iron & copper & the charge if different according to the direction of the Current. This explains the discrepancy noticed by Cable Electricians between a[b] P & N charging current=[f]

TAE   J. F. Ott

X, NjWOE, Lab., N-85-12-08:3 (*TAED* N313003). Document multiply signed and dated. [a]Marginalia written by John Ott. [b]Obscured overwritten text. [c]Marginalia and drawing made by John Ott. [d]Circled. [e]Multiply underlined. [f]Followed by dividing mark.

1. See Doc. 2804 (headnote). At some time after Edison wrote this document, he and John Ott went through it making pencil notations (transcribed as marginalia) specifying dimensions and other details of the apparatus that Ott was to construct.

2. Figure labels are "X," "S′," "S′," "S″," "S″," and "XYZ."

3. Figure labels are "to XYZ," "C," "X," and "1 to 5 ohms Res."

4. Edison referred to a darkened box he devised in 1875 for visually observing the sparks he attributed to an "etheric force." His spelling of the name varied. See Docs. 668, 670, 764, 766, 840, and 864.

5. Figure labels are "to XYZ," "G," "G," "N," "X," and "C."

6. A thermopile is an acutely sensitive device for generating weak thermoelectric currents from small temperature differentials between pieces of different metals (traditionally, antimony and bismuth). Edison had tried the device in a telephone receiver in 1877 (see Doc. 1062). Smith 1883, s.v. "Thermopile"; Tyndall 1977 [1895], 277–79.

7. Figure labels are: "T," "X," "C," "⟨12 inches long,⟩" "4 inch dia," and "covered with felt."

8. Figure labels are "⟨1 to 5 ohms Res,⟩" "Bifilar Suspension," "mirror," "X," "M," "iron wire," "iron wire," and "N."

9. Figure label is "beam."

10. It is not clear how Edison intended this instrument to work as an XYZ detector. Presumably, the wires at lower left would convey energy that would move the columns at the center, perhaps by thermal expansion. Each column appears to be connected by a filament to a mirror whose slightest motion would deflect a beam of light.

11. Figure labels are "Bifilar suspension," "Liquids," "mirror," "gold plated," "metal—," "metal," "Rubber," "Electric Eng[ine]," and "metal."

12. Figure labels are "torsion wire" and "mirror."

13. Figure label is "weight."

14. Figure labels above are "zinc" and "zinc."

15. Figure label is "Bifilar   metal also silk fibre," "X different substances   S wire of different metals," "S," "mirror," and "X."

16. British physicist John Tyndall (1820–1893) succeeded Michael Faraday as director of the Royal Institution in 1867. He devoted much of his early career to studying the effects of radiant heat on gases. Before 1860 he published *Heat: A Mode of Motion* (Tyndall 1893 [1883]), based on a series of Royal Institution lectures, in which he propounded a mechanical theory of heat. It went through multiple editions under both the original and a subsequent title (Tyndall 1863). *Oxford DNB*, s.v. "Tyndall, John"; Doc. 1662 n. 4.

17. Figure label is "sand bath."

18. Figure label above is "sand bath."

19. In this paragraph and the next, Edison implicitly followed Michael Faraday's explanation of the "retardation" in submarine cables that caused the slowing and blurring of signals sent through them. Faraday interpreted the phenomenon as evidence of an electromagnetic field, arguing that an interval of building up a "strain" in the field around the conductor must precede the flow of current. As historian Bruce Hunt explains:

> a cable, consisting as it did of a central wire separated from an outer conductor of water or damp soil by only a thin layer of gutta-percha insulation, was in effect a huge condenser or Leyden jar; its capacitance was enormous. When a battery was touched to the central wire of a cable, it took an appreciable time for the electrostatic strain to spread laterally through the gutta-percha along the entire length of the cable, so that the rise of the current was retarded. When the battery was disconnected, it took additional time for the strain to relax, and the consequent discharge along the cable greatly prolonged the fall of the current . . . [Hunt 1991, 62–63]

Oliver Heaviside further explored the phenomenon to advance his own interpretations of field theory, including the January 1885 *Electrician* article that may have influenced Edison's thinking about electrostatic induction (Hunt 1991, 66–67; Doc. 2804 [headnote]; Heaviside 1885).

–2874–

Samuel Insull to Alfred Tate

[New York,] Dec. 9 1885.

Dear Sir:—

I confirmed receipt yesterday of your telegram[1] reporting successful working from Toronto to Port Hope[2] to which Mr. Edison instructed me to reply as follows:

"Put up apparatus at more distant stations continued until no results obtained. Have system used for actual service if possible. Report very fully."

You will understand from the above that Mr. Edison desires

to see how far beyond Port Hope you may work successfully. From his experiments here over the Boston line concerning which he wrote you,[3] you will understand that he considers that you ought to be able to work right through to the last "way station" on the line to Montreal. He desires that you should push this work through as quickly as possible.[4] You can well understand that the delay in successfully operating the apparatus is very detrimental to the ~~apparatus~~ invention and it must materially affect the minds of telegraphic people here when entering upon a consideration of the value of the invention. Furthermore, our success here, means the rapid pushing of the invention in London which is just as important territory as is this country. A last and by no means the least important consideration is that the expenses are running up quite heavily and my impression is that Mr. Edison's mind is beginning to be agitated on this point. We are all extremely glad to hear of your success and hope that it may continue. You want to be careful to look for any funny business on the part of these friends (?) who are supposed to render you assistance.[5] Very truly yours,

S. Insull   Private Sec'y

TLS (carbon copy), NjWOE, Lbk. 21:162D (*TAED* LB021162D).

1. Although neither Tate's telegraph nor Insull's confirmation has been found, Tate wrote at length on 9 December about his test of new phonoplex receivers in accordance with Edison's directives in Doc. 2870:

> I wired you last night stating that new instrument worked splendidly between Pt Hope and Toronto and the result of my work this morning confirms the announcement . . .
>
> This morning opened with a heavy rain and when I got to the Telegraph office about nine o'clock I found the Morse lines working miserably.
>
> At 10.30, the rain still continuing, I called Pt Hope up on the Phone and from that time until 12 oclock was in uninterrupted communication with him. I sent him a lot of regular business which was received just as well as on Morse instruments and the opr stated that it came almost as loud as any sounder in his office.

Tate reported equally good reception on his end, with no induction except from an adjacent wire from Port Hope. Tate to TAE, 9 Dec. 1885, DF (*TAED* D8546ZCT).

2. Port Hope, Ont., a Canadian port of entry on Lake Ontario, was situated on the Grand Trunk Railway and another rail line east-northeast of Toronto. Tate arranged on 22 October to ground one of the Toronto-Ottawa-Montreal telegraph lines at Port Hope in order to make ongoing tests over the sixty-three miles from Toronto. Crossby 1881, s.v. "Port Hope"; Tate to TAE, 22 Oct. 1885, DF (*TAED* D8546ZBM).

3. See Doc. 2870.

4. The next day, Tate planned to install the new induction coils with a single primary winding recently received from Edison. He then worked the phonoplex between Toronto and Port Hope on 11 December for Harvey Dwight, who reported "very fair" results to Erastus Wiman. Tate extended the test circuit eastward to Belleville, Ont., on 16 December and the next day was working "splendidly" with the new coils over 113 miles of wire between there and Toronto. (Insull promptly relayed this news to Edison in Akron.) Tate planned to continue another 55 miles to Kingston and then on to Prescott, Ont., 225 miles from Toronto. Tate to TAE, 10, 11, 13, 16, 17, 28 Dec. 1885, all DF (*TAED* D8546ZCV, D8546ZCX, D8546ZCY, D8546ZDA, D8546ZDD, D8546ZDE, D8546ZDG, D8546ZDN); Insull to TAE, 17 Dec. 1885, Lbk. 21:183 (*TAED* LB021183).

5. Tate had been hoping for assistance with his tests but was wary of engaging with potential rivals. His most immediate concern was about Benjamin Birdwood Toye, manager of the Great North Western's Toronto office, who had been assigned to help. Toye was, according to Tate, "somewhat of an inventor (?)" who had patented a telegraphic repeater and claimed to have experimented several years ago on a system like the phonoplex. After coming to suspect Toye of subtly undermining the trials, Tate diplomatically avoided him. Tate to TAE, 20 Oct. and 6 Nov. 1885; Tate to Insull, 23 Nov. 1885; all DF (*TAED* D8546ZBG2, D8546ZBY, D8546ZCN); Insull to Tate, 10 Nov. 1885, Lbk. 21:82 (*TAED* LB021082); "New Patents," *Telegrapher* 7 (17 Dec. 1870): 131.

Two other clouds on the horizon were George Black of the Great North Western's office in Hamilton, Ont. (and subsequently its manager), and Abner Rosebrugh, a Toronto physician. Black, esteemed by Erastus Wiman as "one of our best operators and most expert electricians in Canada," had complained to Wiman that the phonoplex infringed on patents issued and pending to himself and Rosebrugh for a system of condensers that permitted telephone communication over Morse telegraph wires. Wiman alerted Edison (as did Tate), whose reply (evidently not copied into a letterpress book), came as a "relief" to him and seems to have closed matters with respect to Black (Black to Wiman, 17 Oct. 1885; Wiman to TAE, 21 and 30 Oct. 1885; Tate to TAE, 22 Oct. 1885; all DF [*TAED* D8546ZBI, D8546ZBH, D8546ZBR, D8546ZBM]; "Canadian Electrical Association," *Western Electrician* 25 [8 July 1899]: 23; U.S. Pat. 212,433). Rosebrugh, however, who had related patents in his own name and with Thomas Rosebrugh, also of Toronto, was rumored to be a "shark" and several times called on Tate, who (in his words) kept his "mouth tightly shut." Rosebrugh, too, may have had a connection to Wiman; he later made a detailed report on the February 1886 demonstrations of Edison's railway telegraph system, with which Wiman was closely involved (and which Rosebrugh argued depended—as did the phonoplex—on the condenser circuits devised by Black and himself). Rosebrugh's attorneys offered a licensing arrangement in early November, as he himself did a few weeks later, which Edison declined (Lazier, Dingwall & Monck to TAE, 4 Nov. 1885; Abner Rosebrugh to TAE, 17 Nov. 1885; Tate to TAE, 6 and 9 Nov. 1885; all DF [*TAED* D8546ZBV, D8546ZCH, D8546ZBY, D8546ZCB];

TAE to Rosebrugh, 21 Nov. 1885, Lbk. 21:132 [*TAED* LB021132];
Rosebrugh 1886; Preece 1886 [Silvanus Thompson discussion], 296;
Guillemin 1891, 741–42).

–2875–

Notebook Entry:
Miscellaneous[1]

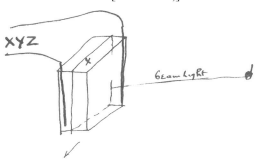

glass cell.[2] x transparent porous septum say parchmentized
paper   Each liquid has different regrangibility. passing beam
concentrated arc Light rapidly across cell make XYZ   may
be standing still   ⟨Base 4 × 5½⟩[a]

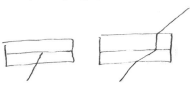

Proposition. iron alters magnetism by heat, nickel more so;
hence surround iron core by wire; balance in bridge[3] & use
current to magnetize core, then by a beam or hot copper bar
heat end iron cores suddenly= if we get a jump on the mirror, [we?][b] ok= In practice hot & cold water could alternately
be passed through steel magnets covered with Coils & with
Commutators get a constant current.

Thermic reverser phone.[4c]

[A][5]

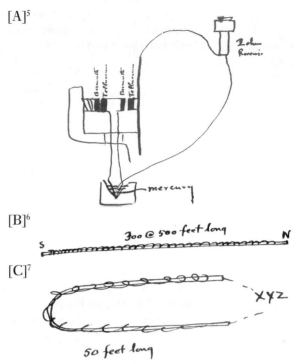

[B][6]

[C][7]

No 4   Iron wire wound uniformly with say 22 wire—& the iron wire magnetized—

[D][8]

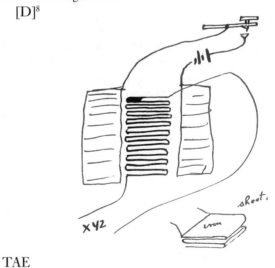

TAE   J. F. Ott

X, NjWOE, Lab., N-85-12-08:49 (*TAED* N313049). Document multiply signed and dated. [a]Marginalia written by John Ott. [b]Illegible. [c]Followed by dividing mark.

1. See Doc. 2804 (headnote) regarding Edison's investigation into forms of energy. John Ott later made pencil notations of dimensions and other details, as he did in Doc. 2873.

2. Figure labels are "XYZ," "x," and "beam Light."

3. That is, a Wheatstone bridge.

4. Figure label is "Hg." Cf. Doc. 1062 n. 2.

5. Figure labels are "1 ohm Receiver," "Bismuth," "Tellurium," "Bismuth," "Tellurium," and "mercury."

6. Figure label is "300 @ 500 feet long."

7. Figure label is "50 feet long."

8. Figure labels are "XYZ," "iron," and "sheet."

–2876–

Notebook Entry:
Railway Telegraphy

[New York,] Dec 12 85[a]

Coils   Primaries— ⟨Switch to throw 2 3 4 5 & 6 cells quickly—⟩

1.[b] 7 ohm coil.

2 cells   3 cells   4 5[c] & 6 cells—[d]

1[b] 3 ohm coil

2 3 4 5 & 6 cells[d]

1[b] 2 ohm coil

Same cells.

1[b] 1 ohm coil

1[b] .40 coil wound so as to be same size as the 7 ohm coil in diameter & length

1[b] 14 ohm coil[d]

Reg 7 ohm length

2.[e] 3 ohm coils   multiple arc[d]

2[b] 7 ohm multiple arc

2[b] 14 "   "

all tried with the 2 3 4 5 & 6 cells—[d]

2[b]— 2 ohm in series

2[b]   3 "   "

2[b]   7 "   "

2[b] 14 "   "

~~Secondaries~~ Condenser across the vibrator to be varied [~~from 8?~~][f] first 8 then 16 32. 48. 72. 100. Each experiment.[d]

~~after trying these~~ Secondaries. Regular telephone coil, varying battery & varying condenser[g] then, The three large coils, then, The several varieties of small coils with low resistance secondaries & also high resistance primaries—[d]

Try all these coils, etc at the station on[c] Table having two or 3[c] sets set up—and while experiments are being tried on one set another is being got ready   Thus with 3 sets or even two the Extpts can be made Continuous=

During the Experiments the Condenser devices etc for getting rid of induction & also for connecting to the line must remain undisturbed[d]

[~~Thoro?~~][f] A <u>thorough</u> understanding must be had with the train man so he records the Experimenting.[c] another man should be with him & have a telephone & he should record his impressions independent of the other man—

The nature of every change made at the sending station should be teleghd to train so the two persons each of whom have a telephone to Ear Can record them & giving them time to record their impressions as to loudness etc=

One man should watch the reed at the sending station [& ~~be?~~][f] keeping it at same tone, which he can do by adjusting & using [~~another f?~~][f] a tuning fork—

After the best form of Coil & condenser & battery is obtained—another trip should be taken using these forms only and variations in the reed & condensers etc gone through—to ascertain what improvement can be made in this line—

Each[c] experiment of coils etc should be numbered & these numbers teleghed so all these persons involved record the same experiment.—

After all these experiments & the best form of coils batteries condensers note in reed— connection to line— etc you might make ½ doz receiving telephones of different resistances & test them[1]

J. F. Ott   M N Force

X, NjWOE, Lab., 85-10-01:23 (*TAED* N310023). Document multiply signed and dated; miscellaneous calculations not transcribed. [a]Date written by John Ott. [b]Preceded by checkmark in left margin. [c]Obscured overwritten text. [d]Followed by dividing mark. [e]Followed by checkmark. [f]Canceled. [g]"varying . . . condenser" interlined above.

1. John Ott continued Edison's notes in this book on 7 January, when he listed and described two dozen "Induction Coils to be tried." A week later, he began to make and test various telephone receivers. Related work continued intermittently through 30 January, when he delivered "one complete cabinet" to the Railway Telegraph and Telephone Co., likely for use in the demonstration described in Doc. 2890 (N-85-10-01:41–61, Lab. [*TAED* N310041, N310057, N310059, N310059A, N310061]). In December, Hugh De Coursey Hamilton made a number of comparative receiver tests, mostly involving the diaphragm (N-85-10-03:49–55, Lab. [*TAED* N311049, N311051, N311053]).

From Samuel Insull

My Dear Edison

Your telegram asking for municipal lamps[1] came to hand yesterday & I immediately communicated with the Lamp Co who expressed the lamps last night.[2]

I enclose you two letters from Tate.[3] I have kept copies of them in case you may want to refer to them for reply on your return. I have written Tate that you are away.[4]

I was sorry to miss you Sunday. I wanted to see you about the machine works consolidation.[5] It is necessary that we should get the Illuminating Stock [($13 000)?][a] off the books and also the Shafting Co[6] Stock owned by the machine works ($8,800) total [($21 300)?][a] so Batch and myself agreed that a dividend should be declared of 10% ($22 000) to clear these two amounts off.[7] You will remember that I told you the Machine Works assets figured up [$251 000?].[8a] Of course this dividend will reduce it to [$229,000?].[a] Then the profits for this quarter should be at the least $12,000 so that would bring it back to $241,000. Then I think that when we come to value the Stock of material on hand we can increase it a further $10,000. Our effort will be to keep the assets up to $250,000 & yet distribute between you & Batch the Illuminating Co & Shafting Cos Stock.

I am arranging with Batchelor for the Machine Works to buy from you all that Stuff that came in from menlo Park.[9] It will amount up to quite a figure I think. Of course I cannot get cash for it but I can get stock in the new organization. It is necessary that all these things be fixed before Batch & myself go to see D. M. & Co.[10] about Tube Co.[11]

After talking it offer with Batch we have about come to the conclusion that the best way will be to organize a Co. of $750,000—issue $600,000 to buy the three concerns (machine works, Tube Co & Shafting Co) & leave $150,000 in the Treasury. The investment in the three concerns today amounts roughly to [$320,000?][a] of which the Stock owned by you represents about $235,880 so that of the $600,000 of the new concern you would get about $442,300 of the $600,00 issued.[12]

This I will show you is a first class deal for you. To day you have you have the following interests in the three concerns

Machine works	80 per cent
Tube Company	36 " "
Shafting Company	68 " "

In the new organization your total interest will be 75 per cent. In other words you will give away 7% of your interest in the Machine Works business in order to get 37% more interest in the Tube Company business & 5 per cent more interest in the Shafting Companys business. I call this a good deal for you of course always assuming that the Tube Co & Shafting Cos business will be good from now on. Yours Sincerely

Saml Insull

ALS (letterpress copy), NjWOE, Lbk. 21:175 (*TAED* LB021175). [a]Illegible.

1. This refers to the lamps used in Edison's municipal street lighting system. Little is known about the development of this system, for which Edison filed patent applications in October 1884 (U.S. Pat. 328,573; 328,574; 328,575). He likely applied about the same time for at least one other patent (on cutouts for the lamp) that never issued; an undated drawing shows several electromechanical shunt cutouts like those described in Jenks 1887 (p. 10) as the first form of cutout (Undated Notes and Drawings [c. 1882–1886], Lab. [*TAED* NSUN08]). The first trial installation was put into service at Lockport, N.Y., in March 1885 and a second plant began operating in Portland, Maine, in the fall. At the time Insull wrote this letter, a third plant was being installed in Lawrence, Mass.; it began operating on 24 December, and a fourth plant started at Jacksonville, Fla., in January (Jenks 1887, 14–15). These plants likely provided the spur to renewed work on the system by Edison and Batchelor in the fall of 1885, including work on a high-voltage dynamo (Cat. 1235:271 [item 36]; Batchelor Journal Cat. 1305:9, 15–18; both Batchelor [*TAED* MBN012270, MBN013009]; Cat. 1151, Lab. [*TAED* NM020AAQ images 49–51]; Voucher [Laboratory] nos. 608, 617, 621 [1885]).

The requirements of the municipal (street lighting) system, with lamps strung over sizeable distances and an essentially constant electrical load, differed from those of a distribution network for a dense area of residences and businesses, whose demand (and consequent losses in the conductors) would peak for a brief period each day. To reduce the amount of copper needed in the lengthy conductors, Edison planned to use small currents at a relatively high voltage, eventually finding in practice that ordinary wire could carry 4 amperes at 1,200 volts as far as 10 miles with acceptable losses. Recent calculations at the Edison Machine Works predicted that 18 pounds of copper wire in the municipal system would carry current for the same number of lamps—with less loss—than 1,800 pounds of conductors in the conventional three-wire arrangement (Jenks 1887, 8; Memorandum, 12 Oct. 1885, Unbound Documents [1885], Batchelor [*TAED* MB172]).

In order to maintain a sufficient resistance over the entire circuit to employ such low current and high voltage, Edison used low resistance lamps connected in series. (Aggregate resistance of elements in series is the sum of every individual resistance. In a parallel circuit it is related to the sum of the inverse of every individual resistance; it falls as the number of resistors rises. Edison reportedly had built a small network of low-resistance lamps wired in series to demonstrate the inadequacy of that principle for interior lighting in 1879; the following year, British inventor Joseph Swan put forth his own version of a low-current, series-wired system for indoor use; see Docs. 1705 n. 1 and 2022 n. 6). The Edison municipal lamp had a thick filament, and lamps of various intensities (from 16 to 50 candlepower) could be produced by using filaments with the same cross-section but different length. Placing the lamps in series called for an automatic switch (termed a "cutout") to maintain the circuit around any one that failed (see, for example, Docs. 2916 and 2984). The cutout and the lamp's electrical characteristics meant that the municipal system could also operate with a mix of incandescent and arc lights in each circuit ("The Edison System of Municipal Lighting," *Electrical World* 9 [12 Feb. 1887]: 78; McClure 1889, 186–90; Whipple 1888, 143–44); Edison Electric Light Co. brochure, n.d. [1886?], p. 4, PPC (*TAED* CA001D).

2. Edison left New York for Akron on the evening of 13 December. The editors have not found his telegram but Insull instructed the Edison Lamp Co. on 15 December that "Edison wires have shipped quick by Express to Aultman Miller and Company Akron Ohio ten fifty Volt thirty two Candle municipal Lamps also ten same but lowest Volts made   must be sure they are same amperes   hurry." Insull to Erastus Wiman, 14 Dec. 1885; Insull to Edison Lamp Co., 15 Dec. 1885; both Lbk. 21:170, 173 (*TAED* LB021170, LB021173).

3. The editors have not identified the two letters among the several that Tate sent to Edison about his work with the phonoplex. Among the recent extant ones are those written from Toronto on 10 December (of which there is also a copy) and 13 December, as well as one sent from Belleville, Ont., on 16 December (which Insull may not have received yet). Edison responded to this stream of correspondence in Doc. 2879. Tate to TAE, 10, 13, 16 Dec. 1885, all DF (*TAED* D8546ZCX, D8546ZCV, D8546ZDA, D8546ZDD).

4. Insull's letter to Tate to this effect was dated 18 December. Lbk. 21:186 (*TAED* LB021186).

5. This interest in "consolidation" marks the first step toward organizing what would become the Edison United Manufacturing Co.

6. The Edison Shafting Manufacturing Co. was incorporated in June or early July of 1884 to fabricate shafting, pulleys, hangers and related transmission equipment under contract with the Edison Machine Works (abbreviated as the Edison Shafting Co. in *TAEB* 7 chap. 5 introduction; Edison Shafting Manufacturing Co. agreement with Edison Machine Works, 14 July 1884, DF [*TAED* D8432A]). Plans were in motion for the Machine Works to absorb both the Shafting Manufacturing Co. and the Electric Tube Co., an arrangement that evidently was to take effect on 31 December although it was not ratified until the end of January 1886 (Electric Tube Co. to TAE, 31 Dec. 1885, Lbk. 21:162J [*TAED* LB021162J]; Cat. 1336:7 (item 9, 30 Jan. 1886), Batchelor [*TAED* MBJ003007B]; Edison Machine Works draft agreement with Edison Electric Light Co., n.d. [Jan. 1886?]; John Tomlinson to Harry Livor, 16 Feb. 1886; both DF [*TAED* D8627ZAW, D8627D]). Henry ("Harry") Livor, general manager of (and investor in) the Shafting Manufacturing Co., remained in charge of what became the Shafting Dept. of the Edison Machine Works. The George Place Machinery

Co., which had acted as sales agents for the Shafting Manufacturing Co., continued in that role, although the Shafting Dept. opened its own New York showroom at 21 Chambers St. by February 1887 (Doc. 2681 n. 2; Livor agreement with Edison Machine Works, 1 Jan. 1886; letterhead of Livor to Insull, 26 Aug. 1886; Place Machine Agency agreement with Edison Shafting Co., 9 Dec. 1885; all DF [*TAED* D8627A, D8627Z, D8532M]; "Manufacturing and Trade Notes," *Elec. and Elec. Eng.* 9 [Feb. 1887]: 80).

7. The Edison Machine Works had recently declared a dividend of 2.5 percent on 1 October 1885, payable in shares of the Edison Illuminating Co. of New York instead of cash, with Edison receiving forty shares. It had previously announced dividends of 3.75 and 7 percent in November 1882 and June 1885, respectively (Batchelor to TAE, 14 Nov. 1885; Ernest Berggren to Insull, 28 Oct. 1885; both DF [*TAED* D8531ZAC, D8531Z]). In annual reports to New York City's tax commissioner, however, Insull stated both the amount of surplus earnings and the annual dividend as "none" (Insull to Commissioners of Tax and Assessments [New York City], 2 Apr. 1885 and 9 Apr. 1886, both DF [*TAED* D8531C, D8627L]).

8. Account summaries drawn up in October 1885 listed the "book assets" of the Machine Works at $300,811.67 on the first of that month, up from $277,102.79 at the start of 1885, with machinery and tools accounting for nearly half the total in each case. In addition to stock of the Edison Illuminating Co. of New York and the Edison Shafting Manufacturing Co., the Machine works held $27,450.00 in shares of the Edison Co. for Isolated Lighting in October, an increase from $19,288.66 over the same interval. The value of raw materials on hand went from $15,880.56 to $20,783.84. The "total actual assets" that the Machine Works reported for tax purposes at the start of 1885 and 1886 were $173,768 and $231,108, respectively. Edison Machine Works memorandum of "Book Assets," Oct. 1885; Insull to Commissioners of Tax and Assessments (New York City), 2 Apr. 1885 and 9 Apr. 1886; all DF (*TAED* D8504ZZA, D8531C, D8627L).

9. The editors have found no specific information about this transfer of property. Possibly the items included a chalk press, steam engine, and boiler taken from Menlo Park to New York sometime prior to 1887. Insull to Alfred Tate, 3 Mar. and 29 July 1887, both DF (*TAED* D8736AAW, D8736ADB).

10. Drexel, Morgan & Co., the famous international banking house in New York, though principally involved with railroad finance in this era, played a major role in funding and launching Edison's electric lighting enterprises, including the Edison Electric Light Co. and the Edison Electric Illuminating Co. of New York. It also controlled Edison's lighting patents in Great Britain. In addition, several of the firm's current or former partners were personally involved in Edison's lighting concerns in the United States and abroad, notably J. Hood Wright, Egisto Fabbri, and Charles Coster. The bank remained connected with Edison's electric lighting affairs until its organization of the General Electric Co. in 1892. *TAEB* 5–7 passim, esp. Docs. 1612, 1648, and 1649; Carosso and Carosso 1987, 271–73; Strouse 1999, 181–83, 230–34; *American Electrical Directory* 1886, 322.

11. The editors have found no further information about this meet-

ing. Drexel, Morgan & Co. apparently had a financial interest in the Electric Tube Co., having advanced it $75,000 in February 1885. John Tomlinson to Electric Tube Co., 10 Dec. 1885; Charles Berggren to Insull, 25 Feb. and 19 June 1885; all DF (*TAED* D8533X, D8533D, D8533R).

12. That is, to preserve the existing ratio of Edison's $235,880 share in the $320,000 aggregate.

–2878–

Memorandum: Edison Lamp Co. Agreement with Edison Electric Light Co.

[New York, c. December 21, 1885[1]]

Basis of New Contract.[2]

Lamp Co have exclusive right mfr during life patents—

Present prices—life guaranteed 1000 hours of 10 Lamp [--][a] per Electrical horse power[b]

Pay Light Co for handling accts & guarant'g bills   3 cents per lamp on Lamp used N & S america only—

Light Co have orders filled first.

Lamp Co Carry stock 50,000 Lamps

Edison[c] give all his improvements in filiment Lamps for 10 yrs from date to Lamp Co—[b]

In case Lamp Co do anything prejudicial to interest of Light Co they receive notice & must correct or call for Arbitrators to ascertain if it is so. if they decide it is & Lamp Co fail[d] to stop it within 2 months than Exclusive Character Contract fails & they may license other parties, paying us however 3 cents for Every Lamp made by others.

AD, NjWOE, DF (*TAED* D8528ZAA). [a]Canceled. [b]Followed by dividing mark. [c]This paragraph enclosed by a brace in left margin and annotated by a question mark. [d]Obscured overwritten text.

1. Not having found a contract between the Edison Lamp Co. and the Edison Electric Light Co. on the provisional terms outlined in this document, the editors have conjectured this approximate date based on John Tomlinson's 18 December bill for preparing such an agreement. Tomlinson billed at the same time for consulting with Francis Upton about a new contract between the Lamp Co. and the Edison Electric Light Co. of Europe, a related subject likely dealt with sometime after Upton's return from Europe in the third week of August. DF (*TAED* D8528W).

2. The mutual obligations between the Edison Lamp Co. and the Edison Electric Light Co. had been the cause of much uncertainty and occasional contention for several years, and the Lamp Co. had not been included in the complex set of agreements that redefined relations between the parent Edison firm and the other manufacturing shops in September 1884 (see Docs. 2039 esp. n. 1, 2395, 2638, 2661 esp. n. 6, and 2725 n. 6). The editors have not found a completed or signed version of a contract, but some of the terms outlined in this document were

included in an undated twenty-four page typed draft. Some of these terms also appear in brief memoranda (also undated) by John Tomlinson (Edison Lamp Co. draft agreement with Edison Electric Light Co., n.d. [1885?]; Tomlinson memoranda, n.d. [1885?]; all DF [*TAED* D8528ZAB, D8528Y, D8528Z]).

–2879–

To Alfred Tate

[New York,] 23rd Dec [188]5

Dear Sir,

You did quite right in going out on the line to ascertain how far the phonoplex will practically work. I desire to call your attention to a fact which I think will be greatly to the advantage of the phonoplex. Instead of working through from Toronto to Montreal it will be of more benefit to the Telegraph Co to work the line through Morse and to work from Toronto half way phonoplex and from Montreal half way phonoplex thus giving them three circuits on that wire instead of two.[1] The local business is of course from about one hundred miles to one hundred fifty miles from each terminal. [------][a] Through business from the stations on one section to that of the other can be done on the Morse. All you have to do is put in the middle of the line a relay and shunt both ends to the grounds with a condenser as you have been doing. You had better talk to Mr Dwight about this and discuss the advantages of having three circuits instead of two. I think he will agree with me that it will be far better to divide the line for the phonoplex into two circuits. If he does agree then the phonoplex can always be worked in this way and will doubtless work on all circuits because if you cannot work from Toronto to Montreal phonoplex you can certainly divide the line, as above suggested, into two parts. Do you see my point? Telegraph men here with whom I've talked say that this capacity of the phonoplex to be divided into two circuits on a through wire is one of its greatest advantages and it would always be used this way instead of trying to work through.

Please box up all the apparatus sent you which is not of the latest pattern and which you do not want to use such as the induction coils motographs &c & ship them back to New York. You should get them through the U.S Custom House free of duty as returned articles of US. manufacture & I think you may be able to get back the duty already paid to the Canadian Custom House. If you cannot get them through to New York free of duty please hold them & advise me.

Let me know immediately by wire what additional apparatus you require to fix up the Toronto–Montreal line & I will ship it to you.[2] As soon as you have the wire working for phonoplex on regular business satisfactory to Mr Dwight I want you to come to New York as I have several wires that I want to put in operation in the United States.[3] I think you should thoroughly post one of Dwights men how to put up the phonoplex apparatus so that in case Mr Dwight desires to add it to their other circuits ᵃcircuits all we would have to do would be to ship him the apparatus and the man you teach can put it up   Yours truly

Thos. A. Edison   I[nsull].

LS (letterpress copy) NjWOE, Lbk. 21:192 (*TAED* LB021192). Written by Samuel Insull. ᵃCanceled.

1. As Edison explained in a patent application executed at the end of November, he wished "to utilize sections of the line only or independently of other sections for the induction-signals [phonnoplex], while the entire line is employed, as heretofore, for the ordinary Morse signals." This configuration was "based upon the discovery made by me that by grounding the line at any point through a condenser of sufficiently large capacity the line becomes practically divided at that point for the induction-signals, which will pass to ground through the condenser from either section of the line . . . while the line remains intact from end to end for the ordinary Morse signals." U.S. Pat. 435,689.

2. Tate telegraphed in reply that "Dwight wishes line Toronto to Kingston one hundred sixty three miles prepared as per your letter to test system for Commercial practicability   this includes toronto port hope Cobourg Belleville & Kingston." He explained in a letter that Harvey Dwight was "getting very impatient" to have the line put in commercial service and likely would "want the system extended to Montreal so as to give three separate circuits." He requested condensers to set up the phonoplex at Cobourg, "an important office on the line [just east of Port Hope] which does a great deal of local business." Tate to TAE, both 28 Dec. 1885, DF (*TAED* D8546ZDL, D8546ZDN).

3. Tate was still working in Ontario in early January but returned to Edison's New York office by 6 February 1886. Tate to TAE, 6 Jan. 1886, DF (*TAED* D86390); Tate to George Markle, 5 Feb. 1886, Lbk. 21:247 (*TAED* LB021247).

−2880−

To Theodore Miller and John Miller

New York Dec 24 1885

John[1] and Theo.[2]

I send you by Express today an electric shocking coil which one of the boys in the Laboratory made from old material.[3] The two wire clamping posts marked H are for the two handles. The other two are for the wires that lead to the battery. by turning the handle the electric circuit is broken and closed rapidly on the little wheel where the ivory pieces are set in. Be sure the spring that presses on the surface of the wheel does not get pulled away so it doesnt touch the surface at times. You will notice a screw, whereby the pressure on this spring may be regulated. The wheel should be turned about 200 times per minute for a black cat and 199½ for a cat with a sanguine temperment. You will notice that in the center of the spool containing the wire there is a rod made up of very fine wires   by pulling this out, the strength of the shock is greatly diminished and its strength increases as you gradually shove the rod into the coil.

I send you one cell of battery also a supply of Bichromate of potash to make the battery fluid with.[4]

place the Zinc in the glass cell. Fill the cell with water so it comes within ¾ inch of the top of the Zinc. Then put the Carbon rod in the earthenware cell and fill it with the red fluid then put this cell in the water within the [annode?]ᵃ zinc. Then pour into the water surrounding the zinc an amount of sulphuric acid equal to 5 thimble full   Then the battery is ready for business

To make the red fluid, get a glass or earthenware crock— (dont use metal) and put into this about 2 quarts of water and two double handfulls of the red material which must be pounded up as fine as wheat, then pour into this ½ pint of Sulphuric acid which you can get at the drug store, then stir it well with a wooden stick until the liquid is very red. Be sure in pouring the sulphuric acid that you do not let any of it spatter into your Eyes   I think your Brother Robert[5] understands batteries and he will doubtlessly show you how to make the fluid etc.—

This coil is very powerful. I tried it on a Dutch Carpenter today and it Knocked him down instantly, but didn't hurt him   This was when the rod was in the coil and with 2 cells of battery, so there is no danger from one cell. It will shock 25 persons joining hands, and make a cat knock off the plastering

I sent you on the 22d a Telescope which [shows?]ᵇ the rings of Saturn Etc. it [is?]ᵇ not too large for Land views & I hope it will suit you[6]   I also ordered Messrs Bunnel & Co[7] to send you a pair of telegraph instruments with battery and wire so you can set up a complete telegraph line from one part of the house to the other and learn to telegraph. Hoping you will have a merry christmas and not watch me and Mina so closely when I come again. I ~~remain~~ am with the most distinguished consideration   yours Truly

T A Edison

ALS, NjWOE, DF (*TAED* D8514ZAG). ªCanceled. ᵇPaper damaged.

1. John Vincent Miller (1873–1940) was the penultimate of Mina Miller's siblings. He eventually attended both St. Paul's School in Concord, N.H., and Yale University concurrently with their youngest brother, Theodore. After graduating from Yale in 1897, he enrolled in mechanical engineering at Cornell but interrupted his studies to enlist in the Navy during the Spanish-American War. Miller entered Edison's laboratory at Orange in 1899 as an experimental assistant and remained closely associated with him in a variety of engineering, administrative, and financial positions that encompassed mining enterprises, the Edison Storage Battery Co., and the Edison Portland Cement Co. He later served as secretary of Edison's estate. "Miller, John Vincent," Pioneers Bio.; "John Miller Dies; Was Edison's Aide," *NYT*, 17 Aug. 1940, 15; Vincent 1899, passim; Jeffrey 2008, 170–74.

2. Theodore Westwood Miller (1875–1898), slightly more than a year younger than John Vincent Miller, was Mina's youngest sibling. He graduated (with John) from Yale University in 1897 and studied law at New York University until the following May when, soon after the declaration of war against Spain, he enlisted in Theodore Roosevelt's new Rough Rider regiment. Miller was fatally wounded in Cuba two months later. Vincent 1899, passim; collection of obituaries [July 1898] in MFP (*TAED* X018D5AP); Jeffrey 2008, 170–74.

3. The induction "shocking coil" was a common enough device for both therapeutic and amusement purposes. The intensity of the induction could typically be controlled by sliding one of the coils in or out relative to the other. Edison and a partner had manufactured a medical "inductorium" in 1874 so powerful that "it is seldom that a person can stand one half of the current generated from the one cell." The phrase also pertained to a similar instrument consisting of a coil that could be pulled through a magnetic field so as to induce a current without need of a battery (Doc. 434; see also Doc. 435 [and headnote]; "A Pocket Shocking Coil," *Amateur Mechanics* 1 [May 1883]: 155–57; Preece 1878, 213–14). The crank-operated coil given to Theodore and John was likely the one characterized in a Miller family memorial as the "small dynamo" that distracted the boys from the routine of publishing their own newspaper, which ceased on 24 December (Vincent 1899, 21–22).

4. Bichromate of potash, a reddish compound and efficient oxidizer, was frequently used as a depolarizing agent in batteries. Prescott 1877, 71–72; Seiler 1884, 8.

5. Robert Anderson Miller (1861–1911), fourth of the surviving Miller children, was a native of Canton, Ohio. He worked there in the family business of C. Aultman & Co. until 1889, when he was appointed postmaster at Ponce, Puerto Rico, a position he held until his death. He married Louise Igoe in 1887. Obituaries collected by a clipping service are in Unbound Clippings (1911) (*TAED* SC11); see also Jeffrey 2008, 170–74.

6. On 23 December, Edison purchased a telescope from Joseph Rob-

inson, a New York optician and "Importer and Manufacturer of Optical and Mathematical Instruments." He paid $76.35, including express charges to Akron. Voucher (Laboratory) no. 638 (1885).

7. J. H. Bunnell & Co., a major manufacturer and dealer in telegraph and telephone equipment in New York, was formed in 1878 by Jesse H. Bunnell and Charles McLaughlin. A Union military telegraph operator during the Civil War, Bunnell (1843–1899) and Edison met as fellow operators in Tennessee. Doc. 378 n. 8; "Jesse H. Bunnell Dead," *NYT,* 10 Feb. 1899, 7; "Jesse H. Bunnell," *Electrical Engineer* 27 (16 Feb. 1899): 209.

-2881-

From John Farrell

NEW YORK, Dec 31st 1885[a]

My dear Mr Edison,

Your favor with check enclosed of 10$ duly recd with many thanks for the "Brady testimonial"[1]   Yours Truly

Jno Farrell[2]   Secy

ALS, NjWOE, DF (*TAED* D8503ZEU). Letterhead of Hoffman House. [a]"NEW YORK," and "188" preprinted.

1. Friends of the pioneering photography professional Mathew B. Brady (1823?–1896) of Washington, D.C., were organizing a raffle to sell a portrait from his gallery of the recently deceased General George B. McClellan. The proceeds were meant to assist Brady through his most recent financial hardship, due to the grave health of his wife (Juliet Handy Brady died in May 1887). Telling Edison that Brady had been "honored with yr patronage & friendship years past," John Farrell solicited him to contribute $10 for twenty tickets (*ANB*, s.v. "Brady, Mathew B."; *DAB*, s.v. "Brady, Mathew B."; Doc. 1297 n. 4; Farrell to TAE, 23 and 30 Dec. 1885, both DF [*TAED* D8503ZEO, D8503ZET]). In 1878, Edison and several associates sat with the new phonograph at Brady's Washington, D.C., studio. Brady, hoping that the photographs made by Levin Corbin Handy (his nephew) would sell widely, sought exclusive access to Edison as a portrait subject (Docs. 1297 n. 3, 1308 nn. 1–2, 1309 n. 2).

2. Possibly John J. Farrell (1864–1932), a native New Yorker, who became a political journalist for the *Newark Evening News* during the 1890s and then served from 1913 to 1932 as the executive clerk to New Jersey's governors. He is mentioned as a friend of Brady in Meredith 1974 (238). Obituary, *NYT,* 12 May 1932, 19; "'Fourth Estate' Honored," *Trenton Evening Times,* 24 Mar. 1899, 1.

3

January–April 1886

Edison spent more of the winter and early spring away from his New York home, shops, and offices than at them. When he returned to the region in early May, it was as a newly re-married man and the owner of a country estate in suburban Llewellyn Park, New Jersey. During this interregnum, he at-tended to the details of refashioning his personal life, to cre-ative inventive work (particularly on the phonoplex telegraph system), and to speculation about the nature of force and energy in the physical world.

Before embarking on his marriage trip, Edison also minded a range of inventive and business affairs at home. On the first of February, he attended a successful public demonstration of his railway telegraph system on Staten Island in the company of Leland Stanford, Henry Seligman, and Erastus Wiman, among others. He may have taken inspiration from a bliz-zard that hit the region a few days later for a burst of mis-cellaneous technical notes and drawings, including a design to make hand-held kersosene lamps (like those used by rail-road brakemen) resistant to strong winds. He also sketched a machine to clear snow from the streets and compress it into blocks for summertime cooling.[1] In mid-February, his work with Charles Batchelor was formalized with the founding of the Sims-Edison Electric Torpedo Company, for which Batchelor was developing an electric motor that Edison hoped would be manufactured by the Edison Machine Works. The Electric Tube Company and the Edison Shafting Company were both merged into the Edison Machine Works about that same time. According to account records, he also seems to have made some experiments on the phonograph, though the nature of that work is not known.[2] In personal matters, he moved to expunge a blot on his reputation by finally agreeing to pay the judgment against him in a long-running lawsuit over a disputed debt. And, shortly before the wedding, he ordered a caretaker to stop placing flowers on the grave of his late wife.[3]

Edison and Mina wed on 24 February at the Miller family home in Akron, Ohio. Joining several of Edison's long-time professional associates in attendance was Frank Toppan, a U.S. naval officer who, despite having met the groom little more than a year earlier (apparently through Ezra Gilliland), served as best man. In preparation for his new life with Mina, Edison in January bought a large furnished home on nearly twenty-four acres in Llewellyn Park, an exclusive community and one of the nation's first planned suburbs. Edison's purchase of such a large and expensive estate, even below its full value (in order to liquidate the property of the prior owner, an embezzler) generated some curiosity about his personal finances and some spurious speculation about his own financial dealings.[4] Having within the last year acquired land in Florida, where he was now building and equipping a new home and laboratory, Edison started the year "loaded up to the muzzle with notes which I must meet."[5] To acquire the home that became known as "Glenmont," he took a mortgage of $85,000 and apparently cashed out at least $10,000 of government bonds (held in the name of his late first wife).[6]

As the newlyweds traveled to Florida, Mina, not yet twenty-one, expressed some wistfulness at leaving her family for this new stage of life, which was already being chronicled by newspaper updates of their southward journey.[7] The couple dallied in eastern Florida before crossing the state to Fort Myers, on the Gulf Coast. Along the way, Edison was thinking about the phonoplex telegraph and designing a major modification that essentially made it a triplex system, capable of sending and receiving three messages in each direction simultaneously.[8] In Fort Myers, they joined Ezra Gilliland, his wife and sister-in-law, several domestic servants, and Marion Edison, who had just turned thirteen.[9] The new Edison and Gilliland homes were not yet finished, so the entourage piled into one, at least for a while.[10] Edison's new laboratory, which he envisioned as a place of active research rather than vacation puttering, was not ready either, though shipments of supplies and equipment continued to arrive from the North, as duly noted in the local paper.[11]

Edison, for his part, seems to have settled into an expansive and relaxed frame of mind.[12] Hardly had he arrived in Fort Myers than he began pouring a torrent of notes and drawings into his notebooks, eventually filling six of them. From the perspective of the twenty-first century, Edison seems more alive and more fully present during the Fort Myers weeks, at least on paper, than in much of the peripatetic prior year, when so much evidence of his activities was filtered through associates and subordinates. His Fort Myers books are remarkable in the range of his ideas, from the practical (lamp filaments), the impractical (a "Larynaxial piano"), the speculative (an "XYZ" force), to the grandiose ("Our solar system is a Cosmical Molecule").[13] It is also remarkable that Mina participated, at least in writing or recopying some of the entries. Even Marion made it into these books, copying into them a few news articles and a telegram message. Edison wrote persistently about physical forces and energy, nurturing a speculative strain of thinking that had emerged in prior periods of relative leisure. Yet like Michael Faraday, whom he deeply admired and whose *Experimental Researches* he may well have been rereading on this trip, Edison's mind was never far from the laboratory; he included on these pages a number of instruments and experiments he might use to test his ideas and, he hoped, turn them to some practical use.[14]

Edison somehow also found time to make detailed notes for Eli Thompson, whom he hired to supervise the Florida grounds and staff. His instructions show a wide knowledge of local crops and decorative plants, though he did not indicate the sources of this information. They reveal an equally remarkable ambition to transform the rough ground into an estate pleasing to both the eye and the palate. He wished, he wrote, "to Carry Everything to Extreme Excess down here." At some point, he seems to have shipped north to his sons several alligators as pets.[15]

1. See Doc. 2892.
2. Ledger #5:588, Accts. (*TAED* AB003; image 291). In December 1885, an article in the *New York World* (republished elsewhere) reported that Edison had resumed work on the phonograph. Perhaps confusing the phonograph with other acoustic research and with the aerophone, an instrument on which Edison had experimented in 1878, the report stated that he was developing a large steam-powered device with a thirty-foot funnel capable of projecting sound from atop a tall building. Edison was quoted as explaining, however, that such a machine was merely to prove the principle of the improved phonograph that he intended to produce for office dictation ("Edison Photographs

a Sound," *Wheeling Register,* 23 Dec. 1885, 2; "Edison at Work on the Phonograph," *Washington Post,* 26 Dec. 1885, 4).

3. See Docs. 2893 and 2894.

4. A *Washington Post* article titled "Inventor Edison's Income" purported to give "The Unvarnished Truth Concerning His Money and His Company." It used the fluctuations of the price of Edison Electric Light Co. stock over several years, in connection with Edison's purchase of Glenmont, to suggest that such "an amount of wealth not hitherto accorded to the inventor" was the result of speculation or worse (*Washington Post,* 14 Mar. 1886, 4). The article also appeared as "Edison's Wealth" in the *Chicago Daily Tribune* of 14 March (p. 12) and in altered form in numerous other papers.

5. See Doc. 2883. Possibly Edison was thinking of three notes to the Ansonia Brass & Copper Co. totaling $7,344.65 that he assumed from the Edison Construction Dept. They were due at intervals of four, six, and eight months from 21 November 1885. The notes were transferred to Edison's books and recorded in his voucher system in a single transaction on 28 February. Voucher (Laboratory) no. 102 (1886).

6. TAE statement of withdrawal, 19 Jan. 1886, DF (*TAED* D8603R); Doc. 2886 n. 5.

7. See Doc. 2903.

8. See Docs. 2902 and 2911; Edison's phonoplex system made its commercial debut on the Baltimore & Ohio lines by the end of February.

9. Marion Edison evidently left her New York school summarily without explanations to her teachers. Her young brothers remained in their classrooms at least through 1 February but did not travel to Florida. Their whereabouts during the honeymoon are unknown, but they looked forward to seeing their "new Mamma" afterwards. See Doc. 2895; Thomas Edison, Jr., to TAE and Mina Edison, 1 Mar. 1886, CEF (*TAED* X018B1AA).

10. See Doc. 2907

11. Shortly before his wedding, Edison was quoted in a newspaper interview on his plans for the new laboratory: "I shall be governed by circumstances in my experiments, but it looks now as if we were only on the threshold of electrical knowledge. The next twenty years will see great strides. I shall take five or six men to Florida with me and keep them busy." *New Orleans Times Democrat,* 2 Feb. 1886, Cat. 1140:116a, Scraps. (*TAED* SB017116a).

12. Cf. Doc. 2609 (headnote).

13. See Doc. 2937 (headnote).

14. See Doc. 2912 (headnote).

15. Doc. 2946; Theodore Miller to Mina Edison, 16 May 1886, FR (*TAED* FS001AAA).

–2882–

*George Bliss to
Edison Lamp Co.*

CHICAGO, Jan. 2, 188[6][a]

Gentlemen:

As I mentioned the other day in a letter to you,[1] the breakage which we have been experiencing with lamps out here during the past two months has been something very excessive. We received the following letter to day from the Manitou Iron Springs Hotel:[2]

"Western Edison Light Co., Chicago.[3] The last lot of lamps which we received from you are very short lived, some only burning from 50 to 60 hours, and as they are used in the same circuit with lamps that have been burning many months it is very evident that there must be some defect in the new ones as the old one still survive. Can you give any explanation? I have endeavored to keep the current as uniform as possible, and am always careful to avoid overcrowding." Very truly yours,

Western Edison Light Co.   Gen'l Supt.[4]

⟨Edison   Hows this? or do you believe it?   EHJ⟩[b]

⟨Upton— How would it do for you to personally learn the Lamp business I̶e̶ ie[c] The Carbonization— You are a Scientific man like myself Batch Etc. It seems to me if I was running the Lamp factory that there wouldnt occur any such thing as losing the art.

As the financing is rather Easy I suggest you do like the rest of us learn the business thoroughly & not be dependent on others= you are degenerating into a mere business man— Money isnt the only thing in this mud ball of ours   TAE⟩

TL, NjWOE, Upton (*TAED* MU086). Letterhead of Western Edison Light Co., George Bliss general superintendent; "Defect report" typed in preprinted subject line. [a]"CHICAGO," and "188" preprinted. [b]Marginalia written by Edward Johnson and followed by "Over" in Edison's hand to indicate page turn. [c]Circled.

1. Not found.

2. The Iron Springs Hotel, near the Pike's Peak Cog Road depot in Manitou, Colo., had recently been built anew after a fire in about 1882. Its modern features included a 161-light Edison plant and steam heat. Daniels and McConnell 1964, 28; Atchison, Topeka and Santa Fe 1898, 15; Hooper, Bell, and Wood 1890, 24; *American Electrical Directory* 1886, 175.

3. The Western Edison Light Co. was formed in May 1882 in Chicago to sell and install electric lighting plants in the surrounding region, principally Illinois, Iowa, and Wisconsin. Doc. 2299 n. 1.

4. A longtime Edison associate in Chicago, George Harrison Bliss (1840–1900) had been involved in promoting and commercializing Edison's electric pen, phonograph, and telephone. With Edison's intercession, he established the Chicago office of the Edison Co. for Isolated Lighting and then became superintendent of the Western Edison Light Co. Docs. 861 n. 8, 2278 n. 8, 2426 n. 1.

–2883–

*Draft to James
Connelly*

[New York, c. January 4, 1886][1]

My Dear Connolly[2]

You have struck me at a time that I couldnt raise $200. Im loaded up to the muzzle with notes which I must meet[3]   I am very sorry

TAE

ADfS, NjWOE, DF (*TAED* D8603B1).

1. Edison drafted this reply on the back of the first page of James Connelly's 4 January letter, which was mailed from New York. Connelly to TAE, 4 Jan. 1886, DF (*TAED* D8603B).

2. James H. Connelly (1840–1903) was a freelance writer and newspaper reporter whom Edison seems to have met in 1884. Connelly and a business partner, David Curtis, formed the Cook Publishing Co. and began publishing *The Cook: A Weekly Handbook of Domestic Culinary Art for All Housekeepers* in 1885, for which Edison loaned them $500. Now, with the business foundering in the wake of Curtis's personal misfortune, Connelly implored Edison to buy $500 or $1,000 in stock shares of Cook Publishing to carry the enterprise a few weeks until "we will be put firmly upon our feet by the advertising already promised." Doc. 2749 n. 1; Voucher (Laboratory) no. 677 (1885); Cook to TAE, 4 Jan. 1886, DF (*TAED* D8603B).

3. Possibly Edison was thinking of three notes to the Ansonia Brass & Copper Co. totaling $7,344.65 that he assumed from the Edison Construction Dept. They were due at intervals of four, six, and eight months from 21 November 1885. The notes were transferred to Edison's books and recorded in his voucher system in a single transaction on 28 February. Voucher (Laboratory) no. 102 (1886).

–2884–

To Willis Stewart

[New York,] Jan. 8th. 1886.

Dear Sir:—

In re yours of November 28th, 1885.[1]

You know I am all out of business now, and could not take the contract personally, but perhaps I could get the Light Company to do so. Please send me all the data of every kind, the area to be lighted, and every species of information that will enable me to draw out a general plan to see if it is feasable.[2]

I have made a new improvement so I can take the present three wire system, which saves sixty-two and one-half per-cent, and save one-half as much again.[3] Very truly yours,

T. A. Edison, per Mc.G[owan].

TL (carbon copy), NjWOE, Lbk. 21:162H (*TAED* LB021162H).

1. Edison had evidently given Stewart an estimate (not found) for converting the Santiago station to water power. Stewart wrote on 28 November to see if Edison would enter a construction contract backed fi-

nancially by Grace & Co.; Edison's notes on that letter formed the basis for his reply (Stewart to TAE [with TAE marginalia], 28 Nov. 1885, DF [*TAED* D8534W]). Stewart referred again to Chile's abundant water power in a letter of 20 January 1886, when he advised that W. R. Grace & Co. might soon "invite you to send on a No. 2 (50 l[igh]t) dynamo, to be used in connection with a Sprague motor to show the transmission of power" over long distances. This demonstration was intended for an upcoming Chilean exhibition of North American machinery, an affair potentially useful for Grace's expansion as an American machinery exporter. Edison wrote on that letter (referring to the Edison Machine Works): "Does Batch want to take the risk of sending the Dynamo and Motor"; it is not clear what, if anything, came of the request (Stewart to TAE [with TAE marginalia], 20 Jan. 1886, DF [*TAED* D8630H]; Coe 1919, 9363; U.S. House 1887, 301, 303–305).

2. Changes in the Santiago plant and its operation would need approval from the Edison Electric Light Co., whose officers (notably Charles Coster and Frank Hastings) tended to side with Kendall & Co. (the appointed agent for Edison lighting in Chile) and the current superintendent of the Santiago station, George Wellington Waters. Stewart to TAE, 26 Mar., 17 and 28 Apr. 1886; Stewart to Insull, 28 Apr. 1886; W. J. Clark to Insull, 8 May 1886; all DF (*TAED* D8630W, D8630ZAD, D8630ZAI, D8630ZAJ, D8630ZAM); regarding Kendall & Co. see Doc. 2602 n. 5.

3. It is not clear what improvement Edison had in mind. He had made about a score of rough sketches of electrical distribution systems in recent days. Some appear to show variations on his three-wire feeder-and-main system, including several with dynamos in series, implying a relatively high voltage on the feeder lines. Their features are generally consistent with Edison's municipal system, specifically to an adaptation made before the end of 1886 to "light groups of several small towns or suburban districts from one station," as an Edison Electric Light Co. brochure claimed. Among these sketches are two showing a scheme for carrying current at 800 volts through an undefined distance from "water power & Dynamos" (a location perhaps relevant to Stewart's interest in waterpower) to an unspecified "station" of some sort. Although the editors have no clear evidence of Edison's role at this time in modifying the municipal system, it is conceivable that he did so in response to practical experience at the several locations in which it was operating. N-85-05-28:237–65, Lab. (*TAED* N309237, N309240); Edison Electric Light Co. Central Station Catalogue, n.d. [1886], PPC (*TAED* CA001D).

-2885-

From Grace & Co.

Valparaiso, Chile. Jan. 9 1886.[a]

Dear Sir:—

The Electric Railroad patent granted to yourself by the Government of Chile expires on the 24th of April of this year, unless a demonstration of the system is made in the country before that date.[1] We consider the patent worth saving, and before you receive this will cable for a Sprague 1½ HP motor[2] (for which we have code word) to be used at the mill of Mr. Enrique Lanz[3] for the purpose of making a demonstration. Mr. Lanz has a 50-light dynamo, and will arrange car, rails, &c., for a distance of 1000 ft.

Having already been to the expense of taking out the Patent, we ask you to furnish this motor for the purpose of saving the privilege, on condition that we shall remit its value whenever the same may be sold. Mr. Lanz will probably keep the motor in case it works well.

The Certificate of Railroad Patent is now in possession of Mr. G. W. Waters,[4] of the Santiago Co.[5] Please advise him to turn it over to us, and oblige, Yours Truly,

pp[6]   Grace & Co[7]   Mr. Catton[8]

⟨Insull   arrange with sprague if possible

Insull   write him to put a 50[b] Light dynamo on a Car and belt up so the speed of the surface[b] of the car wheel is 3[b] to 5 times slower than the speed of the surface of the pulley on Dynamo—

He can rig up the jack shaft[9] Easy Enough then if unsuccessful he has his 50[b] Lt Dynamo to sell   also tell Waters to deliver RR patent to Stewart   E[dison]

I see that patent Expired April 24 so he cant save it⟩[10]

D, NjWOE, DF (*TAED* D8630B). Letterhead of Grace & Co. [a]"*Valparaiso, Chile.*" and "*188*" preprinted. [b]Obscured overwritten text.

1. The editors have not specifically identified this patent but in late 1883, Willis Stewart, in conjunction with Enrique Lanz, was attempting to obtain protection for Edison's electric railway system in Chile. By August 1884, Stewart was making preliminary plans to apply it on a mining railroad and was asking Edison for updated information on conducting current through the rails. Stewart to TAE, 14 Dec. 1883 and 8 Aug. 1884, both DF (*TAED* D8372M, D8435ZAX).

2. The cable has not been found. In March, Stewart asked Insull to "punch up" Frank Sprague to send the drawings (presumably of motors) he had requested. He had advised Edison in 1884 that "The subject of motors in general is an important one here, & I suffer from the lack of a sample motor to show this application of electric currents." Since about May 1884, cable correspondence with Stewart had been encrypted according to a list of code terms (not found). Stewart to Insull, 27 Mar. 1886; Stewart to TAE, 8 Aug. 1884; TAE to Stewart, 9 May 1884; all DF (*TAED* D8630X, D8435ZAX, D8416BOC).

Sprague designed and patented in 1883–1884 a heavy-duty motor that, because of the distinctive compound winding of its field coils, could run at constant speed under varying load as well as at variable speed and power. It also featured self-adjusting commutator brushes to reduce sparking (Doc. 2575 n. 9; "The Sprague Motor," *Electrical World* 5 [25 Apr. 1885]: 168–69; Passer 1953, 238). After Sprague exhibited the motor to acclaim (including Edison's) at the 1884 International Electrical Exhibition in Philadelphia, he incorporated the Sprague Electric Railway and Motor Co. with Edward Johnson and John Tomlinson at the end of that year. By virtue of its connection with Johnson and other Edison associates, the firm later contracted its manufacturing to the Edison Machine Works, and the Edison Electric Light Co. promoted sales of the motors (Dalzell 2010, 68–75; Sprague Electric Railway and Motor Co. incorporation certificate, 24 Nov. 1884, Sprague [*TAED* X120CAL]; Edison Electric Light Co. circular, 25 May 1885, DF [*TAED* D8526S]; see Doc. 2998).

3. Enrique Lanz was a founder and director of the Sociedad de Fomento Fabril, an industrial development organization. In addition to his financial support for Edison's electric railroad in Chile, Lanz had installed an isolated electrical lighting plant at his flour mill near Valparaíso. Doc. 2602 n. 8; Tupper 1887, 9, 13, 17–18.

4. George Wellington Waters (1860–1932), a native of New Jersey, was the superintendent of the Santiago central station. Doc. 2711 n. 6; General Records of the Dept. of State, RG59, Entry 205, Box No. 1307 (1930–1939, Chile A–Z), record located in *Reports of Deaths of American Citizens Abroad, 1835–1974* (online database accessed through Ancestry.com, 16 Jan. 2013).

5. La Compañía de Luz Eléctrica Edison-Santiago was formed in 1882 as the local Edison illuminating company in Santiago. The object of an ongoing and muddled struggle for control, the firm was apparently still the only functional Edison electric lighting interest in Chile. Doc. 2602 n. 4; Stewart to TAE, 28 Nov. 1885 and 17 Apr. 1886; Stewart to Samuel Insull, 26 Dec. 1885; W. J. Clark to Insull, 8 May 1886; all DF (*TAED* D8534W, D8630ZAD, D8534X, D8630ZAM).

6. An abbreviation of "per procurationem," meaning "by proxy." *OED*, s.v. "p.p." in initialisms listed under "P."

7. Grace & Co. established its agency in the Chilean port of Valparaíso in 1881 to stabilize the Grace family's shipping interests along the west coast of South America amid the War of the Pacific. Among its six founding partners were John W. Grace in Lima, Peru, and William R. Grace, Michael P. Grace, and Charles Flint in New York. Clayton 1985, 135.

8. Not identified.

9. A jack shaft is an intermediate driven shaft in a mechanism for transmitting power. *OED*, s.v. "Jack" compounds.

10. The editors have not found a formal reply from Edison. At the end of March, Stewart advised Insull that "The electric railway privilege is probably lost." Stewart to Insull, 27 Mar. 1886, DF (*TAED* D8630X).

*From Edward P.
Hamilton & Co.*

New York, Jan'y 12th *1886.*[a]

Dear Sir.

We would respectfully submit our services, learning you consider the purchase of Real Estate And would suggest the property recently occupied by Mr. H C. Pedder[1] In Llewellyn Park Orange N.J.[2] The photographs[3] & full particulars we would be pleased to submit. This property has cost some $400,000 furnished. It can be bought fully at half its cost either furnished or unfurnished[4]

Orange & that section of Country being our specialty, Can offer other properties should you not feel inclined toward the Pedder place.[5] Kindly advise & oblige   Yours Very truly,

Edw P. Hamilton & Co[6]

L, NjWOE, DF (*TAED* D8603L). Letterhead of Edward P. Hamilton & Co. [a]"*New York,*" and "*188*" preprinted.

1. Henry C. Pedder (b. 1841?) had been a confidential clerk with the fashionable New York department store of Arnold, Constable & Co. In 1879, he and his wife purchased about thirteen and a half acres (to which they later added) in Llewellyn Park. Early the next year, they commissioned a house by architect Henry Hudson Holly, who had already built in the Park. The three-story house, in Holly's signature Queen Anne style, was completed by 1882 at a cost of $70,000. Scarcely two years later, having made some $20,000 in additions to the structure, Pedder was accused by his employer of embezzlement. To make restitution, Pedder sold the house, its furnishings, outbuildings, and land (now almost twenty-four acres) to Arnold, Constable & Co. for one dollar and then reportedly left the country (U.S. Census Bureau 1970 [1880], roll T9_781, p. 261D, image 0164 [West Orange, Essex, N.J.]; Yocum 1998, 1:9–10, 14–18, 20–22, 30; Herron 1998, 1:11–12). The construction materials were described in an 1881 issue of *American Architect and Building News* ("House for Mr. Henry C. Pedder," 10 [27 Aug.]: 98); Yocum 1998 and Herron 1998 describe the house in full detail.

2. Llewellyn Park was a planned residential community in the township of West Orange, N.J., about twelve miles from New York City. Though politically independent since 1863, West Orange was served by the post office and train station in nearby Orange and, throughout Edison's lifetime, portions of it were often referred to simply as "Orange" (Doc. 2559 n. 3; Yocum 1998, 1:9–10; *Ency. NJ,* s.v. "West Orange"; *U.S. Official Postal Guide* 1886, 483; *Am. Cycl.,* s.v. "Orange [N.J.]"; Hill 2007, 37–38). According to Mina Edison's recollection years later, Edison offered her the choice of a "beautiful" house on Riverside Drive in New York City or the one in Llewellyn Park. In her telling, she selected the more rural location she believed he wanted because "the confusion and excitement of a large city were deadening to experimental research" (Coman and Weir 1925, 11).

3. Among the photographs sent was probably one of the earliest to be taken of the house, in 1882 or 1883, showing the dwelling much as it would have appeared in early 1886.

On Edison's first inspection, the former Pedder residence probably appeared much as it did in this photograph taken in 1882 or 1883.

4. According to 1885 estimates by Henry Holly, the cost of the improved land, buildings, and furnishings of the entire Pedder estate, which included a greenhouse, stable, and other outbuildings on separate lots, was somewhere between $235,000 and $306,000. Holly to J. Asch, 12 June 1885, DF (*TAED* D8540A); cf. Yocum 1998, 1:21–22.

5. Seemingly unknown to this broker, Edison had agreed only the day before to buy the property directly from Arnold, Constable & Co. He signed the contract on 20 January, paying $125,000 for the twenty-three room house, outbuildings, and land. A separate contract for $1 included all the Pedders' furnishings and personal possessions, itemized in an inventory running to thirty-six typed pages. He took a mortgage from the seller for $85,000 that he paid in two installments: $10,000 in July 1886 and $75,000 in April 1890. Yocum 1998, 1:32 esp. nn. 70–71; TAE agreement with James Constable, Richard Arnold, Frederick Constable, and Hicks Arnold, 20 Jan. 1886, Miller (*TAED* HM860278); Israel 1998, 248–49; "Inventory of Contents of House at Llewellyn Park," 20 Jan. 1886, NjWOE.

6. A dealer in the real estate of the New Jersey suburbs of New York, the firm of Edward P. Hamilton & Co. specialized in properties in and around Orange, where they kept their New Jersey office. Hamilton had been in business since 1867. *New York's Great Industries* 1885, 93.

[New York,] January 16th. [188]6

To Frazar & Co.[1]

Gentlemen:—

Referring to the letters which I have addressed to your houses authorizing them to act as my representative, I beg to state that so long as your agency lasts, any inquiries which may come to me for electric light apparatus will be turned over to your firm, providing that said inquiries come from the territory covered by your agency.[2] It is just possible that some business may be done through my European friends, although I think it very improbable that such will be the case. This letter must not be taken as in any way referring to any business that may be transacted through London or any of the other European centres.[3] My intention is that you shall act as my agent, and every inquiry that comes to me will be sent immediately to your good selves, but you can well understand that I cannot control such business as may be secured through European agencies. You will have every facility to get material as cheaply as it can be obtained by other parties for use in the United States or other parts of the world, as my desire is to foster your agency and assist you to do business as far as lays in my power. Yours very truly,

T.A.E

TL (carbon copy), NjWOE, Lbk. (*TAED* LB021162N). Initialed for Edison by Alfred Tate.

1. Everett Frazar established Frazar & Co. in 1854 as merchant and commission agents in Shanghai, and the firm later added branches in Nagasaki, Yokohama, and Hong Kong. Its New York house had sought to introduce Edison's electric light in Japan in 1882, and since at least April 1884, Everett Frazar had been guiding Edison's application for an electric light and telephone concession in Korea. The Yokohama branch became the sole agents for Edison's light in Japan and Korea in May 1885. Doc. 2678 n.1.

2. Frazar had been using Francis Upton as an intermediary to try to make changes in the arrangements of his Yokohama branch with Edison. His wishes included adding the Straits Settlements (with Singapore) to their territory and a request, in Upton's words, "to extend the agency of his firm for another year and from year to year so thereafter long as the business makes satisfactory progress under them." Edison made a cursory reply to "Extend 1 year & include the places asked," but Samuel Insull thought that doing so could "run afoul of Mr Edisons existing contract obligations" in the region. Edison sent a letter to Frazar's Shanghai branch on 16 January. Frazar acknowledged it (and others not found) as "extending the time of the China, Japan and Straits agencies" for a year (until May 1887). Upton to TAE, 25 Sept. and 7 Nov. (with TAE marginalia) 1885, and 12 Jan. 1886; Frazar & Co. to Upton, 7 Jan. 1886; Frazar & Co. to TAE, 21 Jan. and 25 Feb. 1886; all DF (*TAED* D8535ZBS, D8534V, D8630D, D8630C, D8630J, D8630N); Insull to

Upton, 11 Nov. 1885; TAE to Frazar & Co., 16 Jan. 1886; both Lbk. 21:86, 162L (*TAED* LB021086, LB021162L).

3. One possible competitor in the Straits Settlements was Edison's Indian and Colonial Electric Co., Ltd., the London-based firm that held rights (since 1882) to Edison's lighting system in British possessions of the Indian Ocean basin; however, it had done little business and was soon absorbed by the Australasian Electric Light Power and Storage Company, Ltd. Another potential entanglement was with the Compagnie Continentale Edison, whose rights to license Edison's electric light patents in the French colonies would have included nearby Indochina. Doc. 2624 n.1; Insull to Edison Lamp Co., 23 Mar. 1885, Lbk. 20:199D (*TAED* LB020199D).

–2888–

Draft Patent
Application: Railway
Telegraph

[New York,][1] January 21 1886

Dyer

Take out following patent.[2]

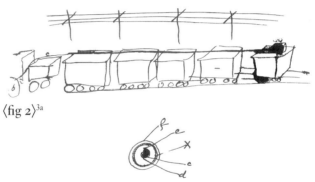

⟨fig 2⟩[3a]

on the caboose I have a reel on which I coil a cord about ½ inch diameter   This cord contains either one or two conductors. there are several hundred feet of it sufficient to reach the whole length of the longest freight train. it is to be used as one plate of the condenser on the RR telegh of which the RR telegh wires form the other plate.

The cord shewn in fig 2 consists 1st of the small wire c insulated by d—then e is rope; over this is wound spirally copper wire making a flexible cord—over this again is tape & then the whole is braided with hemp—[4]

The center wire runs to Locomotive to get a good ground [~~under?~~][b] but the Caboose wheels may be fixed to make a good ground & thus the center wire need not be used—

We could use a flat band—

Lead Covered wire etc.

We have ascertained that it is length of conductor is what is wanted[5]   instead of 2000 feet of sufface on two or 3 cars, 75 feeet one 30 or 40 is better. greater length paralell with the telegh wires is the ticket= we can possibly use a ~~telegh~~ telephone instead of morse on long freight trains   I think you might mention this.

get strong claims— take patent in my name—

Edison

PS   Charge to RR Telp & Tel Co[6]

ADfS, NWOE, Lab., Cat. 1151 (*TAED* NM020AAR). Letterhead of E. T. Gilliland; several faint and incomplete sketches on reverse of second page have been omitted. [a]Marginalia probably written by Richard Dyer. [b]Canceled.

1. Although Edison wrote on Ezra Gilliland's Boston letterhead, the editors have no evidence that he was away from New York, where, the previous day, he signed papers for the mortgage on his new home. TAE agreement with Arnold, Constable & Co., 20 Jan. 1886, Miller (*TAED* HM860278).

2. Edison's memorandum became the basis of a patent application filed on 16 February (Case No. 663) under the title of "Railway Signalling Apparatus." It covered "a more simple and efficient form of the train conductor than heretofore proposed" and contained six drawings (three of them similar to Edison's sketches here) and six claims. The Patent Office swiftly rejected it for, among other reasons, a lack of specificity and the existence of prior patents, including one to Ezra Gilliland. The application was amended and rejected a number of times and eventually abandoned sometime after January 1894. TAE Patent Application Case 663, 16 Feb. 1886; U.S. Patent Office to TAE, 6 Apr. 1886; Dyer & Seely to TAE, 31 Jan. 1894; all PS (*TAED* PT014AAB, PT014AAF, PT014AAS); cf. Doc. 2588.

3. Figure labels are "f," "e," "x," "c," and "d."

4. Figure labels are "Loco," "inductor," "[inner?] wire," and "signalling apparatus."

5. The proposed patent (see note 2) explained that "the increase of length in the train conductor is far more beneficial than increased width although there may be the same area of condensing surface in either case. This would seem to be due to the fact that the effective opposing condensing surface of the line wire is dependent upon the length rather than the width of the train conductor."

6. That is, the Railway Telegraph and Telephone Co.

Notebook Entry:
New Laboratory

[A][2]

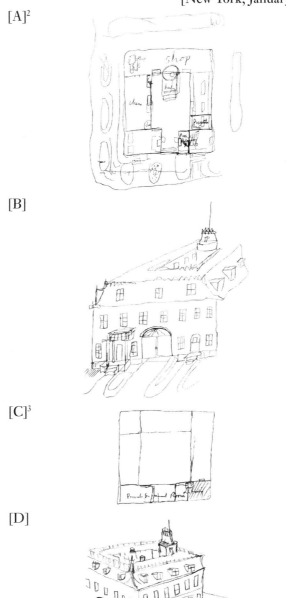

[B]

[C][3]

[D]

X, NjWOE, Lab., N-85-10-01:63 (*TAED* N310062).

1. These drawings were made in the notebook between an entry by John Ott dated 20 January 1886 and one by Edison on 3 February 1886 (Doc. 2891). They appear to be Edison's earliest ideas for a new laboratory that would include a machine shop, a chemical laboratory, a private experimental room, and a library.

2. Figure labels running clockwise from top left: "Shop Eng[ine]," "Shop," "boiler," "Private," "Book Keeper   Lib[rary?]," "[stair?]," "chem," and "[batteries?]."

3. Figure labels are "Private Experiment Room" and "Library."

–2890–

*Charles Batchelor
Journal Entry*

[New York,] Feb. 1st 1886

10.[1] <u>R.R. Telephone and Telegraph:</u>—[2] Was present at an exhibition of this system at Staten Island. it worked well.[3]

AD, NjWOE, Batchelor, Cat. 1336:7 (*TAED* MBJ003007C).

1. Charles Batchelor consecutively numbered each entry in this journal.

2. Batchelor's designation of this system as "Telephone and Telegraph," in contrast to Edison's customary references to the railway "telegraph" alone, more accurately reflected the origins and flexible uses of wireless telegraphy. See Docs. 2780 and 2800 (headnotes).

3. Edison participated in this choreographed demonstration of his railway telegraph system. According to press reports, the operator at Clifton, Staten Island, transmitted messages to a train en route from there to Tottenville. The messages, sealed in envelopes to guarantee their secrecy until the train left, were addressed to visitors on board, including Edison, Erastus Wiman, Leland Stanford, and Henry Seligman, among "a large number of railroad and telegraph officials." During the course of the ride, some thirteen and one-half miles, Edison telegraphed to his daughter Marion and to the *New York World* offices, both in New York City, and Seligman directed a message to his brother David, also in New York, and received a reply. The exhibition evidently continued in some form on 2 February, when William Wiman sent his congratulations from a "train going from Tottenville via Clifton" to Edison in New York. "Messages from a Moving Train," *NYT,* 2 Feb. 1886, 1; "Telegraphing Through Air," *Washington Star,* 2 Feb. 1886; William Wiman to TAE, 2 Feb. 1886; Cat. 1140:116b, 134; both Scraps. (*TAED* SB017116b, SB017134); a more detailed account of the event, including the names of sixteen of the attendees, appeared in the *New York World* and subsequently in the *Railway Conductors' Monthly* ("Telegraphing from Trains," Vol. 3, Mar. 1886, 177–79).

[New York,] Feby 3rd 1886

Regulation temperature House by Expansion[1]

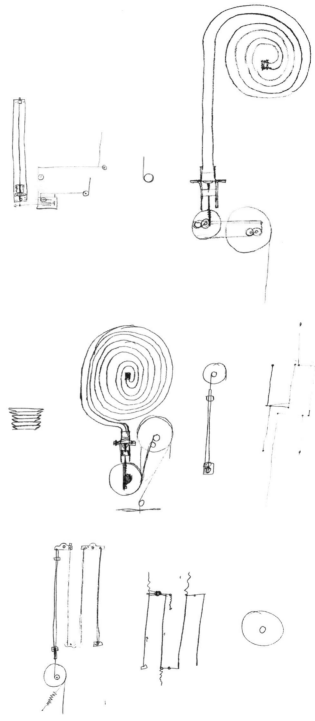

[A][2]

Compounding Hard Rubber Rods so Expansion regulate hot air grate in house[3]

Expansion Zinc 1 in[ch] 336. iron—1 in 846. glass 1 in 1248 water increases $\frac{1}{9}$ in bulk from 32° to 312   mercury $\frac{1}{65}$—

12—	336
6—	168
3—	84
$1\frac{1}{2}$—	42.

TAE

X, NjWOE, Lab. N-85-10-01:79 (*TAED* N310079). Document multiply signed and dated; miscellaneous calculations omitted.

1. In his early work on electric lighting Edison focused his efforts on thermal expansion regulators to prevent platinum filaments from reaching their melting point. See Docs. 1424, 1426–27, 1432, 1454, 1456, 1462, 1466, 1469, 1472, 1485, 1491, and 1503.

2. Figure labels are "iron," "Rubber," "iron," "Rubber," "iron," and "Rubber."

3. In one of his first electric light regulators in 1878, Edison used the expansion of rubber in a manner related to that in his tasimeter, an instrument he was developing contemporaneously. See Doc. 1289.

Notebook Entry:
Miscellaneous

Gave instructions to John Ott to make another wick between the 2 wicks of the Regular sperm oil hand Lantern used on Railroads[1]   This central wick dipping into a reservoir of Kerosene within the oil can= This will more than double the brilliancy of the Lamp, cannot blow out & is very cheap[a]

also to use in reg Kerosene Lamp a capilliary space flat—& adjustable so Kerosene is burnt in thin sheet without wick by capilliary action[2]

Thermo   Theory is that one junction blackened absorbs heat, other polished reflects heat—[3]

[A][4]

[B][5]

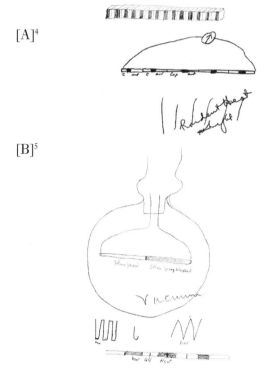

Lamp factory Expmnts

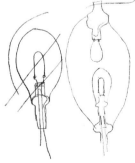

one Lamp within the other

hard rubber polished balls to roll around & Electrify the inside of the globe—[6]

[C][7]

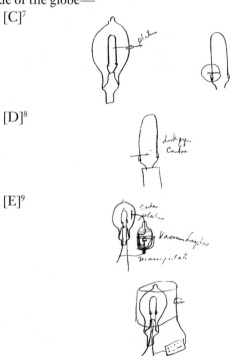

[D][8]

[E][9]

1000. storage bat or Chl[oride] Silvr batteries[10]

glass ring foil inside for Earth outside Electrified.

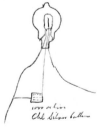

1000 or less   Chl Silver batteries.

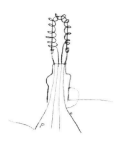

manipulate[11]

Try fibres cut from ferns stalks, they are palms also orchid flower stocks

Roll Licorice out in sheets and Carbonize— dip paper in hot liquid Licorice, scrape &dry & cut regular grind the fine Carbzd Lampblack up with Licorice; roll in sheets & Stamp out.—

Borelled[12] Hydrogen deposit Boron in filiment—

Fibres boiled in Licorice=[a]

Torpedo wire[13a]

Try thin Celluloid Elastic as possible—fine relay wire & central phos bronz[14]—wh[i]p cord braid—[a]

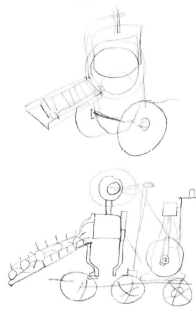

St Cleaning— Machine for gathering snow & Compressing it to ice & dropping definite blocks in gutter as horses advance apparatus[15]   Blox ice being used for Radiated cold in Summer

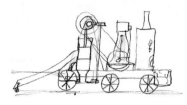

TAE

X, NjWOE, Lab., N-85-10-01:93 (*TAED* N310093). Document multiply signed and dated; miscellaneous calculations omitted. ªFollowed by dividing mark.

1. Hand lanterns were used by a variety of railroad employees for giving signals and inspecting equipment. While signal oil and kerosene were replacing sperm oil by the 1870s, the ready availability of sperm oil led to its continued use by eastern railroads into the 1880s. Sperm oil lamp burners had either one or two tubes through which the wick descended into the oil reservoir. Edison apparently planned for Ott to add a central third tube for the kerosene wick and to put a small kerosene reservoir within the sperm oil reservoir (Barrett and Gross 1994, 1–5, 303–12, 322–26; Hobson 1991, 17–19, 31–35; Maril 1989, 11). Account records indicate that work on "Oil Lamp Experiments" took place during the second half of March (Ledger #5:590, Accounts [*TAED* AB003 (image 292)]).

2. Edison made several drawings on 5 March in a pocket notebook of designs for kerosene lamps that would not be affected by wind. PN-86-3-04, Lab. (*TAED* NP021B).

3. This note and the following three drawings on thermoelectricity were written upside down in this book; the editors have placed them in the order they were most likely written. In this note, Edison proposes that rather than produce thermoelectric currents in the usual manner from a small temperature differential between two pieces of distinct metals, he might generate them using the same metal treated in such a way that it would respond differentially to heat.

4. Figure labels are "C[opper]," "ant[imony]," "C," "ant," "Cop," "ant," and "Raident Heat & Light." One way in which Edison could generate or measure small electrical charges produced by radiant heat or light was by using a thermopile.

5. Figure labels are "Silver polished," "Silver Lamp blacked," "Vacuum," "heat," "heat," "heat," "cold," and "Heat."

6. Edison had hypothesized that the problem of "electrical carrying" (or "carbon carrying"), in which fine particles of carbon migrated from the filament to the lamp globe, was caused by static attraction between the carbon and the glass. In two patent applications filed in October 1882 and a third filed in November 1883, he proposed methods for electrically neutralizing that attraction. This appears to be the purpose of this drawing and those following. He returned to the problem of static attraction in a notebook entry made during his Florida honeymoon. See Docs. 2025 and 2346 esp. n. 2; U.S. Pats. 268,206; 273,486; 425,761; N-86-08-18:199, Lab. (*TAED* N314199).

7. Figure label is "platina."

8. Figure label is "disk paper Carbon." Here and in one of the drawings below, Edison is likely referring to the silver-chloride battery developed by Warren de la Rue in 1868. Each cell consisted of a zinc electrode immersed in a solution of sodium chloride or zinc chloride and a silver electrode in the form of a wire imbedded in a stick of fused silver chloride. A series of such cells (de la Rue used as many as 11,000) could be made into a relatively light, compact, and constant battery. Hospitalier 1883, 1:29–30; Fownes and Watts 1883, 324–25.

9. Figure labels are "carbon," "platina," "Vacum Leyden," and "manipulate."

10. Figure label above is "tin."

11. Figure labels are "P[ositive]" and "N[egative]."

12. The editors have not determined what Edison meant by this term.

13. Figure labels are "Steel wire," "Linen braid & Linseed Lead very thin—," and "finest linen, Balata." Edison presumably had in mind a cable for powering and controlling the Edison-Sims torpedo (see Doc. 2779 n. 2).

14. Phosphor-bronze alloy was invented by Dr. Künzel of Dresden and manufactured by the Phosphor Bronze Co. Among its uses was a telephone wire invented by Lazare Weiller, who was associated with the company and first displayed his wire at the Paris Electrical Exhibition of 1881. Nursey 1885, 130, 136.

15. Charles Batchelor experimentally compressed snow in February 1888, finding that a pressure of ten atmospheres would reduce the volume of "very wet snow" by half (Cat. 1337:42 [item 541, 20 Feb. 1888], Batchelor [*TAED* MBJ004, image 23]). Edison's interest in snow removal was likely prompted by two recent storms. One in January dumped twelve inches of snow on New York City and occupied the entire force of the Dept. of Street Cleaning. While laborers hauled off about 6,000 loads the first day, side streets and those in the upper part of the city remained uncleared. The New-York Snow Melting Co. made an unsuccessful effort to melt the accumulation and allow it to run off into the gutters. The *New York Times*, noting this attempt at mechanized melting, editorialized, "It really seems as if some means might be devised of preventing the blockade of business that a snowstorm enforces upon New-York. Even the enormous expense of carting the snow bodily from the principal thoroughfares of the city and dumping it into the rivers would probably be a true economy, if the enormous expense entailed by the presence of the snow until nature takes it out of the way could be accurately ascertained and assessed." A massive snowstorm struck much of the eastern and southern United States on 3–4 February. New York City received several wind-whipped inches, but rail travel was paralyzed south of Washington, D.C. "Snow in the Streets," *NYT*, 10 Jan. 1886, 8; "Carting Off the Snow," ibid., 11 Jan. 1886, 5; "The Heavy Snowstorm," ibid., 5 Feb. 1886, 4; "Hovering Around Zero, ibid., 5 Feb. 1886, 5; "The Great Snow Storm," *Atlanta Constitution*, 5 Feb. 1886, 1; "Snow-Melting Machine," *American Machinist* 7 (1884): 7; on the history of urban snow removal see MacKelvey 1995 and "Snow Removal," National Snow and Ice Data Center (http://nsidc.org /cryosphere/snow/removal.html), accessed 28 Aug. 2012.

[New Brunswick, N.J., February 8, 1886][2]

N. Jersey Supreme Ct.

Lucy F. Seyfert[3]

v.

Thomas A. Edison[a]

In Case. Statement.

Judgment Feby. 20. 1884—Damages & Costs[4]	$5348.64
Interest Feby. 20. 1884 to Nov. 11. 1884	236.23
Sheriffs fees on fi. fa.[5] including costs of resale.	81.22
	5666.09
Proceeds of sales under first writ	2750.00
Deficiency on— " "	2916.09
Sheriffs fees on alias writ[6]	35.94
	2952.03
Proceeds of sales under alias writ	11.00
	2941.03
Costs of alias writ & writs to Essex & Hudson ($1.18 each)[b]	3.54
	2944.57

Interest from Nov. 11. 1884 to Feby 10. 1886		
91 d[ay]s	44.01	
1 yr.	176.72	220.73
Costs of proceedings supplementary to execution as taxed by clerk exclusive of commissioners fees	22.16	
Add Comrs fees—	21.60	43.76
Total due Feby. 10. 1886		$3,209.06

D, NjWOE, DF (*TAED* D8603ZAB). [a]"Lucy . . . Edison" enclosed in right brace pointing to "In Case. Statement." [b]"(1.18 each)" interlined above.

1. This document summarizes Edison's costs related to the settlement of a long-running lawsuit against him. The case arose from a $300 note that Edison gave in 1874 to an investor in the Automatic Telegraph Co. The note passed to William Seyfert, who later conveyed it to his wife, Lucy Seyfert. When Mrs. Seyfert attempted to collect, Edison disowned the note, claiming it had been part of a personal loan to her husband. See Doc. 2662 esp. nn. 2–3, 6; see note 4 below.

2. The place and date are taken from the cover letter enclosing this statement, which Lucy Seyfert's lawyers sent to Edison's attorney. Woodbridge Strong & Son to John Tomlinson, 8 Feb. 1886, DF (*TAED* D8603ZAA).

3. Lucy Fisher Hunter Seyfert (1829?–1887) was the daughter of

Jacob V. R. Hunter, a wealthy ironmaster, who in 1846 acquired a one-quarter interest in the Reading Ironworks, owned at the time by the firm of Seyfert, McManus & Co. Lucy's husband, William M. Seyfert, who took over the management of the property in 1848, was the son of Simon Seyfert, a founding partner in Seyfert, McManus & Co. When Jacob Hunter died in 1861, his estate of several hundred thousand dollars passed to Lucy and her nine siblings. Hunter had amassed his fortune through ownership of the Sally Ann Furnace in Rockland Township, Pa., and his investments in mines, timberland, and the Reading Ironworks. Doc. 2662 n. 5; Philadelphia, Pa., *Death Certificates Index, 1803–1915,* FHL film no. 1003713, online database accessed through Ancestry.com, 13 Feb. 2013; "Sally Ann Furnace, in Rockland; Owned by Jacob Van Reed Hunter," *Reading Eagle,* 3 Mar. 1918, 18.

4. The Seyfert case was the subject of two trials (and the object of numerous delaying tactics). The first trial, before the Middlesex County Circuit Court in December 1882, resulted in a judgment of $5,065.84 against Edison. In 1883, his attorney obtained a new trial on technical grounds. In the second contest, in February 1884, the New Jersey Supreme Court again decided against Edison and assessed him court costs in addition to the damages to Seyfert. The Middlesex County sheriff partially satisfied the judgment by the sale of some of Edison's Menlo Park property, but the death of Edison's wife interrupted the process. The state Supreme Court then twice ordered Edison to appear for discovery to determine what other assets he might have, and Edison twice failed to appear. In both instances, his attorney argued that Edison, having moved to New York, was out of the court's jurisdiction. These motions were denied. The second time, in December, the Court also cited Edison for contempt. Edison's purchase of a large house in Llewellyn Park, N.J., and his upcoming wedding likely motivated him to settle the claims now. The day that Edison signed the mortgage papers on his future home, Samuel Insull asked attorney John Tomlinson to "please find out <u>immediately</u> the amount still due on Seyfert Judgment & we will arrange to pay." It is not clear when the matter was fully resolved but probably by mid–April, when Tomlinson arranged to pay the sheriff's fees. Docs. 2662 nn. 2, 6 and 2698 n. 1; *Seyfert v. Edison* (1880), Lit. (*TAED* QD011); Vroom 1886, 18:428–34; New Jersey Supreme Court opinion, 4 Dec. 1885; Abraham Schenck to Tomlinson, 14 Apr. 1886; both DF (*TAED* D8503ZEG, D8603ZAS); Insull to Tomlinson, 20 Jan. 1886, Lbk. 21:216 (*TAED* LB021216).

5. Short for *fieri facias,* a Latin term meaning "cause to be made." It is a judge's writ of execution permitting an officer of the court to seize and sell a debtor's property in satisfaction of a judgment. Rapalje and Lawrence 1888, s.v. "fieri facias."

6. An alias (Latin for "otherwise") writ is issued when the original writ did not produce the intended effect. Rapalje and Lawrence 1888, s.v. "alias."

–2894–

To William Mawer[1]

[New York,] 9th Feby [188]6

Dear Sir,

Referring to your memorandum on foot of account I do not want you to place any more flowers on grave[2]   Your truly

Thos. A. Edison   I[nsull]

L (letterpress copy), NjWOE, Lbk. 21:260 (*TAED* LB021260). Written by Samuel Insull.

1. William Mawer (b. 1848), a native of England, was the proprietor of a florist shop at 376 Belleville Ave. in Newark. U.S. Census Bureau 1970 (1880), roll 778, p. 176D, image 0014 (Newark, Essex, N.J.); ibid. 1982? (1900), roll T623_964, p. 11B (Newark, Essex, N.J.); Holbrook 1885, 622; ibid., 1888, 723.

2. The grave was that of Edison's first wife, Mary, who died in August 1884 and was interred at Mount Pleasant Cemetery in Newark (see Docs. 2712 [headnote], 2717–2718). On 1 February, Edison paid Mawer $29 due on account and inquired whether he was still putting flowers there. Mawer replied that he had not done so recently because of severe winter weather and wished to know if he should resume when conditions improved (TAE to Mawer, 1 Feb. 1886, Lbk. 21:239 [*TAED* LB021239]; Mawer bill, 6 Feb. 1886, Voucher [Household] no. 51 [box 11; 1886]).

–2895–

A. C. Mears to Samuel Insull

[New York,] Feby 10, 1886.

Dear Sir,

I hope Miss Edison's absence for about two weeks is not due to illness. I regret this interruption to her studies, as she has been making very satisfactory improvement.

You will greatly oblige me by informing Mr. Edison that I would like to know when to expect her again in her classes. Her Music lessons are also interrupted, which is a real loss to her.

I sincerely wish she could be with me altogether in boarding-school, as the regularity of her study would tend greatly to her progress.[1] Yours Respectfully,

Mme A. C. Mears   by S.[2]

L, NjWOE, DF (*TAED* D8614B).

1. Alfred Tate replied to this letter two weeks later, attributing the delay to the absence of both Edison and Insull. He stated only that Marion Edison "has gone to Florida to spend the balance of the winter." In reply to another inquiry in March from Mears (not found), Insull explained that Marion would live in the family's new home in Orange when she came back, and he did "not think she will return to your school." When Mears billed Edison $118.40 for tuition in May, she allowed a deduction of $77.50 for Marion's extended absence (Tate to Mears, 25 Feb. 1886; Insull to Mears, 11 Mar. 1886; both Lbk. 21:299,

384 [*TAED* LB021299, LB021384]; Mears to TAE, 7 May 1886, DF [*TAED* D8614C]; Voucher [household] no. 232 [box 33; 1886]). Marion's brothers seem to have attended Mme. da Silva and Mrs. Bradford's school at 15–17 W. Thirty-eighth St. as day students through 1 February. The institution advertised itself as a coeducational "English, French, and German Boarding and Day School" (Voucher [household] no. 233 [box 33; 1886]; Advertisement, *The Churchman* [8 Aug. 1885]: 168).

2. Not identified.

–2896–

*Notebook Entry:
Phonoplex and
Railway Telegraph*

[New York,] Feby 12 1886

Phonoplex

Wind a 7 ohm or less coil with bare wire with hard thread very much spiraled so as to give nearly all air space between wire so coil will charge & discharge quicker—

To obviate induction arrange condenser thus

graduate this Condenser so it will allow regular Current to lift weight but not the induction Current.[1]

Try this for vibrator for RR telgh

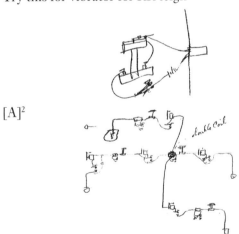

[A][2]

another line   Repeating into by induction

Phonoplex Repeating[3]

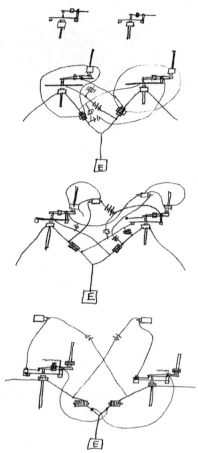

Induction coil & condenser Combined   try Expmnt[4]

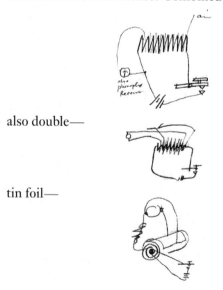

also double—

tin foil—

paper with foil pasted on side— wound[a] spirally & double

TAE

X, NjWOE, Lab., N-85-10-01:115 (*TAED* N310115). Document multiply signed and dated; unrelated and undated doodles, sketches, and miscellaneous calculations on several facing pages omitted. [a]Obscured overwritten text.

1. Cf. Doc. 2902.
2. Figure label is "double Coil."
3. Although Edison had signed a patent application for a phonoplex repeater in November, it was not filed until 19 February (see Doc. 2867 n. 7). The drawings in this document seem to have only a general relationship to the devices described and illustrated in that patent.
4. Figure labels are "also phonoplex Receiver" and "air."

–2897–

*Charles Batchelor
Journal Entry*

[New York,] Tuesday, Feb. 16 1886

19.[1] <u>Sims-Edison Fish Torpedo</u>:[2] [~~Scot?~~][a] Edison, I, and Gardner Sims[3] signed as incorporators of this company today.[4]

AD, NjWOE, Batchelor, Cat. 1336:9 (*TAED* MBJ003009D). [a]Canceled

1. Charles Batchelor consecutively numbered each entry in this journal.
2. The Sims-Edison torpedo was based on the original design developed by Winfield Scott Sims with funding from the Army (see Doc. 2779 n. 2). It consisted of a cylindrical hull of copper 28 feet long, with the one-mile model having a circumference of 21 inches and the two-mile version a circumference of 28 inches. It weighed from 3,000 to 4,000 pounds and carried a load of 250 to 400 pounds of dynamite. When submerged it was supported by a float attached by an upright steel stanchion. The torpedo, which was constructed entirely of copper and brass to avoid rust and corrosion, was designed so that no section weighed more than 800 pounds and the whole thing could be taken apart and put together in less than fifteen minutes. Sims-Edison

Electric Torpedo Co. circular, [1886], PP (*TAED* CA017B); "Highly Destructive Torpedoes," *New York Tribune*, 31 July 1886; "Torpedoes for Uncle Sam," *New York Sun*, 8 Aug. 1886; both Cat. 1057:46b, 46d (*TAED* SM057046b, SM057046d).

It is unclear what contribution Edison himself made to the torpedo. What little evidence there is of the experiments for its new electric motors (tested at the Edison Machine Works during 1885) seems to indicate that Charles Batchelor had primary responsibility for this work, with the assistance of Henry Walter (see Doc. 2779; Unbound Documents [1885], Batchelor [*TAED* MB159, MB169]; also Batchelor to Edison Machine Works, 2 Feb. 1888, DF [*TAED* D8818ACN]). It is likely that Edison's primary role was advisory, thus his position as consulting engineer in the Sims-Edison Electric Torpedo Co. (see note 4) and for promotional purposes.

3. Gardiner Sims (1845–1910) formed a partnership in the late 1870s with Pardon Armington to manufacture steam engines in Lawrence, Mass., where he had been superintendent of the J. C. Hoadley Engine Works. In 1881, they developed a high-speed engine for the Edison electric system, for which they became a major supplier. They relocated to Rhode Island and incorporated as Armington & Sims Co. in 1882 or 1883. Docs. 2078 n. 1 and 2131 n. 7; Obituary, *ASME Transactions* 32 (1910): 1501–2; Obituary, *Power and the Engineer* 32 (5 Apr. 1910): 656; Greene 1886, 260–61.

4. The incorporators of the Sims-Edison Electric Torpedo Co. included Edison, who became consulting electrician; Batchelor, who served on the executive committee; Gardiner Sims, who became vice president and consulting engineer as well as a member of the executive committee; Winfield Sims, general manager; John Anderson, president; Lewis May, treasurer; Frank W. Allin, secretary; William M. Deen, who served on the executive committee; and George H. Stayner. Sims-Edison Electric Torpedo Co., Certificate of Incorporation and By-Laws, 16 Feb. 1886; Sims-Edison Electric Torpedo Co. circular, n.d. [1886]; both PP (*TAED* CA017A, CA017B).

The company was likely formed in the expectation that large orders would be forthcoming from the Army. A recent report indicated that

> the United States Government have recently purchased five Sims torpedoes, and that the War Department has contracted for five more, which are being constructed by the inventor. The Government have also taken money for seven others which are being made at the Edison Machine Works. It would seem, therefore, that in America at any rate this weapon is looked on as having successfully passed that experimental stage at which so many promising inventions fail, and may now be regarded as a recognised means of defensive and offensive warfare. ["The Sims Torpedo," *Engineering* 41 (5 Feb. 1886): 139]

According to the report of Henry Abbot, who had overseen all of the Sims torpedo tests, the Edison motor proved superior to the Weston motors Sims had used previously and to the Siemens Bros. of London motor he adopted in 1885. In comparison to the latter, it generated the same power but at a much lower armature velocity that enabled it to work efficiently without gearing. These results led Abbot to conclude that the "torpedo may now be regarded as perfected in its essential details; and its practical value as a war weapon can probably be estimated with some precision after the trial runs of the five new boats are completed and the detailed reports thereon are submitted to the Board [of Engineers]." However, by the time the last three motors made under Sims contract of 25 June 1886 were delivered, they showed "such a liability to unexpected defects and delays as to hold them still in the experimental stage and prevent a general acceptance of this weapon as a safe and reliable one for harbor defense." The company made efforts to improve the torpedo over the course of the next decade but these efforts proved futile as controllable torpedoes using cables were abandoned before World War I. *Report of the Chief of Engineers, U.S. Army* 1886, 51; idem 1887, 418; idem 1889, 481; "Electrical Means for Harbor Defense," *Electrical World* 17 (25 Apr. 1891): 305–6; "Bettering Their Torpedo," *NYT*, 9 Dec. 1892, 8; Armstrong 1896, 86–87; Bishop 1918, 54–55.

–2898–

Notebook Entry:
Arc Lamp

[New York,] Feby 17 1886—

Carbon for[a] arc Light= Experiments[1b]

Liccorice disolved in water is to be used in every proportion. It is to be mixed thinly with the different Carbon powders put in the mixer & the mixer heated up to 220 to 250 to drive off all the water   you can continue this for different times until the licorice is solid & will just bind the carbon together in the press—

Make samples of every kind of proportion of Lampblack— with hard Carbon also with the petroleum Carbon also Lampblack alone— also hard Carbon alone also petroleum Carbon alone[b]

The do ~~this~~ all the Experiments over again using Coal tar as a base instead of Liccorice

The do them all again with Sugar as a base.

also with Wood Tar as a base.

also with Copal varnish as a base.

O

also take some of the unburnt petroleum Carbon powder it & without mixing anything with it press it Hot & squeeze it hot. It softens—[b]

also try Experiment after all the different Carbons have been fully Carbonized at the Lamp factory and after burning a couple to get their burning time to rig up a place in Laboratory where you can bring up a mould just to a dull red or little below—

Then take one or two of the remaining carbons of that batch

& soak in Licorice water, heating the Carbon to 2 or 300 ϴin sand bath before plunging in Licorice syrup— then after[c] it has remained in for 2 or 3 hours take out put in mould & bring gradually to 8 or 900 fahr— then take it out again resoak & recarbonize   see how many times you have to do this to get 12 hours burning

also try soaking in sugar syrup   also in thick Coal Tar. also pitch—

J F Ott   M N Force

X, NjWOE, Lab., N-85-12-08:59 (*TAED* N313059). Document multiply signed; miscellaneous calculations not transcribed from facing page. [a]Interlined above. [b]Followed by dividing mark. [c]Obscured overwritten text.

1. Edison had spoken to Francis Upton in January about "experimenting on Carbons for arc lights." Upton suggested enlisting John Lawson, a capable chemist and knowledgeable "insider" whom Upton feared losing from the Edison Lamp Co., and indicated that the firm was "willing to pay a good portion of his salary in case you can use him" (Upton to TAE, 22 Jan. 1886, DF [*TAED* D8626B]). Edison's interest in the subject was likely spurred by a proposal from George Bushar Markle, Jr., part of a powerful family of bankers and anthracite coal operators in Hazelton, Pa., and an early investor in the Edison central station lighting plant there (Docs. 2424 [headnote] esp. n. 39 and 2617 n. 2). The parties discussed some business arrangement, possibly for a new company in which Edison would have a large minority interest; Edison assented but did not sign the contract before leaving for Florida and apparently never did so (Markle to TAE, 22 Jan. 1886; Markle to Samuel Insull, 10 Feb., 3, 12, and 16 Mar. 1886; Insull to John Tomlinson, 19 Apr. 1886; all DF [*TAED* D8633B2, D8633D1, D8622C, D8622D, D8622F, D8618D]; Alfred Tate to Markle, 5 Feb. 1886; Insull to Markle, 9 Feb. and 8 Mar. 1886; Lbk. 21:247, 262, 352 [*TAED* LB021247, LB021262, LB021352]). Edison did, however, continue to experiment periodically on arc light carbons until October, when he wrote to Markle's brother Alvan: "I have not met with the success which I had hoped to in producing arc light carbons. I can make a better carbon than those ordinarily sold, but my results are not sufficiently in advance of those obtained by other people to satisfy me." At the start of 1888, Samuel Insull decided to wipe off Edison's books several old accounts he considered "worthless." The largest among them was $10,648.82 in experimental expenses for "arc light, arc light machines and municipal machines" (TAE to Alvan Markle, 18 and 25 Oct. 1886, Lbk. 23:19, 51 [*TAED* LB023019, LB023051]; Insull to TAE, 26 Jan. 1888, DF [*TAED* D8835AAP]; see also Doc. 2928).

Manufacturers of arc light carbons at this time used a variety of carbon substances mixed with a carbonaceous liquid into in a paste, which was then molded or squirted under hydraulic pressure through a die to form the proper shape. The Markles evidently offered to sell Edison equipment suited to this purpose, including "the hydraulic press, the squirter, and a big roller grinder" (TAE to Alvan Markle, 19 and 25 Oct.

1886, Lbk. 23:19, 51 [*TAED* LB023019, LB023051]). Formed carbons were baked to the desired hardness. Controlling the baking temperature was extremely important and is the likely reason that Edison used a sand bath, a common piece of laboratory apparatus in which a container filled with sand is heated so as to supply even heating to substances or chemicals in another vessel within it. In some cases, after the original baking, manufacturers would soak the carbons in a carbonaceous liquid and bake them a second time to increase their purity and hardness, key qualities of durable carbons that would run at consistent temperatures and emit steady light (Alglave and Boulard 1884, 46–55; Atkinson 1889, 163–73; Maier 1886, 147–52; Houston and Kennelly 1902, 307–20).

–2899–

To Mina Miller

Jersey City NJ[1] FEB 18 *1885*6. 6:21 pm[a]

Miss Mina Miller—

Better not send the girl[2] with Mr. Gilliland but take her with us if agreeable to you   they left tonight   six train

Edison

L (telegram), FFmEFW, EMFP (*TAED* X104GA). Message form of Western Union Telegraph Co. [a]"FEB" stamped and "*1885*" preprinted on form.

1. This message was sent from the Erie Railroad depot in Jersey City.
2. Probably Louisa Rittersbaugh (b. 1838?), a domestic servant in the household of Lewis and Mary Valinda Miller. She was the daughter of George Rittersbaugh, a saddler in Canton, Ohio, and his wife Elizabeth. The Millers hired Rittersbaugh as early as 1860 to help care for the children and perform other duties in their Canton home, and she seems to have moved with the family to Akron about 1863. U.S. Census Bureau 1963? (1850), roll M432_730, p. 519B, image 390 (Canton, Stark, Ohio); ibid., 1967? (1860), roll M653_1037, p. 127, image 259 (Canton, Stark, Ohio); Hendrick 1925, 83, 102, 114; Robert Miller to Mina Miller Edison, 27 Feb. 1886, N(ChaCI) (*TAED* X477B); Mary Valinda Miller to Mina Miller Edison, 14 Mar. 1886, FR (*TAED* FI001AAE).

–2900–

To John Ott

[New York, February 20, 1886?[1]]

John—

Have Hamilton set up the sulphate Zinc tube[2] artificial line & use condensers, and arrange to get good phone & good 7 ohm coil & arrange it so you can just get it to knock the weight up with 3 new cells; put an ampermeter in the 3 cell circuit (use strap on Kinneys galvanometer)[3] so tas to always be sure you have the same amperes through the coil. The you will be able to make any tests of Coils & other devices I may send you by mail—

Try this:[4]

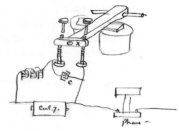

X is a yoke that works on the end of the sounder lever & is held by friction   when the lever comes down the points on the yoke strike bothe of the screws at the same[a] time and leave at the same time   Thus the spark on each point is $\frac{1}{4}$ what it would be on one point the circuit being broken in two places at the same time. What I want to find out is if I can dispense with the Condenser around the points if I break the circuit in two <u>or more</u> places at the same time— If I cant perhaps I can reduce the amount required of Condenser. Don't you remember down at Goerck st we made a sounder that broke the circuit in 4 places at the same time   It was made like this so[5]

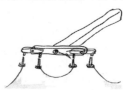

If the two points shows up better without a condenser than a single point— you better make one with 4 points top & bottom[b]

also make a 7 ohm coil Thus=[6]

where the wood is there are no lines of force and the wire does[a] no good

Try and work up a good low Resistance Callaud[7] for phonoplex—

T A Edison

⟨Finished Experiment Monday March 1, 86⟩[c]

ALS, NjWOE, Lab., Unbound Notes and Drawings (1886) (*TAED* NS86ADC). [a]Obscured overwritten text. [b]Followed by dividing mark. [c]Marginalia written by John Ott.

1. John Ott wrote this date on the letter, presumably when he received it.

2. Edison referred to a recent modification of the Daniell cell, a two-fluid battery widely used to provide a steady electromotive force.

In this adaptation, each electrolyte (typically copper sulphate and zinc sulphate) was placed in a separate test tube and the top of each tube connected with the other through an inverted U-shaped piece of tubing. This arrangement minimized the mixing of electrolytes that occurred through the porous septum in conventional forms of this battery. Sloane 1887.

3. Edison meant Patrick Kenny. The editors have found no information about this particular instrument.

4. Figure labels are "coil. 7." and "phone—."

5. Figure label is "insulation." In 1881, Edison experimented with and applied for patent protection on a similar device for breaking a dynamo circuit simultaneously at several points in order to reduce sparking at the commutator. His use of this principle for spark suppression dates to May 1879, when he adapted telegraph sounders to this purpose, predating the existence of the Edison Machine Works and the Testing Room at Goerck St. in New York. Docs. 1735 n. 12, 1745 first figure and n. 1, 2122 n. 5.

6. Figure labels are "wire" and "wood."

7. Developed by Armand Callaud of France prior to 1863 and subsequently used on French telegraph lines, this gravity cell was a less expensive and more durable form of the Daniell battery. Doc. 358 n. 6; *KNMD*, s.v. "Callaud Battery"; "The Sulphate of Copper Battery of M. Callaud," *Electrician* 3 (13 Feb. 1863): 167; "Pile de Callaud," *Journal of the Telegraph* 3 (15 May 1870): 141.

–2901–

*Charles Batchelor
Journal Entry*

[New York,] Saturday Feb 20th 1886

23:[1] <u>Noctes Edisoniana</u>: Farewell dinner to T.A.E. prior to his marriage.[2] Present. C Batchelor, E H Johnson, S. Insul, A O Tate, J. C. Tomlinson, G. C. Sims, Lieut Toppan,[3] S. Bergman, C. E. Chinnock.

General regrets that Mr Kreusi & Mr Upton had somehow been overlooked.

AD, NjWOE, Batchelor, Cat. 1336:11 (*TAED* MBJ003011C). Written by Charles Batchelor.

1. Charles Batchelor consecutively numbered each entry in this journal.

2. Batchelor noted in a journal entry dated 23 February under the heading "Edison's marriage" that he left "with party by special car for Akron and attended on 24th   returning on the 25th." Johnson, Insull, Tomlinson, Sims, Bergmann, and Charles Bruch (whose family was friends with the Millers) were among those traveling in the private car. Cat. 1336:11 (item 25, 23 Feb. 1886), Batchelor (*TAED* MBJ003011E); "Edison-Miller," *Akron Enquirer*, 25 Feb. 1886, Ser. 7, EP&RI (*TAED* X001N1); "Mr. Edison's Wedding," *NYT*, 25 Feb. 1886, 5.

3. Frank Winship Toppan (1855–1922), a native of Newburyport, Mass., was a U.S. Navy officer. Toppan graduated from the Naval Academy in 1879 and received the rank of ensign in 1881 (U.S. Naval

Academy Cemetery Inventory Form, 2005 [usna.edu/cemetery/lookup
.htm, accessed 12 Nov. 2012]; "Died," *NYT*, 22 Apr. 1933, 9). According to his testimony in a later patent interference, he became associated with Ezra Gilliland in November 1884 and soon afterward entered
into discussions with Gilliland and Edison about forming a company to
exploit the railway telegraph patent of William Wiley Smith in which
Gilliland had a one-half ownership interest; an Edison account record
for Toppan suggests that his unspecified work on behalf of the railway
telegraph continued intermittently into September 1885. In spite of his
relatively recent association with Edison, Toppan served as best man at
the wedding. In 1888, he became the manager of Edison's new phonograph factory in Orange, N.J. Toppan left the Navy in 1887 but came
out of retirement to serve in the Spanish-American War. He later attained the rank of lieutenant commander and received a private burial
in the cemetery of the Naval Academy in Annapolis (Toppan testimony,
Edison and Gilliland v. Phelps, 25–26, MdCpNA [*TAED* W100DKC];
Ledger #5:709, Accts. [*TAED* AB003, image 344]; "Mr. Edison's Wedding, *NYT*, 25 Feb. 1886, 5; "Edison-Miller," *Akron Enquirer*, 25 Feb.
1886, EP&RI [*TAED* X001N1]; "The Phonograph Works," *Herald*
[Orange, N.J.], 1 Sept. 1888, Unbound Clippings [*TAED* SC88097c];
Callahan 1901, 548).

–2902–

*Memorandum to
Richard Dyer:
Phonoplex*

[Jacksonville,] Feby 28 1886

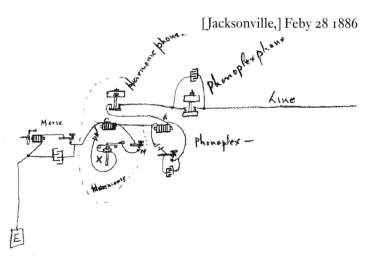

⟨(Triphonoplex)   Hows that?⟩[1]
Dyer—

Take out patent on this   In addition to regular phonoplex
I add another receiving phone just the same as the phonoplex
phone except it has a light weight and has an upward limiting
nut so the diaphragm cant throw it up more than $\frac{1}{1000}^a$ of an
inch— This weight responds to the rapid vibrations thrown
on the wire by a local self make & break telephone diaphragm X
C is a 7 ohm primary coil— when key N is closed the dia-

phragm vibrates as its lever self mks & bks. This causes the coil to send [-]^b waves— The phonoplex phone is prevented from responding by being shunted with a small Condenser which nearly wipes out the rapid [arriving?]^c but weak vibrations of the harmonic & in addition is provided with so heavy a weight that while the strong phonoplex waves lifts it no single harmonic wave has the strength. Thus I get 3 messages over a way wire.[2]

T A Edison

ADS, NjWOE, Lab., Cat. 1151 (*TAED* NM020AAS). ^aObscured overwritten text. ^bCanceled. ^cIllegible.

1. Figure labels above are "Morse," "Harmonic," "X," "N," "Harmonic phone," "d," "phonoplex phone," "phonoplex—" and "Line—"

2. Edison's description of this system became the basis for a patent application that he executed on 11 May, after his return from Florida, and the drawing above was the template for the first of its two figures (U.S. Pat. 370,132). The patent was an extension of his basic specification on the phonoplex, which doubled the capacity of way lines. He intended the new arrangement to triple way line capacity by differentiating among two types of induction signals in addition to regular Morse signals, effectively permitting it to work as a triplex telegraph. Except for the Morse instruments, the system was based entirely on the phonoplex principle, unlike the hybrids with duplex or quadruplex circuits between terminal stations that he had devised some six months earlier (see Doc. 2849).

The patent described the three types of matched transmitters and receivers, each pair working with a distinctive signal. Ordinary Morse instruments constituted one pair. Another set was the "simple induction apparatus," like the transmitter used in all phonoplex arrangements (and described in the original patent), and its matching receiver. The third class of transmitter, identified in this document as the "self make & break telephone diaphragm," produced an effect similar to that of the "circuit controller" in the railway telegraph. The vibrating diaphragm controlled not only the contacts to make or break the line circuit, but also the circuit to the electromagnet whose consequent on-and-off operation caused it to move in the first place. Edison indicated that it would generate "rapidly-occurring induction waves or vibrations which form a musical note, such note being divided at the transmitter into dots and dashes for producing Morse harmonic signals." It worked in conjunction with a "harmonic sounder" similar to the ordinary induction receiver but with a weight lighter and more limited in its range of motion (cf. Doc. 2896). As described above, the diaphragm would vibrate the weight in response to the rapid and relatively weak pulsations of the harmonic transmitter; the regular phonoplex sounder, largely isolated from those signals by a condenser, would not respond. To increase the volume of sound, Edison prescribed using "a metal weight constructed to be resonant and made as a ring standing on edge, and preferably open at one side." By the latter part of 1887, Edison was using an actual "circuit controller" to attain the same result.

U.S. Pat. 370,132; "Edison's Improved 'Phonoplex,'" *Electrical World* 10 (8 Oct. 1887): 192.

–2903–

Mina Edison to
Mary Valinda Miller

St. James Hotel[1]   Jacksonville Florida   Feb. 28/86
My dearest Mamma.

On the last day of the month which made me a bride I thought nothing would be a more fitting close than a letter to you. Also this is Sunday and I now know right where you are, while I write. I see you all at church in the hopes of course that all are able to be there. Some singing has been heard from the parlor, gospel hymns, etc. Mr. Edison found in his coat pocket the Discipline[2] which he has been looking through— Some parts of it is quite obscure and he intends to study it well having all the points he cannot understand explained[a] by Papa when he sees him again. He wanted to know the other day if I married him to convert him, he is willing and wants to believe but he says he has such an inquiring mind that he cannot believe what he cannot see the whys and wherefores of but he does not want me to be influenced in the least by him nor believe anything he says on the subject for he thinks he must be wrong but can't help it. No talking or arguing on the subject will do any good everyday life is what must be the convincing power.

Yesterday as we left Atlanta I saw Mr. and Mrs. Whitney[3] (I think there names are, two <u>very large</u> persons that are <u>always</u> together and have been at Chautauqua, perhaps the name is Whitestone) anyway you may know whom I mean from above discription; Homely as mud fences both of them.

People stare at us so; all knowing that the person with me is Mr. Edison who has just been married and of course the curosity of seeing him and then somewhat increased by his recent marriage causes quite a stir. Arrive at a hotel and hardly leave my room until we leave again, that is I do not go to my meals but Mr. E. has taken a drive about to see the sights. We saw very little scenery coming from Atlanta here as we travelled last night but this morning we saw peach groves in bloom. The orange trees all seem to have been killed this far down at least.[4] Also we saw the yellow pine tree from which is derived the turpentine we use. It looked so queer to see a large forest of tall pines all hacked away just a little distance from the ground. One thing peculiar to the trees here is that they are all so very straight. that is due to the fact that there is so little wind it is perfectly flat here[b] in Florida   We have seen nothing yet resembling the jungles we read and see illustrated in books that is still in store for me. You would or will be surprised when you see how they live here, towns! nothing to them what ever and a dozen or so of colored people living in a house with probably but one room and no windows. They keep out in this way as much heat as possible as the sun is intensely hot in the day time. All the men of color are so black here; nearly every one of the darkest shade.[c] Excuse blot—

Upon opening my trunk this morning for the first time since I left home found my nugs[5] and every thing all safe. I feel some what refreshed from the change of clothes made. This is a miserable[d] hotel and a worse city. Tomorrow we leave for St. Augustine and as I have stated before will remain there some time. I am not as well as when I wrote before[6] but nothing serious a few days will bring me out aright again. I wonder whether all the beautiful decorations have been removed.[7] I wish I might see them again. Were they not beautiful? I shall never forget what a beautiful sight you all made under[c] the awning when we left nor can I ever forget when we as a family were in the Sitting room— It all looked so lovely to me.

I want to write a few lines to Helen[8] so will say goodbye once more, writing again soon.

Love to all— Your truly loving daughter—

Mina.

Mr. Edison is lying on the bed trying to sleep but we have come into the flies home. He sends love and telegraphed this morning the message of safety.[9] Lovingly   Mina.

ALS, FFmEFW, EMFP (*TAED* X104FAA). [a]"x" interlined above. [b]Interlined above. [c]Obscured by ink blot. [d]Obscured overwritten text.

1. The St. James, reputedly one of the state's finest tourist hotels, opened in 1869 and was enlarged several times, most recently in 1881 when four stories were added to the original three. The Stick-style structure could accommodate five hundred guests by this time and boasted, among other features, telegraph offices, a passenger elevator, steam heat, electric lighting, an orchestra, and "plenty of agreeable society." Varnum 1885, 46; Braden 2002, 64, 81; Albion 2008, 19–20.

2. The *Book of Discipline* sets forth the laws, plan, polity, and process by which Methodists govern themselves. It contains sections on the history, theology, and social teachings of the church. Initially published in 1784, the *Discipline* undergoes quadrennial amendments. *MEAR*, s.v. "United Methodism;" *Book of Discipline* 2008, v, 11.

3. Not identified.

4. Arctic air swept through much of the United States in mid-January, producing what the *New York Times* reported as the "coldest weather ever known in the South" and threatening much of Florida's orange crop. By March, reports indicated that most of the state's or-

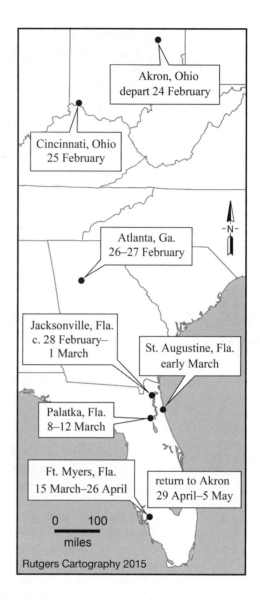

Akron, Ohio
depart 24 February

Cincinnati, Ohio
25 February

Atlanta, Ga.
26–27 February

-N-

Jacksonville, Fla.
c. 28 February–
1 March

St. Augustine, Fla.
early March

Palatka, Fla.
8–12 March

Ft. Myers, Fla.
15 March–26 April

return to Akron
29 April–5 May

0   100

miles

Rutgers Cartography 2015

ange trees had recovered. "The Trail of the Storm," *NYT,* 12 Jan. 1886, 1; "The Frostbitten South," ibid., 15 Jan. 1886, 4; "Florida's Orange Groves," ibid., 8 Mar. 1886, 3.

5. Mina seems to have used an obscure word for a type of medicinal distillate, possibly of balsam. *OED,* s.v., "nug 1."

6. The editors have not found this letter.

7. Mina and Edison were married under a canopy of palms and in front of a fan-shaped design consisting of calla lilies and, according to one account, several varieties of roses. They received congratulations in the Miller library beneath an elaborately decorated floral wishbone of roses. "Edison-Miller," *Akron Enquirer,* 25 Feb. 1886, Ser. 7, EP&RI (*TAED* X001N1); "Mr. Edison's Wedding," *NYT,* 25 Feb. 1886, 5.

8. Mina's friend Helen Storer (1866–1917) came from a prominent

Akron family. Storer's father, James, was a business owner and the city postmaster, and her mother, Maria, was a community organizer. Mina and Helen were high school classmates and continued their studies in Boston after graduation, Mina at Miss Johnson's School and Helen at Wellesley College. Enders 2006, 124–25; U.S. Census Bureau 1982? (1900), roll T623, p. 3A (Akron Ward 1, Summit, Ohio); "Alumnae News," *Wellesley Alumnae Quarterly* 2 (Apr. 1918), 189.

9. Not found.

–2904–

*From Thomas
Edison, Jr.*

Glenmont[1] March 1st 1886.

My dear Papa and Mamma.

I suppose you are in[a] your southern home having a good time. we kiss our new Mamma every night be fore going to bed   Willie[2] and I goes to school every day, I was promoted[a] in higher room in[a] the fifth reader and willie is in[a] the second reader and I am in long[a] division and willie is in subtraction[a]

Willie and I hav each got a rabbit and each got a pigeon. yesterday[a] Willie[a] and I let are pigeons out on the lawn[a] to play   Oh papa we have head a ferful storm has blown down trees and taken[a] roofs off of stores in orang   uncle[a] and antie[3] gives us all the candy   we are all well   love and kisses to you[a] Mamma good papa and Mamma

T A Edison Jr

ALS, CEF (*TAED* X018B1AA). [a]Obscured overwritten text.

1. The Edisons' new home in Llewellyn Park, Orange, N.J. It is not clear how the name became associated with the property, though a number of homes in the area already had names, and architect Henry Hudson Holly, who designed this one, endorsed the practice. (The naming of houses was common enough and certainly familiar to Mina, whose family home in Ohio was called "Oak Place.") "Glenmont" seems to have originated during the Edisons' ownership, possibly from the house's location on a rise or "mount" near the "Glyn Ellyn" ravine. Yocum 1998, 1:19–20.

2. William Leslie Edison (1878–1937), the youngest child of Edison and his first wife, Mary.

3. This was probably Edison's brother (William Pitt Edison) and sister-in-law (Ellen J. Edison), who took the boys to their farm near Port Huron, Mich., during the summer.

–2905–

*Notebook Entry:
Miscellaneous*

[St. Augustine,] March 2 1886

Pass strong currents through Solutions & then test with their proper reagents. also ditto in field of powerful magnet   see what change if any[1]

Phonoplex X lifts weight nearly up—reg throws it open ckt when whole power of weight makes down sound[2]

X, NjWOE, Lab., PN-86-03-03 (1886) (*TAED* NP021A).

1. This appears to be instructions for a set of XYZ experiments; see Doc. 2872 (headnote).

2. Figure label above is "X." Edison intended this device as a highly responsive sounder. He more clearly illustrated and explained it a few weeks later at the end of Doc. 2911.

–2906–

Notebook Entry: Miscellaneous

[St. Augustine,] March 4 1886

Deaf[1]

Vacuum—also air—Hydrogen CO_3 under varying pressures also Ether—

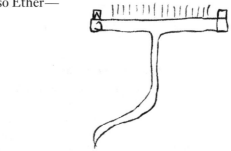

Bristles etc prevent rebound of sound wave   felted— Velvet.

[A][2]

Try Microphone again—[3]

Good   To disolve out the resin of Bamboo etc use boiling linseed oil ie[a] hot & 24 hours. This is Duncan resin process[4] nothing else will do it= E

TAE

X, NjWOE, Lab., PN-86-03-03 (1886) (*TAED* NP021B). [a]Circled.

1. Edison had apparently begun working on "Ear Trumpet Experiments" in February 1886 but this is the first evidence of his recent designs for a hearing aid (Ledger #5:572, Accts. [*TAED* AB003, image

283]). He had worked on an acoustic hearing device (named the "au-riphone") in 1878; in the rush of publicity around his invention of the phonograph, he had received numerous inquiries about the device and announced plans to market it commercially (see Docs. 1228, 1298 esp. nn. 1–2, 1326 esp. nn. 2–3, 1361, and 1464). Edison was spurred to work on such a device by the hearing difficulties he had experienced since his early teens. However, by 1884, he was quoted as saying that he did not wish to be "cured" of impaired hearing: "I am not very deaf. There are lots of things I don't want to hear. Now I don't have to hear 'em" ("Edison Prefers Being Deaf," *Chicago Daily Tribune,* 30 Aug. 1884, 10). His renewed interest in the subject coincided with his remarriage, and there are several entries in the Fort Myers notebooks related to hearing aid designs.

2. Figure labels are "dia[phragm]" and "dia."

3. Edison referred to a form of his carbon transmitter. In 1878, he and David Hughes had waged a bitter priority dispute over the microphone principle. Hughes and Evans 2011, chap. 7; *TAEB* 4, chaps. 3–5 passim.

4. Edison may have had in mind the work of Andrew Duncan the younger, a Scottish physician and a professor at Edinburgh University. Duncan included a discussion of methods for distilling extracts and resins for pharmaceutical uses in his popular *Edinburgh New Dispensatory* (Duncan 1819, 592–605).

–2907–

Ezra Gilliland to
Samuel Insull

Ft Myers Fla Mch 7 [1886]

My Dear Sam

We arrived here safely about a week ago[1] and found our houses not completed[2]   The work is being pushed as rapidly as possible and we will probably get into one of the houses about the middle of this month both famalies will then live together until the other is completed say 2 weeks later   The Laboratory is not yet completed and probably will not be in time to do any work this season.[3]

I left in such a hurry that I forgot many papers which I need badly   Park & Tilfords bill particularly[4]   I have no means of knowing whether we get whats charged or whether we leave a portion of it at Key West where its transferred— We ought to have a letter press and a copy book   there are one or two in the (books) laboratory at 292[5] which I bought at about the same time you bought one or two which left an Extra or two   I wish you would ~~by~~ buy a medium price[a] letter press and pack one or two Copy books with it and send it by Mallory Line soon as convenient—[6] send our mail to Tampa Fla

Care Steamer Manatee—[7]

In this way we get it nearly a week earlier or in say 5 to 6

days & by regular mail to Ft Myers it takes 2 to 3 weeks   I
have not heard a single word from New York from any body
since the day I left there 3 weeks ago— [8]

TAE reached Jacksonville last Sunday and sent a message
asking how houses getting along since which time I have heard
nothing from him[9]

I was sorry I had to rush off so fast had no opportunity to
say good bye—

I told Johnny[10] to have Galley send the organ[11] by Mallory
Line with the pictures which Toppy[12] was going to buy and
send   I hope he has done so as there is not even a jews harp
down here and we need ~~money~~ music[a] badly—[13] With regards
to all   I am   Yours very truly,

E T Gilliland

. Write me all news   let me know how you get on with Insurance—[14] Dont fail pay these last bills out of Ins money

ALS, NjWOE, DF (*TAED* D8603ZAI1). [a]Interlined above.

1. Ezra Gilliland's party included his wife Lillian and her sister,
Jeanette (Nettie) Phelps Johnson, Edison's daughter Marion, and two
maids, Nora and Helena (Lena) McCarthy. They reportedly arrived on
25 February and lodged at the Hotel Keystone on First St. *Fort Myers
Press,* 27 Feb. 1886; ibid., 1 May 1886; both Clippings, EMFP (*TAED*
X104S006A1, X104S007D); U.S. Census Bureau 1970 (1880), roll
T9_295, p. 299C, image 0299 (Indianapolis, Marion, Ind., enumeration
district 115); *Marion County, Indiana, Index to Marriage Record 1886–
1890,* Book 495, included in the Indiana Marriage Collection, 1800–
1941, online database accessed through Ancestry.com, 24 May 2012.

2. Gilliland referred to the two identical residences that he and Edison had agreed to build there. According to a recent press account, the
two-story houses facing the Caloosahatchee River each had a fireplace,
large kitchen, and piazza. The structures, assembled from pre-cut
materials, were largely completed but not yet habitable by late February. *Fort Myers News Press,* 20 Feb. 1886, Clippings, EMFP (*TAED*
X104S005D).

3. The *Fort Myers Press* reported on 13 February that the workmen
had begun construction on the laboratory, which, like the two nearly
finished homes, had been prefabricated in Maine. It was a one-story
wood-frame building with a gabled roof and wood shingles. One estimate put the cost of building and equipping the laboratory at about
$16,000. The *Fort Myers Press* reported on 3 April that the construction
workers would soon be dismissed; however, given the time that would
be needed to unpack and set up the voluminous supplies and equipment
that Edison and Gilliland ordered in February and March, it seems unlikely that the laboratory would have been in a functional state before
Edison returned north at the end of April. *Fort Myers Press,* 13 Feb.
1886; ibid., 3 Apr. 1886; both Clippings, EMFP (*TAED* X104S005C,
X104S007B); Rosenblum 2000, 1:5–7, 2:27; Fort Myers Journal Book
(1885–1887): 20–30, Accts. (*TAED* AB022); Albion 2008, 12–18.

4. Park & Tilford operated several grocery stores in New York City, and Edison often ordered provisions from the firm. The founders, Joseph Park and John M. Tilford, commenced business in the city in 1840. The headquarters store, at 59th St. and Fifth Ave., had an Edison electric lighting plant in the sub-basement powering 500 incandescent lamps. Woolley 1910, 243; "Messrs. Park & Tilford," *NYT*, 1 Jan. 1886, 6; see Vouchers (Laboratory) for Park & Tilford, 1883–1885.

Writing in reply to this document on 13 March, Insull enclosed a duplicate (not found) of the Park & Tilford bill, the original having already been sent to Gilliland. A ledger of Edison's expenses at this time reflects two recent bills from the firm. The former was paid before Gilliland left New York; the latter, for $265.45 on 19 February, was paid on 24 March. Insull to Gilliland, 13 Mar. 1886, Lbk. 21:390 (*TAED* LB021390); Fort Myers Ledger (1885–1887): 200, Accts. (*TAED* AB021).

5. Bergmann & Co.'s factory at 292 Ave. B in New York, where Edison had a laboratory on the top floor.

6. C. H. Mallory & Co. of New York and Mystic, Conn., established a line from New York to Key West and Galveston, Tex., as early as 1873 but did not begin direct service from New York to Florida's west coast until 1908. Freight destined for Edison and Gilliland presumably came via this route to Key West, where it would have been transferred to another vessel for Fort Myers. Mallory also operated (since 1876) a weekly "Florida Line" with intermediate stops between New York and Fernandina or nearby Jacksonville, Fla., from which passengers or freight to the Gulf Coast were dispatched by rail (Browne 1912, 81; Baughman 1972, 157). Insull promised in reply that he would send the press and copybooks by the next Mallory steamer on either 16 or 17 March (Insull to Gilliland, 13 Mar. 1886, Lbk. 21:390 [*TAED* LB021390]).

7. The *Manatee* was a side-wheel steamboat launched in 1884 which operated at this time along Florida's Gulf Coast between Tampa, Sarasota, and Fort Myers. It brought Edison and his new bride to Fort Myers on 16 March. Soon afterward, owner Samuel Stanton sold the ship, and by late April, it was running between Jacksonville and Sanford on the St. Johns River. That sale may have been what prompted Edison to wire alternative instructions about mail on 19 March (Doc. 2913). *Johnson's Steam Vessels* 1917, 158; Stanton 1967, 3–4; *Garner et al. v. The Captain Miller*, 33 F 585 (Circuit Court N. D. Florida 1888); Tate to Insull, 19 Mar. 1886, Lbk. 21:444 (*TAED* LB021444).

8. Insull replied that he had written to Gilliland "several times" since returning from Edison's wedding. Insull to Gilliland, 13 Mar. 1886, Lbk. 21:390 (*TAED* LB021390).

9. Edison's telegram has not been found.

10. John Randolph.

11. Merritt Gally (1838–1916) was a clergyman turned inventor. Although he was chiefly known for his invention of a platen job-press, Gally later patented a multiplex telegraph system. He manufactured organs and self-playing musical instruments in New York and patented a number of improvements for them (*DAB*, s.v. Gally, Merritt; Ord-Hume 1986, 143). Gally shipped to Edison in Florida a "double set Orchestrone" (purchased for $150) on 4 March; he had already sold an organ to Edison on 31 December 1885 for $56.81, as well as Orchestrone music on two occasions that month. The first such organs, which used

perforated metal or card discs or perforated cardboard books to trigger the notes, were successfully marketed in the 1870s. Instruments using perforated paper rolls began to appear in the 1880s, and it was one of this type that Gilliland and Edison had shipped to Fort Myers. In November 1887, Gally wrote Edison that he had "in stock a large reed instrument of similar size to yours, with full keyboard organ added" (Gally invoices, 11 and 18 Dec. 1885, 4 Mar. 1886, Cat. 1165 (Fort Myers accounts): 7–8, 27, Accts., NjWOE; Voucher [Laboratory] no. 659 [1885]; Ord-Hume 1986, 108; Gally to TAE, 23 Nov. 1887, DF [*TAED* D8704AFA]; Fort Myers Journal Book [1885–1886]: 8, 26, Accts. [*TAED* AB022, images 6, 15]; Gellerman 1985, 80).

12. That is, Frank Toppan. Insull enclosed with his 13 March reply a bill of lading for "some of the pictures that Toppan ordered for you." The remaining pictures were sent on or before 15 March, but the editors have no additional information about these items. Insull to Gilliland, 13 and 15 Mar. 1886, both Lbk. 21:390, 421 (*TAED* LB021390, LB021421).

13. In addition to the organs mentioned above, Edison and Gilliland may also have planned on having a piano. Frank Toppan purchased one on Edison's "house" account in October, but the bill was filed with the Fort Myers invoices. Wallet & Cumston invoice, 1 Oct. 1885, Cat. 1165 (Fort Myers accounts): 1–2, Accts., NjWOE.

14. Gilliland referred to insurance on his and Edison's goods lost aboard the schooner *Fannie A. Milliken*. After some delay in settling the claim, Insull expected a check on 16 March. He calculated that $2,956.25 was due to Gilliland, including $1,900 for lost furniture, plus an undetermined amount owed on the account of Sigmund Bergmann, presumably equipment he had shipped on the schooner. Insull promised to deduct from the total $1,458.42 charged against Gilliland "on the Florida Books." At the end of March, Edison's own account for the Florida laboratory was credited with $1,720 for furniture. Among the items lost on the *Milliken* for which Edison and Gilliland purchased replacements were books worth $935.76 and six large bolts, each more than three feet in length, destined for the laboratory. Insull to Gilliland, 13 and 15 Mar. 1886, both Lbk. 21:390, 421 (*TAED* LB021390, LB021421); Voucher (Laboratory) no. 164 (1886); Fort Myers Journal Book (1885–1887): 21, 28, 37, Accts. (*TAED* AB002, images 12, 16, 20).

–2908–

To Samuel Insull

Palatka— [March] 11 1886.

Your letter about 200 shares and tel[e]g[ra]m previously recd I spose refer to same Transaction—[1] Don't sell <u>less</u> than 25= Unfortunately Gill teleghs that both his & my cert[ificate]s are locked up in our safe at Bergmann #2   has sent you the Key— you can have them transferred if you like signing as my attorney—[2a]

Tell John Ott & Jim Gilliland to keep me posted every two or 3 days what progress they are making— also Tate=[3]

Hamilton at Lab has more time let him do they Writing—tell me the [--]^b bugs they find— How is Hamilton getting on with the Condensers—

Has Batch wound the 125 volt <u>new</u> dynamo (no spark)—[4] when does J Hood Wright[5] return. ⟨th19⟩^c

Get the exact figures for Kruesi's bid for up & down town station.[6]

Is Batch Making a model of the Railway Truck I had designed before I left=[7]

Send me a North Amer Review to Ft Myers—[8]

AD, NjWOE, DF (*TAED* D8603ZAL). ^aFollowed by dividing mark. ^bCanceled. ^cMarginalia written by Insull and followed by Edison's dividing mark.

1. Insull was trying to sell some of Edison's stock shares in the Railway Telegraph and Telephone Co., the certificates for which were in Ezra Gilliland's possession (Insull to Gilliland, 3 Mar. 1886, Lbk. 21:336 [*TAED* LB021336]). As he explained to Edison on 3 March:

> Dr Crowell is dickering with me for 200 shares of Railway Telegraph. The old gentleman I think has an enquiry for some & wants to make a turn on it. I offered it to him at 25 (you told me to sell at 24) & he would not take it. I think tomorrow he will offer me 22½ & I shall of course refuse it as I have no alternative after your instructions to me. Please <u>wire</u> me immediately on receipt of this what I am to do in case I get an offer of 22½ .... Remember we need funds badly here. [Insull to TAE, 3 Mar. 1886, Lbk. 21:332 (*TAED* LB021332)]

Three days later, Insull wired Edison (in care of the San Marco Hotel, St. Augustine): "Sold one Hundred and forty shares Grasshopper at twenty five. Instruct Gilliland send me stock quickly" (Insull to TAE, 6 Mar. 1886, Lbk. 21:347 [*TAED* LB021347]).

2. Edison had wired from Palatka the day before that Ezra Gilliland had sent Insull the key to the safe. TAE to Insull, 10 Mar. 1886, DF (*TAED* D8603ZAK).

3. Insull wrote Edison on 15 March that Jim Gilliland "has nothing to report. He is making a new Dynamo & Batchelor is helping him get it out." (This may have been the "new headlight dynamo" that Insull mentioned in a draft telegram to Edison; according to account records, work continued on this project from late 1885 to July 1886). At Insull's prompting, Alfred Tate wrote to Edison the same day about the phonoplex, largely concerning its generally satisfactory use on Baltimore & Ohio Telegraph Co. lines in New York City, and particularly about battery problems and operators' complaints of unexpected changes in the tone of the receiver. John Ott reportedly also made a summary of his activities for Edison on the 15th, but his letter has not been found. He sent another letter on 20 March about various experiments, including efforts to prevent the loose weight in the phonoplex receiver from turning on its axis, which Tate thought caused the alteration of

its tone. Ledger #5:585, Accts. (*TAED* AB003 [image 289]); Insull to TAE, 15 Mar. 1886; Tate to TAE, 15 Mar. 1886; both Lbk. 21:408, 424 (*TAED* LB021408, LB021424); Insull to TAE, 18 Mar. 1886; Ott to TAE, 20 Mar. 1886; both DF (*TAED* D8639B, D8636F).

4. Batchelor had recently sent Edison a brief description and a tracing (not found) of a modified standard armature "made on your new principle" that was intended to produce 125 volts. On 15 March, Insull reported that Batchelor was "now winding the new 125 volt Dynamo (no spark)   He expects to get a test on it either this or next week. He has sent you a Blue print of the windings to Fort Myers." The editors have not found more information about this design. Batchelor to TAE, 3 Mar. 1886, EP&RI (*TAED* X001J4C); Insull to TAE, 15 Mar. 1886, Lbk. 21:408 (*TAED* LB021408).

5. A founding partner of Drexel, Morgan & Co., James Hood Wright (1836–1894) held a variety of interests in Edison electric light and power enterprises. These interests included stock shares in the Edison Electric Light Co., the Edison Machine Works, and the Electric Tube Co. Doc. 2566 n. 2; TAE agreement with Wright, Edison Electric Light Co., and others, 19 Feb. 1885, Miller (*TAED* HM850242); TAE agreement with Wright, Edison Electric Light Co., and others, 18 May 1886, DF (*TAED* D8627U).

Insull replied that "J. Hood Wright returns on 19th. Shall Batchelor & myself go to see him or shall we leave it till you get back to New York. I think the latter is the safest plan   He is more likely to agree if you ask him." The editors have not discerned the purpose of this proposed meeting. One possible reason for seeing Wright, either as a major investor or as a representative of the Morgan firm, was the ongoing effort to reorganize Edison's manufacturing shops. Another rationale could have been financial planning for another New York City central station. Insull to TAE, 15 Mar. 1886, Lbk. 21:408 (*TAED* LB021408).

6. Preliminary planning for a second central station district in New York City had started as early as 1882. Initially, it was to encompass the area around and just north of Madison Square. Its projected northern boundary was pushed north to the theater district and, eventually, as far as the edge of Central Park, between Madison and Eighth Aves. With such a large area under consideration, the prospective Second District was divided into two sections, each to be served by its own central station plant (Docs. 2223 n. 3, 2445 n. 3, 2599 n. 8, 2669 esp. n. 7; New York Edison 1913, 108–11). John Kruesi sent Insull "the estimates that were got out for Mr. Edison just before he went away. . . . We are now engaged in making accurate determinations for the same districts and it will take at least a week longer before we can make a close estimate on the new determinations." Insull forwarded Kruesi's letter and estimates (not found) to Edison (Kruesi to Insull, 13 Mar. 1886, DF [*TAED* D8627G]; Insull to TAE, 15 Mar. 1886, Lbk. 21:408 [*TAED* LB021408]).

7. The truck was intended to accommodate the Sprague railway motor. Insull replied that Charles "Batchelor is not making a model of the Railway Truck you designed before you left. Johnson says that there are a number of details which no one can get out but you that the Truck cannot be ordered till those details are gotten out. . . . Batchelor of course cannot go ahead till he gets an order from Johnson." He added that a truck based on the design of Frank Sprague "will be finished very

shortly" at the Edison Machine Works. He telegraphed essentially the same information to Edison at Fort Myers a few days later. In Sprague's own distinctive 1885 design, the motor hung astride the axle underneath the car; his truck was tested before the end of the month. Insull to TAE, 15 and 23 Mar. 1886, both Lbk. 21:408, 449 (*TAED* LB021408, LB021449); Dalzell 2010, 77–78.

8. The March issue of the *North American Review* contained an article ascribed to Edison on "The Air-Telegraph: System of Telegraphing to Trains and Ships." Edison 1886.

<table>
<tr><td>–2909–

From Samuel Insull</td><td>[New York,] March 13th [188]6</td></tr>
</table>

[New York,] March 13th [188]6

My Dear Edison:[1]

I wired you immediately on receipt of your letter from Palatka asking you if the documents which I sent you to St Augustine had been remailed;[2] I got no reply from you and am now wondering what has become of the stuff I sent to St. Augustine—amongst other things was an agreement providing for the issue of the Electric Railway Stock—[3] It is Essential that your signature should be got to this—until this is received everybody is kept waiting for their stock.[a] If you have not got these letters I hope you will send for them to the San Marco Hotel, execute the papers and return to me with the greatest possible despatch.

Railway Telegraph:

I wired you that I had sold 140 shares of stock a $25. I am in hopes of selling some more but I cant exactly understand whether I have your authority to sell over 200 shares. If you have received some of my letters sent to St Augustine I read your telegram dated Palatka as meaning that I can go ahead and sell above the 200 shares.[4] If you[a] have not received my letter I read your telegram as meaning that I cannot sell the difference between 140 and 200 shares at less than $25.— I may say that it is not my intention at the present moment to sell at less than $30., although a great deal will depend on the likely purchaser desiring to buy as badly as I desire to sell!

I wish you would drop me a line in reply to the various letters I have sent you to St Augustine as I should like to hear from you definitely what I am to do after you have read my previous letters[a] to you.[5]

The experiments from Chicago to Milwaukee are turning out successfully altho' up to the last letter received from there by Dr Crowell, perfect communication had not been obtained between the train and stationary point;[6] the trouble had something to do with leaks on the cars due to bad insulation, but are considered as mere detail troubles and they will be put right very shortly.

Phonoplex

The two circuits on the B&O are running Successfully[7a] and Bates[8] has given instructions for a circuit to be put in between Baltimore and Harpers Ferry—a distance of 95 miles—[9] He is going to send a man with Tate so as to get him posted with a view to their putting in their own circuits. After this will follow the putting in of the Postal Telegraph Co's circuits— Johnson[10] has this matter in hand and is waiting the return of Mr. Chandler[11] who is out of town at the present time. I shall be able to deal with the foreign business without any trouble whatever. I think[b] I have got one man in view for part of the $25,000[a] which we need for foreign phonoplex and I am inclined to think there will be no trouble in raising the whole of it as the perfect working of the invention in actual practice makes it a comparatively easy matter. I have been thinking it over since you went away and I believe that you would get more money out of France and Germany if you would put these two countries into the Company we propose forming. My idea is that in as much as the expense is merely that of an expert and the negotiations with the[a] Government will not require a wonderfully high order of business ability, that you will not require to make the big division of the royalties with a business representative that you have to in the case of other inventions. What I mean is that if you send a man to France and Germany on salary; if he exhibits the apparatus and it proves a success, the adoption of it by the Government is a certainty—in fact it is a necessity— now could not this work be done by an expert of the Foreign Company to better[a] advantage than if you placed each of these two countries in the hands of a capitalist; made a big divvy with him because he supplies the money and undertakes the negotiation? This is only an idea of mine and of course I am working on the assumption that France and Germany will not be included in the Foreign Company

Electric Light:

There is nothing much to tell you in Electric Light. Nothing has occurred in connection with the Uptown Station since you left with the exception of a meeting Johnson had to talk over preliminary plans[a] with the Managers of Shops and Heads of Departments. I asked Johnson whether I could report anything to you and he said there was absolutely nothing to say. Central Station business is rather dull at the moment. Clarke,[12] who used to be at Baltimore is here negotiating on behalf of a Syndicate who propose to put up money for five or six stations in upper New York state right away. Whether anything will come of this I cant say at the moment.[a] I cannot give you the exact state of our work at Goerck Street, as the temporary confusion consequent on uniting the books of the three corporations has thrown us a little back [there?].[c]

I shall be able to write you a full account of the Machine Works business this day week as Everything will be in order then—[13]

I have got a letter from Renshaw[14] in relation to English phonograph. He is looking into the matter and has made a claim[a] on Nottages Executors[15] who recd £600. from the United Telephone Company[16] for their rights under the Phonograph agreement.[17] Renshaw says that Nottage will probably claim that the disclaiming of the phonograph was a voluntary act on the part of your assignees (The Telephone Company) and that the amount paid Nottage was in liquidation of his contract rights. Renshaw says that as this point is a doubtful one he would suggest that if Nottage disputes your claim[a] that he be asked to refer the question informally to the arbitration of any practical man patent agent or Counsel to be agreed upon. What do you say to this.[18]

I have been delayed in Sending this letter to you as my joints are all tied up with what Chadborn[19] advises me is rhuematism—in fact I would not be able to send this to you this morning but for Tate's kindness in taking it from dictation. I shall wire you tonight asking what address will reach you if letters are mailed Monday as I think I shall have quite a good deal to send you by then[20]   Yours Sincerely

Saml Insull

L (letterpress copy), NjWOE, Lbk. 21:396 (*TAED* LB021396). Written by Alfred Tate. [a]Repeated as page turn. [b]"I think" interlined above. [c]Illegible copy.

1. This letter was addressed (by Alfred Tate) to "Thomas A. Edison Esqr   Florida!!"

2. On 11 March, Insull wired Edison at Palatka: "Have documents mailed St. Augustine been returned   sent letters Sandford." The editors have not found Edison's letter to Insull. Lbk. 21:381 (*TAED* LB021381).

3. Insull referred to an agreement drafted on 2 March to distribute 15,000 shares of the Electric Railway Co. of the United States. Insull mailed the agreement in triplicate to Edison on 3 March, and Edison signed and mailed the copies back with his letter of the 11th. Charles

Coster of Drexel, Morgan & Co. sent word on 17 March that he had obtained the shares that Edison and his associates "were anxious to get hold of." Edison received 1,334 shares, with much smaller allotments to Charles Batchelor, Francis Upton, and Insull. TAE agreement with Stephen Field, Simeon Reed, and Sherburne Eaton, 2 Mar. 1886, Miller (*TAED* HM860282); Insull to TAE, 3 and 15 Mar. 1886, Lbk. 21:332, 408 (*TAED* LB021332, LB021408); Coster to Insull, 17 Mar. 1886, DF (*TAED* D8602D).

4. See Doc. 2908 n. 1. Edison wired from Palatka on 10 March: "Send letters to Palatka for two days then to Sanford   Sanford House. Gilland has sent key safe where stock is. better borrow meantime   sell none less than twenty five." TAE to Insull, 10 Mar. 1886, DF (*TAED* D8603ZAK).

5. Insull had written to Edison at least twice in the last two weeks. In a four-page letter on 1 March, he inquired what terms Edison might already have offered for the use of the phonoplex in Canada (see Doc. 2913). Two days later, he wrote again for guidance on the prospective sale of railway telegraph stock, as discussed above and in Doc. 2908. Insull to TAE, 1 and 3 Mar. 1886, both Lbk. 21:325, 332 (*TAED* LB021325, LB021332).

6. Eugene Crowell had succeeded John Tomlinson as president of the Railway Telegraph & Telephone Co. Instruments and men (including Charles Rudd, Samuel Dingle, and George McGregor) had been sent to Chicago before 1 March for a trial installation of the railway telegraph system on the Chicago, Milwaukee & St. Paul Railway. According to Insull, Crowell, who evidently remained in New York, expected the equipment to be ready a few days later. Based on early success along forty miles of line, the railroad issued a thousand invitations to railroad officers and newspaper reporters to a public trial of the system between Chicago and Milwaukee on 19 March. After the event, Insull telegraphed that it was a "big success." Insull to TAE, 18 Mar. 1886, DF (*TAED* D8639B); Insull to TAE, 1, 18, and 23 Mar. 1886; Insull to Ezra Gilliland, 3 and 13 Mar. 1886; all Lbk. 21:325, 438, 449, 336, 390 (*TAED* LB021325, LB021438, LB021449, LB021336, LB021390).

7. These circuits were on lines of the Baltimore & Ohio Telegraph Co. in New York City, at least one of them up Broadway from the company's main office in the financial district to about Thirty-first St. They were working before 1 March, when Insull wrote to Gilliland (and separately to Edison) that there had been a "big gale here the other night & all the B&O wires were so badly mixed up that they could get no messages through from the Main to the Uptown Offices. Peculiar to say that although they could get nothing through their Regular Sounders the Phonoplex worked right along all right." Alfred Tate to TAE, 15 Mar. 1886; Insull to Gilliland, 1 Mar. 1886; Insull to TAE, 1 Mar. 1886; all Lbk. 21:424, 321, 325 (*TAED* LB021424, LB021321, LB021325).

8. David Homer Bates (1843–1926), a native of Steubenville, Ohio, was the president and general manager of the Baltimore & Ohio Telegraph Co. He began his career as a telegrapher with the Pennsylvania Railroad and was working in Altoona, Pa., at start of the Civil War when he was one of four elite operators selected for the War Dept. Telegraph Office in Washington, D.C. Bates served there as a cipher operator from 1861 to 1866, managing the office most of that time. President Lincoln

appeared there almost daily, and Bates later published two book-length accounts of the president's visits, including *Lincoln in the Telegraph Office* (1907). In 1879, he became president of the new American Union Telegraph Co. and, when that firm was acquired a few years later, assistant general manager and acting vice president of Western Union. He left there by 1884 to head the B & O Telegraph Co., where he remained until it, too, was bought up by Western Union in 1887. "David H. Bates Dies; Knew Lincoln Well," *NYT,* 16 June 1926, 25; "Western Union in Possession," ibid., 4 Feb. 1881, 2; "Swallowed Up at Last," ibid., 7 Oct. 1887, 1; "David Homer Bates," *Telegraph Age* 24 (1 Oct. 1906): 478; Bates 1907, 14–16, 45; Bates to TAE, 30 Mar. 1886 and 11 July 1888; both DF (*TAED* D8637H, D8801AAV).

9. Alfred Tate advised Edison of this development by letter two days later, as did Samuel Insull by telegraph on 18 March. Tate to TAE, 15 Mar. 1886, Lbk. 32:424 (*TAED* LB021424); Insull to TAE, 18 Mar. 1886, DF (*TAED* D8639B).

10. Presumably Edward Johnson, who had been in charge of automatic lines for the Atlantic & Pacific Telegraph Co. from 1875–1876, during the tenure of Albert Chandler as the company's vice president. *NCAB* 33:475.

11. Albert Brown Chandler (1840–1923) was president of the United Lines Telegraph Co. and, as one of the receivers of the Postal Telegraph and Cable Co., that company's functional administrator. After guiding the Postal Co. through its reorganization, he was elected president in 1886. A Vermont native, Chandler went to Washington, D.C., in 1863 to serve as a cipher operator in the War Dept. Telegraph Office. Later that year, he was given the additional duty of disbursing clerk for Gen. Thomas Eckert, superintendent of the Department of the Potomac. (Like Bates, he later wrote about his wartime experiences and acquaintance with Abraham Lincoln.) Chandler spent nearly a decade after the war at Western Union, where his responsibilities included trans-Atlantic cable traffic. He left in 1875 to become assistant general manager (later president) of the Atlantic & Pacific Telegraph Co. When that firm was absorbed by Western Union in 1882, Chandler resigned to become president of the Fuller Electrical Co. Two years later he became counsel to the Postal Telegraph Co., which led to his appointment as receiver in 1885. Ullery 1894, part 3:30, 32–33; Insull to Chandler, 28 July 1886, LM 12:80 (*TAED* LM012080); Ross and Pelletreau 1905, 3:212–13; "Albert B. Chandler Dead," *NYT,* 4 Feb. 1923, S5.

12. Henry Alden Clark (1850–1944) was an agent of the Edison Co. for Isolated Lighting (Docs. 2474 n. 10, 2725 n. 3; "Henry Alden Clark," *NYT,* 17 Feb. 1944, 19). His responsibility for Washington, D.C., and western Pennsylvania overlapped that of other agents; consequently he had by this time been induced to give up Pennsylvania and take upstate New York instead. He refused to relinquish Washington, however, and continued to hold that territory (as well as his original assignment of Maryland), leading to a long-running dispute with Uriah Painter (Edward Johnson to Painter, 9 and 26 Jan. 1885; Painter to Johnson, 20 Dec. 1885; Painter to Insull, 21 May 1890; all UHP [*TAED* X154A4AQ, X154A4BD, X154A4FM, X154A9AT]; Charles Hughes to Insull, 11 June 1886, DF [*TAED* D8633M]). Clark apparently succeeded in forming a syndicate to construct central stations in Amsterdam, Co-

hoes, Troy, Syracuse, Utica, and one other town in upstate New York (Insull to TAE, 15 Mar. 1886, Lbk. 21:408 [*TAED* LB021408]).

13. Insull reported on Edison Machine Works activities two days later in response to Edison's queries in Doc. 2908, which he received on 15 March. He left for Chicago on unspecified Machine Works business on 18 March. Insull to TAE, 15 and 18 Mar. 1886, both Lbk. 21:408, 438 (*TAED* LB021408, LB021438).

14. Alfred George Renshaw (1844–1897), a London solicitor, looked after legal affairs in Great Britain for Edison. He acted as the trustee for Edison's interest in the defunct Edison Telephone Co. of London, Ltd., and its successor, the United Telephone Co., Ltd. His 2 March letter to Edison is discussed in the notes below. St. Mary Magdalene, Holloway Rd., Register of Baptism, p. 45, LMA, *London, England, Birth and Baptisms, 1813–1906*, online database accessed through Ancestry.com, 4 Oct. 2012; *England & Wales National Probate Calendar (Index of Wills and Administrations), 1858–1966*, online database accessed through Ancestry.com, 4 Oct. 2012; Docs. 1933, 1954, 2118 n. 5.

15. The late George Swan Nottage (1822–1885) had been the managing principal of the London Stereoscopic and Photographic Co., which held the license to manufacture and sell Edison's phonograph in the United Kingdom. When he died in April 1885, his right to the license passed to heirs who may have included his son as well as his business partner and cousin Howard John Kennard; the estate was administered by his wife, Martha Christina Nottage. Nottage was also an influential political figure in London, having served as alderman from 1876 until his death, sheriff from 1877 to 1878, and lord mayor in the last year of his life. Doc. 1237 n. 7; Boase 1892–1921, s.v. "Nottage, George Swan"; "Howard John Kennard," *Minutes of the Proceedings of the Institution of Civil Engineers* 126 (1 Jan. 1896): 408–9; "The Right Honourable Nottage, George Swan," 19 Apr. 1885, *England & Wales, National Probate Calendar (Index of Wills and Administrations), 1858-1966*, p. 555, online database accessed through Ancestry.com, 11 Oct. 2012.

16. The United Telephone Co., Ltd., was formed in London in 1880 by the merger of the Edison Telephone Co. of London, Ltd., with the rival Telephone Co., Ltd., which controlled Alexander Graham Bell's patents.

17. Renshaw's 2 March letter (enclosing two others) was part of a skein of correspondence going back at least to mid-January. At that time, George Gouraud asked Insull to give him a letter demanding payment from the Nottage estate of monies allegedly due Edison, a matter he had evidently already discussed with Edison. In February, Edison asked Renshaw to look into a report (presumably from Gouraud) that Nottage had received a sum of money to liquidate his interest in the phonograph license, half of which Edison explained should come to him as a royalty (Renshaw to TAE, 2 Mar. 1886; Gouraud to Insull, 16 Jan. 1886; both DF [*TAED* D8603ZAF, D8603P]; TAE to Renshaw, 9 Feb. 1886, Lbk. 21:269 [*TAED* LB021269]). One of the enclosures with Renshaw's reply was the answer to his own inquiry to Waterhouse and Winterbotham, attorneys for the United Co., who confirmed that the company had paid Nottage £600. That payment was a consequence of the company having disclaimed (or excised) the phonograph from Edison's fundamental British patent on the telephone, which it did to

protect its telephone interest from a technical legal flaw in the description of the phonograph that otherwise could have invalidated the entire patent. Although as Waterhouse and Winterbotham pointed out, Edison's 1878 agreement with Nottage expressly provided that such a disclaimer would terminate the phonograph license, an arbitrator recommended that Nottage receive some compensation for his lost interest (TAE agreement with Nottage and John Kennard, 22 Mar. 1878, Miller [*TAED* HM780039]). Armed with this information, Renshaw sent a letter to Nottage's heirs, a copy of which he enclosed to Edison (Waterhouse and Winterbotham to Renshaw and Renshaw. 1 Mar. 1886; Renshaw and Renshaw to London Stereoscopic & Photographic Co., 2 Mar. 1886; both DF [D8603ZAG, D8603ZAH]).

18. The question of royalties became even more complex than was immediately apparent. Under a side agreement in 1878, Nottage and his partner had promised to pay Edison a £1,500 advance on future royalties; after the amount due Edison reached this threshold, he was to be paid according to the fifty-percent formula in the primary contract (TAE agreement with Nottage and Kennard, 22 Mar. 1878, Miller [*TAED* HM780040]). Contrary to Renshaw's expectations, Nottage's heirs now claimed that the royalties due Edison on phonograph sales had reached only £1,200, and that consequently they had overpaid him £300, the exact amount that Edison said was owed him. Renshaw explicated the matter in a letter to Edison on 15 April, across the top of which Edison queried: "Insull on this basis is anything due" (Renshaw and Renshaw to TAE, 15 Apr. 1886, DF [*TAED* D8603ZAT]). The controversy over the phonograph licensing rights continued for several years, and it is not clear how—or if—the royalty issue was resolved (see Gouraud to TAE, 8 Dec. 1887 and 31 Jan. 1888; both DF [*TAED* D8754AAP, D8850AAM]; *Edison United Phonograph Co. v. London Stereoscopic and Photographic Co.*, Misc. Legal [*TAED* HX91096A]).

19. New York physician Edwin Ruthvin Chadbourne (1853–1909) attended the Edison family. U.S. Census Bureau 1967? (1860), roll M653_437, p. 18, image 707 (Bridgton, Cumberland, Me.); *Biographical Review* 1896, 482; *Directory of Deceased American Physicians, 1804–1929*, record for Edwin Ruthvin Chadbourne, online database accessed through Ancestry.com, 23 Oct. 2012; Doc. 2718 (headnote n. 5).

20. The telegram has not been found but Insull's eight-page letter of Monday, 15 March, is referred to in notes above.

–2910–

To Samuel Insull

Fort Myers Fla 18 Mch. 1886[a]

Samuel Insull

Arrived yesterday takes mail seven to nine days[1] use telegraph[2]

Edison

L (telegram), NjWOE, DF (*TAED* D8603ZAM1). Message form of Western Union Telegraph Co. [a]"1886" preprinted.

1. Mail service to Florida's west coast was spotty before completion of the Plant System railroad from Jacksonville to Tampa in 1887. Un-

til about this time, mail went by train from Jacksonville to Cedar Key, then by steamboat down the coast to Fort Myers. This route apparently was interrupted quite recently by the sale of the Tampa-based steamer *Manatee* (cf. Doc. 2907). An alternative, which Edison seems to have used for freight, was via the Mallory shipping line from New York to Key West and then by steamboat to Fort Myers. Browne 1912, 81; Pettengill 1952, 47–49, 81–82; "The Florida Railroads," *Railroad Gazette*, 15 Apr. 1887, 253.

2. By coincidence, Western Union notified Insull the same day that his telegram of 15 March (not found) to Edison at Sanford, Fla., had not been delivered because Edison was not at the hotel there. Acknowledging receipt of the present message from Edison at Fort Myers, Insull wrote that he had immediately "sent you some long letters which I have had waiting on my desk for an address for some days past. It has been impossible to keep you posted as you have not kept me well supplied with addresses I have wired San Augustine, Palatka and Sanford to send any letters for you to Fort Myers as both Dyer & myself have written you to one or the other of those places." Western Union to Insull, 18 Mar. 1886, DF (*TAED* D8633G5); Insull to TAE, 18 Mar. 1886, Lbk. 21:438 (*TAED* LB021438).

–2911–

To John Ott

Fort Myers March 18 1886

John

Tate says the operators on phonoplx buckle the diaphragm, that they cannot tell when the magnet is the right distance or even when it is touching the diap'm[1] they Keep on turning until the diaphragm is badly buckled. please try some Experiments to stop this. I make one or two suggestions thus[2]

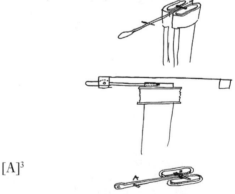

[A][3]

A is a lever which has a handle c outside of telephone case B is an end on it of brass. & fits down in a slot in the iron magnet it is flush with the top. When the magnet is say $\frac{1}{32}$ from face of the diaphragm you work the lever from the outside up and down & this tells you how far it is away of Course when the diaphragm touches the face of the magnet you cant work the lever at all— This lever serves to tell you just where you are & the distance—perhaps a single ~~pi~~ lever & piece will answer[4]

Another way is to make the diaphragm one pole & the head of the magnet the other pole or contact point & when the magnet touches the diaphragm it closes the circuit around the <u>sending</u> Key that works the sounder

The operator can leave his key open & adjust until it closes the sounder and then he knows that the diaphragm and magnet are in contact— I don't think this as good as the lever

Another way is an index lever

This I don't think so good as the lever as the diaphragm may be buckled & then the index wouldnt show proper

Doubtlessly you will think of other ways—

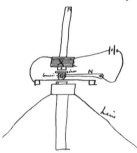

Here is an experiment I want you to try[a] & see if it wont work farther or louder[b] than the others[5]   X is a one or 2 ohm Bell Receiver magnet, G is the regular weight secured to lever N. a local battery is used which is broken when the weight leaves the diaphragm   This tends to make a <u>self vibrator</u> but I want it so adjusted and[c] the weight so heavy that it wont act as a self vibrator. The weight being too heavy or the magnet adjusted so far away or the Current being too <u>weak</u> by putting in Resistance in the local Circuit. Now when the wave comes over the [main?][d] line it throws the weight up as the Local magnet <u>almost</u> does it   the weak line current <u>added</u> to it is sufficient to throw it up. but it now <u>falls</u> back with the <u>full</u>

power of the weight as the Local magnet is open   thus you get a loud sound due to the falling of a big weight, which without the local magnet the <u>weak</u> current on the main line would be unable to <u>lift</u>   ⟨Made and tested this April 12 86   J. F. Ott⟩[e]

AD, NjWOE, Lab., Unbound Notes and Drawings (1886) (*TAED* NS86ADC2). [a]Obscured overwritten text. [b]"or louder" interlined above. [c]Repeated at bottom of one page and top of the next. [d]Obscured by ink smear. [e]Marginalia written by John Ott.

1. The editors have not found a letter from Tate describing the diaphragm problem. In a letter of 15 March, he did describe a problem with the movement of the weight causing inconsistent sound. Tate to TAE, Lbk. 21:424 (*TAED* LBo21424).

2. Ott received Edison's letter on 5 April and promised to try the suggestions given below. Ott to TAE, 7 Apr. 1886, DF (*TAED* D8636I).

3. Figure labels are "A" and "B."

4. Figure label is "dia."

5. Figure labels above are "brass," "X," "s," "G," "iron," and "Line." This instrument appears to be like the one that Edison sketched in Doc. 2905.

EDISON'S FORT MYERS NOTEBOOKS   Doc. 2912

Three days after arriving in Fort Myers on his honeymoon with Mina, Edison started a series of notebook entries that, during their six-week stay, nearly filled six standard notebooks, a copious output even by his standards. This torrent of wide-ranging ideas included practical experiments and contrivances as well as speculations about the fundamental nature of matter and energy and some ways he might explore them.[1] He used a seventh book (Doc. 2946) for detailed instructions on planting the grounds of his new Fort Myers home, displaying both considerable ambition for the estate and a working botanical knowledge of the region. Doc. 2944 is the last entry selected from this period.[2]

The heavy use of these books reflects several unusual circumstances (in addition to the central fact that Edison was away from his assistants and from daily business, although he was with his new bride). Edison made entries nearly every day, but beginning in April, he produced considerable overlap by drafting entries in one book and reworking them more carefully in another.[3] Mina, in addition to signing most of his entries as a witness, also wrote several in her own hand. Some she obviously copied from Edison's work in another book, while others she probably copied from papers that are no longer extant.[4] Edison's daughter Marion, who was with the couple in Fort Myers, created a few entries of her own with items that she seems to have transcribed from newspapers and a telegram to Edison.[5]

These notebooks were truly idea books for Edison. A number of his topics follow in a general way those in the third volume of Faraday's *Experimental Researches,* a book slender enough to carry easily. His thoughts moved organically from one topic to another across several broad themes. Theoretical or speculative questions about energy and the structure of matter form one theme. Edison outlined experiments for discovering a new type of energy that he provisionally called "XYZ" and for exploring the conversion of light and heat into electricity and magnetism, a subject he first took up systematically in December 1885.[6] Near the end of the Fort Myers visit, he made extensive notes on his theory that gravitation is an electromagnetic phenomenon and proposed several experiments for exploring that idea (see Doc. 2937 [headnote]). Edison's ongoing experimental projects—practical efforts on which he had been engaged in New York—are well-represented in a second category. Subjects include electric lighting, particularly incandescent lamps and lamp manufacturing processes, generators, arc lights, storage batteries, and the new municipal lamp; wireless, phonoplex, and other telegraph systems; and telephones.[7] A third broad theme consists of new practical inventions, represented by a few notes and drawings for such projects as a cash carrier system, a cotton picker, and a depth gauge for ships, as well as renewed interest in developing a hearing aid. Then there are the notes and drawings for impractical-looking machines scattered throughout these books. Like the "Expansion Engine," which would work by the enlargement of heated copper rods, a telephone transmitter based on the "Multiple pulley principle," or the "Larynaxial piano," such devices may seem fantastical but were real enough in Edison's mind for him to draw with considerable detail.[8]

Although the Fort Myers laboratory was not yet equipped to try most of these ideas, Edison and Mina did carry out at least one experiment—shocking an oyster to see if its shell would open (a "Dead failure").[9] It is possible that they tried others, such as with the hearing aid devices described in Doc. 2921, but the comments (like "good") in these pages generally read more like a winnowing of suggestions worth trying later than the results of actual tests. Edison did not take up most of these ideas until he returned to New Jersey, and in some cases, not until he opened his new Orange laboratory in November 1887.[10]

Because of the distinctive character of these notebooks, the editors have taken the unusual steps of selecting most of Edison's original entries and generally presenting them in their entirety (as defined by date). Alternate versions of these entries, either by Edison or Mina, have not been selected, but the notes identify their locations.

1. N-86-03-18, N-86-04-03.1, N-86-04-03.2, N-86-04-03.3, N-86-04-05, N-86-04-07, N-86-08-17; all Lab. (*TAED* N314–320).

2. Edison wrote Docs. 3039 and 3041 soon after his return; see also Doc. 3038.

3. The only dates on which Edison did not make entries were 27 and 31 March and 1–2, 10, 19, 20, and 24 April. It is possible that he postdated other entries.

4. N-86-03-18:157–67, 199–207; N-86-04-03.3:29–39; all Lab. (*TAED* N314157, N314199, N317029, N317035, N317037).

5. N-86-04-03.1:118, 124, 140–43, 162–65, 268–70; Lab. (*TAED* N315, images 60, 63, 71–72, 82–83, 118–19).

6. See Docs. 2804 (headnote) and 2872.

7. Edison probably had telephones in mind when he sketched a variety of arrangements for producing lampblack, the form of carbon widely used in transmitters and which he sometimes supplied to the Bell companies. N-86-03-18:209–15, Lab. (*TAED* N314209).

8. See, e.g., Docs. 2925, 2932, and 2934.

9. See Doc. 2915.

10. For example, see Docs. 2921 n. 5, 2947, 2949, 2950, 3109, and 3111.

–2912–

Notebook Entry: Miscellaneous

[A][1]

[Fort Myers,] March 18 1886

100 Volts

Carbon

175 volts,

[B][2]

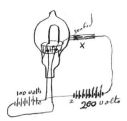

X is 1st sealed vacuum obtained—then wire is shoved in X & 200 Volts charge given it & while charging X is again sealed & wire drawn away leaving the platina cylinder inside of globe permanently charged 200 volts contrary

[C][3]

Run Curve on 6 of these[4]

Carbonize in 4 days in reducing solution. also water absorbing solution.

Linseed oil. also Linseed oil & sulp Iron Proto[5] Saturated solution. Chloride Zinc. Protochloride Tin.

City gas passed through several tubes of finely divided copper to absorb O— also then thro[ugh] phos[phorous] anhyd[ride] to get H_2O[6]

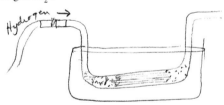

Slow carbonization 3 days—

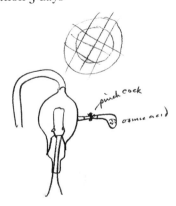

get vac fair—[7] open pinch cock then get another good vac & work Lamp then Let in Osmic acid & bring Lamp up high to deposit Osmium—

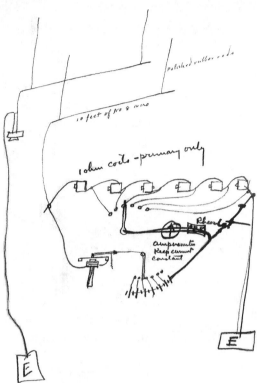

Experiments for baloon teleghr[8] see if a measured Emf give greater jump—

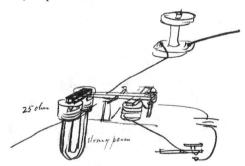

See if sudden pulling off armature gives good phonoplex Jump—[9]

Carbonize 64 hours slow in Linseed oil= proto[a] salts that are reducers in Linseed & alone if can get heat also Chloride

Cal & Chlo Zinc glacial phos acid[10] in sealed tubes to get pressure— also in parafine—Napthaline—& high melting point Hydrocarbons

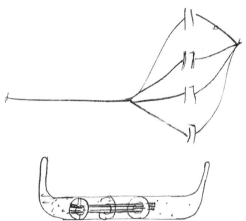

Sealed tube with glass or nickel washers—in center of which pass the filiments to hold them in position— Then Carbonize with tubes sealed and filled with powdered metallic Lead—fusible metal[a]—Zinc—etc.

Process for burning natural gas into Telephone Lamp black & then to book ink— Experiment—[11]

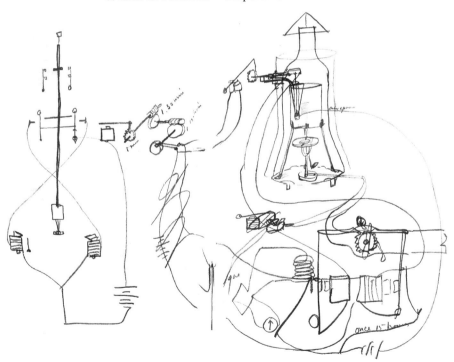

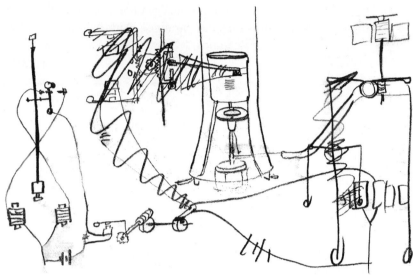

100 min. 1 Rev min[12]
TAE   Mina

X, NjWOE, Lab., N-86-03-18:3 (*TAED* N314003). Document multiply signed and dated; several miscellaneous calculations omitted. [a]Ob-
scured overwritten text.

1. Figure labels are "100 volts," "carbon," and "175 volts." The following set of three drawings appears related to Edison's efforts to prevent electrical carrying by neutralizing what he believed to be the static
attraction between heated carbon particles from the filament and the
glass globe. Cf. Doc. 2892.

2. Figure labels are "sealed," "X," "100 volts," "c," "Z," and "200
volts."

3. Figure label is "300 volts."

4. Lamp "curves" graphically represented the range of lifetimes in a
sample of test lamps. See Doc. 2177 (headnote).

5. That is, proto-sulphate of iron (ferrous sulphate).

6. Figure labels are "Hydrogen" and "Sand."

7. Figure labels above are "pinch cock" and "osmic acid."

8. See Doc. 2807. Figure labels above are "Polished rubber rods,"
"10 feet of No 8 wire," "1 ohm coils—primary only," "Rheostat," and
"ampermeter   Keep current constant."

9. Figure labels above are "25 ohm" and "strong perm[anent]."

10. That is, metaphosphoric acid, called "glacial" for its glasslike
crystalline character. *OED*, s.v. "glacial" adj., 2b.

11. Figure labels are, from left to right, "1 min," "1.30 min," "1 15
min," "air space," "gas," and "once 15 hours." Edison made three
drawings of similar apparatus on 19 March. The first of these is the
clearest (see p. 369) (N-86-03-18:23, Lab. [*TAED* N314023]).

12. Edison wrote this line of text on the page opposite the drawing.

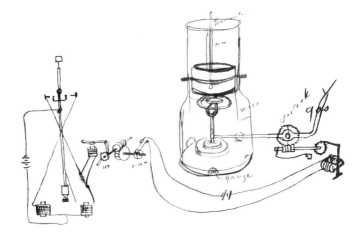

One of Edison's 19 March drawings of a combustion chamber with thermostatic "auto pressure regulator" for producing lampblack. The motor regulating the gas inlet (at right) was controlled by switches (at left) in response to pressure in the vessel.

–2913–

To Samuel Insull

Ft Myers Fla   Mch 19 1886[a]

Saml Insull[1]

Canada six dollars   try get fifty dollars circuit instead[2] have Hamilton post me progress laboratory work[3]   send mail direct here   Tampa steamer discontinued[4]   get number Gillilands patent[5] from Lockwood Boston & send power Atty here for signature[6]   have Bergmann ship Gouraud[7] more transmitters[8]   safe key sent mail

Edison

L (telegram), NjWOE, DF (*TAED* D8603ZAN). Message form of Western Union Telegraph Co. [a]"1886" preprinted.

1. Because Insull had left for Chicago the previous day, Alfred Tate transmitted this message to him there. Tate to Insull, 19 Mar. 1886, Lbk. 21:444 (*TAED* LB021444).

2. Insull had queried Edison in a letter on 1 March: "What proposition did you make to Wiman as to Royalty in Canada   Tate says you told him that you made no proposition. Wiman told me that you pro posed six dollars a station for Canada" (cf. Doc. 2810). Insull expressed some urgency about negotiating with Erastus Wiman, having learned through Alfred Tate that the Canadian parties liked the phonoplex and wished to use it on more circuits. Insull to TAE, 1 Mar. 1886, Lbk. 21:325 (*TAED* LB021325).

3. Insull instructed Tate on 20 March to have John Ott "wire Edison fully" (in Insull's name) about the work at the lab and to have Ott and H[ugh?]. de Coursey Hamilton "mail Edison full details." The same day, Hamilton wrote a "Record of work completed & in progress at laboratory to be embodied in telegrams." Ott made his own more succinct reports that day and again a week later. Insull to Tate, 20 Mar. 1886; Hamilton memorandum, 20 Mar. 1886; Ott memoranda, 20 and 27 Mar. 1886; Insull to TAE, 21 Mar. 1886; all DF (*TAED* D8636B, D8636D, D8636F, D8636G, D8603ZAP).

4. See Doc. 2907 n. 6.

5. See Doc. 2848 n. 7.

6. Insull later requested John Tomlinson to draft a power of attorney from Gilliland to Gouraud, but it has not been found. Insull to Tomlinson, 19 Apr. 1886, Lbk. 22:33E (*TAED* LB022033E).

7. George Edward Gouraud (1842–1912), a native of New York and a former Union army colonel (and the son of pioneering promoter of photography François Fauvel-Gouraud), was a longtime business agent in London. He had dealt with Edison since 1873 and played significant roles in efforts to commercialize the phonograph, telephone, and electric light in Great Britain. Gouraud was preparing to embark on a world tour, for which Edison gave him a general letter of introduction to Edison electric lighting companies in Europe and the U.S. Doc. 159 n. 7; Welch and Burt 1994, 104; Gouraud to TAE, 18 Dec. 1885; TAE letter of introduction for Gouraud, 18 Dec. 1885; both DF (*TAED* D8503ZEM, D8503ZEN).

8. Gouraud had wired from London on 13 March: "Gilliland cable numbers English patents   more telephones   power attorney   quick" (Gouraud to TAE, 13 Mar. 1886, DF [*TAED* D8640F]). Insull relayed this cable to Ezra Gilliland with a request for instructions, and he evidently retransmitted it to Edison as well. When Insull received this reply from Edison, he wired Alfred Tate from Chicago to get "for particulars gillilands english patent on transmitter" and to have Bergmann ship six of those instruments to Gouraud (Insull to Gilliland, 15 Mar. 1886; Insull to TAE, 18 Mar. 1886; both Lbk. 21:421, 438 [*TAED* LB021421, LB021438]; Insull to Tate, 20 Mar. 1886, DF [*TAED* D8636B]). The transmitters were manufactured by Bergmann & Co., but when James Gilliland tested them he found them "not in shape to do business with." He hoped his brother could sell the patents because, he noted, "They commence Taking out the Bell Telephones March 30th   if any thing is done it will have to be done right away" (J. Gilliland to Insull, 18 Mar. 1886, DF [*TAED* D8640G]).

–2914–

Notebook Entry: Incandescent Lamp

[Fort Myers,] Mch 19 1886

Experiment quantitatively & acurately on bringing up metals in Vacuo— Iron—Steel—brass—German Silver Nickel Tin   Aluminium—Zinc—Lead—Magnesium—Cadmium, ~~Nick~~ Cobalt, Copper—Silver—gold—platinum— Read Res[istance] & amperes & volts from 100 to melting point— curve it—[1a] Try these metals in Hydrocarbon Vapor to deposit & afterwards eat out[b]

paint enamel over all the metal & part of widened Carbon[2]
& bake in sand[3] up to hardening point as in <u>pottery</u>
TAE   Mina

X, NjWOE, Lab., N-86-03-18:29 (*TAED* N314029). [a]"from 100 . . .
curve it—" interlined below. [b]Followed by dividing mark.

1. That is, to produce a graphical representation of the distribution
curve for the changes in resistance, amperage, and voltage.

2. Filaments had broadened ends to improve the copperplated connection to the lead-in wires.

3. Edison most likely meant a sand bath.

–2915–

Notebook Entry:
Miscellaneous

[Fort Myers,] March 20 1886

Thermo-sensitive—for measuring fine line ray as in spectrum & still have advantage of best couple.[1]

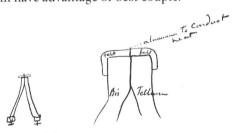

Spectroscopic Slit

Cash Carrier[2a]

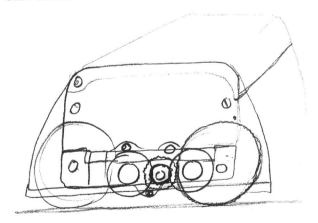

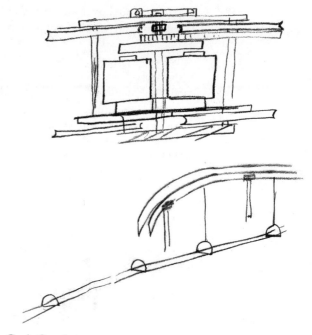

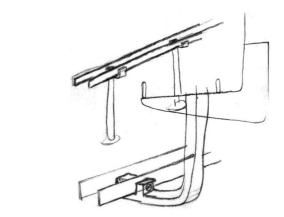

Cash Carrier

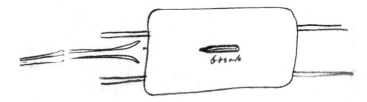

Cash Carrier[3]

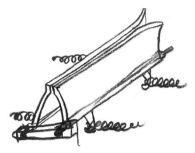

break—proper length

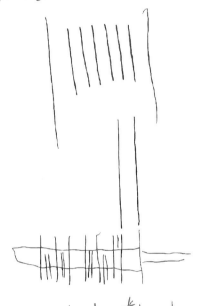

[A][4]

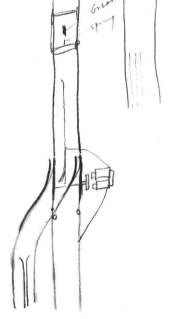

Cash Carrier[5]

Heat Translation into E[lectricity]

Core of wood

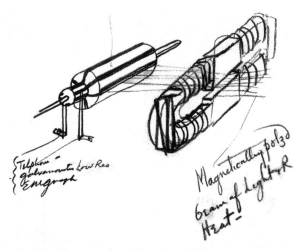

use bands of copper[6]— Zinc Lead Tin—Cadmium—Silver nickel Iron—German Silver Antimony—platina—Carbon—Sulphides of Lead—& other conducting Sulphides phosphides   also tubes filled with[b] Conducting Solutions, with non–pol[ari]z[a]ble Electrodes.[a]

also all above expmts try for XYZ—[c]

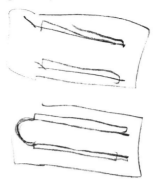

Now magnetize the beam of light at right angles[7]

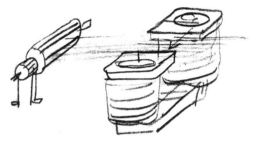

Try all the above   Now use polarized light—and magnetically effect it straight & at right angles

Heat Trans

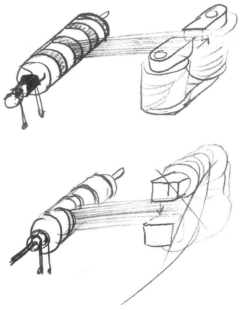

Same as tother

Heat Trans

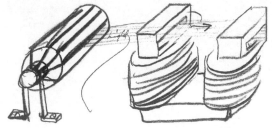

also—cylinder wound Spirally[8]   also used polarized Light—
If it gives reverse currents shield ½ of them
get Continuous ones
Try filtering the beam through Various substances.—
Electrify the beam—by continuous Holz[9d] machine[a]
pass it beam[e] through a filter containing a conducting Solution with Electrodes arranged to send powerful current
through the Solution[d]— Same direction & at right angles— o[a]

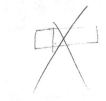

Cash Carrier—[10]

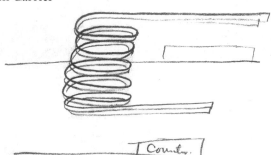

Method of getting on Second floor with motor. Same as a
Mountain Railroad Sou[thern] Pacific[11]

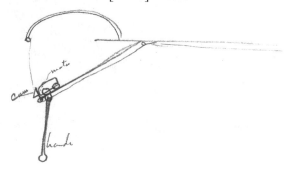

Cash carrier[12] showing Cam & lever, to give the motor an initial start so armature wont get stuck on Center

Shock an Oyster see if it wont paralyze his shell muscle & make the shell fly open—[a] ⟨Dead failure   TAE⟩

TAE   Mina

X, NjWOE, Lab., N-86-03-18:31 (*TAED* N314031). Document multiply signed and dated. [a]Followed by dividing mark. [b]Followed by "over" to indicate page turn. [c]Circled and followed by dividing mark. [d]Obscured overwritten text. [e]interlined above.

1. Figure labels are "felt," "aluminum to conduct heat," "felt," "Bis[muth]," and "Tellurium." Edison presumably referred to a thermocouple, a sensitive device for measuring heat.

2. Edison meant a means of delivering cash or papers through the interiors of buildings. At this time Samuel Insull and a partner, Edwin G. Bates, were eastern agents for the Rapid Service Store Railway Co., a Detroit firm that sold a wire carrier system for delivering cash and parcels in shops. The system had been invented by Edison's boyhood friend Robert McCarty, who sent Edison twenty-five shares of stock in the company as repayment of a $100 debt. Another Edison friend, Milton Adams, was connected with Insull and Bates's agency. Insull likely had made Edison aware of some of the technical problems plaguing the system, which he described in a 31 January telegram to the company. In 1887, Rapid Service was acquired by the Lamson Co. Insull agreement with Rapid Service Store Railway Co., 7 Aug. 1885; McCarty to TAE, 13 Sept. 1885; Rapid Service Store Railway Co. circular, n.d. [c. 1885]; Bates and Insull to Rapid Service Railway Co., 31 Jan. 1886; all DF (*TAED* D8541ZAA1, D8503ZCO, D8541ZAO1, D8634E); TAE to McCarty, 27 Oct. 1885, Lbk. 20:59 (*TAED* LB021059); Buxton 2004, 13.

3. Figure label is "break."

4. Figure labels are "break" and "spring."

5. Figure labels are "hand" and "Rubber."

6. Figure labels above are "Telephone— galvanometer Low Res E[lectro]M[oto]graph" and "Magnetically polzd beam of Light & R[adiant] Heat—."

7. Edison is referring to the Faraday Effect in which a magnetic field will rotate the polarization of a beam of light moving through a transparent medium.

8. Figure label above is "Light."

9. German physicist Wilhelm Holtz (1836–1913) designed a self-charging electrostatic generator that produced a notably long-lasting and powerful discharge. It worked by induction between two glass plates in motion relative to each other, not (as in earlier machines) by friction, and the induced charges were collected in a Leyden jar. Holtz invented it between 1865 and about 1870; by this time, the machine was manufactured in various forms by Ritchie & Sons of Boston, Chester & Co. of New York, and James W. Queen & Co. of Philadelphia. Atkinson 1896, 108–10; Warner 1993, 1:xli; Turner 1983, 192; Hayes 1872, 667; Barnard 1870 [1869], 548.

10. Figure label is "Counter."

11. Edison referred to the method used by engineer William Hood

of the Southern Pacific Railway to build a line (completed in 1876) through the Tehachapi Pass in California as the railroad expanded eastward towards New Orleans. Hood created a path through the Tehachapi Mountains involving a rise of 2,754 feet in just 16 miles by using a series of switchbacks and tunnels to reduce the grade to only 2.1 percent. In the most famous section of the road, known as "the Loop," the line crosses over itself, which appears to be the model for Edison's proposed solution for moving the cash carrier between floors. Daniels 2000, 59.

12. Figure labels above are "Cam," "motor," and "hand[l]e."

–2916–

To Edward Johnson

[Fort] Meyers, [March] 21 1886[1]

Dear Johnson

Yours re—municipal received Just in time to answer by mail today.[2] I will think it over further but think I can give you want you want. Here is the rough Idea ask John Ott to let <u>Hamilton</u> make the mixtures for you[3]

<u>Note</u>   Ordinary charcoal does not conduct to any appreciable Extent   the conducting power of charcoal depends up to a certain Limit to the temperature it is brought when 1st Carbonzd— Thus the final heat used in making ordinary charcoal is insufficient to render it a conductor for ordinary purposes. were the heat Carried a little higher it would become a better conductor & so on   I have used ordinary charcoal from which I have cut sticks that were 1 inch long by ⅛ diameter that had 5000 ohms resistance. ~~Now~~ it varies greatly in resistance but if it will answer Upton can make you plenty that will have an Even resistance & be high or low resistance just as desired as he can stop at any temperature he desires—

Now spose you put a peice of charcoal across the wires thus[4]

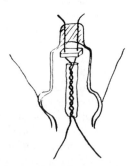

and this peice of charcoal had 3000 ohms resistance. Then with 50 volts it would have only 36 foot pounds of Energy[5] & this would scarcely warm it—but should the Carbon break There would be 800 volts & 9386 foot pounds this would heat it up to incandescence & bring it to such a temperature that its resistance would come down to less than 2 ohms probably less than 1 ohm

Another way would be to powder as you suggest Tamp it in & then put a wad of asbestos etc to hold it well tamped. Then with a large body it would come down to a very low resistance[a]

Ordinary Lampblack is also a practical non conductor. There are various qualitities ~~the~~ of varying resistances. this should be tried

Again you can make a mixture of good conducting Lampblack & oxide of magnesia in varying proportions so the resultant powder well ground & mixed will have varying resistances according[b] to the amount of Oxide of magnesia— The proportion of magnesia might be increased to such an Extent as to make the resultant tamped or moulded peice an insulator yet the 800 volts would jump from particle to particle & contract the peice instantly & produce an Enormous decrease of resistance   This would establish an arc ~~but would m~~ & the twisted wire would have to be depended on to make the cut off— The resistance of this compound would probably never come down so low as to act as a cut off itself like the piece of Charcoal

Now here is a lot of substances which are partial Conductors.[6]

1	Amorphous phosphorus (dry)	very high Res'
2	Sulphate of Lead (Galena)	Like Carbon.
3	Anthracite Carbon—	Very high comes down to Conduct like a metal by heat.
4	Peroxide of Lead	more than Carbon
5	Peroxide of Manganese	more than Carbon
6	Calcopyrites (Double sulphide Iron & Copper)	High[b]
7	Iodide of Copper dry	

No 1 could be used pure

2   mixed with Magnesia Oxide

3   powdered or in block like the charcoal in the drawing— Anthracite the size of the charcoal shewn would come down in Res to less than $\frac{1}{10}$ ohm— This would be the bass material if you could secure it or drill it, but perhaps the powdered would do—

4   should be mixed with magnesia[c]

5   ditto.

6   pure.

7   mixed

you might have something outside of the Lamp to act as Cut out using anthracite. Thus[7]

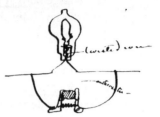

3000 @ 5000[8] ohms res but when Carbon breaks becomes brilliant for second & then comes down to $\frac{1}{5}$ @ $\frac{1}{10}$ ohm res & acts as Cut off—

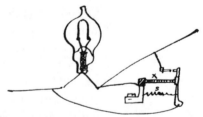

powerful spring—you could use rough broken pieces of anthracite— also try charcoal in here—

Anthracite has its initial resistance changed ~~by~~ to any degree by simply heating in Bunsen flame[9] by ~~pair~~ holding it in with pair pinchers— another way which may work as a Cut off is to mix conducting Lampblack with protoxide of Lead (Litharage)—so the powder will be a high resistance one— now tamp it between wires—or mould it & use it in Exterior apparatus. The moment it comes to incandesce—The Oxygen of the Lead unites with the Carbon & metallic Lead is instantly formed—

note)[d]   No amount of Lead in glass makes it Conduct as the lead is in the form of an oxide & is a non conductor.

Here is an idea—[10]

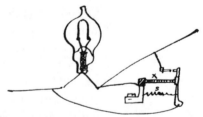

X is a stick of moulded material having a stiff gum like Licorice—Sealing wax Asphalt Etc mixed with Conducting Lamp black so it have say 5000 resistance. This shunts the Lamp— The Spring S Constantly tends to close the cut off lever & point but cannot as the rod being Cold is stiff— The moment the Lamp bks the 800 volts warms the stick up instantly it softens & the spring pulls the points together <u>powerfully</u> & keeps them there[e] Im getting short of Ideas & mail is closing. will send some more soon   god bless you my[b] boy   amen in haste

TAE

ALS, NjWOE, DF (*TAED* D8621L). [a]Followed by dividing mark. [b]Obscured overwritten text. [c]"with magnesia" interlined above. [d]"Note)" interlined above and enclosed with following paragraph by a large open parenthesis. [e]Multiply underlined.

1. The editors' conjecture of the month is based on related correspondence and laboratory records discussed in the notes below.

2. Not found. Johnson had apparently proposed an alternative to the lamp cutouts used in the first commercial installations of the Edison municipal system, which were electromechanical shunts, probably similar to those shown in an undated technical note. Jenks 1887, 10; Undated Notes and Drawings (c. 1882–1886), Lab. (*TAED* NSUN08 [image 108]).

Edison's undated drawing of an electromechanical cutout like that used in early installations of the municipal system.

3. The following discussion is very similar to a set of notes, dated 22 March (possibly retrospectively), that Edison made in one of his Fort Myers notebooks (N-86-03-18:121, Lab. [*TAED* N314121]). Preceding the notes on the municipal lamp is a drawing of a machine to scrape lampblack from a kerosene lantern; Edison drew other methods for obtaining lampblack on 28 March (N-86-03-18:209–13, Lab. [*TAED* N314209]). Johnson received this letter on 5 April, and two days later John Ott wrote Edison that he would make these compounds and that he was also making another one whose resistance would change. Ott also reported making "19 lamps and all worked well   Mr Batchlor tried his best to make them miss, even put 6¾ amperes through them and could not make them miss, these were with Peroxyde Lead and Musilage made into a paste and aplyed with brush this compells it to arc, positive, but your idea serves another purpos bringing the resistance down, to say that of lamp   this would dispence with regulating device in station" (Ott to TAE, 7 Apr. 1886, DF [*TAED* D8636I]).

4. Edison sketched a "cutout," an automatic switch for bridging the circuit around a failed lamp in the series-wired municipal lighting system (see Doc. 2877 n. 1). A similar drawing appears in his notebook (p. 123; see note 3), where he drew a line from the piece of charcoal to the following text: "ordinary non-cond[u]ct[in]g ie only partially [conducting] charcoal which conducts with the 800 volts & practically none at 55."

5. One watt (determined by resistance times the square of the current) is equivalent to approximately 44.3 foot–pounds per minute. The charcoal in the circuit described here would absorb about 0.833 of a watt. Thompson 1904, 229–30; cf. Doc. 1795 nn. 2–3.

6. A similar list appears with the first two items linked by a line to the powdered carbon in the associated drawing (N-86-03-18:123, Lab. [*TAED* N314123]):

1 powder of non-condctg Lampblack with or without a non-condctg oxide   2 Conducting Lampblack with various portions

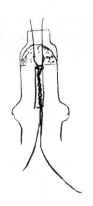

Edison's 22 March drawing of a cutout made using a powdered non-conductor.

of non-conducting oxides. so great an amount of oxide can be used to make it an insulator yet the 800 volts will jump from particle to particle.

The list continues:

3 Dry amorphous phosphate 4 Sulphide Molybdenum— 5 Sulphide Lead— 6 Powdered Calcopyrites 7 ordinary pyrites 8 powdered arsenic 9 [ordinary] arsenic 10 Peroxide Lead & also Per Ox Manganese mixed with non-cndctg oxides— 11 Iodide Copper mixed with non-condctg oxides.

7. Figure labels are "twisted wire" at top and "anthracite" at bottom.

8. Edison sometimes used the "@" symbol to indicate a numerical range.

9. Over the course of the 1850s, Robert Bunsen (1811–1899), renowned professor of chemistry at the University of Heidelberg, improved gas burners to produce a nonluminous flame useful in his analytic work. The so-called Bunsen burner rapidly replaced the blow pipe in much laboratory analysis. *Complete DSB*, s.v. "Bunsen, Robert Wilhelm Eberhard."

10. A similar drawing and description, next to which Edison wrote "OK," end the notebook entry (see note 3).

–2917–

Notebook Entry: Miscellaneous

[Fort Myers,] Mch 21 1886

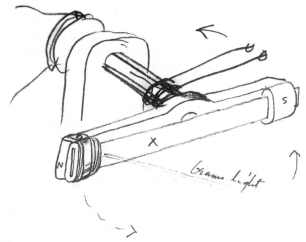

magnetized in the two directions   polzd Light & various changes[1]

X is ⁵⁄₁₀₀₀ sheet steel magnetized N & S. beam light strikes near end. Whole rapidly rotated   use telephone, Mirror etc. Theory is that Heat will disturb the magnetism & induce a current[2]

Translation Heat Light into E[lectricity] or XYZ

[A][3]

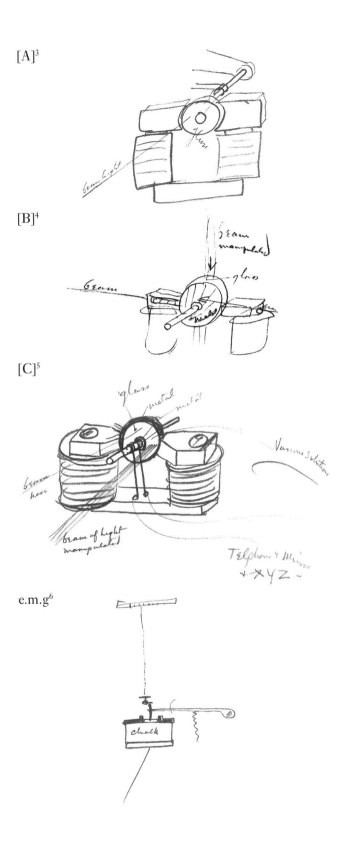

[B][4]

[C][5]

e.m.g[6]

Theory— The water will raise the end of lever resting on chalk up & work mirror— Try paper & other porous substances which swell by water.[a]

How would it do to make Zinc amalgam—& then mould Zincs for batteries by Hydraulic press— always be amalgamated   beat Fuller—[7]

For Storage battery plates— mould powdered Lead & dry Salt or other large crystal soluable salt—by hydraulic pressure then disolve out—[a]

Make collection of rare metals large quantities. Tellurium— Selenium—Vitrous & metallic, Cadmium— Thalium— Bismuth Antimony Tin, Arsenic, Molybdnum— Silicon & others— also selenides phosphides, Tellurides Sulphides, Arsenides.

Phonoplex   Prolongation of the wave—for long circuits.[8]

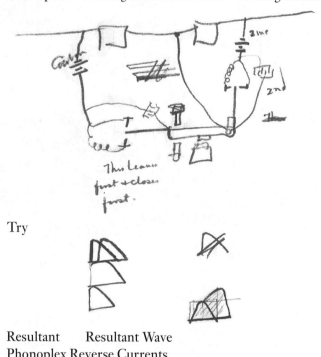

Try

Resultant   Resultant Wave
Phonoplex Reverse Currents

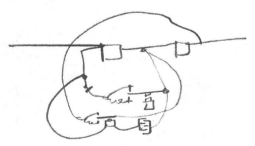

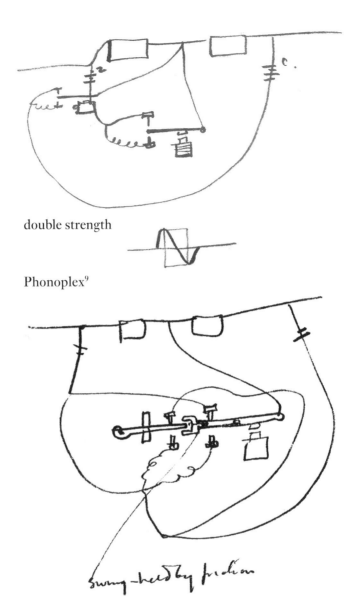

double strength

Phonoplex[9]

Phonoplex— If get cheap condenser work phone thro a condenser around magnet thus preventing change of adjustment by variations in strength line current—[a]

XYZ experiments   Electrodes to be immersed & weighed in various solutions—[b]

100	Zinc	100 Hard Rubber
"	Copper	Sulphur
	Lead	phosphorous
	Iron	Amorp[hous] phos
	Carbon	Parafine

Antimony	Asphalt
Tin	Sulphide Iron
Cadmium	" Lead
Molybdnum	Tin
Nickel	Zinc
Thallium	Copper
Silver	Antimony
Bismuth	Arsenic[c]

phosphide Iron
Lead
Tin
Zinc
Copper
Antimony
Arsenic[d]

all the Borides—peroxides—Iodides fluorides, Selenides, Tellurides—[a]

Translation Raident H[eat] & Light into E[lectricity][10]

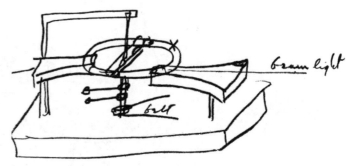

X circle of glass tubing ¾ @[11] 1 inch dia—filled various conducting fluids. rotated to Cut beam light various angles—[12]

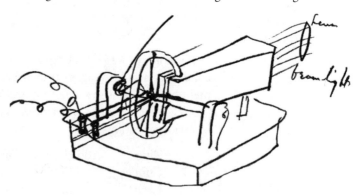

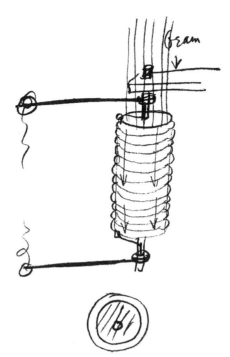

Beam Light circular passes through tubes   Coiled[e] Tubes
filled with various solutions

[E]¹⁴ᶠ

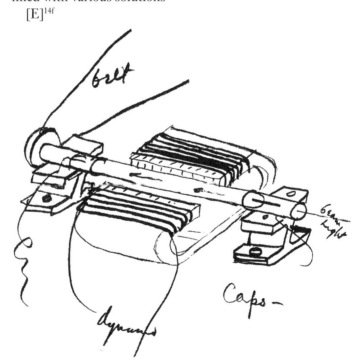

XYZ—use Iodide K   Sensitive & accumulative—

Try some expmts with vibrator primary Coil one ohm to 7 powerful battery plates of various metals in liquids insulators immersed like meter plate.[15]

Straightening filiments in Lamps—[16]

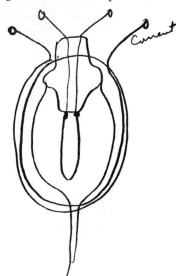

Straightening filiments in vacuo—[17]

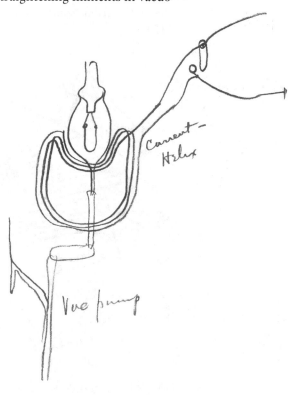

Mutual attraction between Current in wire & filiment straightens Carbon

Straightening filiments in lamps—[18]

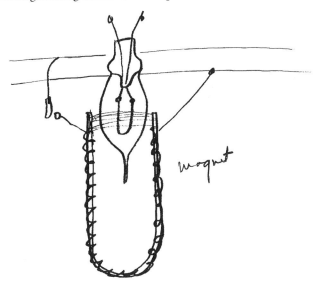

use Tellurium electrodes as dehydrogenizer—saving powdered Tellurium—[a]

Try Carbonize thin Celuloid in Linseed oil hot—[a]

See if Licorice can be bleached if there are more than one substance that can be seperated by solvents or precipitated—[a]

Licorice as clamps to enlarge   pasted on filiments before Carbonization

[G][19]

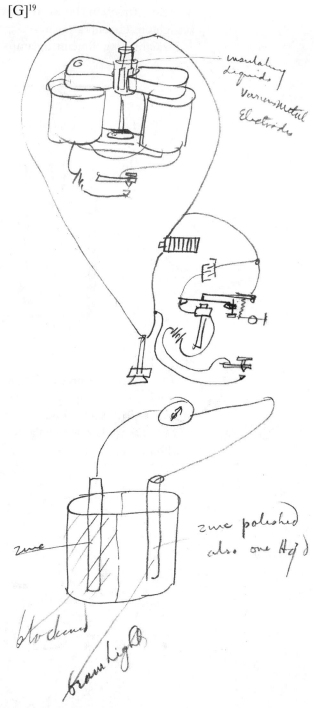

insulating liquids

Various Metal Electrodes

zinc polished
also. one Hg)

zinc

blackened

beam light

Sol Sul[phate] Zinc[20]

Pass beam of light for Spectroscope between poles powerful
Electromagnet notice any displacement Lines—with dif Sub-

stances[g]   also magnetize glass prism—also the Bisulphide prism also a glass cell prism   use <u>various liquids</u>—[21]

Phonoplex

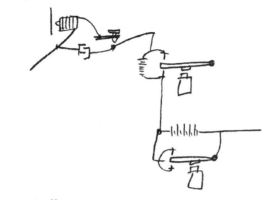

Phonoplex[22]

[H][23]

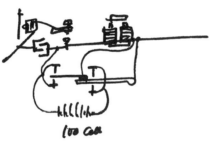

Phonoplex[24]

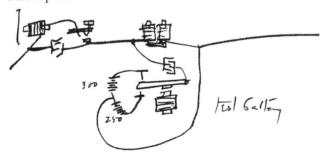

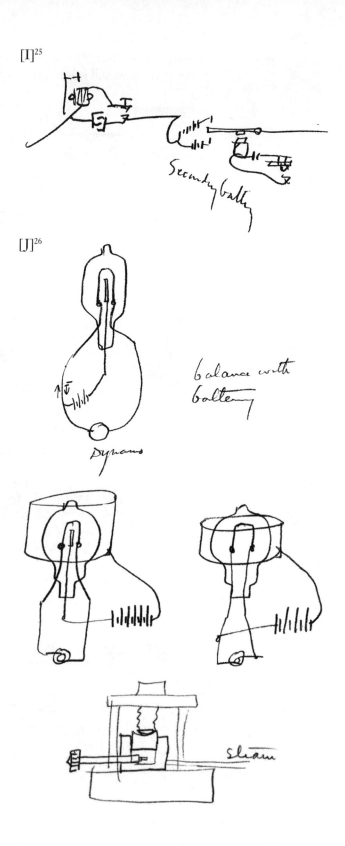

[I]²⁵

Secondary battery

[J]²⁶

Dynamo

balance with
battery

Steam

squeezing out Licorice filiments with broadened Ends[27]
XYZ Solutions for immersing the various substances

Cyanide K
Ammonia Conc
Bisulphide Carbon
Benzol
Benzine
Nitro Benzol
Aniline oil
Bromine
Chloral Hydrate
Absolute Alcohol
Turpentine
Oil Cassia
Linseed oil
Pure Sulphuric acid
Fuming   "
Idiform
Phenylamine
Ether
Chloroform
Tar—wood
    "   Coal
Carbozotic acid
water glass
Hydrocyanic acid
Butter Antimony
Bichromate K
Iodide methyl
    "   Potassium
Iodine H_2O
    "   in Bisulphide Carbon
Chloride Carbon
Napthaline
Ammonia Citrate Fe
Chlorine Water
Peroxide Hydrogen
Permanganic acid
Permanganate K
Fluoride Ammonia

Phosc Anydride
Sulphide K
Nitric acid
SO_4—[h] 3 proportions
NCl—strong
Chloride Zinc
Acid Pot Sulphate
Caustic Soda Potash
Lime—
Sulphate Iron
Ferri—& Ferro Cy Iron
[Nitroprusside?][i] Sodium
Sulpho Cy—K
Chl ammonium
Nitrite Soda
acetic acid gloc
acetate Lead
Chl Calcium
Borax
Sulphate manganese
Molybdic acid

bamboo   clamp to compress fibre

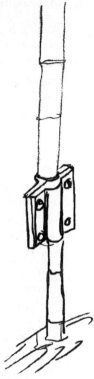

getting Turpentine by vacuum[28]

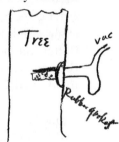

Spin out Balata for silk—also balata & a shining resin for ditto[29a]

Make ±20 $\frac{1}{2}$[h] m[icro]farad Condensers of copper & iron— with parafine paper— then make artificial zinc Sulphate Line—& make this Equivalent to Atlantic Cable   Then with a small emf from a section of a resistance [30]

& sensitive motograph & [vibrator?][i] try rapidity transmission & correct success by substituting Reg foil Condensers—[a]

Telephone

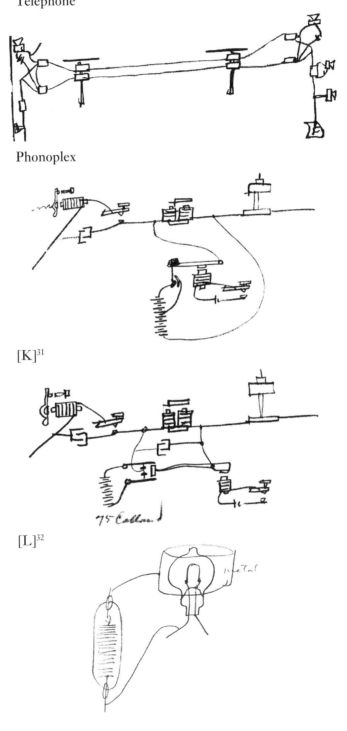

Phonoplex

[K][31]

[L][32]

DuLuc Dry pile 1000 volts in Vac[33]   also not in vacuum—

TAE   Mina

X, NjWOE, Lab., N-86-03-18:61 (*TAED* N314059). Document multiply signed and dated. [a]Followed by dividing mark. [b]Columns below divided by a vertical line. [c]Each column continued by five large "X" marks. [d]Column continued by four large "X" marks. [e]Added in left margin. [f]Drawing followed by dividing mark. [g]"with dif Substances" interlined above. [h]Obscured overwritten text. [i]Illegible.

1. Figure label above is "beam light." Edison's writing is small and unclear at this point, and he may have written "charges"; the editors are unsure of his meaning in either case.

2. The first page of the entry, up to this point, is undated. The editors have included it here because it bears a closer relationship to what follows than with the preceding entry (Doc. 2915).

3. Figure labels are "beam light" and "glass."

4. Figure labels are "beam," "metal," "beam manipulated," "glass," and "beam."

5. Figure labels are "glass," "metal," "metal," "beam here," "beam of light manipulated," "Various Solutions," and "Telephone & Mirror & XYZ—."

6. Figure label is "chalk."

7. The Fuller battery, used primarily in Britain, was a bichromate of potash battery in which an amalgamated zinc plate was placed upright in the porous jar. Niaudet 1884, 218.

8. Figure labels are "Carbon," "zinc," "2nd," "~~This~~," and "This leaves first & closes first."

9. Figure label is "Swing—held by friction."

10. Figure labels are "X," "beam light," and "belt."

11. Edison sometimes used this symbol to indicate a numerical range.

12. Figure labels below are "Lense" and "beam light."

13. Figure label is "beam."

14. Figure labels are "belt," "dynamo," "Caps—," and "beam Light."

15. That is, Edison's electric meter. It operated on the principle of electrochemical deposition of a metal from a solution onto an electrode that could be removed and weighed. See Doc. 2163 (headnote).

16. Figure label is "current." Edison had patented the use of a spring to compensate for the bending of the filament by thermal expansion and contraction (U.S. Pat. 317,633). The drawing here seems to represent the possibility of using the magnetic field generated by loops of wire to straighten filaments in the lamp. Edison made a set of drawings in July incorporating several of these designs, from which Edward Rowland made a measured drawing for a patent application (Case No. 670,

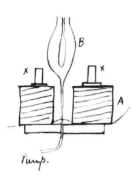

Charles Batchelor's 1887 drawing of methods that Edison had been "using for some time" to straighten filaments.

subsequently abandoned). Rowland dated Edison's drawings as 20 July; however, he had traced a copy of his own measured drawing on 12 July. It is likely that 20 July was the date on which Edison executed the application (Cat.1151, Lab. [*TAED* NM020ABA]; App. 2.A). In May 1887, Charles Batchelor described a similar method of straightening carbons that Edison "has evidently been using for some time. He finds that a magnet attracts the current passing in the carbon and he utilizes this for the purpose" (Cat. 1336:209 [item 366, 18 May 1887] Batchelor [*TAED* MBJ003208B]).

17. Figure labels are "Current—Helix" and "Vac pump."

18. Figure label is "magnet." Another version of this design can be seen in a drawing of 11 March (*TAED* B037BB).

19. Figure labels are "insulating Liquids" and "various metal electrodes."

20. Figure labels above are "zinc," "zinc polished   also one Hg'd," "blackened," and "beam Light." A zinc sulphate solution was the standard electrolyte in Edison's chemical electric meter. Doc. 2163 (headnote).

21. Certain liquids when used in a hollow glass produce better dispersion than a standard flint glass prism. Bisulphide of carbon was particularly favored because it gave a longer spectral line and could also be used for projecting spectra on a screen. Lockyer 1873, 24; Browning 1883, 41, 43–44, 62–64.

22. Figure label is "100 to 500 cells—."

23. Figure label is "100 cell."

24. Figure labels are "300," "250," and "test battery."

25. Figure label is "Secondary battery."

26. Figure labels are "Dynamo" and "balance with battery."

27. Figure label is "steam."

28. Figure labels are "Tree," "vac," and "Rubber gasket."

29. Balata is a latex similar to gutta percha that retains its elasticity better and was often mixed with gutta percha. Brannt 1900, 303, 309–11.

30. Figure label is "$^1/_{10}$ volt."

31. Figure label is "75 Callaud."

32. Figure label is "metal."

33. The battery developed by J. A. de Luc in the early 1800s provided a long-lasting, high-voltage power supply. It consisted of a large number of small discs of zinc foil and silver foil electrodes separated by paper, which acted as the electrolyte, packed tightly together on glass rods. Ord-Hume 2005, 167–68; Miller 1855, 1:375–76.

–2918–

Notebook Entry: Miscellaneous

[Fort Myers,] March 23 1886

Make a set of long magnets of ½ inch norway iron[1] of following lengths— 1, 2, 3, 4, 5 6 8 12 20 40 feet in length   wind Evenly after covering with paper No 20 well insulated wire from end to End   Then magnetize with same amperes in each Experiment   and move an astatic needle fibre suspended[a] with or without director magnet at different distances—& make <u>Curves</u>, opening & closing the Long mag ckt— I want to get the projecting power for the lines of force of very long magnets—[2] EMf of magnetism analogous to Electricity

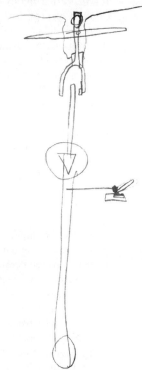

at same time get the charging & discharging times of these magnets[3]

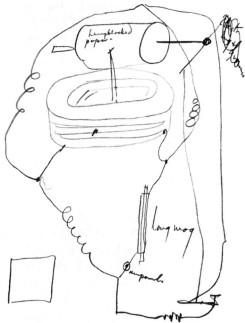

Recording self induction of magnets—[4]

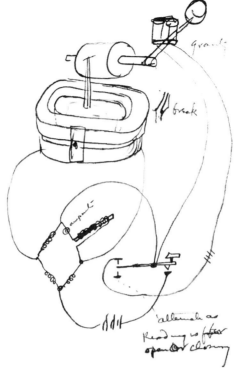

instead of allowing the cylinder to Rotate[b] by dropping of a lever by gravity Have the cylinder Connected by friction to a Constantly rotated Elec Engine with governor—release by ~~Clock w~~ magnet= this will give true Curve on blackened paper—perhaps only a Torsion wire galvanometer a la W.U Phelps-style=[5] bristle to mark—[c]

~~If the~~ If the ~~m~~Long magnet Expmts shew any abnormal results. Make following   500 feet smooth soft No 4 iron wire—cover carefully with paper spirally—wind its whole length with single layer No 22 wire   Then wind this on wooden Helix like magnet.

also make one of Lead Zinc Tin & also copper same length instead of No 4 iron

Get some sheet Tin & have Taylor the foundryman[6] cast some thin iron on them   if successful get some fancy stamped tin ware & cast thin plates   if good have some fine samples made & arrange with Troy Stove man.[7]

also Try brass[b] work tinned & brass cast on it at Bergmanns[c]

Try that Condenser we made with sheets of oxidzed Linseed oil. see if it still acts as Storage battery

[A][8]

Make XYZ Thermo piles of following 10 pair series—[9] [–][d]

using Radiant heat Light, Electricity & magnetism one end Cold on [--][d] other   also, Radiant heat one End Light filtered other End   also R Heat & Electricity op[posite] Ends— also R Heat & magnetism. also mag & E   also Light & E   Light & mag.

Sulphur—	Copper
Sulphur	glass
Sulphur	Hard rubber
Sulphur	Arsenic
Sulphur	Parafine
"	Asphalt
Sulphur	Sulphate[b] Iron
Sulphur	Melted Chlorides[b] & Other Salts.
"	Tin Zinc Iron Lead, Carbon

Parafine. ~~Arsen~~ & the above

Arsenic & the above

Hard Rub & the above

Melted Salts & the above & one with the other

[B][10]

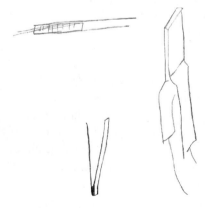

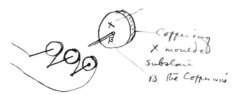

Thermo disc—[11]

The EMf bet Cop & an insulator as[b] a Thermo must be great. Try lime—water etc— cell being of sides of Copper.[12]

Various Insulators. Hard Rub. Sulphur gutta, etc

square foot surface   touch one side quickly with flame— then discharge thro sen[sitive] Gal[vanometer]

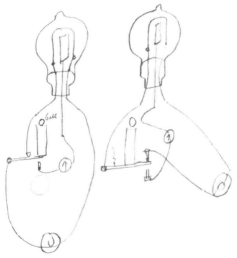

See if there is a Current after the Current is disconnected but there is still a brilliant Incandesce[13]   Use 100 cp Lamp

Try life

[E][14]

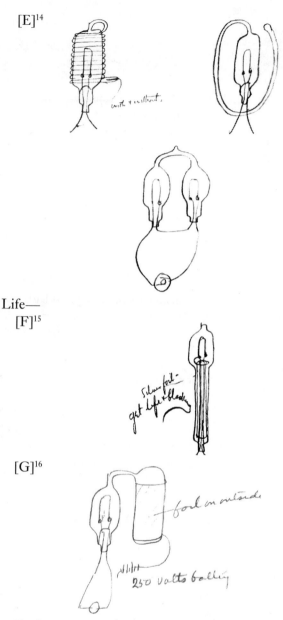

Life—
[F][15]

[G][16]

Try battery to both poles—also Holz mac[hine]—also reverse battery   also 1000 cell test battery.[17d]

May be vapor Hg plays important ~~par~~ part. Try jump spark for couple hours after Lamp finished before getting life= (to make ozone

Make 10 Lamps for Life with dry phosphorus in—

also Sulphur—also Zinc—Tin, Lead, Bismuth, Iron—Antimony—Arsenic—Iodine.[c]

Carbonize some filiments with long cotton fibre on them

& put in Lamps so filiments hang down & shew static attraction.[18]

XYZ

Make self Vibrator but with points that can be made to clamp the Ends of Zinc Tin etc wire so whole ckt will be of proper material to Conduct XYZ

also a Telephone where every kind of change of material can be made—perhaps 40[b] or 50[b] cheap telephones Every one an anomaly of XYZ devices would be better & run the scale on Everything with vibrators in all XYZ[b] Experiments
TAE—   Mina

X, NjWOE, Lab. N-86-03-18 (*TAED* N314131). Document multiply signed and dated. [a]"fibre suspended" interlined above. [b]Obscured overwritten text. [c]Followed by dividing mark. [d]Canceled.

1. Norway iron (also known as Swedish iron) was considered the highest grade of wrought iron, and its purity and softness made it the preferred form for electromagnets. Lilienberg 1915; Gordon 1996, 176–79; Lockwood 1883, 53; Fleming 1886, 62.

2. Edison had long correlated the size of a magnet, particularly its length, with the strength of its field of force. This association seems to have been an intuitive one based on his reading of Michael Faraday, and he had experimented in 1873 on the relationship between the proportions of magnets (some of them very long) and their field strength. Edison used long magnets again in the 1879 design of his first commercial dynamo, a distinctively tall machine. He held to that basic pattern until 1883, when the British electrician John Hopkinson, guided by more sophisticated theory, used short, squat magnet cores to produce a more economical magnetic circuit. Docs. 363, 375, 386, 1735 n. 14, and 2419 (headnote); Friedel and Israel 2010, 55–57.

3. Figure labels are "Lampblacked paper—," "long mag," "ampmeter," and "~~Electr Eng & dynamo~~."

4. Figure labels are "gravity," "break," "ampmeter," and "alternate as Read mag is ffor open or closing." Although Edison dated this page "Mch 24 1886," those pages preceding and following it are dated 23 March, and together they appear to constitute a continuous set of notes and drawings regarding these magnet experiments.

5. Edison referred to a form of torsion-wire galvanometer manufactured by George Phelps. The torsion-wire galvanometer, first devised by William Ritchie in 1830, was based on Charles-Augustin de Coulomb's torsion balance, which used the force required to twist a thread or wire to measure feeble electric and magnetic forces. Edison may have acquired what he called "Coulombs balance Gal[vanometer]" in 1873 (see Doc. 353). Bud and Warner 1998, s.v. "Galvanometer"; Buchanan 1846, s.v. "Torsion."

6. Zachary Taylor had an iron foundry in Brooklyn, which he (with Charles Batchelor and John Kruesi) incorporated as Taylor & Co. in February 1886. Cat. 1336:9 (item 20, 17 Feb. 1886), Batchelor (*TAED* MBJ003009E).

7. The editors have not identified this individual. Troy, N.Y., had been a center of stove manufacture since the 1820s, and at this time there were seven foundries employing about 2,000 workers. Descriptions of these foundries can be found in Weise 1886, 281–85; see also Groft 1984.

8. Figure labels are "Sulphur" and "copper."

9. Figure labels are "Sulphur" and "copper."

10. The following drawing shows the actual form of the thermocouple for the series of pairs listed above.

11. Figure label in drawing above is "Copper ring   X moulded substance   B the Copper wire."

12. Figure labels are "Cop" and "Cop."

13. Figure label is "ball."

14. Figure label is "with & without."

15. Figure label is "Silver foil—get life & blackening."

16. Figure labels are "foil on outside" and "250 volts battery."

17. Edison is probably referring to the 1,000 volt de Luc dry pile mentioned in Doc. 2917.

18. See Doc. 2892 n. 6.

–2919–

Notebook Entry:
Miscellaneous

Make Lamps of all kinds of glass and test conductivity.[1]

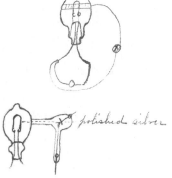

Also one of polished hard rubber[2]

[A][3]

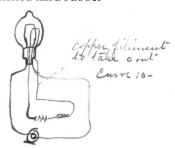

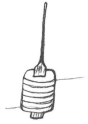

Try various liquid with capilliary index see if Magnilia[4] or E[lectricity] contracts or expands. Same—

Put under water & break capilliary x off,[5] let the H_2O & SO_4 remain in. Take out and melt x solid then decompose & burst it.

Radiant matter in vacuum.

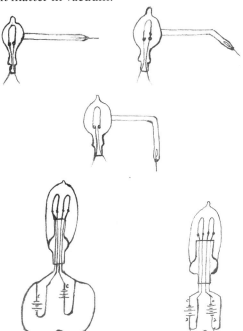

Vary incandescence of one, other constant

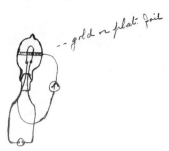

Make measurements.[6]

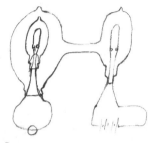

10 Lamps each[7]

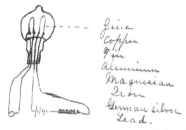

Work these slowly after Reg[ular] Lamp, finished & good vacuum—

Afterwards work them gradually up, slowly keeping vacuum running until bursted—

Run life curve—

Mina

X, NjWOE, Lab. N–86–03–18 (*TAED* N314157). Written by Mina Edison; document multiply signed and dated.

1. See Doc. 2892 n. 6.
2. Figure label is "polished silver."
3. Figure labels are "copper filiment to take O out" and "Curve 10—."
4. Mina probably copied this entry from a set of loose notes made by Edison; the original word was probably "magnetism."
5. Figure labels are "x" and "[Seat?]."
6. Figure label above is "gold or plat foil."
7. Figure labels are "Zinc," "Copper," "Tin," "Aluminum," "Magnesian," "Iron," "German silver," and "Lead."

[Fort Myers,] March 23 1886

If high incandescence of carbon filiment causes electrification & conduction in vacuo then perhaps platina heated by flame (blow pipe) to high incandescence will give Current to Central Electrode[1]

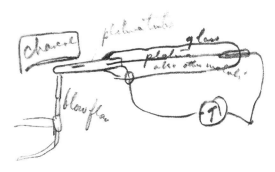

TAE

X, NjWOE, Lab., N-86-03-18:169 (*TAED* N314169).

1. Edison referred to the phenomenon known by this time as the Edison Effect and to the distinctive lamps with a third electrode by which it could be produced (see Docs. 1898 [headnote] and 2756 n. 3). Figure labels are "charcol," "blow flow," "platina tube," "platina also other metals," and "glass."

[Fort Myers,] March 25 1886

Hang ten lamps up by small wires from center of ceiling of a room or rooms get life also 10 Lamps surrounded within several inches with metal grounded. get curve of life— Think the former will show very much better—[a]

screen of mica geneva cross see if matter purely radiant[1]

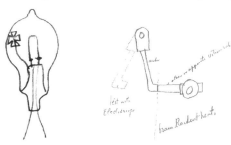

Theory is that the two vitrous substances rubbed together are positively & negatively Electrified by the heat acting on surfaces as Thermo pile.[2] They are thermo piles of very high EMF same as Bismuth rubbed on antimony are charged very Low Emf—[3a]

The Centre wire guarded with thin glass [melted on?][b] see if conduction.

see if conduction through glass

The oxidation on surface of a filament by hot sulphuric acid throws it into all kinds of Contortions by relieving the strain of the surface. in Experiments on slow carbonization Suspend in sand bath in tube Single filiment, & let it contort as it will, then try one with quick carbonization[4]

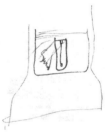

Put several of the finest fibres together with Licorice[c] with broadened End of Licorice— Sand bath auto regulated preliminary—[5a]

See if Licorice disolves in any of the Hydrocarbons etc.[a]

Roll out Licorice hot—also using oil—[6a]

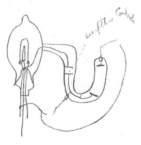

In the auto prelim sand battery[7] put several sets of Carbons taking each set out at different g degrees, weigh—[d] Carbonize fully & Run curves on finished Lamps[a]

perhaps paper soaked in Licorice calendered Hot stamped filiments can be carbzd by auto Prelim so perfect as to beat bamboo, then a million cld be Carbzd at once— use uncallendered paper so Licorice fill up spaces[a]

perhaps $^{25}/_{1000}$ uncallendered can be cut with revolving knife so there is no cavities & perfectly Even quality obtained. if impossible do this natural state it could be soaked glue— Rosin etc making it hard & thus cut down to size—

if vacuum doesnt conduct sound then diapm & compressed air[c] ought to be the thing[8]

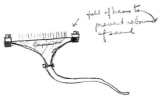

Try different gases compressed—[9]

Holes made deep

prevent rebound of sound   slightly bevelled or Cone shaped holes

Deaf apparatus

Screen of cone holes over big dia[10a]

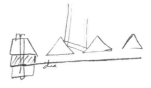

⟨no good⟩

⟨Little better⟩

distance of dia from underside[c] distance of bevel from straight line but try all distances—[11] ⟨OK⟩

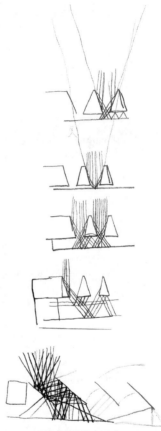

Coned rings seem to fulfil Every condition for concentrating against a diaphragm without rebound of the air wave[12]

Coned rings—[13] May[c] want partition t think have something like this in Laboratory

TAE

X, NjWOE, Lab., N-86-03-18:179 (*TAED* N314179). Document multiply signed and dated. ªFollowed by dividing mark. ᵇIllegible words interlined above. ᶜObscured overwritten text. ᵈInterlined above.

1. Edison's drawing resembles the apparatus used by British chemist William Crookes in a well-known experiment made public in 1879. When Crookes directed a cathode ray beam in an evacuated tube at a solid object in the form of a Maltese cross, he noticed that the cross blocked the rays, effectively casting a shadow. Crookes argued from this result that the rays consisted of particles traveling in straight lines, an interpretation vigorously disputed by others who argued that they were electromagnetic or some other type of disturbance in the ether. Crookes contended that the result also supported his controversial earlier claim about the existence of what he called "radiant matter" (borrowing Michael Faraday's phrase) under certain conditions—a fourth state of matter that was neither solid, gas, nor liquid (see Doc. 2582 n. 1; Crookes 1881; Heilbron 2005, s.v. "Cathode rays and gas discharge"; Brock 2008, 231, 238–41). The dispute over radiant matter continued to this time, and Edison had called it "a humbug" in 1884 (see Doc. 2589).

2. Figure labels above are "test with Electroscope," "amber," "other or opposite vitrous sub[stance]," and "beam Raident heat." The electroscope, an instrument for detecting differences of electric potential, was originally developed by William Gilbert to prove that substances other than amber could be electrified by friction. *Ency. Brit.* 11, s.v. "Electroscope."

3. On 1 October 1887, Edison wrote below this note: "Rotate collectors to take off E use fine wire primary & coarse secondary to reduce tension."

4. Figure label is "Sand."

5. This note contains the first known reference to a carbonization process that Edison called "auto preliminary." It may have evolved from his notes of 18 March (Doc. 2912) in which he described both a longer carbonization process and methods for putting raw filaments under pressure by molten metal, or its genesis may go back to the slow carbonizing experiments described in Doc. 2898. As he developed the idea, the term "auto preliminary" seemingly refers to methods of treating filaments (often by soaking) with carbonaceous materials that would penetrate the fiber and then be subjected to a preliminary carbonization process, all with the goal of producing a more homogenous structure and more even resistance (cf. Doc. 2346 esp. n. 10). In order to prevent these filaments from becoming distorted and to ensure that they would be heated evenly, Edison planned to carbonize them under pressure using radiated or convected heat. It was for this reason that he proposed in this entry to use a sand bath. It is not clear why Edison used the term "auto regulated," but it likely refers to the greater control he would have had over temperature and pressure from using a sand bath. Edison's experiments led to the filing of several related patent applications by the end of the year (see Docs. 2987, 2996, and 3006; U.S. Pats. 484,184; 484,185; 485,615; 490,954; and 534,207). By 1887 Edison often referred to this process for treating and carbonizing filaments in the laboratory with the shorter term "preliminarize" and to filaments thus treated as "preliminarized" (see, for example, Docs. 3039 and 3041).

6. Text is "see if this Condr."

7. Edison may have meant to write "bath," or he may have intended the meaning of "battery" as a series of similar instruments or devices. *OED*, s.v. "battery" 13a.

8. Figure labels above are "Compressed aid" (presumably intended as "air"), "dia[phragm]" and "dia."

9. Figure labels above are "compressed air" and "full of hairs to prevent rebound of sound."

10. Figure label is "dia." The previous day, Edison sketched a "Table scale & slide with music box to test deaf apparatus." Somewhat resembling an instrument for photometric measurements, this device could place a source of sound over a range of distances up to forty or fifty feet. N-86-03-18:171, Lab. (*TAED* N314171).

11. Figure labels are "angle ok" and "same distance."

12. Additional rough sketches similar to those here for "deaf apparatus" have not been reproduced. They appear on pp. 194–95.

13. Figure label is "½ section."

–2922–

Notebook Entry:
Miscellaneous[1]

[Fort Myers,] March 28 1886

Try Rubber parchment wet partchmnt & dried under the pressure   Hard Rubber thin & shaped   spun brass   oiled silk or cloth.[2]

wire gauze & oiled cloth, Linseeded paper laid 'on' inside—

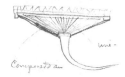

Try Ether, & other very light spec gravity Liquids   also try heaviest gases under pressure—[3]

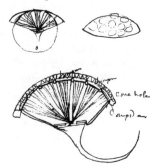

⟨good⟩[4]

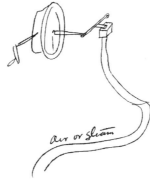

[--][a] open & close spectroscopic ground slit by Emg[5]   make powerful sound

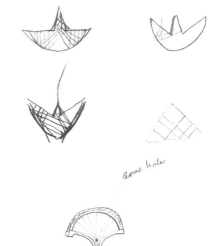

Deaf[6]

moulded hard rubber   10 inch diaphm   no compression also with Compressed air

It may be that radiant heat should be magnetized in a certain way before striking the thermo pile & the heat in this form would nearly all be turned into E[lectricity].[b]

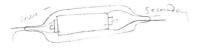

Try Regular Edison Coil in Vacuuo—heat tube well—see if sharper al[7]

Try Reg 7 ohm Coil in Vacuuo against one in air on phonoplex.

Try ¼ or ⅛ inch spark coil with High Vacuo—with Vibrator outside—

Try in our small condenser box Lead—Linseed paper dried so still little sticky—Lead—Then parafine paper then Lead Linseed paper Lead—connected in series.[8]

charge & see if constant Volts like test battery—
Remember that condenser that ~~gave a~~ acted as storage.[9]

Auto preliminary Carb[oni]z[a]t[io]n[10]

Siemens Regenater Lamp[11] for Heating using Bunsen flame or Even a row of Bunsen burners with glass chimney using heated air instead of direct flame on sand bath

Note= How would copper 15 @ 20 mesh do instead sand—Conduct heat well—[c]

Lamp[b]

Important= I think there is no doubt that if a coating of infusible oxide can be put on the Carbon filiment, That the oxide will be carried by the static charge and not the Carbon=[12] The attempts that have not been successful have been with the uncarbzed filiment. The filiment should be ¾ carbonzed at least if not fully carbonzed and then coated and the coating go through the preliminary stage[d] the same as if it was a filiment.

Coat carbonzed filiments with syrupy—Acetate, chloride, [~~C~~-][a] of Aluminum—Magnesium, Calcium,[d] also gelatinous silica, Boracic acid— If single coating not Enough recoat after Each preliminary say 500 deg[e] until proper thickness secured—[b]

Perhaps If the filiment fully carbonzed and suspended in a bath of chloride or acetate of magnesium—aluminum—Chloride Calcium Etc. and connected with the Current and the temperature of the solution kept high that the Oxide if Current is weak will be deposited in coherent state on the filiment=

Note— get the Enamals that used on pottery—& mix infusible oxides with them—

see if addition of the 500 Volts makes any change in the current going to the Central wire—[13]

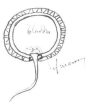

bladder or metal, Rubber etc   Compressed air also not thin dia membrane, to Ear tube   Completely Enveloped with coned circle— ⟨good!⟩[14f]

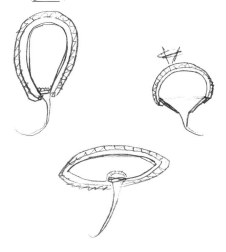

TAE

X, NjWOE, Lab., 86-03-18:217 (*TAED* N314217). Document multiply signed and dated. ªCanceled. ᵇFollowed by dividing mark. ᶜPartially enclosed by curved line at right. ᵈObscured overwritten text. ᵉ"500 deg" circled. ᶠMultiply underlined.

1. This entry is preceded by several rough drawings from the same day showing devices for the manufacture of lampblack and others possibly related to the hearing aid. N-86-03-18:209–16, Lab. (*TAED* N314209).

2. Figure labels for this drawing, presumably a hearing aid, are "Compressed air" and "dia[phragm] here or here."

3. Figure labels above are "compressed air" and "wire—."

4. Figure labels above are "diap[hrag]m," "diapm," "cone holes," and "compsd air."

5. Figure label above is "air or steam." Edison is probably referring to the metal rod with a slit that was attached to the ground plate of a Browning automatic spectroscope, around which the prisms could be moved. Schellen 1872, 94–95.

6. Figure label is "Cone holes."

7. Figure labels are "prim[e?]—" and "secondary." Although the coil tests Edison described here would have intrinsic interest for his ongoing study of electromagnetic energy, they also had potential practical applications to the railway telegraph and especially the phonoplex, where the "sharpness" of signals induced in the line was crucial to the system's operation.

8. Figure labels are "parafine," "Lead," "Linseed," and "Lead."

9. That is, like a storage battery.

10. Figure label is "Sand bath."

11. Edison meant the regenerative gas lamp devised by Frederick Siemens in 1879. The air used in combustion was warmed by waste heat recovered from the combustion process, thus producing a more intense light with less gas. "Regenerative System for Gas-Light," *Journal of the Franklin Institute* 107 (1879): 343; "A New Form of Regenerative Gas-Lamp," *Nature* 40 (23 May 1889): 82–83; "Société des Ingénieurs Civils de France: Presidential Address of M. Emile Cornuault," *Journal of Gaslighting* 97 (22 Jan. 1907): 208.

12. A filament treated this way might reduce the darkening of the glass bulb caused by carbon carrying. In an October 1882 patent application, Edison proposed "covering the flexible carbon filament . . . with a coating of insulating material not decomposable by carbon, and fusible at the highest temperatures only. For this purpose I prefer to use one of the earthy oxides. . . . The effect of this coating of the filament is that the static attraction will draw over particles of the oxide or other material instead of particles of carbon" (U.S. Pat. 492,150). He returned to this idea in his 1884 Florida notebook (Doc. 2609 [headnote]) where he proposed various methods for coating filaments with oxides. He speculated that "the oxide will not be reduced and that the organic material will be carbonized as a lace work the oxide staying there hence when carrying commences ½ or more will be white oxide hence the carbon will last twice as long and the globe will be no blacker besides I will have a high resistance Carbon" (see Doc. 2635).

13. Figure label above is "500 volts."

14. Figure labels are "bladder" and "if necesry." On preceding pages

also dated 28 March, Edison drew a number of arrangements for using compressed air or other gases to transmit sound in a hearing aid. Those devices included a number of semi-circular collectors of sound, similar to those shown here. Edison drew alternative versions of this hearing aid design the next day. N-86-03-18:214–25, 243, Lab. (*TAED* N314209 [images 107–12]; N314243).

-2923-

From Samuel Insull

[New York,] 29th March [188]6

My Dear Edison

My absence in the West on Machine Works business has prevented my writing you much: although the telegrams I have had sent you have kept you well posted I think.[1]

With relation to the Electric Railway I wired you yesterday of the success with Spragues motor. Batchelor informs me today that he has written you fully on this subject & as he is better posted than I am I will not try to supplement his letter[2]

Tate is now writing you as to phonoplex.[3] I went with him this morning to see Wiman. I am sure from Wimans manner that he was much pleased with what he saw of the phonoplex in Canada   I also gathered that Dwight was much pleased as while we were with Wiman he dictated a letter to Dwight giving details of prices &c as to Royalty. Tate & myself talked $50 a Circuit & said nothing about $6 per station.[4]

Railway Telegraph at this moment is somewhat stagnant Rudd[5] Dingle[6] & McGregor[7] are at Wadsworth[8] on the Chicago Mil. & St Paul.[9] They are experimenting with a car in the Yard. Rudd is troubled a great deal by induction which he says is much worse than at Staten Island. He says "the margins must be raised"—I presume he means that the Receiver must give the sound out louder in order before the system can be placed safely in the hands of the ordinary operator. He predicts however entire success & says it is only a matter of a little experimenting. Have you any suggestions. Stock is dull. I was obliged to you for your permission to sell 200 shares at 35 & also disappointed at it. I thought I had done wonderfully well when I sold 140 shares at 25. I then wired you my most that I hoped to sell more between 30 & 35. The latter figure was my most sanguine idea but you seem to consider, from your telegram that 35 would be low.[10] Any way the market is off— I failed to sell the 200 which I had hoped to & today I do not think I could sell at above 20 or possibly 25. I am on the lookout however nobody is offering any stock & the first purchaser who comes along I shall try to catch for you. My strong belief is that Railway Telegraph will ~~never~~not amount to a great deal in a long time. I do not think Railway Co.'s will adopt it with much of a "rush." If your belief is the same you should let me sell some of your stock even if I cannot get over 20. Dont think that because I write thus that I am not talking the thing up. I simply give you my opinion & something tells me that your opinion is somewhat akin to mine. The Doctor[11] is still painting glowing pictures of the business. He may be right, but still I cannot bring myself to think so.

Isolated business has improved a little the last week. Since I wrote you last the Co has sold plants amounting to[a] about 1000 lights. This is not much still it is better than nothing

The central station business looks promising. I wired you yesterday that with relation to Uptown New York the law had been appealed to. Johnson tells me that a mandamus has been applied for to compel the authorities to give us permission to lay our conductors.[12] Johnson has been agitating that the 27th St property should be secured any way & a meeting of the Illuminating Co directors is to be held for this purpose.[13]

The Boston Block plant is running about 2000 lights have been contracted for & overhead lines (short) are being run for this work as at present the permit to put wires underground has not been obtained.[14]

At last Stuart of Cincinnati has raised the Capital for a 1000 Light Block Plant. The people who are interested in the Union Depot[15] (now lighted by us, are putting up the money. They have plenty of capital & this Block is said to be only the start of a large Co.[16]

The Syndicate for putting in half a dozen stations in Upper New York State is going along all right. The first plant Amsterdam 1600 lights will soon be in course of construction[17]

Humbird[18] of Pennsylvania claims to have about 12 or 14 Cos all of which he will have installing plants before the Summer arrives. Whether he will be able to do this you are as well able to judge as I am.

Stewart of Chili writes me that Grace & Co have issued a circular asking for $100,000 Chilian currency ($50,000 gold) for a Valparaiso Station of 1200 lights.[19] Grace & Co in the subscription circular guarantee themselves 6% on this capital so I presume they will be sure to raise the money. Stewart still wants you to take up the question of running the Santiago Station with water power. He says until Santiago is put on a good basis that business will lag.

I have received the information from Lockwood about Gillilands English Patent & Tomlinson is drawing the papers. Bergmann has shipped half a dozen more Transmitters.

That key of safe has never reached here so I still owe 140 shares of Railway Stock which I gave my word I would return inside of a week & I do not like breaking my word in such matters.[20]

I continue to wire you very fully as I read your first telegram from Fort Myers[21] to mean[b] that it is next to useless to depend on letters   Yours very Sincerely

Saml Insull

The weather has been very uncertain here & Spring seems a very long time coming

ALS (letterpress copy), NjWOE, Lbk. 21:464 (*TAED* LB021464). [a]"plants amounting to" interlined above. [b]"to mean" interlined above.

1. Insull left for Chicago on 18 March. Because he instructed Alfred Tate to communicate with Edison in his (Insull's) name from New York, evidence for the date of his return is uncertain. Insull to TAE, 18 Mar. 1886, Lbk. 21:438 (*TAED* LB021438); Insull to Tate, 20 Mar. 1886, DF (*TAED* D8636B).

2. Charles Batchelor's letter has not been found but he reported the result in his journal as "excellent." Insull's telegram read, in part: "Johnson says tested twenty fourth Street Sprague Motor success   exceeded expectations in power ease of control non sparking   further tests with Car this week." Referring to Henry Rowland, professor of physics at the Johns Hopkins University, Insull added: "Rowlands report highly endorses Spragues methods." A few days later, Batchelor noted that another test "in which Proff. Rowland assisted was a perfect success at 24th Street   Now it is proposed to fit up a car." The tests were made along a short track on the grounds of a sugar refinery on Twenty-fourth St. Cat. 1336:15, (items 33, 36; 29 Mar. and 1 Apr. 1886), Batchelor (*TAED* MBJ003015C, MBJ003015E); Insull to TAE, 28 Mar. 1886, Lbk. 21:486 (*TAED* LB021486); Dalzell 2010, 78.

3. Tate reported in detail about preliminary tests he made of lines on which the Baltimore & Ohio Telegraph Co. planned to install the phonoplex. These were a 106-mile wire from Baltimore to Hagerstown (instead of to Harpers Ferry, as originally planned), and 85 miles from Baltimore to Harrisburg. Tate discovered no difficulties except the discharge from relay magnets; finding no troublesome induction effects, he did not expect to need condensers on adjoining wires. By May, the B&O reportedly was using the phonoplex from Baltimore to Harrisburg and on to Pittsburgh. Tate memorandum, 27 Mar. 1886, DF (*TAED* D8636H); Tate to TAE, 28 Mar. 1886, Lbk. 21:475 (*TAED* LB021475); "The Edison 'Phonoplex' or 'Way-Duplex,'" *Electrician* 16 (7 May 1886): 517.

4. Alfred Tate gave Edison a similar account of this 28 March meeting. Because Wiman had forgotten about the royalty proposal, Tate and Insull did not bring it up and instead offered the phonoplex at $50 per year per circuit, or one-half the U.S. price. Tate pointed out that Harvey Dwight of the Great Northwestern Telegraph Co. wanted to have

more circuits installed. Tate to TAE, 28 Mar. 1886; TAE to Dwight, 11 May 1886, Lbk. 21:475, 22:67 (*TAED* LB021475, LB022067).

5. A capable inventor with several patents to his name, Charles H. Rudd (1842–1894) had worked for the Gold and Stock Telegraph Co. until about 1880, when he reportedly joined the Western Electric Manufacturing Co. in Chicago (U.S. Pats. 55,541; 68,577; 259,589; 191,887; U.S. Census Bureau 1970 [1880], roll 201, p. 321B, image 0636 [Evanston, Cook, Ill.]; Scribner 1919, 12). He and Samuel Dingle had participated in the tests and demonstration of Edison's railway telegraph system on Staten Island in 1885 and early 1886. Rudd died of injuries caused by the accidental detonation of a high explosive compound he had invented and was testing for Western Electric (Rudd 1887, 269; U.S. Patent Office 1896, 314; "Condition of Charles H. Rudd," *Western Electric News*, 15 [11 Aug. 1894]: 69; "The Late C. H. Rudd," *Electrical Engineer* 18 [3 Oct. 1894]: 272).

6. Samuel K. Dingle (b. 1850?), originally from New York City, participated with Rudd on the earlier railway telegraph experiments on Staten Island. An experienced telegrapher, Dingle had worked for the Western Union Co. for nine years before leaving to become an assistant electrician with the Electro-Graphic Co. He held similar positions with the Postal Telegraph Co. before returning to Western Union in 1884. Dingle later worked for the Railway Telegraph and Telephone Co. and its successor, the Consolidated Railway Telegraph Co. In 1892, he formed the Phelps & Dingle Manufacturing Co. of Passaic, N.J., to manufacture pneumatic tires patented by Lucius Phelps. Whittemore 1889, 342–43; "Messages from a Moving Train," *NYT*, 2 Feb. 1886, 1; New Jersey Dept. of State 1914, 558; U.S. Pat. 482,487.

7. George C. MacGregor (b. 1849?), a native of Elizabeth, N.J., was the manager of the Railway Telegraph and Telephone Co. After earning a degree in civil engineering from Rensselaer Polytechnic Institute in 1871, he joined the Pennsylvania Railroad as a construction engineer, resigning in 1879 as a principal assistant of maintenance of way. MacGregor later was chief engineer of the Raleigh & Western Railroad. Nason 1887, 419; "Telegraphy on an Express Train," *NYT*, 6 Mar. 1886; Rensselaer Polytechnic 1875, 33; "Personal," *Engineering News* 6 (9 Aug. 1879): 249; "Personal Mention," *Railway Age* 38 (9 Sept. 1904): 352–53.

8. The small town of Wadsworth, Ill., about forty-one miles north of Chicago's Loop, was created in 1874 by the construction of the Chicago, Milwaukee & St. Paul Railway line and named for one of the company's principal investors. *Ency. Chgo.*, s.v. "Wadsworth, Ill."

9. The Chicago, Milwaukee & St. Paul Railway originated in 1847 as the Milwaukee & Waukesha Rail Road Co. After a number of name changes, it became the Chicago, Milwaukee & St. Paul Railway in 1874 on completion of its line between Milwaukee and Chicago. After the turn of the century, the railroad reached Seattle and added "Pacific" to its name. *Ency. Chgo.*, s.v. "Chicago, Milwaukee, St. Paul & Pacific Railway Co."

10. The telegrams from Insull and Edison have not been found, but on 18 March, shortly before the Chicago exhibition of the railway telegraph, Insull wrote that although he had not sold any shares beyond the initial lot of 140, "I am trying to sell 200 shares more & I have put the

price at 35 but if I cannot make a sale at that price I shall come down to thirty." Insull to TAE, 18 June 1886, Lbk. 21:438 (*TAED* LB021438).

11. That is, Eugene Crowell.

12. The New York state legislature had created the Board of Commissioners of Electric Subways, known as the Electrical Subway Commission, to enforce the placement of telegraph, fire alarm, and electrical wires underground, as a new law required. The commission proved to be an obstacle to the creation of a Second District. As Spencer Trask, president of the Edison Electric Illuminating Co. of New York, reported in 1886, the Commission had refused to issue permits for the company to open the streets, apparently in the hope of creating a uniform underground grid through which all types of wires could be run. The Edison company, which had its own system of underground conductors in use for the Pearl St. plant downtown, viewed the delay as an unwarranted hardship. It is not clear what legal redress the Edison Illuminating Co. sought; it finally gained the needed permits in 1887. New York Edison Co. 1913, 108–11; Doc. 2568 n. 1.

13. The Edison Illuminating Co. decided on 9 April not to purchase the unspecified Twenty-seventh St. property. Instead, it authorized Charles Coster to acquire "the Racket Club on 26th St. and 6th Ave." for $80,000; the property was probably that of the Racquet Club (variously the Racquet Court Club) at 55 W. Twenty-sixth St. The station was eventually built nearby, at 45 W. Twenty-sixth St. Cat. 1336:19 (item 39, 9 Apr. 1886), Batchelor (*TAED* MBJ003019A); Martin 1922, 172; King 1893, 549; Carpenter 1909, 439.

14. The Boston block plant was the first project of the Edison Electric Illuminating Co. of Boston, formed on 26 December 1885. Charles Batchelor visited the plant, in a converted stable and tenement building near the city's theater district, on 10 February. Its first commercial service, on 22 February 1886, was for a performance of *Iolanthe* at the Bijou Theater. The plant opened with one Armington & Sims engine and two Edison H dynamos, augmented in short order by a similar engine and dynamos from the World's Industrial and Cotton Exhibition at New Orleans and by similar equipment from an Edison plant in Rochester. These additions gave the plant a capacity of 2,400 16-candlepower lamps and made possible the expansion of service Insull referred to here. Cat. 1336:9 (item 14, [10?] Feb. 1886), Batchelor (*TAED* MBJ003009A); Mansfield 1901, 798.

The term "block plant" designated a facility in one building that provided electric power to others on the same block so that the wiring need not cross a city street. This was not precisely the case with the Boston plant because the Bijou Theatre was located across the street, but Insull used the term more loosely to differentiate the station from the much larger central stations powering entire city districts and also from the smaller isolated plants that normally provided electricity to a single building. "Block Plants," advertisement of the Engineering Supervision Co. of New York, *Isolated Plant* 8 (Sept. 1916): back cover; G. W. Bromley & Co., "Map of the City of Boston," 1886, Norman B. Leventhal Map Center, Boston Public Library, http://maps.bpl.org/id/12259 (accessed 13 Sept. 2012).

15. The Cincinnati, Indianapolis, St. Louis, and Chicago Railroad built the massive Central Union Depot (also known as Grand Central

Station) to accommodate a number of railroads in Cincinnati; it opened in April 1883. The Ohio Edison Electric Installation Co. of Cincinnati created an electric plant (with two Edison H dynamos) for the depot in a cramped space under the sidewalk. The plant apparently was the source of problems. In February 1885, Archibald Stuart reported that Ohio Edison had "had very bad luck" with it. "New Station in Cincinnati," *NYT,* 10 Apr. 1883, 2; Cat. 1336:27 (item 56, 6 May 1886), Batchelor (*TAED* MBJ003027); Stuart to Insull, 5 Feb. 1885, DF (*TAED* D8523J).

16. According to Archibald Stuart, the Edison Construction Dept. had made a canvass in Cincinnati for a 1,000-light block plant on the basis of which Insull, in March 1884, had provided a construction estimate to the Central Edison Light Co. of Cincinnati. On that company's behalf, Stuart asked Insull on 20 March 1886 to locate the old canvass and forward it to him. Insull advised him to contact the Isolated Co., which had taken custody of the Construction Dept.'s records. The editors have found no follow-up correspondence. Stuart to Insull, 20 Mar. 1886, DF (*TAED* D8622G); Insull to Stuart, 27 Mar. 1886, Lbk. 21:461 (*TAED* LB021461).

17. The Amsterdam central station (1,600 lights) was to be the smallest of six plants in upstate New York projected by the syndicate arranged by Edison agent Henry Clark. The plant may be the one that started operating about this time as a joint venture of the Amsterdam Arc Light Co., the Amsterdam Street Railroad Co., and the Edison Electric Light & Power Co. Stuart 1909.

18. James S. Humbird (1853–1922) was the agent in Maryland and Pennsylvania for the Edison Isolated Co. and the Edison Electric Light Co. An alumnus of Washington College (now Washington & Lee University) in Lexington, Va., Humbird served as president of the Association of Edison Illuminating Companies in 1884 and 1885 and as vice president in 1886. In 1885, he formed a partnership with Frank Gorton, secretary and treasurer of the Western Edison Light Co. of Chicago, to act as Western agents for the Edison concerns. George Westinghouse was reportedly so impressed with Humbird's promotion of the Edison system that he hired him in 1888 as a general selling agent, and Humbird attained prominent positions in the Westinghouse organization. Humbird was later among the incorporators of the National Carbon Co., a consolidation of eleven U.S. and Canadian companies that constituted the majority of the world's carbon industry. Jordan 1878 [1911], 1:1500; "James S. Humbird Dies in Atlantic City," *Cumberland (Md.) Evening Times,* 2 Sept. 1922, 9; Insull to TAE, 19 Apr. 1885, DF (*TAED* D8523ZAO); Washington and Lee University 1888, 138; Association of Edison Illuminating Cos. 1887, 7; *Edisonia* 1904, 196; "Obituary" [Frank Gorton], *Electrical World* 56 (22 Dec. 1910), 1505; "Chicago," *Electrical Engineer* 7 (Jan. 1888): 27; "The London Westinghouse Company," *Electric Power* 1 (Sept. 1889): 293; "Miscellaneous Notes," *Western Electrician* 6 (1 Feb. 1890): 62; "Big Carbon Firms Combine," *NYT,* 10 Jan. 1899, 1.

19. Circular not found.

20. Ezra Gilliland apparently sent the key on 21 March. Gilliland to Insull, 21 Mar. 1886, DF (*TAED* D8634K2).

21. Doc. 2910.

Cotton picker[1]

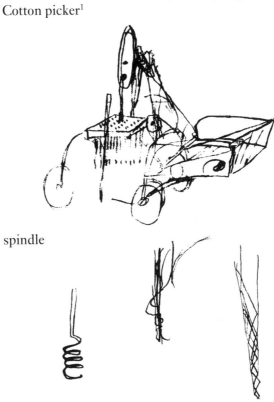

spindle

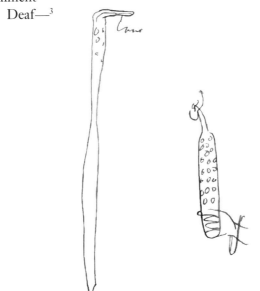

Compound filiment   Magnescan[2] filiment   Duplex filiment   Insulated filiment   Enameled filiment   Magneso-
filiment
Deaf—[3]

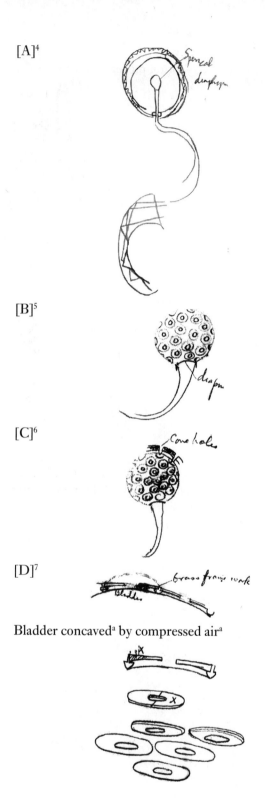

[A][4]

Spherical
diaphgm

[B][5]

diaphm

[C][6]

Cone holes

[D][7]

brass frame wk

Bladder

Bladder concaved[a] by compressed air[a]

x

x

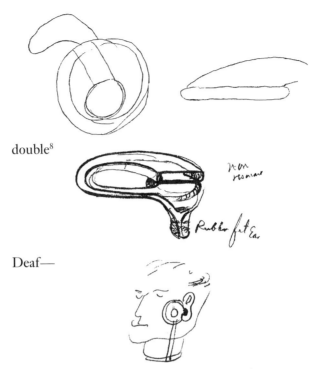

double[8]

Deaf—

Ear piece. spring inward to clasp side head[9]

[E][10]

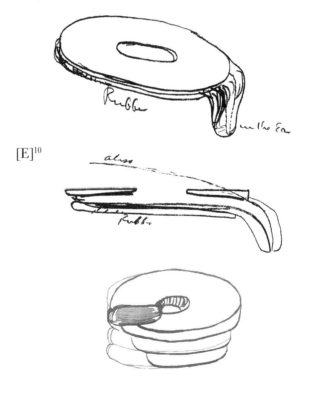

Deaf   Get Edison Telephone—also Gilliland Telephone[11] also The[12]

and Test thoroughly the Effect of Mouth pieces & effacacy of the chambers of all sizes & shapes also if chamber must be tight at edges— use a pipe to sound   also the voice.

Deaf— for Compressing the air use water[13]

or lift like Hg in Giesler[14]

TAE

X, NjWOE, Lab., N-86-03-18: (*TAED* N314245). Document multiply signed and dated. [a]Obscured overwritten text.

1. According to a newspaper notice of the Edisons' southbound journey (republished in mid-March), "While passing through Georgia the other day Thomas A. Edison, the electrician, conceived the plan of a machine for picking cotton, and showed sketches of his new invention to persons in a hotel at Atlanta" (*Waterloo [Iowa] Courier*, 17 Mar. 1886, 6). Edison reportedly was working on a cotton picker in late 1887 and thought that it might be one of the first things developed at his new Orange laboratory ("Edison's Perfected Phonograph—His New Laboratory Nearly Completed—Other Electrical Work," *Electrical Review* [New York], 11 [5 Nov. 1887]: 8–9; "A Talk with Edison," *Mechanical Engineer* 14 [5 Nov. 1887]: 105). Efforts to develop an automatic cotton picker dated to the 1850s but were not commercially successful until the early twentieth century (Holley 2000, chap. 26; *EH.net Encyclopedia*, s.v. "Mechanical Cotton Picker" [http://eh.net/encyclopedia/article/holley.cottonpicker], accessed 4 Dec. 2012).

2. Edison perhaps meant "magnesane," a rare term for magnesium chloride ($MgCl_2$). *OED*, s.v. "magnesane."

3. Figure label is "[wire?]."

4. Figure label is "Sperical diaphragm."

5. Figure label is "diap[hrag]m."

6. Figure label is "Cone holes."

7. Figure labels are "Bladder" and "brass frame work."

8. Figure labels are "non resonant" and "Rubber   fit ear."

9. Figure labels are "Rubber" and "in the Ear." On the page facing the first page of this entry (p. 244), Edison made doodles of several heads in profile (though none with ears) and several rough sketches that could be related to diaphragms.

10. Figure labels are "also" and "Rubber."

11. Probably the Gilliland transmitter discussed in Doc. 2848 n. 7.

12. The editors have not positively identified the specific device that Edison apparently had in mind, but his drawing below is reminiscent of ideas he planned to try in October 1884 (see Doc. 2750 esp. figs. B and M).

13. Figure label is "[Res----l?]."

14. Edison referred to the simple but effective and relatively rapid mercury pump devised by Heinrich Geissler (1814–1879) about 1855. It operated by raising a reservoir of mercury so as to force air from an attached chamber, then lowering the reservoir so that the mercury flowed away from the evacuated chamber. Edison acquired a Geissler pump in 1879, and its principle was incorporated into the design of novel vacuum pumps at his Menlo Park laboratory. Thompson 1888a, 6–8; see Docs. 1714, 1786 n. 2, and 1816.

–2925–

Notebook Entry:
Miscellaneous

Fort Myers Fla April 3 1886

Ideas[1]

Lyrnaxial piano

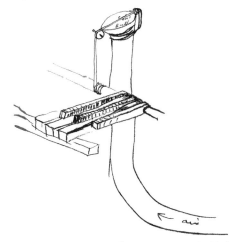

Helmholz artificial Laryrnx[2] with [streatchable?][a] Cords on ends of rubber[3]

In the gelantenous Silica for compound carbons[4] mix minimum quantity Caustic Potash to cause it slightly to fuse at highest temperatures to hold Silica together—[b]

Lampblack— Drop on hot plate (red hot to ignite it) Crude petroleum & revolving cylinder—

With Natural gas Dont use outside Oxygen but mix[c] previous to issuance from burner with the gas— ascertain by experiment smallest quantity Oxygen that will take all the Hydrogen of the gas & leave no excess

Regulate admission air[5]

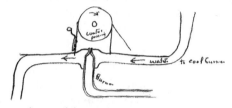

or mix air previous with gas—

T A Edison

X, NjWOE, Lab., N-86-04-03.3:1 (*TAED* N317001). Document multiply signed and dated. [a]Illegible. [b]Followed by dividing mark. [c]Obscured overwritten text.

1. This heading appears on the notebook's first page to designate its purpose. Edison made many of the same entries, in what seems to be a rougher form, in another notebook, where this entry appears on the first five pages. N-86-04-03, Lab. (*TAED* N315).

2. Hermann von Helmholtz, perhaps the most eminent German physicist of his generation, experimented and theorized extensively on acoustics. He devised electromagnetic instruments with resonators to replicate vowel sounds but did not build an artificial larynx. His well-known *Sensations of Tone* (an 1875 English edition of which Edison owned) illustrated an "artificial membranous tongue" whose action Helmholtz described as analogous to that of the elastic vocal chords of the human larynx. Helmholtz 1875, 146–47.

3. Figure labels above are "cord," "Rubber," and "air."

4. It is not clear if Edison referred to arc light carbons compounded chemically by adding salts (as described in du Moncel 1883 [135–37]) or to the "compound carbons" made of many smaller carbon rods, such as were used in well-publicized 1884 experiments at the Trinity House on the English Channel ("The Trinity House Experiments on Lighthouse Illuminants," *Nature* 30 [14 Aug. 1884]: 362).

5. Figure labels are "water passing," "water to cool burner," and "Burner."

[Fort Myers,] April 4 1886

I propose to rotate the cotton picker spindles by a blast of air acting on a wheel like Sturtevant blower—[2] The direction being reversable as the spindle goes up or down Towards the plant.

Multiple armatures, each armature wound Complete in regular way   Commutator 1st 1 block to wire to first armature 2nd block to " " 2nd armature & so on[3]

On the filiment of carbon coated with Silica etc, The thickened ends can be freed from the coating by immersing in an acid—Hydrofluoric, Sulph'c etc but this need not be originally coated if each one done by by hand seperately.

The increase in the number of candles per horsepower with white radiating surface will be greatly increased[4]

Make a ¹⁄₁₀ M[icro]F[arad] condenser (parafine) in form Cylinder also one exactly same Capacity or near   test both outside then enclose one in Vacuum—see if Vacuum increases capacity or sharpness—

TAE—

X, NjWOE, Lab., N-86-04-03.3:9 (*TAED* N317009). Document multiply signed and dated.

1. Another version of this entry can be found in N-86-04-03.1:9–19, Lab. (*TAED* N315009).

2. Edison referred generally to the centrifugal air blowers designed and manufactured in the United States by Benjamin Franklin Sturtevant, who had invented machines to produce wood veneers and pegs for boots and shoes. To remove dust created by these processes, he designed pressure blowers and rotary exhaust fans by 1864 and became the major manufacturer of this equipment. By 1869, Sturtevant was also manufacturing a system of forced-draft ventilation for buildings; by the end of the next decade, he was operating a large factory in Jamaica Plains, Mass. *DAB*, s.v. "Sturtevant, Benjamin Franklin"; Petroski 2008, chap. 8; Landau and Condit 1996, 32; Rosenberg 2007, 140–41.

3. In the other version of this entry (see note 1) Edison drew several arrangements for using three Gramme armatures connected together which he believed would be non-sparking. Although he made a note to obtain a patent, he apparently did not do so. At this time, Charles Batchelor was working on a non-sparking dynamo design for Edison (see Doc. 2908), a problem to which he and Edison returned in November (Cat. 1336:139 [item 241, 4 Nov. 1886], Batchelor [*TAED* MBJ003139B]).

4. In the other version of this entry (see note 1) Edison thought the increase would be ten to fifteen percent.

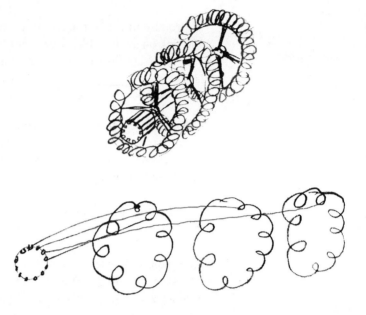

–2927–

Notebook Entry: Incandescent Lamp

[Fort Myers,] April 5 1886

Carbonize less dense material such as paper, wood (pine willow etc) Licorice mixed with MgO by auto preliminary to get high resistance then if it stands well—compound the surface by MgO—

I think with perfect Carbzn The lighter materials will answer fully as well as bamboo & be exceedingly high resisistance   Try parchmentized paper—white Holly Lampblack & Licorice & MgO—Lampblack Tar & MgO—Licorice & MgO. punch out of rolled sheets of this material[1]   Municipal 32s[2] will answer

Mixtures for filiments which are soft can probably be rolled down between tin foil several layers of material & foil one over the other & rolled together   thus obtained even

Several might be stamped out simultaneously bent in loop shape & Carbzd together   The foil melting or could be eat out by acid Thus making it easier to make fine filiments

With Auto Preliminary Soak original filiments with Licorice, prelim & Carbz reg— Then soak Licorice & prelimin & resoak 2 or 3 times, then Carbz regular   This has never been fairly tried

TAE

X, NjWOE, Lab., N-86-04-03.3:13 (*TAED* N317013). Document multiply signed and dated.

1. Edison had made earlier attempts to punch or cut filaments from rolled sheets; see Docs. 2057 n. 3, 2291, 2306, 2308, 2346, 2632, 2634, and 2635.

2. Edison's reference to "32s" is uncertain but may be to the candle-power rating of municipal system lamps. In his designs for a municipal lamp cutout, Edison had proposed using some of the high-resistance materials listed here; see Doc. 2916.

–2928–

Notebook Entry:
Miscellaneous[1]

[Fort Myers,] April 6th 1886

Arc Light Carbons— before baking make sheets of pottery mixtures and roll outside covering on Carbon— make mixtures so as ~~exp~~ contraction will be the same put it on very thin, perhaps 2 or 3 exceedingly thin $\frac{1}{1000}$ Coats best. Alumina— Magnesia, clays etc[2a]

Grasshopper battery—[3] make the size of the cells as small as possible so as to prevent slow discharge by surface— [~~Osm~~---][b] Iridisomine points so there is no fine metal points to follow & prevent instantaneous discharge

Lamp   Make a mouse mill like Thompsons recorder[4] or perhaps the Gas Lighter Mouse mill[5] run by motor will answer. use this in static experiments on Lamps for counter charges—[6a]

Speaking tube— Try greatest distance with inch gas pipe diaphragm on ends and 35 lbs to square in pressure inside you can speak— Try lead pipe instead $\frac{1}{4}$ inch inside bore—

Try at different pressures   if increasing pressures causes greater distance Carry pressure up to 100 lbs or more to square inch   Try glass Tubing but on & sealing wax poured on joint.[7]

Deaf—

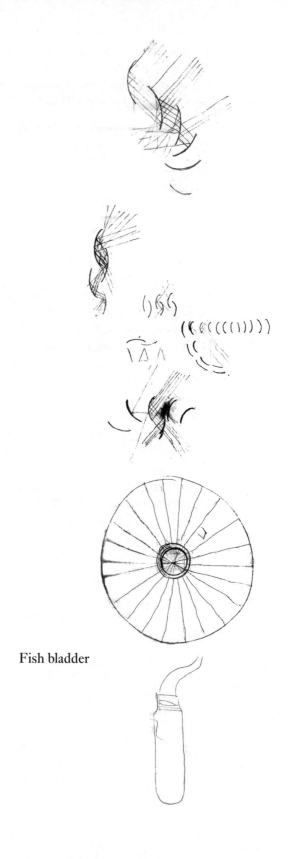

Fish bladder

Deaf

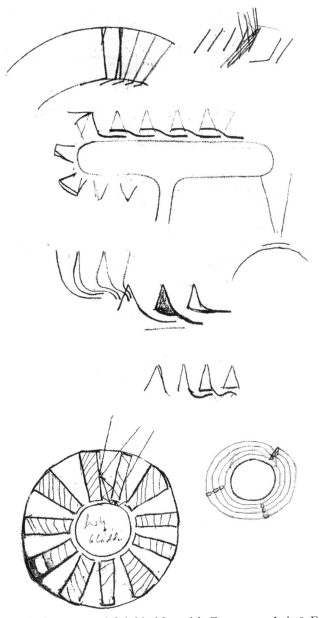

30 mesh sieve around fish bladder with Compressed air & Ear piece.[8]

Telephone—

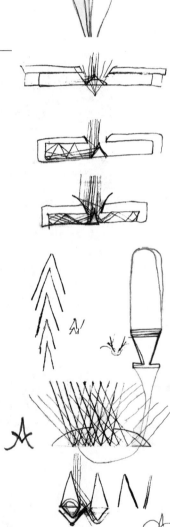

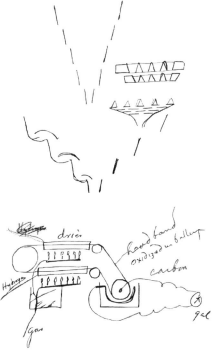

Conversion of heat into E[lectricity] by ox[i]d[a]t[io]n & reduction of O of Lead on continuously moving band of lead,[10]
passing into liquid close to Carbon in proper liquid Lead is
Oxidized—thence through drier tube thence through Hydrogen reduction tube to battery again & so on continuously

Experiments to get a porous material whose pores will let O
through & not N[11]

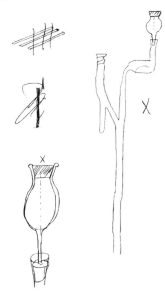

Use plaster paris—Cork—Lime—natural—Meerschum—pressed Chalk every deg pressure, Cocoanut charcoal—dif charcoal—dry clay moulded oxides phosphates etc.[12c]

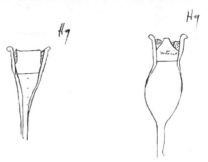

Leather bladder   Alligator Leather. parchment paper   Electrify the surface of the porous material.

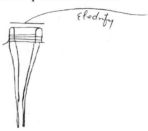

parchment paper[13]—Leather etc.
  Thermo   Thermo Couple Rotation

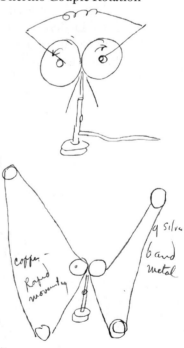

Try dif speed.[14]

Thermo

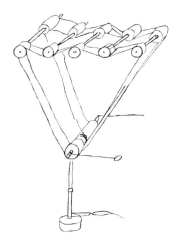

air Telegh— Baloon Telegraphing[15]

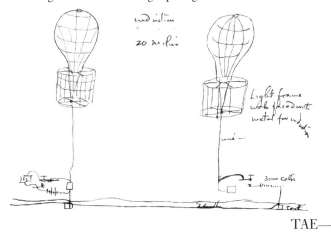

TAE—

X, NjWOE, Lab., N-86-04-03.1:27 (*TAED* N315027). Document multiply signed and dated. [a]Followed by dividing mark. [b]Canceled. [c]"dry clay . . . etc." written opposite preceding text.

1. An alternate version of this entry appears in N-86-04-03.3:19–39, Lab. (*TAED* N317019, N317029). The notes from pages 29–39 of that book are dated 7 April and were written by Mina Edison, presumably based on this notebook entry; drawings and related text that she copied there are marked in this original entry with a large "X."

2. To improve the conductivity of arc light carbons, some manufacturers coated them. Charles Brush is credited with being the first to use copper, which became the most common coating, although nickel, iron, or iron oxide were used as well. By the twentieth century carbons were sometimes impregnated with minerals such as magnesia. "Charles F. Brush," *Electrical Review and Western Electrician* 57 (5 Feb. 1910): 715; "Carbons for Arc Lights," *Teleg. J. and Elec. Rev.* 16 (27 June 1885): 570–73; "Arc Light Carbons," *Electrical Engineer* 41 (1908): 773; see also the references in Doc. 2898 n. 1.

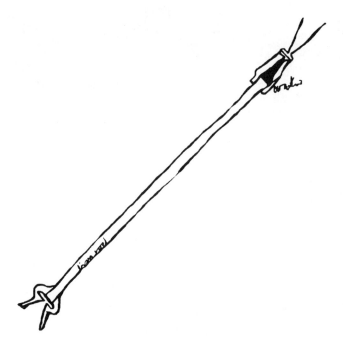

3. Presumably a new battery for the grasshopper (railway) telegraph.

4. William Thomson's mouse mill was an electrostatic generator designed originally for electrifying the ink in his siphon recorder. Prescott 1877, 1130–39.

5. This recent adaptation of Thomson's electrostatic generator was used to ignite gas in a lamp. It consisted of an ebonite cylinder with strips of tinfoil that was rotated within a larger cylinder having two tinfoil inductors. *L'Ectricité dans la Maison* [review], *Teleg. J. and Elec. Rev.* 16 (21 Feb. 1885): 166; Gray 1890, 195–96; Guillemin 1891, 223.

6. Presumably to reduce carbon carrying. See Doc. 2892 n. 6.

7. Figure labels are "dia" and "Tube." In the version in the other notebook, Edison made a drawing labeled "Molecular Speaking" that used an iron rod, presumably vibrated by the movement of water.

8. Faint figure label above is "fish bladder."

9. Edison doodled "Sanibel e Sanibel" on the following drawing.

10. Figure labels above (clockwise) are "~~Hydrogen~~," "drier," "Lead band oxidized in battery," "carbon," "gal[vanometer]," "gas," and "Hydrogen."

11. Here and in the following notebook entry, Edison explored methods of isolating atmospheric oxygen. In this entry, he considered materials that could act as a filter to physically separate oxygen from nitrogen in the air. The use of a membrane to separate gases derives from the work of Thomas Graham on the diffusion and effusion of gases, with which Edison was familiar (see Doc. 2324); a copy of Graham's *Elements of Chemistry* is now in the library at the Orange laboratory. On Graham's work and the history of research on membrane transport of gases, see Mason 1991 and Maxwell et. al. 1986.

12. Figure labels are "Hg," "material," and "Hg."

13. Figure label above is "electrify."

14. Figure labels are "copper—Rapid movement." and "g[erman] silver   band metal."

15. Figure labels are "induction   20 miles," "Light frame work faced with metal for induction," "wire—," "3000 cells," "Earth," and "Earth." See Doc. 2807.

–2929–

Notebook Entry:
Miscellaneous

[Fort Myers,] April 8 1886

For seperating O from air—draw into a vacuum through immense surfaces of iron or other metal   perhaps those metals which have an afinity for O. greatly.[1]

Experiment[a] on loops of all metals in Vacuo for Paper in Sillimans Journal—[2]

Cooling molten metals various kinds in Vacuo—also with current (strong) passing through metal while liquid in Vacuo—also within powerful magnetic field—ditto with Current at various angles

Siren Phonograph[3]

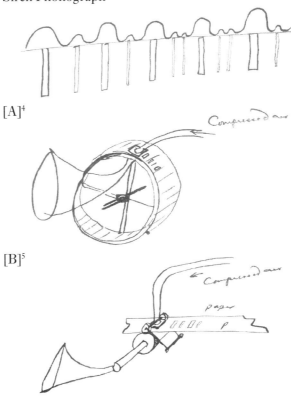

[A][4]

[B][5]

Siren Phonograph[6]

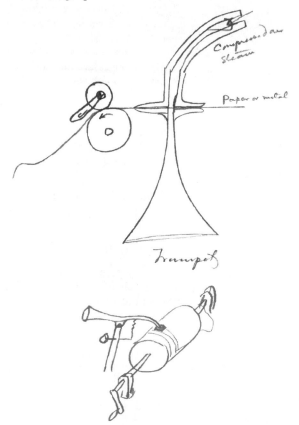

Indenting directly in phonogh by direct impact of the sound wave   Think[a] it will reproduce from same device by exhaustion of air in turning in reg phon use air Compsd[7]

TAE

X, NjWOE, Lab., N-86-04-03.3:41 (*TAED* N317041). Document multiply signed and dated. [a]Obscured overwritten text.

1. Gaseous oxygen could be produced by heating metallic oxides, which readily give up their oxygen, but the high cost of these oxides made their large-scale use impractical. Many experimenters sought to develop cheap methods for isolating it from the atmosphere on an industrial scale. Among such processes were those that involved heating air with either moist cuprous chloride or a mixture of manganese peroxide and potassium chlorate in a cast-iron retort. In 1886, a new method using barium oxide was commercialized by two French brothers, Ar-

thur and Leon Quentin Brin, who formed Brin's Oxygen Company in London. The Brin Process was the most successful commercial method for several decades. Edison had likely read an article by the Brin brothers about their process that was reprinted from *La Nature* in the same October 1884 issue of *Scientific American* that featured an article about Edison's exhibit at the Philadelphia Electrical Exhibition. In it, the authors described the use of suction pumps rather than a vacuum to suck air into the large iron retorts where the reaction between the oxygen and barium took place; however, another pump did create a vacuum to suck up the oxygen and remove it from the retort. Almqvist 2003, 65–69; Payen and Paul 1878, 19–20; Kolbe and Humpidge 1884, 9–12; *Encyclopædia of Chemistry* 1877, s.v. "Oxygen"; Jensen 2009; Molinari and Pope 1920, 196–97; Brin and Brin 1884, 243–44.

2. Edison used a common alternate name for the *American Journal of Science*. He did not publish such an article.

3. Edison experimented with and patented a similar device in 1878 known as the aerophone but never developed it into a commercial technology. See Docs. 1210, 1250, 1266, 1341, 1362, and U.S. Pat. 201,760.

4. Figure label is "Compressed air."

5. Figure labels are "Compressed air" and "paper."

6. Figure labels are "Compressed air   steam," "paper or metal," and "Trumpet—."

7. Figure label is "air."

Notebook Entry:
Miscellaneous

[Fort Myers,] April 11 1886

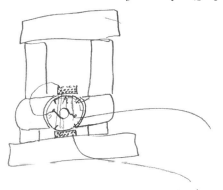

Wind as shewn heavy wire so ampere spires[1] same as on armature then put it in main Circuit[2]

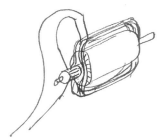

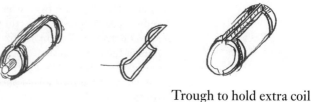

Trough to hold extra coil

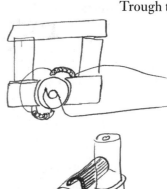

Extra prolongation to shift lines force to keep neutral point constant[3]

⟨good⟩

Wind armature thus, one turn around, then before going to Commutator, start end on another turn round but at right angles & after this turn is made go to Commutator & so on winding whole armature in this way. ½ of the wire is available but the other ½ prevents the armature itself from sending out lines of force ~~the~~ field is increased & the lines Concentrated so we get about same volts as in regular way if not more.[4]

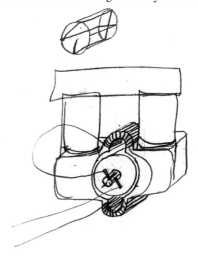

⟨good⟩[5]

Phonoduplex or Quad[6]

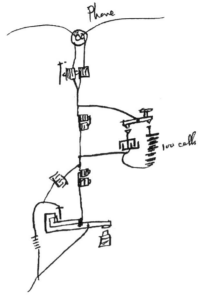

Sextuplex by phonoplex[7]

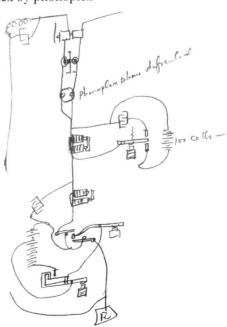

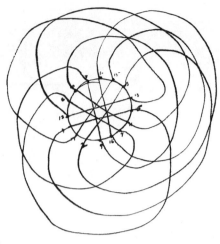

⟨OK⟩[8]
  Phonoplex

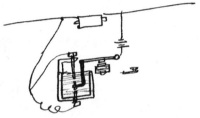

water & dif Liquids—[9]

To get rid of the Condenser

Phonoplex= to prevent battery running down[10]

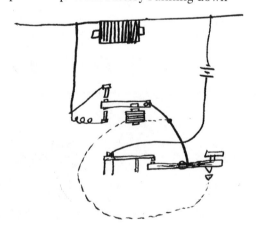

⟨good⟩

Phonoplex[11]

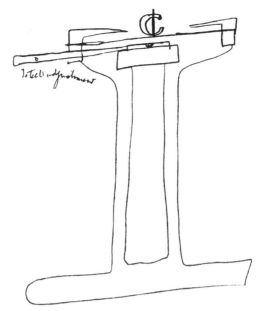

Permanent Magnets[12]

steel washers[13]

[A][14]

Perm Magnets

steel plates as heads to soft iron shoe.[15]

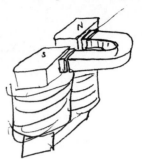

steel plates   glass hard
　　[B][16]

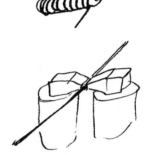

steel wire hardened & pulled thro & wound
　　Condenser Safety Catch   Dont forget about making Zincs
of batteries[17] of Compressed amalgam of Zinc—[18]

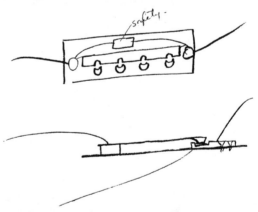

sheet of thinnest paper tissue unparafined— High volts will
jump & short ckt condr
　　Telephone Receiver—
　　A current passing through an ~~iron~~ steel[a] wire makes ½ of
one side N other S. Same in band[19]

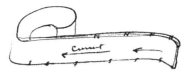

Hardened & Magnetized

Hence

paper between the coils

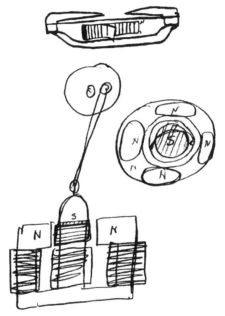

Reciprocating Dynamo[20]

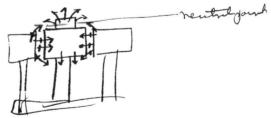

Quick discharge— Phonoplex 7 ohm overhanging coil—[21]

overhanging coil. see if better

7 ohm Coil with steel wire core magnetized

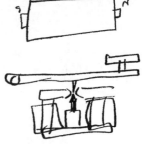

break Current between mag
Phonoplex quick break   Beak ckt in Vacuo

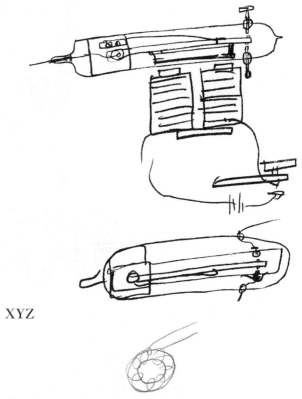

XYZ

Continuous ring no magnetic lines outside yet big induction   see if where the Current comes of ie[b] the wire poles there isnt a momentary magnetism on closing & opening. investigate this ring biz Thoroughly

PS   Try closed iron magnet 7 ohms on phone[22c]

Try XYZ things in this

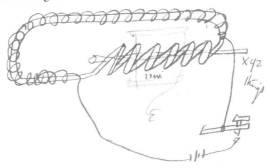

also[23] see if E^d reconverted.

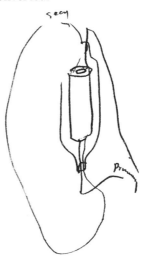

This[24] is in other book[25]   make this 200 ohms secondary
& 7 ohm primary[c]   7 ohm phono coil in Vacuo high   see if
charge & d[ischarge] qkr—

TAE

X, NjWOE, Lab., N-86-04-03.3:49 (*TAED* N317049). Document multiply signed and dated. [a]Interlined above. [b]Circled. [c]Paragraph partially enclosed by line at left. [d]Obscured overwritten text. [e]"make this . . . primary" interlined between preceding and following sentences.

1. That is, turns of wire on the armature.

2. In an alternative version of this note, dated 7 April with related drawings, Edison elaborated: "Wind each turn round then before going to com[mutator] wind at right angles then each loop will be practically 2 but only ½ will cut field lines of force & there will be no mag due to [coil?]." N-86-04-07:7–13, Lab. (*TAED* N319007).

3. What appears to be another version of this design, dated 8 April, is in N-86-04-07:15, Lab. (*TAED* N319015). Edison appears to be seeking a way to maintain a fixed neutral point so as not to have to rotate the commutator brushes. In order to minimize sparking at the commutator bars, the brushes needed to be at the neutral (or non-sparking) point where current in the armature windings reverses direction as the rotating coils pass into a region of opposite magnetic polarity. However, current flowing through the armature wires has the effect of shifting the magnetic lines of force—and consequently the neutral point—in the direction of the armature's rotation, making it necessary to move the brushes accordingly. See Doc. 2420 n. 14 regarding Edison's earlier work on this fundamental problem of dynamo design.

4. An earlier version of this note, dated 8 April and with the note "Copy this," is in N-86-04-07:27, Lab. (*TAED* N319015). In a memorandum to patent attorney Richard Dyer, also dated 11 April, Edison described a design that he wanted to patent in which the armature would have no magnetic effect:

> part of the face of the field is cut away and large wire wound parallel with the armature wire and with a carrying capacity equal to the armature and so arranged as to resistance current and turns of wire that its magnetizing effect on the armature is equal to the wire on the armature.
>
> The current from the armature is passed through this fixed coil in such a direction that it magnetizes the armature just the opposite to the magnetization due to the wire on the armature. This causes the armature to have no magnetic." [Cat.1151, Lab. (*TAED* NM020AAD)]

Edison asked Charles Batchelor on 11 May to experiment with a "non polar armature" so that "In order to have no poles in the armature wind every wire that passes over top wind one horizontally to neutralize the magnetic effect" (Cat. 1336:31 [item 66, 11 May 1886], Batchelor [*TAED* MBJ003031B]).

5. An earlier version of this design, dated 8 April and marked "OK Copy," appears in N-86-04-07:25, Lab. (*TAED* N319015); also see pages 29 and 31.

6. Figure labels are "Phone" and "100 cells." An earlier version of this design, dated 8 April, is in N-86-04-03.1:127, Lab. (*TAED* N315127).

7. Figure labels are "phonoplex phone difrential" and "100 cells—." An earlier version of this design, dated 8 April, is in N-86-04-03.1:129, Lab. (*TAED* N315127).

8. Drawings related to the armature winding above, some dated 8 April, are in N-86-04-03.1:130–39, 143–45, Lab. (*TAED* N315127).

9. Edison drew a line to associate this phrase with the shaded area inside the vessel.

10. Another version of this design is in N-86-04-07:149, Lab. (*TAED* N319015).

11. Figure label is "To [tel?] adjustment."

12. Similar drawings appear in N-86-04-07:151, Lab. (*TAED* N319015).

13. Figure labels above are repetitions of "N" and "S."

14. Figure label is "washer."

15. Edison drew a line to associate this descriptor with the junction between the pole pieces and the U-shaped magnet. Figure labels are "S" and "N."

16. Figure label is "wood."

17. That is, the batteries' anodes, or positive electrodes.

18. Figure label is "safety."

19. Figure labels are "current" and repetitions of "N" and "S."

20. Figure label is "neutral point." Rough sketches of this design are in N-86-04-07:33–34, Lab. (*TAED* N319015).

21. Figure labels are "coil" and "iron Core."

22. Figure label is "iron."

23. Figure labels are "Iron" and "XYZ things."

24. Figure labels are "sec[ondar]y" and "primary."

25. See Doc. 2922.

–2931–

Notebook Entry:
Miscellaneous

[Fort Myers,] April 12 1886

Piano wire Tuned to respond to reed on[a] grasshopper

worm & wheel to adjust

[A][1]

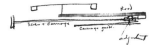

grasshoper Call[2]

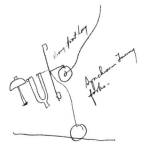

Grasshopper— Receiver with adjustable chamber so that column air can be adjusted periodic with the vibrations of the reed[3]

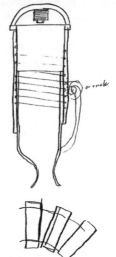

I think Goerck should build a cheap gramme[4] which will use the air cooling method of winding like that shewn in head light dynamo[5]

E   wound either with field distortion or right angle winding on armature

New cheap gramme   non sp[ar]k[in]g dynamo

$^{10}/_{1000}$ sheet iron rings built up with tissue paper   The rings are split and can be sprung enough to get coils on

[B][6]

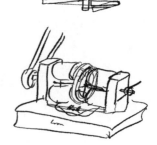

It is doubtful is splitting the ring to get coils on is as good as a winding machine

Split bobbin run by belt slowly & fast—fast to fill it & slow to unwind on Iron[a] core—[b]

Have I patented split disks of all put together fat one common opening then after Coils put on shift them promiscously[7]

TAE

X, NjWOE, Lab., N-86-04-03.3:91 (*TAED* N317091). Document multiply signed and dated. [a]Obscured overwritten text. [b]Followed by dividing mark.

1. Figure labels are "screw & carriage," "carriage guide," "Reed," and "adjustment."

2. Figure labels are "strong—foot long" and "Synchronous Tuning forks—." In 1876 Edison worked on an acoustic transfer telegraph that used synchronously vibrating tuning forks to control the switching of circuits on a telegraph line in order to send multiple messages simultaneously. He also used a similar arrangement to control the movement of printing telegraph typewheels at distant stations. Here he appears to be combining such a device with a telephone call bell to act as a receiver for the grasshopper telegraph. See Docs. 749 and 782.

3. Figure label is "or rack." The adjustable chamber would act in fashion similar to a Helmholtz resonator.

4. Frenchman Zénobe Gramme developed a ring-armature dynamo in 1870 that became the first commercially successful dynamo. Edison had experimented periodically with Gramme machines and ring armatures of this type while developing his electric light system. See Docs. 1489, 1611, 1621, 1626–28, 1634 nn. 1 and 3, 1641, 1646, 1716, 1719 n. 1, and 1760.

5. The editors have found no information about this machine, but Edison's new father-in-law, Lewis Miller, was president of the American Locomotive Electric Headlight Co. That company's product, if it was like other headlight dynamos, probably had a ring armature, which facilitated air cooling (though drum-wound armatures, including Edison's, could also be designed to promote air cooling). *American Electrical Dictionary* 1886, 375, 385, 407–8, 419; "The Electric Headlight for Locomotives," *Brotherhood of Locomotive Engineer's Monthly Journal* 18 (1884): 336; "The International Electrical Exhibition at Vienna," *Electrician* 11 (27 Oct. 1883): 569–70; Schellen 1884, 172, 299, 302, 322, 335, 361, 368, 373, 383–84, 396, 418, 425, 435, 451.

6. Figure labels are "[Hole?]" and "iron."

7. Edison does not appear to have patented this design.

Notebook Entry:
Miscellaneous

Device for automatically indicating depth water on steamers continuously   also an alarm by gal needle swinging to a stop & closing bell ckt[1]

The device is put in one side of wheatstone Bridge
    Testing quality & homeogenity & for flaws of steel & Iron shafts[2a]

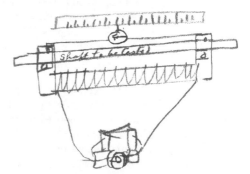

Compass need[le] on a Torsion slides along near shaft. if shaft homegenios The Zero[b] point will be constant   if a flaw There will be a <u>Consequent</u>[b] pole[3] & produce a dip in the curve   There are many variations possible here—

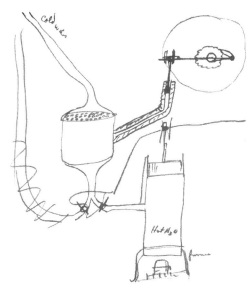

Expansion Engine[4]   powerful Hydraulic[b] press big Cylinder full Copper[b] tubes very thin & $\frac{1}{32}$ @ $\frac{1}{64}$ $\frac{3}{16}$ bore cylinder 15 inches ~~long~~ dia[c] of ~~iron~~ copper heavy=Copper Ends— 2 feet long— Hot water 212 let up in then run back into hot boiler & Cold let thro & so on all auto= Bi Sul Carbon or other good Expandable liquid to be used

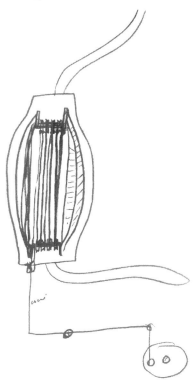

Expansion Engine[5]   Hot & cold H_2O passed alternatively through the multiple pulley Engine

On phonoplex Try 2 7 ohm Coils in Series with 8 cells instead of 4 so same amperes passes through both as one— see if this don't do better if not make one sounder with 4 points so they will leave & close Exactly & use separate battery & Seperate Condensers around the sparking points.[a]

perhaps the 2 coils in Series requiring 8 cells causes such high Emf at points that to get rid of spark too much condenser must be used & this takes it from the line but with one Sounder & separate points OK

Multiple phonoplex[6]

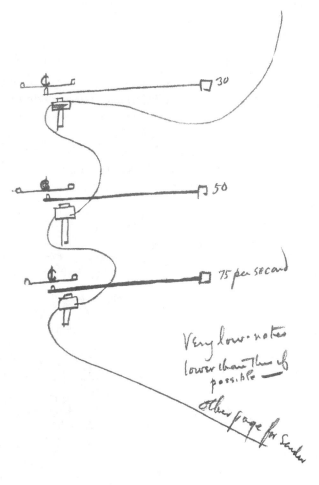

Multiple phonoplex

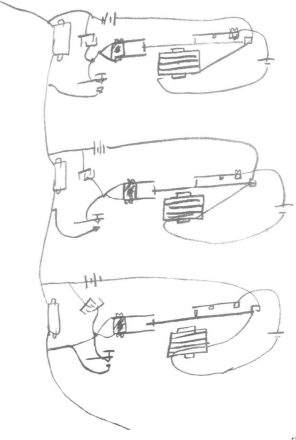

TAE

ADDENDUM[7d]

[Fort Myers,] April 13 1886

Try dissolving Licorice in hard boiled Linseed oil—then coat tin[b] foil with thin layer dry [---][e] naturally. Then recoat till get it ${}^{10}/_{1000}$ then put in hot oven & keep 48[b] hours—

also Try linseed films compound ie[f] many films in one & use hot oven   The reason previous Experiments with films didnt work well was that they were only naturally dried while they should be dried by hot oven slowly raising heat.[8] also Try tissue paper uncallendered & callendered no holes dip & dry naturally & then bake slowly with high final in oven

TAE—

X, NjWOE, Lab., N-86-04-03.3:101 (*TAED* N317101). Document multiply signed and dated. [a]Followed by dividing mark. [b]Obscured overwritten text. [c]Interlined above. [d]Addendum is an X. [e]Canceled. [f]Circled.

1. Figure labels are "ocean st[eame]r," "water," "insulated wire 1000 feet," "copper bouy," and "Bottom ocean."

2. Figure label is "Shaft to be tested."

3. A consequent pole is one formed between the poles normally found in magnetized iron. When Edison identified consequent poles in dynamo field magnets in 1883, he attributed them to flaws in the iron. See Doc. 2577.

4. Figure labels above are "cold water," "Hot H$_2$O," and "furnace." In this device, it appears that as the copper tubes in the cylinder were alternately subjected to hot and cold water they would cause the "expansible liquid" in the cylinder to expand and contract, creating hydraulic pressure that would move the ratchet mechanism. What may be related undated sketches, including one labeled "Tasimotor," are in N-86-04-07:45, 114, Lab. (*TAED* N319, images 21 and 52).

5. Figure label above is "wire." In this device the hot and cold water would presumably cause bands of hard rubber or other heat sensitive substances to expand and contract through a system of pulleys to produce mechanical motion. Edison drew a telephone based on this multiple pulley system two days later (see Doc. 2934). What may be a side view of the pulley arrangement is in N-86-04-07:119, Lab. (*TAED* N319, image 54).

6. Figure labels are "30," "50," "75 per second," and "Very low notes   lower than this if possible—." At the bottom, Edison wrote: "other page for sounder," referring to the next figure. This design appears to use phonoplex instruments set to respond to different frequencies as a sextuplex.

7. Because it is unclear if this page (which appears after several intervening pages dated 14 April) is misdated or copied from another version that Edison made on 13 April, the editors have chosen to present it as an appendix to this entry.

8. For Edison's earlier work on films, see Docs. 2291, 2307–8, 2623, 2630, 2632, 2634–35, and 2750.

–2933–

Notebook Entry: Miscellaneous

Deaf—[1]

[Fort Myers,] April 14 1886

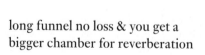

long funnel no loss & you get a bigger chamber for reverberation

[A][2]

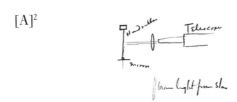

25 foot scale—
Dips in one then the other[3]

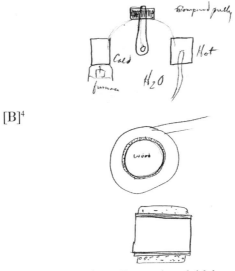

[B][4]

7 ohm phonoplex   Iron wires laid lengthwise on 2 inch block wood   Same quantity iron wire as in reg & the Cop[per] wire wound   Length same

TAE

X, NjWOE, Lab., N-86-04-03.3:115, 119–23 (*TAED* N317115). Document multiply signed and dated.

1. Figure labels are "Compressed air   Bladder or Rubber" and "dia[phragm]."
2. Figure labels are "Hard rubber," "mirror," "Telescope," and "beam light from star."
3. Figure labels are "furnace," "Cold," "H_2O," "mCompound pulley," and "Hot."
4. Figure label is "wood."

–2934–

Notebook Entry: Miscellaneous

[Fort Myers,] April 15 1886

Try some fundamental experiments on lines of force—
The present Theory don't seem to explain certain expmts.[1]

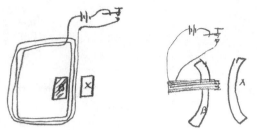

B & X Iron   see if X magnetized—if so if its proportionate—

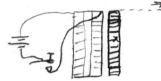

X Iron   see if magntz if so the proportion between it and
when in Center. also move it by ¹⁄₃₂ outwardly as the arrow
measure relative strength

Fundamental magnetic Expmts

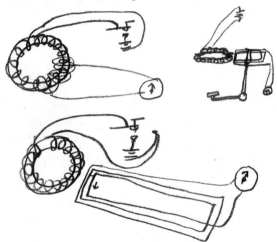

draw to & from   whirl around

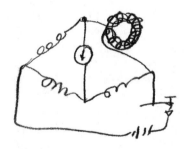

Fundamental Magnetic Expmts[2]

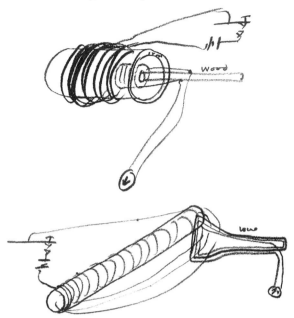

move in every direction & plot the strengths of induction[3]
Fundamental Magnetic Expmts[4]

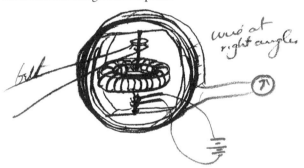

Phonoplex Coil[5]

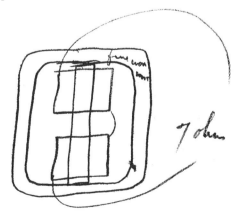

Same length as regular

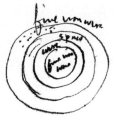

This strengthens lines force & more powerful on theory that lines force drawn inward[6]

Fundamental Magnetic Expmts

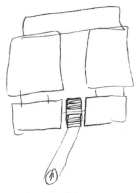

nothing

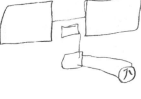

strong

stronger

Telephone Receiver

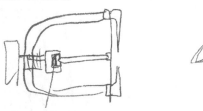

dif substances, pressed chalk Lime & every other salt, dry & also moist. Emg action.

Telephone Receiver[7]

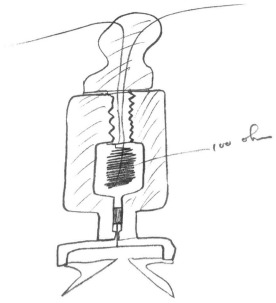

Liquid greatest Expansion   Hydraulic principle

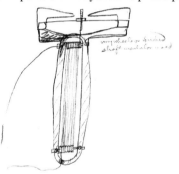

Multiple pulley principle[8]
Telephone Receiver[9]

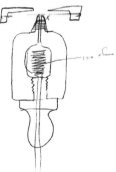

Telephone Recvr

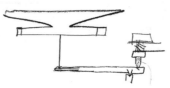

The Thermo Receiver. Telurium etc. all that is necessary is leverage on this class of receivers—[10]

XYZ[a]

If the molecules of an element like a metal is effected when a line of magnetic force strikes one side of the wire, giving a Current in one direction & when merely allowing the line of force to strike the other side a reverse Current, Then it is very likely that Compounds (no elements) should be effected different—hence dynamo mac wound with Hard rubber,—tubes containing liquids instead of wire—& moulded solids in wires or rods like peroxide & etc—

XYZ[11]

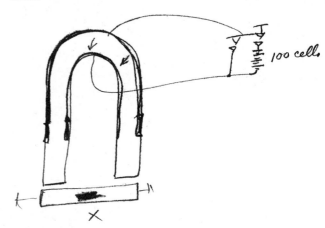

glass tube filled with diaelectrics   also try other liquids—

Cutting with E lines of force at right angles ought to produce another form Energy along the Compound & thus produce an attraction. X is trough containing some liquid or others—also all Solids & metals. Try dif Ends f as some may be non condrs of XYZ— Try metals a better device next page

XYZ[12]

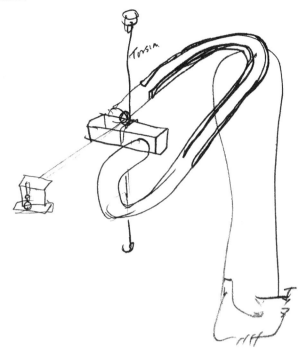

Fundamental Magtc Expmt

Ascertain when a fair Vac is obtained if guage shews diference when bulb magnetized[13]   Try at diferent rarifacations

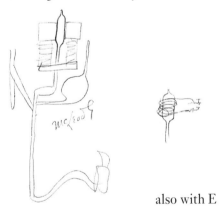

also with E

TAE

X, NjWOE, Lab., N-86-04-03.3:125 (*TAED* N317125). Document multiply signed and dated. [a]Multiply underlined.

1. Figure labels in both drawings are "B" and "X." What may be related undated sketches are in N-86-04-07:53, Lab. (*TAED* N319, image 25).

2. Figure labels are "iron" and "wood."

3. Figure label above is "wire."

4. Figure labels are "belt" and "wire at right angles."

5. Figure labels are "fine iron wire" and "7 ohms." What may be related undated sketches are in N-86-04-07:59, Lab. (*TAED* N319, image 28).

6. Figure labels above are "fine iron wire," "space," "wire," and "fine iron wire."

7. Figure label is "100 ohms." Edison seems to have intended a 100-ohm resistance coil to heat a liquid, which would expand and move the piston against the diaphragm. A similar undated drawing is in N-86-04-07:61, Lab. (*TAED* N319, image 29).

8. Figure label above is "ivory wheels or bushed shaft insulator used." It appears that the motion of the diaphragm would cause a pawl-like device to rotate the ivory wheels and thus the wires. It is unclear what the electrical effect of this action would be. See Doc. 2932 for another instrument using the "multiple pulley principle."

9. Figure labels on the large figure are "x" and "100 ohms"; in the diaphragm detail, they are "X," "dia[phragm]," and "Rubber." Presumably the rubber would move the diaphragm by responding to the expansion of a liquid heated by the 100 ohm resistance coil as shown in the figure above.

10. As the metal ("Tellurium etc") was heated by the passage of the current, it would cause the lever to pivot.

11. Figure labels above are "X" and "100 cells."

12. Figure label is "Torsion."

13. Figure label is "McLeod G[auge]." The McLeod gauge, invented in 1874, is a Torricellian device capable of measuring extremely low gas pressures. Edison began using the gauge in lamp research in 1879. Doc. 1816 n. 3.

–2935–

Notebook Entry:
Miscellaneous

[Fort Myers,] April 16 1886[1]

Trough containing diferent conducting solution— Expmt is to see if with dilute sols whether the magtc lines will not

cause the Attraction[a] of the Electric lines & with it the conducting matter   concentrating the liquid nearest mag & leaving only pure water in top of sell

XYZ— Reversion of E[lectricity] & M[agnetism][2]

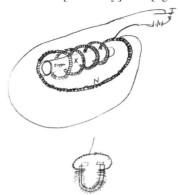

A copper horseshoe wound with several layers of softest iron
wire ⅛ thick over the whole length of which is wound insulated wire & charged powerfully with battery.[3] Expt for XYZ
bet Copper poles[4]

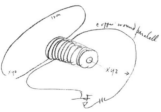

put in dif things

XYZ   Copper wire wound spirally with insulated Iron
wire—[5]

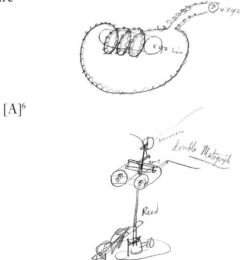

[A][6]

Tasimotor[7]

Air Telegh— Barometer at Geneva being 27.08 on top Mt Blanc 15 732 feet above sea was ~~16.008~~ 16.08. correct Geneva by Sea Level. Then[a] experiment on long Vacuum tubes for conducting vibrations from 7 ohm phonoplex coil to teleghone[8] at different pressures   use regular Hg Guage as McLeod not required—

On Top Mt Blanc difficult to make yourself heard   The Barom 16.08, hence The Compressed air for Deaf device & in long speaking Tubes is OK—

Try alcohol or other liquid instead Compressed air in fish bladder—

=Important put man on this & work up system of Iron underground telephone—No electricity= In water a bell is heard 45 000 feet which in air can only be heard 656 feet. Vel of sound in Water 49[00?][b] per sec Iron 17,500

iron wire suspended by strings in a box in the Earth only transmit the sound molecularly not move the whole wire— also just bury in the Earth & see how far can Transmit—

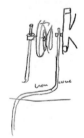

Try dif forms of transmitter & receiver[9]

Iron[a] wire Non-E Telephone[10]

Perhaps the buried iron wire could be wraped with[a] loose cotton laid lengthwise, & slightly braided over to hold it in its place— This would prevent transmission of vibrations to outside matter.

Experiment to find a non-condr of sound from an iron wire to matter around it—

glass tubes or even lead tubes with fine piano wire in 200 feet lengths could have a vacuum in it permanently & the joint be made an elastic one so vibrations would be carried for miles

Iron wire Non-E Telephone

The lead pipe or glass pipe[a] or even copper with fine wire could have a diaphragm joint[11]

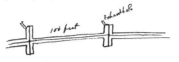

Telegraphing at sea[12]

as sound will travel 50 times further in water than air—I propose to run from a steam whistle[a] are large funnel & tube to water under the steamer   the End is provided with a diaphragm, or not as Expmt will determine—by cutting whistle into breaks[a] we have dots & dashes which are composed of vibrations. these are communicated to the water. on another ship a tube runs up from the bottom of vessel with water in in[13] or a diaphm is on the End and a thread passes up to another diapm to receive the impolses   Thus each vessel being provided with sending and receiving apparatus communication can be had a distance probably of 25 miles. a siren, or heavy vibrating fork may be arranged to give vibrations to the water.[14]

Expmt in water telephone transmission[15]

Marine Telgh

x funnel 50ᵃ feet long closed on end but full water (closed by parchment etc.

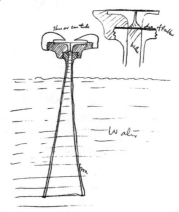

also diapm in side of ship deep down & long funnel inside ship to Concentrate hole 2 foot or 4 feet in area [composed?]ᵇ of number holes & diapms

Water Telephone Receiver[16]

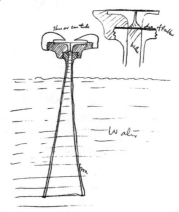

perhaps a column of air in receiver in exact tune with the whistle tone on Steamer would help—

Flexible Ear tubes of metal[17]

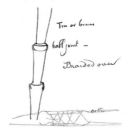

Explosive Siren[18]

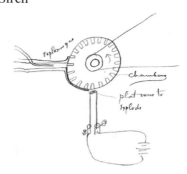

small one for projection— See if sound most powerful in
direct line—[19]

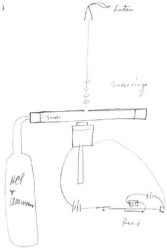

Try gas & use plat spiral to ignite
Molecular Receiver from iron wire   impact telephone
Non-E—[20]

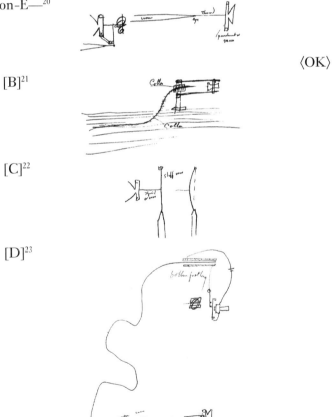

⟨OK⟩

[B][21]

[C][22]

[D][23]

Molecular recvr[24]

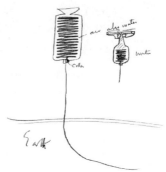

True photophone[25]

This under pressure coated with colloidon & silver salts—also Bichromatzed Gelatines   Effect Light to produce charge which must Either increase or decrease tension on surface & be transmitted to Carbon button & telephone at each flash

glass jar filled with chlorine & Hydrogen[26]   weak light ought to effect it without causing it combine

Mfr of Silk[27]   Try   Gum Chicle[c]

glue mixed with Bichromate Potash proper proportion & made in cylinder & placed in larger body of Hydrocarbon solid—soften & pull out. afterwards disolve HCarbon—Expose fine filament of gelatine to Light.

Bases—. Collodion—Cupric ammonium paper—gutta percha—Balata—Rubber disolved in Benzol or Bisulphide Carbon into which Rosin & various other Hydrocbns disolved until ~~Rub~~ Resultant film Rubr has slight Elasticity.— Exterior can be glue—Licorice—Molasses candy— Fibres not in continuous length be made by working over & over the Candy with Rubber Rosin core & then disolving in hot water

MCQueens[a] man[28] says strong [gloceral?][29b] acetic acid changes[a] glue to ropy mass drawn out silky threads   if so action[a] surface with Bichrom K & [Copper?][b]

Seperation of solid matter in milk[30]

Milk forced through fine hole with great velocity again Roughened[a] surface   This breaks globule & unbreakable ones pass down incline quickly

Seperation Cream from Milk   Try Grahams Dialysis—[31] use also electricity thro porous diaphms.

Use porous diaphms various thicknesses & substances & use vacuum underneath

use air pressure on one reciprocation to free pores of globules & next recipctn use Vacuum to draw water of milk through—[c]

Evaporate in Vacuum—

Freeze the H_2O—

Try filtering through long tube filled with fine particles use vacuum to draw— Centrifugal through porous substance

Fundmtal Magnetic Expt[32]

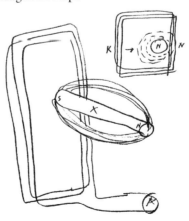

X steel mag—move it from N towards R but not enough so K will come within range of lines of force   want to ascertain if its the <u>drawing</u> in towards magnet of lines of force or not.

Phonoplex Coil, sheet wound as shewn

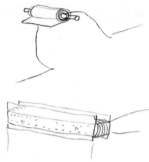

use foil copper with parafined tissue between   wind it 2 inch thick core being ¾—fine iron wire.

[E][33]

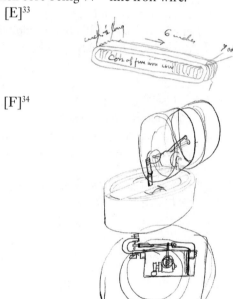

[F][34]

Fundamental Mag Expt   Faradays—[35]

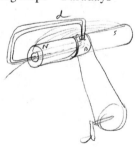

If the iron dont act as guard—bush[36] with thick bush of copper insulated from everything through which wire passes   also not insulated

Revolved—[37] see if B is good guard.

Fundamental Magnetic Expmt   <u>XYZ</u>[38d]

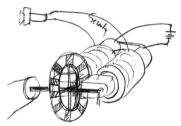

a powerful permanent magnet would be better, or The primary only in one side of a bridge & telephone in balance wire. The wheel is rotated with great velocity & at diferent spots has a piece of substance or cell with liquid to see if there is anything that will retard magnetism   See[a] 215[39]

<u>XYZ</u>[d]   If any substance will disturb the lines of force and the wheel is turned sufficient fast to give a musical note it will be heard in telephone. now if this disturbance is created without the production of electricity or magnetism, then we have a <u>new form of energy</u>

Fundamental Expmt   Pull thro with & with heavy copper guard closed—

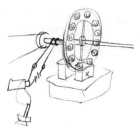

Electric pen spools no mag[40] X liquids & substances to see if the electric field is disturbed.

Milk—[a] Try disolving out butter oil with Bisulphide Carbon[e] Etc—then Evap BiSul[c]

E[lectric] R[ail] R[oad][41]

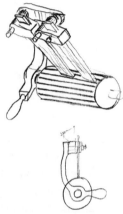

Brush device for motors that reverse direction of rotation

TAE

X, NjWOE, Lab., N-86-04-03:2 (*TAED* N317155). Miscellaneous doodles not reproduced; document multiply signed and dated. [a]Obscured overwritten text. [b]Illegible. [c]Followed by dividing mark. [d]Multiply underlined. [e]Interlined above.

1. Edison misdated the first page of this notebook entry as 1883. However, the preceding entries and the rest of the pages in this entry are all dated 1886. Some entries were copied from another book. One that was not, dated 16 April, is a drawing of a ship, labeled "Archimedes Screw to propel Steamer." N-86-04-07:127, Lab. (*TAED* N319127).

2. Figure labels are "copper," "X," and "N."

3. What may be related undated sketches are in N-86-04-07:103-4, Lab. (*TAED* N319, images 46–47).

4. Figure labels below are "XYZ," "iron," "copper wound parallel," "XYZ," and "put in dif[ferent] things."

5. Figure labels are "XYZ here" and "& XYZ." Edison's suggestion here may have some relation to a notebook entry of 24 March "to see if eCopper conducts faster than Iron." He proposed to test the idea by braiding bare copper and iron wires and placing a galvanometer in circuit between them. N-86-03-18:177, Lab. (*TAED* N314171).

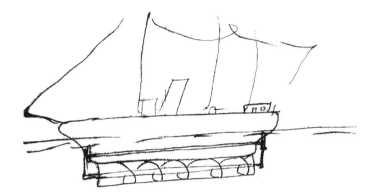

A ship propelled by an Archimedean screw, one of the 16 April drawings in another notebook that Edison did not copy into this entry.

6. Figure labels are "mirror," "double Motograph," "spring," and "Reed."

7. Edison dated this note 15 April, but the editors have included it here because it comes in the midst of a series of entries dated 16 April. He apparently coined this compound word from the Greek root for "tension," the same source as the name of his tasimeter. The intended operation of this device is unclear (*OED*, s.v. "tasimeter"; Docs. 1289 n. 3 and 1329 n. 1). Another rough sketch of this design is in N-86-04-07:120, Lab. (*TAED* N319, image 55).

8. Edison presumably meant "telephone."

9. Figure label above is "iron wire."

10. Edison dated the page starting with this heading and ending with "would be carried for miles" as 15 April. Because the subject fits seamlessly with the pages immediately preceding and following, all dated 16 April, the editors conjecture that he was mistaken and have accordingly included the page in this document.

11. Figure labels are "100 feet" and "exhaust hole." Edison turned the book upside down, probably at a later time, and made several drawings on the even-numbered pages (174, 180, 182, 184) between the entries "Iron wire Non-E Telephone" and the "Flexible Ear tubes of metal." One is very rough and the others are unrelated to surrounding text; they are not included here.

12. Figure label is "diaphragm." Sketches that may be related to Edison's designs for telegraphing at sea are in N-86-04-07:128–29, Lab. (*TAED* N319, image 59).

13. Edison turned the page here and wrote the heading "Marine telegh" at the top of the new page.

14. Figure label is "water."

15. Figure labels are "Listen," "Water," and "Talk." Edison drew a distinctive clock doodle upside down on page 178 facing this text. He made another notable but seemingly isolated sketch (on page 172, again with the notebook upside down) of a book lying on a table.

16. Labels on the large figure are "[This?] or Ear tube," "water," and "iron." Figure labels on the small drawing (showing a closeup view of the action on the diaphragm) are "dia of Rubber" and "water." A related drawing is in N-86-04-07:121, Lab. (*TAED* N319, image 55).

17. Figure labels are "Tin or brass," "ball joint—," "Braided over," "cotton," and "cotton." Similar sketches as well as another design for a flexible ear tube are in N-86-04-07:121–23, Lab. (*TAED* N319, images 55–56).

18. Edison dated this page 15 April but the editors have included it here as the following drawing of another design for an explosive siren is dated 16 April. Figure labels are "explosive gas," "chambers," and "plat wire to explode." A rough sketch of the revolving chambers is in N-86-04-07:124, Lab. (*TAED* N319, image 57).

19. Figure labels are "Listen," "Smoke rings," "smoke," "Hcl & ammonia," and "Reed." Another version of this drawing indicates the use of "explosive gas." N-86-04-07:125 Lab. (*TAED* N319, image 57).

20. Figure labels are "iron," "Eye," "Thread," and "parchment or mica."

21. Figure labels are "Cotton" and "Cotton."

22. Figure labels are "thread or wire" and "stiff wire."

23. Figure labels are "bobbin foot long" and "iron wire."

24. Figure labels on the left drawing are "air," "cotton," and "Earth"; figure labels on the right drawing are "also water" and "water." What may be a related sketch is in N-86-04-07:131, Lab. (*TAED* N319, image 60).

25. Edison presumably referred to the photophone of Alexander Graham Bell and Charles Sumner Tainter, a device for transmitting sound wave patterns on a modulated beam of light (see Doc. 2792 n. 6). Figure labels are "solution used for negatives   decomp by light" and "beam flashed   each flash produces a decomp & vibration to X."

26. Figure label above is "Listen."

27. Another version of this entry appears in another notebook and contains two drawings, one of which is reproduced; the other shows the box at center with a collodion base on top of which is licorice with the description "Heat & draw out to silk." N-86-04-07:134–36, Lab. (*TAED* N319, image 61–63).

28. Not identified.

29. Edison probably meant "glyceral."

30. For Edison's interest in the separation of cream from milk, see Docs. 2361 and 2622 n. 5. The first drawing in another version of this entry is similar to the one below except that the label in center rectangle is "milk." Two additional drawings appear at the end. The first is labeled "centrifugal through a porous ~~dia~~ substance"; the second is labeled "Lighter globes stay near center" and "Rotate." N-86-04-07:139–43, Lab. (*TAED* N319, image 64–66).

31. Thomas Graham (1805–1869) was a British chemist best known for two major contributions. His work on the diffusion and effusion of gases led him to what is known today as Graham's Law; namely, that the rate of effusion of a gas is inversely proportional to the square root of its mass. Graham also established the foundation of the study of colloidal chemistry. In addition, while attempting to separate colloids and crystalloids, he invented the "dialyzer," the precursor of the modern dialysis machine. *Complete DSB,* s.v. "Graham, Thomas."

32. Figure labels for drawing on the left are "S," "X," "N," and probably "K" at the bottom. Figure labels for the drawing on the right are "K," "N," and "N." Related sketches are in N-86-04-07:145, Lab. (*TAED* N319, image 67).

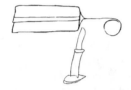

Edison's 16 April drawing of a process for producing silken fibers from a heated mixture of glue and bichromate potash.

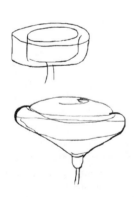

Edison's alternative 16 April drawings of centrifugal processes for separating milk solids.

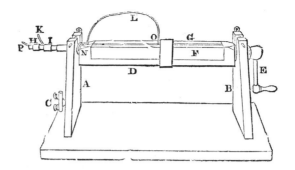

33. Figure labels are "inch ½ long," "6 inches," "7 ohm," and "core of fine iron wire."

34. Edison made this drawing on the facing page; it appears to be related to several drawings above on pages 180, 182, and 184 (see note 11).

35. Figure labels are "N," "d," "C," "B," and "S." What may be related sketches are in N-86-04-07:145–48, Lab. (*TAED* N319, images 67–69). Edison seems to have drawn a variation of the rotating apparatus used by Michael Faraday for the experiments presented in a long paper in 1852 and subsequently published in his *Experimental Researches* (Faraday 1965 [1855], §34). Faraday's device included two bar magnets (**F** and **G** in the drawing), laid side by side along the axis of a rotating shaft; covered wire **L** could be positioned in various ways, including down the axis of the magnets. The aim of this set of experiments, he explained at the outset, was to define more carefully his longstanding conception of "lines of magnetic force":

> the idea conveyed by the phrase should be stated very clearly, and should also be carefully examined, that it may be ascertained how far it may be truly applied in representing magnetic conditions and phaenomena; how far it may be useful in their elucidation; and, also, how far it may assist in leading the mind correctly on to further conceptions of the physical nature of the force, and the recognition of the possible effects, either new or old, which may be produced by it. [Faraday 1965 (1855), para. 3070]

36. That is, to create a bush or bushing.

37. Figure labels above are "B" and "iron."

38. Figure label is "Secondy." Related sketches are in N-86-04-07: 149–50, Lab. (*TAED* N319, image 69–70).

39. The following page and paragraph.

40. Figure label on magnet core is "X."

41. Figure label on the small drawing is "spring." Another version of these drawings is in another notebook and labeled "Copy" along with several related drawings. N-86-04-07:156–67, Lab. (*TAED* N319, image 73–78).

Notebook Entry:
Miscellaneous

Elec RR[1]

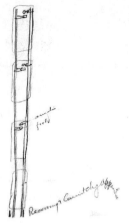

Work up idea of accelerating bobbin to work a centrifugal clutch throwing in device

XYZ[2]

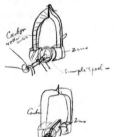

[A][3]

Wind wire on hard rubber & other substances.[4]

cylinder ½ Zinc ½ Carbon or Copper   wire wound perpin & at right angles   The particles between Z & C become charged & ought to give something

XYZ   If Electricity passing in a wire is given off as heat, Then a wire under[a] some kind of polar environment give off Electricity when the same is conducted along it from high to low temperature

Applcn of heat to a wire & Allowing[a] it to Conduct to & from a magnetic field ought to give Electricity

XYZ[5]

wind wire also sheet both directions—Rubber & other cylinder
Fundamental Magnetic Expmt[6]

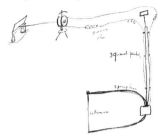

Determine accurately relative Emf with the poles at diferent distances in $\frac{1}{1000}$s plot curve— also substitute magnets
with exactly same poles but diff lengths & make curve— also
curves of Emf at diferent distances apart with dif degrees of
saturation Try Cast iron magnets also Reg mag & Cast
iron pole pieces[7]

Try pole pieces dif kinds & shapes Also[a] to ends etc
Fundamental Magtc Expmt Magnetic bridge[8]

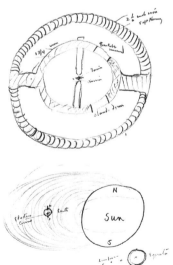

The Earth[a] revolving around the sun cut the lines of force
so the current tends to pass around both sides of it[9]   This

would neutralize & give no Effect were it not that one side of the Earth is heated & this gives direction to the Current around the Equatorial belt. This makes the magnetic poles and The electric Current & Consequent magnetism causes the attraction[a] to the sun   the orbit of the Earth is the point where the tendency to go off in a straight line is balanced Exactly by the mutual attraction of the Electric Current—magnet of the Earth & magnet of the sun

TAE

X, NjWOE, Lab., N-86-04-03.3:225 (*TAED* N317225). Document multiply signed and dated. [a]Obscured overwritten text.

1. Figure labels are "armature," "field," and "Reversing & Circuit chg apparatus." Another version of this drawing from the same date and labeled "Copy" is in N-86-04-07:169, Lab. (*TAED* N319, image 79).

2. Figure labels are "Carbon & other metals," "Zinc," and "Simple Spool—."

3. Figure labels are "Carbon," "Zinc," and "Siemens Rev Circuit windg."

4. Figure label on left drawing is "Zinc."

5. Figure label is "100 cells." Another version of this drawing labeled "wind wire both ways   use Hard rubber cylinder" is in N-86-04-07:175, Lab. (*TAED* N319, image 82).

6. Figure labels are "20 000 ohm," "39-inch pendulum," "spring base," and "cutting wire." Another version of these drawings and text is in N-86-04-07:181, Lab. (*TAED* N319, image 85). In that version, the phrase "Pendulum should only make a single swing" appears at the bottom of the page.

7. Figure labels are "[brass?]," "Res," and "ampermeter."

8. Figure labels are "2½ inch iron   soft Norway," "1¾ iron," "Iron to be measured," "S," "N," "Torsion," "Mirror," and "Standard iron." This appears to be the genesis of Edison's "magnetic bridge or balance for measuring magnetic conductivity," which he described in a paper presented at the August 1887 meeting of the American Association for the Advancement of Science (see Doc. 3059; Edison 1888a; McClure 1889, 212–15). Another version of this drawing from the same date and labeled "Copy" appears in N-86-04-07:183, Lab. (*TAED* N319, image 86).

9. See Doc. 2937 (headnote). Figure labels above are "Electric Current," "N," "S," "Earth," "N," "Sun," "S," "Lines force," and "Equator."

ELECTROMAGNETISM AND GRAVITATION
Docs. 2937, 2936, 2938, 2940, 2944, 2947, 2950, and 2957

Toward the end of his honeymoon trip, Edison's thoughts turned to physics and astronomy, subjects about which he could only speculate. He hypothesized that gravitation is an electromagnetic phenomenon, and that electromagnetism could explain the workings of the solar system. With a long-standing (if passive) interest in astronomy,[1] Edison had only to take a short step from thinking about dynamos and conducting wires to imagining "The Earth revolving around the sun cut[ting] the lines of force so the current tends to pass around both sides of it" in a way that could explain a number of phenomena. In the same breath, he hypothesized that both the Earth and Sun act as powerful magnets on each other, with the Earth's orbit defined as "the point where the tendency to go off in a straight line is balanced Exactly by the mutual attraction of the Electric Current."[2] Having proposed this alternative explanation of celestial mechanics, he then considered the general possibility that "gravitation is an mutual attraction of all the atoms," each of which must have some polarity.[3] Like many contemporaries, he intuited a correspondence between the worlds of the very large and the very small. "Our solar system is a Cosmical Molecule,"[4] he wrote, inverting what would be called the "astronomical view of nature," the more conventional approach in which the conceptual framework of celestial mechanics was used to explain terrestrial physics.[5] Edison summarized and codified these speculations in an undated notebook entry (Doc. 2957), probably after he returned to New York. Among the numbered items in that entry is the blanket statement that "gravitation attraction has no existence . . . all weights are due to Static or Current attractions." Seemingly a summation of many of his astronomical speculations, the entry concluded with a partial cosmogony based on electromagnetic principles. In it, Edison's handwriting progressed (as it often did) from precise and deliberate at the beginning to a hasty cursive near the end.

By taking up the vexing problem of gravitation, Edison of course was stepping onto intellectual ground contested for centuries. Waves of competing hypotheses and philosophical speculations had come and gone. Michael Faraday, with an unshakeable belief in the explanatory power of the electromagnetic field, was an early skeptic of action-at-a-distance theories, and by the middle of the nineteenth century, explanations positing some sort of gravitational medium had come into vogue, at least in the Anglophone world.[6] That medium often had putative properties like the field (or its associated ether), and it was not unusual for explanations to be couched in similar terms. William Thomson had proposed in 1854 that "the potential energy of gravitation may be in reality the ultimate created antecedent of all the motion, heat, and light at present in the universe."[7]

Like so much else in the notebooks from his honeymoon, Edison's conjectures echo the third volume of Faraday's *Experimental Researches*. In the introduction to the paper "On the possible relation of Gravity to Electricity," Faraday wrote of the presumed unity toward which Edison was groping:

> The long and constant persuasion that all the forces of nature are mutually dependent, having one common origin, or rather being different manifestations of one fundamental power . . . , has made me often think upon the possibility of establishing, by experiment, a connexion between gravity and electricity, and so introducing the former into the group, the chain of which, including also magnetism, chemical force and heat, binds so many and such varied exhibitions of force together by common relations. [Faraday 1965 (1855), para. 2702]

Faraday's negative experimental results did not diminish his conviction. In a passage that Edison would have found heartening, Faraday defended his search: "It is not to be supposed for a moment that speculations of this kind are useless, or necessarily hurtful, in natural philosophy. They should ever be held as doubtful, and liable to error and to change; but they are wonderful aids in the hands of the experimentalist and mathematician."[8]

Edison's interest in these subjects, unlike in electromagnetism more generally, did not rise to the level of a genuine research program. He did, however, propose at least one experiment inspired directly by Faraday to use a "Simple dynamo to determine Earths Magnetism."[9] Shortly after returning from Florida, he spent about a hundred dollars for nearly a score of books on astronomy, magnetism, or meteorology, and his bookkeeper started an "Astronomical Experiment" account record that included payroll and small equipment purchases from May to July.[10]

1. Edison had joined Henry Draper's 1878 solar eclipse expedition, during which he tried out the new tasimeter as a thermal observing instrument (which he also hoped to use to measure the heat of stars). He took an interest in the transit of Mercury the same year (see Docs. 1289 n. 3, 1312, 1405, and *TAEB* 4 chap. 4 introduction). In 1877, he had corresponded with Samuel Langley about measuring the heat of stellar spectra (see Docs. 1095A and 1135) and with Draper about the discovery of solar oxygen (see Docs. 989, 992, 998, 1010, and 1018). However,

Edison made no pretense of keeping up with astronomical research. He was unaware of study in recent decades of correlations between sunspots and disturbances of the Earth's magnetic field that showed the connection between solar and terrestrial magnetism to be far more complex than he thought (North 2008, 545–48; Hufbauer 1991, 46–49).

2. Doc. 2936.

3. Doc. 2937; see also Doc. 2941.

4. Doc. 2937.

5. Van Lunteren 1988 (p. 162) attributes the phrase to historian John Theodore Merz.

6. Van Lunteren 1988, 161–65. Faraday justified his skepticism, in part, by recourse to Newton himself (Faraday 1965 [1855], 507, 532, 570–71). Geoffrey Cantor, a modern biographer, provides a lucid explication of Faraday's gravitational conjectures and experiments. Cantor believes that Faraday was able (in part by a selective reading of Newton) "to bring gravitation at least partially within his framework of lines of force and thus partially bridge the apparent divide separating the phenomena of gravity from those of electricity" (Cantor 1991, 245–58, quoted 253).

7. Of course neither Thomson nor anyone else could fully explain the relation of gravity to other forces, and one historian concludes that "gravitation came to pose one of the greatest obstacles to the electromagnetic view of nature" that dominated physics by the end of the century. Van Lunteren 1988, 168; Thomson quoted in Smith and Wise 1989, 520.

8. Faraday 1965 (1855), para. 3244.

9. See Doc. 2947.

10. Newton's *Principia* was among the titles Edison purchased. Vouchers (Laboratory) nos. 225–36 (1886) for D. Van Nostrand; Ledger #5:592, Accts. (*TAED* AB003 [image 293]).

–2937–

*Notebook Entry:
Planetary and Atomic
Magnetism*[1]

[Fort Myers, April 17 1886?[2]]

The rotation of the Earth around the sun Cut the lines of force eminating[a] from the sun. ~~The rotation of~~ This causes one side of the Equator to be N & the other side P but not Current Now the rotation of the earth on its own axis causes the tendency in one half of the Earth to be neutralzed & increases the other thus a strong current circulates around the Earth East[a] to west. This causes a north & south magnetic pole The mutual attraction of the whole Equals Exactly the tendency of the Earth to leave its orbit[3]

porous cell filled with Sulphate Zinc & amalgamated Zinc Electrode   test if there is a current from East & west or north & south.[a] 2500[a] feet apart.

The motion of the Earth due to the seasons has to be explained

The sun must have an orbit then which produces a disturbance or displacement of the lines of force

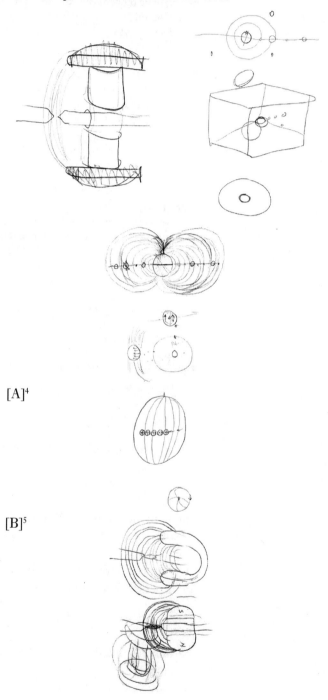

[A]⁴

[B]⁵

All atoms[a] have a n & s polarity   The N of the atom points equatorially & not towards the South Earth pole— gravitation is an mutual electrical attraction of all the atoms   The reason of the diference of weight of diferent substances is that each substance has a diferent number of atoms or molecular groups. The atoms of each molecule are closer together being greater in number in each molecule in dif substances   The Molecules may all be the same distance apart. The Total Attraction is greater for the Earth as the molecular atoms are greater— all ~~substances~~

All substances we call Elements are composed of molecules of diferent atoms   all atoms are primal hence matter is composed primarily of one substance, the primal molecule. Our solar system is a <u>Cosmical Molecule.</u>

different Molecules say of Iron are composed of as many atoms[a] as there are lines in[a] the spectrum   ~~Th~~Each atom has a diferent motion in the molecule. Each atom rotates on its axis with inconceivable velocity

X, NjWOE, Lab., N-86-04-07:191 (N319190). [a]Obscured overwritten text.

1. See headnote above.
2. This entry appears between entries dated 17 and 18 April.
3. Figure labels are "west" and "east." Edison's drawing appears after the word "neutralzed," from which he drew a line to the remainder of the paragraph below it.
4. Figure labels are five iterations of "N/S."
5. Figure labels are "S" and "N."

–2938–

Notebook Entry:
Miscellaneous[1]

[Fort Myers,] April 18 1886

The rotation of the earth around the sun (which is an immense magnet or spherical mass with polarity one $\frac{1}{2}$ N other half S Throwing out lines of force through space) cuts these lines of force so one side of the earth is N the other S but this does not produce a current but the rotation of the earth on its axis causes the circulation of a tremendious current around the earth in direction and paralell with the Equator clear up to the two poles which it creates. The mutual attraction of the lines of force of the Earth with its poles and the lines of force of the sun is at the present distance of the Earth from the sun just sufficient to balance the tendency of the earth to go in a straight line, hence its orbit— The lines of force eminating from the sun are eleptic as produced by the counter lines of force from bodies outside of our planetary system, hence the Earth to always be in an Exact balance between its tendency to proceed[a] in a straight line & the attraction of the sun must move in this eleptic orbit—

The rotation of the earth or rather oscillation of the earth to produce the[a] seasons must be due to the fact that the sun oscillates in the same manner—[b]

Expmt   To ascertain the strength of Electric polarizatn of matter on the earth subject to conduction & in current form Take two wires each [-][c] 500 feet[d] mile long one east & west the other north and south, No 0000. Sink a several[e] large porous pot at one end in good damp earth fill with solution of sulphate zinc & use amalgmated electrode zinc for ground plate— Measure daily morning noon & night Current & volts on both wires.

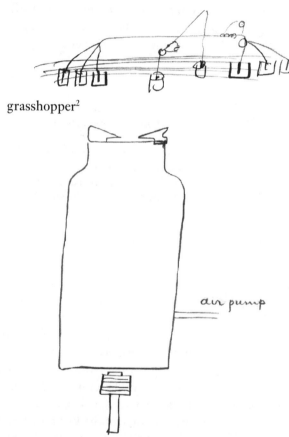

grasshopper[2]

air pump

ascertain if by compression of air the column of air which is in tune with a certain note can by increased compression respond to higher note ie[f] as you compress the column may be shorter

If above[a] correct changes in Barometer throws Resonators out of order— ditto velocity of sound in air—

Fundamental Magnetic Experiments[3]

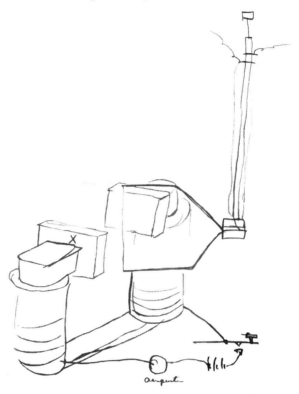

Put poles of ~~each~~ magnet inch apart & then make a great variety of substances metal & insulators X through which lines of force must pass & see if there is a specific Magnetic Condctg capacity— plot curves— also see if current is thrown in X metals

XYZ[4]

Spectrum— pass tubes filled with liquids also wires— paralell & at right angles to the beam of light   pass a beam of one kind of light through a tube lengthwise & move this paralell & at right angles to another beam of diferent Colored light & also the spectrum & also white light— Try these Expmts also in magnetic field both directions—

Make a jump spark spectrum also flame spectrum of iron

nickel and other metals & then The same in a <u>powerful</u> field paralell & at right angles to the <u>L of F</u>   also in a partial[a] Vac tube & this between poles in various directions of a powerful magnet. also use static from holz[5] to change it in vacuum.

   [A][6g]

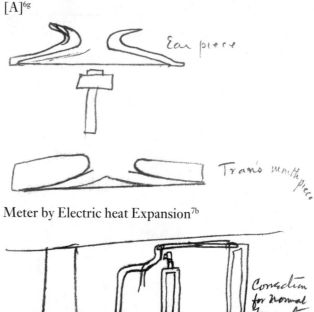

Meter by Electric heat Expansion[7b]

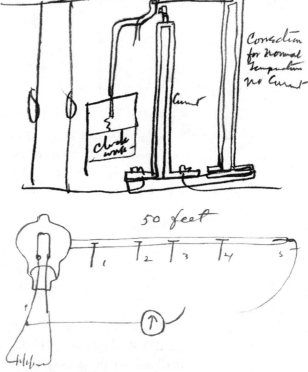

See if it will conduct in straight line & get Res Every 10 feet—[8b]
   disc[a] glass Coated in Vacuo with iron & white[a] wax melted & flowed over before Vac broken— See Effects magnests letting light through[h]

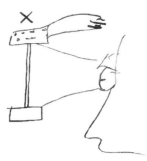

also put secondary coil around X & see if beam <u>light</u> through alum gives Electric wave in Mirror gal—also heat

Expt[9]

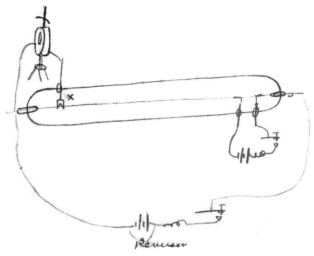

Experiment to determine if heat [–][c] conducted along a wire is hastened and retarded by Electric Currents to & fro. part of the wire is kept incandescent until by[a] conduction heat researches the thermo metal X when mirror Gal detects it   Conduction time 1st ascertained with C off Then with current P & then N from C   also vary strength.

Fundamental—[10] Theory that gravitation is due to the circulation of an electric current around the earth due to the rotation of the earth on its axis cutting the lines of force thrown out into space by the sun which is a magnet whose[a] polar[a] center line is paralell with that of our earth— The current circulates around the Earth paralell with the Equatorial belt clear to both poles ie[f] like the lines of latitude.

The weight of all matter diminishes ie[f] electric attraction diminishes until we reach the center of the earth— matter insulated from the conducting crust of the earth is attracted by so called[i] static attraction, hence the weight of such matter should diminish as we receede from the earth   a pound of glass[a] weighed by a <u>spring</u> balance [at?][a] the sea level should weigh less on the same balance in a baloon 5 miles high.

Here is an Experiment to determine if matter where the E has a current form is attracted greater that in static form[11]

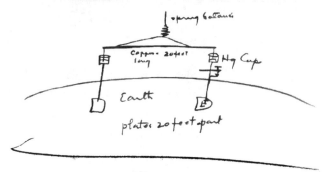

Test in New Lab—[12] Comcl availabilty of Cobalt Nickel & Iron for heat Effecting magnetism[h]

Make a filament about same as regular lamp but straight & 10 inches long ½ dia— wind fine wire around outside primary & secondary— Then heat fil with Current, bal in side of[j] bridge the wire & use another tube same amount of wire on same size tube so as to get same self induction— Now when Iron white [bal?][13]—then gradually lower temp by res in current & take throw of gal each time—afterwards put Iron one side Nickel other & get curves—

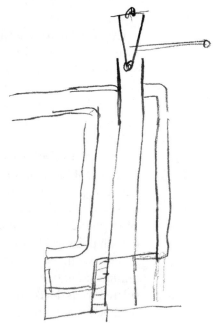

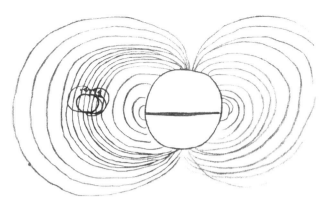

The reason why the earth rotates on its axis while going around the sun is because the lines of force cutting the earth on its face nearest the sun are the strongest gradually weakening as you go farther out from the sun[14]   The weakening of the line in 8000 miles is sufficient to produce rotation by mutual attraction between the strongest earth lines and the strongest[a] sun lines—

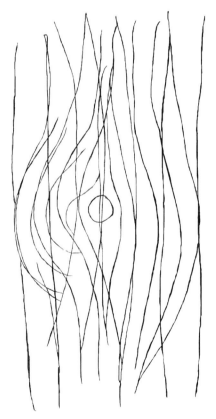

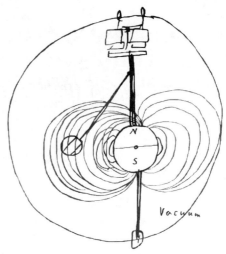

apparatus to prove the theory[15]

TAE

X,NjWOE, Lab., N-86-08-17:3 (*TAED* N320003). Document multiply signed and dated; miscellaneous doodles not transcribed. [a]Obscured overwritten text. [b]Followed by dividing mark. [c]Canceled. [d]"500 feet" interlined above. [e]Interlined above. [f]Circled. [g]Drawing followed by dividing mark. [h]Followed by index pointing to following paragraph, written on the facing page. [i]"so called" added in right margin. [j]"side of" interlined above.

1. See Doc. 2937 (headnote).

2. Figure label is "air pump." A variant of this note in another book is preceded by a circuit diagram marked "To get rid of heavy inductive wave." Edison apparently intended to use the frequency of a resonating air chamber to create inductive impulses in the transmitter (at bottom) of the railway telegraph, perhaps with the intention of overriding inductive interference. N-86-04-07:205, 207, Lab. (*TAED* N319, images 97–98); cf. Doc. 2948 n. 2.

3. Figure labels are "X" and "ampmeter."

4. A draft version of this note from the same date is in N-86-04-07: 211, Lab. (*TAED* N319, image 100).

5. That is, a Holtz electrostatic generator.

6. Figure label for the top drawing is "Ear piece" and for the bottom drawing "Trans[mitter] mouthpiece." A draft of these drawings from the same date is in N-86-04-07:213, Lab. (*TAED* N319, image 101).

7. Figure labels are "clock work—," "Current," and "Correction for normal Temperature   No Current."

8. Figure labels above are "50 feet" and numerals 1 through 5. A draft version of this note from the same date is in N-86-04-07:217, Lab. (*TAED* N319, image 102).

9. Figure labels are "X" and "Reverser." A draft version of this note from the same date contains a more detailed drawing and description of the experimental circuit, as well as a closeup of part of the right side of the circuit. N-86-04-07:223, Lab. (*TAED* N319, image 105).

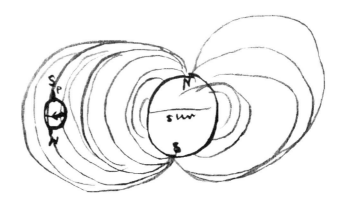

10. A draft version of this note from the same date includes a drawing after the first paragraph to illustrate this hypothesis. N-86-04-07: 225–27, Lab. (*TAED* N319, images 106–7).

11. Figure labels are "spring balance," "copper 20 feet long," "Hg Cup," "Earth," and "plates 20 feet apart."

12. Edison had started to think about a new laboratory during the winter (see Doc. 2889) but did not begin making definite plans until spring 1887. Although several of the experiments outlined in the Fort Myers notebooks were among the earliest ones Edison undertook at the new Orange laboratory, the editors have not found evidence that he took up this particular one.

13. Edison likely meant to write "hot."

14. Above the small circle (presumably representing the Earth) on the left side of the figure above, Edison wrote the numbers "5," "6," "7," "8," and "9" on succeeding lines of force. Read from right to left, as the lines issue from the Sun, they would indicate the lines' successively diminishing force.

15. Figure label above is "vacuum."

–2939–

Notebook Entry: Miscellaneous

[Fort Myers,] April 19 1886

Telephone Receiver   ⟨Copy⟩[1]

See which is strongest end

XYZ   If Iron loses its peculiar property of magnetization at a red heat, probably non magnetic metels may gain the property, hence Try all with Core & armature at all temperatures up to bright red—[a]

Iron at a point about dull red say 900 fahr loses it magnetism hence by manipulating the temperature causing a slight rise & fall it will magnetize & demagnetise a field magnet and thus by winding an extra low resistance coil can be thrown in Current & Commutated thus getting heat into Electricity   ⟨good!!!⟩[b]

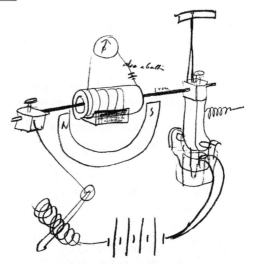

Experiment to bring iron up to 900 ie[c] the critical point where it loses its magnetism & see if can get induced Currents.[2]

If Iron loses its magnetism slowly up to 900 fahr. Then and change in heat of a magnet ought to throw induced Currents in a coil

Therefore hot & cold water oscillated through magnet composed of thin[d] tubes one within the other ought to be a way of getting heat into Electricity

TAE

X, NjWOE, Lab., N-86-04-03.1:173 (*TAED* N315173). Document multiply signed and dated; two unclear rough sketches not included. [a]Followed by dividing mark. [b]Multiply underlined. [c]Circled. [d]Obscured overwritten text.

1. The editors have not found another version of this drawing.

2. Figure labels above are "also a [battery?]," "iron," "N," and "S." A few days earlier, Edison had speculated along a different line about the possibility of recovering electrical energy from melted steel: "If steel

is a definate combination of carbon and iron, then when cast iron is molten it should decompose with a strong current say 1000 amperes per 3 square inch section." N-86-08-17:57, Lab. (*TAED* N320047).

–2940–

Notebook Entry:
Miscellaneous[1]

[Fort Myers,] April 21 1886—
Perhaps our solar system is rotating as a whole and the sun rotating cuts its own L[ines of] F[orce] & Thus the heat is accounted for—[2a]
Fundamental[3]

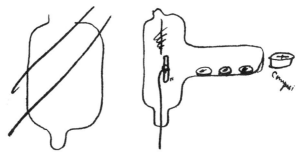

also compasses inside—
ascertain the strength of deflection in a Compass needle 12 inches away from a small ½ size telephone perm mag   then proceed to exhaust air & see if lines of force spread out in Vacuo watching for increased deflection of needle as exhaustion proceeds
Phonoplex   To Silence phone when Sending & also same battery[4]

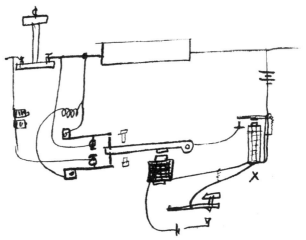

X has Copper Core around iron core   Cop ¹⁄₁₆ thick act very sluggish

[A][5]

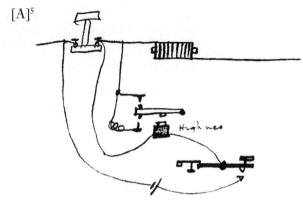

Not good
Phonoplex[6]

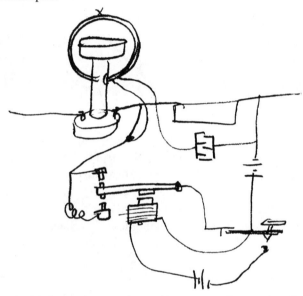

X Right angles to other coils & neutralizes— low Res—
Fundamental[7]

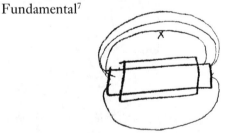

How are L of F conducted. can it be that the actual iron molecule goes round via X   put in path of lines of force a gas flame
to oxidize it and a wet sheet of Ferri also ferrid potassium also
look through Spectroscope[b] at Bunsen burner through which
L of F pass see if get iron spectrum

XYZ Vibrate a flame in path of L of Force   see if it effects it & give sound in telephone

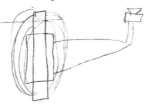

also galvanometer

Reed[8] vibrating diaphragm give vibration to chamber thus which gas passes throu flame in & out L of F. ⟨Dec 10 87   Ordered made by John Ott.⟩

Fundamental. As Iron gives a jerk when[b] heated by Elec at about red heat and expands & contract & also loses its magnetism or power to become magnetic about same point & I think reverses direction of Current in Thermos at about same point (ascertain)—And[b] other metals dont, This shows that its molecules or atoms are closer together than other metals, & ~~therefore~~ and have to be seperated by heat before they become like the other conductors not subject to magnetism, hence a slight Electric Current in Iron makes powerful attraction & scarcely none in other metals. perhaps Certain Metals if cooled down to 75 or 100 below Zero & put in a powerful Coil would shew magnetism   Try   Use rotating disk & solid carbonic acid & ether[c]

TAE

X, NjWOE, Lab., N-86-08-17:31 (*TAED* N320031). Document multiply signed and dated. [a]Followed by dividing mark. [b]Obscured overwritten text. [c]"Use . . . ether" written in left margin.

1. See Doc. 2937 (headnote).

2. The source and quantity of heat energy within the Sun had been the subject of dispute for decades. The question became entangled with passionate debates about biological evolution after Charles Darwin advanced a theory of natural selection requiring an enormous geological

time scale. The principal hypotheses were that an inflow of meteors provided enough kinetic energy to sustain the Sun (or at least prevent it from cooling too rapidly), or that the gravitational collapse of gas clouds would do the same. William Thomson notably championed some form of both general ideas at various times, though in such a way as to present a direct challenge to Darwinian theory, an opposition that he still firmly maintained. North 2008, 539–42; Lindley 2004, 165–77; Smith and Wise 1989, 505–38; Burchfield 1975, chaps. 2–3; Hufbauer 1991, 55–57.

3. Figure label of drawing on the right is "compass."

4. A draft of this drawing and note from the same date is in N-86-04-07:233, Lab. (*TAED* N319, image 110).

5. Figure label is "High res."

6. Figure label at top is "X." Another version of this drawing from the same date is in N-86-04-07:239, Lab. (*TAED* N319, image 113). That notebook contains several other phonoplex sketches on pages 230–52 (images 110–20). Similar designs were tested in early May. Unbound Notes and Drawings (1886), Lab. (*TAED* NS86ADC5, NS86ADC6, NS86ADC7); see also N-86-04-28:15, Lab. (*TAED* N321015).

7. Figure label is "X."

8. Figure label above is "Reed."

–2941–

Notebook Entry:
Molecular Magnetism

[Fort Myers,] April 22 1886—

It is probabl~~ye~~ that an electric current cannot get the atoms of non magnetic metals completely rotated so N & S are exactly opposite[a] each other. They rotate with difficulty while the[a] iron atoms rotate easily and when a piece of iron is saturated the atoms are all opposite in ~~o~~polarity ie[1b] 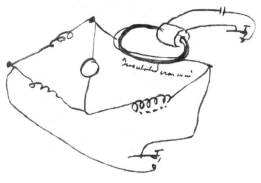 in Copper even by a current heated[c] to the melting point they are only partially rotated ⟨drawing⟩ The effect of heat must be to rotate the atoms in the opposite direction as at red heat magnetism is lost, while electric conduction which can only take place by rotation is diminished by heat.

~~Independent~~ now the question is does[a] iron become a better conductor of E if magnetised   Try[2]

measure it magnetized and demagnetized— If not then as iron outside of its magnetic[a] property acts to a Electric Current like another metal then iron must be have[a] a Compound atomic arrangement   one system of atoms rotating Easily gives us magnetism while the regular atomic system Causes it to act like another metal to E.

I must carefully investigate all phenomena Connected with the sudden Jerk in red hot iron noticed by Barrett,[3] see if it does it without current but by heat alone without Electricity.[d]

It may be iron has no magnetic polariztion in air hence it shows magnetism   perhaps all metals no magnetism but air is polazed by them & cant show it[4e]

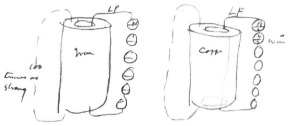

Hence the air seems conduct lines of force from[a] Iron several hundred times better than the[a] same lines of force from copper etc

This would seem to show that there was either a surface polarization of the copper or the end atoms on Iron were rotated exactly north & south with a given [Line?][f] current while[a] with other metals there was only partial rotation ie[5b]

If rotation of Copper atom was $\frac{1}{100}$ part of that of iron atom with given current The polar pull or attraction at the Ends of Copper would be only $\frac{1}{100}$ part of that of Iron   when Iro because The total surface of active atom surfaces would only be $\frac{1}{100}$— Iron very probably is fully rotated when saturated.

If conduction is rapid occillation[a] of atoms[a] (from the great earth pole line or internal attraction due to earth cutting lines of force from the sun) Then the amplitude is due to the amperes, and rapidity to the number hence a one ampere cell will give a Current which will give a certain amplide of axial rotation say 100 M in second, 2 cells added together & still same current & twice resistance will give [---------][f] ½ ~~the~~ same[g] amplitude but twice the number ie[b] 200 M in the whole length[h] or what is probably more correct keep up the same amplide in the first ohm   The increased pressures preventing a fall of amplitude in the first ohm by the addition of the second ohm

X, NjWOE, Lab., N-86-06-18:47 (*TAED* N320047). [a]Obscured overwritten text. [b]Circled. [c]"by a current heated" interlined above. [d]Followed by a dividing mark. [e]Paragraph written sideways in right margin, continuing upside down at top of page. [f]Canceled. [g]Interlined above. [h]"in the whole length" interlined above.

1. In this and the following drawing Edison indicated the polar orientation of molecules with the letters "N[orth]" and "S[outh]."

2. Figure label is "Insulated iron wire."

3. Physicist and psychic researcher William Barrett (1844–1925; knighted 1912) was born in Jamaica but educated in Manchester and in London, where he became an assistant to John Tyndall at the Royal Institution in 1863 and was influenced by Michael Faraday. He was appointed professor of physics at the Royal College of Science in Dublin in 1873. About that time, Barrett began a long-running research program into the magnetic, thermal, and electrical properties of metals, and Edison referred to one of his early studies, "On Certain Remarkable Molecular Changes Occurring in Iron Wire at a Low Red Heat" (Barrett 1873). Barrett was also known by now for research into psychic phenomena and as one of the founders (1882) of the Society for Psychical Research. *Oxford DNB*, s.v. "Barrett, Sir William Fletcher."

4. Figure labels on the left drawing are "100 times as strong," "Iron," "LF" and on the right drawing are "Copper," "LF," and "wire." In both drawings, Edison indicates the polar orientation of molecules in the wire with the letters "N" and "S."

5. Figure labels on the left drawing are "all north Exposed" and "iron"; the label on the right drawing is "¼ north Exposed." Letters "N" and "S" again indicate the polar orientation of molecules.

–2942–

Notebook Entry: Miscellaneous[1]

XYZ

[Fort Myers,] April 22 1886

See if a beam of Light paralell & at right angles effects lines of force— also heating chamber[2]

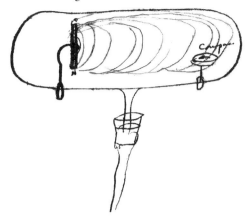

XYZ   Vacuum— run curve of EMf as Exhaustion proceeds also work beam Light[a] both directions   heat case dif temperatures & run curve— also statically charge[3]

[A][4]

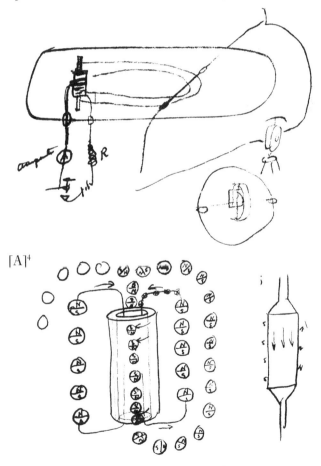

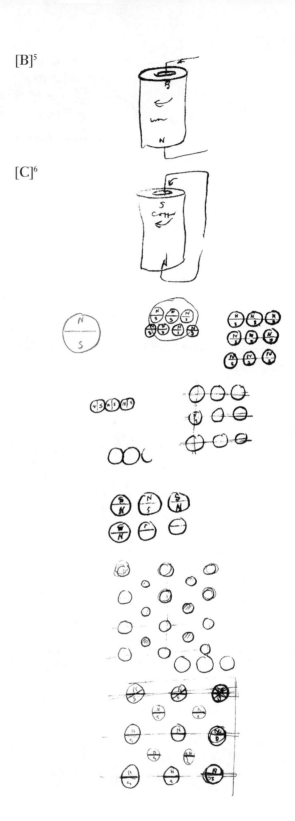

If steel is a compound of carbon Then when cast iron is molten It should decompose with[a] a strong Current ie[b] a current proportionate as its conductivity is greater than a liquid. using wrought[c] iron poles The Carbon should go to one pole & pure Iron to other. <u>This is an experiment worth Trying!!</u>

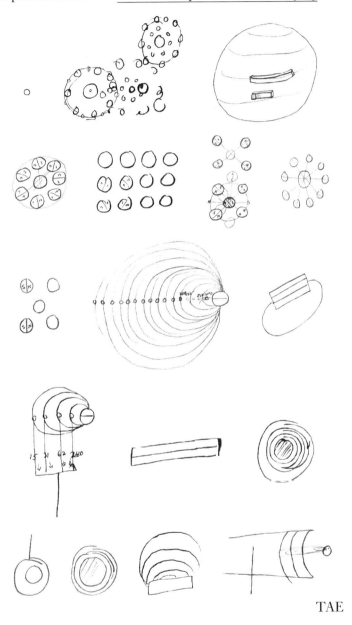

TAE

X, NjWOE, Lab., N-86-04-07:261 (*TAED* N319261). Document multiply dated; miscellaneous doodles and calculations not reproduced. [a]Obscured overwritten text. [b]Circled. [c]Interlined above.

1. Among Edison's doodles on these pages are the name "Lorenzo De Medici" and famous lines from Thomas Gray's "Elegy Written in a Country Churchyard," which Edison misquoted slightly: "The boast of heraldy of pomp and power/All that beauty off that wealth ere gave/Alike await the inevitable." Edison recited the same lines in 1878 for an early public demonstration of the phonograph. See Doc. 1277.

2. Figure labels are "S" and "N" (on the bar magnet) and "Compass."

3. Figure labels are "[ampmeter?]" and "R."

4. Figure labels are repetitions of "N[orth]" and "S[outh]," a pattern that Edison repeated in several of the drawings below.

5. Figure label is "iron."

6. Figure label is "copper."

–2943–

*Notebook Entry:
Electromagnetism*

[Fort Myers,] April[a] 23 1886

Fundamental.[1]

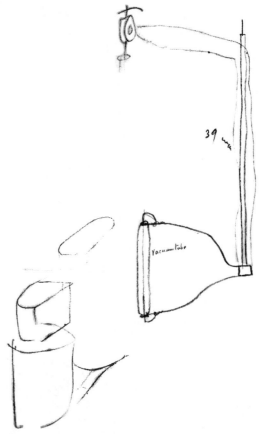

See if the EMF is exactly same in tube without & with a <u>very</u> high vacuum—

Fundamental[2]

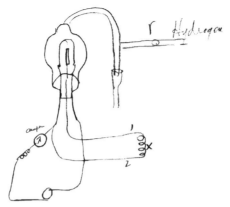

Exhaust fill with H   Exhaust & fill sevl times. Then ~~at atmo-spheric pressure or near there~~ with ordinary Barometer guage, start pump & take Emf at 1 & 2 with X open until pump has run hour beyond reading by ordinary guage   break Vac by letting in H & Commence again   make curve with X closed say 500 ohms— also repeat at different resistances   Keep current same ie[b] 20[c] c[3] at high Exhaustion—[d] also keep CP same 20 during all exhaustions & run curves with X open & closed— also try dif gases   Try a platinum Loop & Center plat wire— also—Iron Loop & Center platinum wire   Use a regular lamp & make two center wires exactly alike in length & size & relative position one plat one zinc   see if diference

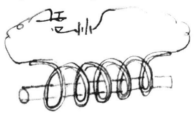

glass tube with wire in 10 or fifteen turns, Vacuum in tube, see which is strongest with or without vacuum—[4]

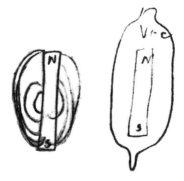

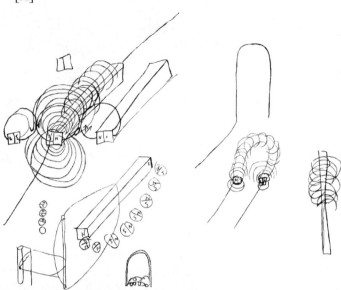

Fundamental   Ascertain if with a given Vacuum manipulating L[ines] of F[orce] increases or diminishes pressure—⁶

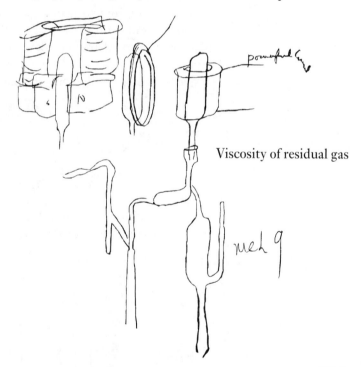

Viscosity of residual gas

TAE

X, NjWOE, Lab., N-86-04-03.1:199 (*TAED* N315199). Document multiply signed and dated. [a]Obscured overwritten text. [b]Circled. [c]Obscured overwritten text. [d]Followed by "over" as page turn.

1. Figure labels are "39 inch" and "vacuum tube"; cf. Doc. 2936.

2. Figure labels are "r," "Hydrogen," "[ampter?]," "1," "2," and "X."

3. That is, 20 candlepower.

4. Figure labels below are "N" and "S" on the left drawing and "Vac," "N," and "S" on the right drawing.

5. In the following drawings Edison indicates polar orientation with the letters "N[orth]" and "S[outh]."

6. By "pressure," Edison meant voltage. Figure labels are "S," "N," "powerful Current," and "McL[eod] G[auge]."

–2944–

Notebook Entry:
Miscellaneous

[For Myers,] April 25 1886

Cosmos[1]

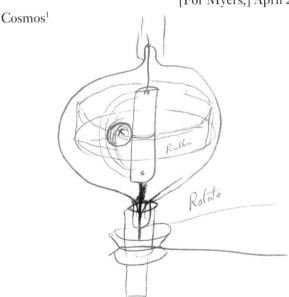

X Copper Ball.[a]
Cosmos

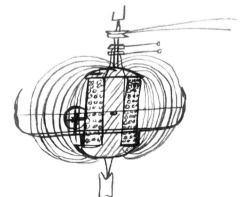

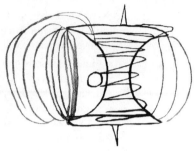

X Copper ball— This ball cant get out   will rotate on its axis
& in contact with nothing
    Cosmos   Try variety Expmts with this[2]

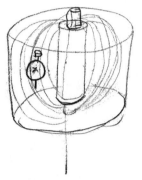

Copper just balanced in Kerosene   Magnet rotated see if rotation on its axis— also rotate the jar & hold magnet stili.
    Make a lot of hollow balls ½ inch dia[meter] of Iron Lead
Tin Zinc & other metals, balanced exactly by themselves in
Benzine also[a] water with salt in— Then Throw in an Iron ball
& several Copper see if they arrange themselves definitely—
also cut water with lines of force, pass Current through with
variety of balls in liquids— see if any definitiveness comes of it.
    New dodge   Arc Lights[3]

Rotate the arc & keep it in center   stops flickering— ⟨Good
work it up⟩

                                  TAE—

X, NjWOE, Lab., N-86-08-17:67 (*TAED* N320067). Document multiply signed and dated. ªObscured overwritten text.

1. See Doc. 2937 (headnote). Figure labels are "X," "N," "S," "Rubber," and "Rotate." On or after 22 April, Edison used several preceding pages for a chart and illustrated notes on the comparative size, mass, orbital period, and strength of presumed lines of force of various planets. N-86-08-17:61–65, Lab. (*TAED* N320047).

2. Figure label in drawing on the left is "X."

3. Figure labels are "S" and "N."

–2945–

From Samuel Insull

New York, April 28th. 1886.[1]

My Dear Edison:—

You will have received from Dr. Crowell a letter requesting you to go to Chicago in connection with the Railway Telegraph Co's. affairs. The Doctor feels very strongly on this matter, and was very much disappointed at your telegram to me stating that you could get through all the troubles as well in the Laboratory as by going to Chicago.[2] The fact is that Dr. Crowell does not believe such to be the case. He reasons that a trip to Chicago on your part, so that you can personally study the operation of the apparatus, will do more good than the minutest investigations by your assistants, or the most thorough study of the matter in the Laboratory. Inasmuch as the Doctor feels this way, I think that even if your opinion is exactly the contrary, you should humor Dr. Crowell by complying with his wishes, as he justly observes he has a very large amount of money invested in the Railway Telegraph Co.,[3] and it is but natural that he should feel somewhat concerned if everything does not go exactly as rapidly as he would wish. I promised the Doctor that I would write you and urge you in the strongest possible manner to comply with his request.

Business at Shops has been comparatively good during the last month or so. The Machine Works has shipped $136,000. worth of goods during the first three months of this year, divided up as follows: January, $42,000.; February, $43,000.; March, $51,000. The shipments for the month of April will I think be just about as [----]ª large at the Machine Works as March. The orders coming in, however, at the present moment are comparatively small, and we have got quite a stock of machines on hand. We shall doubtless feel quite a tightness down at Goerck Street during the months of May and June, but after that, unless the actual business belies entirely the predictions of everybody, we shall have more work than we can possibly do.

Bergmann's business is very brisk. In February he shipped $28,400. worth of goods and in March $33,000. In April he will doubtless ship from $35,000. to $40,000. worth of goods. He has just got an order for 100 more printers of an entirely new type for a new Chicago Co. This order will be followed by others. He says that the "losses" on this order will be very great ???. His chandelier business is somewhat slack just now, but this will doubtless improve as the season advances.

At the Lamp Co. they shipped during the month of February 51,626 lamps, bring $20,306.15   Their home orders for the same month were 28,547 lamps and for foreign account 17,250 lamps, making a total of 45,797 lamps. During the month of March they shipped 45,881 lamps, bringing $21,081.67, and received orders for home account for 42,022 lamps and for foreign account for 3,946 lamps, making a total of 45,968 lamps in all, being practically the same as the orders for the month of February.

With relation to the business here at 65 Fifth Avenue,[4] the enclosed statement[5] showing the business done by the Isolated Department from February 1st. till last Saturday explains itself. You will see that the plants aggregate 6,785 lamps, and total $79,882.42 on contracts. Business in this department is somewhat paralyzed by the anticipation of the new arrangement by which the isolated business will be taken over by the shops. Batchelor wired you the details of this arrangement,[6] and there is therefore no necessity for me to recapitulate. Mr. Coster is preparing a basis of contract, and by the time you reach home the matter will be in shape for immediate action. There is no doubt but what the arrangement proposed is a good one for everybody,—that is, alike for the Light Co's. business and the shops. The difficulty which Mr. Batchelor and I foresee, and which has caused a great deal of discussion between Batchelor and myself, as we cannot devise any way of overcoming it, [-][a] is that whereas now we are able to get payment for our machines immediately our dynamos are shipped (I mean from the Isolated Co.), when this new deal takes place we shall be compelled to carry our accounts for at least four months before we can expect to get any cash from the sale of the dynamos for isolated plants.[7] This introduces the question of how the Machine Works can stand such a heavy draft on its finances extending over such a lengthened period. In addition to this, it will be necessary to put some capital (probably about $10,000. from each shop) into the new organization. When we remember the length of time above referred to that we shall have to carry our dynamos, it seems almost impossible for the Machine Works to raise this $10,000. I mention these subjects simply with the idea of suggesting that you should think over them by the time you get home. There is no doubt but what we must find some way of dealing with this difficulty, as I feel confident that you will agree with the rest of us (as your telegram to Batchelor would seem to indicate) that the proposed arrangement will prove advantageous to everybody in the long run.[8]

With relation to central station business, I would state that Markle[9] has formed a Company in Detroit with $250,000. capital, the contract for which with the Light Co. will be signed in a day or two.[10] Hastings told me this morning that a premium is already offered in Detroit for the stock of this Company. Coming further East, we find that a Company has been formed in Reading with $60,000. capital, and Hastings goes to-morrow to Reading to get the contract signed, immediately after which a 1600 light station will be installed. The same evening that Hastings is at Reading, Humbird will in all probability be at Wilkesbarre closing a contract for a central station there. The Amsterdam Co. with a capital of $50,000. is now erecting a 1600 light station. The Rochester Co., with a capital of $150,000., is now raising their capital, and have only $14,000. more yet to get. Atlantic City, with a capital of $50,000. to start on, have I understand ordered the greater part of their plant. Topeka, Kansas, is forming a Company with $100,000. capital, which has all been subscribed, although the contract with the Light Co. has not yet been formally signed. Laramie, W[yoming]. T[erritory]., [-][a] is erecting a 1600 light plant, on what capital I am not aware at the moment. Hix[11] has raised a Company for Philadelphia with a $1,000,000., and all that he is waiting for is the passage of the Ordinance by the Board of Aldermen to put its wires underground.[12] He told me the other evening that he felt confident this Ordinance would go through all right. Humbird tells Johnson that he expects to get twenty companies started and erecting stations before the 1st. August. With relation to New York, nothing can well be said, except that things are in practically the same shape as they were a month ago. The only modifications that has occurred being that Commissioner Loew, of the Underground Commission, is dead, and no Commissioner has as yet been appointed in his place.[13] I cannot give you any real information with relation to this matter. You know the subject is very little spoken of, and I think you better wait until you return and then Johnson can tell you all about it. I may mention, however, that three houses have already been purchased for the erection of the first station in 27th. Street.

We have just got into our new offices at 40 and 42 Wall St.[14] We are not quite settled there yet, but by the time you get back everything will be in good shape. With relation to "Foreign Grasshopper," it is absolutely impossible to do anything until things look more hopeful in connection with the home company here.

Tate has written you fully with relation to the "Phonoplex."[15] I have some ideas which I think you will consider good as to the exploitation of the "Phonoplex" in Europe. I think that I can show you a way by which we can insure its immediate introduction into several of the European Continental Countries.

Hoping that you have had a pleasant vacation, and that you and your wife are in good health, believe me, Very sincerely yours,

S[amuel]. I[nsull].

P.S. Will you please wire me on receipt of this as to when you will be in New York. I am daily expecting important papers from England in relation to Indian & Colonial affairs. That reorganization so long talked of will take place after all, and some money will be got out of the Company for you.[16] Gouraud is back here again.[17]

TLS (carbon copy), NjWOE, DF (*TAED* D8603ZAX). ªCanceled.

1. Edison had previously telegraphed Insull his intention to leave Fort Myers on 26 April and arrive in Akron, Ohio, on the 29th. TAE to Insull, 20 Apr. 1886, DF (*TAED* D8603ZAT1).

2. Eugene Crowell's letter has not been found. On 22 April, Insull wired Edison in Fort Myers that Crowell was "very anxious you and Gilliland should go to Chicago on your way home   Rudds progress very slow," since his initial success. Edison replied: "Know whats wanted with going   can do better in Laboratory." The trustees of the Railway Telegraph and Telephone Co. passed a unanimous resolution on 30 April calling for "concentrated, immediate and continuous efforts" in Chicago and specifically, for Edison "to proceed to the scene of action." Insull reiterated Crowell's anxious request the next day, but that was after Edison intended to arrive in Akron (see note 1). Edison and Mina left Ohio for New York on 5 May. Insull to TAE, 22 Apr. 1886, Lbk. 22:47B (*TAED* LB022047B); TAE to Insull, 23 Apr. 1886; Railway Telegraph and Telephone Co. resolution, 30 Apr. 1886; Insull to TAE, 1 May 1886; all DF (*TAED* D8603ZAU, D8639H, D8633J); Jane Miller to Mina Edison, 9 May 1886, FR (*TAED* FM001AAM).

3. With 2,259 shares (of 10,000 issued) as of November 1885, Crowell was the Railway Telegraph and Telephone Co.'s largest investor. List of stockholders, 21 Nov. 1885, DF (*TAED* D8546ZCK).

4. The long-time headquarters of the Edison lighting companies, including the Edison Co. for Isolated Lighting; cf. note 14.

5. Enclosure not found.

6. Telegram not found.

7. The Edison Machine Works had been a supplier to the Isolated Co., which installed plants. With the shops planning to take over this installation work collectively (see note 8), the Machine Works would be acting instead as a contractor and would not be paid for equipment until each plant was completed and paid for by its customer.

8. The heads of the Edison Machine Works, the Shafting Manufacturing Co., and the Electric Tube Co. met on 30 January to "confirm the consolidation of all three into 'the E.M.W.'" At a related meeting the next week, Edward Johnson proposed that the Isolated Co. maintain a small construction company in the short term "with the object," according to Charles Batchelor, "of getting entirely out of construction at some future time." At another conference on 6 April, Johnson, Insull, Francis Upton, Sigmund Bergmann, and, presumably, Batchelor agreed that a merger of the manufacturing shops was "impracticable" at the moment, but they adopted a plan by which the shops could form a selling department to take over the business of the Isolated Co. Edison reportedly wired his assent to this scheme. Cat. 1336:7, 13, 17 (items 9, 29, 38; 30 Jan., 4 Mar., and 6 Apr. 1886), Batchelor (*TAED* MBJ003007B, MBJ003013B, MBJ003017).

9. John R. Markle (1845–1921), a steam and electrical engineer, was the manager of the Michigan Department of the Edison Electric Light Co. and the Edison Co. for Isolated Lighting in Detroit, in which roles he had helped organize the Edison Illuminating Co. of Detroit. Markle was a Civil War veteran who had enlisted at age sixteen and served in Iowa and Ohio regiments. After the war, he worked first in cigar manufacture and general merchandising in Denver and then in the grain and produce business in Chicago. As a representative of the Edison lighting interests, he was credited with helping to establish almost 100 isolated plants and seven central stations in Michigan between 1881 and 1891. U.S. Census Bureau 1982? (1900), roll 248, p. 2A (Chicago Ward 4, Cook, Ill.); Letterhead of Markle to TAE, 23 June 1884, DF (*TAED* D8425F); Record for John R. Markle, *Iowa Civil War Soldier Burial Records*, online database accessed 19 Dec. 2012 through Ancestry.com; Miller 1957, 22; Mitchell 1891, 119–20.

10. The Edison Illuminating Co. of Detroit was organized on 15 April 1886. Markle brought together local business leaders George Peck, Edward Voigt, Simon Murphy, and J. E. Scripps, who capitalized the company in return for one-half of the first stock issue. Under its patent license agreement with the Edison Electric Light Co., the Detroit Illuminating Co. paid twenty-five percent of the stock issued and a cash payment of five percent of the original capitalization to the New York firm. (Markle received one-tenth of the stock issued to the Light Co. as a commission; he also served as sales agent.) The Detroit Illuminating Co. built its first plant on Washington Blvd. on land rented from Scripps and his brother. It went into operation on 8 November 1886 with a 4,000-lamp capacity. National Civic Federation 1907, 667; Miller 1957, 9–11, 27, 435 n. 10.

11. William Preston Hix (1836–1911) was a general agent of the

Edison Co. for Isolated Lighting in Philadelphia, having until recently had a similar role in St. Louis. Hix also organized the local illuminating company for Topeka mentioned above. Doc. 2470 n. 2; Letterhead of Hix to Insull, 6 Aug. 1885; Letterhead of Hix to TAE, 16 June 1886; both DF (*TAED* D8541ZAA, D8621G); *Hix v. Edison Electric Light Co.*, Supreme Court [N.Y.] Appellate Div. 1897, 3, 40–41.

12. After meeting with Edward Johnson in January, Hix enlisted the aid of prominent Philadelphia banker B. K. Jamison as a silent partner to help raise $1 million in capital for an Edison company in that city. Johnson formally hired Hix to organize the company on 23 February 1886, granting him a commission of 15 percent of the original stock offering, two-thirds of which would be secretly paid to Jamison. Hix obtained a charter for the Edison Electric Light Co. of Philadelphia on 13 December 1886, and the firm entered into a patent license agreement with the Edison Electric Light Co. in February 1887. *Hix v. Edison Electric Light Co.*, Supreme Court [N.Y.] Appellate Div. 1897, 3, 14, 18–20, 35–39, 134–35.

Philadelphia does not seem to have passed a private bill in 1886 or 1887 authorizing the local Edison company to lay underground wires. However, in August 1886, the city did enact a general ordinance governing the laying of all underground conductors, cables, and tubes. It required companies to secure a permit from the Board of Highway Supervisors, after which the chief of the city electrical department would supervise all work. Philadelphia 1887, 243–46.

13. Charles E. Loew (1827–1886) died on 21 April of acute peritonitis. A native of Alsace, Loew was president of the Iron Steamboat Co. and an influential Tammany Hall politician who served as New York City alderman and a collector of revenue, as well as New York County clerk. He was appointed in July 1885 to the Electric Subway Commission, the body with oversight responsibility of underground conductors (see Doc. 2923 n. 12). Loew's vacant seat was filled on 25 May by the appointment of Roswell Pettibone Flower. "The County Clerk," *NYT,* 22 Apr. 1870, 2; "Obituary: Charles W. Loew," ibid., 22 Apr. 1886, 5; "Experienced in Laying Wires," ibid., 26 May 1886, 1.

14. Edison's personal office moved from 65 Fifth Ave. to 40–42 Wall St. (the headquarters building of the Manhattan Bank, between William and Nassau Sts.) sometime after 23 April, when he telegraphed to the old address. Edison's attorney John Tomlinson and patent attorneys Dyer and Seely shared the rented space at 40 Wall St., and the Edison Electric Light Co. of Europe began using that as its business address. The Edison Electric Light Co. relocated in July to 16–18 Broad St., as did the Edison Electric Illuminating Co. of New York. The Edison Co. for Isolated Lighting remained at 65 Fifth Ave., which became the headquarters of the new Edison United Manufacturing Co. TAE to Henry Villard, 24 July 1886, Lbk. 22:265 (*TAED* LB022265); Letterhead of Dyer and Seely to Samuel Insull, 24 June 1886; Charles Batchelor to Tomlinson, 27 Aug. 1886; TAE to Insull, 23 Apr. 1886; John Vail to TAE, 28 Dec. 1886; all DF (*TAED* D8637S1, D8603ZAU, D8624O, D8627ZAA); Cat. 1336:95 (item 149, 8 July 1886), Batchelor (*TAED* MBJ003095A); *Trow's* 1888, 543.

15. Alfred Tate had just sent Edison an eight-page report, largely concerning his efforts (not as yet successful) to isolate the phonoplex

receiver from outgoing signals, problems with burned-out condensers on B&O lines, and negotiations for expanded use of the phonoplex in Canada. Tate to TAE, [27?] Apr. 1886, Lbk. 22:52 (*TAED* LB022052).

16. Edison's Indian and Colonial Electric Co., Ltd., negotiated a conditional agreement with the Australasian Electric Light Power and Storage Co. by the end of May. Under the proposed terms, the Indian and Colonial would wind up its affairs and transfer its interests and business to the Australasian Co. Any cash or securities remaining after the firm discharged its liabilities was to be distributed among the stock-holders, who were also to receive shares in the Australasian Co. The stockholders ratified this agreement on 11 June. Edison's Indian and Colonial Electric Co., Ltd., report, 2 June 1886; Samuel Flood Page to TAE, 11 June 1886; both DF (*TAED* D8630ZBA, D8630ZBT).

17. George Gouraud had apparently been in London as recently as mid-March, when he cabled Edison from there. Gouraud's address in New York was 74 Madison Ave. Gouraud to TAE, 13 Mar. 1886, DF (*TAED* D8640F); Insull to Gouraud, 17 June 1886, Lbk. 22:191B (*TAED* LB022191B).

–2946–

Memorandum to Eli Thompson: Fort Myers Grounds

[Fort Myers, March–April, 1886][1]

Mr Thompson[2]—

It will require 280 Lighter Loads of muck to cover 8 acres 4 inches deep— Each Lighter load 30 000 lbs or 8000 cart Loads 1000 lbs Each—or about 1500 5 ton loads on wood track[3]

We shall want a Bannana[a] Bed about 20 feet square. you can probably buy these. I noticed up by the wind mill on the Island[4] up in the narrow channel of the Calahousehatchie, lots of bananas bushes.[b]

Shall want about 1000[a] pineapple. I believe there are two ways of planting: one of which is longer but ~~surer~~ sure, plant [--------][c] this[d] kind—[b]

in garden across road.[e] ⟨ordered them of Montgomery[5] from Key Largo—⟩

~~Set out towards woods two ponciana Trees~~.[f] It might be best in cases like this to set out[a] duplicates very close together at the end of a year if both should grow the poorest one can be cut down. This will insure having at least one.

I think we should have a lemon hedge inside fence towards woods. The lemons procurable here seem to be a poor variety—if you cannot procure the regular Italian Lemon shoots elsewhere or raise the slips yourself from seeds found in the Lemon we have north then you will have to use the lemon & lime (The Limes seems to be good) procurable here. ⟨Use Lime altogether except few Lemon Reg Sicily Lemon if you can⟩

I saw near Jacksonville some very fine pecan trees. please

see if you can propagate slips from the nuts procurable from the stores. also, propaigate also[a] soft shell almonds Brazil nuts, dDate palm From Dates procurable in stores, English walnut, Filberts,—

Procure some fig tree slips, and set out about ½ dozen— put them out towards the old house as they are an uguly tree. I dont know if there are any fig trees around here if not perha[ps?][g] you will have to get them from Jacksonville where there is a nursery— Major Evans[6] will tell you about the nursery there and what you can get there.

As peaches grow in perfection down here, ascertain what period of the year when the <u>season is not</u> backward that the peaches ripen   if they ripen <u>as early</u> as April 25 or May 1st set them out. perhaps there is an <u>early</u> variety   ½ dozen trees will answer.

What time do grapes ripen here   if during our stay which will be hereafter from January 15 to May 1st—set some out

We want a strawberry patch. about 20 by 100 of best early bearing strawberries— also about 25 current bushes, also a bed of Red and black Raspberries mixed— I think a row a bed about 20 × 100 will do for both. ⟨garden⟩

⟨House Landscape⟩[a]

    ⟨Wild Coral plant[h]
    Lavender Bushes[i]
    4 Castor beans
    5 Olives.
    5 Jamaca apple
    5 Egg Fruit.⟩
About 10 mangos
    "   10 alligator pears.
    10 Sapidillos
    2 Spanish gooseberry[b]

⟨Landscape—⟩

Frierson had one tree last year
4 paw paw trees
6 pomegranates
⟨2 mulberry native.
2 custard apple.⟩

I think to ⟨as many⟩[j] orange trees of best variety ⟨as will go on end of the House plot will answer⟩[k] will do us on the grounds as I propose to plant some across the road— ⟨plant a lot in garden in case we want to transplant⟩[b]

½ doz Guavas[a] if they bear during our season if not one will do—[b]

I think you better plant some peanuts so as to obtain about 3 bushels for our use— ⟨garden⟩

I wonder if plum pear & Siberian Crab will grow here if so put out Couple of each ⟨~~H~~Landscape—⟩

I will send you from NYork on my arrival several good books on floriculture & Horticulture etc. also as soon as possible Ship you a full ~~catalogue of~~ assortment of seeds and bulbs. I think about 1800 varieties— also will send you a supply of good garden truck seeds. the later you[a] Can plant at such time as will insure us a good & regular supply for the table during our stay—

I propose to have 4 acres cleared across the road and a board fence put around it thus.

The garden truck & propogating beds can be placed over here. you should[a] set out plenty of Extra slips over there so in case of a failure on the River side lot of a shoot you can transplant

I suppose Cotton seed will sprout & grow in this part of the Country any time if so plant so we can see some while here say 10 or 15 hills—

I think perhaps the strawberry bed might go across the road—

You better use some of the space across road to Experiment with on diferent fertilizers—[b]

You are authorized to have ~~4~~about 4 acres cleared which I understand a man who has charge of the place next to old house will do for $35 per acre

Now I want it perfectly clear[a] of everything ~~Except the large trees~~[f], Even if it costs more— you are also authorized to build a board fence around the whole[7]

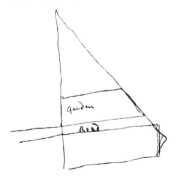

I shall send you every kind of grass seed so you can experiment in small patches in the garden—

Also procure 4 cabbage palmanto about 8 feet from the ground to where the leaf[a] rods come out   Set them out on grounds—[8]

8 feet

Lattice work complete from ground to top
[I think?][c]

What I desire in the flower line is a <u>few nice ones</u>—of <u>Every variety</u> that I can procure.

You should carefully study the books I send relating to the care & planting of the flowers.

I will send you about 1000 yards of common print cloth, which you can place around the more tender shoots when a freeze approaches   This cloth will prevent radiation   If run through boiled linseed oil & hung out until dry it will prevent radiation almost entirely—

will also ship barrel of boiled oil—

We propose to have our ground the best manured in florida   Therefore you may order for the River grounds, 6[a] ton of oil cake,[9] two ton of phosphates, and two ton of guano— and for the garden 4 ton of Oil Cake—2 ton[a] phosphates and 4 ton of guano, providing the guano does not cost more than $50 per ton=

I think you should go back from the river & look for <u>black muck</u> fresh water muck. There are plenty of <u>ponds</u> back of Myers which appear to have black ground   you can hire a cart or use the lighter & yacht and get it up the river wheeling it to light⟨er⟩ by wheelbarrow & line of <u>boards</u>

It evidently wants some fine decayed fibrous spongy matter like they are putting in the Coconut holes to hold the <u>manure</u> & prevent it going clear through to <u>China</u> — You can doubtlessly discover a good spot to get it which is accessable dont stint it but procure plenty of it— I want to <u>Carry</u> Everything to Extreme[d] <u>Excess</u> down here   You will probably have to pay the owner of the ground for it.—

I think you better get some better shoots of Bamboo from Major Evans and plant them keeping them well <u>manured</u> and watered.[b]

If Cherry trees will grow put out 2 or 3. oxheart,[a] Red etc
2—grapefruit trees[10]

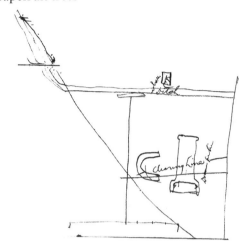

[A][11]

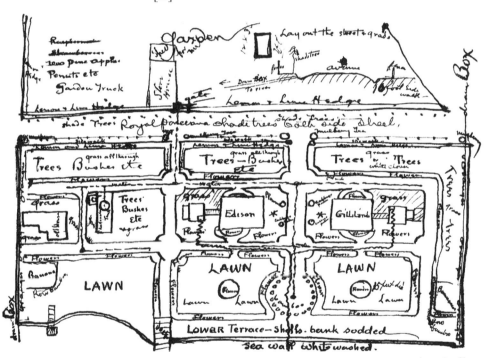

Plant in garden about 20 ft square of tobacco= ⟨garden⟩[b]
4 Trees of Japaneese plums in House plot—
½ dozen gooseberry bushes. ⟨House plot.⟩

2 apricote trees   House plot.
2 persimmons   "
2 Tamarind   "

6 ⟨2⟩ mulberry trees ~~outside in street as shewn on map~~[f]
Any tropical fruits you hear of get shoots, if you can.
¥[–][c]
Among the Oranges get at least 2 Mandarin Orange trees.
2 Green gage plum in House plot
In garden set out a double[a] row sugar can 50 foot long—[b]
Plant in garden dozen hills of poor mans dish rag—its a species of squash   plenty at St Augustine.
Want two[a] hives of ~~beess~~ bees ⟨over in garden⟩[l] for our flowers. you can purchase these about 2½ miles from Myers on the river road to Parkinsons[12] or up the river above Parkinsons
    Edison House.

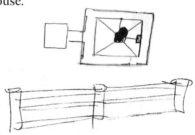

15 ton for Lighter Load—
25 lbs per square foot.
4300 tons.
280 Lighter Loads[13]

[B][14]

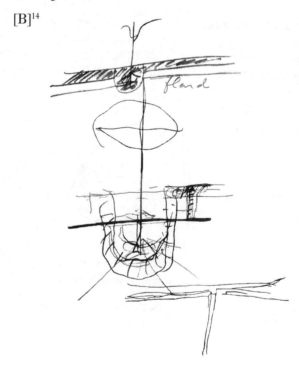

AD, NjWOE, Lab., N-86-04-05:5 (*TAED* N318005). Miscellaneous calculations not reproduced. ᵃObscured overwritten text. ᵇFollowed by dividing mark. ᶜCanceled. ᵈInterlined above. ᵉ"in garden across road" added after dividing mark. ᶠStrikeout in pen. ᵍObscured by ink smear. ʰ"Wild coral plant" written in right margin. ⁱ"Lavender Bushes" written in right margin with line to indicate insertion. ʲ"as many" interlined above. ᵏ"as will . . . will answer" interlined above. ˡ"over in garden" interlined above.

1. Edison wrote these instructions over an unspecified period of time. His initial draft was in pencil; he later made corrections and additions in pen. Those emendations are transcribed as marginalia to differentiate them from the original text. Edison added a number of subject headings such as "garden," "landscape," "house landscape," and "House plot," presumably to direct Thompson where various plants and trees should be planted.

2. Eli Thompson (1831–1895), a native of Wayne Township, Indiana, was a carpenter and former chief of police (1871–1874) and city marshal (1875–1877) in Indianapolis. Edison likely knew him from his own days as a telegraph operator there. With his two sons, Thompson oversaw the assembly and completion of Edison's house and laboratory at Fort Myers and remained as a caretaker of the property at a reported salary of $70 per month. Thompson contracted malaria, however, and left Edison's employ as early as September 1886. He returned to Indianapolis, where he later served as a policeman in Union Station ("Death of Eli Thompson," *Indianapolis News*, 7 Dec. 1895, 7; TAE to Thompson, 5 Oct. 1886, Lbk. 22:467 [*TAED* LB022467]). Records of expenses (mostly payrolls) paid on behalf of the property from December 1885 to April and from April to September 1886 (during Thompson's tenure) are in two Fort Myers Cashbooks, the earlier one not yet accessioned in June 2014 (Fort Myers Accounts, Accts., NjWOE; *TAED* AB023).

3. Edison seems to have added the salutation and first paragraph, both in a different style of writing from the text below, in the available space above the body of the first page.

4. The editors have not identified this island.

5. Likely Alex Montgomery (b. 1829?) of Fort Myers, a neighbor of Maj. James Evans (see below). Montgomery's occupation was listed as a "boatman" in the 1885 Florida Census. Florida State Census 1971 [1885], roll M845_9, p. C (Monroe County).

6. Major James S. Evans (1824–1901) is considered the founder of the town of Fort Myers. A native of Suffolk, Virginia, Evans came to Florida as early as 1842 and soon after joined a government surveying crew. He seems to have spent subsequent winters in Florida while returning to Virginia in the summer. In 1859, he was part of a team assigned to survey the Caloosahatchee region. When this crew disbanded in December 1859 or early 1860, Evans stayed in the area to establish a claim and to develop a farm for coconuts, coffee, and other tropical plants. With the backing of an association of investors in Suffolk, he brought twelve slaves to Fort Myers who planted several hundred coconuts and other plants. With the outbreak of the Civil War, Evans returned to his native state and served as an officer of Virginia troops, attaining the rank of major. He reappeared in Fort Myers in 1873 and made good his legal claim to the land. Evans had the town plat drawn up in a manner that allowed

squatters to remain on their claims, which he sold to them for trivial sums. From 1886 until his death, Evans looked after Edison's interests in Fort Myers. U.S. Census Bureau 1982? (1900), roll 172, p. 5A (Fort Myers enumeration district 0077, Lee, Fla.); "Died in Florida," *Suffolk (Va.) Herald,* 18 Jan. 1901, 2; Grismer 1949, 93–96, 275; Florida Land Association, Letters of Association, 1 Oct. 1859, Ser. 1, Box 2, Folder 5, Kilby; U.S. Census (1860) database of Slave Schedules, Monroe County, Fla., accessed through Ancestry.com, 5 Mar. 2013.

7. Figure labels are "garden" and "Road."

8. Figure label is "8 feet."

9. Oil cake is a fertilizer made from the fibrous material left over after seeds have been pressed for oil. Cottonseed oil cake was commonly used in the South. *OED,* s.v. "oilcake"; *Chambers's Ency.* 1887, s.v. "oil-cake."

10. Figure labels are "Clearing Line," "wagon," "shed," "Ditch," and "Drain pipe."

11. Edison's map of the property is reproduced close to actual size. In the central row from right to left are the Gilliland house, the Edison house, the laboratory building, and, at the far left, the "old House," known as the Summerlin Cottage, which was to become the caretaker's residence. Two undated scale drawings of the structures and grounds (one showing the general plan for landscaping) are in Seminole Lodge & Fort Myers, DF (*TAED* D8745AAH).

12. Edison is probably referring to the site of the ferry that A. T. G. Parkinson operated on the Caloosahatchee about twelve miles upriver from Fort Myers, near what is now Olga. Parkinson, an Englishman, owned and managed citrus groves on the Caloosahatchie near the ferry landing. Beehives could be purchased from Reinhold Kinzie, who had recently introduced them on his farm a few miles upriver from Parkinson's landing. Dodson 1972, 406; Norton 1892, 266; Grismer 1949, 109.

13. This list preceded by related calculations on facing page.

14. Figure label is "florid[a?]."

4

May–September 1886

Edison and his bride Mina returned from their Florida honeymoon about 4 May after a stopover with her family in Akron, Ohio. When they reached New Jersey, they took up residence in the large furnished home he had purchased in suburban Llewellyn Park. Like the honeymoon itself, their new home represented the start of their life together and the close of an unsettled period of twenty-one months. Even his thoughts seemed to return, at least symbolically, to firm ground as the extravagant speculations about gravitation and the structure of the cosmos faded from his notebooks, replaced by down-to-earth work on telephones, electric lamps, and other familiar lines of experiment.

The restoration of some accustomed order to Edison's domestic and intellectual life did not, however, translate to tranquility. Two unrelated events disrupted his work within six weeks of returning to New Jersey. One was a strike at the Edison Machine Works, the contention being over hours, overtime pay, and shop rules. Charles Batchelor, negotiating for the company, acceded to the workers' demands in the first two areas but would not yield any of his managerial prerogatives. The men walked out, and Edison and his partners in the business promptly reassessed their standing intention to expand the shop in the New York area. They decided instead to relocate it far from the city in order, as Edison would put it, "to get away from the embarrassment of the strikes and communists to a place where our men are settled in their own homes."[1] The men came back to work on Batchelor's terms, but in the meantime a suitable site had been found in Schenectady, some one hundred seventy miles up the Hudson River valley from

New York City. Edison arranged a personal loan of $45,000 to enable the Machine Works to buy the property in June.[2] The move took place during the fall and would require a significant reshuffling of some of Edison's principal lieutenants and their duties.

The other significant change arose from reports reaching Edison about the premature failure of his incandescent lamps in some cities. He responded that "a gradual deterioration [in manufacturing quality] may take place without anyone knowing exactly the 'why or wherefore,'" as had happened on other occasions. Referring to the Edison Lamp Company's factory, he promised to personally "look through the various processes at East Newark & if I can make any improvement I shall most certainly do it."[3] That very day, 11 June, he went to the factory and launched what became an extensive experimental campaign to improve lamp life and the manufacturing process. Starting, as he often did on such projects, with a search of the scientific literature, he quickly narrowed his attention to moisture and residual gas in the bulbs. His efforts to remove them led immediately to three patent applications and a busy summer of related research. Mina helped as a copyist (and witness) for his notes, perhaps in the evenings at home.[4] Edison spent so much time at the East Newark factory that by the end of June he had established his personal laboratory there. Giving up his rented laboratory space in New York further detached his affairs from the metropolis, though he still maintained an office there, as did the Edison Electric Light Company and related businesses.[5]

Other changes were also in the offing, though these were more administrative and financial matters in which Edison took less personal interest than the output of the Machine Works or Lamp Company. The Edison United Manufacturing Company was incorporated in early July as an organization to coordinate and streamline the sale and installation of goods made by the Edison manufacturing shops: the Machine Works, the Lamp Works, and Bergmann & Company.[6] Francis Upton, traveling in Europe for his health, made preliminary arrangements with financier Henry Villard, a longtime Edison backer, to try to harmonize the Edison lighting companies in France and Germany and also reconcile the competing Edison and Siemens interests in Germany. Villard envisioned a comprehensive European arrangement that would provide cash for building Edison central stations in the United States, such as the "uptown" or "second" New York district, long-delayed by the lack of such capital.[7] Edison made his own approach to Siemens & Halske, asking Germany's preeminent electrical manufacturer to become his commercial agent for the phonoplex telegraph system in Europe, but the firm declined, citing its rivalry with Edison in electric lighting.[8]

The phonoplex in the United States, however, was a notable bright spot in Edison's business affairs. Convinced that he could soon "bring the invention to perfection," he experimented during the late spring in conjunction with practical tests of the system made by his assistants in and around Chicago.[9] By early July, he was sufficiently satisfied to put Samuel Insull in charge of developing a commercial business based on renting the phonoplex equipment. Insull's efforts found a willing (and profitable) market, primarily among railroad companies.[10] During the summer, Edison also arranged for the publication of a well-illustrated article about his system in the British journal *Engineering*.

With the Machine Works strike behind him and immersed in the challenge of improving lamp manufacture, Edison seems to have returned to a familiar equilibrium during the summer. He was ensconced in experimental work at his East Newark laboratory, mostly on electric lighting but also some related to telephones. His children (at least his two sons) appear to have been away for much of the season. Young William Leslie was at the farm of his uncle, William Pitt Edison, and his wife in Michigan; Thomas, Jr., visited the Miller family in Akron, Ohio. Both boys then went with the Millers to Chautauqua, where they were reunited with their sister Marion in the latter part of July. Mina, too, got away, perhaps to Chautauqua and definitely to Akron in late August, while Samuel Insull took up residence with Edison at Llewellyn Park in the latter part of that month. It is more difficult to discern the motives and attitudes of Edison's young wife. Shortly after the honeymoon, Mina seems to have expressed to her family some anxiety, perhaps about the responsibility of running a household. Older sister Jane reassured her that from "What I have seen of Mr. Edison seems to me he is very patient, loving, & kind—and I think he is very fond of you." She also recommended hiring good servants to manage the house.[11] Edison's father, meanwhile, addressed a request for money to Samuel Insull, concluding with the wish: "I would like to See Mrs T.A.E very much."[12]

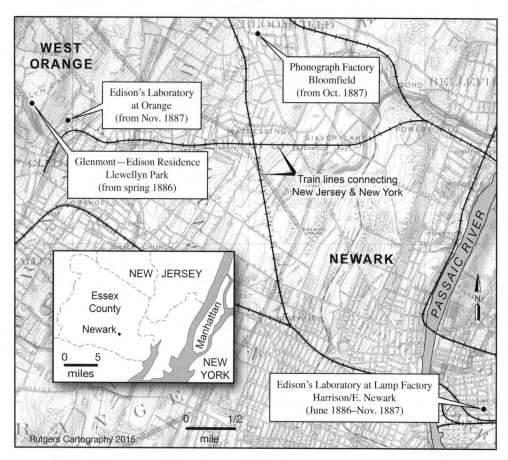

Points of significance to
Edison's life and work
in Newark region,
1886–1887.

1. "Edison Machine Shops Going," *New York Tribune*, 24 June 1886, 1.

2. See Doc. 2956.

3. See Doc. 2960.

4. See Doc. 2962 (headnote). Mina filled a book copying Edison's lamp notes in August and September. N-86-08-03, Lab. (*TAED* N324AAA).

5. Though Edison paid rent to Bergmann & Co. through mid-July, he probably was working exclusively in East Newark well before then. Voucher (Laboratory) no. 355 (1886) to Bergmann & Co.

6. See Doc. 2958.

7. See Docs. 2959, 2965, 2973, and 2975.

8. See Doc. 2955.

9. See Doc. 2948.

10. See Doc. 2971.

11. Jane Miller to Mina Edison, 12 May 1886, FRS (*TAED* FM001AAN).

12. Samuel Edison to Insull, 22 June 1886, DF (*TAED* D8614K).

Small hand dynamo driven by Electromotor[2]

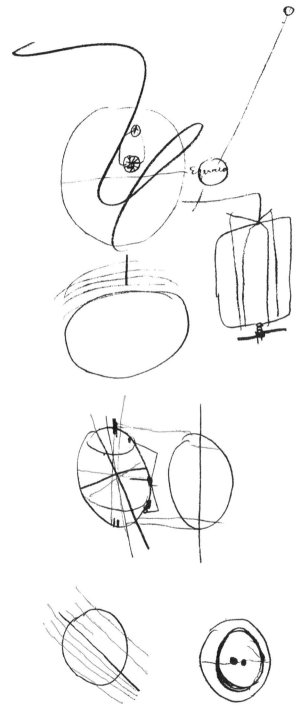

Simple dynamo to determine Earths Magnetism[3]

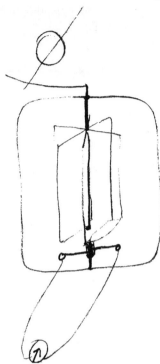

Use pendulum & E[a] Motor
[A][4]

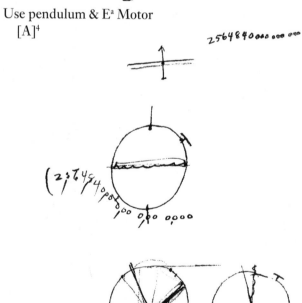

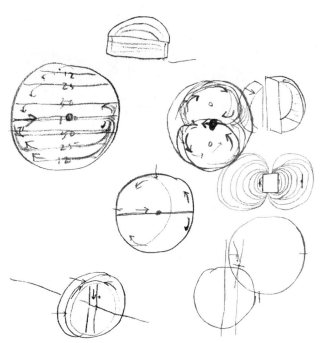

Sept 22—Sun on Equator   June 21 long   Dec 21 short.[5]

TAE

X, NjWOE, Lab., N-86-04-28:43 (*TAED* N321043). Document multiply dated. ᵃInterlined above.

1. See Doc. 2937 (headnote). On a page preceding this entry, Edison made a rough graph of the mean diurnal variations in the intensity of the magnetic dip at Singapore and St. Helena, using data from tables in Volume Three of Michael Faraday's *Experimental Researches* (Faraday 1965 [1855], pp. 318, 320). The magnetic dip (or inclination), indicated by the position of the needle in a compass held vertically, reflects the orientation of lines of force relative to the Earth's surface, and it is subject to both global and local influences, such as air temperature. In another notebook, probably also in the period immediately after his return from Fort Myers, Edison created a set of monthly tables related to atmospheric magnetism that likely was influenced by his reading of Faraday's long article on atmospheric magnetism in Volume Three. In that article, Faraday analyzed the Earth's magnetic dip, especially in relation to seasonal changes, and speculated about its effects on the lines of force emanating into space (N-86-04-28:41, N-85-10-01:184–87, 190–203; both Lab. [*TAED* N321 image 20; N310 images 79–80, 82–88]; Faraday 1965 [1855], §32 esp. paras. 2872–91).

2. Figure label is "Equator."

3. Edison made a related drawing about 19 April of an experimental arrangement to "Cut the lines of force of the Earth & break it up into waves of [revolu?] as waves are too gradual   Then [take?] make same Vol sound from Current & Read amperes." N-86-04-3.1:183, Lab. (*TAED* N315173).

Experimental device to "Cut the lines of force of the Earth" and make them audible with a vibrating "Reed" (near bottom) and telephone receiver.

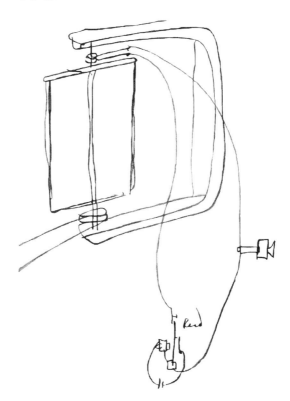

4. Adjacent to this drawing, Edison performed the long division of 33,000 into 8.464×10^{21}, which he wrote out in full across the facing page. When he recopied the quotient (2.56484×10^{17}) at the top of the page, he inadvertently left off three zeroes. One horsepower is equivalent to 33,000 foot-pounds per minute.

5. Following this entry is one probably made by Arthur Payne containing calculations based on the formula for calculating the centrifugal force on an object, expressed here this way: "Equal to its weight— multiplied by the number of feet in the radius of the circle and this produced by the square of the number of revolutions per second. and this multiplied by 1.2275." This formula was published in a number of reference works including Loomis 1870 (p. 24), an 1885 edition of which may have been in Edison's library by this time (Cat. 208503, NjWOE).

–2948–

To Charles Rudd

[New York,] May 13th. [188]6

Dear Sir:—

I have just been trying some experiments with the "Phonoplex."[1] I find that with one primary coil and two cells of battery, I get a certain result through an artificial line with condensors, and by putting two coils of the same resistance and size in series and double the battery, so that they[a] are the same amperes, I get precisely the same quickness of discharge, but double the electro-motive-force, hence I think that putting a number of primaries in series, say two or three, and connecting the secondary in series, you will derive great benefit. I know you certainly would if you use primary coils only, if you were to try using excessively small condensors at the receiving end of the wires   On the "Phonoplex," when I use the musical sound diminishing the size of the condensors did not seem to effect the musical sound, while it constantly diminished the inductive sounds.[2]

I do not see why you cannot use the cable, for you can cut it up as much as you please and can easily make a splice by using tape.[3] I am trying a great number of experiments to help you out, and I certainly believe that between the two of us we will bring the invention to perfection. Yours very truly,

T.A.E.

TL (carbon copy), NjWOE, Lbk. 22:81 (*TAED* LB022081). Initialed for Edison, probably by Alfred Tate. [a]Obscured by smudge.

1. This letter was addressed to Rudd at Wadsworth, Ill., where he was testing the railway telegraph. When Edison returned his attention to the phonoplex in early May, he had Rudd's problems with the

railway telegraph fresh in his mind (Unbound Notes and Drawings [1886]; N-86-04-28:15, 43; all Lab. [*TAED* NS86ADC5, NS86ADC6, NS86ADC7; N321015, N321043]; see Doc. 2945). With both systems using induction signals and telephone receivers, there was a good chance that a means of overcoming inductive interference in one would work in the other. The approach described here was apparently intended to increase the signal's strength without muddying the characteristic vibration that made it recognizable to the receiver. Throughout May, Edison continued to design circuit arrangements for the phonoplex; the intent of many of his drawings is unclear but see also note 2 below (Unbound Notes and Drawings [1886], Lab. [*TAED* NS86ADC8, NS86ADC9]). This work included John Ott's 20–21 May designs of instruments for "determining the curvature of discharge" (what Edison elsewhere referred to as the signal's sharpness) by using a pendulum as a time-keeping device (Unbound Notes and Drawings [1886], Lab. [*TAED* NS86ADD, NS86ADE]; cf. Doc. 2947). During May and June, Edison received a number of reports on the railway telegraph trials from Rudd and S. K. Dingle (see their correspondence in Telegraph, DF [*TAED* D8639]).

2. By the term "musical sound," Edison likely referred to the sound produced by the oscillating diaphragm described in Doc. 2902 n. 2 (also cf. Doc. 2938 esp. n. 2). That device, which he was incorporating into the phonoplex system just at this time, produced "rapidly-occurring induction waves or vibrations which form a musical note." Edison used it to create a third channel or type of signal that allowed the phonoplex to operate as a triplex system. He executed a patent application for this idea on 11 May. His drawings of the phonoplex from this period are unclear, but he sketched what may be a similar arrangement that was marked "Tried" on 7 May (Unbound Notes and Drawings [1886], Lab. [*TAED* NS86ADC6]). Other drawings that may be related include those from 21 April and 18 May (Unbound Notes and Drawings [1886]) and notebooks N-86-04-28:15 and N-86-04-07:233–37; all Lab. (*TAED* NS86ADC5, NS86ADC9, N321015, N319233).

3. This may have been an insulated wire cable like the "flexible cord" for railway telegraphy described in Doc. 2888.

<table>
<tr><td>–2949–

*Notebook Entry:
Miscellaneous*</td><td>[New York,] May 15 1886</td></tr>
</table>

Telephone[1]

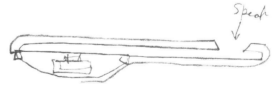

Is iron merely a non polorizable material whereby contact is made with the air to produce stress, [if?][a]   Try this[2]

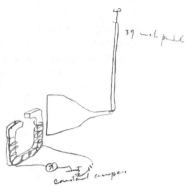

Hollow core pour various liquids in see if change lines force—
Note[b] Try if alloys of iron with other metals are magnetic   investigate their properties.

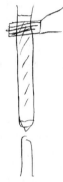

put layered sheet iron round carbon & magnitize it   keep
arc central

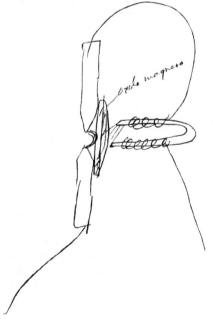

Cause magnet attract arc to plate Lime or Magnesia Zircon etc.[3]

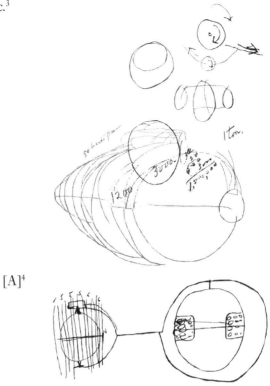

[A][4]

Notes— Comet dilatate as they recede & contract as they approach the sun p 102 Vol 6 Astron Society—Herschel[5]

Tail of Comets always point away from Sun—

in sun spot period Corona extends great distance from Sun Spectroscope shows its gas— non sun slight Corona & shown in Spectroscope that light comes from solid— Color white while Extended Corona pink—are

My theory is that Electric arc occur on surface Equatorial part sun reduce EMf on parts & give great dif EMf where arcs are, & lines of force from sun attracts arc out into space, as magnet attracts arc of arc light. one picture shows several Corona streamers [---][a] Knife shaped. These are matter[c] from arc's attracted out by lines of force—

TAE

X, NjWOE, Lab., N-86-04-28:89 (*TAED* N321088). Document multiply signed and dated. [a]Canceled. [b]"Note" written outside parenthesis enclosing following sentence. [c]Obscured overwritten text.

1. Figure label is "speak."
2. Figure labels are "39 inch [pendulm?]" and "constant ampers."
3. Figure label is "oxide magnesa—."

4. Edison's intent for this drawing and its associated text and calculations is unclear. He may have been trying to represent lines of force or other forces that the experimental arrangement shown above was designed to observe.

5. Edison paraphrased John Herschel's paper in the *Memoirs of the Royal Astronomical Society* describing how comets appear to show "dilatation of volume as they recede from, and concentration within a smaller compass as they approach, the sun." Herschel 1833, 102.

–2950–

Notebook Entry: Electromagnetism and Gravitation[1]

[New York,] May 17, 1886

Apply magnet also Coil also current to increase sensitiveness to Astronomical observation[a]

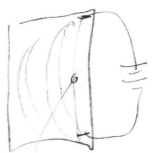

Hydrometer—See if specific grav increasing[2]

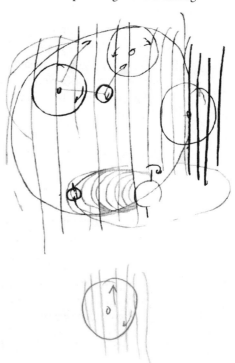

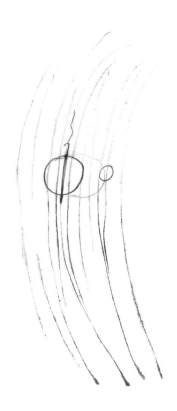

TAE

X, NJWOE, Lab., N-86-04-28:100 (*TAED* N321100). Miscellaneous calculations omitted. [a]Followed by dividing mark.

1. See Doc. 2937 (headnote).
2. On scores of the following pages until page 167 (Doc. 2957), Edison made numerous sketches and calculations, many of which appear related to planetary magnetism.

–2951–

From Eugenie Stilwell

Somerville N.J.[1]   May 18/86

Dear Al—

My teacher Miss Parsons[2] requested me to write and ask you if you had received the last two bills for my tuition; and if so she has not received any checks yet and is very much in need of the money.[3]

I hope you not will not think this is intruding as I have been waiting until I thought you were fully rested after your trip. Of course you enjoyed it and I[a] hear your are looking extremely well.

With love and best wishes I remain   Yours Sincerely
Jennie L. Stilwell[4]

ALS, NjWOE, DF (*TAED* D8614F). [a]Interlined above.

1. Located on the north bank of the Raritan River (roughly thirty-five miles southwest of Orange), Somerville became the county seat of Somerset County in 1783. *Ency. NJ*, s.v. "Somerville."

2. Emma L. Parsons (1842–1914) was co-principal (with Laura H. LeFevre) of the Somerville Seminary, founded in recent years (possibly to fill a void created by the closure of the Somerset Young Ladies Institute in 1878). Parsons studied mathematics at Ingham University, in her home town of Le Roy, N.Y., and after graduating in 1861, she may also have earned a master's degree there. She embarked on a variegated career that included teaching mathematics at Ingham, and by 1883 she was president of the Ossining Institute (Ossining, N.Y.), of which LeFevre was secretary. Parsons reportedly joined the English Department at Fisk University in Nashville in 1892 and subsequently taught physical geography and bookkeeping at Fisk. She remained there for twelve years and died in Pasadena, Calif. U.S. Census Bureau 1967? (1860), roll M653_756, p. 661, image 46 (Le Roy, Genessee, N.Y.); ibid. 1882 (1900), roll 1563, p. 15A (Nashville Ward 3, Davidson, Tenn.); "Obituary," *The American Missionary*, n.s., 6 [Apr. 1914]: 36; "Death of Miss Parsons," *Batavian* (Batavia, N.Y.), 1 June 1894, n.p.; Snell 1881, 668; Woman's Board of Missions 1883 [*Fifteenth Annual*], 101; "College Directory," *Old and New* 6 (1872): 5.

3. In June, Edison's staff paid two bills for board and tuition to the Somerville Seminary, one dated 9 March and the other 18 May (which brought forward the unpaid March balance). Tuition was $87.50 for a half semester, plus charges for music and painting lessons and other fees such as rental of a church pew and laundry services. These bills and at least one other from the school in 1886 are in Vouchers (Household), nos. 125, 296, 509 (box 33; 1886).

Edison had paid a similar bill at the end of 1885 only after Emma Parsons and Margaret Crane Stilwell, the mother of Eugenie and the late Mary Stilwell (Edison's first wife) both urged him to remit. Mrs. Stilwell pointed out that the school had "very few pupils" and limited resources. Edison had instructed Samuel Insull to "Write to Miss Parsons   say Edison lost the bill   to forward another & you will pay it= please attend to this promptly." Below his signature he added, "Write Mrs Holzer you will attend to it," referring to Alice Stilwell Holzer, another Stilwell daughter and the wife of William Holzer, one of his manufacturing superintendents. Despite Edison's avowed good intentions, however, subsequent bills from the school were paid only after reminders from Parsons or her student. In November 1886, young Jennie told Insull that "I honestly think it would be better for me to discontinue going there to school. I do not feel as though I ought to stay there and impose on Mr. Edison." Edison reportedly prevailed on her to remain (paying $131.19 in December 1886) and she did stay despite her ongoing embarrassment, at least through the spring term of 1887, when she and Parsons again pleaded for payment. Parsons to Margaret Stilwell [with TAE marginalia], 14 Dec. 1885; Margaret Stilwell to Insull, 18 Dec. 1885 and 18 Jan. 1887; Eugenie Stilwell to Insull, 26 Nov. 1886, both 7 Feb., and 16 June 1887; Margaret Stilwell to TAE, 29 July 1887; Parsons to TAE, 15 Jan., 9 Mar., and 18 June 1887; all DF (*TAED* D8514ZAB, D8514ZAD, D8714AAB, D8614P, D8714AAD,

D8714AAE, D8714AAN, D8714AAT, D8714AAA, D8714AAF, D8714AAP); Vouchers (Laboratory) nos. 660 (1885), 601 (1886).

4. Eugenie L. "Jennie" Stilwell (1868–1942) was a younger sister of Edison's late first wife. Edison had paid her educational expenses since at least 1882, when she attended (with his two oldest children, Marion and Thomas, Jr.) an English and French school in New York City. In 1884, Edison paid Jennie's fees at the Bordentown Female College, in Bordentown, N.J. See Docs. 2344, 2710 n. 1, and 2723.

–2952–

Charles Batchelor
Journal Entry

[New York,] May 19th 1886.

71[1]   Labor troubles. Men have all struck on 17th & after consultation for two days we have decided to stand up against them in regard to running the shop as we see fit— We have conceeded the 9 hours work and 10 hrs pay— The contract clause we cannot allow.[2]

AD, NjWOE, Batchelor, Cat. 1336:33 (*TAED* MBJ003033C).

1. Charles Batchelor consecutively numbered each entry in this journal.

2. On 1 May, a workers' committee presented to Batchelor, general manager of the Edison Machine Works, a set of demands that included ten hours' pay for nine hours' work, time-and-a-half overtime pay until 8 p.m., and double pay for overtime thereafter. Management agreed to these terms but rejected three additional demands: an end to piecework and subcontracting; that one man should supervise no more than one machine; and that all workers belong to a union. The workers waived the demand for a union shop but struck over the remaining two issues. In all, about 350 men from the Edison Machine Works and the Electric Tube Co. (a division of the Machine Works), including laborers, pattern-makers, and machinists, left their places. Cat. 1336:23, 43 (items 49 and 79, 1 and 26 May 1886), Batchelor (*TAED* MBJ003023, MBJ003043); "Edison's Men on Strike," *New York World,* 20 May 1886, in Cat. 1336:35 (item 72, 20 May 1886), Batchelor (*TAED* MBJ003035A), reprinted as "Strike at the Edison Machine Works," *Elec. and Elec. Eng.* 5 (June 1886): 238.

Batchelor claimed to have arranged with customers and suppliers so that the Machine Works could shut down "very easily" for three months. The company was unable, however, to complete three torpedo boats for the U.S. government on time, resulting in forfeiture of the contracts and what a trade journal called "a heavy loss." "Strike at the Edison Machine Works," *Elec. and Elec. Eng.* 5 (June 1886): 238; "The Edison Machine Works to Locate at Schenectady," ibid. 5 (July 1886): 279; "Edison Machine Shops Going," *New York Tribune,* 24 June 1886, 1.

On 26 May, Batchelor indicated that about forty-four men reported to work. Only four of these were returning strikers, joined by twelve replacements, according to a news report. The remainder may have been employees of other Edison shops, who, according to the *New York Tribune,* helped to break the strike. Batchelor and Henry Livor met with the strike committee on 31 May and acceded only to the pay demands. "In every other respect," Batchelor noted, "the shop [is] to be run just as the managers decide & no interference whatever [to] be tolerated." The men returned to work the next day. Cat. 1336:43, 47 (items 79 and 85, 26 and 31 May 1886), Batchelor (*TAED* MBJ003043, MBJ003045C); "Strikes and Labor Notes," *NYT,* 27 May 1886, 5; "Edison Machine Shops Going," *New York Tribune,* 24 June 1886, 1.

The strike evidently precipitated a permanent rupture in the operations of the Machine Works. Edison had been planning to move the shop to more spacious quarters, and the company had purchased land at 10th and Berry Sts. in Williamsburg, Brooklyn, in March for this purpose. The labor difficulties, however, seem to have convinced Edison and Batchelor to look for an entirely different location. Batchelor was quoted in *Electrician and Electrical Engineer* warning that when the shops reopened it might "be in some other place away from this city." Soon after, he and Livor inspected two recently built but unfinished shop buildings in Schenectady, N.Y. "I'm satisfied," Batchelor told the *New York Times* later that month, "that we can do our work cheaper at Schenectady than here, and that we will have no trouble there from strikes." Edison, commenting that fall on the impending move to Schenectady, noted that it would be "done to get [a]way from the embarrassment of the strikes and communists to a place where our men are settled in their own homes." According to the *Tribune,* only "the most skilful and most loyal" workers would be transferred; the rest (about 300) would be dismissed. Edison had remained in the background during the strike and seems to have made no public comment until it ended. He was, however, president of the Machine Works and its controlling stockholder, and the announcement of the shop's move to Schenectady was attributed to him. "The Edison Machine Works to Locate at Schenectady," *Elec. and Elec. Eng.* 5 (July 1886): 279; "Driven Away By Strikes," *NYT,* 24 June 1886, 8; Cat. 1336:13 (item 27, 2 Mar. 1886), Batchelor (*TAED* MBJ003013A); "Edison's Men on Strike," *New York World,* 20 May 1886, in Cat. 1336:35, 43 (items 72 and 79, 20 and 26 May 1886), Batchelor (*TAED* MBJ003035A, MBJ003043); "Editor's Note-Book," *Chautauquan,* 7 (Oct. 1886): 54; "Edison Machine Shops Going," *New York Tribune,* 24 June 1886, 1; cf. Jehl 1937–41, 1009; also Insull 1992, 46–49.

-2953-

To Samuel Flood Page[1]

[New York,] 16th [20] May [188]6[2]

Dear Sir,

Your favor of the 31st March came to hand whilst I was still absent at my Florida Laboratory, & in as much as it proposed a settlement differing considerably from that I had talked of with Mr Insull, it was impossible for him to reply without personal consultation with me.[3] I found myself unable to go into the matter until a few days back. I then cabled you as follows:—

"Proposal concerning patent account unsatisfactory If you desire prevent patents lapsing you should instruct Hanford pay fees your account pending your making acceptable agreement with me."[4]

To which you replied

"Will pay fees on your account subject to arrangements being made"[5]

I have replied to this as follows:

"Will not incur further expense patent account. You should pay fees your account pending agreement between us. letter mailed"[6]

My reason for sending this last cable is that I have already paid out such a large amount in connection with these patents that I do not feel disposed to further increase the amount. The risk you take is but small in paying the fees as there can be no doubt of your being able to acquire the patents providing you are willing to reimburse me the amount I have spent in obtaining same. If the patents are worth anything they are certainly worth far more than the mere cost of obtaining them.[7] I took them out in perfect good faith understanding at the time that there would be no question of my being unable to obtain from your company (as the Successors of the Edison Co) their bare cost. You argue from a wrong basis altogether in your letter of 31st March. You simply take the patents which you propose to maintain reckon the expenses on them & forget altogether that I have spent a large sum on those which you propose should be allowed to lapse. If I were to render you an account in strict conform with the clause of the agreement which covers the question of future patents the cost to you even of the patents you desire to acquire would be far greater than the total cost of obtaining the whole of the patents.[8] Furthermore if you recognize the whole of the account you will acquire the whole of the patents. If some of them still remain in my hands it is but natural that I should endeavour to dispose of them to other parties. This course would be unfortunate and far from my wishes as it might prove greatly to your disadvantage to have patents on Electric Lighting bearing my name held by any English company other than your own.

It is hardly equitable to take a proposition made by me nearly two years ago (I refer to my letter of August 22nd 1884)[9] as applying to the case at this late date. In 1884 I could have disposed of Central Station Dynamos at prices which would have about equalled your[a] contract price with the Edison Machine Works but it is impossible to do this today. Mr Insulls proposition of 31st March[10] was suggested by a desire to relieve you of all responsibility with relation to those Dynamos. The manager of the Edison machine works has repeatedly consulted with me on the subject and again & again expressed a desire to make delivery to you of these machines which would necessitate you paying between $1600 & $[1700?][b] the final payment on each machine. I have opposed this as I knew you had no immediate use for these machines & I did not want to involve you in any further <u>immediate</u> payments in connection with them.

In the event of your accepting the proposition made you by Mr Insull on my behalf (in his letter of 8th March) I should simply negotiate with The Edison Machine Works with a view to their cancelling your[a] obligation to them in connection with the order for Central Station Dynamos. I should not expect to make one cent of profit on the arrangement and should consider myself fortunate if I obtained from the Machine Works sufficient to reimburse me the amount I have now invested in the English patents in question. You must remember that the improvements made in Dynamo construction have caused a very great depreciation[a] in the value of Central Station Dynamos in fact so much so that even if it is possible to sell them[a] at all the price obtainable will be <u>very small</u> indeed as compared with the price easily obtainable here and abroad four or five years ago—or in other words at[c] the time at which you gave the order. To-day with one half the weight of material we get the same product from our Dynamos & with very much better efficiency. This is owing to a long series of experiments conducted by me at very great expense to myself

With relation to future patents I think you should undertake to pay the cost of preparing cases here for the English Patent Office then my American Patent Attorney can forward the cases to you & you can decide whether you will file them or not. The question of publication is a matter which can readily be arranged between my American Patent Attorney & your Patent Solicitor.[11] I would suggest that you place your portion of this business in the hands of Mr Hanford who has already a perfect understanding with my Patent Attorney here with a view to the prevention of the invalidating of a patent from publication both here & in England.

If I were to agree to your suggestion of forwarding all cases to you for decision as to whether you would accept them or not I should be at the expense of the preparation of the cases by my American Patent Attorney. Of course I desire to save myself all expense in this connection as whatever advantage there may be in the matter accrue to your Company not to myself.

Awaiting your speedy reply[12]   I remain   Yours truly

Thomas A. Edison

LS (letterpress copy), NjWOE, Lbk. 22:86 (*TAED* LB022086). Written by Samuel Insull; a copy in an unknown hand is in DF (*TAED* D8630ZAV1). [a]Obscured overwritten text. [b]Illegible; text from copy. [c]Interlined above.

1. This letter was addressed only to the secretary of the Edison & Swan United Electric Light Co., Ltd., in reply to one sent by Samuel Flood Page (see note 3). Flood Page (1833–1915), former manager of London's Crystal Palace, became the United Co.'s secretary and defacto manager in December 1883. Doc. 2572 n. 7.

2. Both the letterpress and freehand copies of this document are unambiguously dated 16 May. However, the letter quotes cable messages dated subsequently, on 18 and 20 May (see nn. 4–6). In another letter addressed to Flood Page on 20 May, Edison referred to this document as having been written that day. Lbk. 22:96 (*TAED* LB022096).

3. Flood Page wrote after the United Co.'s board met to consider Samuel Insull's 8 March proposition regarding ownership of Edison's electric light and power patents in Great Britain, an offer that essentially restated terms Insull had delivered in London more than a year before (Flood Page to TAE, 31 Mar. 1886, DF [*TAED* D8630Z]; Insull to Flood Page, 8 Mar. 1886, Lbk. 21:364A [*TAED* LB021364A]). The question of Edison's patents, wrapped up with his claim for more than $11,000 in experimental expenses, was the subject of dispute as long ago as 1883 but had taken on new urgency, as some of the patents were coming up for renewal and Edison threatened to allow them to lapse (TAE to Edison Electric Light Co., Ltd., 10 Dec. 1883, DF [*TAED* D8316BQE]; see Docs. 2572, 2719, and 2738). In his letter, Insull reiterated earlier proposals that the company reimburse Edison for those expenses directly related to obtaining the patents, in return for which Edison would transfer the patents to the firm and waive his contractual right to repayment of experimental expenses. Insull pointed out that Edison had "left off taking out patents in as much as he can get no satisfactory arrangement either as to those he has already taken out or as to future cases." As part of the deal, Edison also wished to acquire the "C" central station dynamos ordered—but not fully paid for—by the antecedent Edison Electric Light Co., Ltd., several of which remained at the Edison Machine Works (see note 9). According to Flood Page, the United Co. board, acting on the advice of its own technical experts, determined that it wished to acquire only fifteen of the thirty-four existing patents in question. Further, it proposed to count toward the patent expenses the £1,700 it had advanced on each of the six dynamos and pay the balance of the patent fees by consigning one of the machines to Edison.

4. This is the text of Edison's cable to the United Co., dated 18 May (DF [*TAED* D8630ZAQ]). On his own authority, Insull had already arranged to pay fees on the patents expiring in March but refused to do so

for those coming due in April (Insull to Flood Page, 8 Mar. 1886, Lbk. 21:364A [*TAED* LB021364A]).

5. The full reply from the United Co. on 18 May explicitly acknowledged receipt of Edison's cable that day. DF (*TAED* D8630ZAR).

6. Edison cabled this message to the United Co. on 20 May, referring to another letter of that date (see note 2). Lbk. 22:84 (*TAED* LB022084); DF (*TAED* D8630ZAT).

7. Despite recent reforms, British patents cost substantially more than American ones. Parliament lowered the fees in 1883 to £154 (from £178) for fourteen-year specifications (equivalent to approximately $770) and introduced the option of shorter lifetimes of four years (£4) or seven years (£54). A new act in 1884 allowed the payment of renewal fees in annual installments: £10 for years four to seven, £15 for years eight and nine, and £20 for years ten to thirteen. By comparison, a U.S. patent valid for seventeen years cost $35, of which $15 was paid upon filing the application and the remainder due when the patent issued. MacLeod and Nuvolari 2010, 25–26; Abbott 1886, 2:593–94.

8. Edison referred to the ninth article (not the sixth, as incorrectly stated in Doc. 2719 n. 1) of his 1882 contract with the principals of the Edison Electric Light Co., Ltd. The clause required the company to reimburse Edison for "all expenses incurred by him in experiments" that resulted in patents, as well as for the costs of taking out and maintaining the patents. TAE agreement with Drexel, Morgan and Co., Egisto Fabbri, Edward Bouverie, and Grosvenor Lowrey, 18 Feb. 1882, CR (*TAED* CF001AAE1); see also Doc. 3007 esp. n. 6.

9. Flood Page had quoted from Edison's offer (Doc. 2719) to buy from the United Co. as many of the central station dynamos whose value would equal the amount Edison had expended on experiments since his original contract with the British company. Of the six "C" dynamos ordered in May 1882 by the British Edison company, one had been consigned to the Nederlandsche Electriciteitmaatschappij; four others, paid for only in part, remained at the Machine Works despite sporadic efforts to resolve their fate. The editors have not been able to account for one other. See Docs. 2270 esp. n. 4, 2374 esp. nn. 4–5; 2572 esp. n. 4, and 2674.

10. Edison presumably meant Insull's 8 March letter to Flood Page (Lbk. 21:364A [*TAED* LB021364A]), to which he refers in the next paragraph.

11. In his 31 March letter (see note 3) Flood Page proposed "That as soon as you are ready to take out a patent in America, you should send to us forthwith as complete a specification as you can and that we should elect within one month and let you know by cable whether we care to take out a patent for ourselves in England. It is of course essential that nothing should be done in America which could in any way be held to be publication" in the meantime.

12. Flood Page replied on 11 June, shortly after the United board considered Edison's letter (see Doc. 3007). The company evidently made no decision at that time other than to ask Edison to calculate the amount due him, according to a strict interpretation of the ninth article of his 1882 contract, "for the patents now under consideration." Edison took no formal action on that request until November. Flood Page to TAE, 11 June, DF (*TAED* D8630ZBS).

To William Marks[1]

Dear Sir,

I have your favor of 18th inst[2]   Since I wrote you last I have started to move my Library from 65 Fifth Ave to my House at Orange. It may be two weeks before I get my Books sufficiently in order to find the works you desire. Will this very greatly inconvenience you   Yours truly

Thos. A. Edison   I[nsull].

L (letterpress copy), NjWOE, Lbk. 22:100 (*TAED* LB022100). Written by Samuel Insull.

1. William Dennis Marks (1849–1914) was the Whitney professor of dynamic engineering at the University of Pennsylvania since 1877. A native of St. Louis, Marks graduated in 1870 from Yale, where he earned a degree in civil engineering the following year. He served as general superintendent of the International Electrical Exhibition in Philadelphia in 1884, by which time he had two technical books to his credit. Marks became involved that same year as a consultant with the Edison Electric Light Co. of Philadelphia. He left his academic position in 1887 to act as the company's chief engineer and eventually its president (1892–1896). Doc. 1923 n. 5; Chamberlain 1901, 1:376; "William D. Marks," *NYT,* 8 Jan. 1914, 11; Franklin Institute 1885c, 10.

2. Marks's letter was part of a chain of correspondence that he initiated on 10 April. Having been asked to write articles on telegraphy and telephony for an appendix to the *Encyclopaedia Britannica,* Marks asked Edison to loan any papers or books that would help him highlight the contributions of American inventors. Edison wrote instructions on that letter to "Loan him ~~all~~ the testimony (2 large books) on telephone interferences   also tell him to look at Prescotts work on telephone old & new editions." These marginal notes became the basis of a 13 May typed reply referring generically to George Prescott's explications of the telephone, which were based on text written by Edison (Doc. 1232 n. 1) and published under at least three titles by this date (Prescott 1878, Prescott 1879, Prescott 1972 [1884]). The reply did not specify which volumes of the Telephone Interference cases, a multi-part record of intertwined Patent Office interference proceedings, Edison proposed to lend but he most likely meant the two volumes of testimony and exhibits for Edison (see Docs. 1270 n. 1 and 1792 n. 1; TI1–2). The 18 May letter from Marks thanked Edison and requested information about "your recent work in induction telegraphy." Later, Marks graciously stated that he was in no hurry and, as it turned out, he did not receive some of the material for at least six months. Samuel Insull, telling Edison in November that he had promised the interference volumes to Marks "some time ago on your behalf," asked him to look for them in his library. Marks to TAE, 10 Apr. (with TAE marginalia), 18 and 24 May 1886; all DF (*TAED* D8606B, D8606E, D8606F); TAE to Marks, 13 May 1886; Insull to TAE, 9 Nov. 1886; Lbk. 22:146, 23:90 (*TAED* LB022146, LB023090).

To Siemens and Halske[1]

Dear Sirs

I am desirous of opening negotiations with you with a view to your representing me throughout Europe (The United Kingdom of Great Britain and Ireland excepted)[2] in the exploitation of telegraphic inventions. The representation would involve the manufacture of the apparatus and the business of disposing the same. If agreeable to your goodselves I would propose making a Contract with you which would not only refer to the invention which I have more particularly in mind in writing to you but also to any telegraphic inventions which I may make for a certain period of years in the future. I would agree on such a basis of remuneration to myself in the shape of Royalties as would not tax the Inventions to any great extent. My desire is to obtain good representation of a permanent character and thus insure the working of my inventions to the very best possible advantage both from a technical and commercial point of view.

The reason which has prompted me to write you is my desire to immediately exploit in Europe my Way Duplex or Phonoplex System which is a device for doubling the capacity of a wire which is tapped by "Way" or "Intermediate" Offices. I enclose you cuttings from "The Electrical World" of this City which will give you some idea of the nature of the invention.[3] Up to this writing I have not tested the Invention on circuits of more than one hundred miles in length.

I also expect to produce other telegraphic inventions in due course.

Should you be willing to take the matter up I will send my Business Representative to Berlin to make the necessary Contract with you.[4]

Awaiting your reply I remain   Yours very truly

Thomas A. Edison— R[andolph]

LS (letterpress copy), NjWOE, Lbk. 22:107 (*TAED* LB022107). Written by John Randolph.

1. Founded in 1847, Siemens & Halske was the Berlin manufacturing firm for the constellation of Siemens family enterprises and a major manufacturer of telegraphic, telephonic, and lighting equipment. Doc. 1851 n. 1.

2. Edison's phonoplex was represented in the United Kingdom by Michael Moore (see Doc. 2865).

3. Edison referred to "The Edison 'Phonoplex' or 'Way-Duplex,'" an illustrated explanatory article from the 17 April 1886 issue of *Electrical World* (7:177). The same issue also contained a brief but laudatory

editorial notice of the phonoplex. Untitled clipping, *Electrical World* 7 (17 Apr. 1886), Cat. 1044 (No. 30): 53, Scraps. (*TAED* SM044087c).

4. Werner von Siemens responded (as translated in New York) that he could accept Edison's offer only if it were "a general and mutual representation" because "We cannot be Agents of Edison in single matters being entangled in law-suits with the Edison Companies in Berlin, Paris & Brussels." Siemens to TAE, 11 June 1886, DF (*TAED* D8630ZBQ).

Although he and Alfred Tate had discussions with other prospective agents, Edison seems to have had no representation on the continent for the phonoplex system as late as 1888. Sometime in 1887, he had a contract for Norway drafted on behalf of Knud Bryn, general manager of the Christiana Telephone Co., but it is not clear whether the agreement was consummated. The next year, Armin Tenner sought the agency for Russia and Germany. Tenner, though, advised that the phonoplex could not be adopted in Europe unless it included a recording device. Edison did some experiments with an ink recorder but reported "having considerable trouble with it" in April 1888. Through Alfred Tate, he asked Tenner whether the regular phonoplex could be used on the railroads in Russia and Germany for business conversations while reserving "the Morse side of the circuit for messages which must be recorded." Edison vowed to keep working on a phonoplex recorder, but the editors have not determined the outcome of his experiments. Tenner to TAE, 23 Feb., 6 Apr., and 3 May 1888; Tate to Tenner, 19 Apr. 1888; TAE agreement with Knud Bryn, n.d. [1887]; all DF (*TAED* D8853ABL, D8853ACE, D8853ACJ, D8818AIX, D8753AFS); "Personals," *Electrical World* 9 (11 June 1887): 286.

–2956–

Charles Batchelor
Journal Entry

[New York,] May 25th 1886.

78.[1] New Shops. The E.M.W.

At a meeting of the Edison Machine Works Board today it was decided to give to Geo Place[2] the right to negotiate for the Schenectady Locomotive Works[3] at a price not to exceed to us (including all commissions) $42,500—[4] Meeting was held at Laboratory and adjourned to Edison's house at Orange, N.J. Livor[5] to go up with him & bring more detailed information of the property.

AD, NjWOE, Batchelor, Cat. 1336:41 (*TAED* MBJ003041B).

1. Charles Batchelor consecutively numbered each entry in this journal.

2. Edison had done business since the late 1870s with George Place (1835–1914), proprietor of the George Place Machinery Co. at 121 Chambers St. in New York. Place was married to Iphigenia, the socialite sister of Henry Livor, and he was an early and prominent member of the exclusive Union League Club of New York. U.S. Census Bureau 1982? (1900), roll 1111, p. 9A (Manhattan, New York, N.Y.); "Obituary Notes," *NYT*, 18 May 1914, 9; Docs. 951 n. 1, 2681 n. 2.

3. Batchelor meant the McQueen Locomotive Works, a factory complex situated on nine acres in Schenectady, N.Y., near the New York Central and a branch line of the Delaware and Hudson Railroad. According to his notes, the principal buildings measured 350×80 feet and 325×125 feet. Only one was "floored" and both lacked roofs. There was also an engine house, 30×50 feet. Construction of the plant began after Walter McQueen, the chief locomotive designer and superintendent of the Schenectady Locomotive Works, split with his employer in 1885 and planned to form his own company. He initially had the financial support of state senator Charles Stanford, entrepreneur Nicholas I. Schermerhorn, and banker George G. Maxon, but when their backing fell through (see below), McQueen left the buildings unfinished and returned to the Schenectady Locomotive Works. Hammond 1941, 112; Cat. 1336:37 (item 74, 22 May 1886), Batchelor (*TAED* MBJ003037); Howell and Munsell 1886, 153–54, 198, 209.

4. Financing for the McQueen Locomotive Works collapsed with the deaths of Charles Stanford (August 1885) and Nicholas Schermerhorn (April 1886). John A. DeRemer, the company secretary, was appointed receiver, and the remaining stockholders agreed to sell the property at a price between $45,000 and $90,000. In the meantime, the strike began at the Machine Works, and Edison began seeking a new location for his consolidated shops. Henry Livor, Batchelor, and Samuel Insull seem to have inspected a potential site in Williamsport, Pa. It may have been Livor who discovered the availability of the McQueen works. George Place reported that the Schenectady property might be had for as little as $35,000, and on 22 May, Livor and Batchelor were dispatched to inspect it. According to contemporaneous press accounts, Edison offered $37,500 but the McQueen stockholders stuck to their minimum of $45,000. A committee of local professionals and businessmen, concerned lest the property remain vacant, raised $7,500 by subscription to cover the difference. This amount was apparently given to Edison or his associates before their purchase of the property, which now included about twelve acres. Place and DeRemer completed negotiations in New York City by 5 June. The next day, Edison arranged for a personal loan from Drexel, Morgan & Co. of $45,000, which he then loaned to the Machine Works. The transfer of title took place on 25 June at George Place's office. Crowley 1890, 158; "Obituary Notes," *NYT*, 25 Apr. 1886, 9; Schermerhorn [1928?], 4; "The Two Original Buildings" [typed compilation of newspaper excerpts], Box 1, Folder 11, Rice; Hammond 1941, 113; Cat. 1336:37, 45, 61, 71, 87 (items 74, 83, 94, 96, and 126; 22, 29 May; 5, 6, and 25 June 1886); all Batchelor (*TAED* MBJ003037, MBJ003045B, MBJ003061B, MBJ003071A, MBJ003087B); TAE promissory note to Drexel, Morgan & Co., 23 June 1886, Miller (*TAED* HM860285A); Insull to Alfred Tate, 9 Feb. 1887, DF (*TAED* D8736AAC).

5. Henry "Harry" M. Livor (1846–1904), a long-time associate of Edison, was at this time a sales agent for the Machine Works and a stockholder in the firm. Doc. 2681 n. 2; Edison Machine Works agreement with Livor, 1 Jan. 1886; TAE et al. agreement with Edison Electric Light Co., 18 May 1886; both DF (*TAED* D8627A, D8627U).

[New York, May 1886][2]

1st   The sun is a solid globe of molten matter revolving on its axis once in 25 & 8 days[3]

2   A powerful Electric current circulates around the whole of its body from west to East parallel with its Equator and to the poles.

3   This Electric current is sufficiently strong to keep all its matter except at the poles and center at[a] vivid incandescence

4— ~~breakage in~~ Lines of force ~~are~~ pass through its body in the same manner as a Cylinder[a] of wire wound so the wire would be in the same direction as the Currents Circulating in the incandescent matter. ~~These lines form a vortex ½ passing throug one ½ passing through the sun via north pole~~ half of these lines pass through the sun the other through space from pole to pole forming a vortex

5— Exccentricity of the lines of force due to the attraction of the sum total of the lines of force of all the planets & stars causes the sun in the its rotation to Cut its own lines of unequal strength & thus produce its current, & heat.[4]

6. planets are held in their orbit & propelled by the neutral attraction of closed circuit Currents

7. ~~The~~Except the shell all planets are molten & conduct Electricity [---][b]

8   The lines of force weaken as they recede from the sun at the square of the distance hence The ½ of the planet nearest the sun passes through lines of force stronger than the ½ [furthermost?][c] from the sun   This difference produces powerful Currents in the molten matter & causes rotation of the body on its own axis.

9— These currents causes lines of force to proceed out of the ~~earth~~ planet and produce a field ~~dys~~ disymetrical with the axis so the magentic poles are never the astronomical poles

10   This[a] system shifts ~~completely~~ at Every revolution of the planet hence there must be a diurnal variation of the needle, & since the axis shifts there must be an annual, & since the ~~sy~~ symetry of the lines of force are disturbed by planetary conjunction there must be secular variation.[5]

11. One ½ of the planet is powerful Electrified N & the other S. ~~by~~ [there?][b] passing through the lines of force The poles of which are in the[d] equatorial regions— The signs should change once in 24 hours and have 2 max & 2 minimum in potential for the Earth.

12— ~~Incandescent~~ matter in vacuo or space conducts electricity like a metal ~~when the~~ between an <u>incandescent</u>

substance like the sun and other incandescent matter as pro or cold[e] conductor as proved by the Centre wire in filiment Lamps[6]

13. gravitation attraction[a] has no existence but [--------][b] all weights are due to Static or Current attractions

14 Lines of force proceeding from the N pole of the sun to the south pass along the circumference and by curved lines tooutwardly to the End unlimited space. The Body of lines of ½ section of the ring a s of the ring of the Vortex is the line of balance. magnetic or conducting matter on this line is neither attracted or repelled by the sun[f]   if any[a] conducting matter having a current in it or electrified is in space between this line & the sun it will be repelled if on the other side, attracted.

15— Sun spots are due to faults in the molten surface of the sun   an electric arc forms across the fault the two side will have the[a] current & diminish in temperature, & appear dark by comparison.[g] The arc being mobile & conveying a current will be attracted outwardly to the center of strength of the lines of force just as an ordinary Electric Arc tends to go to the center of the L of F of a magnet. The arc breaks long before it reaches this point hence the incandescent matter surrounding the sun—[7]

16 prominence[a] are arcs of small difference of potential but with great current strength due to great mass of matter

17 The tails of comets are due to the breaking of the arcs from repellent[a] action as the approach the sun

18 Difference in Conducting power of diferent planets instead of the specific gravity must be correct.

19 all planets were once projected from the sun by faults in the equatorial strata. The region from which they were projected[a] was at dif temperature velocity[a] & the matter must have been necessarily different. The moment the curren general current of the sun ceased to pass through the detached matter attraction nearly ceased but the powerful arc still keep it in electrical & it was thrown out into space where getting a rotary motion soon adopted a very eccentric orbit & rotated on its axis & finally the eccentricity diminished & it took its regular orbit

19— Comets are matter recently thrown from the sun The reason comet can never fall in the sun is that they have powerful Currents in them & if they pass Mercury towards the sun they pass the Line[a] of no attraction to the line of the repellant

20 periodicity of sun spots is produced by shifting farther

from the sun the neutral or non attractive center of L of F by approach of Jupiter or other planetary conjunctions.[a] [Ar?][b] more powerful attraction of the arcs cause many to break & thus produce a disarrangement of the equatorial surface diminishing great[a] number small arcs & creating few <u>large</u> arcs & consequently spots by comparison & by depriving the poles or Edges of faults of current by conduction

X, NjWOE, Lab., N-86-04-28:167 (*TAED* N321167). Miscellaneous drawings not reproduced. [a]Obscured overwritten text. [b]Canceled. [c]Illegible. [d]"in the" interlined above. [e]"or cold" interlined above. [f]"by the sun" interlined above. [g]"& . . . comparison" interlined above.

1. See Doc. 2937 (headnote).

2. This entry follows Doc. 2950 and several pages of sketches and calculations, many of them pertaining to Edison's astronomical speculations. It is followed first by several additional pages of similar sketches and calculations and then by Doc. 3005. The notebook contains no dated entries after May relating to Edison's astronomical speculations.

3. Edison wrote the "8" above the "&," presumably to indicate an implied change of units. The Sun's rate of rotation was given as 25 days 8 hours in many reference works, including the 1862 edition of Richard Green Parker's *A School Compendium of Natural and Experimental Philosophy* (p. 360), Edison's first science book (see Doc. 1 n. 5).

4. See Doc. 2940 n. 2.

5. Edison presumably meant the word "secular" in its scientific sense of a change perceptible over a long period of time. In astronomy, it refers particularly to "changes in the orbits or the periods of revolution of the planets." *OED*, s.v. "secular" 7, esp. 7a.

6. That is, the tripolar or so-called Edison Effect lamp with a third electrode between the legs of the filament. Docs. 1898 (headnote) and 2756 n. 3.

7. Recognition of the periodic variability of sunspots, their migration on the Sun's surface, and their correlation with terrestrial magnetic effects all occurred around mid-century. There were as yet no satisfactory explanations of these phenomena. North 2008, 545–48; Hufbauer 1991, 46–49.

–2958–

*Charles Batchelor
Journal Entry*

[New York,] June 3rd 1886.

89.[1] United Edison Shops[2]

E.H.J, T.A.E., Bergman, Hutchinson[3] Kline[4] and I met at laboratory to discuss ways and means to get over the difficulties that put themselves in the way of starting this business. It was decided that the E.MW., Bergman & Co[5] & the Lamp Co should each put up $10,000 cash. As the E.M.W. has so many obligations to meet this summer the money should be advanced (so Mr Johnson said) to it by the Isolated Co. All goods delivered to the United Edison Shops should be paid for by note of the United Edison Shops endorsed by the <u>three companies</u>.

It was generally agreed by all that if Sypher & Cos[6] place could be got for $19,000 per year that it would be well to do so. Bergman[a] & Co sell their fixtures to the United Edison Shops at 50% off.

AD, NjWOE, Batchelor, Cat. 1336:59 (*TAED* MBJ003059). [a]Obscured overwritten text.

1. Charles Batchelor consecutively numbered each entry in this journal.

2. Batchelor referred to one of many meetings about the formation of the Edison United Manufacturing Co., an organization intended to coordinate and streamline the sales and installation of goods made by the Edison Machine Works, the Edison Lamp Works, and Bergmann & Co. The idea for such a company was being discussed by early March as a means for the Edison Co. for Isolated Lighting to get out of the central station construction business. The Edison Electric Light Co. and the Isolated Co. authorized this move on 24 June and the new entity was incorporated in New York soon after, with Edison, Batchelor, Sigmund Bergmann, Edward Johnson, Joseph Hutchinson, and Francis Upton as trustees. The Edison United Manufacturing Co. was capitalized at $150,000, to be raised through the distribution of stock at a par value of $100 per share. The three shops maintained their separate identities but assumed responsibility under the United Co.'s broad charter (which permitted it to own and share patents) to sell and install equipment for both isolated and central plants in direct competition with other manufacturers. The United Co.'s role as a contractor for isolated plants was unclear, at least initially, even to Edison illuminating companies in other cities. The Isolated Co., stripped of its functions, was effectively absorbed into the Edison Electric Light Co. by the end of 1886. Cat. 1336:13, 87 (items 29 and 125, 4 Mar. and 24 June 1886), Batchelor (*TAED* MBJ003013B, MBJ003087A); Edison United Manufacturing Co., certificate of incorporation, 2 July 1886, NNNCC-AR (*TAED* X119PA); Edison Electric Light Co. annual report, p. 13, 26 Oct. 1886, CR (*TAED* CA001B); TAE agreement with Edison Electric Light Co., Edison Co. for Isolated Lighting, and others, 27 Oct. 1886, *Edison Electric Light Co. v. United States Electric Lighting Co.*, Defendant's Depositions and Exhibits [Vol. IV], pp. 2320–25, Lit. (*TAED* QD012E2320); Association of Edison Illuminating Cos. 1887, 40.

3. Joseph Hutchinson (b. 1853?) was manager of the Edison Co. for Isolated Lighting. He became general manager of the Edison United Manufacturing Co. upon its formation. Doc. 2420 n. 23; U.S. Census Bureau 1982? (1900), roll 1105, p. 16B (New York City [Manhattan], New York, N.Y.); Insull to Hutchinson, 25 Oct. 1886, Lbk. 23:49 (*TAED* LB023049).

4. Philip Henry Klein, Jr. (b. 1862) was the secretary of Bergmann & Co. He was to have been a director of the Edison United Manufacturing Co. but was dropped from consideration on 24 June in favor of Harry

Livor (who also was dropped later). Klein instead became secretary of the new firm. In 1892, he became an incorporator of the new General Incandescent Arc Light Co., along with Sigmund Bergmann and William Meadowcroft. U.S. Census Bureau 1982? (1900), roll 1120, p. 10A (New York City [Manhattan], New York, N.Y.); Doc. 2579 n. 2; Klein to Samuel Insull, 1 Apr. 1884; Klein to TAE, 5 Oct. 1886; both DF (*TAED* D8424ZAF, D8629A); Cat. 1336:87 (items 125 and 127, 24–25 June 1886), Batchelor (*TAED* MBJ003087A, MBJ003087B); "Telegraphic Brevities," *NYT*, 5 Oct. 1892, 5.

5. Established in 1881 as a partnership among Sigmund Bergmann, Edison, and Edward Johnson, Bergmann & Co. made fixtures, switches, and other small electric devices in its factory at Seventeenth St. and Ave. B, New York. Doc. 2420 n. 27.

6. Sypher & Co. was the preeminent New York City dealer in antique furniture, tapestries, art, porcelains, and other furnishings. It specialized in antiques from the palaces and great houses of Europe. In 1886, the company occupied what was described as "immense ware and salesrooms" at 860 Broadway between Seventeenth and Eighteenth Sts. Batchelor reported on 24 June: "Sypher place lost to us but we agreed to take 65 5th Avenue." *Finance and Industry 1886,* 150; Advertisement, *NYT*, 18 May 1886, 8; Cat. 1336:87 (item 125, 24 June 1886), Batchelor (*TAED* MBJ003087A).

–2959–

Telegrams: From/To Francis Upton

June 5, 1886[a]
Berlin 12:33 pm

Edison NY

After making workable Contracts between European interests Villard[1] is promised large Capital towards building stations Everywhere[2] this with your full powers will obtain minimum and increasing guaranteed lamp royalty[3] rau Coming here I stay[4] quick favorable action imperative[5]

Upton[6]
[New York]

Drexels say Siegel knows nothing of Villards scheme If Siegel agrees will give power attorney[7]

L (telegrams), NjWOE, DF (*TAED* D8630ZBE). On Western Union Telegraph Co. message form; second message written on reverse, probably by Samuel Insull. [a]Date from document; form altered.

1. Financier Henry Villard (1835–1900) was an early and enthusiastic backer of Edison's development of electric light and motive power in the United States and abroad. Villard was a director of the Edison Electric Light Co. and had chaired the company's special committee on manufacturing and reorganization in early 1884, when he helped Edison reorganize his domestic lighting enterprises. After a financial failure, he began a lengthy stay in Europe that summer and, by Au-

gust, had settled in Berlin, where his further efforts as an industrial promoter were shaped in association with Georg Siemens, a founding director of the Deutsche Bank and a cousin of Werner Siemens. By early June 1886, Villard had reviewed correspondence in Paris related to the various Edison companies (with cooperation from Oscar Siegel), and concluded that harmonizing the French and German companies could result in a well-financed scheme for the construction of central stations (see Doc. 2965). He was also already planning his return to New York, where he resumed full-time residence in September 1886. By then, Villard was the American investment advisor and manager for Deutsche Bank, a financial force in the later formation of Allgemeine Elektricitäts-Gesellschaft (1887) and Edison General Electric (1889). See Docs. 2152, 2439 n. 1, 2599 n. 4, 2638, 2685 n. 1, 2690; Wilkins 1989, 433–34; Hughes 1983, 45, 75–77; Siemens 1957, 1:99; Villard 1904, I:x, II:319; Kobrak 2007, 35–37, 45–50; "Henry Villard's New Enterprise," *Washington Post*, 6 Oct. 1886, 1; "Gleaned from Passenger Lists," *NYT*, 25 Oct. 1886, 8.

2. As Upton explained in letters on 3 and 7 June, Villard expected that better relations between the Edison firms in Paris and Berlin would enable him to obtain financing for a new construction company that would build central stations throughout Europe. Villard also envisioned this entity building plants (or at least financing them) in the United States. Upton to TAE, 3 and 7 June 1886, both DF (*TAED* D8630ZBD, D8630ZBI).

3. According to Upton, Villard expected to secure a fixed lamp royalty on lamp sales (with a guaranteed minimum) that could bring as much as $20,000 annually to the Edison Electric Light Co. of Europe. Upton predicted that European sales, presently 400,000 lamps per year, would soon reach one million. Upton to Insull, 7 June 1886; Upton to TAE, 7 June 1886; both DF (*TAED* D8630ZBH, D8630ZBI).

4. In a draft of this cable, Upton noted, "Villard brings [Louis] Rau & [Emil] Rathenau together" (Upton to TAE, c. 5 June 1886, Upton [*TAED* MU097]). During March 1886, Rau had been negotiating with Rudolf Sulzbach, president of the German Edison company (DEG); he planned to meet with Villard on 11 June. Upton pointedly cabled Edison that the meeting would be "futile" unless someone, such as Villard, had authority to act for Edison (Oscar Siegel to TAE, 5 Mar. 1886; Upton to TAE, 7–8 June 1886; all DF [*TAED* D8625555, D8630ZBI, D8630ZBM]).

5. Upton had pleaded for immediate attention on 3 June, and Edison had led him to expect that the European company directors would confer authority upon Villard at a meeting on 7 June (Upton to TAE, 3 June 1886, DF [*TAED* D8630ZBB]; TAE to Upton, 4 June 1886, Lbk. 22:166 [*TAED* LB022166]). While he awaited for further information, Upton wrote additional letters to Edison and Insull, describing the situation in which the French and German companies thwarted each others' success, the dire consequences that could result if they continued "drifting apart into an open rupture," or if the Compagnie Continentale Edison entered bankruptcy. He warmly recommended Villard as one who "understands the situation here and will work to save something for America." Upton was sure Edison had "never had so fine a chance to have an able man on this side of the water. . . . Unless

some such measure is taken nothing will come from Europe to America. [Oscar] Siegel is in Paris and busy in a bank. He has neither time nor inclination to carry on negotiations. I am too young and have not time." He also hoped to use the prospective European reorganization to "get a slice for the Lamp Co and [I] am going to try to get hold of the French Lamp Factory. This will give us a stand in Europe and will allow us to avail ourselves of cheap labor and ~~we~~ enlarge our market" (Upton to Insull, 7 June 1886; Upton to Edison, 7 and 10 June 1886; all DF [*TAED* D8630ZBH, D8630ZBN, D8630ZBI]).

6. When Upton left New York on 15 May, he told Edison that while the trip was necessary for his health, he hoped to do some business on behalf of the lamp factory. He also used the trip to gather information (from James Hipple) about the German company's lamps, experimental expenditures, and competition with Siemens & Halske. During his absence, Upton delegated management of the factory's business to Samuel Insull. Upton to TAE, 15 May and 3, 7, and 24 June 1886; Upton to Insull, 7 June 1886; all DF (*TAED* D8626G, D8630ZBZ, D8630ZBD, D8630ZBI, D8630ZBJ, D8630ZBH).

7. The Edison Electric Light Co. of Europe approved Villard's power of attorney on 14 June, granting him wide authority to act on his own judgment. However, the board still wanted Oscar Siegel's opinion about it and wished to know whether Siegel's power of attorney should be revoked or shared with Villard. The question of joint powers lingered after Siegel indicated his approval of Villard. Close to the date of his planned departure to New York from France (on 26 June), Upton counseled: "best hold powers my return." TAE to Upton, 14 June 1886; TAE to Philip Dyer, 22 June 1886; TAE to Villard, 24 July 1886; Insull to Charles Coster, 22 Nov. 1886; all Lbk. 22:183, 211, 265; 23:116 (*TAED* LB022183, LB022211, LB022265, LB023116); Upton to TAE, 24 June 1886; Drexel, Morgan to Insull, 23 June 1886; both DF (*TAED* D8630ZBZ, D8602M).

–2960–

To Edward Johnson

[New York,] 11th June [188]6

Dear Sir,

I return you your memoranda of 7th with reference to Lamps & indicators.[1]

I desire it to be distinctly understood that there is no attempt on the part of the Lamp Co to Keep your Company in the dark. It is curious that reports adverse to our lamps come from Cincinnati whereas from Chicago we get just the opposite opinion.[2] However I am going to assume that the adverse reports are the correct ones & this morning I start in at East Newark to try & correct the trouble. You are yourself sufficiently acquainted with Lamp manufacture to know that a gradual deterioration may take place without anyone knowing exactly the "why or wherefore." This has occured before & may be the case on this occasion.[3] Anyway I am going myself to look through the various processes at East Newark & if I can make any improvement I shall most certainly do it[4]

With relation to Indicators I heard yesterday that John Howell[5] denies the correctness of Hammers[6] conclusions. This shall be looked into today & you shall be advised about it by tomorrow or Monday. If the Lamp Cos Indicators are defective I think they should correct them[7]   Yours truly

Thos. A Edison   I[nsull]

L (letterpress copy), NjWOE, Lbk 22:180 (*TAED* LB022180). Written by Samuel Insull.

1. Not found.

2. After a trip West in early May, Charles Batchelor noted in his journal that "in Cincinnatti the short life of lamps was complained of bitterly, whilst at Chicago I find them making contracts every day and guaranteeing 1000 hours." Remarking that lamps in both cities were operated to give about the same intensity of light, Batchelor attributed the difference to Chicago having "better men and larger plants also their plants are put up to have very little variation in [voltage] drop," all of which could, at least in principle, produce more even voltage regulation. Cat. 1336:31 (item 64, 9 May 1886), Batchelor (*TAED* MBJ003031A).

3. See, e.g., Docs. 2183 and 2882.

4. This date marks the start of a period in which Edison spent a large amount of time at the Edison Lamp Co. factory where, within a few weeks, he set up his laboratory. Batchelor remarked in his diary of this date that, according to Samuel Insull, Edison "had to go out there & take charge of the manufacturing to get it back on to a solid basis. He claims that [Francis] Upton has known the lamps were bad all the time but has generally put off all complaints untill they could not be put off any longer." On 9 July, following his return from Europe, Francis Upton noted in a letter to his mother that "Edison has been at the factory while I was away and did some resetting of processes and much talking." Cat. 1336:73 (item 102, 11 June 1886), Batchelor (*TAED* MBJ003073); Upton to Lucy Upton, 9 July 1886, Upton (*TAED* MU091).

5. John White Howell (1857–1937) headed the testing department at the lamp factory since 1881, shortly after he graduated from the Stevens Institute of Technology. Doc. 2129 n. 4.

6. William Joseph Hammer (1858–1934) became the Edison Electric Light Co.'s chief inspector of central stations in 1884, having been associated with Edison's lighting enterprises since 1879. Doc. 1972 n. 7; "Hammer, William J.," Pioneers Bio.

7. Hammer had written a set of recommendations based on his inspection of the voltage indicators at the Harrisburg, Pa., central station, against which Howell submitted to Insull four handwritten pages of criticism. The Lamp Co. had replaced the indicators with a new type, probably the one of Howell's own design described at the end of Doc. 2538 (headnote). Hammer apparently found some fault with the accuracy and utility of both sorts of instruments. Howell vigorously defended both types, concluding that "we have done our full duty by the Harrisburg Co, especially as the indicators they returned are hard to get

rid of owing to their being old fashioned." Howell to Insull, 8 June 1886; Howell memorandum, 6 June 1886; both DF (*TAED* D8626I, D8626J).

-2961-

From William Leslie Edison

Farm [Port Huron, Mich.,] June 13th 1886[1]

Dear mama and papa

O I am[–]ª g-havinng so much fun I wish Tomyᵇ was ɵOnly here   uncle Pittᵇ gave me a pare ofᵇ mules I drive them every morning.ᵇ O I have a lovely colt I named it Tom   i have a pig I will have ten little pigs. I cant write a long letter for I have to go to work   I hope you are all wellᵇ   good bye love and kisses to all, your loving son

willie Edison

ALS, NjWOE, FR (*TAED* FD001AAA). Monogrammed "E" letterhead. ªCanceled. ᵇObscured overwritten text.

1. Young William was with Edison's brother and sister-in-law in Port Huron for an undetermined period of time. At the end of June, replying to his new stepmother, he pleaded: "I don't want to come home so soon   Cant I stay all summer." By 21 July, however, William joined his siblings and some of the Miller family in Chautauqua, N.Y. Older brother Thomas, Jr., spent the first part of the summer with the Millers in Akron. See Doc. 2974; William Edison to Mina Edison, 28 June 1886; Mary Valinda Miller to Mina Edison, 15, 18 and 22 July 1886; all FR (*TAED* FD001AAB, FI001ABL3, FI001ABL4, FI001ABL5).

SUMMER LAMP EXPERIMENTS AND "SOAK PATENT" APPLICATIONS   Docs. 2962, 2964, 2966, 2967, 2968, 2969, 2972, 2976, 2978, 2979, 2980, and 2988

Concerned that reports of disappointing lamp life signaled manufacturing problems at the Edison Lamp Company's factory in East Newark, Edison resolved to spend his time there "to try & correct the trouble."[1] He decided in mid-June to move his laboratory operations to the factory and to launch an extensive experimental campaign to improve lamp life and the manufacturing process.

Edison undertook a brief literature search on 18 June regarding chemicals he might use both for removing vapor from the glass bulbs and for cleaning the glass or clamps. Ten days later, he and John Ott began making tests along these lines. Their experiments continued for about three months, with Edison making most of the records in two notebooks;

the numbered lamps in this series run to 222.[2] In addition, Edison's wife Mina made a copy, under his supervision, of records for lamps 1–80. The entries copied and witnessed by Mina were also signed variously by Edison and assistants Ott, Martin Force, and Albert Keller.[3]

Edison had experimented for only a few days when he drafted three related patent applications on 1 July. These "Soak" patents, as he termed them,[4] covered several paths to a common goal: the elimination of vapors, particularly of water and mercury, inside the lamp bulb, usually by chemical absorption. The process for removing gaseous mercury described in Doc. 2966 was incorporated into Edison's U.S. Patent 438,307. Doc. 2967, which described the use of sulfuric acid in place of mercury in the vacuum pumps, became Edison's patent application Case No. 668 but was subsequently abandoned. The third draft (Doc. 2968) describes a complex process in which the bulb was flooded with hydrogen gas to displace most atmospheric oxygen.[5] Any remaining oxygen would combine with hydrogen, under the influence of the filament's heat, to form water, which would be absorbed by a desiccant such as phosphorous anhydride. Edison also drafted two other applications on 7 July. One was related to Doc. 2966 and was incorporated into his U.S. Patent 438,307. The other pertained to removing water vapor from the lamp; it was developed into U.S. Patent 411,019.[6]

In many of his summer experiments, Edison used an outside heat source, such as a kerosene lamp, to heat the lamps and drive off gases. His standard procedure had been to bring the filament to high incandescence by a strong electric current, a practice that risked shortening the filament's life. However, Edison found that "the amount of gases and vapors contained in the pores of the filament" was inconsequential; the real culprit was the gas residing in or on the surface of the bulb itself. Heating the glass from the outside attacked the problem directly while safeguarding the filament. Edison mentioned this new practice in Doc. 2968 and specifically claimed it in a patent application (of which no draft has been found) that he executed on 15 July.[7]

In some of the lamp experiments, Edison also put a drying agent such as phosphoric anhydride into the bulb to absorb moisture (see Doc. 2968). He realized at some point that this substance was a good conductor of static charge and therefore a possible remedy for the persistent problem of "electrical carrying" of carbon from the filament to the glass by static

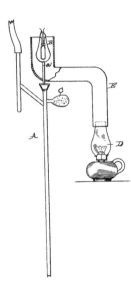

*Drawing from Edison's U.S. Patent 411,018 showing a method of "heating the globe (**B**) from an external source during the process of exhaustion" by an oil- or alcohol-burning lamp (**D**).*

attraction. Following the test of one lamp he noted that "if phos anhy could be got in so as not to give off water & in exact quantity scarcely any blackening would occur."[8] Edison drafted another patent application on 1 August (Doc. 2976) in which he claimed the use of a transparent film of phosphoric anhydride (or similar substances) to equalize electrical potential within the bulb and diminish electrical carrying.[9]

Edison also began coating the filaments themselves with an oxide as a means of increasing resistance and possibly also to prevent carbon carrying. These experiments, started in mid-August, involved soaking the filament in a carbonizable material to lock the oxide to the carbon. This process had the ancillary benefit of filling gaps in the filament's structure so that it would heat evenly under an electric current. Edison included this soaking process among the claims he gave his patent attorney Richard Dyer on 26 September for several patent applications on treating filaments.[10]

For all of Edison's work it is unclear what effect, if any, his experiments and patent applications had on manufacturing at the Edison Lamp Company. Francis Upton, returning from Europe in early July, found that Edison had done "some resetting of processes and much talking" but he did not elaborate.[11] It is also possible that Edison's experiments led to the design and installation of new carbonizing furnaces at the lamp factory.[12]

1. See Doc. 2960.

2. N-86-06-28 and N-86-07-07, both Lab. (*TAED* N322, N323).

3. That Edison supervised Mina's work is evident from the additional information about the preparation of many of the lamps that she added; Edison also wrote some additional notes in his own hand. N-86-08-03, Lab. (*TAED* N324).

4. Edison likely used "soak" in the sense of drawing out, draining, or exhausting something. *OED*, s.v. "Soak" v. 8.

5. No application based on this draft has been found; possibly it is among the missing applications from this period; see App. 2.A.

6. Cat.1151, Lab. (*TAED* NM020AAV, NM020AAY). As an authoritative *History of the Incandescent Lamp* (1927) put it in reference to the first decade of lamp manufacture, "The chief difficulty was then and still is getting the moisture, in the form of water vapor, out of the lamp bulb." From the early 1890s until well into the next century, General Electric pursued extensive research into chemical "getters" (often phosphorous compounds) to "get" moisture from the bulbs during production. Howell and Schroeder 1927, 124; see also 125–31.

7. U.S. Pat. 411,018.

8. N-86-08-03:27, Lab. (*TAED* N324AAA, image 14). Regarding the use of absorbents see also Docs. 2969, 2970, and 2972.

9. This application issued in 1889 as U.S. Patent 406,130.

10. Memorandum to Dyer, 26 Sept. 1886, Cat. 1151, Lab. (*TAED* NM020ABF). Edison did not execute the application for this soaking process to produce filaments of "even density and homogeneous structure" until 26 November; it issued in 1893 as U.S. Patent 490,954.

11. Upton to Lucy Upton, 9 July 1886, Upton (*TAED* MU091).

12. Edison Lamp Co., "Drawing of Chimney for Carbonizing Furnace," 9 Oct. 1886, Oversize Notes and Drawings (1879–1886), Lab. (*TAED* NS7986BAT); "Specification for the Construction of two Brick Chimneys," 4 Mar. 1887, DF (*TAED* D8734AAF).

–2962–

Notebook Entry:
Incandescent Lamp[1]

[East Newark,?] June 18 1886

Notes for Lamp Expmts—

Ammonium Nitrite present in the breath in large quantities after eating especially— J Chem S Vol 25—p 35—[2]

Rain water 2 milgm per Litre Nitric acid & ammonium Nitrite

Absptn Coconut charcoal see J Chem Soc March 1870—[3] also Vol 25—p 649[4]   latter gives table of absptn Cynogen & NH at dif temps pressure 7760 mm[5]

Nitrous oxide decomp by Copper spiral also iron—[a]

Silver Reduced from Oxide by Hydrogen in Lamp absorb mercury

Try Potassium amalgam—

Try Lead Reduced by H, for absorbing O hot—

Alcohol cleans glass splendid—

other cleaners Bisulphide Carbon Cynide potassium, Benzine, Nitric acid— Alkali hot, Heat— This leaves porous residue charcoal & salts, hence better use alcohol remove it, first hot water, then Alcohol.[a]

TAE

X, NjWOE, Lab., N-86-08-17 (*TAED* N320077). [a]Followed by dividing mark.

1. See headnote above.

2. Edison referred to H. Struve's "Researches on Ozone, Hydrogen Peroxide, and Ammonium Nitrite" among the "Abstracts of Chemical Papers: Inorganic Chemistry" in the *Journal of the Chemical Society* 25 (Jan. 1872, p. 35). Edison's discussion below of concentrations of nitric acid and ammonium nitrate in rainwater comes from R. Angus Smith's "On the Composition of Atmospheric Air and Rain-Water" in the same set of abstracts (p. 33).

3. Hunter 1870.

4. Hunter 1872.

5. That is, atmospheric pressure.

[New York] 21st June [188]6

Dear Sir,

I have your letter as to the Municipal System.[1] I have the matter of the lamp in hand. Mr Howell is working on it under my directions.[2] We propose making a lamp of three amperes & an economy of 160 candles to the horse power—or as close to that as possible

I think your idea of putting a man in charge of the Municipal System a good one. If properly worked it should bring us big business[3]   Yours truly

Thos A. Edison   I[nsull].

L (letterpress copy), NjWOE, Lbk. 22:208 (*TAED* LB022208). Written by Samuel Insull.

1. In a letter of 14 June, Johnson informed Edison that the "municipal system is <u>demanding</u> immediate Attention. [John] Vail will send you notice of a meeting to be held at my office some day this week to definitely settle upon a proper Cut out." He also complained that "the Lamp Factory is Sending these new Lamps out (to Portland) in advance of instructions so to do, & that complications have arisen in Consequence. This should not be permitted" (DF [*TAED* D8621F]). The municipal lighting system in Portland, Maine, had been operating since the fall of 1885 with older style lamps. John Vail wrote Edison the same day as Johnson to inform him that the meeting would be held on 17 June (DF [*TAED* D8621E]).

2. No documentation of this work has been found by the editors. Edison eventually took personal charge of the experiments. In mid-July, he asked patent attorney Richard Dyer about his existing patents and applications on lamp cutouts as he had "some new ideas & I want to see if they are already secured broadly" (TAE to Dyer, 15 July 1886, PS [*TAED* PT032AAA2]). Beginning in late August, he and John Ott began extensive experiments on municipal lamps, focusing especially on cutouts (see Doc. 2984).

3. William Jenks was appointed manager of the Edison Electric Light Co.'s new Municipal Department in July. Jenks testimony, *Edison Electric Light Company vs. F. P. Little Electrical Construction and Supply Company*, Equity Case File 6204, NDNY, pp. 26–28.

[East Newark,] June 28 1886

Expmts on Lamps[1]

No 1   Gasolene washed   ⟨80 c[andlepower]   480 m[inutes]⟩
2   Alcohol & KO—broken accid[entally]
3   "   "   ⟨115 min⟩
4   Nitric A then rinsed Alcohol   ⟨90 m⟩

5   Hot Bi Chrom K & SO_4 Rinsed H_2O ⟨5ml⟩
6   HC then alcohol
7   Alcohol twice, hot.
8   Bisulphide Carbon[a] only
9   Soaked ½ hr alcohol clean Clamps scarcely any air come off Lamp on pump
No 10   Alcohol afterwards Bisulphide
No 11   Alcohol then bulb heated Continuously by Kerosene Lamp while on pump ⟨lasted 195   rather black⟩

⟨cp 80—90—82—80—72—68—65 62 60 56—Evy 10 min⟩
No 12=[b] Lamp washed alcohol then run off with Kerosene Lamp under it— gave spectrum H. <u>Very strong</u> Hg & faint CO or CO_2—

177 on[a] scale shews very strong violet line, this I thought was always a CO Line ~~but~~ 125 being the strongest CO Line but this 177 is 3 times as strong as 125 in this lamp hence it must be due to something else. The spark g[auge] has only plat points   2 lines at 130 & 131— Barker says 177 is H line[2]

No 11— at 16 cp—155 cp hp[3]— 418 at 80 cp— drop in cp— [--][c] 20 min[d] 84 40[a] min[e]—65—60m 64 80m 52 100m [-][c]—51. 120m 47   quite black   Lasted 135 min

No 10— Bisulphide C & first Alcohol— 149 c per hp at 16. 378 at 80— Lasted 365 min

moderately black—	1st hour	2		3rd	4	5	6th
drop 80—			72	64	55—	48	44	40

Cpower—[f]

No. 13. No alcohol Heated by Kerosene Lamp— Fused KO in bulk— Spectrum Strong Mercury— Line 177 Sodium Line. ~~Trace~~ Feaint[e] of Red H line ⟨Line—⟩ 106½— No signs of CO— arcd at 40 min—didnt vary in cp   economy 152 at 16   411 at 80

No 12   Blackening on glass absorbed gases so couldnt get spark after heating Lamp   Vac[a] Low— H brilliant, CO moderate Hg very strong sodium moderate.[f]

14— Phos Anhy[dride] & K— Lamp nothing in Lamp heated Kerosene Lamp— K melted get air out finally   Kept low heat absorb CO. Sealed Lamp off cold being cold   5 min previously hot both Lamp & filiment gave plenty Hg= afterward cooling Lamp & slightly heating K no Hg— Couldnt get a spark through sealed off spark guage set Lamp up for Curve—heated guage—got no Hg—but <u>very</u> strong H— also Lines 133½ 160

101 115 96 93¾ 91¼ 86½ <u>158 156</u>—
No 14 lasted 80 m— 151 c per hp   406 at 80

80 c 88—   74   65   62 cp—[f]

Rather black—[f]

No 15   Exhausted with Kerosene Lamp not heated on vac —sealed off and set up—[f]

No 16— $^1/_{500\,000}$ of atmosphere, lots Mercury—also CO feint Hydrogen moderate & N. fooled with it incipient arcs due to spark being on when Lamp lighted— Vac got so perfect couldnt pass spark at first, heated & made incipent arcs   this knocked vac down slightly saw H, Hg & CO— afterwards by absorption of glass or clamps got so high couldnt get spark through   Sent[a] it to get curve   man reports takes 126 volts to get 16 candle .67 ampere, hasnt volts nuff to set at 80— ~~Pu~~[f]

17   another Lamp not heated by Kero or fil by current, absorption K & Charcoal in Tube, heated filament after seal-ing— Knock vacuum down & charcoal after laying all night[4] didn't appear to absorb— Hydrogen & Nitrogen strong pink on clamps— CO very faint, No Hg— Set it up for curve—

18   Iodine in Lamp with Copper Hgd—[5] to absorb it & prevent going in ~~bulb~~ pump— Arcd 75 min 161 cp hp at 16   438 at

min   20m   40m   60 min

80   80°   82   68   60 whitish deposit globe at first[f]

17[6]   Iodine in bulb— Lasted 60 min— 168 cp hp at 16— 462 at 80.

            20 m   40 min

80 cp[e]   78   70— whitish Iridescence deposit near clamps.[f]

            122— intensity 12— sharp

            108.5   "   9   "

Hg—   114.5   "   8.   slowly increasing

    fluted spec nitrogen

    CO strong

    122 gradually disappears

    H Line thin

    New pump

X, NjWOE, Lab., N-86-06-28 (*TAED* N322000A). [a]Obscured over-written text. [b]Multiply underlined. [c]Canceled. [d]"20 min" interlined above. [e]Interlined above. [f]Followed by dividing mark.

1. See Doc. 2962 (headnote). Detailed descriptions of the prepara-tion of these lamps as well as experimental results are in N-86-08-03, Lab. (*TAED* N324).

2. Because chemical elements and compounds emit unique spectral lines when heated, Edison was able to use a spectroscope to determine

the composition of any chemicals remaining in the lamp after it was treated and evacuated. In conducting his spectrum analysis, Edison would have used a standard photographed micrometer scale placed in one tube of his spectroscope. That scale was based on the millimeter scale developed for chemical analysis by Robert Bunsen and Gustav Kirchhoff in the early 1860s in order for chemists "to be able to observe quickly and easily, especially when lines have to be recognized which only flash for a moment." The scale produced a simple qualitative measurement suitable for chemical analysis, in contrast to those developed by physicists and astronomers, who sought absolute quantitative measurements in order to develop a complete catalogue of the spectrum of wavelengths. However, imprecision in the scale and imperfect standardization of the prisms used in spectroscopes could create some uncertainty in reading the scale, as Edison seems to have experienced (Hentschel 2002, 45–55, quoted 51; McGucken 1969, 24–28; Kohlrausch 1883, 119). Following this entry there is a related but undated note, probably by John Ott, describing the process Edison and his staff used for producing a "spectrum of a tube taken off of a washed Hydrogen pump" (N-86-06-28, Lab. [*TAED* N322000B]). For standard descriptions of spectrum analysis from the period see especially Roscoe 1873, which Edison had acquired along with a Browning spectroscope in 1873 (Doc. 544 n. 2) as well as Browning 1883; Fresenius 1876; and Kohlrausch 1883, 118–121.

3. That is, the amount of candlepower per horsepower. Throughout these experiments Edison sought to determine the efficiency of the lamps by conducting measurements of candlepower like the ones in this entry.

4. Edison dated only the first two pages; the editors have not determined how many days he spent conducting these experiments.

5. Edison probably meant this as shorthand for "amalgamated."

6. Possibly misnumbered. This may be lamp 19.

–2965–

Henry Villard to the Edison Electric Light Co. of Europe, Ltd.

Berlin, June 29, 1886.

Confidential

Gentlemen:

As you have been advised by Mr. F. R. Upton, I have been engaged for some time past in an effort to bring about a revival of the Edison electric light interests in Europe and elsewhere by

1. Settling the prevailing differences between your Company and the French Edison Companies, on the one hand, and between the French Companies and the German Edison Company, and between the latter and Messrs. Siemens and Halske, of this city.[1]

2. The organization of a new strong Company with ample capital, that would make a special business of promoting and

financiering central station enterprises in Europe and in the United States.

The differences between the parties in interest mentioned, were, as you are aware, of such a character that they prevented the development of the existing Continental Companies, kept you out of all income from them and were about resulting in a general destructive war between them and Messrs. Siemens and Halske, in absolute disregard of, and to the certain ruin of, your interests.

I take pleasure in informing you that I have succeeded in securing the acceptance by the management of the German Company of the terms of the memorandum drawn up here by Messrs. Rau, Upton and myself, and communicated to you,[2] except as regards the amount of the minimum guarantee per annum of lamp royalties, which will have to be reduced.

To bring about an understanding between the German Company and Messrs. Siemens and Halske was a most delicate and difficult task, but a basis was finally found according to which the Company is to confine itself to promoting and exploiting electric light enterprises, while the firm is to take charge of their technical execution.

I am also able to advise you that the organization of a new Company for the financial purposes stated is assured. Messrs. Siemens and Halske and leading banks and bankers will be interested in it. Its main object is to be to issue its own obligations secured by mortgage upon central stations.

Only informal agreements have been made so far, but I am confident that formal engagements will be entered into in due season by all the parties concerned. But the definite steps cannot be taken before September, owing to the absence of the principals during the hot season. I will duly report further progress to you at the proper time.

Please communicate this letter to Mr. Upton. It is desirable, however, that the knowledge of its contents should be confined to as few people as possible, as undue publicity will endanger the final carrying out of the programme, which includes as an essential feature a change in the control of the German Company.[3]

Hoping that I have acted to your satisfaction, and that I shall hear from you in due season regarding the power,[4] I am Yours, very truly,

H. Villard   By C. A. Spofford   Attorney[5]

L, NjWOE, DF (*TAED* D8630ZCA).

1. After Francis Upton left Berlin but before he sailed for New York, Villard sent him an update about his (Villard's) efforts to reach an agreement between Siemens & Halske and Deutsche Edison Gesellschaft (DEG), with a promise of more information before Upton sailed. As of 21 June, Werner Siemens had assented in principle on behalf of the former party, and Villard intended to meet with Emile Rathenau of DEG. The settlement ultimately resolved major difficulties that had developed under the firms' 1883 contract, in part by buying out the claims of the Compagnie Continentale Edison on the German firm. Villard to Upton, 21 June 1886, Upton (*TAED* MU090); Siemens 1957, 1:92–93, 99–100.

2. Villard likely referred to an undated and unsigned fifteen-page handwritten (in French) memorandum. TAE agreement with Edison Electric Light Co. of Europe, Compagnie Continentale Edison, and Société Électrique Edison, [1886?], DF (*TAED* D8630ZCV).

3. Villard may have alluded to changes that resulted in the subsequent restructuring of Deutsche Edison Gesellschaft as Allgemeine Elektricitäts-Gesellschaft (AEG) in May 1887. Chandler 1990, 464; Hughes 1983, 76; Siemens 1957, 1:100.

4. Although the Edison Electric Light Co. of Europe approved a power of attorney to Villard on 22 June, the joint power from it and Edison personally was not sent until 24 July. The power granted Villard broad authority to act on behalf of the firm and Edison, with respect to electric lighting in Europe, until 1 January 1887. The letter of transmittal outlining Villard's authority made it clear that the company's foremost goal was to raise money for its bond obligations by November (the bondholders having informally agreed to write down fifty percent of their principal investment). Villard was also instructed to make no restrictions on Edison's future inventions and to keep the French companies from venturing beyond their contracted territories. Neither the fusion of the Paris companies nor disputes between the French and German companies was resolved when Villard returned to New York in September. Power of attorney to Villard, 22 July 1886, DF (*TAED* D8625D1); Edison Electric Light Co. of Europe to Villard, 24 July 1886, Lbk. 22:265 (*TAED* LB022265).

5. Charles Ainsworth Spofford (1853–1921) was Villard's private secretary and associate in numerous enterprises. Doc. 2685 n. 3; "Necrology. Charles Ainsworth Spofford," *Annual Report of the American Scenic and Historic Preservation Society* 26 (1921): 25.

Dyer—

Take out patent and assign to Lamp Co for following. They are a series of <u>Soak</u> patents hence broad claims[2]

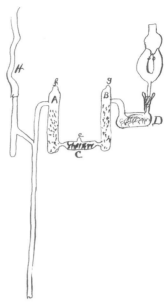

D phosphoric Anhydride or other absorber[a] of water.[3]

A & B are tubes containing an absorbant of Chlorine Bromine or Iodine Vapors such as Quicklime & such metals ~~asin~~ pieces that combines with Chlorine at ordinary temperature or slight increase of temperatures above the ordinary such as Copper Zinc Iron, alkali metal   C is a tube containing solid Iodine, or a[a] salt of Iodine which gradually gives off vaporos of Iodine, or a chloride which gives off free chlorine gradually or a chloride that gives off its <u>heat</u> ⟨(chlorine(?))⟩[b] slowly by a gentle heat ⟨How hot?⟩[b] such as sesquichloride of chromium, or an inert porous material like charcoal etc saturated with Chlorine gas[a]— Bromine & salts acts the same as chlorine—

The object of this trap is to prevent mercury vapors from passing from the drop tube to the lamp. Iodine Chlorine or Bromine ~~all g~~ all attack Mercury at ordinary temperatures forming an Iodide Bromide or chloride. Consequently no mercury reaches the Lamp— The object of tubes A. & B. is on the one hand to prevent these gases or vapors from going into the fall tube where they would combine with the mercury & dirty the tube so it wouldnt work & on the other hand prevent the gas used from entering the Lamp. The metals ~~are~~ in tubes are attacked by the gases & form solid Chlorides Bromides or Iodides. There are absorbant substances that will absorb the gases in the pores of the same that ~~w~~Could be used in A & B such as ~~chalk~~ oxides magnesia, Charcoal & other inert substances that absorb gas. a compound could be used with which the gas or vapor combine, such as quicklime in the case of chlorine.

The object of getting the Mercury out of the globe is to ~~prev~~ diminish the blue and lengthen the life of the filiment ⟨effect of Hg. on lamp?⟩[4][b]

I spose you can make a claim for a chemical substance between the Lamp and the mercury pump which will combine with the Vapors of mercury to make a solid. There are other chemical substances which when mercury passes through or over them combines besides the ones named such as heated sulphur, vapors of nitric acid   in fact any chemist could off hand suggest a dozen things in A B & C that would accomplish the object hence I want to get very broad claims—

T A Edison

ADfS, NjWOE, Lab., Cat. 1151 (*TAED* NM020AAZ). Document multiply signed. [a]Obscured overwritten text. [b]Marginalia written in unidentified hand.

1. See Doc. 2962 (headnote).
2. The other draft "soak" patents from this date are Docs. 2967 and 2968; see also Doc. 2969. Edward Rowland, who likely made the patent drawings, initialed Edison's draft on 9 July. The process described here was incorporated (with parts of another somewhat later draft) into an application that Edison executed on 15 July and filed two days later; it issued in 1890 as U.S. Patent 438,307. Cat. 1151, Lab. (*TAED* NM020AAV).
3. Figure labels are "H," "f," "g," "A," "B," "D," "e," and "C."
4. Since at least 1880, Edison had associated the appearance of a bluish discharge around one of the clamps with the presence of gas or moisture in the bulb, and he adapted the phenomenon as an indication of the state of the vacuum during lamp manufacture. See Docs. 1898 (headnote), 1902, 1996, and 2029 esp. n. 3, 2061 n. 2, and 2139.

–2967–

Draft Patent Application: Incandescent Lamp[1]

Llewellyn Park   July 1 1886

Patent No 2   <u>Lamp Co account   Soak patent</u>   ⟨668⟩[2a]
Dyer—

Take out patent on ~~the~~ a sprengel pump[3] same as we use except the use of Sulphuric acid instead of mercury to get the vacuum   The fall tube is about twenty feet long   we use Lead ~~&~~pot[4] Hard rubber fixtures & mechanical pumps to handle it= The Sulphuric is very strong. get broad claim.

object is to dispense with mercury as its expensive and dangerous while the men can protect themselves by mica masks for accidents, they cant from the insidous Mercury besides it gets the[b] Vacuum free of mercury vapor [---][c] The [vap?][c] very little vapor from Sulphuric acid may be absorbed by a substance which will combine with it such as an oxide of a metals throughs which the vapors must pass before getting to Lamp— it (sulp acid[d] acts also as a drying agent though I prefer to still use the phosphoric anhydride in addition— it doesnt break the tube by pounding & the tube is always clean—

Edison

P.S. No diference in pump from[b] the regular except length of fall tube   E

ADfS, NjWOE, Lab., Cat. 1151 (*TAED* NM020AAX). [a]Followed by dividing mark. [b]Obscured overwritten text. [c]Canceled. [d]"(sulp acid" interlined above.

1. See Doc. 2962 (headnote).

2. This is Edison's patent application Case No. 668. No completed application has been found and no patent issued.

3. Devised in 1865, the Sprengel air pump was the most effective form of pump at this time. Droplets of mercury fell down a long narrow tube (the "fall tube") connected above to the vessel to be evacuated. The droplets entrained air, carrying it to a reservoir at the bottom and gradually producing a high vacuum within the enclosed space. Edison used a form of Sprengel pump in his laboratory and lamp factory. Doc. 1667 n. 2.

4. Edison likely referred here to the use of either leaded or potash glass for the fall tubes.

–2968–

Draft Patent Application: Incandescent Lamp[1]

Llewellyn Park   July 1 1886

Patent acct   Lamp Co— Soak patent—[2]

To effectually absorb all the water within the bulb of the lamp and to cause water to be formed so that it may be absorbed so as to get rid of the Oxygen is object of the patent.

a small after the Lamp has had all the water taken from it by passing pure dry Hydrogen through it from a source of Hydrogen & all the air driven out by displacement due to the inrush of the Hydrogen, a small pellet of an[a] water[a] absorbing substance such as phosphoric anhydride is put in the lamp and while it is full of Hydrogen put on the pump and Exhausted, and Oxygen present combines with the Hydrogen by the action of the incandescent filiment & forms water at[a] a low incandescence & this is absorbed by the pelett [op pecie?][b]

of phosphoric anhydride, which remains in the lamp after it is sealed off and ready for market, it also absorbs any water as water that might have been in the globe. A piece of phosphorious may be used instead of the phosphoric anhydride & this is ignited while Oxygen is still in the globe & before the Hydrogen has been passed the phosphorus burns to phosphoric anhydride   it should be put in a small receptible so that in burning the white cloud of phos anhydride will not rush upward & coat the sides of the lamp— Chloride of calcium & other absorbants may be used but phosphoric anhydride I think is preferable. The pelet when put in the lamp must be made to fall to the bottom of the lamp where it will adhere and where it will remain during the use of the lamp, as even[c] phosphoric anhydride gives off a slight amount water when strongly heated. Instead of ~~Hyd a gass~~

Want a strong claim on this. The use of dry[c] Hydrogen[a] in the lamp as the gas to start with to Exhaust and ~~an~~ a powerful water absorbant Either in the pump or allowed to remain permanently in the Lamp is used, the latter preferable.

Claim heating the filiment at first to lower incandescence than it is to be burned at to cause the Oxygen to combine with the Hydrogen to form water—

PS   I forgot to mention that while the lamp is being exhausted ~~it~~ the bulb[d] is heated continuously by immersion in hot oil bath or from heat rising from a chimney & flame. Thus—[3]

Make a claim for this also for it in combination with the P in Lamp & the dryer in Lamp or pump—

Edison

ADfS, NjWOE, Lab. Cat. 1151 (*TAED* NM020AAW). [a]Obscured overwritten text. [b]Illegible. [c]Interlined above. [d]"the bulb" interlined above.

1. See Doc. 2962 (headnote).
2. No patent application matching this draft has been found, although Edison's U.S. Patent 411,019 does include the use of a "receptacle forming a continuation of the exhaust-tube of the lamp" that would "contain pieces of glass covered with phosphoric anhydride or other moisture absorbent; or such absorbent may be placed therein by itself."

The application covering the process and claims in this draft may be one of the two missing patent applications, Case Nos. 667 and 672, that would have been executed and filed about the same time as Edison's other soak patents (see App. 2.A; also Docs. 2966 and 2967).

3. Figure label is "bunsen burner."

–2969–

Notebook Entry:
Incandescent Lamp

Lamp Expmts[1]

[East Newark,] July 2 1886

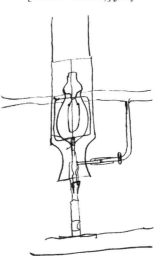

Dyer for patent on process[2a]

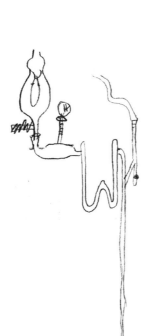

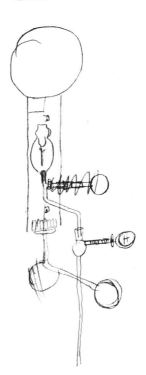

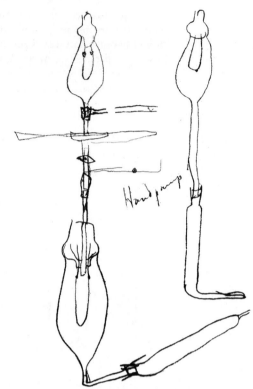

Dyer patent   Phosphorus may be dissolved in Benzine the globe filled & vac got then heat of Lamp makes it PO_3— Oxygen could be used— ammonia & HCl—[4]

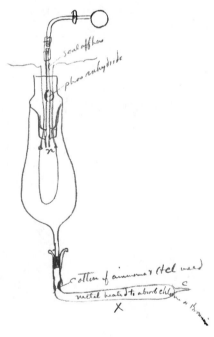

Ammonia absorbed by charcoal or a substance heated giving HCl. gas passed through X cold exit C then sealed also N. The X heated & chlorine absorbed

Try pentachloride phosphorus both to absorb Hg & act as drier The C combining with Hg & Liberating P which forms PO_3 with O, & O of H_2O[a]

Spk to Holzer abt dissolving out those Lamps that PO_3 pumped up into—[a]

~~Hyposulphite~~ Soda Thiosulphite of soda[b] removes gaseous Chlorine from substances it ~~has~~ is absorbed into.[a]

glacial acetic acid melted parafine[c] for pump to replace Mercury also melted Rose metal,[5] or fusible metal

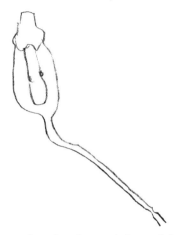

fill with & boil parafine then invert & heat globe to expel parafine & seal off—[6]

TAE

X, NjWOE, Lab., N-85-12-06 (*TAED* N312031). Document multiply signed and dated. [a]Followed by dividing mark. [b]"Thiosulphite of soda" interlined above. [c]Preceding text enclosed by brace.

1. The drawings below may be related to Edison's US Patent 411,018 in which he claimed the use of an external heat source to heat the filament during evacuation in order to remove gases and vapors. See Doc. 2962 (headnote).

2. Figure labels are "H" and "H." The editors have not found a patent application based on the following notes. If one was drafted, it may be one of the missing soak patents (see Doc. 2966 n. 2).

3. Figure label is "Hand pump."

4. Figure labels are "seal off here," "phos anhydride," "N," "cotton of ammonia & HCl used," "metal heated to absorb chlorine [or?] Bromine," "C," and "X."

5. Rose's metal, a mixture of bismuth, lead, and tin, has a low melting point (about 100° C). It was widely used as a solder and also for boiler safety devices. Tidy 1887, 464.

6. Edison sketched what appear to be several vacuum pump arrangements on the next two pages (pp. 42–43). His drawings are partially obscured by doodles, including many iterations of his wife's name. Between those pages and the next dated entry on 4 July (Doc. 2970) are three versions of the famous lines from *Richard III* beginning, "Now is the winter of our discontent. . . ."

–2970–

Notebook Entry:
Incandescent Lamp

[East Newark,] July 4 1886

Lamp Expts

Bichloride of platinum combines with Hg—forming I spose HgCl & amalgam of P & Hg[a]

Try anthracene & other HC compounds to absorb Iodine[a]

pentchlorides are I think prone to give off chlorine— Try these[b]

Try phosphorus vitrious & Red to absorb Hg—

Try pentchloride Antimony for Hg— one atom Cl very loose

mix ~~clay~~ oxides Magnesium[c] roll[d] out cut filiment also oxide others infusible— dry then soak in Licorice or sugar or tar, dry then carbonize— High Res lamps[1a]

Coating Carbon with Boron   claim decomp Chl Boron also Flouride Boron[a]

Might mix Licorice, tar, sugar or Dextrine, or Tragacanth with oxide of Mag etc   roll out & Cut filiments. Carbonze

Try iron filiment & chloride Carbon— also in presence of H. also Cynogen gas from Cynide Mercury or Silver to make filiment disolve out iron[2]

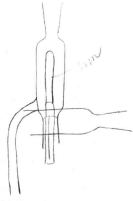

first wash water Then absolute Alcohol. Then on other pump get vac— break & rinse with Benzol— get vac—heat Lamp & Light iron wire.

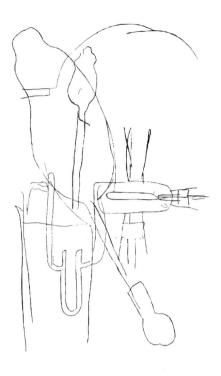

TAE

X, NjWOE, Lab., N-85-12-06 (*TAED* N312049). [a]Followed by dividing mark. [b]Line of text inserted between dividing marks above and below. [c]"oxides Magnesium" interlined above. [d]Obscured overwritten text.

1. This entry appears to mark the beginning of Edison's efforts to produce high-resistance filaments from oxides in a plastic state, a subject on which he drafted several patent applications in the fall. See Docs. 2994 and 3012; Cat.1151, Lab. (*TAED* NM020AB01); U.S. Pats. 411,020, 485,616, and 525,007.

2. Figure label below is "iron."

–2971–

Samuel Insull to
Paul Bossart[1]

[New York,] 7th July [188]6

Dear Sir,

I beg to inform you that Mr Edison has turned over to me the management of the business affairs of the Phonoplex.[2] Of course you will understand that this does not affect your relations with the business Mr. Edisons object being to relieve himself of the details of the business & at the same time to place in control of it someone who can cooperate with yourself & Mr Tate with a view to getting all that is possible out of the invention from a commercial point of view   I feel confident that in informing you of this that I can count upon that cordial

cooperation which must make the invention a perfect success
& thus enable us all to make some money.

Of course your correspondence will from time to time be
submitted to Mr Edison so that you can rely upon his being
well posted as to the results of your individual work. I want you
to feel that at all times I stand ready to forward your wishes &
whenever by any effort here your business in the west can be
facilitated do not hesitate to call me   Yours truly

Saml Insull   Private Secretary

ALS (letterpress copy), NjWOE, LM 12:16 (*TAED* LM012016).

1. Paul W. Bossart started setting up exchanges for the American
Bell Telephone Co. in Kansas and Missouri in 1879. Trading on a prior
acquaintance with Edison, in 1883 he requested the rights to the Edison electric light system in Kansas City. This effort was unsuccessful and he instead received from the Bell company responsibility for
a larger eastern territory. Bossart eventually became superintendent
of the New England Telephone Co. He resigned in February 1886 to
work on Edison's phonoplex, reportedly to "take general charge of the
business." Bossart initially collaborated with Alfred Tate on the Baltimore & Ohio installation and seems to have worked on several other
railroads. He had recently been setting up the system on the Atchison,
Topeka, & Santa Fe, and Insull addressed this letter to him in Lawrence,
Kans. Bossart evidently relocated to Chicago in mid-August. "Paul W.
Bossart," *Telephony* 3 (Apr. 1902): 121; "Paul W. Bossart," *Western
Electrician* 22 (26 Feb. 1898): 118; Bossart to TAE, 24 Nov. 1883, DF
(*TAED* D8349I); *Electrical Review* 7 (20 Feb. 1886): 4; Tate to Bossart,
23 Mar. 1886; Bossart to George Myers, 20 Apr. 1886; Bossart to James
Hill, 20 Apr. 1886; Insull to Tate, 7 July 1886; Lbk. 21:450 and 22:37,
43, 235 (*TAED* LB021450, LB022037, LB022043, LB022235); Insull to
Bossart, 16 Aug. 1886, LM 12:152 (*TAED* LM012152).

2. By the end of July, Insull was evidently mailing out a promotional
pamphlet about the phonoplex that identified him as the manager of the
business; Alfred Tate was named as the electrician ("Edison Phonoplex
System of Telegraphy," n.d. [1886], PPC [*TAED* CA012A]; Insull to
Charles Hosmer, 31 July 1886; Insull to William Hovey, 31 July 1886;
both LM:12:84, 89 [*TAED* LM012084, LM012089]). According to a
list in *Engineering,* the Edison phonoplex had been adopted by October
1886 on the Atchison, Topeka, & Santa Fe and five other railroads (Baltimore & Ohio Telegraph, Kansas Southern, Philadelphia and Reading,
Canadian Pacific, and the Pennsylvania Railroad), as well as the Great
North Western Telegraph Co. ("The Edison Phonoplex," *Engineering*
42 [22 Oct. 1886]: 412).

Notebook Entry:
Incandescent Lamp[1]

General Experiments—

Carbonate & Bi-Carbonate dry ground up with pyrogallic acid absorb Oxygen from the air   use this in pump— Carbonate is better than bi-Carb—

Phos Anhydride can be kept & moulded in any shape under gasalene & of course almost any Hydrocarbon—[a]

Tested cleaned Lamps after going through washing with Bichromate of Potash & SO_4 & distilled water— feint traces SO_4 with Chl Barium but second boiling & washing with distilled H_2O got rid of it entirely=

Mercury is attacked by Sulphuret of Potassium—perhaps it wont be dry— Can use this substance in pump to absorb Hg Vapors—

TAE   J. F. Ott

X, NjWOE. Lab., N-86-07-07 (*TAED* N323AAB). Document multiply signed and dated. [a]Followed by dividing mark.

1. See Doc. 2962 (headnote).

Charles Batchelor
Journal Entry

158[1]   Uptown Station.

We had a meeting at 16 Broad St[2] to discuss the dynamo and conductors for new district. E[dward]. H. J[ohnson]., TA.E. Chinnock, Batchelor Vail,[3] Andrews,[4] & Insull were present Decided to have 125 volts & 500 ampere at the dynamo

Edison proposed leaving off some streets altogether & running feeders to heavy centres & then making mains at such places same size as feeders, other mains of less size

There was much discussion on as to whether it is advisable to use cables as feeders   EHJ. & Chinnock say it is better to have no joints & if a pick cuts into one cable it is only one side that is damaged. Edison maintained that the cable whilst being electrically superior to tubes[a] cannot be as good & well protected from mechanical injury. he illustrated his remarks by showing that all faults to our underground system were mechanical and said that unless you drew the cable through an iron tube it would not be so satisfactory as our tubes. I claim that in large cities like N.Y. it is difficult to lay cables of large diameter and long lengths owing to so many other pipes being now in the ground

AD, NjWOE, Batchelor, Cat. 1336:97 (*TAED* MBJoo3097D). [a]Obscured overwritten text.

1. Charles Batchelor consecutively numbered each entry in this journal.

2. The Edison Electric Illuminating Co. of New York had relocated its offices to this location by early 1886. The Edison Electric Light Co. also moved there, though perhaps not until later in the year. Edison Electric Illuminating Co. annual report, 19 Jan. 1886, DF (*TAED* D8623A); cf. Doc. 2945 n. 14.

3. Jonathan H. Vail (1852–1926), who had been closely involved with central station planning and dynamo construction, was general superintendent of the Edison Co. for Isolated Lighting (since 1881). He took on similar responsibilities for the Edison Electric Light Co. in 1887; he became secretary of the Association of Edison Illuminating Cos. in 1886. *TAEB* 7, App. 4.C n. 7; Doc. 3050 n. 1; *Edisonia* 1904, 196; "Vail, Jonathan H.," Pioneers Bio.

4. William Symes Andrews (1847–1929), a skilled instrumentalist and mechanic, was the chief electrical engineer of the Edison Co. for Isolated Lighting (since 1885) and had recently received three patents for improvements in the three-wire system. Andrews received his early scientific training at Cuzner's Collegiate Academy in his native England, where he was later headmaster. He immigrated to Toronto in 1875 to work for Raybon & Co., a firearms firm, and eventually became superintendent of its factory in Newark, N.J. He began working for Edison in 1879 as an assistant at the Menlo Park laboratory. After a stint at the Edison Machine Works, where he managed the Testing Room, Andrews was appointed the chief electrical engineer of the Edison Construction Dept. in 1883 and supervised the establishment of numerous central stations. In late 1886, Andrews became the general superintendent of Marr Construction Co. of Chicago, which also built central station plants, but he returned to the Edison lighting enterprise as a superintendent and manager for the Edison General Electric Co. and the General Electric Co. Docs. 2223 n. 2 and 2490 n. 7; "Andrews, William Symes," Pioneers Bio.; "W. S. Andrews Dies; An Edison Pioneer," *NYT*, 2 July 1929, 22; U.S. Pats. 317,610; 317,700; 318,157.

–2974–

*Jane Miller to
Mina Edison*

Chautauqua, July 21. 1886

Dear Mina,

We arrived here safely yesterday afternoon— Took dinner at Lakewood[1] and then the yacht for the cottage. Everything seems as usual—quite a number[a] at the hotel. Mr. Firestones[2] are here renting Uncle Jacob's[3] cottage and taking their[b] meals at the hotel. Mrs. Chess[4] and Rosengarten's[5b] also take their meals at the hotel— Mr. Studebacker's[6] came today— They have bought Mr. Frank Carley's[7] cottage. Dr. Vincent has been away for a week or more returns here tomorrow to stay until Friday when he leaves for a week's trip south. Mrs. Vincent[8] &[c] George[9] are very cordial. George called upon us last evening. The Brown girls[10] are still at home ~~yet~~ and are having a good time. Every one asks[b] if you are coming also if Mother is coming   I hardly know how to answer for I don't know really what you intend to do.

Thomas and the little[11] are perfectly well[a] and enjoy every hour. They all have their hair cut close to their heads and are tanned as can be. They seemed very much pleased with the grapes & peaches.

How kind it was of you and Mr. Edison to go to the train with us. I really was ashamed of myself for getting so vexed but I was so surprised when Thomas said we had ~~but~~ a few moments to make the train. We are always so ready to excuse our selves and try to make some one[a] else the cause[b] of all dificulties.[b] I blame myself for not knowing exactly about the train and watch. You did all that you could and you thought of course I knew all about everything which I should have done. As it was everything turned out well for us but how for you I don't know. Everytime I get vexed I am provoked with my-self. I want to be more patient but seems to me I will never reach that point. Everything at your home is so beautiful and you and Mr. Edison made it so agreeable that I regreted very much when the time came for me to pack my trunk. My dear, I enjoyed my visit very much and thank you and Mr. Edison for it. I have thought you so often alone all day long. I hope Mr. Edison returns home early in the evening and you have pleasant drives and visits together. The cottage and tent are as usual and we will be ready to welcome[b] any one of our people when they get ready to come. I hope you will not defer your visit any longer than[b] is really necessary. Grace[12] and Marion occupy the tent. They have a nice company of boys & girls. And I guess will have a good time.

Give my love to Edison and tell him I think he might write[b] a few lines when you write. I wish you would write often dear Mina   It seems lonely here without you girls—

Mrs. Vincent and Miss Harris (Bishop Harris daughter)[13] called and have just left. With ever so much love

Jennie

ALS, NjWOE, FR (*TAED* FM001AAR). [a]Interlined above. [b]Obscured overwritten text. [c]"Vincent &" interlined above.

1. The town of Lakewood, laid out in 1874 on the southern shore of Lake Chautauqua, was a transfer point between trains and steamers for Chautauqua travelers and a popular summer destination in its own right. *Chautauqua* 1884, 58–59.

2. Clinton DeWitt Firestone (1848–1914), born near Canton, Ohio, and his wife Flora (b. 1850?), an Indiana native, lived with their children in Ohio's capital city. Firestone organized the Columbus Buggy Co. in 1875 or 1876, and it quickly became a leading maker of buggies and carriages (and later a stepping stone into the tire business for his younger cousin, Harvey S. Firestone). Active in church and civic affairs, Firestone was a delegate to the Methodist Episcopal general conference in 1884 and to the Republican presidential nominating convention the same year. By 1887, he was a trustee of the Chautauqua Assembly. U.S. Census Bureau 1970 (1880), roll T9_1016, p. 029, image 0536 (Columbus, Franklin, Ohio); "Clinton Firestone Dead," *Motor Age* 25 (26 Feb. 1914): 11; Lee 1892, 1:921; Smith 1898, 1:851–52; "Meeting of the Trustees of the Chautauqua Assembly," *The Chautauquan* 7 (Mar. 1887): 385.

3. Jacob Miller (1827–1889), the brother of Lewis Miller, was Mina and Jane's uncle. He was a co-founder and longtime president of C. Aultman & Co. (which helped launch Lewis Miller's Buckeye reaper), among other business interests in northeastern Ohio. An early financial supporter of Chautauqua, he remained active in the movement. Obituary, *Chautauqua Assembly Herald*, 25 (18 Aug. 1890): 5; Hendrick 1925, 170.

4. Amelia J. (née Carley) Chess (b. 1844?) grew up in southern Ohio before her family relocated to Louisville, Ky., in the late 1860s. Sometime before 1880 she married William Edward Chess, an Indiana native connected with Chess and Wymond, a Louisville cooperage established in 1877. U.S. Census Bureau 1970 (1880), roll T9_423, p. 501C, image 503 (Louisville, Jefferson, Ky.); "Mary Ann Carley," *Western Christian Advocate*, 7 Jan. 1903, 20; Weeks 1886, 432.

5. Anna M. (née Carley) Rosengarten (1850–1934), a Louisville resident like her sister Amelia Chess, regularly attended the Chautauqua Assembly at this time. Around 1882, she married L[eroy?]. T. Rosengarten, who was connected with Chess, Carley and Co., a Louisville oil marketing and shipping firm operating in the Southeast at the secret behest of Standard Oil. (One of the principals seems to have been her brother, Frank Carley; see note below.) By the early 1890s, L. T. Rosengarten was involved with Frank Carley in other enterprises in the region. U.S. Census Bureau 1982? (1900), roll T623_968, p. 20B (East Orange Ward 2, Essex, N.J.); *Kentucky Death Records, 1852–1953* [28 July 1934], online database accessed through Ancestry.com, 14 May 2013; "Mother Carley's Daughter," *Daily Chautauquan*, 8 Aug. 1934, 3; *Western Christian Advocate*, 7 Jan. 1903, 20; *Caron's* 1876, 498; *Poor's Manual* 1894, 201.

6. A friend of the Miller family, Clement Studebaker (1831–1901) was president of the Studebaker Manufacturing Co., incorporated in 1870 to consolidate his family's thriving business of carriage and wagon manufacture. Raised as a German Baptist pacifist, the widowed Studebaker converted to Methodism after he remarried in 1864. A resident of South Bend, Ind., he was a strong influence in state and national Republican politics and a longtime trustee (later president) of the Chautauqua Assembly. *ANB*, s.v. "Studebaker, Clement"; Hendrick 1925, 170.

7. Francis (Frank) D. Carley (b. 1839?), like his sisters a native of Ohio, was a trustee of the Chautauqua Assembly and one of the found-

ing investors (1882) in the Athenaeum, a highly regarded but unprofitable hotel on the Assembly grounds ("Mary Ann Carley," *Western Christian Advocate*, 7 Jan. 1903, 20; U.S. Census Bureau 1963? [1850], roll M432_660, p. 4A, image 12 [Athens, Athens, Ohio]; *Chautauqua* 1884, 77; Vincent 1893, 699). He almost certainly is the F. D. Carley closely associated with various Louisville enterprises, most notably with Chess, Carley and Co., an important front organization for Standard Oil. One historian has characterized him as a "lapsed Methodist minister who set a new standard for pitiless methods" in the oil business (Chernow 1998, 254–55; Carley passport application, 2 May 1885, *U.S. Passport Applications [1795–1905]*, NARA microfilm publication M1372_272, online database accessed through Ancestry.com, 15 May 2013).

8. Sarah Elizabeth Dusenbury Vincent (1832–1909), a teacher from an affluent family in Portville, N.Y., met John Heyl Vincent in Illinois and married him in 1858. "John Heyl Vincent Papers," n.d.; Vincent 1925, 44–47; *ANB*, s.v. "Vincent, John Heyl"; on her formative years and early career, see McGaha 1990, chaps. 5–7, pp. 239–44.

9. George Vincent.

10. Not identified.

11. Jane presumably referred to young William Edison (cf. Doc. 2961 n. 1).

12. Except when away for school, Grace Miller (1870–1952), Mina's youngest sister, lived with her parents in Akron. Jeffrey 2008, 171.

13. Mary Harris (1849–1930) was the sole unmarried daughter of William and Anna Atwell Harris, with whom she apparently had been living in New York or Brooklyn; at her death, she was a resident of Evanston, Ill. Her father, Bishop William Logan Harris (1817–1887) was an educator and missionary. Born near Mansfield, in north central Ohio, he converted to Methodism at age seventeen. He entered the ministry in 1837 and taught for eleven years, eight of them in chemistry, physics, and natural history at the Ohio Wesleyan University. Harris was interested in missionary causes and, after being elected bishop of the Methodist Episcopal Church in 1872, he traveled widely. U.S. Census Bureau 1970 (1880), roll T9_ 892, p. 507D, image 0736 (New York City, N.Y.); *Illinois Deaths and Stillbirths Index, 1916–1947*, online database accessed through Ancestry.com, 14 May 2013; *DAB*, s.v. "Harris, William Logan."

–2975–

To Henry Villard

[New York,] July 24th. [188]6

My Dear Mr. Villard:—

I have had repeated conversations with Mr. Upton on the subject of the Edison Electric Light Company of Europe, and your kind offer to act on their behalf and obtain from the Compagnie Continentale Edison such a satisfactory contract as will enable you to insure to the European Co. a minimum income which will allow of your raising money to enable them to pay off their indebtedness and the bonds on which you have defaulted.[1]

I am personally especially obliged to you for taking up this matter, and on a satisfactory contract being signed, and the money being raised to meet the European Company's liabilities, I shall do myself the pleasure of handing you a certificate for 1,000[2] shares of stock of the Edison Electric Light Company of Europe, as an acknowledgement of your services in the matter.[3]

Trusting that this will be satisfactory to you, I remain, Yours very truly,

TAE

TL (carbon copy), NjWOE, Lbk. 22:264 (*TAED* LB022264). Initialed for Edison by Samuel Insull.

1. The defaulted bonds were those of the Edison Electric Light Co. of Europe, Ltd. (see Doc. 2667 nn. 5, 7–10). Evidently in reply to a suggestion from Francis Upton, Villard indicated his disinclination to acquire such a relatively small sum of bonds either for himself or on behalf of the prospective central station construction company. Villard to Upton, 25 July 1886, Upton (*TAED* MU092); regarding the company's subsequent issue of new bonds, see the Edison Electric Light Co. of Europe statement, [19 Jan.?] 1887, CR (*TAED* CE87041A).

2. This figure was typed as "11,000" and the extra digit struck out by hand.

3. Francis Upton had not directly discussed with Villard compensation for his services. Upton reported that Villard "sees money in the after results and promises to do his best . . . and take his chances." The editors have not found evidence of a stock transfer from Edison. Upton to TAE, 7 June 1886, DF (*TAED* D8630ZBI).

–2976–

Draft Patent Application: Incandescent Lamp[1]

[East Newark?,] ~~J~~August 1 1886

Dyer— ⟨(678)⟩[2a]

Patent— <u>acknowledge Recpt</u> ⟨Ans Aug 3, 86⟩[a]

The object is this invention is to diminish the electrical carrying action which takes place in Electric filiment of Carbon lamps the effect of which is to blacken the globe of the lamp & shorten the life of the filiment. The invention consists in causing a thin transparent liquid ~~filime~~ film Which[b] is a conductor of electricity to adhere to the interior surface of the globe containing the filiment. ~~Th~~Which filim shall be sufficiently viscous to adhere and not to run when subject to gravitation & which will not give off deleterious gases at the temperature which the ~~lamp~~ filiment gives the glass envelope.

Making the interior surface a conductor serves to make such surface about[c] the same potential, electrically as the filiment, hence the tendency to electrical carrying is greatly diminished— The substance which I prefer to coat the globe interior is viscous[d] melted phosphoric anyhdride containing just sufficient water to make it run over the glass— The substance does not give off its water or any gases at[b] the temperature[b] given the globe by the filiment, and[b] is a good conductor for the static Electricity[b] of the globe   where the temperature of the glass is very high, still less water is allowed to combine with The phosphoric anhydride and [---][c] and a transparent gummy mass is obtained. This will not run on the glass, hence to spread it over the interior surface, as a small piece of glass containing iron is placed within the globe with a [pelel?][c] piece of phosphoric anhydride. sufficient water is given it through as narrow glass stem extending within the Exhaust steam. the piece is then heated by holding that part of the lamp over a flame until it becomes glassy— The glass within the iron inside is moved into the mass by means of a powerful Electromagnet on the Exterior[b] and then passed all over the surface interior surface of the globe The glass being brought each time to the source of supply by the magnet   Thus a very thin filim of this substance can be spread over the interior of the glass. The glass-iron piece is then withdrawn from the globe and the lamp is put in connection with the Mercury pump and Exhausted— care sh

where the temperature of the glass is never high such as where the case is large and the filiment small, The phosphoric anhydride can be diluted with sufficient water to make it run evenly over the interior surface of the glass. The Lamp is then put on an Exhauster and heated while being dried, The heat serving to drive off all the aqueous vapor which the substance will give up— The heat should [be at about three?][c] several[c] times greater than That which the globe will afterwards be liable to

I do not wish to confine myself to any particular conducting substance but claim any equivalent substance which shall have the properties enumerated in the first part of this specification. anhydrous[c] Chloride of Zinc may be used. This retains its transparency, & conductivity & gives no vapor at the ordinary Lamp temperature. pure sulfuric acid can be used in low temperature lamps   any chemist will at once suggest other compounds which will serve the purpose

Claim. In an incandencent Electric Lamp a conducting transparent film[b] on the interior of the globe for the purpose set forth

The used of phosphoric acid Chloride of Zinc sulphuric acid or equivalent substances which——
The method of coating by use of the magnet.
and any other strong claims

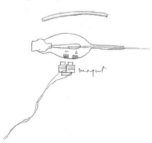

a the glass with iron   b the globule of phosphoric acid.[3]
Bismuth Orthophosphate.

T A Edison

ADfS, NjWOE, DF (*TAED* D8637W1). [a]Marginalia written in an unknown hand, possibly by Edward Rowland. [b]Obscured overwritten text. [c]Interlined above. [d]Multiply underlined. [e]Canceled.

1. See Doc. 2962 (headnote).

2. This draft became Edison's Patent Case 678. He executed the application on 6 August and filed it five days later. It issued on 2 July 1889 as U.S. Patent 406,130. The specification closely followed Edison's draft text and included four claims.

3. Figure labels above are "a," "b," and "magnet." The top drawing, showing a thin coating on the glass, became figure 2 in the patent; the bottom drawing became figure 1.

–2977–

Draft Patent Application: Incandescent Lamp

[East Newark,] Aug 10 1886

Dyer—
    Patent=[1]

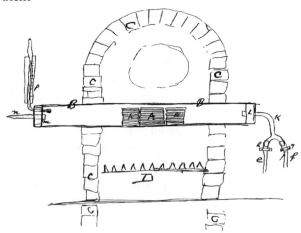

B is a large unglazed porcelain combustion tube passing across the hottest part of the furnace C   in this tube are the forms containing the fibre or filiment of bamboo or other material not yet carbonized. The moulds keep the filiments under strain and pressure. The Combustion tube is closed & luted at both Ends   p is a thermometer to shew the temperature of the Exit gases   n is a tube with fine bore on End to allow gases to pass out. K is a two way tube one connected to a Hydrocarbon fixed[a] gas or vapor The other with Chlorine—both gases or vapors are dried by ~~desicating~~ drying agents to take up the water vapor before entering the tube. The object of the chlorine is to cause the Hydrocarbon gas or vapor to decompose at a lower temperature [~~than?~~][b] The chlorine having an affinity for the Hydrogen of the Hydrocarbon   although the use of chlorine in not absolutely necessary it is better to use it or any other gas which has an affinity for Hydrogen such as Bromine, as Carbon will be deposited in a early stage of the Carbonzation— The gases are passed through slowly the free the tube of air & water vapor. the ~~heat~~ furnace is started at a gentle heat which gradually increases The gases passing slowly through during the process. [~~about 3 more?~~][b] more chlorine than Hydrocarbon should be passed say three of chlorine[c] to one of the Hydrocarbon, and the heat is continued up to the highest heat obtainable[d] by a blower   The fires are then drawn & the tube allowed to cool   The chlorine only being allowed now to pass & this is continued until the tube is below 400 degrees fahr when it is taken out the forms removed and the tube filled fresh forms   of Course several[d] tubes can be put in the same furnace. It is not essential that a Hydrocarbon gas should be used as ~~Sulp~~ Bisulphide of Carbon can be used alone, ~~or with~~ without chlorine.

Seeley note=[a] There have been laboratory Expmts in analysis where flax leaves sticks etc have been put in a combustion tube & carbonzed carbon being deposited thereon.[2] So you will have to make claims in the light of this= I put in a definite article & shews means for accomplishing the result.[e]

TAE

ADfS, NjWOE, Lab., Cat. 1151 (*TAED* NM020ABB). [a]Interlined above. [b]Canceled. [c]"of chlorine" interlined above. [d]Obscured overwritten text. [e]Paragraph from "Seely note=" enclosed by open parenthesis.

1. Figure labels, reading across from left to right and down, are "C," "p," "C," "C," "B," "B," "n," "m," "A," "A," "A," "L," "K," "h," "g," "e," "f," "C," "D," "C," and "C." The editors have not determined whether Dyer filed a patent application on this furnace. An incomplete

drawing for the application (without figure labels) is in Oversize Notes and Drawings (1879–1886; undated), Lab. (*TAED* NS7986BAS).

2. The editors have not found further information about these experiments.

<table>
<tr><td>

–2978–

Notebook Entry:
Incandescent Lamp[1]

</td><td>

[East Newark,] Aug 10 1886—

Made 6[a] lamps with ~~L~~telephone lampblack which is good conductor. an extra platina wire passes in lamp & lays inside lamp down where it is sealed— The Lampblack lays in bulb the lamp being inclined it is heated by Kerosene Lamp to drive off gases & decompose Hydrocarbons & eliminate water— after Lamp sealed off the lampblack is shaken down to bottom of globe where it comes in contact with the Extra platina wire   Lamp is then set up at 80 cp the extra platina wire is connected to positive & some Lamp are connected to negative wire outside lamp   tests show that when wire connected to positive life is longer   it may be that when connected to negative it has a bad effect on filiment & the reason why lamps last longer on p is there is no effect. I have also made several lamps with carbonized Anthracite Coal powder— This gives off scarcely any gas & doesnt dirty the lamp bulb— The p wire to carbon powder gives best results. I notice that the blackening instead of stopping short at clamps extends clear down below clamps to surface of the Anthracite.— ~~I~~

I am trying putting a Coating of glacial phos acid on inside globe & connecting it to the wire that is near the seal so the potential of the globe will be the same as the Clamp— TAE

</td></tr>
</table>

TAE   J F Ott

X, NjWOE, Lab., N-86-06-28 (*TAED* N322AAB). Document multiply signed and dated. [a]Obscured overwritten text.

1. See Doc. 2962 (headnote).

<table>
<tr><td>

–2979–

Notebook Entry:
Incandescent Lamp[1]

</td><td>

[East Newark,] Aug 13 1886

Tried several lamps having ~~fil~~ filiments coated with Alumina, Magnesia— Calcium, Beryllium Zirconia—from their chlorides & acetates—dipping fibre in solution heating of Kerosene chimney to decompose then passing quickly through flame— The Coating holds on filiment up to about 20 to 25 cp then it seems to Jump off although some filiments hold very well— The contraction of the oxide while the filiment doesnt contract but very little causes this Cracking

</td></tr>
</table>

Martin[2] has dipped a dozen carbons in Coal tar in Benzol & Liquorice in water then dipped them in finely powdered oxides of Alumina, & Magnesia, and put them in mould & they are to be run through the carbonizing process, the Theory being that the ɛTar will carbonize & lock the oxide together— the coating is very fair though not complete= I am getting ready a lot of filiments about ½ carbonized ie[a] brought up to 600 deg fahr & these I am going to dip in Tar Liquorice etc &[b] then in infusible oxides & then run through to final carbonization the shrinkage of the Carbon will then be about equal to the tar Carbon— also I am going to dip some bamboo filiments in tar & [Li?][c] infusible oxide & carbonize in regular way[d]

I am also going to soak fully carbonized filiments in tar— Licorice & other Carbonizable liq materials in Liquid shape soaking the filiment before Carbonization also when[e] partially carbonized (this is probably the best period) & when fully carbonized so as to fill up the spaces & breaks due to the initial Carbonization of the bamboo.

TAE   J. F. Ott

X, NjWOE, Lab., N-86-06-28 (*TAED* N322AAD). Document multiply signed and dated. [a]Circled. [b]Obscured overwritten text. [c]Canceled. [d]Followed by dividing mark. [e]Repeated as page turn.

1. See Doc. 2962 (headnote).
2. Martin Force.

–2980–

Notebook Entry:
Incandescent Lamp[1]

[East Newark,] Aug 14 1886

Put two regular Carbons in platina Clamps, then soaked or dipped[a] in Coal tar several times, drying slightly after Each dip— Then gave them to Joe[2] to seal in also another Carbon in platina Clamps dipped several times in liquorice water— gave to Joe I propose to slowly heat lamp from low heat on drier to about 600 fahr put on pump & bring up slowly so as to Carbonize— The O of the Liquorice will probably Oxidize the whole of the Carbon if not sufficiently carbonized in drier—but the Tar will Carbonize in vacuum ok I think

J. F. Ott

X, NjWOE, Lab., N-86-06-26 (*TAED* N322AAE). [a]"or dipped" interlined above.

1. See Doc. 2962 (headnote).
2. John Joseph Force (b. 1860?), a younger brother of Martin, was usually referred to as Joe. He began working as a glassblower at the lamp factory in 1880; this entry appears to mark the start of his work in

Edison's laboratory. U.S. Census Bureau 1965 (1870), roll M593_874, p. 380A, image 154 (Raritan Twp., Middlesex, N.J.); ibid. 1970 (1880), roll T9_790, p. 275C, image 0391 (East New Brunswick [Raritan Twp.], Middlesex, N.J.); Timesheets, NjWOE.

−2981−

To William Wiley [1]

[New York,] August 24th. [188]6

Dear Sir:—

Referring to your favor of the 23rd. inst.,[2] I beg to hand you herewith complete set of cuts referring to my invention known as the Phonoplex System of Telegraphy. I also hand you pamphlet descriptive of same, together with two newspaper notices thereof.[3] Will you do me the favor to forward these to London "Engineering"[4] with my compliments, it having occurred to me that they might like to publish a descriptive illustrative article with relation to the invention, which is entirely novel.[5] Yours very truly,

TAE   I[nsull]

TL (carbon copy), NjWOE, Lbk. 22:350 (*TAED* LB022350). Initialed for Edison by Samuel Insull.

1. William Halsted Wiley (1842–1925), a partner in John Wiley & Sons, the scientific and technical publishing house started by his grandfather, was the New York correspondent for *Engineering* (London) since 1885. A Union Army veteran and a graduate of the Rensselaer Polytechnic Institute, Wiley had also attended Columbia College School of Mines in 1868 and worked as an engineer for about ten years before entering the family business. At about this time, he was on the township committee in East Orange, N.J., an area he later represented for several terms in the U.S. House of Representatives. *NCAB* 14:503; *BDUSC*, s.v. "Wiley, William Halsted."

2. Not found.

3. The pamphlet was probably the illustrated fifteen-page (undated) booklet on the "Edison Phonoplex System of Telegraphy" that Samuel Insull began mailing to railroad officials at the end of July (PPC [*TAED* CA012A]; Insull to Charles Hosmer, 31 July 1886; Insull to William Hovey, 31 July 1886; both LM 12:84, 89 [*TAED* LM012084, LM012089]). The editors have not identified the news clippings, but the phonoplex had been illustrated and described in some detail by the *Electrical World* in its 17 April 1886 issue ("The Edison 'Phonoplex' or 'Way-Duplex,'" 7:177).

4. *Engineering*, an illustrated weekly journal, was founded in London in 1866 by American engineer and journalist Zerah Colburn, with financial assistance from steelmaker Henry Bessemer. Mortimer 2005, chap. 20.

5. *Engineering* published a lengthy original article (with sixteen illustrations) in its 22 October issue ("The Edison Phonoplex," *Engineering* 42:411–13). It also described a novel feature of the system for

the establishment of communication on two separate Morse wires between offices which otherwise are unable to work direct with each other. This is accomplished by what is known as "jumping." For instance, two wires, running in different directions, cross each other at a certain place. The first of these is phonoplexed to the point of intersection with the second, and the induced currents are then thrown into the latter through a condenser . . . [p. 412]

The article's publication may have involved William Wiley's brother Osgood, who placed a different one (albeit with fifteen of the same illustrations) in the *Railway Gazette,* a New York weekly, around the same time. Osgood also inquired about publishing details of Edison's Llewellyn Park home, a request that Edison denied (O. Wiley to TAE [with TAE marginalia], 6 Oct. 1886; O. Wiley to Samuel Insull, 6 and 25 Oct. 1886; all DF [*TAED* D8606G, D8606H, D8606K]; Insull to O. Wiley, 13 Oct. 1886, Lbk. 23:6 [*TAED* LB023006]; "The Edison Phonoplex," *Railway Gazette* 18 [22 Oct. 1886]: 718–19).

–2982–

Notebook Entry:
Telephony

[East Newark,] August 25 1886

Telephone Experiments[1]   Try in telephone powdered—

Platinized light charcoal—
Silver plated
Silicon[a]
Tellurium[a]
Sulphide Lead—
    "   Iron
          Tin
Calcopyrites.[2]

J. F. Ott

X, NjWOE, Lab., N-86-08-25 (*TAED* N326000). [a]Followed by checkmark.

1. Beginning this day, Ezra Gilliland (presumably assisted by John Ott, who witnessed the entries and probably modified the apparatus) experimented for several days with these and other substances in "L[ong] D[istance] transmitter #1" (p. 1). On 28 August they tried double diaphragm arrangements but found those worked no better than the standard design. They then used a "Corrugated Electrode the object being to prevent the granulated material from creeping away from the center of the diaphragm— This worked splendidly and is very decidedly an improvement upon the regular standard form" (p. 10). Pleased with that transmitter's "talking qualities," Gilliland decided to conduct "a long test . . . to determine its staying qualities" (p. 10). During that trial he and Ott discovered that "the new corrugated form of Electrode . . . was somewhat longer and come down near the Diaphragm which partly accounted for its increased loudness" (p. 11). During this time they also

continued to experiment with different substances but failed to find anything that worked better than the standard anthracite carbon. Gilliland
therefore suggested on 2 September the "procuring of highest quality
of anthracite coal   let it be hard and glossy and presume the quality of
carbon will be greatly improved" (p. 15). The following day, he drew
arrangements of corrugated electrodes and dished diaphragms. Also on
2 September, for uncertain reasons, he sketched an induction coil in
a vacuum vessel. N–86–08–25:1–16, Lab. (*TAED* N326001–N326016).

2. Edison presumably meant chalcopyrites; that is, copper pyrites.

–2983–

*Mina Edison to
Samuel Insull*

Oak Place,[1] Aug 30/86

My dear Mr. Insull—

There will be two brothers, Mr. Edison, the children and
myself to take or to have dinner prepared for on Thursday[2]
~~night~~ noon—[a] and also wish you to make one of our happy
number—

Will you please tell Lena[3] that I should like to have her have
everything in readiness and to serve at table—she is the up-
stairs girl—

And to Mary the cook[4] I would like to have you give her this
bill of fare—

Stuffed Duck
Baked sweet Potato
Lima Beans
Sliced Tomatoes with salad dressing
Ice tea if a warm day, coffee, if cold—
Charlotte Russe[5] & fruit—

I hope that you have been well treated while at Glenmont.[6]
Thanking you for your kind service, I am yours sincerely

Mrs. T.A. Edison.

ALS, NjWOE, DF (*TAED* D8614M). [a]Interlined above.

1. The Miller family had occupied Oak Place, their home on twenty-
five landscaped acres in Akron, since about 1870. Hendrick 1925, 103–4.

2. The second of September.

3. Mina apparently referred to Helena McCarthy (b. 1866?). A native
of Ireland, she was working for the Edison family in March 1886, when
she accompanied Marion to Florida (see Doc. 2907 n. 1). She seems to
have married Michael Doyle about 1898 and was still employed by (and
living with) the Edisons as a maid in 1920, although it is not clear if her
service had been continuous. Herron 1998, 1:19–20; U.S. Census Bureau 1982? (1900), T623_968, p. 19B (West Orange, Essex, N.J.); ibid.
1992 (1920), roll T625_1038, p. 1A, image 961 (West Orange Ward 2,
Essex, N.J.).

4. Mary McMahon (variously McMann) was born in England about 1870. She was still in the Edisons' employ as a cook in 1920. Herron 1998, 1:19–20; U.S. Census Bureau 1992 (1920), roll T625_1038, p. 1A, image 961 (West Orange Ward 2, Essex, N.J.).

5. Charlotte russe is a molded dessert of alternating layers of fruit and cake around a center of pudding or custard. Bender and Bender 2001, s.v. "charlotte."

6. Samuel Insull had been staying at the Edisons' home since about mid-August. Insull to Frank Toppan, 26 Aug. 1886, LM 4:026 (*TAED* LM004026).

–2984–

Notebook Entry:
Incandescent Lamp

[East Newark,] Sep 10 86

Cutout[1]

1[2]

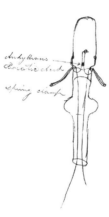

The wires to be plated on same time the carbon is plated on and cover them with rubber tubing

2

To be bound with high resistance compound that will fuse under 1000 Volts and alow spring to close circuit

TAE   J. F. Ott

X, NjWOE, Lab., N-86-08-24:2 (*TAED* N325AAD). Document multiply signed and dated.

1. In late August, Edison and John Ott began a series of experiments on municipal lamps that Ott recorded and Edison initialed in a single

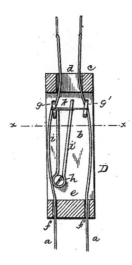

Detailed patent drawing of the cutout mechanism at the base of the Edison municipal lamp. When the high-resistance thread (k) was burned by the current, it released a spring (i′) that closed a bypass circuit around the lamp at g′.

notebook (see also Ott's 28 April notes in N-86-04-28:3–7, Lab. [*TAED* N321001]). The two earliest entries, dated 24 and 27 August, describe efforts to electroplate filaments to the clamps with different metals and metallic oxides. These materials were likely intended to prevent arcing at the clamps when the filament broke. This entry of 10 September marks the beginning of experiments focused on lamp cutouts, which continued through the end of September (N-86-08-24, Lab. [*TAED* N325, images 1–31]; see also John Ott Notebook, Hummel [*TAED* X128B057]). The entries record tests on a variety of high-resistance substances and drawings of different configurations for the cutouts, including those embodied in U.S. Patent 466,400 (which Edison and Ott jointly executed and filed at the end of October). According to the patent, the cutout was designed so that when the filament broke, the full current would pass through a thread of a high-resistance substance, preferably "a mixture of powdered lamp-black with shellac, mucilage, or other adhesive material." The thread would burn immediately, releasing a spring to close a circuit around the broken lamp and maintain continuity to the other lamps in series. If, as sometimes happened, the current formed a destructive arc between the terminals instead of passing through the cutout, the thread would be destroyed anyway. Edison later noted in a patent application filed in June 1887 (U.S. Pat. 476,530) that it was "very difficult" to give the thread "just the right resistance to convey no current when the lamp is in operation and to convey enough to destroy it when the filament breaks."

2. Figure labels are "Anhydrous Boratic Acid" and "spring clamp."

–2985–

Samuel Insull to Margaret Stilwell[1]

[New York,] Sept. 21st. [188]6

Dear Madam:—

The only reason for my not remitting to you is that we have been extremely short of money.[2] My instructions from Mr. Edison are to pay you ~~only~~ $25. per week,[3] and I do not think he has the slightest intention to alter those instructions. I hope to send a check in the course of a day or so. Yours very truly,

Saml Insull   Private Secretary.

TLS (carbon copy), NjWOE, Lbk. 22:421 (*TAED* LB022421).

1. Margaret Crane Stilwell (1831–1908) was the mother of the late Mary Stilwell Edison, Edison's first wife. She was born in New York State but seems to have lived a good portion of her adult life in Newark. After Mary's death, she resided at the Edison home in Menlo Park and took care of Edison's children. Margaret continued to live at Menlo Park in the household of her son-in-law William Holzer until 1888, when she moved to Wakeman Ave. in Newark. She was widowed in 1884. See Docs. 2646 n. 1, 2648; U.S. Census Bureau 1967? (1860), roll M653_688, p. 575, image 573 (Newark Ward 5, Essex, N.J.); Stilwell to Insull, 18 Jan. 1887, 27 Sept. 1887, and 6 Mar. 1888; all DF (*TAED* D8714AAB, D8714AAV, D8816AAD).

2. One of Edison's cashbooks shows regular weekly disbursements of $25 to Mrs. Stilwell from April through May. The editors have not found records of payments from June. In July, she received a single payment of $100 but there is no record of any payment in September or October. In November, she received another $100. There was no payment for December, but she received $200 in January. Cash Book (4 Jan. 1881–31 Dec. 1887): 72–112, Accts., NjWOE.

3. Edison had been providing financial support for his first wife's family since at least 1881. Nicholas Stilwell, his father-in-law, had been able to work only intermittently due to ill health for some time before his death. Mrs. Stilwell seems to have relied on Edison to pay her rent and perhaps other expenses (including tuition for her daughter Eugenie Stilwell) and had to remind Edison's secretaries to send the money promised. Edison was still subsidizing Mrs. Stilwell at least as late as 1897, when she was residing in Asbury Park, N.J. Prior to Edison's second marriage, Mrs. Stilwell often looked after his children and seems to have supervised the housekeeping at Menlo Park. *TAED*, s.v. Stilwell, Margaret Crane; TAE to William Carmen, 26 Apr. 1881, Lbk. 8:251 (*TAED* LB008215); Docs. 2646 n. 1, 2712 n. 4.

–2986–

Draft Patent Application: Incandescent Lamp

[East Newark, N.J.,] Sept 26 1886

Dyer—

Patent= ⟨680⟩[1a]

The object of this invention is to make a filiment obtain a filiment of vegetable matter capable of forming a filiment of carbon after carbonization by heat which shall be free from pith seams and be have all parts of the body of the same relative density   I have discovered that the roots of the palm & many[b] other roots have at their center an extremely hard & homeogenious portion from which filiments can be prepared— after drying especially the scrub palmetto which has a root consisting of a spongy exterior and a central core of like willow in the center of which is a perfectly round cylindrical fibre in certain kinds of the root and [a center?][c] in others a central[d] part from which a cylindrical or flat filiment may be cut. fig 1 shews this kind

⟨Fig 2⟩[2]

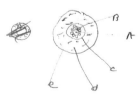

e[d] is the outer shell of the root   a the spongy material   c the large central [-][c] core= A is the solid material surrounded with holes very close to each other   by spitting the [-][c] core c the central part may be obtained nearly round thus.

⟨Fig 3⟩[3]

In other roots from the scrub palm or palmetto which grows abundantly in florida— There are one & sometimes two perfectly cylindrical fibres with small central holes in them which do no harm— These fibres are obtained by splitting the central core

⟨~~Fig 4~~⟩[4]

These fibres have no pithy seems but are ~~the~~ aggregation of an immense number of paralell fibres locked ~~to eath~~ to each other with considerable force—

The ~~wood~~ roots should be gathered while alive the fibres or material taken therefrom and allowed to thoroughly dry before manipulating them to put them in shape for use.

The method of preparing the fibre is by drawing them through a cylindrical cutting die made in two halves the holes being graduated from Large to small the filiment being drawn through successively so as to take a slight shaving Each time

fig 3 shews the apparatus[5]

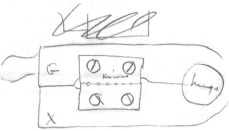

The face of the knives can be ground to keep them sharp— X is fixed G is provided with a handle to open[6]

The fibres are cut the right length and drawn through the first hole   afterwards about ¼ of inch at each end is not drawn through the smaller holes this allows of enlarged ends on the fibres which are handy for clamping   after the last cutting hole the filiment is drawn through a polishing die—

Dick   Must this go in two patents.[7] I want to get a broad patent for making a filiment for carbonization for Lamps made out of the central part of roots of plants or trees— also a specific claim for the ~~fib~~Cylindrical fibres & central part of the roots of palms or palm family— also for the method of forming cylindrical filaments with & without enlarged ends by ~~split~~ cutting dies. split or not, its a <u>cutting</u> draw plate not a draw plate like wire drawers— also for the split dies for allowing enlarged ends—

ADf, NjWOE, Lab., Cat. 1151 (*TAED* NM020ABE). Marginalia written in an unknown hand. [a]Followed by dividing mark. [b]Interlined above. [c]Canceled. [d]Obscured overwritten text.

1. This is the case number assigned by Edison's attorneys, Richard Dyer and Henry Seely. Edison executed the application on 26 October and filed it two days later. The Patent Office examiner initially rejected it for lack of novelty. Dyer & Seely argued that the peculiar characteristics of the plant fibers described were sufficiently different from other patented filament materials. The Patent Office ultimately reversed itself and allowed the application, awarding U.S. Patent 454,262 to Edison in 1891. J. B. Littlewood to TAE, 23 Dec. 1886; Frank Brown to TAE, 15 Jan. 1889; Dyer & Seely to the Commissioner of Patents, 21 Dec. 1888 and 20 Nov. 1890; all Pat. App. 454,262.

2. "Fig 2" was apparently added later, possibly by Dyer or someone in his office. This is figure 2 in the issued patent. Figure labels are (clockwise) "B," "A," "c," "d," and "e."

3. Edison also drew an index pointer toward this figure. "Fig 3" was apparently added later, possibly by Dyer or someone in his office. This is figure 3 in the issued patent.

4. "~~Fig 4~~" was apparently added later and then crossed out, possibly by Dyer or someone in his office. This is figure 1 in the issued patent.

5. Figure labels are "G," "X," "knives," and "hinge."

6. Figure label is "fibre."

7. The text and figures related to the method and apparatus of preparing the fibers were not included in the application. The editors have not determined if Edison filed a separate application.

–2987–

Draft Patent Application: Incandescent Lamp

[East Newark,] Sept 26 1886

Dick[1]

Patent. ⟨685   686 & 687⟩[2]

~~Claim~~ I take the vegetable filiment before carbonization and soak it for several hours in a solution which contains disolved carbonizable matter such as sugar molasses, Licorice, Coal tar [-][a] these materials permeating the interstices ~~being~~ between the microscopic fibres of which the filiment is made up   the filiment is then taken out of the solution the surface cleaned from the adhering solution & then allowed to dry[b]   afterwards it is placed under[b] strain & pressure in the moulds &

then put in closed boxes & carbonized   the extra material being between the fibres Carbonizes & serves to lock them together and thus cause the electric current to pass through all
parts of the filiment equally

 claim— [soa?]ª S̶o̶a̶k̶i̶n̶g̶—

The strength of the solutions should be of a consistency a
little less than table syrup—it may be varied within wide limits and still accomplish the results—

Claim soaking the filiments of c̶a̶r̶b̶o̶n̶ previous to carbonization in a Solution containing a carbonizable material in
solution.ᶜ

Dick— I think some one has soaked wood in sugar etc & afterwards carbonzed same for arc carbons, so draw your claims
accordingly

 2nd patent

Doing the same thing to filiments already carbonized or
partially carbonized & then carbonizing them again & <u>fully</u>

 <u>Dick</u> Carreé makes his Arcᵇ carbons by squirting the compound of groundᵇ carbon & tar through die, carbonizing &
then Soaking in sugar etc & recarbonizing—hence you should
draw your claims to get around this.³

 Prepare these immediately

ADf, NjWOE, Lab., Cat. 1151 (*TAED* NM020ABF). ªCanceled. ᵇObscured overwritten text. ᶜFollowed by dividing mark.

1. Richard Dyer.
2. These are the case numbers of the patent applications related to
this set of instructions. It is likely that Edison executed and filed the
cases together (or nearly so) in the first week of November, when the
only one of them that resulted in a patent (Case No. 686) reached Washington. That application was initially rejected on the basis of earlier
patents for incandescent lamp filaments as well as Ferdinand-Philippe
Carré's process for making carbon arc electrodes (see note 3); it issued
in 1893 as U.S. Patent 490,954. Thomas Autisell to TAE, 12 Jan. 1887;
Patent Office examiner to TAE, 15 Jan. 1889; Gustav Bissing to TAE,
24 Mar. and 2 June 1890 and 18 Dec. 1891; Dyer & Seely to Commissioner of Patents, 4 Jan. 1889, 18 Mar. 1890, 27 May 1890, and 11 Dec.
1891; all Pat. App. 490,954.
3. French engineer Ferdinand-Philippe Carré (1824–1900), a pioneer of mechanical refrigeration, is credited with being the first to freeze
water by artificial means (1863). He investigated the overseas shipment
of frozen meat and explored the use of refrigeration to desalinate sea
water. He was also noted for his carbon electrodes for arc lighting (Alphandéry 1962, s.v. Carré, Ferdinand-Philippe-Édouard; Galiana and
Rival 1996, s.v., Carré (Ferdinand); "Fedinand Carré," *Sci. Am. Supp.*
30 (1890): 12389). His process for producing these electrodes was cited
by the patent examiner in rejecting Edison's application (see note 2).
Edison's patent attorneys replied that the

Carré process consisted in molding sticks of powdered carbon and a binding solution, and then carbonizing the whole. This differs from applicant's process in that applicant begins with an uncarbonized material alone and carbonizes this partially, then soaks it in carbonaceous material and then completely carbonizes the whole. This is a different thing from the use of a mixture of carbon and other material at the beginning of the process." [Dyer & Seely to Commissioner of Patents, 4 Jan. 1889, Pat. App. 490,954]

—2988—

Notebook Entry:
Electric Lighting[1]

[East Newark,] Sept 26 1886

Memorandum—[2]

Put flat platinum inside of filiment & soak in syrupy chloride platinum, or Double Chl of Pt and Ammonia   heat & see if Locks together ie[a] metallic pt deposited   ⟨Tried not very good⟩   also immerse the joints in syrupy solution of chl plat & pass current to bring joint to red heat or use arc decompose the PtCl— also try Hg Amalgam & Cu also Pt amalgam.[b]

Put bamboo filiment in sealed tube along with phos anhy & chloride carbon in cool end   also some Copper dust to absorb the Cl— then treat to red heat   protect Eyes—do it slowly try different thing in tube   Try filiment with PO_5 alone in cool part   also with nothing   also ascertain what temperature great change takes place in carbon— use sand bath & Thermometer   Try Sodium to absorb H_2O instead PO_5.[b]

Put PO_5 Chl Carbon Copper dust & filiments bamboo & get vac   Then carbonize by heat— also fils with CCl only

Paint a plated joint with syrupy PtCl then get vac—(heat the Cu joint in flame before putting on PtCl) then bring fil up so clamp heated & PtCl decomposed— also make a plat socket break shank of carbon & insert tight fit then work syrupy Chl Plat in & decompos by Needle point flame or arc

See if by long running a lamp can be worked up to 100 CP without cleaning clamps[b]

Use spark gauge with carbon electrodes & at low pressure ie[a] point where greatest volume spark—spark must not touch sides glass.—see if carbon deposited   also try in gasolene gas to deposit with spark. Try two carbon electrodes side by side in gasolene & use jump Spark— Try Licorice & sugar on plat to shank & put whole in Mangesia so pt wont melt.— Make 12 plat cups by flattening & spiraling end pt wire   break shanks off carbon insert & plate with Cu get life[b]

Nitrocellulose the surface of reg fibres   immerse in Ether alcohol solution dry & then reduce by sulphydrate ammonia   also cut fils out of nitrocellulose paper— immerse in Ether Alcol Sol dry & reduce to cellulose by Sulphydrate ammonia— try stannite sodium   see if can get solvent for charcoal looking residue from the Oil Cos retorts. See if softens— make some Caramel Carbonize a fil in hot sugar   treat sugar with slaked lime then boil down hard & see if it carbonizes without melting   Soak some paper (smooth) in sugar for 2 days also caramel also starch[c] dry & cut some filiments get life— Draw fibres through die hot see amount lengthening & shrinking

J. F. Ott

X, NjWOE, Lab., N-86-06-26 {*TAED* N322AAI). Document multiply signed and dated. [a]Circled. [b]Followed by dividing mark. [c]Obscured overwritten text.

1. See Doc. 2962 (headnote).
2. A partial draft of this memorandum, labeled "Notes," is in Unbound Notes and Drawings (c. 1887), Lab. (*TAED* NSUN10, image 47).

–2989–

Notebook Entry:
Railway Telegraph

[East Newark,] Sept 30th 1886
Railway T[elegraph] & T[elephone] Experiment.
set up and tested the Edison arrangement of using a singing telephone with cushion contact[1] as a substitute for the Automatic[a] Vibrator—[2] It worked first class— another was made and forwarded to Rudd for practical test on the train on Chi St P & Mil RR[3]   The device[a] was made and set up as shown in the[a] following sketches—[4]

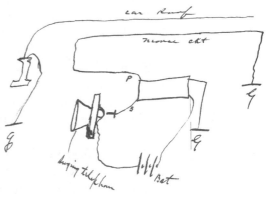

E T G[illiland]

X, NjWOE, Lab., N-86-08-25:17 (*TAED* N326017). Written by Ezra Gilliland. [a]Obscured overwritten text.

1. Edison developed his singing telephone transmitter, a modification of the Reis telephone, early in his telephone researches. The first design dated from July 1876 and the standard form, which emerged in 1877, was used for the transmission of music during public exhibitions. It consisted of "a long tube, having one end covered with a thin sheet-brass diaphragm, which is kept tight by a stretching ring. In the center of the brass diaphragm is soldered a thin disk of platina, and immediately in front of this disk is an adjustable platina-pointed screw secured to a rigid pillar" (Doc. 889 n. 1). In the design shown in Gilliland's drawing, the long tube is replaced with a standard telephone transmitter mouthpiece. The editors have not definitively identified what Gilliland meant by "cushion contact"; he may have had in mind a hydraulic dampening system like those recently patented by Edison and Sigmund Bergmann (see Doc. 2839 n. 3).

Edison's musical telephone, with its signals of relatively high frequency, created powerful induction effects that had been observed well before he began experimenting on wireless telegraphy. In 1877, during a transmission of musical tones from New York City to Saratoga using an earlier version of this device, the sounds were picked up on parallel wires at least thirty feet away and carried to Albany and Providence, R.I. Sewall 1903, 13–15.

2. That is, what Edison called the circuit controller. See Doc. 2780 (headnote).

3. Nothing further is known of these tests on the Chicago, Milwaukee & St. Paul Railway.

4. Figure labels are "car Roof," "Morse ckt," "singing telephone," and "Bat[tery]."

Having spent much of the summer and early fall working to improve his electric lamps, Edison was now forced by circumstances to begin facing a fundamental flaw inherent in his light and power system. The laws of physics governing relationships among voltage, current, and waste heat necessitated an unappealing tradeoff between high electrical losses in the lines or thick (and expensive) copper conductors. Even with every economy Edison had been able to make, including high-resistance lamps and a three-wire distribution network, his direct current system was expensive to build, and construction of a new central station district in New York remained stalled for lack of capital. Those same physical laws, however, suggested an alternative approach: increasing the voltage at which current was distributed would greatly diminish electrical losses. Edison had recognized this fact and made it the basis of his three-wire system of direct current. Other inventors were prepared to take the idea much farther, however, by using alternating current (AC) which, at least in theory, could be transformed easily from low to high voltage and back again. George Westinghouse, intent on building a rival electrical enterprise, was keenly aware of developments in alternating current technology. He supported research in the United States, had acquired the key American patents for transformers developed in Europe, and was starting to build a business enterprise around alternating current distribution.[1]

Edison was aware of emerging AC technologies generally and of Westinghouse's activities in particular. Sometime in October, he outlined in a notebook a handful of "Reasons against an Alternate Current Converter system" (Doc. 3005).

In practice, however, he did not categorically reject the utility of AC. His first inventive step was to design a self-exciting AC dynamo, which would be a significant improvement. Mainly, though, he and Charles Batchelor worked on various electromechanical devices to convert alternating to direct currents or to transform the voltage of direct currents.[2] Between October and the end of 1886, he drafted nearly a dozen patent applications in some way related to high-voltage conversion or distribution. In mid-November, belying the pragmatism of this recent inventive work, Edison wrote a lengthy and uncompromising memorandum to Edward Johnson (Doc. 3008) laying out a comprehensive case against AC on grounds of economics, engineering efficiency, public safety, and an unstated but unmistakable pride in his own system as a stand-alone entity.

Although most of Edison's inventive energies toward the end of the year were focused on electrical distribution systems, he did return in December to a perennial dynamo problem. He designed a new process for the manufacture of dynamo armatures in hopes of reducing eddy currents—and the resultant wasteful heat—in armatures.[3]

Edison faced the prospect of competition in an area in which he took a deeply personal interest: the phonograph. The graphophone, a recording and playback instrument devised at Alexander Graham Bell's Volta Laboratory in Washington, D.C., was receiving favorable publicity and the support of a nascent company. The desirability of a practical and convenient machine for business dictation had been in his mind for years but had always been pushed to the background by more pressing projects. Now, the on-again, off-again discussions between the graphophone's backers and his own Edison Speaking Phonograph Company brought the subject to the fore, and Gilliland proclaimed on 5 October that he had "Commenced work on the standard Phonograph."[4] However, Edison made no serious effort in this direction until May 1887.

Several other activities revolved around the Edison Machine Works. For one, its move to Schenectady was completed in December, necessitating some reorganization of its management. Samuel Insull, Edison's secretary and business manager, went with the company as its treasurer and de facto general manager; his roles in New York were partially taken up by Alfred Tate. With the Machine Works enjoying a much larger plant in Schenectady, Edison tried to expand its production by contracting with inventor Charles Porter to manufacture steam engines he was trying to design.[5] With the aid of Edward Johnson and other associates, Edison also tried to reach a similar agreement with the Sprague Electric and Railway Motor Company (headed by his brilliant former assistant, Frank Julian Sprague) for the manufacture and testing of its heavy-duty electric motors.[6]

The Edison Lamp Company was also looking to expand its markets. Francis Upton returned to Paris in the fall with legal authority to negotiate on behalf of Edison and the Edison Electric Light Company of Europe (based in New York). By late November, Upton secured agreements permitting the Lamp Company to sell its products in European countries previously reserved to the Compagnie Continentale Edison, as part of a broader effort to restructure the fractured Edison lighting business in France.[7]

Edison laid the groundwork for his own (and Mina's) trip to France almost three years hence by accepting an invitation to participate in the 1889 Exposition Universelle in Paris.[8]

Amid all this inventive and organizational activity, so typical for Edison, his personal activities and those of his young wife and children recede from the documentary record. One event that does stand out is the dedication of the Statue of Liberty in October. The occasion was marked by a fireworks extravaganza, for which Edison asked permission to bring his family atop one of the tall buildings in downtown New York.[9] Otherwise, his sons were attending classes at the Dearborn-Morgan School in New York,[10] and his daughter Marion probably resumed her daytime place at Mme. Mears's school, also in New York. Mina's family continued, in various combinations, to be a presence in the Edison home, but Mina seems to have had her doubts about her ability to manage that home. After her brother Edward visited, he chided that she "should be just the happiest young lady in Orange. Why do you let little things worry you so?" Reassuring her that he had felt entirely "easy and at home" there, he advised that she needn't touch anything in the house "more than once a month . . . and still everything would be prim and pretty. Now do not let yourself be worried."[11]

1. See Doc. 3002 (headnote).
2. Batchelor's work with the dynamo-like rotary converters overlapped with his simultaneous effort to address the perennial problem of sparking at dynamo commutators; see Docs. 2999, 3009, and 3010.
3. See Doc. 3019.
4. Doc. 2993.
5. See Docs. 3004 and 3018.

6. See Doc. 2998.

7. See Doc. 3013.

8. Georges Berger to TAE, 3 Dec. 1886, DF (*TAED* D8632D).

9. See Doc. 3003.

10. See Doc. 3015.

11. Edward Miller to Mina Edison, 1 Oct. 1886, CEF (*TAED* X018C2B1).

-2990-

From Arthur Leith

Madison, Wis. Oct. 2, 1886.

Dear Sir:

Preparatory to a debate at the Wis. University[1] this winter on the patent system, an answer from you as an inventor to the following questions would be of great weight and a favor highly appreciated.[2]

1. What do you think is the chief weakness of our present patent system?

⟨The system of declaring interferences—[3] The first man in the office should have the patent and to protect poor inventors a brief description should be filed for one dollar in the inventor's handwriting without any expenses preparing drawings⟩

2. Do you favor the retention of the present system?   ⟨Yes⟩[4]

3. If not, what system would you substitute? I am   Very respectfully,

A. T. Leith[5]

ALS, NjWOE, DF (*TAED* D8603ZCB). Letterhead of Wisconsin Dept. of State.

1. The University of Wisconsin was founded at Madison in 1848. It was reorganized in 1866, under the land grant provided by the Morrill Act, to include an engineering college as well as colleges of law and agriculture, and letters and sciences. Lathrop 1904, 8, 5–17; Curti and Carstensen 1949, 1:37–119.

2. Edison's marginalia was the basis for a slightly expanded typed reply to Leith on 7 October 1886. Lbk. 22:478 (*TAED* LB022478).

3. Under U.S. patent laws, the Commissioner of Patents had authority to initiate an interference proceeding when an application conflicted with another one or an unexpired patent for essentially the same invention. The law provided that the first inventor, not the first to file an application, was entitled to the patent, and the interference itself was a quasi-judicial process for determining who met that criterion. If the commissioner determined that the holder of an unexpired specification was not the first inventor, he was not empowered to revoke the original patent but could issue a new one to the first inventor, leaving the claimants to vie in federal court to enforce their conflicting rights. Smith 1890, 52; Abbott 1886, 2:152–53.

4. Edison wrote over this line of text, partially obscuring it. Insull

evidently misread it so that the typed reply he sent (see note 2) read: "I favor the intentions of the present system."

5. Arthur Tennyson Leith (1867–1955) was a student at the University of Wisconsin. He graduated in 1889 and at some point moved to Washington, D.C., where he was for many years a printer in the Government Printing Office. He co-authored *A Summer and Winter on Hudson Bay* (1912) with his brother C. K. Leith, then the chairman of the geology department at the University of Wisconsin. U.S. Census Bureau 1980? (1900), roll T623_158, p. 14B (Washington, D.C., enumeration district 12); "Necrology," *Wisconsin Alumnus*, 57 (15 Feb. 1956): 34; Florida Death Index, 1877–1998, online database accessed through Ancestry.com, 29 May 2012; "Mourned as Dead," *Washington Herald*, 17 Feb. 1910, 1.

–2991–   [East Newark,] Oct 3 1886

Notebook Entry: With the new clamps John Ott is making use platinum foil
Incandescent Lamp   around carbon & shove into a springy socket of platinum[a]

get some apple wood also Box wood   cut fibres—[b]

Stamp out in die plat wire then in another die make it square thus

fibre

get some electrotypers plumbago[1] put some filaments in— also put some in bottom cup put fibre in & then with plunger press it hard then add more plumbago & also fibre press & so on—

Make Moulds of Dextrine & plumbago press— See if Lamp Co got the dies for forms plumbago a reg carbon mould— Draw Manilla through die after it has been soaked in following Sugar Caramel, Licorice starch tragacanth Linseed Mucillage, Gum Arabic Rosin. dry the fibre then with hot die also through hot oil to soften binding material draw down very hard dry & carbonize   Try all the fibres this way—Also bamboo[c]

Carbonize dozen filiment angle of 20 also angle 45

Soak No 20 carbonze fully & resoak & recarbonize.

Write our man in Fla[2] sending sample to dig & send bbl[3] roots also 2 whole banana stalks green—

get some white lacquer & Lacquer spark gauges with Alumina wire see if it changes   also doz lamps— Try Copal—

pass current through a reg carbon in air under Microscope at dull red & notice oxidations[c]

put bamboo in small bulb with light capilliary tip[d] get vacuum heat it slightly then break under sugar water & while full seal & heat for ½ hour so as to get increased pressure—[c]

put bamboo in sealed tube with sugar & heat to 180 to 220 for ½ hour[c]

Boil 12 Carbons in No 20 solution all day.

Try Manilla for wetting   use aniline & sugar

~~draw~~ soak in hot tragacanth for ½ day Manilla then take out & draw it round through split die of Keller[4]   dry in drying oven then cut in lengths & carbonze in Anthricite[c]

also put bamboo in Liquid Sugar in Vac with U tube of Chl Cal to take up water, color with aniline[c]

X, NjWOE, Lab., PN-86-03-04 (*TAED* NP021D). Document multiply dated. [a]Paragraph overwritten by heavy diagonal lines. [b]Paragraph overwritten by heavy "X." [c]Followed by dividing mark. [d]Obscured overwritten text.

1. Wax molds used in electrotyping were coated with plumbago (graphite) to make them conductive. The plumbago had to be free of grit and was "carefully sifted through muslin or a fine wire sieve." Urquhart 1881, 149–50.

2. Legrand Parish had recently arrived in Fort Myers to act as caretaker after Eli Thompson's departure. Edison wrote to him on 23 October. The letter has not been found, but Parish received it four days later and promptly hired a man and a boy to gather roots. He shipped to Edison on 29 October "a bundle of 50 cabbage Palm roots" with a letter indicating that he could get "cottage Palm roots with less trouble and expense than the scrub palmetto," which was obtainable only along the riverbank. The editors have learned nothing more about Parish beyond correspondence collected in the Edison Papers; in two of his letters, however, he addressed Edison as "Dear Cousin." Parish to TAE, 5 and 29 Oct. 1886, 10 Jan. 1887, all DF (*TAED* D8614N, D8603ZCO1, D8745AAA).

3. Abbreviation for "barrel."

4. Albert K. Keller had been working as an "electric machinist" at the Mechanical Dept. of the Bell Telephone Co. in 1885, where he filed at least one patent application (*Boston Directory* 1885, 608; U.S. Pat. 335,364). He accompanied Ezra Gilliland to New York when Gilliland resigned as superintendent of the Mechanical Dept. to join Edison in October 1885. Keller subsequently took charge of materials being shipped to Fort Myers and stayed in Fort Myers to superintend work on the buildings (Gilliland to Insull, 10 Oct. 1885; TAE to Keller, 29 Dec. 1885; both DF [*TAED* D8503ZDC, D8539N]; Doc. 2860 nn. 3 and 5; "More Men for Edison," *Fort Myers Press*, 9 Jan. 1886, News Clippings

[*TAED* X104S004C]). By August 1886 he was among the staff at Edison's laboratory at the lamp factory, and the following year he assisted with experiments on the phonograph (N-86-08-03; technical note, 6 June 1887, Unbound Notes and Drawings [1887], both Lab. [*TAED* N324, passim; NS87AAK]; Edison Lamp Co. to TAE, 18 July 1887, enclosing timesheet of 9 July, DF [*TAED* D8734AAM]). When Gilliland set up a factory to manufacture phonographs in Bloomfield, N.J., at the end of 1887 he made Keller superintendent. Keller was later involved in the development of coin-in-slot phonographs and patented several additional telephone improvements (Welch and Burt 1994, 32; Koenigsberg 1990, xxxv, 7, 60; U.S. Pats. 644,206; 645,958; 645,959; 645,960; 646,701).

–2992–

Notebook Entry:
Incandescent Lamp

[East Newark,] Oct 4 1886

Boil fil bamboo also carbons in Coal tar until carbzd[a]
also in Asphalt thinned by Turpentin
also Carbons High temptur
Oxalic acid melts 212—C
Boiling p[oint] Benzoic Acid 225 C
Saccharose (cane Sugar) 160 C Melting point
Soak Bamboo fils in Cupric Ammonum few min [–][b]
also dip & allow Sol to dry—[a]
Carbonize[c]
Treat some bamboo fils with HCl— SO_4— KO Ammonia— Carbonize—[d]
also same but wash by Soaking in H_2O—
Have Mills[1] Cut some parchmentized paper filiments reg A—[2] Boil in Molasses 10 hours dry & Carbonize—
also carbonze the paper fils without soaking & then Boil carbons in Molasses also sugar—[a]
Make a shaving Knife[3]

1 pt Cotton wool be disolved in Mixture 24 pts SO_4 & 6 pts H_2O a gelatanous precip is thrown down on adding more water   This is amyloid, same as undercoating of parchment paper—

X, NjWOE, Lab., PN-86-03-04 (*TAED* NP021E). Miscellaneous unrelated calculations not transcribed. [a]Followed by dividing mark. [b]Canceled and followed by dividing mark. [c]Connected by brace to "Soak Bamboo . . . sol to dry—." [d]Obscured overwritten text; followed by dividing mark.

1. William Albert Mills (c. 1856–1927) began working for Edison in 1880 as an assistant in the Menlo Park laboratory. He was currently

employed at the lamp factory in East Newark. *TAEB* 5 App. 2; "W. A. Mills Dies   Electric Pioneer," *Newark Evening News,* 11 Nov. 1927; U.S. Census Bureau 1970 (1880), T9_roll 801, p. 523C, image 0450 (Rahway, Union, N.J.).

2. That is, carbons for lamps designed to produce 16 candlepower at approximately 100 to 110 volts, the standard range for central station service. Doc. 2085 n. 4.

3. This sketch was followed by another one, much fainter and likely made at a later date, showing what may be a blade at an angle to an unspecified horizontal surface.

–2993–

Notebook Entry:
Phonograph

[East Newark,] ~~Sept~~ October 5th 1886

Commenced work on the standard Phonograph—[1] Plan is to make a small compact instrument suitable for office use. It is not expected that it will talk loud but is to be made to be held to the listeners ear like a telephone and to be made to talk about as loud and clear as a good telephone on a short circuit—

Is[a] to be driven by a small motor, probably an electrical motor, and so made that it can be readily stopped and started[b] and backed up or reversed ~~or~~ set back—motors to arranged to run as near as possible a uniform speed and have a simple or convenient regulating device to control speed—

The greatest height of perfection will be to make cylinders or or plates containing the record interchangeable, ie a talking record made in one machine to be transferred and reproduced in another ~~machine~~—although the machine will have great commercial value ~~even~~ if this cannot be accomplished—

The cylinder should be about ~~two~~ 1 inch to 1½ inches in diameter made of glass or polished steel should be about 4 or 5 inches in length and have 40 to 50 threads to the inch   this will give it a capacity of about 10,000 words based on 1½ diameter or 5 in in circumference 50 threads to inch would be 250 inches to inch of cylinder in length. 5 in long would give ~~25~~ 1,250.[a] inches to the cylinder with say about 8 to 10 words to the inch or 10,000 words to the cylinder— I believe that diaphragms and needles or points can be made [---][c] so cheap and simple that they can always be a part of the cylinder and moved from the machine with the cylinder and thereby we accomplish interchangeability—as there is no difficulty in repeating many times the record made if the point and diaphragm are not disturbed   a cylinder to thread and diaphragm & point can be the detachable portion of a machine, and the running gear and motor and all other parts can be contained in the stationary part or balance of the mechanism.

The motor should be connected through the mechanism of a flexable shaft to prevent the buzz or jar being communicated to the Phonograph. This will also make the apparatus more convenient to use providing it is made a size that will admit of its being held to the ear to listen and to the mouth to talk to it, in case it cannot be made light enough to admit of this then a flexable speaking tube can be used for talking into and listening, in this event the jaring sound of the motor will not be as likely to interfere—

The cylinder should be made of polished glass or metal and the substance that receives the record or vibration should be either a shellac, gum or wax something of that nature which can be applied with a brush or by dipping into a liquid solution and allowing it to dry on and can be dissolved off and thereby prevent any scratching or injury to the cylinder and be cheap and require no special skill or devices to accomplish this most important part of the work— Gums or shellacs or substances of that nature will be less likely to produce the scratching sound which has been such a serious trouble in the use of tin foil   If the diaphragm & needle are made to always be kept[a] together then the cylinder can be prepared by the Phonograph Company or their[d] Experts and furnished to the customers and a rental charged and a continuous revenue derived—

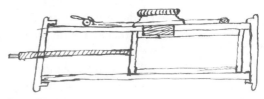

Plan for cylinder diaphragm and point screw and bearings to always to kept together   end of screw has square or other means of connecting it to the motor— the cylinder gets[a] its bearing on a rim or flange located at each end which is made to exactly fit the ~~boar~~ bore of the casting which holds the mouth piece— The mouth piece is made to adjust by revolving   it is also fitted with a device for lifting it when cylinder is to be set back to ~~starting~~ point[2]

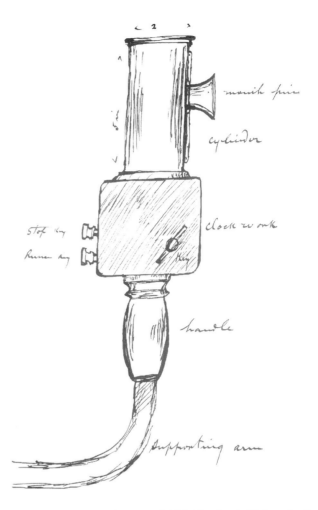

mouth piece

cylinder

clock work

Stop Key

Reverse Key

Key

handle

Supporting arm

X, NjWOE, Lab., N-86-08-25:19 (*TAED* N326019). Written by Ezra Gilliland. [a]Obscured overwritten text. [b]"and started" written at bottom of one page and top of the next. [c]Canceled. [d]Interlined above.

1. Edison had tried unsuccessfully in 1878 to produce a "standard" phonograph that could "be practically applied to numerous branches of commercial and scientific industry" (see Docs. 1276 n. 1, 1397, 1484; *TAEB* 4, chaps. 2 and 6 introductions). Instead, the phonographs in circulation then were all considered exhibition instruments. Gilliland, who had exhibited the device in 1878, probably had that initial aspiration in mind both now and in 1884 when he and Edison first discussed "reducing it to a practical instrument" (see Docs. 1334 n. 1 and 2743 [headnote]). The likely impetus for their renewed interest was the emergence of the graphophone as a credible competitor during the summer of 1886.

In 1881, Alexander Graham Bell and his associates (his cousin Chichester Bell and machinist Charles Sumner Tainter) at the Volta Laboratory in Washington, D.C., had begun making experiments on sound recording on a small Edison demonstration phonograph (Doc. 1195).

They found that using wax as a recording surface greatly improved the fidelity of the sound. By 1884, they had developed the graphophone, a machine that recorded on removable wax cylinders called phonograms; the following year, they filed several related patent applications that issued on 4 May 1886. Edison likely had become aware of the graphophone by 1885 when Edward Johnson, who was a director of the Edison Speaking Phonograph Co., began discussions with Bell and Gardiner Hubbard (who was both Bell's father-in-law and president of the Edison company) about merging the graphophone and phonograph interests. As part of these negotiations, the Volta associates arranged to have several graphophones made at Bergmann & Co., where Edison then had his laboratory. Finding resistance from Edison's allies on the board, Hubbard organized the Volta Graphophone Co. in early 1886. Public exhibitions of the graphophone began during the summer, and *Harper's Weekly* featured it in an illustrated article on 17 July (Wile 1990a; Maguire 1886).

2. Figure labels are "mouth piece," "cylinder," "Stop key," "Reverse key," "Key," "clock work," "handle," and "supporting arm."

-2994-

Draft Patent Application: Incandescent Lamp

New York,[1] Oct 8 1886[a]

Patent. ⟨688[b] 689 & 690⟩[2]

Filiments are formed from oxides got in a plastic state by mixing a small quantity of material which combines with the principal oxide such as an alkaline silicate in small quantity mixed with pure oxide of Alumina, ~~Al~~ Magnesia Zirconia These with water become plastic like clay when finely divided and can be squirted[c] through dies by pressure in the form of cylindrical filiments. These being bent in shape desired are brought up to a full red heat after wards they are taken out & soaked in a carbonizable compound in a liquid form such as sugar—Licorice, Etc[c]

This penetrates all the pores and then the whole is [~~reca?~~][d] put in the furnace in a box with powdered anthracite & brought to a white heat, the walls of [the?][e] pores & surface are coated with carbon. [~~Th---~~][d] as the Carbon does not reduce these oxides, The filiment will stand a high temperature— If it is desired that a lower resistance filiment[c] be obtained a second Soaking & recarbonization can be had or the original porcelanic filiment may have incorporated with it a small quantity of the Carbonizable compound   in this case no alkaline silicate is necessary as the Carbon will act as a binder— The great advantage of filiments made in this manner[c] is that they have <u>very</u> high resistance hence very small copper wires may be used to distribute light over a large area, which is of the highest importance[3]

ADf, NjWOE, Cat. 1151 (*TAED* NM020ABG1). Letterhead of Thomas A. Edison. ᵃ"New York," and "188" preprinted. ᵇFollowed by dividing mark. ᶜObscured overwritten text. ᵈCanceled. ᵉ"walls of [the?]" added in right margin and written off edge of page.

1. Edison drafted this application on old letterhead from 65 Fifth Avenue. The editors have not determined if he did so in his new offices at 40–42 Wall St. in New York or in his laboratory at the lamp factory in East Newark, N.J.

2. These are Edison's patent application case numbers. A docket note on the back of this draft indicates that an application was prepared. The editors have neither found it nor identified any applications associated with those numbers, though it is possible that this draft was the basis for one of the numbered cases. Edison marked an "X" on each page of this draft and designated the two other applications he started on this date (Docs. 2995 and 2996) as "B" and "C," respectively. The latter drafts were the basis of applications assigned Case Nos. 698 and 683, respectively. See App. 2.A.

3. See Doc. 3002 (headnote).

–2995–

Draft Patent Application: Incandescent Lamp

New York,[1] Oct 8 1886—ᵃ

Patent—[2]

I form ~~the~~ aᵇ filiment of clay by forcing it through a hole by a press same as they make arc Carbons etc—~~it is~~ or roll it out in sheets and stamp the filiment out while plastic. if it comes from the die it is bent in shape and slowly baked until nearly all the shrinkage is out— I then put ~~it~~ severalᵇ in a mould mixedᶜ with powdered anthracite Coal & bring them up to a white heat— afterwards they are taken out and put into a tube which can be brought to a white heat a hydrocarbon gas or volatile compound ~~of car~~ containing carbonᵈ is passed through it   this deposits carbon over the whole surface of the porcelain filiment. afterwards the porcelain is eaten away by Hydrofluoric acid or other solvent. The carbon filimentary shell put in a holder on the Ends of the wires running through the inside part of the Lamp & Electroplated thereto—

any oxides or compound which can be moulded or got in shape while plastic & which will stand a white heat will answer—

plumbago may be rubbed down the surface of the non conducting filiment so that its Entire surface becomes a conductor & on passing the Current through the [fil---]ᵉ of plumbago its brot to incandescence & a hard coating is formed on the surface while it is immersed in an atmosphere of a gaseous compound containing Carbon— If an oxide like pure Alumina, Magnesia, or Zirconia be used it is infusible hence it will not be necessary to eat it away by acid or other solvent.

TAE

ADfS, NjWOE, Cat. 1151 (*TAED* NM020ABG2). Letterhead of Thomas A. Edison. [a]"New York," and 188" preprinted. [b]Interlined above. [c]Obscured overwritten text. [d]"or volatile compound . . . carbon" interlined above. [e]Canceled.

1. Edison drafted this application on old letterhead from 65 Fifth Avenue. The editors have not determined if he did so in his new offices at 40–42 Wall St. in New York or in his laboratory at the lamp factory in East Newark, N.J.

2. Edison designated this draft as "B" on each page (see Doc. 2994 n. 2). It became the basis for an application that he executed on 26 November and filed ten days later. The specification issued in September 1889 as U.S. Patent 411,020.

—2996—

Draft Patent Application: Incandescent Lamp

New York,[1] Oct 8 1886[a]

Patent= Carbonizing under pressure.[2]

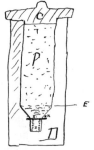

Z are a number of filiments of organic matter. D is a Crucible— C the cover X a piece of Carbon to prevent the filaments from going upwards. The Crucible & Cover is made of Carbon ie[b] plumbago   D is filled with powdered Lead up to E & from there to the cover with large pieces of Lead

The whole is placed in a furnace and gradually brought up to a white heat. when the Lead melts it surrounds the carbons completely and produces a great pressure on them due to the column of Liquid Lead—and as the lead does not boil until the melting point of wrought iron is reached the filiments are perfectly Carbonized & consolidated by the pressure. The Lead being liquid allows them to contract without interposing much[c] resistance— The mould is taken out before the Lead has solidified and poured out leaving a little dross around the Carbons which can be removed by acids— In practice it is preferable to put a small carbon box at the bottom of the

Crucible & cover the whole with a powdered alloy of lower ~~fusing~~ melting[c] point than lead so that the Carbons will be surrounded with a liquid before any great change takes place. Other metals may be Employed such as tin Zinc etc—

ADf, NjWOE, Cat. 1151 (*TAED* NM020ABG). Letterhead of Thomas A. Edison. [a]"New York," preprinted. [b]Circled. [c]Interlined above.

1. Edison drafted this application on old letterhead from 65 Fifth Avenue. The editors have not determined if he did so in his new offices at 40–42 Wall St. in New York or in his laboratory at the lamp factory in East Newark, N.J.

2. Figure labels are "C," "P," "E," "X," "n," "Z," and "D." Edison first considered the idea of carbonizating filaments under pressure by molten metal in a notebook entry of 18 March (Doc. 2912).

Edison designated this draft as "C" (see Doc. 2994 n. 2). He executed the application based on this draft on 26 October and filed it the next day. However, it did not issue as U.S. Patent 484,184 until October 1892. Although the patent examiner rejected the application in early January 1887 on the ground that it was anticipated by the patents of Providence, R.I., inventors Stephen Nicholson and Herbert Bowen (U.S. Pat. 329,670; Brit. Pat. 2,432 [1882]), Edison's attorney did not act on this rejection until January 1889. After some back and forth with the Patent Office and the substitution of new claims, Edison's application was finally allowed on 12 April 1890. Edison failed to pay the fee in time and had to petition the Patent Office the following March to allow him to renew his application. He was allowed to do so and the Patent Office issued the patent after he paid the fee (Pat. App. 484,184).

–2997–

Everett Frazar to Frazar & Co.[1]

[New York,] Oct 12 [188]6.

Dear Sirs—

Mr Edison has handed me the following memo:—[2]

"I want one half doz. complete plants, from which the manilla fibre is taken and used for rope making. The leaves which contain no fibre need not be sent, neither are the roots desired; but if a root will stand transportation, so as to grow in hot house, would like one. The fibre from the stalk is ruined for my purpose by the farmers in their process of getting it, hence I desire the plant whole that I may extract it myself   The stalks should be put in a box" (say ten feet long)

Please send to Manilla for the above, and when procured, ship overland, with memo of all expenses incurred, which I will collect from Mr E. Yours faithfully

Everett Frazar[3]

LS (letterpress copy), NjWOE, DF (*TAED* D8630ZCK).

1. This letter was addressed to the company's Yokohama office. A notation in the upper left corner indicated the ships by which it might be carried there—the *Belgic* on 19 October and the *Peking* on 30 October.

2. Edison's original memo (not found) was evidently dated the same day. The company referred to it in transmitting a copy of this document to Edison nine days later. Frazar & Co. to TAE, 21 Oct. 1886, DF (*TAED* D8630ZCO).

3. American merchant and diplomat Everett Frazar (1834–1901) founded a trading company under his name in Shanghai in 1858 and subsequently expanded his interests to Japan, Korea, and Hong Kong. Frazar had been deeply involved in efforts to bring Edison's telephone and electric light into commercial use in East Asia. Doc. 2678 n. 1; cf. Doc. 2887.

–2998–

Charles Batchelor
Journal Entry

[New York,] Oct 22. 1886

230.[1] Sprague Motors[2]
Meeting between EHJ, Bergman, T.A.E. & myself it was agreed that E.H.J. would accept the following for the future and endeavor to make Sprague accept the same:— The E.M.Wks to do all the ~~experimenting~~ manufacturing. Sprague to have an experimental shop here at New York— We to furnish him with everything in the shape of material and finished parts at actual cost without any profit at all to us[3]

AD, NjWOE, Batchelor, Cat. 1336:135 (*TAED* MBJ003135).

1. Charles Batchelor consecutively numbered each entry in this journal.

2. See Doc. 2885.

3. As the nominal president of the Sprague Electric Railway and Motor Co. (SERM), Edward Johnson provided a personal link to the Edison interests, including the Edison Electric Light Co. (of which he was president), Bergmann & Co. (of which he was a partner), and Edison himself. Charles Batchelor bought 50 shares of the firm's stock in March 1886; by November, when it increased its capital ten-fold to $1,000,000. SERM's stockholders also included Johnson, John Tomlinson, Eugene Crowell, and George Barker. Since obtaining a patent license from the Edison Electric Light Co. in 1885, SERM had built up a modest but promising trade in manufacturing motors for elevators and similar machinery (Dalzell 2010, 67–75; Sprague Electric Railway and Motor Co. agreement with Edison Electric Light Co., 16 June 1885, Miller [*TAED* HM850256]; Johnson receipt to Batchelor, 31 Mar. 1886 and item 249 [20 Nov. 1886], both Cat. 1336:12, 145; Batchelor [*TAED* MBJ003012, MBJ003145B]). In that interval, the Edison Machine Works also built a number of Sprague's experimental railway motors. By early 1887 Sprague had leased factory space in New York City at the Union Lead Works on West Thirtieth St. where, according to a press report, he expected to spend much of his time (Dalzell 2010, 75; "New

York Notes," *Electrical World* 9 [12 Feb. 1887]: 82; regarding recent motor orders and tests see Cat. 1336:117–23 [items 196, 199, 201, 208, 212; 14, 16, 23, and 30 Sept. 1886]; all Batchelor [*TAED* MBJ003117–MBJ003123A]).

–2999–

Charles Batchelor
Journal Entry

[New York,] Oct 23 1886

234.[1] New Armature[2]

Edison proposed an armature as follows:— Put a resistance of German silver wire in between the cup[3] and the bar say equal to the resistance of the armature. This will make the total armature more uneconomical but it will reduce the heating effect of short circuiting each coil to a minimum. It will also reduce the sparking due to the current generated in the coil which ought to be in the neutral point.[4] Am making an armature this way.

AD, NjWOE, Batchelor, Cat. 1336:137 (*TAED* MBJ003137A).

1. Charles Batchelor consecutively numbered each entry in this journal.

2. See Doc. 2885.

3. That is, the portion of the commutator bar shaped to receive the end of an armature induction wire or bar, to which it typically was soldered. See e.g. "American Notes," *Electrician* 22 (19 Apr. 1891): 691.

4. Edison and Batchelor had returned in late spring to dynamo armature heating and commutator sparking, persistent and related problems that limited a machine's efficiency and output (and also had some bearing on the construction of rotary converters or "transformers"). Like other dynamo designers, they recognized the importance in this regard of the neutral point, at which an armature coil passes from one magnetic field to another. In all rotary dynamos, the position of this point is affected by the current flowing in the armature, whose resulting rotating magnetic field distorts the lines of force and shifts the neutral point (see e.g. Doc. 2420 n. 14). In May and June, at Edison's direction, a skeptical Batchelor made a few efforts to construct a "non polar armature" by using a novel compound winding to magnetize the iron core oppositely to the field induced in its windings. When they returned to these experiments in early October, Batchelor noted that counter-magnetizing the core was ineffective when "by far the larger part of the spark comes from self-induction" when the commutator bar broke contact with the brush. He then tried another form of compound winding to produce a counter-electromotive force in each coil at the moment it passed from a commutator brush. After expressing some renewed doubt about the significance of self-induction, Batchelor tried an experiment on 20 October that confirmed it as a major cause of sparking (Cat. 1336:31, 75, 95, 123, 129, 135 [items 66, 106, 148, 215, 218–19, 232; 11 May, 11 June, 8 July, 5, 9–10, 22 Oct. 1886], Batchelor [*TAED* MBJ003031B, MBJ003075, MBJ003123B, MBJ003095A,

Charles Batchelor's 4 November sketch of the extra resistance between the armature induction coils and commutator bars.

MBJ003129A, MBJ003129B, MBJ003135]). William Andrews and a partner, Thomas Spencer, were working along similar lines and filed a patent application in August (U.S. Pat. 406,415).

Batchelor tested the "Non sparker" armature described in this document on 4 or 5 November and "noticed there was always a spark when a heavy change of load but it did not seem to be a cutting spark." A day or two later, he and Henry Walter filed two patent applications intended to address self-induction with this sort of spark-suppressing compound wiring. Batchelor planned a series of variable-load tests to compare the new armature directly against a standard one but seems not to have followed up until May 1887. At that time, when he used a resistance equal to twice that of the whole armature between the induction coil and commutator, he found that the machine ran well, with little sparking, as either a dynamo or a motor. Batchelor indexed his journal entries on this subject by drawing a small circle on each one (Cat. 1336:139, 215 [items 241, 372; 4 Nov. 1886, c. 27 May 1887], Batchelor [*TAED* MBJ003139B, MBJ003215]; U.S. Pats. 360,258 and 360,259). Batchelor also made notes related to sparking and armature heating in another notebook (Cat. 1234:13, 21, 24; Batchelor [*TAED* MBN005013, MBN005018, MBN005024]).

—3000—

From Samuel Insull

[New York,] October 25th. [188]6

My Dear Edison:—

I met Mr. Villard on the street to-day and he told me he would keep an appointment here to-morrow at twelve o'clock, provided you will be here.[1] I have just informed Mr. Upton and have heard the telephone reply which you have answered to his request that you should be here. I certainly think in the case of Mr. Villard you ought to put yourself to the inconvenience of coming to New York. It was very kind of him to take up the European matter, and I do not think it is right to delegate to Mr. Upton and [--][a] myself the work of dealing with a man of Mr. Villard's standing. Mr. Upton will explain this to you himself. Yours very truly,

S. I[nsull].

TLS (letterpress copy), NjWOE, Lbk. 23:58 (*TAED* LB023058). [a]Canceled.

1. Though Edison was in the habit of working at his laboratory in the East Newark lamp factory, Insull addressed this letter to him at home in nearby Llewellyn Park. The editors have not determined whether Edison attended the next day's meeting, although Francis Upton reported a few days later having talked over European business matters with him and Villard. Upton to Louis Rau, 29 Oct. 1886, DF (*TAED* D8625E1).

Samuel Insull to
Joseph Hutchinson

[New York,] October 25th. [188]6

Dear Sir:—

Mr. Edison desires me to inform you that there is a very large watch factory going up at Canton, Ohio.[1] You can judge of the size of the establishment by that fact that they are going to employ 2,000 men. He suggests that you look into the matter as to lighting same.

Mr. Edison desires that you send to Mr. Louis Miller, Oak Place, Akron, Ohio, prices for 100, 200, 300, 400, 500, 600, 700, and 800 light plants, respectively, with and without engine power, and estimate the wiring, plant to consist of sixteen candle power lamps; wiring to be of character suitable for machine shops.[2] Mr. Miller controls the Buckeye Mower & Reaper[3] business and might possibly purchase a plant. Yours very truly,

S. I[nsull].

TLS (carbon copy), NjWOE, Lbk. 23:49 (*TAED* LB023049).

1. John Dueber, principal of the Dueber Watch Case Co., purchased a controlling interest in the Hampden Watch Co. in 1885. With the enticement of $100,000 in donations from residents of Canton, in late 1886 he began building a new plant there for the combined firms. The factory had separate facilities for the manufacture of cases and watch works; after its completion in 1888, the Dueber-Hampden Co. became one of Canton's largest employers. Gibbs 1954, 3–19; Kenney 2003, 54–55.

2. The Edison United Manufacturing Co. promptly acknowledged Insull's letter, noting that they had apprised the Central Edison Co., in Cincinnati, of the new factory. They also promised to send an estimate to Lewis Miller (not found). Edison United Mfg. Co. to Insull, 27 Oct. 1886, DF (*TAED* D8629B).

3. See Docs. 2826 n. 17 and 2853 n. 1.

EDISON AND HIGH-VOLTAGE ELECTRICAL DISTRIBUTION   Docs. 3002, 3005, 3008, 3009, 3010, 3011, 3014, and 3022

In October and November, Edison took steps to adapt his system of direct current (DC) electric distribution to a new competitive threat from high-voltage alternating current (AC) systems. He also outlined the case against high voltage AC on grounds of economics, engineering efficiency, and public safety. He argued at the same time in favor of his own direct current system, emphasizing its fundamental coherence and completeness from dynamo to lamp. The immediate prompts for doing so came from recent technical and organizational developments that were quickly making AC a credible rival in the United States. Always confident of his ability to out-invent competitors, Edison reacted pragmatically, showing little of the dogmatic quality that later characterized his opposition to AC.

Like many lighting engineers, Edison recognized the inherent problems of low-voltage distribution over large areas. As the Scottish engineer Rankin Kennedy, surveying the field in early 1887, put it, "the chief difficulty in the way of wide and extensive [electric] distribution lies in the low potential necessary for the working of the present forms of [incandescent] lamps. . . . The pressure being small a very large volume of current is required to convey a moderate amount of energy for distribution; large currents require large conductors."[1] And large conductors meant a lot of copper.[2]

Transmitting electrical energy at high voltages would reduce the need for so much copper but would introduce the engineering difficulty of lowering the "pressure" to a level usable by incandescent lamps. Edison began looking into this problem even before he had completed the Pearl Street station in New York (his first commercial power plant) in 1882. He designed mechanical contrivances (motor-generators and what he would later call a rotary converter) for reducing voltage.[3] A more fruitful approach was the 220-volt three-wire distribution system he created especially for small cities and towns, where the density of lamps would be lower than in Manhattan. Even these so-called "village plants" had a limited service radius and were still expensive to build, as Edison knew from having financed about a dozen of them.[4]

The problem of voltage reduction was taken up most urgently in Europe, where lighting companies and their systems proliferated but none was yet able to dominate the markets. Electricians tried to exploit the well-known ability of a wire carrying a changing current to induce, through the magnetic field around it, a current in another wire nearby. Some of the day's best electrical minds focused on adapting the induction coil to an AC system. Early forms of such "converters" were inefficient, owing to incomplete magnetic circuits. They also tended to be difficult to regulate, due to electricians' proclivity to connect several coils in series; consequently, voltage fluctuations in one would ripple through and destabilize the other coils. These defects were recognized and corrected in stages. Lucien Gaulard and John Gibbs, working together in London in 1883–1884, made the first crucial advances, following a suggestion by Rankin Kennedy. Then, by about 1885, engineers of the Ganz Company in Budapest—Károly (Karl or Charles) Zipernowsky, Otto Bláthy, and Max Déri—analyzed and improved the Gaulard and Gibbs converter, using a closed iron core to make a strong magnetic circuit and wiring the primary coils in parallel to provide steady voltage. Their design quickly became known as the ZBD transformer and received patent protection in the United States in early November 1886.[5] Referring in 1887 to the Ganz device as a "transfomer," Kennedy claimed that it "leaves nothing to be desired in the matter of simplicity and efficiency."[6]

What might Edison have known of these developments across the Atlantic? Information about the transformers (sometimes called secondary generators) was publicly available, published in English-language journals and patent records. Gaulard and Gibbs exhibited versions of their device in England starting in 1883 (including London in 1885) and won a major prize for their 1884 installation in Turin, Italy. A demonstration of the ZBD transformer in Budapest in May 1885 was noted by the English technical press.[7] Relying mainly on patent records, Kennedy was able to publish a brief but thorough technical history of the transformer in the middle of 1887.[8]

Edison had not made a study of the subject, but a convergence of events drew his attention to it in the fall. Edward Johnson would have alerted him to Frank Sprague's mid-September evaluation of high-voltage AC transmission. Considering engineering and economic aspects, Sprague wrote unequivocally that "this kind of distribution has come to stay, and is going to be a formidable rival to the system of direct supply by continuous currents."[9] Louis Rau, president of the Compagnie Continentale Edison, sent a similarly blunt warning in October that the Edison system was meeting "a tremendous competition" and indeed had been left "entirely by [the] side" by "alternating currents with transformers . . . enabling the current to be carried to a long distance with comparatively little expense."[10] Gaulard and Gibbs received a U.S. patent on 26 October, about the time Rau's letter would have reached Edison. Their specification described an induction transformer, which by now incorporated not only their own improvements but those of the Ganz engineers, and its use in a complete system of electric distribution. Its claims—the document's legally binding portion—were broad, covering the "herein before-described art or method of electrical distribution and conversion." And it was assigned to George Westinghouse, a successful inventor and astute man of business.[11] Westinghouse had recently become interested in electric lighting and, with the help of William Stanley and a few other talented engineers, had been investigating AC. He sponsored Stanley's research at Great Barrington, Massachusetts, that led directly to a successful demonstration of an AC system there in late 1885. After forming the Westinghouse Electric Company early the next year he put up a demonstration system near Pittsburgh in the fall of 1886 and another at Buffalo, New York, that November. Edison quickly identified Westinghouse as the primary high-voltage AC threat. His awareness of the company probably was sharpened by the facts that Henry Byllesby, until recently one of his own lieutenants, was now its general manager, and Franklin Pope, an old antagonist, was among its directors.[12] The status of alternating current systems was such that by the middle of 1887, Rankin Kennedy noted that "the parallel system with the potential transformer has been bodily imported by [Americans] from Britain, in a complete and perfect form." It had caused, he observed, "a somewhat amusing flutter among the electrical fraternity."[13]

One of Edison's first reactions was to begin drafting a patent application (Doc. 3002) for a self-exciting AC dynamo that would obviate the need for a separate DC machine to energize the field windings. Although the specification that resulted in March 1888 described the dynamo in the context of a high-voltage distribution system, it was too limited and too late to forestall the development of AC power by others.[14] Edison put greater emphasis on the improvement of his rotary converters.[15] Before the end of the year, he completed at least eleven patent applications dealing with AC generation and conversion or with the layout and regulation of high-voltage distribution systems.[16]

Over the longer term, Edison gave highest priority to a different strategy. He redoubled his ongoing efforts to improve the efficiency of the incandescent lamp, a goal he would pursue relentlessly in his new laboratory. The great problem, after all, was simply to reduce the amount of copper needed to carry current for a lighting system without disproportionately increasing electrical losses. Edison saw the greatest gains to be made not in the means of generation or transmission of electrical power, but in its consumption. The lamp was an integral part of his entire system, and devising lamps to use only half the power they did at the time, which he was confident of doing, would make it "positively impossible" for AC rivals to compete with his own system.[17]

Edison could summon from lengthy experience reasons to distrust the approach taken by Westinghouse, a newcomer to electricity. He outlined these objections in more technical and economic detail in Docs. 3005 and 3008.[18] Foremost among them was the possibility that alternating current, both by its nature and by dint of the high voltages proposed, could threaten the safety of the public and electrical linemen. Though these concerns may now appear self-serving, and Edison did later push them to incendiary extremes, they were also part of a larger discourse about the safety of electrical wires in cities.[19] Edison did not refer to medical authorities to support his arguments about the outsized physiological effects of AC but his beliefs seemingly were in earnest. Responding in 1882 to an animal welfare advocate's inquiry about humane euthanasia, he unhesitatingly recommended an alternating current generator to "kill instantly without suffering the very largest of animals" (Doc. 2330). Edison also referred to the difficulty of insulating dynamos, which were prone to short circuits even at the relatively low voltages he used; most proposals for AC generation involved producing high voltages at the machines (sometimes by connecting them in series).[20] Obtaining good insulation for conductors was another difficulty, against which he had directed countless experiments. If wires were left uninsulated, they were prone to losses through static discharge. Though that effect would be a small one, Edison knew well how a conductor carrying a changing current could dissipate energy into the surrounding dielectric, a phenomenon he had tried to exploit for wireless telegraphy. Notably, Edison did not refer to one fact that would later become a major obstacle to the adoption of AC: the absence of a practical AC motor.

When the Edison Electric Light Company obtained rights to the United States patents on the ZBD transformer in late November 1886, it did so over Edison's initial objections.[21] Tellingly, when he designed a system some seven or eight months later to light his home and those of several neighbors, he chose his own DC converter technology instead of AC transformers. Edison chose to take power from his laboratory, less than a mile away, presumably to avoid planting the noise and smoke of an isolated plant amidst an exclusive residential community. The homes were wired for an aggregate of 550 lamps, yet the system required an extraordinary pressure of 1,200 volts at the dynamos.[22] Unlike the Westinghouse-sponsored demonstrations at Great Barrington and Pittsburgh, Edison's installation at Orange did little to promote its distinctive plan of electric distribution.

1. Kennedy 1887, 187.

2. Copper was a major part of the cost of an Edison lighting system, although prices for the raw metal were at historic lows in 1886 (Doc. 2424 [headnote, n. 9]; Toole 1950, 319–21; Hyde 1998, 59–60). By mid-1887, Edison explicitly recognized that the proper distribution of copper throughout the feeder and main network was at least as important for determining construction costs as the total amount of copper in the system (cf. Doc. 3050).

3. See Docs. 2242, 2276, 2284, 2440, and 2689. Edison had turned his attention to electric lighting in 1878 after a trip West, where he discussed the possibility of generating electric power at remote rivers and sending it at high voltage over long distances to populated areas (*TAEB* 4, chap. 4 introduction; Doc. 1437 n. 5). One of Edison's 1883 patents covered what was essentially a form of induction transformer with a rotary switching apparatus for changing a direct current into an alternating one. In 1888, this patent was placed in interference with a transformer patent or application of Károly Zipernowsky (U.S. Pat. 278,418; Edison's testimony, pp. 7–9, *Zipernowsky v. Edison*, Lit. [*TAED* W100DMA004]; Edison's U.S. Pats. 265,786; 266,793; and 287,516 also pertain to earlier transformer or converter devices).

4. See Docs. 2424 and 2437 (headnotes) and *TAEB* 7 App. 2.A.

5. Zipernowsky and Déri received U.S. Patent 352,105 on 2 November. It was based on transformer patents obtained in Europe in 1885. That specification and one from 1883 (U.S. Pat. 284,110) were quickly licensed to the Edison Electric Light Co. (see Docs. 3013 n. 9 and 3022). Edison testified later that he personally met Otto Bláthy in October 1886 but "did not exchange fifty words with him." Edison's testimony, pp. 17–18, *Zipernowsky v. Edison*, Lit. (*TAED* W100DMA004).

6. Kennedy 1887, 261. The history of the induction transformer has been written often and well for various purposes, starting with Rankin Kennedy's summary of patent records (Kennedy 1887). John Ambrose Fleming, writing after the course of AC development had become a bit more clear, offered a more theoretical engineering approach (Fleming 1892, vol. 2 chap. 1). Accounts of the broader historical context for the transformer and the development of AC transmission more generally include MacLaren 1943 (chap. 8), Hughes 1983 (chap. 3), Moran 2002 (chap. 2), Jonnes 2003 (chap. 5), Billington and Billington 2006 (chap. 2), and Klein 2008 (chap. 11).

7. Hughes 1983, 88–95; see also Doc. 3010 n. 4. Edison later testified in a patent interference case that he became aware of the ZBD transformer about the time published reports of it appeared in the summer of 1885. Edison's testimony, pp. 11–12, *Zipernowsky v. Edison*, Lit. (*TAED* W100DMA004); cf. the early interest of Elihu Thomson in alternating current (Carlson 2013, 249–51).

8. Kennedy 1887.

9. Sprague's opinion had been solicited by Johnson, and he produced a comprehensive six-page report touching on four main topics. Regarding safety, he concurred with Max Déri's claim that alternating currents posed no greater physiological danger than continuous ones. Second, AC generators would be no more costly, and possibly less so, to build and operate than DC dynamos. He also favorably evaluated the theoretical basis for the efficiency of induction transformers, specifically the design of British-born engineer and inventor Sebastian Ziani de Ferranti, who had been working along lines similar to those of Gaulard and Gibbs and Zipernowsky. Finally, using a hypothetical example, Sprague showed the exponential reduction in the weight of copper conductors needed for high-voltage AC compared with low-voltage DC. Although Sprague offered to give further information if requested, he concluded that "the whole question seems to me to be solved by a comparison where long distances are used between the two systems, and in this case the alternating current distribution unquestionably has the advantage." He closed with a warning to Johnson: "you cannot too soon take steps to prevent some one getting in the field ahead of you." Sprague to Johnson, 13 Sept. 1886, Sprague (*TAED* X120CAN); regarding Ferranti, see Hughes 1983, 97–98.

10. On the other hand, Siemens & Halske in Berlin prepared a forty-six page typewritten report (probably in November) that made its way to Jonathan Vail; it reached the opposite conclusion that "in central stations, there is no room for the transformers of Gaulard and Gibbs and Deri-Zipernowsky." Rau to TAE, 5 Oct. 1886; Siemens & Halske report, Nov. 1886; both DF (*TAED* D8630ZCG, D8624N).

11. U.S. Pat. 351,589.

12. Passer 1972 [1953], 129–38; Bedell 1896, 38; MacLaren 1943, 176–77; Hughes 1983, 102–104; Prout 1921, 113; *American Electrical Directory* 1886, 349; Tate 1938 (150–51) relates Byllesby's decision to join the Westinghouse organization, reportedly for a higher salary not matched by the Edison company.

13. Kennedy 1887, 301. No less an authority than the U.S. Patent Office evidently failed, as late as 1883, to grasp the basic principle of the induction transformer. Fleming 1892, 81–92.

14. By contrast, when the Thomson-Houston Electric Co., which had also been working on AC systems, saw Westinghouse Electric take the lead technologically, it entered into a patent-sharing agreement in March 1887. Carlson 2013, 252–57.

15. See Docs. 3008–3011.

16. These applications included those developed from Docs. 3002, 3011, and 3014. They resulted in eight patents, of which several of the earliest (by execution date) pertained specifically to AC systems: U.S. Pats. 438,308; 524,378; 369,439; 365,978; 379,944; 369,441; 369,442; and 369,443. An application filed on 3 November (Case 699) was abandoned but its lone drawing was later copied into an Edison casebook (Patent Application Drawings [Case Nos. 179–699], PS [*TAED* PT023, image 129]). Another application filed on 19 November was included in the *Zipernowsky v. Edison* interference proceeding; it may have been Case 693 (Edison's testimony, pp. 11–12, *Zipernowsky v. Edison*, Lit. [*TAED* W100DMA]). Case 703 did not issue, and the editors have found no record of it other than a reference in a related specification

for a distribution system (U.S. Pat. 369,443). Case 704, for a system of electrical distribution, was filed on 6 December; it was allowed by the Patent Office but did not issue as a patent (Patent Application, 6 Dec. 1886, PS [*TAED* PT017AAA]). As many as a dozen other applications from the October–December period are unaccounted for and their subjects unidentified (see App. 2.A).

Edison had learned to manage the regulation of his three-wire DC system (see Doc. 2505 [headnote]), but the difficulties would be compounded by inserting another dynamic element—a converter—into local distribution circuits. Several of his patents from this period addressed the integration of converters into three-wire substations (see also his 7 December notes and sketches regarding the layout of a full "Converter System" in Unbound Notes and Drawings [1886], Lab. [*TAED* NS86ADI]). At other times Edison assumed that high-voltage AC current would be carried to each customer and reduced on location, making the system easier to regulate but introducing safety hazards. Edison referred to both distribution models in Doc. 3008. It did not become clear for some time whether the substation or individual transformer model would be more generally adopted (see Fleming 1892, 2:421–22).

17. Doc. 3008. A few weeks earlier, Edison attached the "highest importance" to an improved lamp that would permit the use of small copper conductors over a large area (see Doc. 2994). Writing approvingly of Edison's "new lamp" at the end of 1886, John Vail predicted that it "will be a big thing, even did it save in copper only." Vail calculated significant cost savings that would be possible from using the new lamps in several stations, including an expanded Boston service area. Vail to TAE, 28 Dec. 1886, DF (*TAED* D8624O); see also Doc. 3050.

18. David 1991 (79–91) provides a helpful overview of contemporary concerns about the economic and technical feasibility of AC systems.

19. See, e.g., Docs. 2568 n. 1, 2816 nn. 3–4, and 2945 n. 12.

20. Edison was not alone in maintaining this objection. The eminent French electrician Hippolyte Fontaine reportedly concluded that the limits of dynamo construction made it impossible to transmit 100 electrical horsepower from one machine through 100 ohms over a considerable distance. The alternatives, according to the report, included increasing the size of conductors, using multiple dynamos, or simply accepting high losses in transmission. By coincidence, this account of Fontaine's remarks was published directly opposite the announcement of a London exhibition of the Zipernowsky-Déri transformer and distribution system. "The Electrical Transmission of Power," *Teleg. J. and Elec. Rev.* 17 (15 Aug. 1885): 153.

A different approach to the dynamo problem was suggested by Edison associate William Andrews, who about this time became general superintendent in Chicago for the Marr Construction Co., a subcontractor for building Edison electric plants. Andrews and a partner (Thomas Spencer, also a veteran of Edison's central station construction enterprise who joined the Marr firm) sought to patent a system in which one set of transformers would raise the pressure at the generating station to 5,000 volts and another set would later reduce it for use. As Andrews explained to Dyer & Seeley (Edison's own patent attorneys), this arrangement would eliminate the need for a 5,000 volt generator, one which would be not only "expensive to build by reason of the careful insulation necessary" but also "a somewhat dangerous one to handle." Andrews to Dyer & Seeley, 3 Oct. 1886, DF (*TAED* D8637ZAA); "Andrews, William Symes" and "Spencer, Thomas," both Pioneers Bio.

    21. See Doc. 3013 n. 9.

    22. See Doc. 3056; *Electrical Review* [New York] 11 [1 Oct. 1887]: 1–2.

–3002–

Draft Patent Application: Electric Light and Power

[East Newark?] Oct 25 1886

Dyer—

    Take out patent for this[1]

    Use regular dynamo with regular brushes but they only serve to energize the field which is regulateable by the resistance   No lamp current is taken off of these. On the End of the shaft next commutator are two continuous contact disks[a] secured to shaft & rotated together, on the surface which is continuous a contact[a] brush rests, both ~~wh~~ disks are insulated from each other. I permanently connect one commutator bar to one disk & the commutator bar directly opposite on the other side of the commutator I connect to the other disk then at every revolution the current is reversed in the wires leading from the disks ~~which~~ while the regular Commutator brushes take off a continuous current   The two wires connected to the two disk contact brushes pass to a distant point say a mile etc   There the two primary wires of converters are connected across multiple arc preferably although they can be in series. The secondaries are connected in series to give say 200 volts, the center wire being connected between thus I have a three ~~two~~ wire system, if 2 wire the secondaries are put in Multiple arc   the High tension Machine is say 2000 volts—

    The Converters have large masses of iron in the form of fine iron wire or sheet,[2] hence the slow reversals will not show in lamp, but perhaps if not necessary you neednt say anything about this way of getting around defect of reversals.— I don't want to confine myself to using this machine in connection with converters as the Lamps may be put on direct the tension being reduced of course to the regular required volts.

    also I can in addition to putting on Lamps on the reverse circuit or when using the converter put lamps on the regular brush circuit— Yours

TAE

ADS, NjWOE, Lab., Cat. 1151 (*TAED* NM020ABJ). [a]Obscured overwritten text.

---

1. See headnote above. Alternating current (AC) generators typically used a separate exciter dynamo to supply direct current (DC) to their field windings. Edison wished to patent a current collector that would allow both AC and DC current to be drawn from the same armature (cf. Doc. 3005, possibly written earlier). He signed the patent application based on this draft on 26 November; the resulting patent issued in March 1888. The specification broadly described a dynamo for producing "high-tension [alternating] current to be conveyed to a distance and converted by tension-reducers into a current of low tension adapted for lighting and similar purposes." The claims, however, were written narrowly to cover only the construction of the armature and commutator. U.S. Pat. 379,944.

2. Cf. Doc. 3019.

<table>
<tr><td>–3003–</td><td>[New York,] October 27th. [188]6</td></tr>
</table>

–3003–

Samuel Insull to
Cyrus Field[1]

[New York,] October 27th. [188]6

Dear Sir:—

I have just received a telephone message from Mr. Edison at his house at Orange requesting me to ask you if you could possibly favor him with a permit for himself and family to view the fireworks from the top of the Washington Building[2] tomorrow, Thursday, night.[3] If you can oblige Mr. Edison, he will esteem it a great favor. He would write you himself, but would not have time to reach you during the day, as he will not be in the City. There will be six in Mr. Edison's party. Yours very truly,

Private Secretary.

TL (carbon copy), NjWOE, Lbk.23:60 (*TAED* LB023060).

1. Merchant and capitalist Cyrus Field (1819–1892) had a crucial role in promoting the first successful Atlantic cable in 1866. He later established the American Telegraph Co. and, since 1882, published the *New York Mail and Express.* Doc. 2450 n. 1.

2. Field was the principal builder of the Washington Building (also called the Field Building), named to commemorate the site believed to have been George Washington's headquarters. Located by Bowling Green at 1 Broadway, overlooking the Battery, the original ten-story building was completed in 1885 but already at this time was being expanded by another two floors. Edison proposed (unsuccessfully) in 1883 to supply the building with electric lights. Doc. 2450 nn. 2–3; Landau and Condit 1996, 125–27.

3. The fireworks were planned as the conclusion to a day of festivities for the unveiling of the Statue of Liberty on Bedlow's Island. Despite rain and mist, several hundred thousand people reportedly attended the dedication, with president Grover Cleveland, French engineer Ferdinand de Lesseps, and sculptor Frédéric Bartholdi among the many dignitaries. Persistent bad weather delayed the fireworks until

the evening of 1 November, when the statue and her torch were lighted for the first time by arc lamps and an isolated generating plant of the American Electric Manufacturing Co. "The Statue Unveiled," *NYT*, 29 Oct. 1886, 2; "The Unveiling of the Statue," ibid., 29 Oct. 1886, 4; "Liberty's Torch Lighted," ibid., 2 Nov. 1886, 1.

–3004–

To John Tomlinson

New York, [October 28, 1886?]¹ᵃ

⟨Hurry this up— Edison⟩

Mr Tomlinson—

Draw up a contract between the Edison Machine Works & Chas T Porter,²—Covering following points.

1st   Mr Porter has an Engine—has patents or applications thereon—³

2nd   Mr Porter desires to enter into Contract with E Machine Works to manufacture & sell such Engine⁴

3rd   Mr Porter engages to inspect the manufacture give full instructions in the preparations of Drawings patterns Etc for all sizes of Engines which the ~~public~~ EM WKsᵇ may desire.ᶜ

4   Mr Porter is willing to give Machine Works sole License to manufacture & sell his Engine during the life of the patents on same—

5. [--]ᵈ The consideration given Mr Porter for the Exclusive License is 5 percent on the selling price of the Machine Works—~~it being understoo~~ to the public or to parties furnishing the money—

6th   Mr Porter is to further receive a salary of $2500, per year payable monthlyᵉ while Engaged at the Machine Works in preparing drawings & inspecting & testing Engines— He agrees to remain with Machine works ~~as long~~ until all the standard sizes have been made & tested satisfactorily but has the priviledge in any Event of leaving in 3 years from date if he so desires—but the 5 pct will go on during the life of the patents—

7— To prevent any future misunderstanding as to the duties of Mr Porter state that, the Engine is to be made according to his instructions & ideas only, and that any orders he may give respecting character of workᶠ will be given through the foremaen of the respective departments

8. The Machine Works agrees if the first model Engine is satisfactory to immediately prosecute the business of making a full set of standard sizes & push the business

PS—8[5]   Mr Porter in addition to the $2500. is to be al-
lowed to draw the first year $1500 extra which shall be charged
against his future royalties—

AL, NjWOE, DF (*TAED* D8627ZAF). Letterhead of Thomas A. Edison; each numbered paragraph preceded by a checkmark. [a]"New York,"
preprinted. [b]"EM WKs" interlined above. [c]Obscured overwritten text.
[d]Canceled. [e]"payable monthly" interlined above. [f]"respecting character
of work" interlined above.

1. Edison wrote this date when he added the marginalia below—in a
different ink from the main text—at the top of the first page.

2. A lawyer by training, Charles T. Porter (1826–1910) introduced
the first commercially successful high-speed stationary steam engine
in 1862, one substantially of his own design. Edison adopted it for several of his early dynamos and collaborated with Porter until the latter's
forced resignation from the Southwark Foundry of Philadelphia, which
manufactured the engine, near the end of 1882. Obituary, *NYT,* 30 Aug.
1910, 7; Porter 1908, 1, 321–24; Docs. 1936 n. 2, 2074 n. 1, and 2006 n. 2.

3. The editors have not identified the patents or applications for
Porter's new engine design, whose construction and operation he detailed in a twenty-two page handwritten memorandum (Porter memorandum, 20 Oct. 1886, Miller [*TAED* HM860291]). Draft contracts
referred to a patent allowed (but not necessarily issued) to Porter on
1 June and to three pending applications, one filed on 29 July 1886 (U.S.
Pat. 368,422). According to another undated memorandum by Edison,
Porter expected his new engine to cost the same to build as an Armington & Sims machine but to "be twice as economical" in operation (Edison Machine Works draft agreements with Porter, all n.d. [Nov. 1886];
TAE draft memorandum, n.d. [Nov. 1886]; all DF [*TAED* D8627ZAN,
D8627ZAN1, D8627ZAN2, D8627ZBE]).

4. Porter sent Edison a letter in August 1886 (not found) asking for
employment in the lighting business. Edison replied that he had little
influence in the Edison Electric Light Co.'s operations, but he forwarded Porter's request to company president Edward Johnson with an
expression of "quite a great deal of sympathy for Mr. Porter as I should
imagine from his letter that he needs a position badly." During October,
Edison and Porter jointly worked out terms of an arrangement for the
Edison Machine Works to build the new engine, with Porter communicating directly with Edison's attorney John Tomlinson about Edison's
draft of their agreement (TAE to Johnson, 13 Aug. 1886; TAE to Porter, 13 Aug. 1886; both Lbk. 22:322–23 [*TAED* LB022322, LB022323];
Porter memorandum [with TAE marginalia], n.d. [Oct. 1886]; Porter to Tomlinson, n.d. [Oct. 1886]; both DF [*TAED* D8627ZAG,
D8627ZAE]). In 1883 and 1884, Edison had explored the possibility of
licensing the manufacture of steam engines, notably those of the Armington & Sims Co., to the Edison Machine Works, but he did not make
any definitive arrangements (see Docs. 2608 esp. n. 8 and 2616).

Several versions of a contract were drawn up along the terms laid out
in this document. There was also provision for the creation of a new corporation to buy the engines from the Machine Works and sell them to
end users. According to Edison's notes on that subject, the twenty pro-

spective investors in the new firm included himself and his customary manufacturing partners, as well as Tomlinson, Ezra Gilliland, Samuel Insull, Luther Stieringer, and Charles Coster on behalf of Drexel, Morgan & Co. Despite a negotiating session on 24 November that seemed to produce a final agreement, Porter later raised objections to several points, among them the formation of the new intermediary corporation. The final contract (not found) was signed on 30 November. Cat. 1336:139 (item 238, c. 2 Nov. 1886), Batchelor (*TAED* MBJ003139B); Tomlinson to Batchelor, 3 Nov. 1886; Edison Machine Works draft agreements with Porter, all n.d. [Nov. 1886]; TAE draft memorandum, n.d. [Nov. 1886]; Porter to Tomlinson, 28 Nov. 1886; all DF (*TAED* D8627ZAD, D8627ZAN, D8627ZAN1, D8627ZAN2, D8627ZBE, D8627ZAM); Cat.1336:145 (item 251, 24 Nov. 1886), Batchelor (*TAED* MBJ003145); Alfred Tate to Porter, 21 Oct. 1891, Lbk. 51:374 (*TAED* LB051374).

Porter carried out experiments at the Edison Machine Works in Schenectady and Edison's Orange laboratory for several years. He filed at least one patent application in October 1887 but the following January, Samuel Insull cautioned that despite the Machine Works having spent $13,622 on the project, "no test has been made which would justify us in thinking that he will produce a successful engine." Porter suffered a long period of ill health that hindered his work (and made him "cranky," in Edison's view), but Edison and his laboratory staff carried on. After Insull again warned in July 1888 that expenses were growing out of hand, Edison promised to "shut Porter off" if their joint efforts came to no avail within a month. As it turned out, Edison stuck with the project until June 1891, when he concluded that "Porter is evidently cracked" and therefore "it would be a waste of money" to continue (U.S. Pat. 391,916; Insull to TAE, 26 Jan. and 26 July 1888; Charles Emery to TAE [with TAE marginalia], 20 July 1888; TAE to Insull, 30 July 1888; all DF [*TAED* D8736AAP, D8835ADO, D8805AES, D8818AQB]; Tate to Porter, 7 June 1889; TAE to Insull, 29 June 1891; Lbk. 30:264, 50:188 [*TAED* LB030264, LB050188]). Porter filed two patent applications in April and May 1893, though it is unclear whether the novel engine designs they covered were the subjects of his work with Edison. Both applications issued as patents in April 1894 (U.S. Pats. 517,982 and 517,983).

5. Edison added this postscript on a separate piece of paper.

−3005−

Notebook Entry: Electric Light and Power[1]

[New York?, October 1886][2]

Reasons against an Alternate Current Converter system— <u>Danger</u>

1st— The pressure on the high tension mains are destructive of life—

2nd   Running high pressure mains into stores & houses to connect to converters is dangerous to life.[a]

3rd   Converters act as Condensers and notwithstanding two systems ie[b] high & low are not electrically connected one may get a several thousand volt discharge under certain conditions between low tension system and the ground

4   Impossibility without danger of repairing high line when current on[3]

5   Disruptive action of reverse currents require heavy insulation

6   ~~To~~

6 __

7   to get same economy of conductors The greatest difference of potential [--][c] on high line & ditto low line must be 3 times greater than with continuous currents. to do what 2000 volts continuous will do diference of potential of at least 5000 volts must be used— If on the low line 2 wire distribution system is used & 100 volt lamp then 140 mean must be used or diference of potential of 280 volts alternated current   If three wire system 560 560 volts diference potential & this is fatal—

Continuous currents polarize the points where the current enters the body & prevent its passage— There is no polarization with alternated currents hence full current passes & kills

8— As Alternate Current machines must be worked with continuous or constant field <u>not</u> <u>derived</u> from the main supply circuit, any shortcircuiting of line must burn out machine as it is not saved as in Continuous Current systems[4]

8   The doing away of commutators on Continuous Current machine is no gain as there must be a field energizing dynamo or section on the Alternate Machine & this must have a brush commutator   Experience has shown that small machines spark worse & give more trouble than big machines—

X, NjWOE, Lab., N-86-04-28 (*TAED* N321261). [a]Followed by dividing mark. [b]Circled. [c]Canceled.

1. See Doc. 3002 (headnote).

2. The editors conjecture that Edison probably wrote this outline in the first part of October, given the rather preliminary and general nature of Edison's objections to alternating current, a subject to which he had not yet given much thought (cf. his more detailed critique in Doc. 3008). He probably made this entry before Doc. 3002; the double commutator described there was likely intended to address, at least in part, the defects described in paragraphs numbered "8" below.

3. Edison tried to address this difficulty in a patent application, filed in mid-1887, for a high-voltage system having redundant conductors, any of which could be separately switched off. U.S. Pat. 545,405.

4. That is, because the separate exciter dynamo would not be affected by changing line conditions and would continue to supply current in all circumstances to the main generator's field windings.

*Draft Patent
Application:
Incandescent Lamp*[1]

The object of this invention is to ~~obtain~~ carbonize filiments for incandescent Electric Lamps which will be of even resistance and will not become distorted while heated electrically while in the Lamp— ~~and will be in a bath which process is especially~~ [------ than?][b] ~~the class of Lamps which do not [contain?][b] filiments of carbon coated with carbon a filiment[c] from[d]~~

The invention consists in carbonizing the filiment entirely by radiated & convected[e] heat, and not by conduction. The filiment is suspended out of contact with any body in a carbonizing chamber a small weight serves to Keep it from distorting while being carbonized thus preserving the shape originally given to the carbonizable filiment.[3]

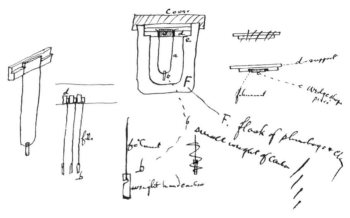

Several hundred filiments in each mould

Soak this patent,[4] make broadest claims.

Claim. Carbonizing carbonzable filiments by radiated & convected heat.—

Out of contact so no heat is conveyed to filiment by conduction—

Suspended filiment, bent or formed to Shape originally & placed under strain while being carbonzed to make it hold its shape——

P.S. The filiment used in Lamp is broken off about ¼ inch from point of suspension so that it ~~be~~ will be carbonzed by radiation at this point & not by conduction from the support.— all carbonzation heretofore of[f] every body The filiment was in contact with matter at some of its parts—our competition without exception use powdered charcoal.

T.A.E.

Where are the other applications??

ADfS, NjWOE, Lab., Cat. 1151 (*TAED* NM020ABL). Letterhead of Thomas A. Edison. [a]"NEW YORK," and "188" preprinted. [b]Canceled. [c]"a filiment" interlined above. [d]Followed by a line of interlined and heavily canceled text. [e]"& convected" interlined above. [f]Obscured overwritten text.

1. This draft became the basis for an application that Edison executed on 6 December, one of three (including one derived from Doc. 3012) related to filament manufacture that he signed that day. It was filed nine days later and issued in 1892 as U.S. Patent 485,615.

2. Edison drafted this application on old letterhead from 65 Fifth Avenue. The editors have not determined if he did so in his new offices at 40–42 Wall St. in New York or in his laboratory at the lamp factory in East Newark, N.J.

3. Figure label at top is "Cover." Clockwise from top right, figure labels are "d—support," "c   wedge shape piece," "filiment," "F. flask of plumbago & clay," "b   small weight of carbon," "b   weight hard carbon," "filiment," and "b   fils."

4. See Doc. 2962 (headnote) esp. n. 4.

–3007–

To Samuel Flood Page

[New York,] November 9th. [188]6

Dear Sir:—

Referring to your favors with relation to my patent accounts, since my former correspondence with you,[1] Mr. Insull and myself had the pleasure of an interview with Viscount Anson,[2] one of your Directors, with relation to this matter. In your letter of the 11th. June, you requested that I should supply you with the amount of the account which would have applied on the two patents, Nos. 2,052 and 2,336 of 1882.[3] The work involved in dividing up my experimental expenses would be so heavy, that I think it hardly necessary to give you the detailed information as to these two particular patents. I find that the amount spent by me, since the signing of the agreement with your predecessors, the Edison Co.,[4] amounts to upwards of $75,000. If the account for patents was rendered in accordance with the agreement, your Company would have to pay double this amount if they took over the whole of the patents, or such part of it as is applicable to the various patents which they may decide to accept. You will therefore see that the suggestion made by me, that your Company should reimburse me simply for the actual amount of cash spent by me for patent fees and the preparing of the patents, is a very reasonable one for me to make. Lord Anson thoroughly understands my ideas in relation to this matter, and I would suggest that you have an interview with him on the subject. With relation to future patents, your Company has the right at the present time to acquire any patents that I may obtain, but the agreement does not provide for the taking out of these patents, inasmuch as I should myself be unwilling to incur the expense in the first case of having such work done. While I have every desire to assist your Company and to give them the benefit of any inventions that I may make, I must confess that I have no desire to invest a large amount of money in English patents, without any certainty of being able to obtain from your Company the reimbursement of such expenditure. I suggested to Lord Anson that it would be greatly for your interests, and certainly far more agreeable to me, that a supplemental contract should be entered into providing for dealing with this matter in the future. I proposed to him that your Company should engage a patent lawyer here acceptable to me, whose duty should be to prepare the patent papers, so far as it is possible to prepare them on this side of the Atlantic, and then to forward the same to your Company, leaving it to you to decide for yourselves whether you desire to apply for the patents or not.[5] The cost per application would be 10-10-0, provided that the same lawyer is engaged by you as the one employed by me to do my American work. If your Company is willing to reimburse me the amount I have already expended in the manner communicated to you in my previous letters, I shall be glad to enter into any agreement as outlined above.[6] Yours very truly,

TL (letterpress copy), NjWOE, Lbk. 23:85 (*TAED* LB023085). A typed copy is in Lbk. 54:403 (*TAED* LB054403).

1. See Doc. 2953.

2. Lord Thomas Francis Anson (1856–1918) was Viscount Anson until he became the third Earl of Lichfield in 1892. A founding director of the Edison Electric Light Co., Ltd., he remained a director of the Edison & Swan United Electric Light Co. until his death. Anson was traveling in Canada in late summer and Samuel Insull invited him to meet Edison in New York in late September. Doc. 2467 n. 11; Cokayne 1887, 5:76; Insull to Anson, 23 Aug. 1886, Lbk. 22:341 (*TAED* LB022341).

3. Edison (or Insull) mischaracterized Flood Page's earlier letter. The secretary stated that the company had paid renewal fees on these two patents so that they would not lapse. He then asked Edison to "kindly inform me for the information of the board what the amount of your account for the patents now under consideration would have been" (referring to all fifteen patents sought by the company) under a strict reading of the original 1882 contract. Doc. 2953 n. 3; Flood Page to TAE, 11 June 1886, DF (*TAED* D8630ZBS).

Edison's British Patent 2,052 (1882) for "Electric Generators and Engines" pertained to a radial bar disk dynamo for producing direct current (Dredge 1885, 2:ccc; see Docs. 2082 n. 2 and 2228 esp. n. 10). Patent 2,336 (1882) applied to lighting railway cars by a combination of batteries and generators driven by the car axles, with switches to keep the dynamo polarity unchanged regardless of which way the armature rotated. That specification was issued in the name of draftsman Henry Byllesby, who at the time was working for the Edison interests in New York, and William Stern, who seems to have been something of an electrical agent. Byllesby and Stern received a corresponding U.S. patent (258,149) about the same time, a half interest in which was assigned to Edison. Dredge 1885, 2:cccviii; *TAEB* 6 App. 5.C.

4. Edison referred to the Edison Electric Light Company, Ltd., organized in March 1882 to control his patents in Great Britain. It merged with the Swan United Electric Light Co., Ltd., in October 1883 (effective June 1883) to form the Edison & Swan United Electric Light Co., Ltd. Docs. 2221 n. 4, 2514 n. 12.

5. Cf. Doc. 2953.

6. The intertwined questions of patent fees and reimbursements to Edison dragged into 1887 (see Doc. 3026).

–3008–

Memorandum to Edward Johnson[1]

[East Newark, c. November 10, 1886][2]

NOTES ON DISTRIBUTION OF ALTERNATING CURRENT. 1886.

Reverse current machines (loss 25%) must have continuous current Dynamo charge field, and as iron in field is small in quantity, a very considerable amount of power is required so that the commercial efficiency cannot be greater than 70%.

Reversing the polarity of many tons of copper in high and low tension wire causes a loss of perhaps[3a]   1%

Loss by static charge of high tension quite small, but in distributing circuits where wire is along and close to walls as in inside wiring it is a large factor, on 1600 Lt plant at least[a]   3%

If conductors are underground[a]   20%

Loss due to converter say[a]   5%

Loss on distributing mains[a]   3%

Loss on high tension wire[a]   2%

Total loss[a]   34%

As it is not practical to work a number of small reverse dynamos in mult arc, single dynamos must be used, hence the necessity of a spare, especially as owing to high tension of the current it is more liable to cross and a short circuit is fatal as its field is made by another machine, and it is not anhilated by a short circuit. The damage will be done before a catch can work in most cases unless the machine is extraordinarily well insulated.

If a network is to be used, and converters be substituted

for feeders then there will be great differences of EMF at the different points due to drop in the converter. This you will understand when we consider a converter a dynamo. There is a drop of EMF between full load and light load. The distribution will have to be made in this case on two wire system, requiring 63% more copper, as the amount of copper in distributing mains must necessarily be great, so that there shall be only 2 or 3% difference between the volts on lamps. This 63% is a large factor compared to feeders when 15% or even any p-c makes no difference between the lamps if feeders are regulated. If double converters are used with 3 wire system this will be saved, but they will infringe our patent.

If the converters are regulated at the points where they are placed, it would require too many people. If the converters are all placed at one Central Station, then they must use feeders, and regulate them, thus again infringing our patent.[4]

If they do not use a network but carry the High tension circuit all over town, putting on converters of various sizes to every consumer's place they must work them in mult arc and if this is the case, the efficiency of the converter will be greatly diminished. The total resistance will be low so as to necessitate very large wires from the distant station.[5]

The drop between the first and last converter would be great. If they attempt to obviate this loss by increasing the volts from, say, 2000 to 4000 the wires on the converter must be 4 times smaller and must cost at least 10 times as much as fine sizes increase greatly in cost. The insulation must be greater, hence greater removal of the wire from the inductive core and less efficiency—again 2000 volt converter to say nothing of 4000 in every consumer's place is not pleasant[6]

In a 1600 light plant there must be several feeders, hence several converters— These must be placed somewhere and parties will undoubtedly exact something for the privileges as they must be at definite points.[7]

The positive wave going through a lamp must be, if it was this shape[b] 100 volts in 100 volts lamp, the negative wave 100 volts the other way making 200 volts difference potential but as the wave can not start instantly and stop instantly it must be thus[8b]

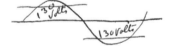

thus it will require 130 volts or thereabouts to produce the equivalent of 100, thus we have for shocking purposes a reversed intermitting current of unlimited amperes as far as the body is concerned (!) and a difference of 260 volts it will certainly be unpleasant.[9]

It will be difficult to get a practical meter as chemical meters cannot be used.[10]

If Ganz agent states that the converter costs $3.00 per lamp then it is equal to cost of our dynamo— Our 500 Light is $1500 I think.

Now having 44% loss suppose we get the power for nothing, then this loss does not count against the system.

But supposing they get the power for nothing (ie) the water or coal let us see how the investment and running expenses are.

The Dynamo at distant Stations will cost more as they must have double capacity in 1600 Lt Stations and ⅓ more in 3200 Lt Stations, and as 1600 and 3200 Lt Stations will probably be the only places where the conditions will be favorable if they have two dynamos and a space they will have to erect two circuits— If they use coal, the boilers, pumps, heaters etc., will have to be 22% larger

The cost of converters will be equal to the cost of our dynamos. The amount of copper in the distributing mains will be 63% more, but there will be at least 3% drop due to reversing and static effects so this must be doubled again, if it is not doubled the lamp breakage will be a large item. They must have someone in the town to regulate the EMF so their labor will be equal to ours.

If they only put the system in and take only the largest consumers put a converter on the premises, no regulators or mains are necessary, the house wiring will only need to be 4 times heavier; but they must neglect the general public if they do so, we come in, taking the general public, placing our feeders near large consumers and make it impossible for them to compete if we pay $4 for coal

The patent that I have that Batch has worked up[11] obviates all the defects and losses of the other system and there are many places where it may be used to advantage.

The fact of the matter is that the moment Capital has confidence and will furnish unlimited capital if they can make 8% on it we will think no more of putting in 20,000 Feeders than the P.R.R.[12] does of spending $100,000 to straighten a curve, that will save $5000 a year, and all systems with their makeshifts will make way for the plain direct system.

You will notice that I give no loss due induction from one side of circuit to the other as there is none, it simply seems to lower the EMF, but it is the induction which would go in a parallel wire having no connection with the induction circuit and the surrounding matter—houses, trees, earth, wall etc. This is all lost as Heat. Why the converter itself is a case in point—look at the tremendous amount of energy that jumps from one circuit to the other— This is lost in the second circuit as heat and light, but is of course useful. The wires are close together it is true but so are the walls of a building in the house wiring, and energy can be lost just as well by static E as Dynamic.[13]

The more I study the converter business, the more satisfied I am that we shall be able to give Westinghouse[14] all the law he wants on this particular subject or any other.

Do you know that the Dynamo Batch made for transforming is a very perfect contrivance, in fact it is perfection.[15] I will bet any amount that you can put several on our circuit, using three wires to distribute low current and they will run for months without requiring attention.

At first thought the revolution of the armature and employment of brushes would seem to be a bad thing as compared to the Zip converter—[16] There is nothing in this. You know the Zip converter makes a great noise owing to molecular movement of the iron which is fully as great as the noise of our revolving converter which runs very smooth, there being no belt and no strain on shaft the bearings don't heat— Again having double coils there is no spark, hence the brushes are put in a fixed position. You know if there is no spark and the brushes are fixed the commutator will last for years, and require no attention— All the converters start up when station starts. Any regulation at station regulates converters perfectly, in fact they can be boxed up, only giving convenience for filling oiler etc.

But here comes the new point, we run a 400 ampere machine at 800. Now there is not the slightest trouble in running armature alone 1600, therefore at 800 Revs it will convert 425 amperes— I suppose you know that it will transfer more than it will run as a dynamo then by reducing the resis. of the armature and running it at 1600 revolutions you can transfer 850 amperes on a 400 ampere machine without any strain for belt &c, in fact do it beautifully 850 amps or 1130 [0].75 amp lamps or 1700 of new lamps. As Batch can probably make them for $150 extra, you have cost for converter for 1.46 per lamp of .75 amp. (present lamp) or 97¢ per lamp (new).

Thus we can convert for half price, use straight currents

and distribute on 3 wire system using two converters in series together.

Just as certain as death Westinghouse will kill a customer within 6 months after he puts in a system of any size— He has got a new thing, and it will require a great deal of experimenting to get it working practically. It will never be free from danger, and there is no condition where we cant go in and make a big dividend where he would lose.

None of his plans worry me in the least, only one thing that disturbs me is the fact the Westinghouse is a great man for flooding the country with agents and travellers.

He is ubiquitous and will form innumerable companies before we know anything about it.[17]

Mr. Wither[18] tells me they employ 30 General Agents and have 1500 Local Agents, and do a business of $2,500,000 a year.

Hutch[19] is going along in the old way employing one agent for a territory that he couldn't visit each factory in 20 years.

I'm making 1400 lamps for Vail[20] 17 pr HP.[21b]

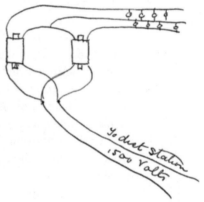

You are dead onto it—when it comes down to dollars and cents, and practicability (ie) constant currents, nothing that anyone else could possibly do could touch us in the least.

The reason our converter has no spark is that as a dynamo the field is distorted in one direction and you have to set your brushes in that direction—but as a motor the distortion is in the opposite direction and you should set your brushes in that direction. Moral—The brush stands still at all loads and no spark— As there is no strain or spark but merely a spindle revolving doubling the speed over that of a dynamo is nothing— In fact they will run beautifully and never give any trouble, thus a 400 amp. dynamo becomes a 800 amp. converter (ie) Any dynamo if rewound and speed doubled will supply twice as many lamps as it would as a dynamo.[22] This seems funny—

For instance— One 400 amp dynamo if run at 1700 revs. and wound lower resistance so that volts would be same at 1700 as it is now with 850 would run 800 ampere but the strain on belt and bearings and the spark etc., would make it unreliable—the defects which limit you in the case of dynamos are absent when double wound and used solely for converting.— Catch on?

My impression is that except in very difficult places we shouldn't use over 1200 volts. This reduces the copper down so that it will be telegraphic in size— We must look out for crosses and such things for if we ever kill a customer it would be a bad blow to the business— When we must use from 3000 to 5000 volts such as lighting Buffalo from Niagara Falls—I should not wind double, but have motors run dynamos, but 1200 volts continuous current will never do greater harm than blister the flesh, and I'll bet any amount that 1000 volts alternate current will kill certain.

Why Zip uses 2000 volts alternate— This gives a difference of 4000 volts (!) (HOLY MOSES)[23] and as it is not continuous he has to have to get a mean which gives 6000 volts dif.

Suppose W. uses 2000 and one leg gets crossed, the first man that touches a wire in a wet place is a dead man.

They use great insulation on their converters, but if they do—their capacity and economy will diminish as the sq. of the insulation (ie) The wires get further away from the core

I'm glad you've caught on to this—to my mind we still control the biz with the only continuous converter system high economy—no danger. Economy of distribution system wire and the fourth=Coming high economy and Res. Lamps and a perfected system—and a converter which costs far less than any other and about 95% efficiency.[24]

Speaking of efficiency of converter—catch on to this:— Test at Franklin Inst.[25] shows, if I remember right, 95% efficiency. 91% commercial leaving out a fraction due to load (ie) strain on bearings due to belt— Now the reason we do not get 100 (leaving out shaft friction) was that there are faucault currents in iron arm. and greatest of all short circuiting of coils by brush even when set at the non-sparking point.

Experiment has shown by running armature without brushes and taking temp. that not more than 1% is lost as faucault currents, hence 4% was lost by short circuiting between commutator blocks due to brush—this is gained in the converter and also does not heat armature. This gives 99% and as we are going to double speed converting twice as much, the faucault currents will be 2% using very thin plates—leaving 98%.

And as the difference between 95 and 91 commercial was 4% it will only now with double speed and load be 2% hence 100—2 lost as Faucault (2 in armature for old load) gives 96%. The mere friction due to the weight of revolving arm., and scarcely anything— it's a mere fly wheel and one man could turn half a dozen (!!!)[26] I don't think that more than .1^c of H.P. would be absorbed— It's the strain on bearings due to load that causes ordinary dyna to show loss of 1 to 2 H.P. in friction.

Now starting with a 91% dynamo— then 96% of 91% is I think 87⅛% say there is an extra 2⅛% loss besides in converter—then we get 85%

Memo. The volts of our converter remain at their highest point and are not dimished by the necessity of advancing brushes due to load —hence very low internal resis. to get volts.

We can run an interlaced high volt 2 wire main all over a city and at convenient parts cut in pressure stations &c. The initial cost of these interlaced mains would not be very great then stations could be started from time to time. The fact is I'm getting enthusiastic on this new plan. The double converting power of a dynamo and its perfection and economy as a converter is the point that makes the system a success— There would be very little danger to our men in repairing a 2000 volt const high pressure main for even if they did get it, it would not produce death, but I cannot for the life of me see how alternate current high pressure mains— which in large cities can never stop, could be repaired.

Note:— [d] The larger the dynamos the better for converting. The only thing that limits it is the centrifugal force tending to throw the wire off the arm. when there is no current to help hold it on— The larger the diameter of arm—the less this is—for instance—a pound weight on the surface of the earth, which revolves, has scarcely any energy to cause it to leave the earth, because while it goes round it is nearly a straight line owing to the enormous diam. but you put that pound weight on a 10 in. arm., it has to change from a straight line twenty times a second—for the same pound on the earth while it travels infinitely faster only changes its direction once in twenty-four hours, so that on a 200 amp machine the centrifugal energy would be very much greater than on a 400 ampere both having the same surface velocity. There is not the slightest difficulty in make a 3000 Lt— (½ amp) converter and the price pr Lt would be way down.

Note:—[d] You must remember that economy by going to

water edge only means saving in boilers and coal expenses. It don't save on engines, dynamos & converters which is the principal investment; for the great diminution in investment we must look to lamps, an improvement of 50% in economy lamps saves coal, but this is a mere flea bite compared to the fact that we use 50% less dynamos, converters, wire, real estate, boilers etc. The mere saving of coal would not pay $\frac{1}{2}$% on the saving in investment.

This investment with the water station earns 15–20% then what does the coal amount to? I am fully alive now that we have a good workable system, that even the gains due to our system is nothing as compared to a lamp that is 50% more economical. The coal it saves is a sum so insignificant as to be no factor when we consider the immensely reduced interest account. Not only have we an infinitely better, and the ultimate system to beat competitors, but it is in the lamps that I hope to make it positively impossible for them to exist in Central station work to say nothing of Isolated.

D (typed copy), Vail (*TAED* ME004). ᵃFollowed by dashed lines to number at right margin. ᵇDrawing taken from handwritten copy of William Andrews. ᶜText from Andrews's copy. ᵈParagraph typed single-spaced and with wide margins.

1. This typed copy was marked "Memo. from T. A. Edison to Edwd H. Johnson" in handwriting not identified by the editors; it ended up in the papers of John H. Vail. William Andrews, who was himself working on AC generation and transmission, also wrote a copy by hand (which he titled "Notes by T.A.E.") in one of his notebooks, now in the E[dwin]. W. Rice, Jr., Papers (Edison EL Coll., Box 1, Folder 1, NScIS [*TAED* X710A]). Andrews annotated his copy with critical comments in the margins, as noted below. Except for trivial details, the body of both copies is the same (including expressions of surprise or disbelief inserted into the text, likely written by Johnson on Edison's original), and it is possible that one was made from the other. The editors have selected the typed version because it provides a cleaner text for transcription; they have inserted Andrews's drawings into the blank spaces provided for sketches.

2. Johnson wrote a direct reply to Edison's memorandum on 13 November. William Andrews's notebook copy of this document (see note 1) was marked "1885," possibly well after the fact. Johnson to TAE, 13 Nov. 1886, DF (*TAED* D8621K).

3. Edison may have been alluding to reactance, a dissipation of energy in the shifting magnetic field around the conductor of an alternating current. The term seems not to have come into general use until several years later. William Andrews, in his copy (see note 1), dismissed Edison's critique as using "an unfair assumption as not many Tons will ever be employed." In the twentieth century, the length of AC transmission lines was effectively limited in part by reactance losses, and some

power suppliers turned instead to high-voltage DC for longer lines. Denny 2013, 61–65, 77–78; *OED*, s.v. "reactance."

4. Andrews commented here: "No regulators required."

5. Andrews commented here: "wrongly assumed   counter EMF prevents this."

6. Andrews commented here: "wrongly assumed."

.7. Andrews commented here: "They are placed on poles in the streets."

8. Figure labels are "130 Volts" and "130 Volts."

9. Andrews placed four large exclamation points down the margin along this paragraph.

10. Andrews noted here: "Ferranti has a meter." Sebastian Ziani de Ferranti, a British electrical engineer and pioneer of alternating current systems, invented a meter of superior accuracy in 1885, but it used a motor that apparently worked only with direct current. Ferranti does not seem to have had an AC meter until 1888. Wilson 1988, 26–27, 60; Gooday 2004, 239–44.

11. Edison probably referred to his U.S. Patent 287,516 for a double-wound armature to be used as a converter, similar to the devices Charles Batchelor and Henry Walter had recently been working on.

12. Pennsylvania Railroad.

13. Electrical manuals and textbooks commonly distinguished between static and "dynamic" electricity to differentiate an electrical charge on a body from an electrical current flowing through a conductor. Edison's 1885 experiments on wireless telegraphy had dramatically demonstrated the discharge of considerable static energy into the seeming insulating medium of air.

14. The name of American inventor and manufacturer George Westinghouse (1846–1914) was already synonymous with his railroad air brake. A veteran of both the Union army and navy, Westinghouse gained his earliest technical education in his father's shop in Schenectady, N.Y. He briefly attended Union College in that city before moving to Pittsburgh in 1868. There he formed the Westinghouse Air Brake Co. in 1869 to manufacture his inventions in that field, and he successfully fought the railroads for control of his air brake patents. About a decade later, he began purchasing patents for railroad signals and switches and synthesizing their features into his own pneumatic system, for which he formed the Union Switch & Signal Co. in 1881. His work on a third transmission system, one also based on pressure differences, is at least suggestive of what he later did with electric distribution. In about 1884 Westinghouse began to develop a means of harnessing natural gas by taking it from the wellhead at unusably high pressures and conveying it through a network of successively larger pipes so that it reached consumers at a safe pressure. Westinghouse became interested in electric lighting around the same time and may have used the gas system as an analogy. (Edison had explicitly modeled his electrical conductors on a network of gas pipes, but those were for gas manufactured from coal and distributed at low pressures.) Westinghouse hired William Stanley (a young electrical inventor who was already experimenting with AC) in March 1884 and set him up in a new laboratory, and he acquired several Gaulard and Gibbs transformers in late 1885. When he organized the Westinghouse Electric Co. at the start of 1886, it took over the manu-

facture of incandescent lamps already being done by the Union Switch & Signal Co. The new firm would also develop and manufacture inventions for AC electric systems (*ANB*, s.vv. "Westinghouse, George" and "Stanley, William"; Jonnes 2003, chap. 5; Usselman 2002, 133–37; Carlson 2013, 87–89; Hughes 1983, 100–103; Passer 1953, 135; Leupp 1918, chaps. 8–9; Prout 1921, chap. 4).

15. Edison had patented some form of rotary converter for changing the voltage of direct current in 1883 (see Doc. 3002 [headnote]; U.S. Pat. 287,516). Charles Batchelor had recently been working on such a device with Henry Walter, assistant superintendent of the Edison Machine Works, who had assigned Batchelor a half interest in a converter patent issued on 26 October. That specification covered a converter whose armature carried separate motor and generator windings, each with its own commutator. Such a device (which Batchelor called a "transformer") could induce a current in the generator having a different voltage than the current coming in to the motor, and the specification noted that "Machines of this general class have been proposed for the transmission of power electrically." Walter's invention was for equalizing the counter electromotive forces induced in the motor and generator windings. With those forces balanced, the neutral points would remain fixed and the commutator brushes nearly free from sparking. That device, or something like it, would be consistent with Edison's descriptions in Docs. 3009–3010 (U.S. Pat. 351,544; Cat. 1336:79, 121, 141 [items 108, 211, 245; 14 June, 28 Sept., 6 Nov. 1886], Batchelor [*TAED* MBJ003079B, MBJ003121C, MBJ003141]). Walter filed jointly with Batchelor at least two other related patent applications in the next several weeks (U.S. Pats. 360,258; 360,259). Another by Walter alone in January was co-assigned to Batchelor (U.S. Pat. 438,308).

16. Károly (Charles) Zipernowsky (1853–1942), born into a merchant family in Vienna, worked as an apothecary before studying mechanical engineering at the Royal Polytechnic College there. He completed his degree in 1877 and took up electrical work, starting the next year with dynamo design for Ganz and Co., a Budapest engineering firm. He and a Ganz colleague, Max Déri, took out a patent on a "rotating secondary generator" (similar to Edison's rotary converter) in 1884. The two men were soon working on induction transformers and a related system of distribution. With a third colleague, Otto Bláthy, they analyzed and improved the Gaulard and Gibbs transformer, patenting their own design in April 1885. Their ZBD transformer was promptly exhibited in Budapest and, in the summer of 1885, also in London by the Edison and Swan United Electric Light Co. A detailed two-part description of it appeared in the *Electrical Review* in August 1885. See Doc. 3002 (headnote); "Electrical World Portraits.—No. XXXIII. Zipernowsky—Déri—Blathy," *Electrical World* 19 (16 Apr. 1892): 258–59; "The Zipernowsky-Déri System of Distributing Electricity," *Teleg. J. and Elec. Rev.* 17 (1 and 8 Aug. 1885): 92–95, 114–17; Fleming 1892, 2:85–88; Hughes 1983, 95–97.

17. Westinghouse already controlled a formidable electrical manufacturing base that he had incubated at the Union Switch & Signal Co. By the middle of 1887, his lamp works reportedly were larger than Edison's, and the combined shops of Westinghouse Electric, employing some 1,200 men, were busy with contracts for new power plants.

In May of that year, William Andrews wrote to John Vail from New Orleans, where a Westinghouse system was being installed, that it was "idle to scoff at the Westinghouse people as being beneath notice as competitors. They are hard and persistent workers . . . ," and Andrews warned that if the Edison interests did not counter them, "we shall suffer in the future for our negligence." "Business Items," *Manufacturer and Builder* 19 (Aug. 1887): 188; Andrews to Vail, 12 May 1887, DF (*TAED* D8732AAR).

18. Not identified.

19. Joseph Hutchinson.

20. Jonathan Vail.

21. See Doc. 3050. Figure labels are "To dist station" and "1500 Volts." At the end of the year, Vail calculated significant savings from the use of the new 250 ohm lamps. Vail to TAE, 28 Dec. 1886, DF (*TAED* D8624O).

22. See Doc. 3009.

23. These parenthetical comments were probably written by Johnson on Edison's original; a similar annotation appears in the copy of William Andrews (see note 1).

24. In his enthusiastic reply (see note 2), Johnson outlined in some detail how a distribution system might be designed in accordance with the claims made by Edison. "Jesus— This is a big thing you have got onto now," he wrote, one in which "the Extra loss in your converters will be more than made up by the added economy of a River Station." He concluded that "Zip[ernowsky] or W[estinghouse]. must work out a system Or Steal Yours." Johnson wrote another memorandum from which the first page is missing but which he likely addressed to Edison about this time. He discussed his plan to promote a high-voltage direct current system based on two dynamos connected in series and Edison's new high-resistance lamps, such a system being adapted to the distribution of power over long distances. Johnson to TAE, n.d. [Nov. 1886?], DF (*TAED* D9899AAL).

25. The year "1884" was added by hand in the left margin preceding "Test." Edison referred to the competitive dynamo trials at the Franklin Institute. Originally scheduled for late 1884, they did not take place until 1885 (Doc. 2820 n. 10). Of the four Edison models tested, two were rated fractionally above 96% electrical efficiency at full load, and one of these at over 91% "commercial" or overall efficiency (Franklin Institute 1885b, 58 [table 8]).

26. This parenthetical note was probably written by Johnson on Edison's original; a similar annotation appears in the copy of William Andrews (see note 1).

–3009–

To Charles Batchelor

New York, Nov 12 1886[a]

Batch=

It has just occurred to me that we have a better thing in our Converter than I first thought   have your man[1] figure it out—see if I am right—

1st   The armature of a 200 ampere machine running without strain from belt spo[iling?]][b] can easily make twice as many revolutions per minute as when used as Dynamo.

2nd   This being case & with same strength field it can be wound to have 4 times <u>lower resistance</u>, hence at double speed give proper Volts & instead of 200 amperes Convert <u>400 amperes</u>[c] & Even then machine will work better than as[d] Dynamo—& there will be no strain [~~dia?~~][e] on moving parts due to this double load—

Let me know if I am Correct

TAE

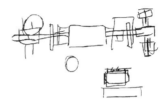

ALS, NjWOE, Batchelor (*TAED* MB218). Letterhead of Thomas A. Edison. [a]"New York" and "188" preprinted. [b]Obscured by ink smear. [c]"amperes" multiply underlined. [d]Interlined above. [e]Canceled.

1. Batchelor's "man" was Henry Walter; see Doc. 3008 n. 15 regarding his collaboration with Batchelor on rotary converters.

–3010–

To Francis Upton

[New York,] November 19th [188]6

Upton, c/o Drexel[1]   Paris

Double wound dynamo used as converter turns out great success, runs double speed splendidly and by reduced armature resistance converts double the amperes over simple dynamo. No sparking or attention as one coil neutralizes magnetism of the other— We adopt it for New York having one station on river and number distributing stations with small areas with regular three wire system, inexpensive cellars used for distributing stations—[2] Call Rau and Ratnaus[3] attention   Get patents and figure them out as against Zipernowsky which is impracticable this country owing lack economy and danger[4]

Edison

L (letterpress copy), NjWOE, Lbk. 83:793 (*TAED* LB023104). Written by William Gilmore.

1. Drexel, Harjes & Co., established in 1868, was the Paris affiliate of Drexel, Morgan & Co. and was one of the leading private banks in Europe. Doc. 2155 n. 2.

2. In his response to Doc. 3066, Edward Johnson enthusiastically described a "river" plant and service district built along these lines. The editors have found no other evidence that the idea was considered. Johnson to TAE, 13 Nov. 1886, DF (*TAED* D8621K).

3. Electrical entrepreneur Emil Rathenau (1838–1915), an early promoter of electric lighting companies in Germany, negotiated the formation of the Deutsche Edison Gesellschaft für angewandte Elektrizät (DEG) in 1883. He remained in charge of DEG at this time. Doc. 2448 n. 1; Hughes 1983, 179.

4. By this date, Károly Zipernowsky had already warranted to Upton (on behalf of the Ganz Co.) that the ZBD transformer would meet certain benchmarks of electrical efficiency, to the satisfaction of engineers of the Edison Electric Light Co. of Europe (Zipernowsky to Upton, 16 Nov. 1886, *Zipernowsky v. Edison*, Lit. [*TAED* W100DMBB]). Upton soon reached an agreement with Ganz for the Edison Electric Light Co. to license the U.S. patents on the ZBD transformer and distribution system (see Doc. 3013).

–3011–

*Draft Patent
Application: Electric
Light and Power*[1]

[East Newark,] Nov 19 1886

Seeley—[2]

Want to get all the claims possible on this method of working from high to low currents   be careful in drawing patent not to call this a tension reducer— It is a motor system driving a Dynamo mechanically   make this distinction broad— so they cannot recite Ayrton & Perrys old suggestion of Motors[a] driving dynamos as a current[a] reducer—[3]

The pith of the invention is

1st— A motor worked by high electromotiveforce driving by direct connection a Dynamo giving out low volts.

2nd   Insulating joints in shaft[b] or means so that the high line is absolutely insulated from the low line & arranged so that no cross can occur [-][c] The Earth intervening would in case the wire on one armature came in contact with the base prevent any Emf from Entering the 2nd ckt in fact the circuits could[a] be miles apart as far a crossing is concerned. ⟨(see?)⟩

3rd   Placing ~~the~~All the motors in Multiple arc—strong claim—or on 3 wire ~~getting one motor to drive~~—system which is practically same thing—

4th— Arranging one motor to drive two machines as well as one— [~~Thus?~~][c] This gives a ~~2~~3 wire unit—

5= Energizing the fields from the dynamos [---][c] and Motor from the low current.

6— also Energizing the Motor field from its own ckt.

Dick[4] will tell you what I want to put this patent in for—[5]

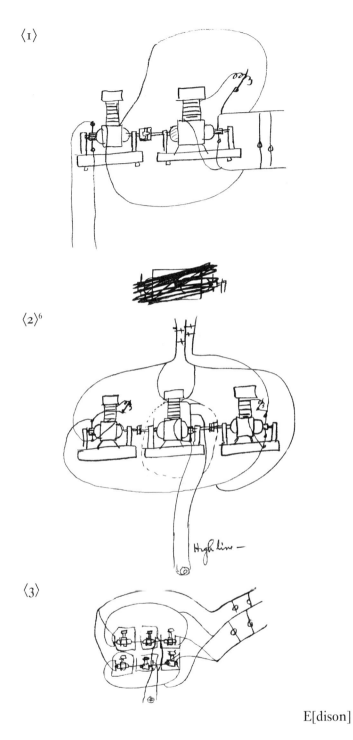

⟨1⟩

⟨2⟩[b]

High line —

⟨3⟩

E[dison]

ADfS, NjWOE, Lab., Cat. 1151 (*TAED* NM020ABP). Document multiply signed. [a]Obscured overwritten text. [b]"in shaft" interlined above. [c]Canceled

1. See Doc. 3002 (headnote).

2. As the partner of Richard Dyer since 1884, Henry W. Seely was one of Edison's primary patent attorneys. Doc. 2429 n. 3.

3. Electrical engineers William Edward Ayrton (1847–1908) and John Perry (1850–1920) were friends and prolific collaborators, first as colleagues at the Imperial College of Engineering in Tokyo and then at Finsbury Technical College in London from 1882 until 1885, when Ayrton took a different post in South Kensington (Doc. 2656 nn. 2–3; *Oxford DNB*, s.vv. "Ayrton, William Edward" and "Perry, John"). Edison likely referred to the suggestion they published in 1883 for regulating the speed of motors under varying load. Ayrton and Perry proposed to have the motor mechanically drive a dynamo which, at certain speeds, would supply a current to the motor field windings counter to that provided by the line, weakening the magnetic field (Thompson 1886, 447–48; Thompson 1904, 1:846–47; Ayrton and Perry 1883, 307–11).

4. Richard Dyer.

5. Edison reiterated some of these points, especially the insulated connection in the rotary converter, in a draft patent application that he addressed to Seely on 28 November. Cautioning that future rivals might "seek to evade" his converter patents, Edison instructed Seely to "cover the only practicable way to do it & in addition Connect it to our 3 wire system & so combine it that I can cover it through combination with other essential features in any system such as feeders, etc—." Cat. 1151, Lab. (*TAED* NM020ABR).

Edison signed the completed application on 10 December. The resulting patent issued in September 1887 with sixteen claims and four drawings. The claims covered permutations on the basic elements outlined above by Edison, especially the coupling of the dynamo and motor and the arrangement of circuits supplied by the dynamo. Three of the figures corresponded to the numbered sketches in this document; the fourth depicted connections with Edison's three-wire distribution system. U.S. Pat. 369,441.

6. Figure label is "High line—."

–3012–

Draft Patent Application: Incandescent Lamp

[East Newark,][1] Nov 21 1886

⟨706⟩[2]

The object of this invention is to obtain homeogenious carbon filiments for incandescent Electric Lamps   The invention consists in the[a] use of The non volitile residues of the resins and bitumens which are Oxygenated Hydrocarbons. With[a] or without infusible conducting or non conducting elements or Compounds.

The invention further consists in the manner of forming the Carbonizable filiment and in the manner of carbonizing the same.

of the oxygenized residues of Resins and bitumens I prefer what is known as Asphaltene which is prepared from common refined Asphalt by heating the same at about 250 centigrade in the open air until the volitile matters are driven off—[3] This is allowed to cool & is then broken up and very finely powdered, it is then put into a mould with a plunger from this mould there is a fine orifice through which the Asphaltene may be forced into a thin filiment the mould being heated to the softening point of the asphaltum.— The filiments are then hung in small Carbon boxes a[a] number of which are placed in a chamber of carbon. The whole is heated for about 15 hours ~~a~~to a heat a little below the softening point ~~all~~ until last traces of[b] the volitile matters are thus driven out & then the heat may be raised slowly without melting or softening the filiment until the whole is thoroughly Carbonized—

If the Asphaltene just before all the volitile matters are driven off in ~~the~~ its preparation is mixed with pure finely divided graphite & then allowed to cool ~~& powders~~ pieces may be put into the filiment forming press without powdering & by a high heat filiments forced out

In this case the filiments may be carbonized by the regular methods without previous drying slowly as the graphite prevents the effects ~~filiment from~~[c] of soften of the asphaltene. ~~Oxides~~ infusible Oxides such as lime, Magnesia Alumina may be substituted for graphite when high resistance filiments are required.

I do not wish to confine myself to asphaltene as it is not an exact compound but claim generally the heating of all Bituminous & Resinous substances[a] Until nearly the whole of their volatile Constituents are driven off Stopping the heat just below the carbonizing or decomposing points. petroleum, Rosin etc may be used—

If filiments are formed out of infusible oxides, They may afterwards be impregnated with asphaltene & accomplish the same purpose The Asphaltene being disolved in Benzol or other good solvent which is volitile at low temperatures

Seeley—

I find asphaltene the best substance for impregnating [------][d] ~~filiments of~~ the clay etc[e] filiments ~~also for soak~~ You may have to divide this application as its for soaking How would it do to put all the claims in & then when we want to can make a division—

Claim formation of filiments from asphaltene or equivalent substance

  2nd   Combined with powdered carbon or graphite

  3rd   Combined with Infusible substances

[~~In f?~~][f] previously formed filiments of carbon or ~~ox~~infusible non-conducting materials impregnated with asphaltene in proper solvent.

~~preparing filiment from asphaltene or equivalent &~~ Etc

TAE

ADfS, NjWOE, Lab., Cat. 1151 (TAED NM020ABO). [a]Obscured overwritten text. [b]"until last traces of" interlined above. [c]"~~filiment from~~" interlined above. [d]Interlined above and canceled. [e]Interlined above. [f]Canceled.

1. Edison wrote this draft on pages from one of his standard notebooks.

2. This is the case number for the application created from Edison's draft. Edison executed the application on 6 December, one of three (including one derived from Doc. 3006) related to filament manufacture that he signed that day. It was filed nine days later and issued in 1892 as U.S. Patent 485,616.

3. Edison planned to adopt similar substances for insulating dynamo armature plates about the same time. In a 26 November draft of a patent application for armatures, he indicated that "almost all of the Bitumens or partially decomposed Resins in proper solvents will answer but I prefer ordinary Asphalt boiled to drive off some of its volitile constituents." Cat. 1151, Lab. (*TAED* NM020ABQ).

–3013–

From Francis Upton

Hotel Chatham, Paris, Nov. 26, 1886.[1]

Dear Mr. Edison:

I send the European Contract to Mr. Coster per instructions from him by cable also the papers belonging to the European Co.[2] The contract is not all I want but is the best I could get. I find that the French law does not allow of an absolute guarantee to the founder's shares so have had to have the guarantee made in accordance with the law.[3] The amount $7782 \times 3 = 23,346$ francs a year is enough to keep the New York Company hardly alive.[4] But there seems to be no doubt but the amount will largely be exceeded in time. The Lamp Co. lamps pay no royalty. I think that the European Co. of New York have some claim on the Lamp Co., but we can submit this question to arbitration if need be.[5]

I signed a clause regarding your ~~giving~~ assigning[a] your patents in the future.

Mr. Siegel ~~was~~ and Mr. Rau were very desirous of having this as they said that it would aid very much in getting new capital. You have only to give up patents in France and from this country we are excluded by the patent laws. You can assign to the Lamp Co and to the Machine Works and we can sell material in Europe under them.[6] This seemed to me to be fair and give you a chance in case the French Company would not do right by you in the future.

The directly and indirectly is modified by the Contract with the Lamp Co. which is ~~also~~ enclosed with this.[7]

This contract gives us the European market so long as we can make the best lamp at a low price.

There are several objectional features such as our giving lists of sales &c and prices of our material. Against these we have the same from the French Co. and we have prohibited them from sending lamps to America or England.

There is no way that I can see for the French Co. to keep us out of Europe[b] so long as we act fairly and make good lamps.

I am perfectly convinced that I never have put time to better use than during the past weeks. It would have been impossible ~~for~~ to have closed the contracts from America and there is now a very strong movement towards putting the Edison light ahead here in Paris. Mr. Siegel could not have made the concessions I did as against concessions on the other side, for he does not understand the whole situation. Whether I have done well will be seen from the results of the next few years. I have worked very hard and have given all the care I could. I have got mad; and argued and asserted and urged everything that could bear on the subject and the contracts as drawn are the results.

I have written Johnson regarding the Zipernowsky patents.[8] I think that the alliance with Ganz & Co[9] is a most excellent one. You are working up the direct current system and Ganz & Co. are working up the alternating system.[c]

Your name is a good one to sail under. I enjoy the good things of life it brings me and think how much pleasure you ~~wol~~ would have, were you here.[c]

The following is the statement of the German royalties to the French Co. made by Mr. Rau to me[10]

1883	2nd six mos.	Marks	18 866.45
1884	1 "	"	9066.40
1884	2 "	"	26 396.60
1885	1 "	"	11 064.79
"	2 "	"	42 867.20
1886	1 "	"	12 454.59
			115,716.03

<table>
<tr><td>Royal[t]y advanced by German Co.</td><td></td><td>350,000.00</td></tr>
<tr><td></td><td></td><td>115 716.03</td></tr>
<tr><td>to be paid</td><td>Marks</td><td>234 283.97^c</td></tr>
</table>

Royal[t]y advanced by German Co.   350,000.00

115 716.03

to be paid   Marks   234 283.97^c

This looks as if in three years^d the founder's shares will have an income from Germany. Yours Truly

Francis R. Upton

ALS, NjWOE, Miller (*TAED* HM860293). ^aInterlined above. ^b"of Europe" interlined above. ^cFollowed by dividing mark. ^dObscured overwritten text.

1. Upton sailed from New York on 6 November with "full powers" on behalf of Edison and the Edison Electric Light Co. of Europe. He also had broad authority to act for the Edison Lamp Co., though that power of attorney did not follow him until 23 November. On the same day that he sent this letter to Edison, Upton wrote similarly to Samuel Insull. Cat. 1336:141 (item 243, 6 Nov. 1886), Batchelor (*TAED* MBJ003141); TAE to Société Électrique Edison, 9 Nov. 1886, Lbk. 23:77 (*TAED* LB023077); Edison Lamp Co. to Upton, 23 Nov. 1886, LM 4:69 (*TAED* LM004069); Upton to Insull, 26 Nov. 1886, Miller (*TAED* HM860294).

2. The contract is a 25 November agreement concerning the consolidation or fusion of the Paris companies; its terms are discussed in the notes below. Upton negotiated for Edison and the Edison Electric Light Co. of Europe, while Louis Rau dealt on behalf of the three Edison companies in Paris. The editors have not found Charles Coster's instructions or his correspondence with Upton on the matter. Samuel Insull had sent Coster the minute books of the Edison Electric Light Co. of Europe on 22 November. TAE agreement with Edison Electric Light Co. of Europe, Compagnie Continentale Edison, Société Électrique Edison, and Société Industrielle et Commerciale Edison, 25 Nov. 1886, DF (*TAED* D8630ZCQ); an undated and unsigned version in French is in DF (*TAED* D8630ZCV); Upton to Insull, 26 Nov. 1886, Miller (*TAED* HM860294); Insull to Coster, 22 Nov. 1886, Lbk. 23:116 (*TAED* LB023116).

3. The contract provided for an exchange of the original promoter's (or founder's) shares for new promoter's shares, and for the annulment of agreements dated 15 November 1881 and 2 February 1882 (p. 4). The new shares were guaranteed an annual income of 3 francs apiece plus thirty-five percent of the profits (determined after deduction of obligations for the legal reserve fund and interest on capital paid in; pp. 11–12), an amount apparently sufficient to enable the Edison Electric Light Co. of Europe to meet its bond obligations. The new shares carried no rights of managerial interference, financial oversight, voting, or attendance at general meetings (p. 12). In addition, Edison and the European Co. surrendered their authority to approve future changes in bylaws that did not affect them or the promoter's shares (p. 13). Through the amendment of numerous bylaws, the contract also enunciated the rights of the Compagnie Continentale Edison to manufacture and sell both Edison and non-Edison apparatus, and to "establish Agencies, Branch Offices and limited partnerships wherever it may think proper" (p. 4).

TAE agreement with Edison Electric Light Co. of Europe, Compagnie Continentale Edison, Société Électrique Edison, and Société Industrielle et Commerciale Edison, 25 Nov. 1886; Edison Electric Light Co. of Europe memorandum, 22 Nov. 1886; Alfred Tate to Bank of Sheldon, 31 Dec. 1887; all DF (*TAED* D8630ZCQ, D8625G1, D8717ACZ).

4. This sum equaled approximately $4,500 at the prevailing exchange rate of one dollar to about 5.15 francs. Upton had tried to get $8,000 annually, an amount he considered the minimum needed by the European Edison company in New York. He commented to Samuel Insull, "It was a hard fight and only upon strong advice of Mr. Siegel did I give up." Denzel 2010, 305; *Ency. Brit.* 9, s.v., "Money"; Upton to Rau, 29 Oct. 1886, DF (*TAED* D8625E1); Upton to Insull, 26 Nov. 1886, Miller (*TAED* HM860294).

5. The new agreement (see notes 2–3) fixed a royalty to the Edison Electric Light Co. of Europe of 20 centimes for every lamp, regardless of what type, sold by the Compagnie Continentale Edison. Lamps supplied by the Edison Lamp Co.'s New Jersey factory were exempted, as were those used in the Continentale Co.'s offices and the Ivry factory. Lamps sold or used in Germany were also excepted until repayment under the assignment contract was complete (pp. 10–11). Upton to Rau, 29 Oct. 1886, DF (*TAED* D8625E1).

6. Upton presumably meant that the U.S.-based companies, in particular the Edison Lamp Co., could not manufacture or sell directly in France under patents licensed to the French firms, one of which had its own factory at Ivry. The contract (see notes 2–3) specified that "Mr. Edison undertakes to offer the Compagnie Continentale Edison in the countries in which it has the right to work patents, patents of new inventions which he may make relating to electric light and to the transmission of motive power," allowing sufficient time for the company to accept or decline. The next paragraph positively stated, however, that "Mr. Edison will assign to the Compagnie Continentale Edison all new patents and all additional patents for improvements . . . relating to the incandescent lamp" (p. 5).

7. On this date, Upton also signed for the Edison Lamp Co. an agreement pertaining specifically to the sale of incandescent lamps in Europe. Among its major provisions was a promise by the French companies to sell those lamps made in the U.S. at the same price as those made in France, allowing the customers to choose freely. (This stipulation did not apply within France.) The Edison Lamp Co., in turn, promised to sell to the French firms at its most favorable prices. It also gained the right to solicit sales for itself in Europe (excepting France), though all orders were to pass through the French companies. Edison Lamp Co. agreement with Compagnie Continentale Edison, Société Électrique Edison, and Société Industrielle et Commerciale Edison, 25 Nov. 1886, Upton (*TAED* MU095).

8. Upton urged Samuel Insull to read the letter he wrote to Johnson; the editors have not found it, but see Doc. 3022. Upton to Insull, 26 Nov. 1886, Miller (*TAED* HM860294).

9. Ganz & Co., a major engineering firm in Budapest, manufactured transformers and other equipment for the alternating current distribution system co-invented by Károly Zipernowsky. The company also owned the key U.S. patents (both issued and pending) for the system,

and Upton signed a licensing agreement on 25 November securing the North American patent rights for the Edison Electric Light Co. The terms were $5,000 paid on signing, a subsequent $20,000 payment, and a ten percent royalty on all goods manufactured under the patents. Upton proudly explained to Insull that without his personal efforts in Paris, George Westinghouse likely would have acquired those rights, "with which he could have made central stations. This alone is in my opinion worth the trip to Europe." Hughes 1983, 96–97; Edison Electric Light Co. agreement with Ganz & Co., 25 Nov. 1886, *Zipernowsky v. Edison*, Lit. (*TAED* W100DMBA); Upton to Insull, 26 Nov. 1886, Miller (*TAED* HM860294).

10. One mark was worth about twenty-four U.S. cents; one French franc equaled about 8.08 marks. *Ency. Brit.* 9, s.v. "Money"; Denzel 2010, 244.

–3014–

Draft Patent Application: Electric Light and Power[1]

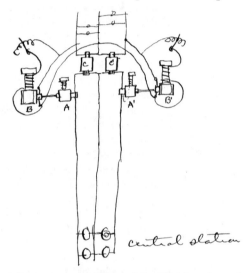

Object of patent regulation of the EMf at the reducing station while that at Central station is constant—[3]

AA′ are two ~~dynamo~~ motors[a] wound so that a ½ speed they give a minimum Counter Electro motor force   B are Motors which are run from the low tension line.— when the Volts are to be increased or diminished on the high line the fields of B are increased or diminished in strength & thus the motor B runs at a higher or lower speed thus causing A to give greater or lesser Counter EMf or back pressure   Thus if the Central station has a Constant of 2500[b] volts & A a Counter pressure of 500[b] volts, The net will be 2000. if now there is a fall of pressure due to load The strength of the field of B is increased, the speed of A diminished & the back pressure diminishes thus bringing up to the required amount.

TAE

ADfS, NjWOE, Lab., Cat. 1151 (*TAED* NM020ABS). ᵃInterlined above. ᵇObscured overwritten text.

1. See Doc. 3002 (headnote).

2. Edison wrote this draft on his New York letterhead, but the editors conjecture that he more likely was at his laboratory in the lamp factory.

3. Figure labels are "B," "A," "c," "c," "A′," "b′," and "central station." Edison signed the patent application developed from this draft on 16 December (cf. Doc. 3011). When the resulting specification issued in September 1887, its single drawing was based closely on Edison's sketch. The patent contained five claims, two of which included his three-wire system of distribution; two other broad claims were erased during the examination process. It incorporated (but did not claim) Edison's "double-wound rotating converters" (marked "C" in his sketch) positioned at a "sub-station" (Pat. App. 369,442). His decision to cross out the word "dynamo" and instead call the driven device in the high-voltage circuit a "motor" may reflect a concern about prior patents on coupled motor-generator pairs (cf. Doc. 3011 n. 3). This method of regulating line voltage took advantage of the principle of back (or counter) electromotive force. Counter-emf, a voltage applied against that of the incoming line current, is generated by the rotation of motor armature windings through a magnetic field. The effect is common to all rotary induction motors and was recognized at the time as essential to their efficient operation (Thompson 1886, 401–407).

–3015–

Bill from Dearborn-Morgan School

Orange, N.J., Nov. 30, *1886*ᵃ

Mr.ᵇ Thos. A. Edison
To Dearborn, Morgan & Co.,[1] Dr.²ᶜ
For Tuition, fromᵈ Nov. 29, 1886,ᵉ to Feb. 4, 1887,ᵉ

of	Thos. A. Jr.,	3 Grade,ᵇ	1 Quarterᵇ		20
"	William P.[3]	1	"	"	12

For Rent of Books, Stationery, etc.,ᶜ   1.30
Injuring Books, Furniture, etc.,ᶜ
*Less Deduction $ if paid beforeᶠ

33.30⁴

*This Deduction is made for *Uninterrupted Attendance* of over two years— See Circular.ᶠ

D, NjWOE, DF (*TAED* D8614P1). Written on ruled billhead of the Dearborn-Morgan School. ᵃ"*Orange, N.J.*," and "*188*" preprinted. ᵇPreprinted. ᶜLine of text preprinted. ᵈ"For Tuition, from" preprinted. ᵉ"188" preprinted. ᶠ"*This Deduction . . . Circular." preprinted.

1. The Dearborn-Morgan School was a non-sectarian institution formed in 1876 by the consolidation of three co-educational schools.

Its billhead stated that it "RECEIVES PUPILS OF BOTH SEXES IN PRIMARY AND ADVANCED GRADES. PUPILS PREPARED FOR COLLEGE OR BUSINESS." Located at 443 Main St. and operated by the legal entity of Dearborn, Morgan & Co. (comprised of J. B. Dearborn, Abby Morgan, C. H. Mann, and D. A. Kennedy), Dearborn-Morgan presently had 227 students enrolled; tuition ranged from $48 to $180 annually. According to an 1896 description, graduates went on to attend elite institutions such as Harvard, Yale, Columbia, Vassar, Smith, and Wellesley. Sargent *Private Schools* 1933 (17): 321–22; Dorflinger and Dorflinger 1999, 54; U.S. Dept. of Interior 1887, 392; ibid. 1899, 2296; Whittemore 1896, 222.

2. A book-keeping abbreviation for debtor, used as a heading for the left-hand or debit column of an account. *OED,* s.vv. "dr," "debtor."

3. That is, William Leslie Edison.

4. The bill was stamped "PAID" by Dearborn, Morgan & Co. on 8 February.

–3016–

Memorandum: Electric Lighting Business Reorganization

[New York, November 1886?[1]]

Mfg Dept. Insull Manager=
Machine Wks   Kruezi. Manager.
Bergmann   Insull. (temporarily)
Lamp Co   Upton.
Sales Dept. Isolated &[a] Cash Stations
Now[b] United Mfg Co. Chinnock Manager.
Engineering work going into Engineering dept.
Chief[b] Constructor[c] for[c] Isolated & cash stations   Field or—[2]

"   Large Central. Henderson[3]   all under Chief Engineer—
Engineering & Construction[d] Dept.   Marks[4] Chieeff
Vail Assistant.
___ Determinator.[e]

Obtaining franchises rights way & general man   Hix—
Supt of Business management of all stations owned[c] or which Co has stock in   Beggs[5]
All offices at 65 5th Ave Except. presidents office & Cos offices—which is down town[6]
Andrews to be taken in, place of Jenks[7] bounced.
All determinations & Complete Specifications of Central stations to be approved by Edison before adoption. No electrical changes
Kruezi   Supt.[8]
Insull   Secy Treasr=
Batch   Vice President &[f] General Manager

Insull to reside at Schenectady

Ditto Kruezi—

Batchelor to go up occasionally[a] for Experiments & Examination accounts & see that things are running satisfactory.

Foremen. ~~Turner~~[9] McDougal[10a] Mach[ine Works?] & Krueusis Tube man[ager?]

Insull to Run Phonoplex from Schenectady also with ~~Batch~~ and Randolph in NYork run my private correspondence[g] Gilmore[11] to go with Insull. Notes to ~~be signed by Batchelor only~~ signed by Insull as Treasr & coun[ter signed?][h] by Vice Prest—or Prest

Batch to come back into Laboratory. Have Walters[12] & draughtsman come with him— Batch also to attend to NY[c] End of any Machine Wks biz with aid of an assistant.

Batch[f] to work up Converters, design for Vail system for Boston[13i] or NY & also get 33 pct more out present dynamo.— plans for & errection new Laboratory—cheap cable for underground[14] —Ore Milling—Head Light Dynamo.—Non Sparking Dynamo. E[lectric] R[ail] R[oad] motors Large 500 hp Dynamo. Compound Gramme. Arc Lt carbons

AD, NjWOE, DF (*TAED* D8629D). [a]Interlined above. [b]Added in left margin. [c]Obscured overwritten text. [d]"& Construction" interlined above. [e]Followed by dividing mark. [f]"Vice President &" interlined above and connected by left brace to "General Manager." [g]"and Randolph . . . correspondence" interlined above. [h]Illegible. [i]"oston" interlined below.

1. Edison's reference below to the proposed role of Charles Batchelor in his laboratory suggests that he made these notes before Batchelor wrote Doc. 3017 on 1 December. Edison likely was planning for the 3 December meeting at which Samuel Insull was, in fact, made secretary and treasurer of the Machine Works at Schenectady and Batchelor became vice president and general manager; see Doc. 3018.

2. Cornelius J. Field (1862–1915), a native of Chicago, was a graduate of the Stevens Institute of Technology. In 1886, he was employed in the engineering department of the Edison Electric Light Co. The next year, he served as chief engineer of the Edison United Manufacturing Co. in New York. Pierce 1901, 899–900; "Cornelius J. Field Dead," *NYT,* 21 Sept. 1915, 12.

3. John Carlos Henderson (1844–1907), a native of Scotland, was a mechanical engineer who worked initially for Henry Villard. In the early 1880s, he was the construction engineer of the Oregon Railway and Navigation Co. and also helped with the installation of an Edison lighting system on the ship *Columbia*, on which he became chief engineer. In 1884, he was a member of the Board of Trustees of the Edison Co. for Isolated Lighting. It is not clear whether Henderson took a position with the short-lived Edison United Manufacturing Co. but by September 1889, he was the chief engineer of its successor, the Edison

General Electric Co. Headstone photo for Henderson, Fairview Cemetery, New Albany, Ind., Find A Grave memorial no. 74709910, online database accessed through Findagrave.com, 9 Sept. 2013; "Obituary Notes," *Electrical Review* 51 (10 Aug. 1907): 231; see Doc. 1892; Edison Co. for Isolated Lighting annual report, 18 Nov. 1884, PPC (*TAED* CA002D); Batchelor to Henderson, 4 Sept. 1889, Lbk. 32:223 (*TAED* LB032223).

4. William Marks remained in Philadelphia.

5. Philadelphia-born John Irvin Beggs (1847–1925), an insurance executive in Harrisburg, Pa., entered the electric lighting business in 1884 as an investor in the Harrisburg Electric Co., which employed an Edison plant. He took over personal supervision of the company and served as its treasurer, secretary, and general manager. Under his control, the firm was, according to the *Western Electrician*, "the most profitable electric light plant in the United States." He became president of the Association of Edison Illuminating Cos. in 1886, holding that position until 1892. It is not clear that Beggs took a post with the Edison United Manufacturing Co. The Edison Illuminating Co. of New York hired him in 1887 as vice president and general manager, and he took responsibility for developing three new central stations, installing isolated plants, and overseeing a rapid expansion of the company's business. He was reassigned to Chicago in 1890 as manager of the Western District of the new Edison General Electric Co. When General Electric was organized in 1892, Beggs became assistant to the president of the new company. "Beggs, John I.," Pioneers Bio.; "John I. Beggs, President of the Association of Edison Illuminating Companies," *Western Electrician* 7 (20 Sept. 1890): 1; *Edisonia* 1904, 196–97.

6. For the new locations of various offices, see Doc. 2945 n. 14.

7. William J. Jenks (1852–1918) had been manager of the Edison central station in Brockton, Mass., and installed the Edison municipal plant in Portland, Maine. Only recently appointed as manager of the Municipal Department of the Edison Electric Light Co., Jenks remained in that position for the time being. Doc. 2475 n. 1.

8. This line marks the top of the other side of Edison's paper and the start of notes about the Edison Machine Works.

9. William B. "Pop" Turner (1843–1914), an Irish immigrant, was the first superintendent of the Edison United shops at Schenectady. Turner is said to have been a skilled millwright and as such was well suited to supervise the setting up of machinery brought from Goerck St. and Brooklyn. He later played a key role in developing a street railroad for Schenectady. Turner remained as general superintendent of the Schenectady works through the consolidation that created General Electric in 1892 but left the company's employ by January 1893. In 1895, Turner was in Melrose, Massachussetts, where he designed machinery for the Turner Tanning Machinery Co. Sometime after 1910, he moved to Wilmington, Del., and operated the W. B. Turner Machine Co. An inventor in his own right, Turner received a number of patents. Turner death certificate, *Delaware Death Records, 1811–1933,* online datebase accessed through Ancestry.com, 23 Sept. 2013; Edward Winters to Francis Jehl, 29 July 1937, MiDbEI (*TAED* X001E18A); Hammond 1941, 150; Yates 1902, 196; "Personal," *Electrical Age,* 11 (21 Jan. 1893): 47; see, e.g., U.S. Pat. 529,220.

10. William M. McDougall (b. 1846), an electrician by trade, became the superintendent of the Machine Works in 1884. He was working in Schenectady by 5 January 1887, but Kruesi, Insull, and Batchelor decided to discharge him in April for unspecified reasons. McDougall was later one of the incorporators of the Acme Storage Battery and Manufacturing Co. of Jersey City, N.J. He received at least two patents for secondary batteries and one for an electric car. McDougall passport application, 8 Mar. 1894, *U.S. Passport Applications, 1795–1925*, online database accessed through Ancestry.com, 19 Sept. 2013; Jehl 1937–1941, 987; Cat. 1336:163A, 185 (items 278 and 328, 6 Jan. and 8 Apr. 1887), Batchelor (*TAED* MBJ003163A, MBJ003185); U.S. Pats. 537,474; 537,475; 627,133.

11. William Edgar Gilmore (1863–1928) was at this time Insull's stenographer and clerk. He went to work in Schenectady in that capacity in December. "Gilmore, William E.," Pioneer Bio.; Doc. 2585 n. 17.

12. Henry Walter.

13. John Vail was in the early stages of planning an expansion of the Edison central station service in Boston. The first station there started in February 1886; the second opened in November 1887. Vail to TAE, 28 Dec. 1886 and 24 May 1887, both DF (*TAED* D8624O, D8732AAT); "The Edison System of Incandescent Lighting from Central Stations" [Edison Electric Light Co. booklet] p. 30, 1888, PPC (*TAED* CA042A).

14. In May 1887, Batchelor "designed a new joint for Edison for connecting together the copper rods in three wire tubing." His method consisted of stamping a metal thread onto the end of one conductor, which would be wound and soldered onto a spiral cut in the end of another conductor. Cat. 1336:203 (item 360, 14 May 1887), Batchelor (*TAED* MBJ003203).

–3017–

*Charles Batchelor
Journal Entry*

[Llewellyn Park, N.J.,] Dec 1st 1886

255.[1] Experimenting

T.A.E. at Shop this morning. Very anxious about his experiments and wants me to arrange to do them for him, to leave E.M.W. as much as possible to itself & come with him— This I told him would be fatal at present   We must try and get them out as we are

AD, NjWOE, Batchelor, Cat. 1336:147 (*TAED* MBJ003147C).

1. Charles Batchelor consecutively numbered each entry in this journal.

–3018–

*Edison Machine
Works Minutes*

[New York,] December 3, 1886[a]

Minutes of the meeting of the Board of Trustees of The Edison Machine Works, held at the house of Mr. Thomas A. Edison, Llewelyn Park, Orange, New Jersey, on the evening of

December Third 1886. Present, Messrs. Thomas A. Edison, Charles Batchelor and Samuel Insull.

Mr. Batchelor tendered his resignation as Treasurer of the Company, which on motion of Mr. Insull was accepted.

On the motion of Mr. Batchelor, Mr. Samuel Insull was unanimously elected Treasurer of the Company, with full powers to transact its business, and perform the functions ordinarily performed by the General Manager.[1]

On motion of Mr. Insull it was decided that no Notes should be issued by the Company, unless signed by the Treasurer, and countersigned by either the President or Vice President.

Mr. Batchelor reported the sale of the property in Brooklyn originally bought for the purposes of erecting shops for the Company and the purchase of the property in Schenectady in its place.[2]

On motion it was agreed to approve Mr. Batchelor's action, and record same on the minutes of the meeting.

On motion of Mr. Insull, seconded by Mr. Batchelor, the following resolution was adopted:

AGREED: That the office of The Edison Machine Works, be moved from #104 Goerck Street, New York, to the new shops at Schenectady, on Monday, the nineteenth day of December 1886.

On motion of Mr. Batchelor the meeting adjourned.

TD, NjWOE, DF (*TAED* D8627ZAO). [a]Date from document, form altered.

1. According to Batchelor's journal notation the same day, Insull also took the office of secretary; he was working from Schenectady by early January, when he used the new letterhead of the Edison Machine Works. Batchelor was elected vice president and general manager. Cat. 1336:149 [item 256, 5 Dec. 1886], Batchelor (*TAED* MBJ003149A); Letterhead of Insull to John Randolph, 5 Jan. 1887, DF (*TAED* D8719AAA1).

2. Samuel I. Hunt, original owner of the Brooklyn property at 10th and Berry Sts., had previously agreed to take back the lot so long as the Edison Machine Works paid the assessments, amounting to about $1,000. Batchelor noted in his journal the next day that a decision to move the shops as quickly as possible had been made after consultation with the Edison Co. for Isolated Lighting, the Sprague Electric Railway and Motor Co., and the Edison United Manufacturing Co. The office furniture and books were dispatched to Schenectady the night of 18 December, when Batchelor declared that "From this day the business is done from there." Within two days, all work at Goerck St. and Bridge St. in Brooklyn had ceased "entirely" (according to Batchelor) and the new Schenectady plant was "beginning to turn out work." The Goerck St. property, which the Machine Works had leased for $9,000 a year, was sold in January 1887. "Real Estate," *Brooklyn Eagle,* 5 Dec. 1887,

2; Cat. 1336:85, 149, 153 (items 119, 257, 265, 271–72, 299; 19 June, 4, 18, 20 Dec. 1886, 28 Jan. 1887), all Batchelor (*TAED* MBJ003085B, MBJ003149B, MBJ003153A, MBJ003153F, MBJ003169A); Insull 1992, 46–49.

–3019–

To Charles Batchelor

[East Newark,] Dec. 18[a] 1886.[a]

Batch=

Have made a machine for coating the plates it does it <u>perfect</u> can get thick or thin coats on <u>perfectly even</u> a good coating is only ¼ of a thousands on each side—[1] The machine makes five revolutions a minute— one hundred spindles with place to put plates on each side & one boy will do about 5 No 20 machines $^6/_{1000}$ plates per day—[2] I will design machine & Gilliland will come over with samples— I find that the burs on the large & 4 small holes are the cause of our troubles to a great extent   am designing a machine to emery the burs off at one stroke— will also send that—

I am making 50 plates to go under our <u>press</u> to test for crosses— I find those ~~telgrap~~ telephone diaghragm plates used for tin types are rolled beautifully and you can get them $^5/_{1000}$ easy   where are they made— Here is a good point— The magnetism of ~~han~~ forged ion cores are increased 20 per cent by annealing after forging—and 30 percent on cast iron some[a] Itallian has tried it— Cant you get the Bridgeport people[3] will little Extra payment to anneal the cores well— also Taylor[4] to anneal his field pieces. This would reduce the amount of iron required 20 @ 25 per cent   you might try the Experiment on 2 cores exactly alike—as regards weight dia & amperes spires—

What about test on dynamo with Resistance in wires leading to Commutator.[5] Yrs

Edison

ALS, NjWOE, Batchelor, Unbound Documents (*TAED* MB219). [a]Obscured overwritten text.

1. This process was probably related to a patent application that Edison drafted on 26 November for a way to "increase the quantity of iron in the rotating armature of a dynamo machine or rotating converter and at the same time not increase the strength of the Focault currents in the iron." No patent resulted from this draft. The iron disks comprising the armature core were ordinarily separated by paper but Edison proposed insulating them with a layer of varnish instead. The material used would have to "be a good insulator of Electricity and a good conductor of heat," as well as resistant to cracks, shrinkage, or being cut by burrs on the stamped plates. Edison intended to dip the plates "in a hot solution of Asphalt or Asphaltene in Turpentine Benzol or other volitile solvent," preferably "ordinary Asphalt boiled to drive off some of its volitile constituents." (Edison was also trying these substances in lamps at the time; cf. Doc. 3012.) Batchelor was making related dynamo experiments on 14 December, when he wrote Edison that he planned to try thin plates coated "with your solution instead of putting in papers" and asked for "a can of the stuff." The machine was built at the Edison Machine Works in early 1887, and Batchelor noted that it "works to a charm" on 12 February. Cat. 1151, Lab. (*TAED* NM020ABQ); Batchelor to TAE, 14 Dec. 1886, DF (*TAED* D8627ZAP); Cat. 1336:177 (item 313), Batchelor (*TAED* MBJ003177A).

2. Standard dynamo models of the Edison Machine Works were designated by numbers at this time, and their armature cores were constructed of thin metal plates placed longitudinally on a shaft. The magnetically active length of the No. 20 armature core was about 30 inches. That machine was rated for 400 amperes and 125 volts at 1,000 rpm. The editors have not found further details about Edison's machine for coating plates. Hering 1886, 424; Franklin Institute 1885b, 54–55; Edison Machine Works price list, 1 May 1889, PPC (*TAED* CA021).

3. Edison may have meant the Bridgeport Forge Co., a large manufactory formed in 1883. A few days later, Batchelor went to Bridgeport and Hartford, Conn., "and talked drop forgings." Orcutt 1886, 768; Batchelor to TAE, n.d. [1887?], DF (*TAED* D8755ACN); Cat. 1336:153 (item 269, 24 Dec. 1886), Batchelor (*TAED* MBJ003153D).

4. Probably Zachary Taylor (1848?–1915), a Brooklyn foundryman who (with Batchelor and John Kreusi) incorporated Taylor & Co. in February 1886. U.S. Census Bureau 1982 (1910), roll T624_972, p. 1A (Brooklyn ward 23, Kings, N.Y); "Annual Report of the Secretary-Treasurer," *Transactions of the American Foundrymen's Association* 24 (1916): 51; Cat. 1336:9 (item 20, 17 Feb. 1886), Batchelor (*TAED* MBJ003009E); Batchelor to Taylor & Co., 21 Sept. 1887, E-6757-2; Batchelor to Taylor & Co. 10 Oct. 1887, Cat. 1247:5; both Batchelor (*TAED* MBLB5009, MBLB6005).

5. See Doc. 2999.

*Draft Patent
Application:
Artificial Materials*

Patent— D[yer]. &. S[eely].

The object of this invention is to produce Mother of pearl surfaces artificially—[1]

The invention consists in forming ~~by inlaying~~ a flat plate of mother of pearl formed of one or more peices   the whole resting on a hard flat surface

Impressions being taken from the face of the mother of pearl on plastic sheets or material by pressure ~~similiar~~ the apparatus used[a] being similar to a Lithograph press—

The colors[a] of mother of pearl being due to minute wavy lines formed by alternate[b] layers of lime & animal matter the depression between the lines being quite sufficient to causes a perfect impression in the plastic surface of the sheet or material pressed on it

I have discovered that not only are the many thousands of lines accurately transferred but that the lime like appearance is transferred as well—

If a [~~pi?~~][c] mirror like sheet of tin foil be impressed on the mother of pearl it will have the exact appearance of the same and its metallic ~~Character~~ appearance is entirely destroyed

Metallic foil, Celluloid ~~hot~~ and other plastic material may be printed as above & used for ornimenting Picture frames fans and in fact take the place of gold foil— Brittania Metal table ware & innumerable other articles can be given a mother pearl surface by this means—[2]

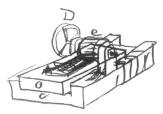

e is the steel roller—B the travelling platten   C the frame   D the wheel   A the Mother pearl in frame— The tin foil being laid in M pearl block is printed just like lithograph[3]

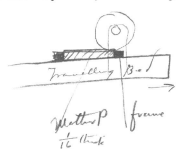

Cant you claim the art of printing from Mother pearl also a new article manufacture Mother pearl surfaces or foil—

The process—ought get Broad claims as this is new art—it

Mem— It works splendid—have made machine—

It is possible to rule fine lines in a wavy manner by diamond tool and a dividing Engine all on steel or other hard metal which will imitate M pearl— should you speak of this, in this applcn I intend to take out another patent on it—

TAE

⟨Bill to me personally⟩

ADfS, NjWOE, Lab., Cat. 1151 (*TAED* NM020ABT). ªObscured overwritten text. ᵇInterlined above. ᶜCanceled.

1. Edison filed a patent application based on this draft in February 1887 (Case 717), but it was rejected and amended multiple times, and eventually abandoned (PS [*TAED* PT032AAB]). He had experimented intermittently on artificial materials for years, beginning with celluloid in 1875 (see Docs. 579, 583, and 586). More recently, he had worked with artificial materials for lamp filaments and contemplated making artificial silk (see Docs. 2634 and 2812). An account entry labeled "Pearl Experiment," possibly related to work on the process for making mother of pearl, shows $900 in expenses between March and October 1887 (Ledger #5:408, Acct. [*TAED* AB003, image 212]).

2. Figure labels are "D," "e," "A," "B," and "C."

3. Figure labels (top to bottom) are "Travelling Bed," "Mother P 1/16 thick," and "frame."

–3021–

Charles Batchelor
Journal Entry

[New York, December 30, 1886[1]]

275.[2] Edison very sick at Orange. N.J.[3]

AD, NjWOE, Batchelor, Cat. 1336:155 (*TAED* MBJ003155).

1. Batchelor wrote this entry at the bottom of the page, following one dated 30 December. The next entry was 31 December.

2. Charles Batchelor consecutively numbered each entry in this journal.

3. See Doc. 2799.

January–May 1887

An illness that afflicted Edison in late December continued into the start of the new year, when he seems to have developed a severe lung infection, most often labeled pleurisy but also described as pneumonia. Confined to his bed for most of January, it was only near the end of the month that he was able to sit up for much of the day and come to the table for meals.[1]

While Edison was sick, he was not idle. He managed some oversight of preparations for an upcoming visit to his winter home and laboratory in Fort Myers, Florida. He arranged, for instance, a shipment of municipal system lamps and a dynamo for lighting the buildings and grounds.[2] He arranged for provisions and equipment, including groceries, distilled water, fishing gear, photographic instruments, and even a gas machine.[3] He also made some experimental notes on treated lamp filaments.[4]

Edison's condition had improved enough by early February for him to travel, but then his daughter Marion fell ill with "congestion of the lungs," further postponing the trip.[5] He did finally leave on 9 February, under a doctor's care in a special railroad car, with Mina, Marion, and his daughter's tutor, Sarah McWilliams. He had received at least one hypodermic injection of morphine, resulting in his arm and hand being "badly swollen" when his train reached Savannah, about 11 February.[6] When they arrived in Fort Myers, the Edisons were greeted by Ezra and Lillian Gilliland, Mina's sister Mary, her brother Robert, and his new bride, the former Miss Louise Igoe.[7] According to a newspaper interview published on his departure from New York, he expected a half dozen experimental assistants to join him as well.[8]

Edison was optimistic about his recovery and clearly was no longer as sick as he had been, but he continued to experience health difficulties. He endured chills and fever in mid-March and then developed a painful abscess under his ear, which (according to Charles Batchelor) was "operated on" toward the end of the month. Another abscess developed in early April, prompting fears of a grave systemic infection, but Edison largely recovered before returning home at the end of April.[9]

Edison and Mina passed their first wedding anniversary in Florida. She was several months pregnant and had apparently confided to her family some doubts about the marriage and the prospect of having a child, particularly a fear that Marion would resent the baby.[10] When her father arrived in Fort Myers in late March, Mina must have discussed her anxieties with him. Lewis Miller remained in Fort Myers after Mina, Marion, and Miss McWilliams had left for Akron, and he used his extended visit (including a hunting trip) to become better acquainted with his son-in-law. He then wrote to Mina, reassuring her with his own observations that she did have her husband's "full affections."[11]

Despite his hopes for leisure activities such as boating, hunting, and fishing, Edison also intended to "work harder, if anything," in Florida than he did at home.[12] He planned to tackle "at least six or seven different ideas"; among them was ship-to-ship telegraphy, by which he hoped "to extend the distance at which telegraphing by sound through water can be successfully accomplished." He thought the quiet Florida waters ideal for such experiments, but it is not clear that he carried them out.[13] By 1 March, he had thought and experimented enough about his pyromagnetic motor to draft a patent application (Doc. 3029), and he remained preoccupied with the motor (and related generator) until the end of the month. Charles Batchelor arrived in Fort Myers on 8 March to help with this work, staying until the twenty-first.[14] Edison took up experiments in ore separation by early April.[15] He also oversaw the installation of lights in his house and laboratory as well as along the road leading to his estate. He had hoped to illuminate the town with his municipal lighting system, but the dynamo he had been so eager to have shipped arrived only in mid-April, too late for him to make the installation. The local newspaper announced hopefully that the plant would be set up the next year, but Edison did not return in 1888.[16] Edison also mentioned to the *New York World* that he intended to make a brief trip to Havana before returning to New York, but the editors have found no evidence of such an excursion.[17]

As usual when Edison was away, his subordinates (principally Samuel Insull and Alfred Tate) looked after his business affairs and kept him apprised by letter and telegram. One complication was a major flood in Schenectady, New York, about 12 April, that closed down the Edison Machine Works.[18] Although Insull had moved to Schenectady and essentially taken over the management of the Machine Works, he continued to manage Edison's finances behind the scenes. Those finances became increasingly strained by the planning and construction of the new laboratory in Orange and by the continuing reorganization of Edison's business enterprises. Charles Batchelor was entrusted with overseeing the plans for the new lab. Edison was already thinking of experiments he wanted to carry out there, and he reminded Batchelor from Fort Myers not to "forget to prepare laboratory plan."[19]

On their trip back North, Edison and his traveling party stopped at the mineral springs near Bartow, reputed to have healing qualities. Newspapers, which took a keen interest in Edison's personal life, reported that the group (excepting Edison and Mrs. James Gilliland) was pitched into deep water when a pier gave way and had to be rescued, but Lewis Miller's biographer noted that the springs were shallow enough for the surprised travelers to wade out. Miller, who was present, was so amused by the exaggerated reports that he had an illustration made of the incident showing Edison rowing to the rescue in a boat, with the caption "Women and Children First."[20] Shortly after Edison's safe return from Florida, Batchelor hired noted architect Henry Hudson Holly and had preliminary drawings of the laboratory made.[21]

Edison was back in Llewellyn Park by 30 April and returned the next day to his laboratory at the lamp factory in Harrison, New Jersey. He did so with Mina at his side as a copyist, though the editors have not learned what, if any, other roles she may have had there.[22] Among the projects he took up was the design (in which Batchelor played a large part) of a new phonograph, likely prompted by Batchelor having been strongly impressed by the recent exhibition of the rival graphophone machine.[23] During the late spring and early summer, Edison also carried out experiments on incandescent lamps and ore milling, worked on the pyromagnetic generator and motor (for which he drafted new patent applications), and designed a process for making wrought iron directly from molten metal.[24]

1. See Docs. 3023, 3025, and 3027; Cat. 1336:165 (item 290, 18 Jan. 1887), Batchelor (*TAED* MBJ003165C); "Edison Going to Florida," *Fort Myers Press*, 29 Jan. 1887, News Clippings (*TAED* X104S011A).

2. TAE to Bergmann & Co., 11 Jan. 1887; TAE to Edison Machine Works, 14 Jan. and 2 Feb. 1887, Lbk. 23:179, 196, 302 (*TAED* LB023179, LB023196, LB023196).

3. Fort Myers Invoice Book, Cat. 1165 (box 2), Accts., NjWOE.

4. N-87-01-27, Lab. (*TAED* NA005AAA).

5. Mary Valinda Miller to Mina Edison, 2 Feb. 1887, FR (*TAED* FI001ABB); Alfred Tate to Samuel Insull, 2 Feb. 1887, Lbk. 23:302 (*TAED* LB023302B).

6. Cat. 1336:173 (item 308, 8 Feb. 1887), Batchelor (*TAED* MBJ003173B); Tate to Insull, 8 Feb. (misdated 10 Feb.), 10 and 11 Feb. 1887; Lbk. 23:394, 382, 405 (*TAED* LB023394, LB023382, LB023405); "Wizard Edison in Florida," *Fort Myers Press*, 14 Apr. 1887, News Clippings (*TAED* X104S012A); Martha Lake and Mary Valinda Miller to Mina Edison, 21 Jan. 1887, FR (*TAED* FI001AAZ1).

Edison's friend Ezra Gilliland reportedly told the *New York World* in late March that Edison's bronchial condition had precipitated heart troubles, so that "it was found necessary several times to administer hypodermic injections of morphine." The Florida climate proved such a tonic that as soon as Edison reached Fort Myers (again according to Gilliland), he "at once rallied, and he is now as hearty and as well as ever." Edison himself affirmed that he had "never felt better in [his] life," but he also showed the reporter "an ugly little sore" on his arm from his last morphine injection in New Jersey. "Wizard Edison in Florida," *Fort Myers Press*, 14 Apr. 1887, News Clippings (*TAED* X104S012A).

7. Smoot 2004, 41; Mary Valinda Miller to Mina Edison, 26 Jan. 1887, FR (*TAED* FI001ABA).

8. Untitled Edison interview, *New York Journal,* 9 Feb. 1887, Clippings (*TAED* SC87001).

9. See Doc. 3033; Mary Valinda Miller to Mina Edison, 20 Mar. 1887, FR (*TAED* FI001ABE).

10. See Doc. 3030.

11. See Doc. 3036; cf. Doc. 3035.

12. "Wizard Edison in Florida," *Fort Myers Press*, 14 Apr. 1887, News Clippings (*TAED* X104S012A).

13. Untitled Edison interview, *New York Journal,* 9 Feb. 1887, Unbound Clippings 1887 (*TAED* SC87001).

14. Doc. 3029 (headnote).

15. See Doc. 3032.

16. Untitled, *Fort Myers Press*, 24 Mar. and 21 Apr. 1887, News Clippings (*TAED* X104S011E, X104S013A); Smoot 2004, 41.

17. Untitled Edison interview, *New York Journal,* 9 Feb. 1887, Unbound Clippings 1887 (*TAED* SC87001).

18. Doc. 3034 n. 1; "Damage by Floods," *NYT,* 13 Apr. 1887, 3.

19. Doc. 3032.

20. Cat. 1336:193 (item 347, 28 Apr. 1887), Batchelor (*TAED* MBJ003193B); "Thomas A. Edison and a Party of Friends Narrowly

Escape Drowning," *Washington Post*, 30 Apr. 1887, 1; Hendrick 1925, 117–18.

21. See Doc. 3040.

22. See Docs. 3037, 3039, and 3041.

23. See Doc. 3042.

24. The generator and motor applications are Docs. 3047 and 3048; the metalworking experiments culminated in Doc. 3064.

–3022–

Francis Upton to Edward Johnson

At Sea Jan 5, 1887.[1]

Dear Sir:

Regarding the Zipernowsky tests

I arranged on my visit to Rome and Milan that Lieb[2] should make the tests you desire under the contract.[3]

He can do this the latter part of this month if ~~we~~ you notify him

I found at Rome a station running about 600 Edison Lamps and 40 arc lamps.

The building is[a] in the yard of the ~~ga~~ Gas Company and is erected for a capacity of 12,000 lamps. Prof. Columbo[4] told me that the boilers for 6000 lights were on the way and that ~~that~~ the machinery had been ordered.[5]

The lighting is distant about one mile from the station and the light appeared to be successful

I did not make any tests, but was much impressed by the evident careful working out of details shown in the whole installation.

There was a distinct impression made that such a system would be a very dangerous competition to the Edison system, as it could do lighting on a very large scale with a moderate investment in conductors.

The plant in Rome uses the coke from the gas works under the boilers as fuel.

The using of such high tension currents makes the inside wiring of the station very simple

The engines were working badly when I was at the station so that it was impossible to get any tests.

Lieb can make the tests needed better than I could. After what I saw I do not think that we need tests to ~~hold~~ show that the Edison Co should hold to the Zipernowsky system. The station showed [at?][b] such thorough work leading to results wanted that the Edison Company will keep a dangerous enemy out of the field. Yours

Francis R. Upton

ALS, NjWOE, Upton (*TAED* MU101). ªObscured overwritten text. ᵇCanceled.

1. Upton reached New York on or before 10 January. Samuel Insull to Upton, 10 Jan. 1887, Lbk. 23:175A (*TAED* LB023175A).

2. John William Lieb (1860–1929) was chief engineer of the Italian Edison Co. in Milan. A graduate of the Stevens Institute and formerly the head electrician at the Pearl St. central station in New York, Lieb was dispatched to oversee construction of the Milan power plant in late 1882. Doc. 2369 n. 7.

3. The 26 November patent license agreement between the Edison Electric Light Co. and Ganz & Co. of Budapest incorporated Károly Zipernowsky's assurance to Upton that the Ganz alternating current equipment would meet certain efficiency standards. The tests were to be conducted by engineers of the Edison company at a plant in Europe designated by Ganz. Zipernowsky to Upton, 16 Nov. 1886; Edison Electric Light Co. agreement with Ganz & Co., 25 Nov. 1886, p. 7; both *Zipernowsky v. Edison*, Lit. (*TAED* W100DMBB, W100DMBA).

4. Guiseppe Colombo (1836–1921), an entrepreneur, engineering educator, and statesman, was the founder and driving force behind the Edison lighting company in Milan. Docs. 2332 nn. 2–3, 2343 n. 14.

5. Although the Rome plant was not yet complete, New York's *Electrical World* informed readers about this time that the station showed that the ZBD transformer system "has certainly passed beyond the experimental stage, and that practical results can be obtained, and probably with great economy." *Electrical World* 9 (15 Jan. 1887): 25.

–3023–

*Charles Batchelor
Journal Entry*

N.Y. Jan. 6th 1886[7].

277[1]   Edison—

Went up to see Edison tonight. He is very sick with pleurisy but is progressing favorably—[2]

AD, NjWOE, Batchelor, Cat. 1336:163 (*TAED* MBJ003163A).

1. Charles Batchelor consecutively numbered each entry in this journal.

2. See Doc. 3021. The term "pleurisy" referred to an inflammation of the pleura (the membranes lining the chest cavity) attributed to any number of causes, including injury and overwork, most commonly in patients of middle age. Medical science recognized pleurisy to take several forms of differing severity and somewhat different symptoms. Edison may have suffered from an acute form, a serious illness typically manifested by fever, coughing, shallow breathing, and pain; according to one medical authority, these symptoms were often most pronounced in otherwise robust patients. Recommended treatments included leeching of the chest cavity, application of hot poultices, use of plaster straps to limit movement, and opiates to diminish pain (Quain 1883, s.v. "Pleura, diseases of"; Bramwell 1889, 1–11).

To Bergmann & Co.

[New York,] Jan 11/[8]7

Dear Sirs:

Will you please prepare for immediate shipment the Municipal lamps which I require for Florida.

I do not know how many of these should be sent, but your books will show the number forwarded last year, and you can prepare as above a like number of the <u>latest improved</u> Municipal lamps and advise me when they are ready.[1]

There will also be quite a number of fixtures for houses required, list of which will be sent you tomorrow or next day. Yours truly

Thos. A. Edison   T[ate]

L (letterpress copy), NjWOE, Lbk. 3:179 (*TAED* LB023179). Written by Alfred Tate.

1. Bergmann & Co. sent "25 Complete latest style Municipal Lamps" on 7 February; these were the fixtures, complete with brackets and cut-out sockets. The Edison Lamp Co. had already sent seventy-five municipal lamps on the first of the month, twenty-five each of ten, sixteen, and thirty-two candlepower. The Edison Machine Works was preparing a municipal dynamo to go to Florida in mid-February. The editors have no evidence that Edison was experimenting with municipal lights on this Florida trip (cf. Doc. 2916); someone (perhaps Edison) wrote "Lighting the town" on Bergmann's invoice, but the equipment may instead have been for use in the laboratory, which Edison had illuminated, as the local paper reported on 10 March. The paper indicated hopefully that Edison intended to light the village of Fort Myers in the future. Around the same time, the *Electrical World* of New York stated, at least partly in jest, that not only were Edison's house and grounds lighted, but that residents had given money to put up poles and wires and now the small town "is nightly illuminated as brightly as any place in New York." Bergmann & Co. invoice, 7 Feb. 1887; Edison Lamp Co. invoice, 1 Feb. 1887; both Cat. 1165 (Fort Myers accounts; box 2, folders 49 and 47), Accts., NjWOE; Samuel Insull to TAE, 12 Feb. 1887, DF (*TAED* D8736AAF); the narratives of lighting in Fort Myers given by Smoot 2004 (40–43) and Albion 2008 (28–30) are based in part on *Fort Myers Press* articles, many of which are in News Clippings, FFmEFW; "Personals," *Electrical World* 9 (12 Mar. 1887): 137.

*Charles Batchelor
Journal Entry*

[New York, January 18, 1887?][1]

290[2]   Edison improving slightly all the time but still a very sick man   the doctors hope to send him to Florida first week in February[3]

AD, NjWOE, Batchelor, Cat. 1336:165 (*TAED* MBJ003165C).

1. Batchelor wrote this entry immediately after one with this date. It is followed by several undated entries and then one dated 23 January.

2. Charles Batchelor consecutively numbered each entry in this journal.

3. In early February, Alfred Tate noted that Edison had to delay his trip until the fourteenth because "Marion has congestion of the lungs." Tate to Samuel Insull, 2 Feb. 1887, Lbk. 23:302B (*TAED* LB023302B).

–3026–

To Samuel Flood Page

[New York,] Jan 24/[8]7

Dr Sir:

I beg to confirm telegram sent you as follows on 13th inst:–[1]

"Ediswan London   Offer contained my letter November ninth[2] last as to adjustment patent accounts and future patents is hereby canceled   Edison"

Such a length of time has elapsed since I made the proposition to you that in justice to myself I felt compelled to cancel my proposition. I was under the impression that your company had decided not to accept same, owing to your silence, as otherwise I should have sent the telegram at an earlier date.[3] Yours very truly

Thomas A. Edi[son][a]

L (letterpress copy), NjWOE, Lbk. 23:229 (*TAED* LB023229). Written by Alfred Tate. [a]Not copied.

1. This is the full text of Edison's cable, which was addressed impersonally to "Ediswan," a common shorthand for the Edison & Swan United Electric Light Co., Ltd. Lbk. 23:190 (*TAED* LB023190).

2. Doc. 3007.

3. The London firm had not been inactive during this interval. In late December, Flood Page both wrote and cabled Edison that the Italian Edison company in Milan had agreed to acquire several of the central station dynamos held for the British company by the Edison Machine Works, whose fate Edison had linked to any proposed settlement (see Doc. 2953). According to Flood Page, the Italians required only assurance that these machines would be compatible with those they already had, after which he could apply proceeds from their sale to closing out the British company's account with Edison. Edison, however, evidently made no reply either to him or the Italians. Flood Page to TAE, 23 Dec. 1886, DF (*TAED* D8630ZCR).

In response to Edison's withdrawal of the November offer, Flood Page reminded him of the unanswered December missives from London, gently pointing out that "the delay has arisen on your side and not on ours." The British company persisted with the proposed Italian deal despite additional delays occasioned by Edison's illness and convalescence in Florida, while Samuel Insull and Alfred Tate managed this correspondence from Schenectady, N.Y., and New York City, respectively (Flood Page to TAE, 2 and 7 Feb. 1887, 2 Mar. 1887; all

DF [*TAED* D8742AAB, D8742AAC, D8742AAD]). Edison finally provided the assurance sought by the Italians in March, and the two machines were transferred to Milan within about two months. In May, Flood Page sent a detailed proposal by which he would sell the two remaining dynamos to Edison, leaving only about $1,000 of Edison's original claim (some $11,200) to be paid in cash. Insull, however, acting on his own authority, reminded the company that the terms offered in November had been withdrawn and, with Edison out of town, "the matter must remain in abeyance until such time as I can consult with him." There the matter rested, and the editors have not determined its ultimate outcome (Insull to Edison & Swan United, 3 Mar. and 26 May 1887; Tate to TAE, 10 Mar. 1887; Lbk. 24:76, 470, 155 [*TAED* LB024076, LB024470, LB024155]; Flood Page to TAE, 10 May 1887, DF [*TAED* D8742AAI]).

-3027-

Charles Batchelor
Journal Entry

[New York, January 28, 1887?[1]]

300[2] T.A.E.

Visited T.A.E. who is now getting better & sits up all day—[3] Talked on probable increase of Stock of Parent Co. on some such basis as I think it likely they would make namely double the stock, pay a 50% stock dividend to present holders, sell 25% of total at about $150. this would give the whole stock about 13% cash dividend over and above $625,000 in the Treasury— I am not sure that is the project but think so— The Western E. Light has also got to be let in at same time also[4]   this will modify it some

AD, NjWOE, Batchelor, Cat 1336:169 (*TAED* MBJ003169A).

1. Charles Batchelor wrote this entry immediately below one dated 28 January. His next entry was dated 29 January.

2. Batchelor consecutively numbered each entry in this journal.

3. Edison seems to have been confined to bed during much of his illness. On 30 January, his mother-in-law acknowledged that he was again able to dress and worried whether his sickness had "affected his lookes in any way." A few days later, she responded happily to Mina Edison's report that he could come to the table for meals. Mary Valinda Miller to Mina Edison, 30 Jan. and 2 Feb. 1887, both FR (*TAED* FI001ABA1, FI001ABB).

4. A published report stated that the Edison Electric Light Co. was considering raising its "capital stock from 12,942 shares to 16,000, par $100; also consolidation with the Western Edison Company. It has declared a convertible scrip dividend of 7 per cent." Batchelor noted on 3 February that the Light Co. board had voted "to increase the capital sufficient to take in the W. E. Light Co.," but he did not give the final terms. He also secured a promise from Charles Coster "that whatever was done in the Light Co there should be nothing done that should impair the interest of T.A.E." The absorption of the Western Edison Co.

as the new Western Department of the Light Co. took effect on 21 February. The former Western company, which controlled Edison patents in Illinois and surrounding states, was also in the process of securing authority to construct and use underground conductors in Chicago. Cat. 1336:171 (item 304, 3 Feb. 1887), Batchelor (*TAED* MBJ003171D); "New York Notes," *Electrical World* 9 (5 Mar. 1887): 122; "Western Notes," ibid., 19 Feb. 1887, 96; "Western Notes," ibid., 2 Apr. 1887, 173; "Western Notes," ibid., 16 Apr. 1887, 196; Edison Electric Light Co. annual report, 25 Oct. 1887, p. 3, PPC (*TAED* CA019C).

–3028–

*Charles Batchelor
Journal Entry*

[New York,] Feb. 8th 1886[7][1]

308[2]   T.A.E.

Saw him at his house and found him much improved & ready to go away tomorrow night to Florida[3]

Took to him a proposition from the Light Co about raising $100,000 for E.M.W. which was Light Co to sell $25 000 worth of its stock to a party for $225 per share on condition that he loan ~~the EMW~~ $50,000 at 6%. Then Light Co to loan E.M.W. $100,000 for 12 months at 6% & put up $200,000 Machine Works stock on option for Light Co to take it at par any time in the time of loan— No dividend to be declared during loan.

I told Edison we could use the 200 (about) shares of Light Co stock we now have for same purpose if he would sooner instead of paying it as a dividend   He preferred to sell his $80,000 of Edison Illg Stock & loan the money to the EMW. which he authorized us finally to do[4]

AD, NjWOE, Batchelor, Cat. 1336:173 (*TAED* MBJ003173B).

1. Batchelor mistakenly wrote "1886."

2. Charles Batchelor consecutively numbered each entry in this journal.

3. Edison had planned to leave earlier in the month but put it off when Marion fell ill with what Alfred Tate described as "congestion of the lungs." Edison was himself still under the care of a doctor on the day of his departure but did get away in a private rail car late on 9 February. Edison's party reached Richmond, Va., on the morning of 10 February. The next day, Tate received a telegram from Savannah, Ga., announcing Edison's arrival there. Edison "feels well," Tate reported, "but his arm and hand are badly swollen, from the effects of an injection of morphia administered a few days before he left." Tate to Insull, 2, 8, 10–11 Feb. 1887, Lbk. 23:302, 382, 394, 405 (*TAED* LB023302B, LB023382, LB023394, LB023405).

4. Batchelor's proposition was to raise working capital for the relocated Machine Works. On 31 January, he and Edward Johnson had visited Edison, who agreed that the Machine Works could raise $100,000 on

its own stock. The money was to be loaned by the Edison Electric Light Co., which would "have the option to take the stock after a year." Cat. 1336:171 (item 303, 31 Jan. 1887), Batchelor (*TAED* MBJ003171B).

At an unspecified time probably not long before this date, Edison had personally loaned $45,000 to the Machine Works that he, in turn, had borrowed from Drexel, Morgan & Co. In March, Samuel Insull raised $18,600 for the Works by selling, with the aid of Charles Coster, 200 stock shares of the Edison Illuminating Co. of New York, nearly enough to repay a short-term debt to Sigmund Bergmann, whom Insull regarded as "the meanest man I know of" in such matters. Insull vowed to be guided by the Drexel, Morgan partner and not "pursue any course which might arouse the ire of Mr. Coster." Insull to Tate, 9 Feb. and 5 Mar. 1887, both DF (*TAED* D8736AAC, D8719AAM).

PYROMAGNETIC GENERATOR AND MOTOR
Docs. 3029, 3031, 3047, and 3048

Doc. 3029 is the start of Edison's practical development of an idea for converting heat energy directly to either electrical energy (in a generator) or mechanical energy (in a motor) without the use of rotary dynamos, steam engines, or other intermediate machinery. The generator was a suggestion he had made in one of the Faraday-infused notebooks he kept at Fort Myers a year earlier, and it could have originated only from a clear vision in his mind's eye of the lines of force emanating from magnetic bodies and curving through the surrounding space.[1] Although he had been experimenting since at least 1882 (and continued to do so) on what he termed the "direct conversion" of coal into electricity,[2] that process took place in a form of chemical battery—in essence an early fuel cell.[3] Now he intended to exploit a fundamental property of magnetic metals to alter a magnetic field around an electrical conductor. The effect would be like that when an armature wire in a generator or motor experiences changing fields, but it would come without the need for rotation in the former case or an energizing current in the latter one. In a paper written for the American Association for the Advancement of Science meeting in August 1887, he described how the method might be applied to a generator:

> It has long been known that the magnetism of the magnetic metals, and especially of iron, cobalt and nickel, is markedly affected by heat. According to Becquerel, nickel loses its power of being magnetised at 400°, iron at a cherry-red heat, and cobalt at a white heat. Since,

whenever a magnetic field varies in strength in the vicinity of a conductor a current is generated in that conductor, it occurred to me that by placing an iron core in a magnetic circuit and by varying the magnetizability of that core by varying its temperature, it would be possible to generate a current in a coil of wire surrounding this core. This idea constitutes the essential feature of the new generator, which therefore I have called a pyromagnetic generator of electricity. [Edison 1888b, 94–95]

Edison first applied this principle, as he pointed out in his AAAS paper, not to a generator but in "the construction of a simple form of heat engine which I have called a pyromagnetic motor."[4] That device was the "Magnocaloric Motor" embodied in his 1 March draft patent application (Doc. 3029); he termed it a "thermo–magnetic motor" in the application filed at the end of May. Edison made additional drawings for it in the early days of March, noting on the second that "At last John Ott starts to make Magnocaloric Motor."[5] Sustained experiments began after Charles Batchelor arrived in Fort Myers on 8 March and continued until he left two weeks later (see Doc. 3031). However, apart from a few sketches Edison made on 25 March,[6] there is no further evidence of work until mid–May, when Batchelor described Edison's new design for the "Heat engine."[7] Over the subsequent few days, Edison embodied this new design in draft patent applications for a pyromagnetic generator and another for a pyromagnetic motor (Docs. 3047 and 3048).

1. Doc. 2939; see also Docs. 2940 and 2941.

2. Edison had a longstanding concern with increasing the efficiency of transforming energy from one form to another. He evidently had used the term "direct converter" in reference to the large direct-connected dynamo that he developed in 1880–1881. Doing away with belts or other energy-wasting transmission devices, Edison had its armature connected through a solid coupling to the crankshaft of the steam engine, and the whole acted as an efficient "converter" of mechanical into electrical energy. His ongoing attempts to raise the output of incandescent lamps, calculated as candlepower of radiant energy per horsepower of mechanical energy, was part of this larger effort to raise the efficiency of energy conversion processes. Ledgers #4:260–62, 340–41 and #5:60, 502, Accts. (*TAED* AB002, images 120–21, 134; AB003, images 44, 248); see *TAEB* 5 chap. 6 introduction and Docs. 2015 n. 3, 2057, 2122 (headnote), 2994, and 3050.

3. See Doc. 2520 (headnote) for Edison's earlier work on fuel cell designs.

4. Edison was not alone in applying this thermal property of iron in the design of a motor. Nikola Tesla applied for a patent in March

1886 on what he called a "thermo-magnetic motor." The device used a Bunsen burner to reduce the magnetism of the motor's magnet; a mechanical device drew the magnet away from the heat to restore its magnetism (U.S. Pat. 396,121; Carlson 2013, 76–77). Charles Batchelor also noted in his journal a September 1887 *Electrical Review* article about a Philadelphia inventor's claim to have made "practically same thing as Edison for a motor" and submitted results of his tests to the Franklin Institute in 1884 or 1885. The inventor, one William Cooper, filed a patent application in 1884 and also reportedly claimed to have made a generator on the same principle but was deterred in both cases by oxidation of the iron (Cat. 1337:12–13 [item 478, 26 Sept. 1887], Batchelor [*TAED* MBJ004012C]; "The Pyromagnetic Generator," *Electrical Review* [New York] 11 [24 Sept. 1887]: 1). In 1888, the Franklin Institute made a claim on behalf of Edwin Houston and Elihu Thomson for the principle embodied in their thermo-magnetic motor almost ten years earlier ("Abstracts from the Secretary's Report," *Journal of the Franklin Institute* 125 [3rd ser., Feb. 1888]: 150–51).

5. Ott may have been with Edison in Florida. N-87-03-02:5–27, Lab. (*TAED* NA008005, NA008015).

6. N-87-03-02:43–51, Lab. (*TAED* NA008043).

7. Cat. 1336:205–7 (item 364, 16? May 1887), Batchelor (*TAED* MBJ003205).

*Draft Patent
Application:
Pyromagnetic Motor*[1]

Magnocaloric Motor[2]

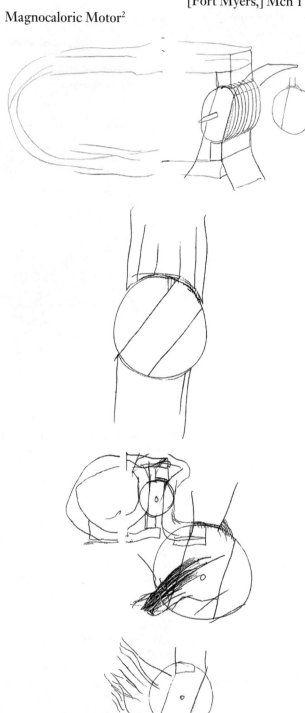

The object of this invention is to produce a simple & economical prime motor

The invention consists in producing a dysymetrical distribution of the magnetic[a] lines of ~~m~~force in a magnetic circuit, by means of heat acting[b] ~~The disturbance~~ on the magnetic substance The disturbance taking place in the moveable part of the magnetic circuit, so that continuous rotation takes place as long as heat is applied to the moveable magnetic part of the magnetic circuit. ~~I have discovered that when Iron is powerfully magnetized that say near its saturation point, that a very small quantity of heat produces powerful changes in its magnetism~~ it is well known that the magnetism of[c] Iron & other paramagnetic bodies is diminished by heat

I have discovered that when the iron is powerfully magnetized say nearly to saturation that its magnetism[a] is extremely Sensitive to heat, a slight increase of temperature producing a powerful diminution of magnetism so that a great portion of the Energy of the heat is transformed into actual motion for various uses

~~The~~ in fig 1 is a simple a[d] apparatus by which ~~I utilize this phenomenon I~~ illustrates the method by which this phenomenon ~~of~~ is utilized

In ~~The~~a prime motor the use[e] of one or mot~~orre~~ magnetic circuits of which one or more parts are moveable & the application of heat to one ~~of~~or more parts of such circuit to produce motion[f]

In a magnetic circuit of which one part is moveable—The disturbance of the magnetic ~~distribution~~ lines of force by heat to produce motion

TAE

ADf, NjWOE, Lab., N-87-03-01:3 (*TAED* NA007003). Document multiply signed and dated. [a]Interlined above. [b]"by means of heat acting" interlined above. [c]"the magnetism of" interlined above. [d]"in fig 1 is a simple a" interlined above. [e]Obscured overwritten text. [f]Followed by dividing mark.

1. See headnote above.
2. For the sake of coherence, the editors have transcribed here the

heading and drawings that Edison made across two consecutive sets of facing notebook pages between "so that" and "continuous rotation" in his text below.

–3030–

Jane Miller to Mina Edison

"Hotel D'Angleterre"[1]   Athens, Greece—Mar. 8 1887

Dear Mina,

Mother wrote me a short time ago about your expectation. I am so sorry you have been obliged to go through so much this winter. I imagine your own trouble was quite enough to go through with patiently without having severe illness to worry you. I hope though my dear you will have an easier time from now on. If you are happy[a] in the thought of having a dear little one I am but if you regret it I am sorry. I think when it comes you will be happy with it. I suppose you have told Marion all about it. Seems to me I would if I were you if you have not already done so. She cannot help but love a little one of yours and I think if you sort of confide in her about it she will feel better. No doubt you have done so all ready. I will try to do a little shopping in Paris for it. Is there anything in particular that you want for it. I thought for my present to it I would try and get a crib like Mrs. Giffords[2] would you prefer all white or blue. Will it not seem strange to have a little one again in our family but that is what we must expect now so many are married. I hope you are[a] happy and cheerful over it. You must go about and be happy for it will tell on your child if you don't— There is no need to regret anything for it is the common thing. Mother said you were afraid Marion would not love it but I am sure she will. I hope I will get home before the eventful time.[3] We expect to sail from Liverpool the 18 of June and I presume will land on the following Saturday. I thought I would spend a day or two with you before going home. I thought I would get for you a pretty summer dress—a house dress or matinee-jacket[4]—and a dress you could wear anytime of the year for a handsome dinner dress. Now if you would like anything different tell me. Also ask Mr. Edison if he[a] would like me to get a seal skin coat for you. From the way he spoke last fall I thought he wanted you to have one. Now don't fail to ask him for I know he would be willing to get you anything you want. I think you can do better here in the way of styles. It may seem a long time a head but it takes so long to get answers to letters. I wish you would write to me. I have had but two letters from you since I left New York it hardly seems right.[5] Why haven't you written? Don't you want to confide in me or tell me how you are getting along &c. I am sure I want you to feel that you can write me anything. Often times when one feels a little blue or a little worried if they can tell one person they soon feel better. Now write me and tell me a little of your plans and expectations. I want to do anything I can for you so don't hesitate to ask me. I wish you would send your measures but I am afraid you will be some what changed— I think I will have Madame Joyeuse[6] just cut the waist and arrange the trimmings as they ought to go and then you can have it finished at home. I think that will be the surest way. Miss Richmond[7] might send me your last measurements though as it would be sort of a guide. We left Naples last Friday evening went to Brindisi by train and then by boat to Corinth and from there by train to Athens. Yesterday we spent visited the Acropolis— ruined Temples— Mars hill[8] &c. Everything in Athens is of white stone and seems so new. They have severe winds here and it is dreadfully dusty— Today we take or make an excursion I don't know just where. We stay here until Saturday noon when we take the boat for Alexandria. I was greatly disappointed that Father failed to come. I felt like giving up the trip too— And when I heard of your condition Mr. Edison's illness and Father not well I felt I ought return[a] home at once and telegraphed saying so. I had about made up my mind to go but I thought perhaps it would be better to find out just how things were at home. Father telegraphed back I should complete my trip.

I am sorry you could not attend Robert's wedding.[9] I suppose[a] though you had a good time with them at your home in the South. How is Mame.[10] I suppose she is Happy with you. Mrs. Studebaker said she looked lovely at the wedding. I am much obliged to you for getting the glass for me   Mrs. S. said that the dishes and glass were both beautiful. You know I suppose Anna's Mother[11] was obliged to come over for her and they return home in April. Anna is not well enough to travel. It was a great disappointment all around. I miss her very much. We left them in Rome.

I will be glad to get home again and see you all. I wonder if you will be glad to see me? Four months or nearly so.

Now write me soon. Tell Mama to write. Love to all—

Jennie

ALS, NjWOE, FR (*TAED* FM001AAW). [a]Obscured overwritten text.

1. The Hotel D'Angleterre accommodated foreign travelers in relative luxury; one guidebook rated it the city's best. It was located near the Palace, a short distance from central Athens. Murray 1872, 128, 132–33.

2. Not identified.

3. Based on scant references in the available correspondence, the editors have not determined when Mina's pregnancy began or ended. Her mother was anticipating a child as late as 15 July, but the pregnancy did not result in a live birth. Jane Miller to Mina Edison, 19 May 1887; Mary Valinda Miller to Mina Edison, 15 July 1887; both FR (*TAED* FM001AAY, FI001ABL3).

4. From about the early 1880s, a matinee jacket was a type of woman's coat considered suitable to wear to the theater. *OED*, s.v. "matinee jacket."

5. Jane had been in Europe since September 1886. Jane Miller to Mina Edison, 4 Sept. 1886, FR (*TAED* FM001AAR1).

6. Mme. C. Joyeuse was a Paris dress shop at 44 Rue du Colisee. Vouchers (Household), Joyeuse bill head, 10 Sept. 1889 (box 2).

7. Not identified.

8. Mars Hill (the Areopagus) is the site of a temple near the Acropolis that was widely (though not universally) believed to be where the Apostle Paul addressed the Athenians. *WGD*, s.v. "Areopagus"; cf. Murray 1872, 200–202; Hackett and Hovey 1882, 203–204; Hurst 1887, 227.

9. Robert Anderson Miller, Mina's older brother, married Louise Igoe on 25 January. Jeffrey 2008, 173; Marriage record for Robert A. Miller, *Indiana Marriage Collection 1800–1941*, accessed through Ancestry.com, 10 Dec. 2010.

10. Mary Emily Miller (1867?–1946), sister of Mina and Jane, was referred to as "Mame" or "Mamie" by family members. Jeffrey 2008, 173.

11. The editors have not identified either Anna or her mother.

–3031–

Charles Batchelor Journal Entry[1]

Myers 19th Mch 1887

323[2]   Mageneto Motors

Since the 8th of present month I have worked making experiment on a new principle of motor for Edison—[3] The principle on which he works is as follows:—'The production of a distortion of the lines of force in the iron of the armature by heat.' and the rotation due to this distortion.[4]

When the armature was made with a number of discs mounted to turn on a pivot and the sections heated at A & B so that no magnetism was noticed there the lines would be displaced & consequently it would endevour to turn—[5]

In the disc form with holes to facilitate heating and cooling the expansion & contraction was great & caused a buckle which made the wheel[a] untrue & consequently gave bad results— He made many kinds of armatures:—

 ring & spiral edgeways.—

 ring & spiral round wire—

flat discs perforated

Flat discs and slotted radially to take care of expansion better as well as heating and cooling quick.

The best result so far was got with a hub all studded with pins like a circular brush— This form allows of quick heating and cooling and does not allow expansion & contraction to affect its shape

AD, NjWOE, Batchelor, Cat. 1336:179 (*TAED* MBJ003179F). [a]Obscured overwritten text.

1. See Doc. 3029 (headnote).
2. Charles Batchelor consecutively numbered each entry in this journal.
3. Batchelor left New York late on 1 March, arrived in Jacksonville on the 3rd at midday, and reached Fort Myers on the 8th. He stayed until 21 March. Cat. 1336:179, 183 (items 322, 324; 1 and 21 Mar. 1887), Batchelor (*TAED* MBJ003179D, MBJ003183A).
4. Figure labels are "N" and "S."
5. Figure labels are "N," "A," "B," and "S."

–3032–

To Charles Batchelor

[Fort Myers,] April 6 87

Dear Batch—

Dont forget to have Laboratory plans prepared—

I want a special or secret part to machine shop for special things I want subrosa—

Upton is going to Europe for several months & thinks he will give up running Lamp Co but may not   I wrote him to see you & let you run finances temporarily—of course there is nothing to do as they have plenty money   I told Upton that I intended in case he left permanently to put Mr Tate in the position— Tate is a rising man is honest smart & ambitious & will be just the man for Lamp factory.[1] See Upton— perhaps it would be well to put Tate in charge immediately Salary 2500 year—

I am feeling better although little dizzy in head.

got train gear in new Engine & it rotates continuously with Considerable power. havent made Cast iron wheel yet   Keller starts tomorrow on it—[2]

Have struck something new & been working on that past 10 days. I think it's going [soon?][a] be immense— We have tried ore milling Expmnt partially— flour falls in Vacuum dead straight while if air is let[b] in Receiver it falls all over the bottom plate of pump—[3] Love to all—

Edison

ALS, NjWOE, DF (*TAED* D8704AAP). [a]Illegibly interlined above. [b]Interlined above.

1. Edison wrote to Upton on the same day:

In case you leave for Europe I think you had better either put the finances of the Lamp Co in hands of Mr Batchelor for the time being or Mr Tate; In case you give up entirely, I intend to put Mr Tate in the position. He is a thorough Bookeeper, is honest, Smart, has good Experience, is young, ambitious, & will in time make a thoroughly suitable and trustworthy man for us= What do you say— [TAE to Upton, 6 Apr. 1887, WJH (*TAED* X098A024)]

The editors have found no reply from Upton, who retained his positions as general manager and treasurer of the Edison Lamp Co. Tate, however, later recalled serving as the factory's acting manager for several months during this (or another) absence by Upton (Tate 1938, 133–34, 139).

2. Edison meant the pyromagnetic (magnocaloric) motor. The editors have not found records of work on it at this time.

3. A series of notebook drawings (undated but seemingly made before 1 May) appears to show this experimental arrangement. Edison included the idea of using a vacuum to improve magnetic separation in a broad caveat in June (Doc. 3059). N-86-08-17:154–65, Lab. (*TAED* N320 [images 64–69]).

–3033–

Charles Batchelor
Journal Entry

[Schenectady Apr. 8th 1887][1]

330[2]   T.A. Edison

There is a telegram from Gilliland[3] at Myers saying that Edison has another abscess below the ear in addition to the one that was operated on three days after I left there and that they fear erysipelas—[4]

AD, NjWOE, Batchelor, Cat. 1336:185 (*TAED* MBJ003185).

1. Batchelor wrote this entry following one on this date. The next dated entry is from 9 April.

2. Charles Batchelor consecutively numbered each entry in this journal.

3. Telegram not found.

4. Medical authorities recognized several forms of erysipelas, all of them characterized by intense swelling and redness of the skin in affected areas, often on or around the face. The illness was the result of streptococcus infection and was known to be communicable between persons and especially from one part of the body to another through the lymph system. In addition to chills, high fever, and accompanying delirium, experts warned of necrosis and possible gangrene. They prescribed a range of treatments, most involving washing the affected area with anti-microbial agents such as carbolic acid, sometimes administered through surgical incisions. Quain 1883, s.v. "Erysipelas"; Whittaker 1889.

–3034–

From Samuel Insull

[Schenectady,][1] Apl 11/ 1887[a]

Dear Sir:

I enclose herewith for signature a limited license in favor of H. H. Unz for use of your patent No. 295,990 on certain conditions as expressed in said license,[2] together with a letter, also for signature demanding from Unz an assignment of his contract with you under date[3]

In explanation of this I beg to say that a few weeks ago I was approached by the A. B. Dick Co. of Chicago[4] who[b] desired to purchase the right to use your patent no. 224,665 which relates to the production of duplicating writing machines (autographic stencils) by use of a corrugated surface.[5] This patent was tied up under the Unz contract for three years, and in such a manner that whether he did anything with it or not we had no means of cancelling the contract until after the expiration of three years.

I hunted up Unz and found that he had gone a certain distance in connection with producing a machine embodying[b] the patent and he stated that he had spent about $1000. in experiments.[6] As soon as we commenced to negotiate with him for an assignment of his contract he attempted to take advantage of his position under the agreement—first demanding a heavy cash premium, then a permanent interest in the patent.

The A. B. Dick people were ready to place upon the market a machine which Mr. Dick had perfected Embodying this patent.[7]

We did not consider it advisable to inform Unz who it was[b] that desired to obtain the patent as he was attempting to play so sharp a game that he would have have discouraged the Dick people and destroyed your opportunity of obtaining a good annuity from an invention which has been unproductive from its birth.

During the course of conversation with him we mentioned that certain Type Writer people had approached us in relation to your patent No. 295,990 for needle pointed type. He immediately jumped at the conclusion that these were the parties who wanted to use the patent he controlled[b] by him and proposed giving them the right to use the latter patent providing he retained the same right himself, was paid $500. and licensed to use also the first named patent on type. We saw he had conceived the idea that this type writer patent was the more important one of the two and we presented to him as our ultimatum this proposition—:[8]

We would give him $500. in cash and if within the present year he could produce a type writing machine Embodying your invention, which worked satisfactorily we would give[b] him a limited license to use the type invention—which license applied only to his make of machine and did not restrict the licensing of other manufacturers by us.

As he had fully impressed himself with the idea that the Type Writer people had found something valuable, which he could'nt just then realize he accepted the proposition, and it is pursuant to this agreement that I send you the within for signature. The license must be delivered to him before the sixth day of May next, and as I have closed with[b] the A. B. Dick Co. you will see the necessity of returning these papers to me at once.[9]

In relation to this typewriter patent the parties who approached us were the Hall people. Hall has a machine but it is really no good. No one Else has spoken of it since Wyckoff Seamans & Benedict[10] sent to your house. I do not consider this a valuable patent—and even could it be ~~made~~ adapted to a type writer I do not think its nature warrants the <u>exclusive</u> licensing of any single manufacturer

I mean by this that if the Remingtons[11] adapted it[b] it would be used where Remington Machines are used— and So with other ~~people~~ manufacturers.

I Express my opinion on this patent in order to explain my reason for having committed you to an obligation. In the first place my conviction is that Unz will not produce any machine atall and next, should he produce a <u>satisfactory</u> one will only induce other Manufacturers of Type Writers to adapt the invention—we can license them to use it, and thus turn Unz efforts to our own advantage and reap pecuniary benefit from the result of his labor.[b]

The A. B. Dick Co. will pay the $500. which we have given to Unz. They also agree to pay a royalty of .75¢ on each machine sold and 5% on renewals: that royalties for the first two years shall not amount to less than $800. in Each year and not less than $1500. per annum thereafter. These provisions are the same as those in the Unz agreement.

We have seen the Dick machine and it is simply <u>perfect</u>. It is without a doubt the best duplicating machine in the market and a dealer who wanted to get the[b] Agency for <u>Railroads alone</u> offered to guarantee the sale of ten thousand machines.

I estimate that your royalties will amount to from $2000. to $5000 per annum. Under these circumstances I felt that it was absolutely necessary that I should get an assignment from Unz of his contract. His attempt at a sharp game threatened our success but I think he overstepped himself, and that our diamond proved to be harder than his. I had not time to consult you[b] by letter as Unz was working in every direction to find out what we wanted with his contract— As Dick perfected his machine not knowing you controlled the patent, and had spoken of it outside, there was a chance of Unz hearing about it so I did the very best that could be done under the circumstances and arranged matters according to the foregoing explanation.

Kindly sign and return the enclosed at once. The Dick contract is being written and when completed will be forwarded to you for signature[12]   Yours very truly

Saml Insull

LS, MiDbEI, EP&RI, Box 2 (*TAED* X001A2D). Letterhead of Thomas A. Edison; written by Alfred Tate. A letterpress copy is in Lbk. 24:332 (*TAED* LB024332); a facsimile copy is in the Samuel Insull Papers, ICL (*TAED* X077AD). [a]"188" preprinted. [b]Repeated as page turn.

1. Although Alfred Tate wrote this letter on Edison's New York letterhead, he was with Insull in Schenectady on his way back from Chicago. Tate reached New York the following afternoon, when he wrote to Edison about his trip and the flooding around the Edison Machine Works. Tate to TAE, 12 Apr. 1887, Lbk. 24:346 (*TAED* LB024346).

2. Edison's patent covered a perforating typewriter for "producing printed impressions in duplicate." The device worked by "pressing upon paper types in succession each of which has a surface of points so as to perforate the paper, and then forcing through the perforations ink upon the sheet of paper." Edison had filed the application in 1878 somewhat surreptitiously, to avoid damaging his electric pen business (U.S. Pat. 295,990; see Doc. 1629 esp. nn. 2–3). The limited license agreement, signed by Edison (and witnessed by Ezra Gilliland) without

a date, authorized Henry Unz to manufacture a typewriter (and only a typewriter) under the patent, provided that he successfully embodied the invention in a practical machine within the calendar year (TAE agreement with Unz, Apr. 1887, Miller [*TAED* HX87006B]).

3. The letter was based on Tate's draft dated 24 March. The finished version, signed by Edison but not dated, was copied into a letterpress book around 25 April, after Edison had mailed it back from Florida. It recapitulated terms of the license and specified that the license was given in accordance with an agreement reached on 23 March (negotiated by Tate in Philadelphia). Tate describes that agreement below; its gist was that Edison would give Unz $500 (and the license) to buy out a much broader September 1885 license on a related patent for autographic stencils (U.S. Pat. 224,665; see Doc. 2843). Unz received $400 on 23 March, and Edison was to fulfill the other terms within two weeks; that period was extended for one month from 6 April (TAE to Unz, n.d. [c. 11 Apr. 1887], Lbk. 24:398 [*TAED* LB024398]; TAE agreement with Unz, 23 Mar. 1887; TAE to Unz [draft], 24 Mar. 1887; both Miller [*TAED* HX87006, HX87006A]; Insull to Tate, 23 Mar. 1887, DF [*TAED* D8719AAO]).

4. The A. B. Dick Co., a Chicago lumber wholesaler, was incorporated in April 1884 by Albert Blake Dick (1856–1934), who had worked for a similar business in Moline, Ill., and also as a bookkeeper for what would become the John Deere Co. Seeking to minimize the inventory he carried, Dick sent around daily inventory sheets to mills and other Chicago lumberyards from which he could procure materials to fill his own orders. To save the effort of writing these routine inquiries by hand, Dick devised a way to make a perforated stencil on wax-coated paper that he could use to produce copies rapidly (see note 7). When he went to patent this invention, he found that the principle was covered by one of Edison's electric pen patents (224,665). After reaching an agreement with Edison, Dick focused his efforts on what he called the mimeograph. The Dick Co. marketed the machine as the "Edison Mimeograph," which it featured on its letterhead by mid-May; by September, the company was also working on means to produce typewritten stencils. It divested its lumber business by year's end and went entirely into the development and sale of the mimeograph and related equipment. *ANB*, s.v. "Dick, A. B."; *BDABL*, s.v. "Dick, A. B."; *IDCH*, s.v. "A. B. Dick Co."; R. G. Dun & Co. report, 17 Mar. 1887; Dick Co. to Tate, 11 May and 26 Sept. 1887; all DF (*TAED* D8702AAC, D8702AAD, D8702AAY).

5. This chain of events seems to have started in mid-February when, as Tate explained to Insull, "two gentlemen from Chicago called to see if they could make an arrangement" for Edison's U.S. Patent 224,665 for autographic stencils. According to Alfred Tate's retelling, one of these emissaries was Dick's friend and associate George Bingham, who may have played a key initial role in the invention. Edison had also recently received at least two inquiries about his perforated type patent (295,990); see below. Tate to Insull, 11 Feb. 1887, Lbk. 23:405 (*TAED* LB023405).

6. Tate later recalled that it was James Russell who identified and located Henry Unz in a Philadelphia hotel. Tate 1938, 116–21.

7. The Edison mimeograph was essentially the same as Dick's origi-

nal version of the machine (see also Doc. 3070). An August 1888 report for the Franklin Institute described the construction and operation of the commercial model:

> There is a hardened steel plate, the surface of which is covered with fine points, to the number of about 200 to the square inch, formed by scoring it with grooves. This plate is made one and one-half inches or three inches wide, and long enough to extend across the width of the stencil sheet, which is to be prepared, and it is set in one end of a convenient wooden frame. A slab of slate, or other suitable material, is set in the remaining portions of the frame, the top surface of all being flush, and making a level bed. The stencil sheet, which is a thin paper covered with paraffine, is laid on this bed, the part of it to be written on being kept over the steel plate. The drawing or writing, which it is desired to duplicate, is then done with a smooth pointed steel stylus, which is used with about the same pressure and ease as a lead pencil. As the stylus passes over the stencil the latter is pressed down on the steel points, so that they are driven through it, making numerous small holes along the lines which the stylus traverses. The stencil thus prepared is placed between the two frames, one of which slides within the other, and is clamped down, so that the stencil is thus stretched smooth and securely held. . . . An ink roller, covered with semi-fluid ink, is then passed over the stencil, pressing the ink through the holes which have been pierced in it, down on to the paper below, thus making the desired impression upon it. Printing may be done in this way very rapidly, at the rate of several hundred an hour, and one stencil may be used for as many as 3,000 copies before it is worn out by the enlarging of the holes. [Franklin Institute 1889, 381–80]

8. Thomas Hall (1834–1911), a Philadephia-born inventor, obtained a patent in 1867 for a ribbon typewriter with individual typebars, and he soon thereafter formed an eponymous manufacturing company in New York. Hall patented a radically different machine in 1881, a portable one with a single key, a stationary platen, and no inking ribbon (*DAB*, s.v. "Hall, Thomas"). The Hall company had contacted Edison by 11 February, when Tate advised Insull that "the invention they refer to is an entirely different one from the above [based on Edison's U.S. Patent 224,665 licensed to Unz] and is covered by Mr Edison's patent No. 295,990 Apl 1st/84. So far as I am aware Mr Edison has no agreement or arrangement with anyone in connection with the latter invention." Hall personally made plans to see Edison and, in early March, while Edison was in Florida, asked to "call to complete arrangements for the use of Patent on perforating type—and to show you the machine which I design to use it" (Tate to Insull, 11 Feb. 1887, Lbk. 23:405 [*TAED* LB023405]; Hall to TAE, 17 Feb. and 4 Mar. 1887, both DF [*TAED* D8704AAK, D8704AAM]).

9. On 10 March, Tate signed on Edison's behalf a receipt for $200 from A. B. Dick Co. as a partial payment under a license agreement then being drafted for Edison's autographic stencil patent 224,665. According to a summary of the proposed contract, the terms would be the same as those in Edison's September 1885 agreement with Unz ex-

cept for minimum royalty payments of $800 the first year and $1,000 the second year. Soon afterward, Tate requested a credit report and other information about the Dick firm. TAE to A. B. Dick Co., 10 Mar. 1887; TAE memorandum of agreement with Dick Co., 10 Mar. 1887; TAE to Joseph Hutchinson, 16 Mar. 1887; all Lbk. 24:143, 149, 196 (*TAED* LB024143, LB024149, LB024196); Tate to William Logue, 16 Mar. 1887, LM 13:408 (*TAED* LM013408).

10. Formed in 1882, Wyckoff, Seamans & Benedict was a partnership of William O. Wyckoff (1835–1895, a former court stenographer), Clarence Walker Seamans (1854–1915, formerly a sales agent for the Remington typewriter), and Henry Harper Benedict (1844–1935, previously an official in the Remington constellation of manufacturing interests). After marketing the typewriters of E. Remington & Sons for several years, the partnership acquired the entire business outright in 1886 and moved its headquarters to New York City (*NCAB* 8:239–41, 517–18; Wyckoff obituary, *Stenographer* 8 [Oct. 1895]: 99–102; *NCAB* A:146; "Henry H. Benedict, Art Patron, Dead," *NYT,* 13 June 1935, 23; *ANB,* s.v. "Remington, Philo"; Beeching 1974, 152; see also note 11). According to Tate, the firm dispatched a representative to Edison's house on 3 February to discuss Edison's perforating type patent (295,990). Edison "offered it for Five Thousand Dollars ($5,000) but nothing further was done" (Tate to Insull, 11 Feb. 1887, Lbk. 23:405 [*TAED* LB023405]).

11. The Remington typewriter name, along with the factory built up by Philo Remington in Ilion, N.Y., was acquired by the Wyckoff, Seamans & Benedict partnership in 1886, who now made and sold typewriters as the Remington Standard Typewriter Co. Philo Remington (1816–1889), famous for making rifles and revolvers, diversified after the Civil War into farm implements and sewing machines and, in 1873, a typewriter based on the original patent of Christopher Sholes and promoted by James Densmore and G. W. N. Yost. (Edison worked with an early version of the Sholes typewriter in 1870–1871 in connection with printing telegraphs and later took credit for having improved it; see Doc. 142 esp. n. 6, *TAEB* 1 chap. 8 intro and App. 1.A.34–35.) Remington sold its models for several years through scale manufacturer Fairbanks & Co., under the guidance of Clarence Seamans, who then took charge of Remington's own typewriter sales department when it was established in 1881. *ANB,* s.v. "Remington, Philo"; Beeching 1974, 152; Current 1954, chaps. 8–10; *NCAB* A:146.

12. The contract was not yet complete on 13 May, when Tate proposed meeting personally with a Dick Co. representative to work out the details. The company reported that sales of the Edison mimeograph were "Excellent, and the machine is giving universal satisfaction." The firm also expressed some reservations about the legitimacy of similar equipment sold by the Broderick Copygraph Co. (a 28 Feb. R. G. Dun & Co. credit report on that firm is in DF [*TAED* D8702AAB]), which Tate advised would "materially interfere with your own sales if it be allowed to remain on the market" (Dick Co. to Tate, 11 May 1887, DF [*TAED* D8702AAD]; Tate to Dick Co., 27 May 1887, Lbk. 24:472 [*TAED* LB024472]). Edison and the Dick Co. signed a contract (not found) on 27 June embracing rights to Edison's autographic stencil patent (224,665). At least part of the delay seems to have been caused by a

prior contract of Edison with Western Electric Co. for the electric pen (Doc. 817, now expired), parties to which included Robert Gilliland as a co-owner of Edison's U.S. Patent 180,857 (issued in 1876) for a process of printing through a punched stencil. Even after completion of the 27 June contract, questions persisted about the status of Edison's U.S. Patent 224,665 (also co-owned by Gilliland) and its legal relationship to the 1876 specification. In early August, Edison's attorney determined that the 1876 patent was unencumbered and could be licensed to Dick. Edison and Gilliland subsequently gave the Dick Co. a separate limited license to use it for the mimeograph, reserving to themselves the continuing right to use or license it for the electric pen. Edison separately agreed with Gilliland in September to divide royalties received from Dick according to Gilliland's three-tenths ownership in that patent (Tate to Dick Co., 20 July 1887, Lbk. 25:44B [*TAED* LB025044B]; Tomlinson to Tate, 4 Aug. 1887, DF [*TAED* D8702AAS]; TAE agreement with Gilliland and A. B. Dick Co., 4 Aug. 1887; TAE agreement with Gilliland, 20 Sept. 1887; both EP&RI [*TAED* X001H1BM, X001H1BN]; a summary record of Edison's mimeograph royalties through October 1887 is in Ledger #5:604, Accts. [*TAED* AB003, image 299]).

–3035–

From Marion Edison

[Akron,] April 24th 1887.

My dearest Papa:

It is Sunday after-noon, everyone mostly is writing and I thought I would follow suit. I have been trying every day since our arrival here to write you as I promised, but there has been so much visiting on hand that it was simply impossiable. We arrived here Thursday[1] after a loud and dusty journey, I realy think that some of our party would have died before we got here if we had not stopped over at Lake Wier[2] and Jacksonville. I suppose Mina must have told you all about our trip so there is little left for me to tell but I will endevor to tell you what wll have been doing since our arrival. Saturday we all went down to Canton to see Robert & Louise.[3] I think their house is very pretty indeed but it does not in estimation come up to Ira's[4] although seven out of nine like it better. Louise made a charming hostess and seems to be very happy.

Auntie Clark (Mrs. Clark you remember her.)[5] is visiting at Robs. she seems very febble and it is realy painful to look at her. I am glad to hear that Mr. Miller makes his little son[6] ~~that~~ take his medicine everyday. You may well be glad that you did not come north with us, it is as cold as ~~and~~ Iceburg here and the trees have hardly commenced to bud yet, I think if it keeps on very long you had better buy your tickets for July.

I have not commenced to study yet but expect to tomorrow

morning if Miss McWilliams[7] gets here in time from Canton where she has been staying since Thursday. She told me before she went that I <u>must</u> have all my visiting with Grace[8] over by Monday and I must try and do as well up here as I had done at Myers. You see darling Papa we in this world always have a little sweet with the bitter. Mina feels rather hurt that you have not written her, she says she has written you four letters and has not had an answer to any of them. please for my sake do not tell her that I said anything[a] about it but I realy think you ought to write her very often if you dont intend having a cyclone soon. Well dear Papa I am afraid my stock of news is nearly exhausted[b] so I will close my hasty letter. Mina is down stairs writing to you so you will just get ours together.[9] All congratulate you~~r~~ upon shooting at a dead aligator. Maime,[10] Mrs. Miller and Grace send love to you and to Mr. Miller   also accept love from me to him and to yourself

Tell Leina[11] to be a good girl and tell Nick[12] to study hard. With[a] bushels & bushels of love and Hoping to hear from you soon I remain as ever you true and affectionate daughter

George.[13]

ALS, NjWOE, DF (*TAED* FB001AAA). On monogrammed letterhead of Marion Estelle Edison. [a]Obscured overwritten text. [b]"h" interlined above.

1. The 21st of April.

2. Lake Weir, in central Florida, was the site of a modest assembly ground built by the Lake Weir Chautauqua & Lyceum Association. The traveling party would have missed the assembly, scheduled for 22 February to 23 March 1887, but could have stayed in the lakeside Chautauqua House hotel. Norton 1892, 304; *Chautauquan* 7 (Oct. 1886): 509; Lamar 1896, 119–20, 127.

3. Newly married Robert and Louise (Igoe) Miller.

4. Ira Mandeville Miller (1856–1934), Mina's oldest brother, had recently married the former Cornelia Wise. After graduating from the Ohio Wesleyan University, Miller entered the family business of Aultman, Miller & Co. He was later involved in other Akron enterprises, including the establishment of a street rail system and the city's first commercial electric lighting system. Lane 1892, 467; Jeffrey 2008, 171.

5. Rachel Kinder Clark.

6. Probably Theodore Miller, Mina's youngest brother.

7. Sarah McWilliams (b. 1843?), an Ohio native known to the Miller family, had traveled with the Edisons to Florida. When Marion was unhappy attending school in Massachusetts a year or so later, she asked to return home to study with McWilliams instead. McWilliams remained connected with the Edisons for years and was listed in the 1900 federal census as a servant living in the household. U.S. Census Bureau 1982? (1900), roll T623_968, p. 20A (West Orange, Essex, N.J.); Mary Valinda Miller to Mina Edison, 6 Jan. and 1 Mar. 1887; Marion Edison

to Mina Edison, n.d. [1888?], all FR (*TAED* FI001AAW, FI001ABD, FB002AAG); Marion Edison to Mina Edison, 1 Nov. 1924, CEF (*TAED* X018A5CM).

8. Grace Miller, Mina's sister.

9. According to Edison's great grandson, David Sloane, the Charles Edison Fund holds letters between Edison and Mina from the period of Volume 8. The Fund has refused several requests by the editors to see and make available any of the correspondence in its possession between Edison and Mina.

10. Mary Emily Miller.

11. Probably Helena (Lena) McCarthy.

12. Probably Nicholas "Nick" Armeda (1869–1955), who had been a deckhand on the *Jeannette* when it brought Edison and Ezra Gilliland to Punta Rassa and Fort Myers in 1885. In 1886, he was hired to look after Gilliland's steam launch at Fort Myers. Armeda was the caretaker of the Gilliland's Fort Myers estate in the 1890s, when it had passed to the ownership of Ambrose M. McGregor. U.S. Census Bureau 1982? (1900), roll 172, p. 9A (Fort Myers, Lee, Florida); *Florida Death Index 1877–1998*, online database accessed through Ancestry.com, 8 May 2014; Gilliland to Eli Thompson, 5 June 1886, Lbk. 22:169 (*TAED* LB022169); Rosenblum 2000, 1:67; payments to Armeda for his services are listed in a Fort Myers Cashbook (1886), Accts. (*TAED* AB023).

13. Marion seems to have acquired this nickname, at least to her father, sometime after her mother's death in 1884. Israel 1998, 234.

–3036–

Lewis Miller to Mina Edison

Myers Apr 26th 1887[a]

Dear Mina

We are preparing to leave for the north in the morning   I have had a most delightful time all have been so[b] kind that I know it comes from genuine friendship and reguard for you. I have had a better opportunity to get acquainted with Mr Edison than ever before. The more I see of him the more I am impressed with his greatness and genuine good hcart   I am thoroughly convinced that he is true to you and true to what he appears to be. And socially he is superior to most any one I know

The many remarks he makes which are seemingly some kind of assinuations or critizisms are genuine innocent wit and are made to Mr Gilliland Mrs Gilliland and any one Else with whom he is well acquainted[b]   the difference you see is the difference of being well acquainted and parcially acquainted   while out on our weeks or four days hunt He was so[b] cheerfull and Entertaining he kept the camp or party in good cheer all the time with his witicisms and stories. he was always ready with some witty remark no matter what was up   And always in such nice language as compared with Mr [Wurl?][1c]

and others   Nothing of low order came from his lips   sometimes he would tell some story that if any one Else but him
would have told it would have been put in such language as
would have been low of cast but not once did[b] he let himself
down   I have purposely watched him to see if any remarks
which might accidentally drop would show any kind of different feelling towards you. He seems to me just as he did a
year ago. Kind a[nd] affectionate in all his ways. You would
naturally Expect that he would visit Mr Gillilands   Since
you left he has been over but very little and never any length
of time   he and I would sit at home of whole evening talking
and reading

I have made these opinions because of your impressions
that you did not have his full affection [--][d]   its so natural
for one after marriage and when Everything must come to a
reality to feel just as you do. This is doubly so in your case
as Mr Edison is a[e] fully matured mind and yours is still formative. You are not prepared to make the allowance that you
should. I see your influence over the family in many ways. The
children have certainly much improved as it appears to me.
Mrs. Wm Edison[2] was very free in giving her opinion, and impressions. She said it was a most fortunate thing that this kind
of a mother got into the family that you seemed to be. She told
me many things that ~~Mr~~ Alvin[3] told her about you which were
very nice indeed. I am convinced that if you will not force a
different impression by your apparent feelling that all is not
right You can have a delightfull home. You have it in your
power to make all around you happy and delightfull. Take Mr
Edisons sayings and looks as witicisms and on his part Efforts
of Entertainment and amusement   You will find that it is just
what he intends. he is so in the habit of those witty remarks to
his men in the works that he cannot & I think ought not try
to suppress them when he comes home   You must try and
meet him with wit have something new Everytime he comes
home. don't think he cares much to hear about house affairs.
Of course you must always Explain to him fully what you desire to do. I am fully satisfied with his future prospects for
making money   he no doubt will ~~have~~ get a very large income
from his light co stock and also from his factories. When you
first went to Llewellyn Park, he had spent so much money
~~that~~ and things just at that time did[b] not look so bright that it
made him a little cautious which was right enough but made
it quite hard for you   he is much pleased with your values of
economy Spoke about a number of times also of your great

desire to make the home the place for entertainments instead of running about.

I think Dott will be very much influenced by you. No doubt she sees your feelling about her and you must win her or Else it will not be just what all hope and desire. I believe that you can win her she certainly takes to your friends and Especially to Mami and Grace[4]   now if she was desirous to throw you off she woul[d] not be interested in them. The boys certainly are much impressed by your influence over them   I am very anxious that you show your true christian influence. Your life not your talks will impress the true character you have   Show by your spirit [~~show~~?][d] what true christian spirit can do   The Miller family are too open with their criticisms of others   its all right to open and not deceptive in your impressions of others   There always to sides to every question one is the side of fault or errors the other the side of good and right. Every one of us have some faults and Errors about us, so there is no time or thing or person about which or whom we cannot find something that is wrong.[b] to grow into a habit to see the errors and not the good is very Easy and natural   it requires an Effort to see the good things and talk about them leave out the Errors and faults   we as a family are faulty in seeing the Errors and holding them up in our view. Jenny is so much troubled in this direction that s[h]e is constantly unhappy. Now Mina I want you to practice on seeing the good qualities of things   Talk and think about them until that side becomes natural. I think this is all you need to change many of your impressions & feel different about many things. I hope you will take these thoughts in same spirit as they are given by a Father

Your Father

ALS, NjWOE, FR (*TAED* FH001AAA). Letterhead of Thomas A. Edison. [a]"188" preprinted. [b]Obscured overwritten text. [c]Illegible. [d]Canceled. [e]Interlined above.

1. Not identified.

2. Ellen J. (née Holihan) Edison (1840–1927) was the wife of Edison's brother, William Pitt Edison. Jeffrey 2008, 157, 165; U.S. Census Bureau 1970 (1880), roll T9_ 605, p. 398A, image 0688 (Port Huron, St. Clair, Mich.); Headstone photo for Ellen J. Eckles, Lakeside Cemetery, Port Huron, Mich., Find A Grave memorial no. 54683102, online database accessed through Findagrave.com, 12 Dec. 2013.

3. Family members sometimes referred to Edison by his middle name, Alva, or simply as "Al."

4. Mina's sisters, Mary Emily and Grace Miller.

Charles Batchelor
Journal Entry

[New York,] Apl 30 1887

349[1]   Edison
   Returned from South much improved but still with ab-
cesses not quite healed.

AD, NjWOE, Batchelor, Cat. 1336:195 (*TAED* MBJ003195B).

   1. Charles Batchelor consecutively numbered each entry in this
journal.

Notebook Entry:
Miscellaneous

[Fort Myers, April 1887?][1]

Experiments New Laboratory—
   Kenny phono motor emg relay on top for repeating WU
Tel—[2a]
   Probable surface of chalk effected by light— use diaphragm
with mirror & heliostat, & condensing reflector or g Lense
make very narrow streak light, talk & turn emg— if beam alters capilarity for instant talking be heard[3]   probably its only
for instant like electricity.[4]

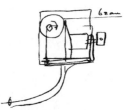

if wks put in Alum Cell   stop all heat & try light—then try
Ultra Violet or Actinic rays only—
   Light Effects chlorine & H—[b] close a tube with Rubber
diaphragm with hearing tube & chlorine & H in proportion
to form HCl—with or without water to absorb HCl which it
does explosively—perhaps saturated Cl water & free Cl with
H OK— let beam light end on enter tube—vibrated from
Mirror on diaphragm—heliostat & sun light—[a]
   Hozion apparatus[5] silent dischg for forming synthetically
sub[stance]s on Com[mer]c[ia]l scale with Res current ma-
c[hine] or high volt dynamos & Silent[c] dischg tubes in Multiple arc

X, NjWOE, Lab., N-86-08-17:185 (*TAED* N320184). [a]Followed by dividing mark. [b]"& H" interlined above. [c]Obscured overwritten text.

   1. Edison likely made this undated entry in Florida or on his return
trip before he reached New York on 30 April. It appears in the notebook about one hundred pages after Doc. 2962 (dated 18 June 1886)

and before Doc. 3039 (dated 1 May 1887). It is immediately preceded by drawings of an electric meter and, before that, by drawings related to ore separation, with which Edison had experimented at Fort Myers in March 1887. Other undated entries on those pages include drawings of electric light systems similar to those from November 1886.

2. The editors have not identified Patrick Kenny's "phono motor," but it may be one of the small electric motors proposed in Doc. 2993 for use with the phonograph. See Doc. 1913 regarding the electromotograph relay.

3. Edison and his assistants had not come up with a satisfactory explanation of the electromotograph principle, a slight variation in surface friction in response to an electric current. One of their earliest explanations was that it was an effect of capillary action. See Doc. 1738 and cf. Doc. 1797.

4. Figure label is "beam."

5. Edison referred to the apparatus devised by Auguste Houzeau to synthesize ozone. Known as an ozonizer, this was one of several devices developed following the pioneer work of Werner von Siemens on what became known as "silent discharge" and is today called dielectric-barrier discharge. In these devices, electrical discharge from an induction coil passes continuously through oxygen gas contained in glass tubes, producing ozone. Dynamos, batteries, and static electric machines could also be used to produce the electrical discharge. The thinner the layer of gas through which the electric discharge passes, the more effective the reaction; when used with rarified gases, as in a Geissler tube, it could produce luminous phenomena. As Edison suggested, silent discharge devices were used for the commercial production of ozone and could be used to synthetically produce other substances, such as ammonia and formaldehyde. Watts 1879a, 727–28; Wiechmann 1906, 133–36; Mellor 1822, 885–89; Preis 2007; Rubin 2002.

−3039−

Notebook Entry:
Electric Lighting

[Harrison,][1] May 1 1887

Electric Lamp Experiments

Dissolve ~~or~~ rather mix as much pure Alumina ~~& any other solution~~[a] also Magnesia ~~Zircon oxide~~ with Asphalt solution possible— dip preliminarized Regular A Carbon[b] & also carbons in solution so surface when Carbonzed will be partially of ~~an~~ a nonconducting infusible oxide   make a order number   15 going to clamp dept[c]   ⟨Martin⟩[2d]

~~In Sugar Solution dissolve as much Lime as it will take up & dip filiments~~ — ⟨No 2 [---][e]⟩   ~~also~~ make a strong acetate of Magnesia—Lime, Alumina etc Solution nearly Saturated [-----][e] & Syrupy[f] then dissolve all the Sugar it will take up & in this solution dip filiments & also carbons[g] on carbonzation. ⟨Martin⟩   this will give a Carbon— Magnesia on Surfaces— make 15 of each Kind   ⟨preliminary⟩

Pack some new stock filiments in powdered asphalt & run up to as high as possible say 600 fahr   take out while hot & pour it out getting filiments out & disolving off asphalt— then dip in regular asphalt solution & reprelimize   also Linseed oil— also Linseed oil Loaded with asphalt. ⟨[Ent?][h] Hamilton⟩[3i]

Dip preliminarized fibres prelim to 500 @ 600 in melted asphalt [---][e] at 600 @ 700 fahr— Try melted sugar   Arrange to duplicate 234 in our new process but putting filiments under strain—[c]

Duplicate [---][e] a good new stock curve But using carbon instead of metal boxes— ⟨Martin⟩[c]

Run some new stock through with powdered fusible metal securing filiments so they wont float. when metal melts. ⟨Hamilton⟩[j]

put a pine stick in metal tube which stick has been soaked in Saturated solution Tannic acid   put some tannic acid in bottom bring up quickly in Stove of pump room   ⟨John Ott⟩[4k]

also put pine stick in tube with Red oxide mercury in bottom to give off oxygen— ⟨John Ott⟩[l]

preliminarize fibres in following Liquids as high as possible— Melted Rosin— Anthracene, Chrysene m[elting] p[oint] 275C Hexabrombenzine mp 300C Hexachlorethane ~~CCl₅~~ C_2Cl_6—mp 182C   The ~~chlorine~~ Haloids may produce beneficial Chemical reaction   Hexachlorbenzine mp—222C Try flour sulphur[5] to prelimize in melts 113C boils 447C   Tribromide Bismuth melts 198 boils at red heat

Fusible metal   4B.—1 Cadmium—2 Lead 1 tin— ~~7 or 8 Bro   1 or 2 Cad 4 Lead & 2 tin~~ 4 Bi—1 Cd 2 Pb 1 Sn—

Antimony Trichloride   melts 73 boils 230 at pressure atmos & 160 in vac—

Stannous chlo melts 249—boils 617—

preliminaze in glycerine, Olive oil, Sugar alone—powdered fine— phenol[m]—187C boiling pt.[c]

Boil some fully Carbonzed filiments Secured in reg forms in Sulphuric acid out doors until they bend—then take out & run Curve—take some [chip?][h] in asphalt sol & run through whole of reg process— ⟨Marshall⟩[6]

put some regs carbons in strong water & SO_4 Sol put current on & oxidize the surface until very black— also put in ~~Strong~~ nitric acid   do same some sets more than others & longer—then treat with asphalt after they have been heated & washed to get acid out   object is to remove the outer scale— ⟨Payne⟩[7c]

Take lot Carbons & pass them through flame so they are partially oxidzed   treat with asphalt

Take reg Carbon mounted on inside part[8] dip in Asphalt Solution 24 hours   seal in Lamp & then gradually bring fil up in Vac while pump running to carbonze asphalt surface stopping each time to permit gas to be evacuated   Make 6 in this manner get curve— ⟨Marshall⟩

Ascertain on a reg set up at 80[9] The increase of Resistance every hour until busted. Then take another and put at 80 selecting one of same economy & volts & insert Res equal each time to increase on the one burning & see what cp will come to   dont burn the standard only during test with Res— object is to see how much decrease in cp is due to increase of resistance— ⟨Marshall⟩

Melt some Zinc ~~& plunge~~ get it red hot plunge secured fibres in it. also try Copper molten in [~~ke?~~]^e white hot.

Run Curve also a curve after being treated with asphalt & run through reg process— ⟨Payne⟩^c

Carbonze ie^n preliminze some filiments in aqua regia weak— Also in melted sulphide—Potassium^o   ⟨Hamilton⟩

Put Lamp on pump with stock Cock under Hg— get Vac & work lamp regular then close cock & let it burn at say 80 cp near as possible for 2 hours keeping pump running—then turn Cock & see if air has accumulated— ⟨Marshall⟩^p

Try passing current through reg filiment secured to inside part dipping in [---ped?]^h molten asphalt also sugar   ⟨Payne⟩

Try preliminzing fibres in Linseed oil, Linseed Loaded with asphalt, asphalt in turpentine—Sugar etc in Sealed glass tubes—up to 5 @ 700 fahr ~~als~~ also try Haloid Combinations with Carbon—Chl Carbon etc— ⟨Hamilton⟩^q

Make Couple Extra brass tubes suspend in them pine pieces. allow them to remain in Lamp heating stove 1 to 3 days ~~see if crack~~ by time   Keep record of time & deliver to me ⟨John Ott⟩^r

Float on surface of asphalt baths & also sugar solution^s best Electrotypers plumbago ~~also~~ [~~Anthra---?~~],^e quite thick so that when solution goes down it will leave a surface of plumbago— you can do this in a small Experimental bath that will do 25 carbons at a time   ~~also dust over filiment when fresh from Bath plumbago powdered~~ [--------]^e   also try finest^m Magnesic Oxide oxide Alumina^m etc. this will give good reflecting surface   make order No of 15 each Kind   ⟨~~Martin &~~ [----] Martin⟩^t

Have box made iron[10]

⟨Payne⟩

put regs in forms ~~up~~ secured   fill with Hg & prelimze various speeds Condensing Hg— Sand bath[11]

TAE

X, NjWOE, Lab., N-86-08-17:193 (*TAED* N320193). Document multiply signed and dated. [a]"~~& any other solution~~" interlined above. [b]"Regular A Carbon" interlined above. [c]Followed by dividing mark. [d]Marginalia written in margin next to line spanning entire paragraph; paragraph preceded by large circled "X." [e]Canceled. [f]"& Syrupy" interlined above. [g]"& also carbons" interlined above. [h]Illegible. [i]Marginalia written inside line partially enclosing paragraph. [j]Marginalia written inside line enclosing paragraph and followed by dividing mark. [k]Marginalia circled and connected by a line to a large circled X. [l]Marginalia circled and connected by a line to a large circled X; paragraph followed by dividing mark. [m]Obscured overwritten text. [n]Circled. [o]This paragraph and the preceding one partially enclosed by line. [p]Marginalia connected by line to large circled "X"; entire paragraph enclosed by line and followed by dividing mark. [q]Marginalia set off and partially enclosed by line; paragraph followed by dividing mark. [r]Marginalia circled; enclosed by line with entire paragraph and large circled "X." [s]"& also sugar solution" interlined above. [t]"Name illegibly overwritten by "Martin"; text from "Float on" to here and a large circled "X" enclosed by a line.

1. Edison likely made the entry at his lamp factory laboratory in East Newark or Harrison, two place names used interchangeably on contemporary maps. The Edison Lamp Co. had been using "East Newark" letterhead, but during April, Francis Upton began writing on a new letterhead with "Harrison" preprinted, and the editors accordingly adopt

that convention. Letterhead, Upton to TAE, 18 Apr. 1887, DF (*TAED* D8734AAJ).

2. Martin Force.

3. H[ugh?]. de Coursey Hamilton.

4. The instructions that Edison assigned to John Ott in this document were included in the first of three one-page memoranda to Ott that Mina Edison wrote out on 8 May, presumably from Edison's drafts or notes. Ott marked the pages to indicate that he had made the requested apparatus. The second directive concerned a covered carbonizing mold; the third concerned construction of an electrified tube arrangement, perhaps for carbonizing experiments. The three sets of directions are the only entries in that notebook, and each was signed by Mina and Ott. PN-87-05-08, Lab. (*TAED* NM030A).

5. That is, in finely powdered form. Powdered sulphur was generally known commercially as the flowers of sulphur. *OED*, s.vv. "flour" noun 2.b; "flower," noun 1.c.

6. John Trumbull Marshall (1860–1909), a Rutgers College graduate, took charge of photometry tests at the Edison lamp factory in 1881. Marshall became a photometric expert and spent his entire working life connected with lamp experiments and manufacture for Edison and General Electric. Doc. 2697 n. 1; Marshall 1931, 51–52.

7. Arthur Coyle Payne (1864–1952), a native of Metuchen, N.J., received an engineering degree from Rutgers College in 1885. It is not clear when he began working for Edison; his name first appears in notebooks about this time. Payne remained in Edison's employ until 1889, working on electric lights and the phonograph and undertaking a South American search for bamboo. He subsequently entered into mining and oil drilling in Mexico and the American Southwest and served four years as a U.S. consul. After 1920, he was actively involved with inventing and promoting display screens for the projection of motion pictures, stock updates, and scientific samples (*NCAB* 39:73–74). Payne started a notebook in early May that he used until mid-summer for recording lamp and filament experiments like those described in this document (N-87-05-09, Lab. [*TAED* NB002]).

8. Edison meant the tapered glass tube carrying the lamp filament and lead-in wires. This entire assembly was inserted into the open base of the lamp globe and sealed there during manufacture. Doc. 2098 (headnote).

9. That is, eighty candlepower.

10. Figure labels are "Thermometer" and "condenser."

11. This entry is continued in Doc. 3041.

–3040–

Charles Batchelor
Journal Entry

[New York,] May 2nd [1887]

350[1]   New laboratory

Saw Edison at Orange & discussed the new plans— Instructed me go ahead & get an architect immediately on it—[2]

AD, NjWOE, Batchelor, Cat. 1336:195 (*TAED* MBJ003195C).

1. Charles Batchelor consecutively numbered each entry in this journal.

2. Batchelor recorded the next day that he had "Put the Laboratory into the hands of H. H. Holly" but did not elaborate on the rather peculiar choice of Henry Hudson Holly (1834–1892). Though a distinguished architect, the author of several books, and an original member of the American Institute of Architecture, Holly specialized in the design of American churches and country homes (including Glenmont, the Edisons' home in Llwewllyn Park). He was a lifelong resident of New York and had his office at 111 Broadway. Holly prepared some form of plan by 9 May that Batchelor took to Llewellyn Park for Edison's review. Cat. 1336:195, 199 (item 350, 356; 3 and 9 May 1887), Batchelor (*TAED* MBJ003195C, MBJ003199); Withey 1970, s.v. "Holly, Henry Hudson"; Obituary, *NYT,* 7 Sept. 1892, 5; TAE agreement with Fairchild & Co., 1 July 1887, ECB (*TAED* X184A2).

Batchelor had assigned a draftsman to make some preliminary drawings on 16 April and visited the site himself a few days later. Cat. 1336:187, 191 (items 336, 341; 16 and 22 Apr. 1887), Batchelor (*TAED* MBJ003187E, MBJ003191C).

–3041–

Notebook Entry:
Electric Lighting[1]

[Harrison,] May 2 1887

Lamp Expts   preliminaze in wood tar in sealed glass tube wood tar being same nature as that from fibre— I suppose a <u>Thermometer</u> can be sealed in tube— the End should be drawn fine so air can be Let out Easy—[a]

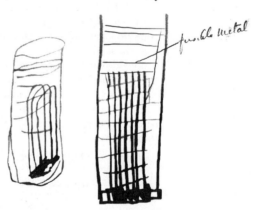

fibre will be kept seperated & also have a tension on them in Carbonizing[2] ⟨Payne⟩

It is probable that with wood tar in sealed tubes and 10 hours run or less that 800 degs fahr can be reached & that there will be no more tar left in filiment to make a crust on surface to cause strain & cracks   it is also doubtful if they will crook if not this will be an Evidence there is little Rupert drop action on surface—[3]

[A][4]

Sealed tube— Wood Tar Thermom in 68 hours reaching— ⟨600⟩

ditto— quickly as possible— a bunch of [---][b] new stock 8 [--][b] round—about 100 in bundle, weighted with say Copper[c] to hold in solution ⟨Order nos for Hamilton⟩[d]

same as above only powdered Rosin in tube which melts.
Same as above [Anthracene?][b] flowers[c] sulphur
ditto. finely powdered sugar—
   "   phenol.
   "   Linseed oil.
   "   glycerine.
   "   mercury
       Linseed Loaded with asphalt
       powdered asphalt.
       powdered Sulphide Potassium
       aniline oil
       acetic acid.
       Satured Sol Sugar.
       Parafine.
       Idiform.[5]

2 pct Sol of Caustic Potash
2   "   Sul acid—
pure water.
Chlorine water—saturated
Pentachloride phos 1 pct Sol[a]

M Force takes Hamilton order nos—

25 are dipped in 62 sugar solution[e] & run through preliminary to 8900 in Carbon boxes. [---- e][b] carbon [laid?][f] in anthracite 10 mesh but $\frac{1}{16}$ of pow 80 mesh comm[er]c[ia]l coal
on bottom. then given to Lawson to run through his final but
not preliminary instead of $4\frac{1}{2}$ inches [or more?][f] make equivalent to $5\frac{1}{2}$ inch to get 100 volts.

25 are to be run to 900 without dipping

25   "   "   put in Lawsons forms in regular manner &
dipped previously being dipped in 62— 25 to be saved   Then
run through pump room as many as possible of the 25. afterwards what comes out to best or all are to be tested for spots
etc notes made about number bent & otherwise inferior & 10
of best are to be selected for curve ie[g] order number[a]

Marshall [--][b] take 30 Lamps Regulars but with new clamps[6]
10 are to be run up on pump in regular way but now so high
up not on highest with exception that instead of running peg
up[7] as high as it will stand to run peg up little less than they
do on regulars

The next 10 are to be run up[c] just as high as it is possible to
go without arcing or melting clamp

The remaining 10 are to be run up as high as possible and[c]
in addition are to be allowed to burn [several?][b] 10 minutes or
so at say 100 candles.

The object of the experiment is to consolidate[c] the carbon
on the pumps and prevent the change that portion of the drop
in candle power due to increase of resistance & which appears
to take place in the first 2 or 3 hours at 80 candles   I think
that bringing the filiment up much higher than usual which
with new clamp is probably possible & holding the peg there
for some time that the filiment will have the change taken out
of it so when it goes to photometer room it will get a reading
which will not change much after[8]

you will probably find that arcing will destroy some of your
lamps so you better [---][b] keeping making putting[h] lamps in
pumps until you get the requisite 10 for a curve keeping account of the number broken & how broken it took to get the
10— you will select Reg 16 cp lamps for this test with the shank

broken off & deposited by new process to inside wires— If you find that arcing bothers you too much you can stop it but putting about 100 ohms resistance in circuit when you have the peg at the highest, the arc ~~never sprin~~ springs with greater difficulty when there is a resistance in circuit somewhat near that of the lamp— It is probable that the deposit at the clamp should be somewhat thicker than would be necessary with the present way of working

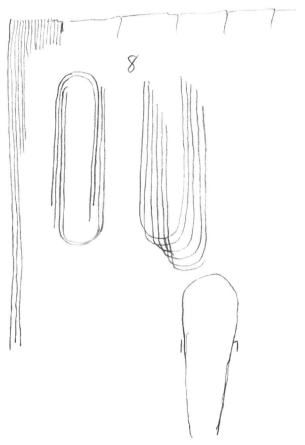

As a magnet will attract an arc I^c have a strong impression that it would be difficult to spring an arc if one of the magnets using on pump in Laboratory were used & powerfully energized—perhaps you better use my pump for the experiments as they have magnets already on—

The great point is to bring the filiment up on the pumps far higher than is now usual in fact the highest attainable limit that is practicable with the use of every device to prevent arcing—

These 3^c sets of lamps should then be set up at ~~the~~ 80 candles each and the drop in candle power taken every hour acurately   this will give us the value if any of bringing them up high on pumps & it may also effect their life for good or bad— End Marshall[a]

~~Deshler~~[9a] Howell   5 Regular lamps to be put up at 80 candles a reading of candle power to be taken every hour & the change in resistance of the filiment taken <u>acurately</u> Every hour   The object of the experiment being to ascertain what proportion of the drop in candle power is due to a change in the mere resistance of the filament—by using a fresh lamp & putting in circuit ~~va~~ varying resistance   The fall in cp from resistance can I suppose be ascertained   I suppose 5 Lamps will be sufficient to give a correct result.[a]

Experiment No 2

Select 5 bulbs which have been blackened   They should be of various shades from a slight blackening to a very Much blackened on— have Holzer carefully remove the inside part & broken carbon, then Taken from a Regular lamp which has been test for Volts cp & ampere the inside part & Carbon & seal this in the blackened bulb— ~~re-ex~~   tell him to be careful not[c] to heat[c] the bulb up where light comes through as this would remove some of the blackening   the have them rexhausted & not run up <u>very high</u> on pump but just enough to get air out. Then ~~you can~~ set them up at same volts [& -][b] & get cp— The object of the experiment is to determine what proportion of the drop of cp is due to blackening— These Lamps should after you are through testing be mounted on a board & the loss on each plainly marked & kept in your museum for further reference[10]   by looking at the[c] tint of a bulb & comparing one can pretty acurately determine the loss of cp by the blackening which will be useful in future Experimenting   if you dont think 5 bulbs enough select such a number as will give us an ~~gamut~~ octave   End Howell[a]

Martin Force—[c] ~~525~~ fibres of each kind 8 round of the different kinds of stock sent me by Payne & Hamilton Including Reg new stock mangrove & bamboo[i] are to be [p--?][b] bent round the hot bender in packed in about 40 mesh anthracite in Carbon boxes and these are to be given Lawson to be run though the full process   after receiving them from him they are to be secured to ~~clamp~~ inside wires by deposit   each filiment is to be cut 2 inches long after it is <u>fully carbonized</u> thus[11]

filiment as it comes from Lawson   Each is to have an order number   ~~allow the~~ out of those of each order number that gets through pump room select 10 for a curve but have accounts kept of number of filiments that are good that comes out of the Carbonzing boxes number broken in depositing ditto pump room   an important Element to determine value of a new fibre is the breakage as well as the life— These filiments are not to be treated in any[c] manner[c] but just run through in the old way   this will determine their value independent of any treatment. I suppose you could string them on your paper sieve [peices?][f] for convenence & thus get the 25 in two boxes. These boxes should all be mixed & placed in various parts of Lawsons Mould[c] [------][b]

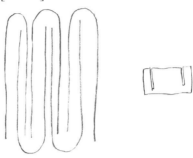

Hamilton[12]   You are to try the following Experiments all of which will require several weeks time & very careful manipulation otherwise they will not be of very[c] much value. The experiments are the bringing up bamboo filiments quickly & also gradually in liquids etc in sealed glass tubes thick enough to withstand the pressure which will issue from a temperature of 600 deg fahr— These glass tubes should be about $\frac{1}{8}$ of inch thick & $1\frac{1}{2}$ to 2 inches in diameter closed at one end by fusion like a test tube   the other end should be open to permit the insertion of the filiments—~~after about~~ which must be secured in carbon or metal[c] receptible so filiment do not touch sides of glass tube & so they will not float to top in the liquid ~~or~~ when a solid in powdered form becomes a liquid— enough material should be used to cover the filiments entirely when the tube is at angle of say 30 degs—

~~The End~~ after the liquid or powder etc is put in tube & everything ready ~~the End~~ it is to be taken in to Joe Force & sealed the End being drawn out to a point so that when the Experiment is finished[c] the tip can be broken off to allow of the Escape of the gases slowly—then the tube is to be cracked at point where it has full diameter the filiments taken out and soaked in a warm <u>solvent</u> of the material used   for instance if asphalt was the material used then The filiments can be washed in considerable quantity of Benzol slightly warmed on sand bath. If Linseed oil a large quantity of^j Turpentine will dilute it so that filiments when taken out will have practically no Linseed oil on them & the turpentine will evaporate   if sugar then^c water will dissolve it— always use a considerable quantity of water or other solvent to wash the filiments

a chamber should be used which will hold say 63 ~~tubes~~ to 6 tubes & heated by a gas stove or other means. a Fahr Thermometer should Enter the chamber The bulb being amid the tube so it will get same heat as they do   the column should appear outside the chamber when mercury stands at 150 degrees   The Thermometer should be capable of indicating up to 600 degress   a very long one is not necessary one about 12 inches long is sufficient & are not so expensive   The chamber should be arranged so that the temperature can be regulated very nicely [----]^b so that starting at 100 fahr it will go gradually up to 600 deg in 8 hours   also be able to go to 600 inside of one hour as I desire to make some experiments rapid & slow— should Any^c tube Explode the chamber will prevent the ~~glass~~ broken glass from coming in contact with the Eyes—

The carbons to be used are regular 8 × 13½ with the shanks on same as now. about 25 ~~carbons~~ filiments^h should be put in Each tube Except in the case where you^c put regular forms in then 3 can be put in each form & 3 forms used making 9 carbons. You can get^c instructions from me as to best method of putting the filiments together for placing in the tube

after washing & drying the filiments they are to be placed in the little paper boxes with the Number marked on the box & also on a slip in the box—each ~~box sho~~ Experiment should be given a number. Thus Hamilton Experiment No 1—Variation No 6   The tube Experiments being called No 1 Experiment & Each tube a Variation with a number—

[~~Each?~~]^b There will be duplicate tubes in Every Experiment   one of the tubes containing the same material as the other will be gradually run up in temperature from the temperature of the atmosphere to 600 degrees Fahr ~~duri~~ gradually during 8 hours— While the other tube ~~contain~~ which is a duplicate will be run up to 600 degrees in one hour or as near that as possible

~~you should b~~ I shall require about ~~25~~ 30 pairs of tubes ~~most~~ all ~~of~~^{k} of which will be 1½ inches diameter inside measurement ⅛ thick—

Have Mr Holzer order it immediately so as to have it by

time your furnace is ready— He will obtain for you a Thermometer   anything relating to making furnaces speak to me
& I will have made—

All of the paper boxes containing order^c numbers are to be
delivered to me—

The following are the materials to be placed in the tubes
please see if we have them all if not make list so they can be
ordered immediately[1]

No 1— Boiled Linseed Oil—

No 2— Pure^c undiluted glycerine

No 3   Mercury—

No 4   Phenol—ie^g carbolic acid

No 5   Parafine

No 6   Wood Tar

No 7   Powdered Rosin 60 mesh   There should be enough
   powder put in tube so that when it melts it will cover the
   filiments fully

No 8— acetic acid   glacial—

No 9— 50 ~~percent~~ parts by weight^m of asphalt disolved in 50
   parts by weight of Linseed oil—

No 10   powdered asphalt 60 mesh

No 11   Aniline oil—

No 12   powdered sugar   this is sold in grocery under name
   of pulverized sugar & is very fine—

No 13   ~~75~~50 parts by weight of sugar & 50 parts water

14   pure water

15   Saturated chlorine water

16— ~~35~~ percent solution of Caustic Potash in water

17   35 percent solution of sulphuric acid in water

18— Idioform—

19— ~~Flower~~ Flour of sulphur

20   powdered fusible metal 10 @ 20 mesh— Formula 4 parts
   Bismuth 1 part Cadmium 2 parts of Lead^c 1 part of Tin
   These are previously fused together & then powdered   I
   think it is brittle if not you can cut it up   it melts about 175
   Fahr— See that we have a supply of material sufficient to
   make 4 or 5 times the quantity needed

21— Tricholoride of Antimony

22— Anthracene

23   powdered sulphide of Potassium

~~20~~ 24— 5 percent solution of chromic acid in water—

25— 5 percent solution of Nitric acid—

26— water 93^c parts   Caustic Potash 5 parts   pyrogallic acid
   2 parts.

27. Chloride of Lime fresh 20 parts   water 80 parts

28. water 90 parts   Hydrofluoric acid 10 parts—be very careful in handling the Hydrofluoric acid not to get any on your fingers or body it makes ulcers but after dilution with water it is not[c] dangerous

~~29~~

I will probably give you other solutions in addition before the End of the Experiment[13]

E[dison]

X, NjWOE, Lab., N-86-08-17:219 (*TAED* N320219). Document multiply signed and dated; miscellaneous doodles not transcribed. [a]Followed by dividing mark. [b]Canceled. [c]Obscured overwritten text. [d]Marginalia linked by line to brace enclosing previous two paragraphs. [e]"in 62 sugar solution" interlined above. [f]Illegible. [g]Circled. [h]Interlined above. [i]"Including Reg . . . bamboo" interlined above. [j]"a large quantity of" interlined above. [k]"all ~~of~~" interlined above. [l]"please see . . . immediately" added between list and preceding paragraph. [m]"parts by weight" interlined above.

1. This entry is a continuation of Doc. 3039.

2. Figure label is "fusible metal."

3. A Prince Rupert's drop is a bead of glass (or other ceramic) with a long tail, formed by rapidly cooling the molten material in water. The drop maintains high internal tension and will shatter explosively if the surface is damaged or the tail broken. *OED*, s.v. "Prince Rupert's drop."

4. Figure labels are "Thermom" and "metal box in sand bath."

5. Edison meant iodoform, a medicinal compound of iodine similar to chloroform, typically forming scaly yellow crystals. *OED*, s.v. "iodoform."

6. Edison and John Ott had experimented with alternative clamps in late summer and fall 1886, but the editors have not identified the "new clamp" mentioned here. See, e.g., N-86-06-28 [14 Aug. 1886], N-86-08-24 [24 Aug. and 16 Sept. 1886], PN-86-03-04 [3 Oct. 1886]; all Lab. (*TAED* N322AAE, N325AAA, N325AAG, NP021D).

7. Edison likely referred to some sort of rheostat or similar device for varying the current in a circuit.

8. Edison issued several sets of instructions for related experiments to laboratory assistant Frederick Saxelby between mid-May and mid-June. Unbound Notes and Drawings (18 and 28 May, 7 and 10 June 1887), Lab. (*TAED* NS87AAH, NS87AAI, NS87AAM, NS87AAO).

9. Charles D. Deshler (1864?–1943) joined Edison's lamp factory laboratory in 1887, two years after graduating with a scientific degree from Rutgers College. Although Deshler was at first criticized by Edison for an apparent lack of initiative, he spent his career in Edison's laboratories at the lamp factory and in Orange. He received the Charles A. Coffin medal in 1893 for work on photoelectric cells. A native of New Brunswick, N.J., Deshler was the son of a local educator and editor who also was an organizer of the antebellum American ("Know-Nothing") Party in New Jersey. Obituary, *NYT*, 16 Apr. 1943, 22; U.S. Census Bu-

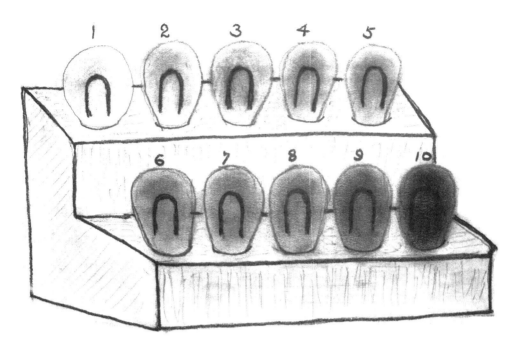

William Hammer's undated drawing of bulbs arranged by gradations of darkness, much like Edison's request for a visual reference.

reau 1970 (1880), roll T9_ 790, p. 147C, image 0135 (New Brunswick, Middlesex, N.J.); Upson 1885, 115; Alfred Tate to Edison Lamp Co., 20 Apr. 1888, DF (*TAED* D8815ABC); Richardson 2004, 261; Hallerberg 1939.

10. William Hammer, who subsequently organized a comprehensive exhibition of Edison lamps (shown most famously at the 1904 World's Fair in St. Louis), sketched at some indeterminate date a display of ten bulbs mounted for display in order of increasing darkness. Hammer labeled his drawing "Standards of Blackening of Edison Incandescent Lamps 1880–1881   Edison's Menlo Park N.J. laboratory and at 1st Lamp Factory at Menlo Park," and he represented it as "The origins of the standard designed & used by me at Edison's Menlo Park Laboratory in 1880." At the time he wrote the caption, Hammer stated that the actual display was in the possession of General Electric, which was formed in 1892. The editors have not determined what, if any, relation the display sketched by Hammer may have to that requested by Edison in this document. Series 2, Box 24, WJH (*TAED* X098HC04).

11. Figure label is "2 inches."

12. Mina Edison transcribed the remainder of the document (from "Hamilton") into a different notebook. Her version, dated 7 May, followed Edison's corrected text except for minor variations of paragraphing and spelling, and it included several additional numbered experiments (see note 13). Miscellaneous Notebook (1887): 3, Lab. (*TAED* NM030003).

13. Mina's 7 May transcription continued with "Additional experiments with tubes—." After a brief introductory paragraph, she extended the list through experiment number thirty-two. Miscellaneous notebook (1887): 15–17, Lab. (*TAED* NM030003, images 7–8).

Technical Note:
Phonograph[1]

Phonograph

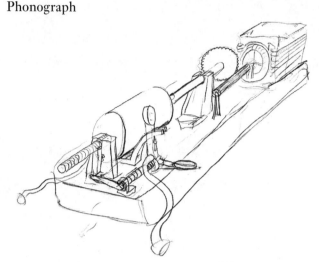

C[harles] B[atchelor]

X, NjWOE, Unbound Notes and Drawings (*TAED* NS87AAC). [a]Date
written by Charles Batchelor.

1. This drawing on looseleaf paper appears to be by Edison, with
Charles Batchelor having dated and signed it as a witness. It coincides
nearly exactly with Edison's resumption of sustained work on the phonograph after a lull since October 1886 (see Doc. 2993). But the date
is suspect. Batchelor recorded in his journal that he was in Schenectady on 7 May, when he "Spent the day <u>at shop</u>" conferring with John
Kruesi and inspecting a building under construction at the Edison Machine Works. On 8 May, however, he "Saw Bell's Graphophone today at
Parlor 2 St James Hotel" in New York City, where the instrument developed by Charles Tainter, Alexander Graham Bell, and Chichester Bell
was exhibited by James Clephane (an Edison acquaintance and former
electric pen agent). Batchelor sketched and described their machine in
some detail; it was powered by a foot treadle, and the mechanism for
tracking the recording and playback styli through a wax cylinder's spiral
groove differed fundamentally from that shown in this document (Cat.
1336:197–98 [items 354–55, 7–8 May 1887], both Batchelor [*TAED*
MBJ003197A, MBJ003197B]; Doc. 639 n. 11; *TAED*, s.v. "Clephane,
James"). It is clear from Batchelor's journal and a newspaper account
of a similar demonstration given the same week by Tainter that the Bell
group aimed to market their machine as a tool for business correspondence, even claiming (as Tainter reportedly did) that a recorded cylinder could be mailed in lieu of a written letter ("A Stenographer Turned
with a Crank," *Newport [R.I.] Mercury*, 7 May 1887, 6). The New York
demonstration, more than this particular document, may mark the start
of Edison's renewed interest in the phonograph, and as he began work,
he did so with prospective business users fixed in his mind (see Doc.
3093 n. 1).

[New York, May 9, 1887?[1]]

<u>557</u>.[2] Municipal lamp.

Jenks, Steiringer[3] and I talked with Edison on the municipal lamp.[4] What will prevent the 'arc' from running down the wires. Edison suggested a number of ways of preventing it amongst which were the following:—[5]

Differential Cut out— two wires run up one conductor insulated except at the top where they are connected by safety fuse— When fuse goes or either wire is burned the magnet will close circuit round the lamp.

This is purely mechanical, if either side wire arcs off the lever will close the circuit by the spring pulling it down.[6]

AD, NjWOE, Batchelor, Cat. 1336:199 (*TAED* MBJ003199).

1. Charles Batchelor wrote this entry after one dated 9 May; the next dated entry is 13 May.

2. Batchelor consecutively numbered each entry in this journal.

3. Luther Stieringer (1845?–1903), a former gas engineer, started what would become a distinguished career as an illuminating engineer in 1881 by designing interior lighting for Edison. In particular, he invented and patented an insulated joint to prevent electrical contact between lighting fixtures and the grounded supports (often gas lines) to which they were attached, and this type of insulated fixture quickly became standard in insurance requirements. In 1886, he supervised construction of a large illuminated fountain on Staten Island. In the middle of that year, he was also connected with the Edison United Manufacturing Co.; by February 1887 he was identified as an "expert" of the Edison Electric Light Co. Doc. 2298 n. 11; U.S. Census Bureau 1963? (1850), roll M432_448, p. 310A, image 14 (Newark West Ward, Essex, N.J.); Obituary, *Chicago Daily Tribune*, 19 July 1903, 4; Dyer 1897, 411–12; "Next to the 'Wizard' Stands Stieringer," *Los Angeles Times*, 21 Dec. 1902, C10; "The Great Illuminated Fountain on Staten Island," *Electrical World* 7 (16 June 1886): 292; Association of Edison Illuminating Cos. 1887, 35, 48.

4. Edison testified in an 1888 patent interference case that he took up work on the municipal lamp about this time at the request of William Jenks because the Edison Electric Light Co. wanted "a more perfect plan of cutting out lamps for the municipal system, as the plans they

had were not satisfactory." Jenks gave a similar account. Edison's testimony, p. 12; Jenks's testimony, p. 72; both *Edison v. Thomson (TAED* W100DLA).

5. Figure label is "Safety catch."

6. In early June, Batchelor noted that he, Jenks, and Stieringer had "tried a lot of lamps of different styles," all apparently intended to address this same possibility of internal arcing. He sketched four basic designs, a few of which showed promise. The design that seemed to work most consistently unfortunately also "made a great deal of flame inside" before a spring could break the circuit, "which would be objectionable only if used in the house." Edison received at least one patent based on cutout designs from this period. Cat. 1337:225 (item 377, 2 June 1887), Batchelor (*TAED* MBJ003225); U.S. Pat. 476,530.

–3044–

Draft Patent
Application: Wrought
Iron Manufacture

[Harrison,] May 14 1887

⟨719⟩   Patent[1]

The object of this invention is a new process ~~for~~[a] to Make wrought iron direct from Molten Cast iron.

The invention consists in running the molten metal into pigs [–][b] in the sand mould or other receptable and the Ends of which are two iron rods connected to a powerful dynamo Electric machine of sufficient Ampere Capacity to Cause[c] the iron ~~to remain liquid &~~ to boil & remain liquid for such as time as the carbon is nearly ~~burnt~~ or all burnt out, ~~As the melting point is gradually raised as the Carbon is burned~~ [with?][b] ~~the iron rods The rods are~~ [p--][b] The Current is taken off the rods pulled & the liquid allowed to Cool. The ends of the bar are slightly larger than the bar itself to prevent the contact rods from melting to improve the grain & quality I cause the mould to be placed in a powerful magnetic field so that when it is cooling the particles of iron shall be given a directive force, and thus give[c] a grain & toughness to the iron

I am aware that Electric Currents have been passed through molten[c] iron, and I am also aware that molten iron has been allowed to cool in a magnetic field,[2] but I claim as new the use of a sufficiently powerful Current to maintain such iron[c] in a molten state for an indefinite time, and also the use of a sufficiently powerful magnetic field as to Cause a directive force to be given the ~~iron~~particles of iron when nearly set. ~~The~~ No[c] Effect will be produced with the small & [--][b] magnetos hitherto used as iron at the moment of hardening from a molten state has very little magnetic capacity hence the[d] Exceedingly powerful ~~fiel~~ magnets of the large modern Dynamo machine will alone produce any effect.

Claim. the passing of Current to keep in liquid state.
Claim powerful field etc
Claim both in Combination[3]

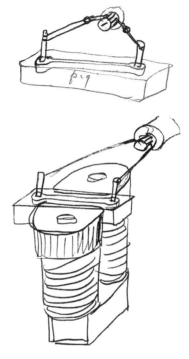

TAE

ADfS, NjWOE, Lab., Cat. 1152 (*TAED* NM021AAA). [a]"a new process for" interlined above. [b]Canceled. [c]Obscured overwritten text. [d]Interlined above.

1. This draft became the basis for a patent application that Edison signed on 24 May and filed on 4 June. The application had nine claims and made reference to four figures (not found). In November, the Patent Office gave notice that, under a longstanding policy, it would not act until the application was divided into separate specifications, one dealing with apparatus and another with processes. Edison did not divide the application but amended it in stages to meet the examiner's objections. The examiner then rejected the revised claims (in April 1890) on several grounds: first, anticipation by a number of earlier patents; second, deviation from the text of the original application; and third, that the claimed process for oxidizing the carbon from molten metal could not work according to the scant description provided. Edison appealed to the Board of Examiners, which heard the case in December 1890. The board issued a mixed verdict, upholding the examiner on most points, notably the precedent of an 1872 British patent for the purification of iron by heat, which was held to cover Edison's claims for the oxidation of carbon. The board did allow three of the revised claims to stand, but Edison effectively abandoned the application (in December 1892) by instructing his attorneys neither to pursue an appeal nor have the al-

lowed claims issued in a patent. Patent Application Files, Case 719, PS (*TAED* PT032AAC).

2. About 6 April, Charles Batchelor pasted in his journal (under the heading "Cast wrought iron") a few printed lines he ascribed to the *New Orleans Times Democrat* of 29 March: "Preliminary tests have shown that iron cooled while a strong current of electricity was passing through it was increased fully one-half in tensile strength and ductility." Cat. 1336:183 (item 327, [6?] Apr. 1887), Batchelor (*TAED* MBJ003183B).

3. Figure label is "pig."

–3045–

Charles Batchelor
Journal Entry

East Newark, May 16th 1887.

<u>363.</u>[1] Phonograph.

Making a phonograph that will be able to lift out cylinder as well as stop and start at will etc   Edison proposes a machine that will cut a groove ahead of the talking needle and the indentation is made at bottom of that.[2] This is reproduced by a point on a hair Spring which just lays in & has a guide to guide it by the wide portion

AD, NjWOE, Batchelor, Cat. 1336:205 (*TAED* MBJ003205).

1. Charles Batchelor consecutively numbered each entry in this journal.

2. On 9 May Batchelor noted they were "Making cylinders of plumbago mixtures, and steatite— The record on this is made by a point and the dust falls away leaving no shaving." Cat. 1336:201 (item 358; 9 May 1887), Batchelor (*TAED* MBJ003201).

–3046–

Technical Note:
Electric Lighting

[Harrison,] May 16 1887

John Ott—

Municipal   About every lamp in 20 doesnt arc when the Carbon breaks—in this case The circuit would remain open— The right of our Co to[a] use of the paper cut out[1] is doubtful therefore I want to get a Cut out (1000 volts, with plates[b] one millimetre apart & one square decimetre in area each produces an electrical attraction equal to 5.7 gramms weight) 1 decimeter is I think about 125 square inches or plate 11 inches square or 45 milligramms to square inch— now if you can put one square inch of surface opposing another of equal area both on delicate springs The Current will cause them to draw together ~~with~~ [f----][c] just the same as if you laid 45 milligramms on one of the plates, I think this is plenty— I propose to make an inside part thus[2]

[A]³

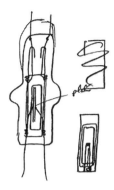

TAE

X, NjWOE, Lab., Unbound Notes and Drawings (1887) (*TAED* NS87AAF). ᵃ"right of our Co to" interlined above. ᵇObscured overwritten text. ᶜCanceled.

1. The editors have identified neither the "paper cut out" nor the competing legal claims to its use.

2. Edison described and explained this device in a patent application he executed on 1 June. It consisted of

> two metal plates placed very close together, but not touching each other, in the lamp-stem, and each connected with one of the wires therein. These plates thus become charged with electricity; but in the normal operation of the lamp the difference of potential between the plates is not sufficient to cause such an attraction between them as to make them approach each other. When, however, the circuit through the lamp is opened by the breaking of the filament, the rise in potential upon the wires causes the plates to be attracted together, so as to come in contact and short-circuit the lamp. [U.S. Pat. 476,530]

3. Figure label is "plates." On the obverse of the first page of numbered notebook paper he used for this note, Edison made two incomplete sketches related to this drawing.

–3047–

*Draft Patent Application: Pyromagnetic Generator*¹

[Orange?]² May 1̶7̶8 1887

Important & Immediateᵃ ⟨721⟩³ᵃ

fig 1⁴

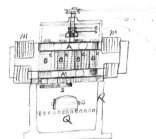

fig 2⁵

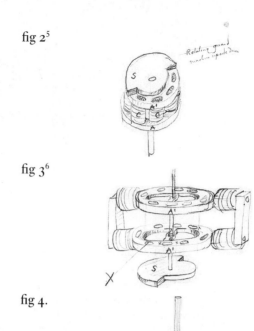

fig 3⁶

fig 4.

nickel plated or enameled iron tubes draw from strips through
a die— 006 thick diameter tubes ⅛ outside measurement.

fig 5⁷

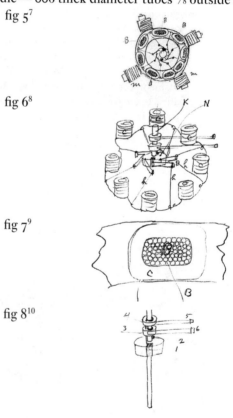

fig 6⁸

fig 7⁹

fig 8¹⁰

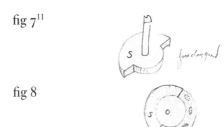

fig 7[11]

fig 8

Dyer—

This is the specification of the Dynamo ~~work~~ I spoke to you about. The Current being derived directly from the heat through its action on magnetism— The f proper name for the motor would probably be <u>Magnocaloric Engine</u> [~~and if this machine?~~][b] or ~~Magn~~ Magnothermic Engine— I prefer the latter name— The name of the Current generator which I now describe—I call a Magnothermic battery

The object of the invention is to Translate heat energy into Electric Energy by the action of heat upon magnetized iron. ~~so~~

The invention consists in acting upon magnetized iron by ~~heat & col rises & fall of heat~~ an increase & diminution of heat at intervals to diminish and increase the magnetism of the iron, The action of which serve to induce Electric Currents in wire wound around such magnetized iron— The intermittent[c] Currents being properly comutated into a continuous one—

Two iron rings A A′ fig 1 2 & 3 form the north & south poles. Electromagnets or Steel magnets m m m in fig 1 & 5 Magnetize[d] the rings— There may be several such magnets Between these two polar rings are bundles of thin iron tubes B fig 1 fig 5 fig 7— These tubes are ⅛ diameter and are composed of iron .006 thick drawn[d] through a die into a tube with ends abutting. The object of using such thin iron is that it shall gain and lose its heat rapidly The power of the generator being proportionate to this property   The~~se bun~~ several bundles of tubes pass through the holes X fig. 3. of the 2 pole pieces and project at both ends. They are packed tightly in these holes by asbestos or other infusible material. The whole of the iron of the tubes are powerfully magnetized by the rings & serve as armatures across the poles— around Each bundle is a bobbin of wire pyroinsulated[c] insulated so that it is Enabled to stand a rather high temperature without changing its insulation. These bobbins are shewn as C[d] in[c] fig 2 fig 1 as C fig 7 fig. 6.

~~In fig 6~~ The whole of the bobbins are connected together electrically after the manner of the Gramme Ring armature. In fig 6 are shewn the commutating devices—as the coils do not rotate it is essential that the equivalent of the usual dynamo brushes must be rotated. There is a disk N[c] on the rotating shaft K   This disk has 2 metallic pieces see fig 8—marked No 1 & 2   Electrical[d] contact with each block is continuously maintained during rotation by the springs 5 & 6 resting on the disks 4 & 3 which are insulated from the shaft but both are in contact with the blocks one to one block & one to the other. There are 8 fixed springs see fig 6 the ends of which bear on the periphery of .N. The blocks coming into contact as they pass around— Each spring is connected to a wire between the coils by an extra wire h[d] as in the gramme ring armature.

Now if the apparatus be arranged as in fig 1 over a furnace R with grate bars Q heat in the form of flame & hot gases would pass upwards and as the only outlet is through the several bundles of tubes B They would be instantly heated, & soon reach a temperature where they would cease to be magnetic. This[d] demagnetization of the iron causes a powerful inducive Current to be thrown into the coils of wire surrounding the tubes. Now to properly collect these Currents into a continuous form it is Essential that while one half of all the tubes and are be-acquiring heat & losing magnetism[e] the other half should be losing heat & acquiring magnetism   to do this I use a rotating guard of fire clay material which serves to close the inlet of ½ of all the bundles of tubes so the flames or hot gases cannot Enter them—while the other ½ are open to the flame— This rotating guard [--][b] S is best shewn in figs 7 8 & in fig 2 the whole apparatus is upside down The guard being shewn more plainly it is also shewn in fig. 3. The shaft being Extended down to bring out the [point?][b] character clearly to the Eye

This guard is rotated by any suitable power & the speed is regulated should be such that with any given fire The maximum electromotive force is obtained— It is obvious that it might be run by an Electromotor receiving Current from the machine itself in this Case it would be self regulating any lessing[d] of the electromotive force due to too great a speed would slow the motor down & thus increase the E.M.F.[a]

The action of the apparatus is very simple—heat enters ½ of all the tubes bundles of tubes. This lessens the magnetism & an inductive Current in one direction is send round the Coils. The other ½ of the Coils are Cooling hence are increasing in magnetism. This sends an inductive Current in the opposite direction—now both Ends Currents meet at one Common point ie[f] the Commutator springs & the result is a multiple arc the 2 Currents Coming together as one as in a gramme ring—

As the guard[d] S is constantly advancing—that[d] Coils which has reached the highest temperature is suddenly [--][b] covered by S[g] instantly it commences to lose heat & acquire magnetism, while a moment before it was ~~losing~~ gaining heat & losing magnetism—at this instant of uncovering & falling of temperature The bobbin of wire is by the action of the Commutator thrown in among[d] the opposite ½ of bobbins—while at the same time [~~anoth?~~][b] a coil from this ½ is ~~covered~~ <u>uncovered</u> & instantly the Current changes sign in its bobbin as it now rises in temperature. Thus by the direction action of heat on iron powerful electric Currents are generated—without the use of Engine & boilers—

The field of force magnets may be permanent steel magnets or electromagnets and be energized by the apparatus[d] itself or by an extraneous source of Electricity, ~~an~~ slight[c] initial magnetism must be given the field magnets to permit them to <u>build up</u>—

~~It is obvious that the~~ As the generative Capacity of the apparatus depends on the diferences of temperature & the rapidity with which the bundles of iron tubes lose their heat it is obvious that The inlet air for the furnace may be made to pass through those Bundles[d] which are guarded [~~f~~][b] by the guard S for the time being & thus rapidly cooled & this heated air ~~with~~ will material assist in raising the temperature of the furnace gases & thus ensure rapidity ~~both by co~~ & at the same time waste heat is utilized.

Claim: The electric generator sub[stantially] as described & other as you may suggest[12]   patent England France & Germany

ADf, NjWOE, PS (*TAED* PT032AAE1). [a]Followed by dividing mark. [b]Canceled. [c]Interlined above. [d]Obscured overwritten text. [e]"& losing magnetism" interlined above. [f]Circled. [g]"by S" interlined above.

1. See Doc. 3029 (headnote).

2. Having also not ascertained Edison's exact whereabouts, the editors have conjectured "Orange," using his preferred term for the township of West Orange (also encompassing Llewellyn Park) served by the Orange post office.

3. This is Edison's patent application case number.

4. Edison drew figures marked 1 through 6 and two apiece marked 7 and 8 on seven numbered notebook pages (pp. 133–45) preceding his text (pp. 147–51, 179–203). For simplicity, the editors have arranged the drawings in numerical order (and, in the case of duplicate numbers, by their presentation in the text). Figure labels are "m," "A," "B," "C," "B," "C," "B," "C," "B," "C," "A'," "S," "Q," and "R."

5. Figure labels are "Rotating guard   machine upside down," "S," "A′," "C," "C," and "A."

6. Figure labels are "A′," "X," "X," "X," "A′," and "S."

7. Figure labels are "m," "B," "B," "B," "m," and "B."

8. Figure labels are "C," "C," "K," "N," "C," "C," "h," "h," and "h."

9. Figure labels are "C" and "B."

10. Figure labels are "4," "3," "5," "6," "2," and "1."

11. Figure labels are "S" and "fire clay guard."

12. Edison executed the patent application derived from this draft on 24 May, and it was filed on 13 June. The Patent Office rejected several of the thirteen original claims in September on the ground of their having been anticipated by three publications (Gore 1868, Gore 1869, and Gore 1873), and the examiner urged Edison to amend the application promptly in preparation for an interference proceeding. Edison's attorneys responded on 26 October by replacing eleven claims entirely and modifying others by changing the phrase "bundles of iron tubes" to "interstitial bodies of magnetic material," the same term used in his pyromagnetic motor application (Pat. App. 380,100). They also replaced the term "electro-thermic battery" with "pyromagnetic generator." The Patent Office notified Edison on 31 October that the amended application was allowable, pending resolution of an interference declared in the meantime with an application filed on 31 August 1887 by Emile Berliner for an "electric furnace generator" based on the changing magnetic properties of iron subjected to heating and cooling. The interference proceeding took place sometime after 18 October 1889. The editors have no other information about the case, but Edison's amended application issued on 14 June 1892 as U.S. Patent 476,983. Berliner's application was divided in May 1891 and resulted in a single specification, U.S. Patent 481,999. A. K. Smith to TAE, 17 Sept. and 31 Oct. 1887; Dyer and Seely to Commissioner of Patents, 26 Oct. 1887 and 18 Oct. 1889; Commissioner of Patents to TAE, 29 Oct. 1889; all U.S. Pat. App. 476,983.

–3048–

*Draft Patent
Application:
Pyromagnetic Motor*[1]

⟨Patent   720⟩[3a]

fig 1[4]

[Orange? c. May 18, 1887][2]

fig 2[5]

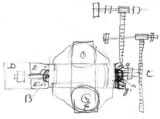

fig 3

The object of this invention is to obtain Motive power from heat.

The invention consists in causing heat to act upon a rotating cylinder formed of iron which is magnetized ~~by the electric current~~ and rotates [--][b] in a magnetic field, in such a manner as to produce a disymetry of the lines of force   The heat reducing or destroying (in (a certain portion of the cylinder) The magnetism for the time being. Advantage of the known fact that ~~iron~~ the capacity of iron for magnetism diminishes by heat & practically disappears at a bright red heat.

Mounted on a shaft is a bundle of thin iron tubes, nickel plated or covered with a thin coating of Enamel to prevent oxidation. The tubes are made thin so as to [~~acquire~~ ---- -- &][b] gain & lose heat quickly   They are placed longitudinally around the main shaft, and form a cylinder ~~The outside being~~ the whole being secured by bands & plates to the cylinder—

~~The~~At both Ends the Cylinder[c] of tubes are open except at the midle where there are at each end two fixed plates. These plates do not rotate with the shaft but are fixed and their inner face towards the cylinder just rest & rub against the ends of the tubes. At the Ends B[c] & C the plate is so arranged that it Covers all the central portion of the tubes as in fig 3.

D is a furnace or fire box. the flame or hot air passing up through the flues E E′   the ends of these flue just touch the ends of the tubes & cover the portion of the tubes not covered by the plates B. & .C.   at the opposite end are also two flues and the flame or hot gases pass from them to the chimney or to the furnace of another similar[d] motor.

The cold air inlet is at the plate .C. The air enters and passes through all the tubes covered[c] by the plates B & C but not by the flues E E′ g g′   when the air reaches the plate B it passes down through a flue to the furnace. This air becomes heated in its passage & serves ~~two~~three purposes   first to reduce the temperature of the iron tubes & thus increase their capacity for magnetism & 2ndly to ~~increase the~~ utilize the waste heat [-][b] & 3rdly to quicken the alternate heating & cooling of the iron & consequently increase the speed & power of the motor

fig 1 shews the lines of force when the two sides of the cylinder is heated

The armature is magnetized by a fixed coil n fig 1 wound and shewn better in fig 4

This coil is formed of ~~bars~~ rods[d] insulated from each other by porcelain or other infusible washers containing such a number of holes as there are convolutions. This coil & the wire on the field magnet are connected in series.

I preferably use a very few turns of large copper wire on both field & fixed coil & energizing the same by a Current of many amperes rather than use a great number of turns & small number of amperes   by this means I am enabled to insulate the whole system by infusible washers or material

The ~~extra~~ fixed coil may be dispensed with and the inlet & outlet flues arranged as in fig 5= ~~in this case the~~ but the motor is not so powerful[6]

The field of force may be a permanent[c] magnet for small motors but for large ones I prefer to use Electromagnets. The fields & fixed coil are to be energized by an auxilliary Dynamo, worked by the motor itself The initial start of this Dynamo being given by hand.

Claim— A prime motor substantially as descried.

Dyer   isnt[c] the above claim a good one in law & isnt it the broadest kind of a claim for Court Construction should[c] anyone use this principle[7]   if not make ~~lot~~ usual claims—

This principle is[c] absolutely new[8]   the power is not very great but I hope to do something with it   patent in US France & Germany

T A Edison

ADfS, NjWOE, PS (*TAED* PT032AAD). Document multiply signed. [a]Followed by dividing mark. [b]Canceled. [c]Obscured overwritten text. [d]Interlined above.

1. See Doc. 3029 (headnote).

2. Although Edison wrote a date of 24 May on this application at the time it was assigned a case number, he executed the completed application that very day. It would have taken a bare minimum of several days for his attorneys to prepare the application and drawings. The editors believe that Edison most likely wrote this draft at about the same time (and in the same location) he drafted the related application for the pyromagnetic generator (Doc. 3047). Both applications were signed on 24 May and filed at the Patent Office on 13 June.

3. This is Edison's patent application case number. The application issued in March 1888 as U.S. Patent 380,100.

4. Edison drew figures 1 through 3 and the unnumbered canceled

sketch together on a single page. The completed sketches in this draft were the basis for the five figures in the final patent. Figure labels on the first drawing are "heat," "B," "heat," "n," and "f."

5. Figure labels are "D," "E," "E′," "B," "f," "g," "g′," "a," and "C."

6. Figure labels are "heat" and "heat."

7. This broad claim was not used in the application. Pat. App. 380,100.

8. Although the principle itself was old, as Edison himself noted in his AAAS paper (Edison 1888b), he believed himself to be the first to apply it practically. However, at this time Nikola Tesla had already filed his own patent application for a motor using the principle (see Doc. 3029 [headnote, n. 4]), and in August, Emile Berliner would apply for a generator patent based on it (see Doc. 3047 n. 12).

–3049–

Charles Batchelor Journal Entry

[New York,] May 19th 1887

367[1]   Laboratory (New)

Edison, I, Gilliland and Insul discussed plans at Holly's office and decided that they were right and that they could complete them.[2] Edison went to Schenectady—[3]

AD, NjWOE, Batchelor, Cat. 1336:209 (*TAED* MBJ003209A).

1. Charles Batchelor consecutively numbered each entry in this journal.

2. Holly laid out detailed masonry plans for the new laboratory on 28 May and reviewed bids for the work over the next several weeks. Edison contracted on 1 July to have the laboratory and library built according to Holly's plans by mason John B. Everett and carpenter P. B. Fairchild & Co., both of Orange. The terms were written incorrectly into the contracts and subsequently amended; Edison was to pay $17,800 to Everett in five installments and $18,609 to Fairchild in six installments, with all work except the library to be completed by 10 October (TAE agreement with Holly, 28 May 1887; TAE agreement with Fairchild & Co., 1 July 1887; ECB [*TAED* X184A1, X184A2]; Holly to Batchelor, 23 June 1887; TAE agreement with Everett, 1 July 1887; C. W. Smith to Batchelor, 1 July 1887; all DF [*TAED* D8755AAE, D8755AAK, D8755AAL]). Four outbuildings were planned and contracted for separately (see Doc. 3067 n. 2).

3. Edison planned to remain in Schenectady until 21 May. Cat. 1336:209 (item 368, 20 May 1887), Batchelor (*TAED* MBJ003209B).

–3050–

To John Vail

Harrison, N.J., 5/20 1887[a]

Vail—[1]

I made 3 lots new lamps about 800 Each came out   The volts of course vary between 98 @nd[2] 110—hence we send you all the 99 & 100 volt lamps we can get in each lot. I think you will perhaps get 250 @ 275 out of the whole—you can set them up at 99½ or 100 volts perhaps 100 Volts would be best=[3] I dont understand How you get 100 volt lamps from Pearl street I thought they used 105 volts— I think you better come right over here go in cellar & take a barrell at Random if you dont get served at Pearl street.— about loss in Build-ings[b]— & mains — I'll you can make an estimate to[b] see if it wouldnt be cheaper to add 33 percent more copper to mains, than to ~~run a~~ two or three Extra feeders. I apprehend that add-ing 33 pct to copper dont increase gross cost of mains laid but very little, the principal cost being Iron tubes boxes laying pav-ing etc. Hence should you lay 2 or 3 more feeders they would call for big outlay in Iron tubes boxes digging etc for instance says mains at 3 pct loss cost 50,000 , feeders, 75 000, now add 33 pct more copper to mains cost probably be $55,000—but if this remains same & 3 feeders are added then feeders would probably cost [t----][c] 85 @ $90 000—[4]

If any losses are to be made let it nearly[d] all be made in feed-ers.

About buildings I think 1 per cent should be greatest loss. The ~~Cost~~ mere cost of Copper in a building must be small compared to labor

What is the use of spending a large sum of money for Cop-per in the mains if the whole thing is going to be nullified by saving of a dollar or 2 in the house wiring

1 pct in House
1½ on mains
10@[e] 12 pct with full load in feeders—

This with[b] old lamp would be with[b] equal number lamps

1½ House
2¼ mains
15 @ 18 on feeders—
Etc Etc

Dont you think that after 500 hours or 22 running days we shall know Enough of new lamp so that we shall be safe in turning over[b] factory to run ~~up stock~~ regular on them, instead of old 10 p[er] hp—[5]

E[dison]

ALS, NjWOE, Vail (*TAED* ME006). Letterhead of Edison Lamp Co. [a]"Harrison, N.J.," and "188" preprinted. [b]Obscured overwritten text. [c]Canceled. [d]Interlined above. [e]"10@" interlined above.

1. By the end of 1886, John Vail had new titles as the general superintendent and chief engineer for the Edison Electric Light Co., with responsibility for the "Department of Engineering and Central Station Construction." Letterhead of Vail to TAE, 28 Dec. 1886, DF (*TAED* D8624O); Edison Electric Light Co. pamphlets, Oct. 1886 (Vail marginalia p. 6) and 1 Feb. 1887, both PPC (*TAED* CA041Q, CA041C1).

2. Edison sometimes used "@" to indicate a numerical range.

3. Vail planned to test a large number of the new high-efficiency Edison lamps against the standard version (possibly at Edward Johnson's request) though he did not have enough power to test as many lamps as Edison had in mind. Edison wrote again to Vail on 1 June (Wednesday) promising to have "125 new Lamps ready for shipment to NYork for test Wednesday morning— It occurs to me you better take from stock on hand over there— 125 Regular 10 per hp lamps without you are willing to trust the Lamp factory" (William Holzer to TAE, 9 Mar. 1887, EP&RI [*TAED* X001A2A]; Vail to TAE, 24 May 1887, DF [*TAED* D8732AAT]; TAE to Vail, 1 June 1887, Vail [*TAED* ME008]). The difficulty of standardizing lamp resistance, which had dogged Edison's manufacturing efforts from the start, persisted with commercial production of the new model. A promotional pamphlet for the new lamps explained the problem:

> It is now fully recognized that the difficulties in obtaining exactness in the manufacture of incandescent lamps—making each lamp the same—are very great, and that only average exactness can be expected. For example: the resistances of the carbon will vary with its cross-section, and a variation which cannot be appreciated by any ordinary gauge, will vary the resistance several ohms. The resistances also depend upon the quality of the carbon, and this will vary slightly from various accidents in the manufacture. ["Catalogue of Lamps Manufactured by the Edison Lamp Company," p. 5, n.d. [1887?], PPC (*TAED* CA041J)]

The company's published tables stated the "ideal" resistance of each type of lamp manufactured for voltages between 88 and 114. The "new" lamps made in 1887 generally had about thirty to forty percent more resistance than older styles of the same candlepower. As confusing as the incomplete standardization may have been for sales agents and customers, the range of voltages permitted a wide range of candlepower in practice by burning lamps above or below their rated voltage. The company presented tables showing the calculated effects on light intensity, power consumption, and lifetime for intervals above and below the voltage rating of each lamp (pp. 6–18).

4. Edison undoubtedly had in mind planning for the uptown New York second district and specifically a detailed cost estimate for the feeders and mains prepared recently by the Edison Machine Works. The proposal for some 90,000 feet of conductors came to $163,265.25; after a discount of about twenty-five percent, the projected cost was $126,198.94, an average rate of almost $1.50 per foot. Construction was well underway by mid-August, when more than a mile of feeders and nearly half a mile of mains had been installed, mostly along Broadway.

Edison Machine Works estimate, 12 May 1887; Vail to TAE, 15 Aug. 1887; both DF (*TAED* D8733AAF, D8733AAM).

5. Calculations by the editors based on the tables published for sales agents indicate that the new 16-candlepower lamp used about 49 watts of power (volts × amps), equivalent to a rate of a little over fifteen lamps per horsepower; the old version used 72 watts, or about ten lamps per horsepower. The new 50-candlepower lamps used about 155 watts, yielding about 4.8 per horsepower; the old ones consumed 210 watts, or about 3.5 lamps per horsepower. "Catalogue of Lamps Manufactured by the Edison Lamp Company," pp. 7–8, n.d. [1887?], PPC (*TAED* CA041J).

–3051–

*Draft Patent
Application:
Phonograph*

[Harrison,] May 21 1887

Dyer—

Patent— Improvement in phonographs[1]

The object of the invention is to improve the articulation of the phonograph—

The invention consists of a cylinder of wax [~~coated?~~][a] or other yielding material such as moulded Kaloin, Camphor Napthalin coated[b] with tin foil or films of gelatin or other smooth Extensible materials, so that the surface will be very yielding & yet none of the material of the cylinder should be cut away—[2]

The invention further consists in causing the reproducing point connected directly to the diaphram to be given motion not by riding in & out of the indentations by by Electric ~~Ele~~ attraction between The point and the Electrified cylinder or point as the case may be. The point does not come in contact with the moving surface but nevetheless ~~The~~ it will ~~fo~~ be given a motion corresponding to Pitch & amplitude of the indentations.[3]

The Electrification of the point or tin foil surface is accomplished by a small disk of hard rubber or glass provided with ~~a~~ rubber & collector— ~~or~~ The disk being turned simultaneously with the phonograph cylinder— ~~instead~~ instead ~~The Mouse Mill Electrifier of Varley[4] may be used instead with commutator~~ Any other Electrifying device may be used ~~Motion~~

The cylinder of wax or other material is connected to the Earth in the usual method in static electric apparatus.

The smooth surface of the cylinder is grooved by a grooving tool like a chaser   This is in advance of the recording point ~~on the~~ [-----][a]   The recording point is of chisel shap & indents the apex of the ridge left by the Threading tool— Thus ~~all~~ the record extends outward beyond the general surface of the cylinder & is very susceptible to attraction for the point on the receiving diaphragm

claim reproducing from[c] ~~articulate speech & other sounds from~~ [-----][a] ~~records~~ phonographic records by electrification of the ~~cyl~~ record or reproducer or both

2nd   Reproducing without contact.[d]

3 The Chisel point—

4— Cylinder of[e] Wax ~~Cyl~~ or other yielding material coated with an Extensible material

5   forming a thread or screw [-----][b] & recording on the apex of the ridge between

6— 6— ~~yielding~~ cylinder of yielding material[f] covered with metallic ~~tin~~ foil—

Etc—

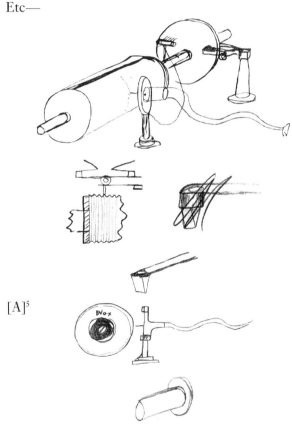

[A][5]

Brass sleve on which wax is cast [---][a] or moulded as a cylinder.

TAE

ADfS, NjWOE, Lab., Cat. 1152 (1887) (*TAED* NM021AAB). Letterhead of Edison Machine Works. [a]Canceled. [b]Obscured overwritten text. [c]Interlined above. [d]Followed by dividing line. [e]"Cylinder of" interlined above. [f]"of yielding material" interlined above.

1. The editors have not found a finished application from this draft, and Edison did not receive a patent based on it.

2. On 17 May, "after trying a large number of compounds of Parafin, Resin, Asphalts, [dies?] etc etc," Batchelor observed that "the best mixture so far for an indentation or even for cutting or scraping is 17 parts of resin and 3 of parafin— We have found that a wax like parafin with tinfoil over it is also very good as it leaves a cushion behind." Cat. 1336:207 (item 365, 23 May 1887), Batchelor (*TAED* MBJ003207A).

3. It was not until 23 May that Batchelor made "a phonograph for Edison in which the receiving needle never touches the surface of the record but is itself electrified and is attracted to the surface more or less as the record is indented more or less." Over the course of the next few days, Batchelor tested at least a dozen soft materials, including silk, hairs, rubber, skins, and hides, for the "phonograph receiver point" and concluded that sealskin hair performed best in producing "a point that gives absolutely no scratching noise." He noted his intention "to get thinner foil and better transmitting diaphragm and point; and much more delicate receiving diaphragms." Cat. 1336:211–15 (item 370, 23–27 May 1887), Batchelor (*TAED* MBJ003211B).

4. The "influence machine," or static charge generator, patented by English electrician Samuel Alfred Varley (1832–1921) in 1860 was the basis for the mouse mill, a combination of charge generator and electromotor adapted for submarine telegraphic printing by several engineers, notably William Thomson. *Oxford DNB*, s.v. "Varley, (Samuel) Alfred"; Prescott 1892, 2:1166–67; Gray 1890, 91–94, 186–87; Thompson 1888b, 588–90.

5. Figure label is "Wax."

–3052–

To S. F. Myers & Co.[1]

[New York,] May 26th. 1887.

My Dear Sir:—

I have your letter of the 18th instant.[2]

I appreciate very highly your kind acknowledgments of my labors, and would be glad if it were in my power to grant the request that you make of me in regard to naming one of your movements the "Edison Watch".

During the past few years a great many [--------][a] articles have been placed upon the market, and my name attached to them by unauthorized parties.[3] In suppressing such actions as these I have always taken the stand that my name is used only in connection with my own inventions. It is necessary that I should maintain this position owing to the number of people who are interested in inventions which I have placed upon the market.

It is with much regret that I am obliged to refuse your request, as I appreciate the feelings which prompted you to make it. It would, however, be inconsistent with the position

which I have always assumed where my name has been connected with inventions other than my own.

Again thanking you for this honor which you have paid me,[b]

TL (carbon copy, fragment), NjWOE, Lbk. 24:463A (*TAED* LB024463A). [a]Canceled. [b]Bottom of page not copied.

1. Founded in 1863, S. F. Myers & Co. was a wholesale jewelry manufacturer and importer in New York City. The firm dealt in a range of
valuables, including watches and watch components. S. F. Myers & Co.
catalogue, Jan. 1885 (no. 22).

2. In a brief typed note, the company sought permission to name one
of its watch movements in honor of "the greatest inventor of modern
times." S. F. Myers & Co. to TAE, 18 May 1887, DF (*TAED* D8716A).

3. For example, on the same day, Edison directed attorney John
Tomlinson to take action against the St. Louis promoters of an "Edison
fire extinguisher," which he had already disavowed (TAE to Tomlinson,
26 May and 14 June 1887, both Lbk. 24:469, 496A [*TAED* LB024469,
LB024496A]). The archival folder "Edison, T. A.—Name Use," DF
(*TAED* D8716) at NjWOE contains several dozen items related to requests for or unauthorized use of Edison's name.

–3053–

To Edward Johnson

HARRISON, N.J., 5-27 1887[a]

E.H.J.

when $^{65}/_{100}$ of all the Lamps die—the ~~averag~~ total hours run
to reach this point will be the average life of Lamps   when the
135th Lamp goes that will be the life in your set[1b]— I should
say that 900 hours will be reached   The Lamps now made
will reach 2000 hours under same Conditions & come down
less in Candle power   I think there is a point about Lamps
not understood Except but by few   Let me record it strongly
on your brain.

I set up a lamp at 16 candles— In 100 hours I test it it is 15
Candles now this loss of a candle with Same Volts cannot be
due to blackening, there is none— I put Lamp in Wheatstone[2]
& find it has changed its resistance so that although there is
same volts there isnt same current[3c]   at End of 200 hours I
find it is[d] down to 13 candles, on measuring I find it has farther
changed its resistance & is slightly blackened. Now I want to
understand how much is due to change of resistance which is
not a loss[c] to the station[c] or to the[c] customer and that due to
blackening which is a real loss to both   I put another fresh 16
cp lamp in circuit & add resistance until there is the same in as[c]
the old Lamp has changed. on reading I Find the fresh Lamp
only comes down to 13½ instead of 13—hence ½ candle is
lost to mankind by blackening— this loss of Candle power by

change of resistance[f] is not <u>costly</u> but only inconvenient & the blackening unsightly— I'm going for these point & have advanced it enormously[e] in the lamp you have even now— if to get 15 per hp you would have to run our regular 10 up to 20 candles ie[g] double c.p. to increase economy 50 pct & at this c.p. they would be black[e] as your hat in 100 hours—

Do you catch on with your <u>intellectual</u> grippers   Thine

Edison

⟨<u>Vail</u>   This is worth bearing in mind   EHJ⟩[h]

ALS, NjWOE, Vail (*TAED* ME007). [a]"HARRISON, N.J.," and "188" preprinted. [b]"in your set" interlined above. [c]Multiply underlined. [d]Followed by a canceled stacked equation interlined below: 65-65=0. [e]Obscured overwritten text. [f]"by change of resistance" interlined above. [g]Circled. [h]Marginalia written by Edward Johnson.

1. The Edison Lamp Co. had a longstanding practice of sampling the lifetime of its production lamps (see Doc. 2177 [headnote]), but Edison seems to have had in mind the tests that John Vail wished to make on the new high-resistance lamp.

2. That is, the Wheatstone bridge (or balance), an instrument used to determine the resistance of a conductor. When the unknown resistor was placed in one side of a two-branch circuit and the two branches electrically balanced so that no current would flow through a "bridge" connection between them, the known resistance on one side could be used to infer the unknown resistance on the other. Maver 1892, 122–30.

3. Cf. Doc. 3041.

7

June–September 1887

Glenmont, the Edisons' home, was a hub of social activity in the early summer, as Edison and Mina hosted her parents during the first week of June.[1] By this time, Mina was entering the later stages of pregnancy and had been busy fixing up a room for the baby. Her sister Jane, on an extended tour of Europe, was preparing to send a special crib from Paris. Jane expected to return to the U.S. in late June and hoped to be present when the baby arrived.[2] Edison himself may have been feeling poorly during at least part of this time. Jacksonville newspapers reported, perhaps spuriously, that he was still afflicted with a painful abscess near his right ear. But Mina's mother found him to be "as happy and as nice as any one could be." Perhaps in conjunction with the visit of his in-laws, Edison purchased five tickets to an open-air charity entertainment at the home of J. Hood Wright, a principal in the New York banking firm Drexel, Morgan & Company.[3]

Edison had long-standing business relations with Wright, whom he would approach later in the summer for start-up capital to develop a large industrial complex in the Orange Valley, near the site of the new laboratory. The factories Edison hoped to construct would turn out highly marketable inventions from the lab. "I honestly believe I can build up work in 15 or 20 years that will employ 10 to 15 000 persons & yield 500 pct to the poineer stockholders," he told Wright.[4]

The construction and furnishing of the new laboratory in Orange was naturally one of Edison's chief concerns during the summer. His determination to build the world's leading private laboratory may have been prompted in part by the opening of Edward Weston's personal laboratory in Newark.

The journal *Engineering* had written earlier in the year that Weston's facility, which included physical, chemical, and electrical laboratories, a machine shop, a library, and an office, was "probably the most complete in the world."[5] This must have been a bitter pill for the competitive Edison, who considered himself the world's foremost inventor and ranked his rival Weston among the "pirates" who had profited illegitimately from his inventions.[6]

Charles Batchelor, deputized to oversee the construction, reported in late June that "We are buying machinery for Edison's laboratory," even ahead of a final agreement with builder J. B. Everett to begin putting up the structures.[7] The buying spree continued during the first week in July when Edison made a trip with Mina and her sister Mary to Philadelphia and probably on to Baltimore and Washington. Philadelphia was home to dealers in electrical and chemical apparatus; while Edison shopped for new laboratory equipment, Mina seems to have gone sightseeing.[8] By the end of October, the laboratory building and furnishing absorbed nearly $56,000 of Edison's funds.[9]

As the laboratory began to rise from its foundations, Edison became increasingly concerned about the quality of the work and what he considered the poor supervision by architect Henry Hudson Holly. At the end of July, he fired Holly and replaced him with the New York architect Joseph J. Taft. By the first week in August, Batchelor was able to report that "The brickwork is done up to the windows of the first floor," and Edison added four small brick outbuildings to the project.[10] But by September, Edison was once again expressing dissatisfaction. "Instead of improving the quality of the work, it is growing worse all the time," Taft complained to Everett, "so much so that Mr Edison is thoroughly disgusted with all the work that you have done, and never will be satisfied with the building."[11]

Edison was also spending heavily to redecorate Glenmont beyond what was already being done to fit up the nursery. In early July, he purchased a long list of items from the New York interior design company Pottier & Stymus, which included Spanish tapestries, paneling, new curtains, fine furniture, and a number of electroliers.[12] His plans included furnishing Glenmont with four hundred electric lights, the power to be drawn from the laboratory. Edison also agreed to provide electric lighting for a number of his Llewellyn Park neighbors, at cost.[13] Other domestic expenses included $300 to his father and $200 to his daughter Marion. He also wrote two $500 checks to Mina, at least one of which, in July, she donated to the Ferry Methodist Church in Orange.[14]

These combined expenditures began to strain Edison's resources and the subordinates responsible for administering them. The first week in July, Alfred Tate, who was now serving as Edison's personal secretary, informed his predecessor Samuel Insull that Edison could raise $10,000 to make the first two payments for the construction of the laboratory, but thereafter was likely to seek the repayment of a loan to the Edison Machine Works of some $15,000.[15] Insull, now in Schenectady managing the Machine Works, had previously kept a close eye on Edison's expenditures, but now reflected that "Mr. Edison does not seem to have anyone with him who urges him to curtail his expenses on his new laboratory. . . . Heretofore, when I have had to provide money, I have always had something to say about how much should be spent."[16]

Edison also continued to reorganize his businesses. At the end of June, he was seeking financing for a new Edison Ore Milling Company that would support a fresh round of experiments.[17] Then, in early July, the Edison Wiring Co. was incorporated to take over the wire manufacturing business of the Machine Works as well as construction responsibilities for the second district of New York.[18]

Edison continued to experiment amid the press of family and financial responsibilities. He worked on a cut-out design for a municipal lighting system in early June. He also continued experiments with materials for phonograph cylinders, and he began to formulate plans for marketing an improved phonograph.[19] Toward the end of the month, prompted by the recent formation of the Welsbach Incandescent Gaslight Company, he turned his attention to incandescent gas lighting.[20] He also drafted a wide-ranging caveat (Doc. 3059) covering, among other things, transformers, leather tanning, ore separation, and his magnetic bridge. At the beginning of July, Edison filed two patent applications, one for a process to cheaply separate fibers from plants for use in lamp filaments and another for a process to heat metals by electricity.[21] Soon after, he succeeded in improving paraffin paper for mimeographs for the new A. B. Dick Co.[22] In late July and early August, he sent draft patent applications to his attorney Richard Dyer for railway signaling, ore milling, a system of electrical distribution, and a method to indicate electrical pressure.[23]

Mina's pregnancy came to an end sometime after mid-July.

In the latter part of that month, her brothers Lewis and Eddie were staying with the Edisons at Glenmont, and her sister Mary, who had recently returned home to Akron from her visit with Mina, reported to her mother that her sister was "getting fat."[24] Edison and Mina appear to have been preparing to go in August to the Miller home in Akron, from where he and his father-in-law planned to go west on a hunting trip.[25] The editors have found no explicit references at the time to a miscarriage or stillbirth, though family physician Edwin Chadbourne submitted a bill on 30 August for $1,000 for unspecified services. One of Mina's sisters made an oblique reference at the end of September to the loss of the baby. "I hope you did not hurt yourself by working," she wrote, before admonishing: "Now, Mina, you must take good care of yourself for you know we want a little nephew and a nice one so don't do anything that will hurt you in the least."[26]

The loss of the baby may partly explain Edison's comparative absence from the documents during September. The editors have found it difficult to track his daily whereabouts, and many of the documents bear a conjectured dateline. It does seem that he spent a good part of his time at Glenmont, close to both Mina and his laboratory under construction. During the last two weeks of the month, however, he shuttled between the lamp factory lab in Harrison, where he was conducting experimental work, and Orange.[27]

On the afternoon of 23 September, the farcical comedy *The American Claimant,* written by Mark Twain with help from William Dean Howells, opened for a single performance at New York's Lyceum Theatre. The star—humorist and raconteur A. P. Burbank—got in touch with Edison through Edward Johnson to offer free tickets. "Burbank wants you and whoever else will go . . . to see a performance of Twain's new play—in which a man who invents everything is the central figure," Johnson wrote. "He says the Phonograph & Megaphone combined make a sensation. Thinks you can suggest some new ideas. What do you say Will you come in?"[28] Having the world's foremost inventor in the audience for a play about an inventor would no doubt have been a publicity coup for Burbank, but there is no evidence that Edison, though he did attend the theater with some regularity, took advantage of this offer.

1. Mary Valinda Miller to Mina Edison, 9 and 12 June 1887, both FR (*TAED* FI001ABH, FI001ABI).

2. Mary Valinda Miller to Mina Edison, 14 June 1887; Jane Miller to Mina Edison, 19 May 1887, both FR (*TAED* FI001ABJ, FM001AAY); Mary Miller to Mina Edison, 15 July 1887, CEF (*TAED* X018C9AC).

3. *Fort Myers Press*, 16 June 1887, News Clippings (*TAED* X104S013D); Mary Valinda Miller to Mina Edison, 14 June 1887, FR (*TAED* FI001ABJ); TAE to Mary Robinson Wright, 2 June 1887, Lbk. 24:483 (*TAED* LB024483).

4. Doc. 3078; regarding Edison's rationales for building the new laboratory, see Millard 1990 chap. 1 and Carlson 1991b.

5. "The American Society of Mechanical Engineers," *Engineering* 43 (14 Jan. 1887): 27.

6. Doc. 2479.

7. Charles Batchelor to John Randolph, 27 June 1887; TAE agreement with J. B. Everett, 1 July 1887; both DF (*TAED* D8755AAI, D8755AAK).

8. Edison's prospective itineraries appear in N-87-00-00.3:1–2 (*TAED* NA003). "Mr. Edison's Experiments," *NYT*, 6 July 1887, 4; Mary Valinda Miller to Mina Edison, 3 and 10 July 1887, both FR (*TAED* FI001ABL1, FI001ABL2).

9. Ledger #5:598, Accts. (*TAED* AB003, image 296).

10. Cat. 1336:253 (item 422; 4 Aug. 1887), Batchelor (*TAED* MBJ003252); see Docs. 3075, 3077, and 3080.

11. Taft to Everett, 2 Sept. 1887, DF (*TAED* D8755ABB).

12. Pottier & Stymus to TAE, 5 July 1887, DF (*TAED* D8746AAE).

13. See Doc. 3056.

14. TAE checks to Samuel Edison, both 3 July 1887; TAE check to Marion Edison, 19 July 1887, TAE check to Mina Edison, 9 and 29 June 1887; all drawn on People's Bank (New York); Bills & Receipts [not selected], D-87-07, DF, NjWOE.

15. See Doc. 3067.

16. Doc. 3087.

17. See Doc. 3061.

18. See Doc. 3065.

19. See Docs. 3055, 3060, 3071, and 3084.

20. See Doc. 3058.

21. See Docs. 3063 and 3064.

22. See Doc. 3070.

23. Docs. 3072, 3074, 3083, and 3089.

24. Mary Valinda Miller to Mina Edison and TAE, 15, 22, and 31 July 1887, all FR (*TAED* FI001ABL3, FI001ABL5, FI001ABM).

25. A Winchester rifle had been purchased at the end of July and sent by express to Akron for this trip. Mary Valinda Miller to Mina Edison, 18 July 1887, FR (*TAED* FI001ABL4); Hartley & Graham Arms and Ammunition to TAE, 22 July 1887, DF (*TAED* D8704ABN); Hartley & Graham Arms and Ammunition to TAE, 4 Aug. 1887, Vouchers (Household), box 11.

26. When Mina was again expecting in January 1888, her sister Mary recalled that she (Mina) had thought she would have twins in 1887 but "there was only one." Family notes for an oral history interview conducted years later indicate that Mina experienced three miscarriages. Edwin Chadbourne to TAE, 30 Aug. 1887, Bills & Receipts [not selected], D-87-07, DF, NjWOE; Mary Miller to Mina Edison, 30 Sept.

1887 and 22 Jan. 1888, both CEF (*TAED* X018C9AD, X018C9AH); Nancy Miller reminiscences, n.d., p. 5, NMC (*TAED* X476D).

27. Tate to W. E. Connor, 14 Sept. 1887, Lbk. 25:91 (*TAED* LB025091B).

28. "Notes of the Week," *NYT*, 18 Sept. 1887, 2; "Amusements. Mr. Burbank's Entertainment," *NYT*, 24 Sept. 1887, 5; Johnson to TAE, 19 Sept. 1887, DF (*TAED* D8704ACW).

–3054–

Technical Note:
Phonograph

East Newark, N.J.,[a] June 7 1887.

Edison's Phonograph[1]

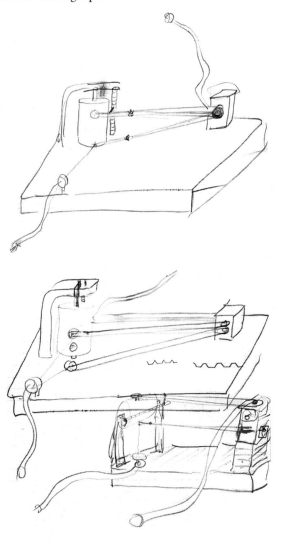

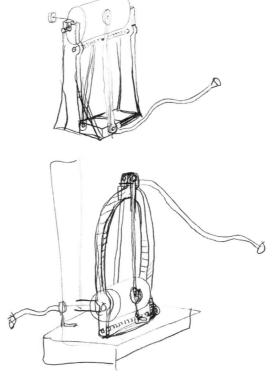

Chas Batchelor

X, NjWOE, Lab., Unbound Notes and Drawings (*TAED* NS87AAL).
Letterhead of Edison Lamp Co. Document multiply signed and dated;
date, heading, and signature written by Batchelor. [a]*East Newark, N.J.,*
preprinted.

1. These drawings on looseleaf paper appear to be by Edison, with
Charles Batchelor having dated and signed them as a witness, likely
before he went into New York for jury duty that afternoon (Cat. 1336:233
[item 388, 7 June 1887]; Batchelor [*TAED* MBJ003233B]). They are the
only complete dated drawings of Edison's phonograph designs from
7 May to mid-November (Doc. 3110). However, additional sketches
from this period are contained in an undated notebook that Edison prob
ably began using about the same time he made these drawings; the first
two drawings in this document closely resemble two drawings on the
third page of that book. According to Batchelor's journal, considerable
work was done on the phonograph during the first week of June, mostly
on the design of needles and diaphragms for reproducing the recording. This work continued over the course of the following week (Cat.
1336:221, 223, 229, 233, 237 [items 374, 376, 381, 387, 396, 397; 1, 3,
6, 11, and 13 June 1887]; Batchelor [*TAED* MBJ003221, MBJ003223,
MBJ003229B, MBJ003233A, MBJ003237A, MBJ003237B]).

Memorandum to
Richard Dyer

Dick

Patent   Municipal—[1]

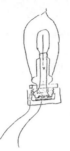 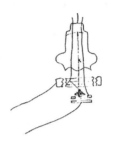 

You see in this device when the arc jumps to Central wire it gets 3 amperes & this is sufficient to fuse the small wire—X This wire connects to the Volute Spiral Contact spring—& holds the point away from the Contact in the socket about $^1/_{32}$ @ $^1/_{16}$—but when it X[b] fuses—the spring socks[2] the contacts together & Cuts lamp out— This is a Electro mechanical instead of magnet—[3]

Edison

ADS (facsimile), MdNACP, RG-241, *Edison v. Thomson*, Edison's Exhibit I, pp. 15–15½ (*TAED* W100DLA; images 14–15). Letterhead of Thomas A. Edison. [a]"NEW YORK," and "188" preprinted. [b]Interlined above.

1. Figure labels in the first two drawings are "X" and "X." Edison did not receive a patent based on the description in this document, although he apparently filed an application that was later entered into an interference proceeding at the Patent Office with an application or patent of Elihu Thomson (*Edison v. Thomson*). Edison's design resembled the so-called Edison Effect lamp, in which a wire positioned in the evacuated bulb near the filament was found to carry a current (a phenomenon that Edison had previously applied to a voltage indicator; see Doc. 2538 [headnote]). Charles Batchelor noted this cutout design in his journal on 2 June, indicating that the lamp was sensitive to the direction of current and could only be set up one way. After suggesting and trying a number of variations over the next two weeks, he tested fifteen of the Edison Effect–style cutouts on 17 June (trying the current in both directions) and found they "all cut out perfectly when the carbon broke" (Cat. 1336:229, 235, 239, 241 [items 380, 394–95, 401–402; 2, 9–10, 16–17 June 1887; all Batchelor [*TAED* MBJ003229A, MBJ003235A, MBJ003235B, MBJ003239B, MBJ003241]). As William Jenks described the cutout in August 1887 to the Association of Edison Illuminating Companies, the third wire was connected with a "fine iron wire holding a phosphor-bronze spring in tension. At the instant of the formation of an arc usually near one end of the carbon, the current divides between the opposite terminal and the middle conductor, the latter taking a sufficient volume to instantly melt the iron wire, liberate

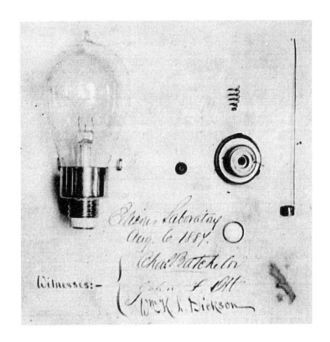

Photograph of Edison's municipal lamp and cutout components, taken about 6 August and pasted in Charles Batchelor's journal.

the spring, and by short-circuiting the lamp within itself cut the arc and even prevent the temporary exhibition of fireworks." (Because of its thermal expansion properties, a short section of platinum was used instead of iron where the wire passed through the glass globe.) The Edison Electric Light Co. adopted Edison's design for commercial use in the latter part of 1887 and had some six to eight thousand of them in service by November 1888 (Jenks 1887, 10; Catalogue of Lamps Manufactured by the Edison Lamp Co., c. 1887, p. 14, PP [*TAED* CA041J]; Jenks's testimony, pp. 72–73, *Edison v. Thomson* [*TAED* W100DLA; images 45–46]; Cat. 1337:4 [item 463, 12 Sept. 1887?], Batchelor [*TAED* MBJ004, image 3]).

2. Edison probably used this word in the American sense of thrusting or driving one thing into another. *OED*, s.v. "sock," v.2.1.c(a).

3. Edison later testified that the Edison Electric Light Co. had asked him in May 1887 to design a cutout apparatus that was inexpensive and completely self-contained; they especially wanted "to get rid of the magnet and all exterior devices as they have proven troublesome." Jenks corroborated the simplicity and relative cheapness of the fusible wire design compared to an electromechanical one. Edison's testimony, pp. 68, 70; Jenks's testimony, p. 75; both *Edison v. Thomson* (*TAED* W100DLA; images 43–44, 47).

–3056–

To William Barr[1]

[New York,] June 14th. 1887.

Dear Sir:—

In reference to your letter of the 7th instant[2] I beg to say that I have only installed sufficient power at Llewellyn Park to supply 400 lamps above what I require for my own use. I have promised to furnish Messrs Auchincloss[3] & Burke,[4] and as yet I am not sure as to the number of lights they will use. How many would you require?

I shall be very glad indeed to supply my neighbors with light so far as the capacity of my plant will permit, and simply charge them with the cost of production.[5] Yours truly,

T.A.E   T[ate].

TL (carbon copy), NjWOE, Lbk. 24:496C (*TAED* LB024496C). Initialed for Edison by Alfred Tate.

1. This letter was erroneously addressed to "E. Barr" but was clearly intended in reply to an inquiry from William Barr, Edison's neighbor in Llewellyn Park (see note 2). William Barr (1827–1908), a Scottish-born dry goods merchant, was dispatched in 1854 by his New York partners to St. Louis, where he incrementally built a hugely successful department store. Barr moved from St. Louis to Orange soon after the Civil War but retained his western business interests and was active in philanthropic affairs in both cities. He purchased "Baronald," his Llewellyn Park home on Tulip Ave. adjoining the property now owned by Edison, about 1868. Obituary, *NYT,* 17 June 1908, 9; *NCAB* 14:522; Whittemore 1896, 319–20.

2. Barr wrote to "make an application— for the Electric light for my house in Llewellyn Park—(just across the ravine from yours—)" and proposed that the contractor putting in Edison's wires should do his own as well. Notes on this letter became the basis for Edison's reply. Barr to TAE, 7 June 1887, DF (*TAED* D8731AAE).

3. Henry Buck Auchincloss (1836–1926) was connected with Auchincloss Bros., a prominent dry goods commission business in New York City and the import and selling agents for J. and P. Coats cotton thread. U.S. Dept. of State n.d., roll 245, passport issued 13 Jan. 1882; Obituary, *NYT,* 10 Sept. 1926, 21; *Trow's* 1881, 55.

4. Irish-born John Burke (1829–1892) made his fortune in the Dublin partnership (with his brother) of E. & J. Burke. The firm was a prominent distillery and, by this time, also a bottler and exporter of Bass's ale and Guinness's porter brands. In 1855, Burke was among the first purchasers of land in the future Lewellyn Park. He moved to the United States about 1859 to oversee the firm's New York operations. Obituary, *NYT,* 5 Feb. 1892, 5; Hall 1895, s.v. "John Burke"; Pierson 1922, 2:308.

5. Edison planned to light his home and several others nearby through underground cables from a generating plant at his new laboratory. According to an October press account, the plant would have two 600-volt dynamos; each home would have some type of converter to reduce the 1,200-volt line current for domestic use but the general outline of the system was still under discussion in early November, when John Vail sent estimates for installing one or two dynamos. At the Edison Machine Works about that time, Henry Walter tested a transformer system for a 1,200-volt distribution network for 325 lamps that almost certainly was intended for Llewellyn Park. Charles Batchelor noted in his journal that Edison's house was lit for the first time on 23 December 1887

("Edison's System of Electrical Distribution Based on Transformers," *Electrical Review* [New York] 11 [1 Oct. 1887]: 1–2; Vail to TAE, 4 Nov. 1887; Walter test report, 7 Nov. 1887; both DF [*TAED* D8733ABG, D8737ABE]; Cat. 1337:35 [item 527, 23 Dec. 1887], Batchelor [*TAED* MBJ004035]). In January and February 1888, the Auchincloss and Burke homes were wired for 100 and 150 lamps, respectively, by the Edison Wiring Co., with Edison expressing irritation at the slow pace of work. Burke's bill for the installation, including his pro-rated portion of the cost of putting in junction boxes and underground conductors, came to approximately $4,100. The bills sent to Auchincloss and Burke for electricity used between March 1888 and February 1889 ($314.85 and $628.70, respectively, at 1.5 cents per ampere hour) referred to the "actual expense of running our engines and dynamos for the accommodation of those who take light from the Laboratory" (Alfred Tate to Auchincloss, 24 May 1888; TAE to Edison Wiring Co., 14 Feb. 1888; TAE to Sigmund Bergmann, 28 Feb. 1888; Tate to Burke, 18 Apr. 1888; all DF [*TAED* D8818ALA, D8818AEA, D8818AFH, D8818AIW]; Tate to Burke, 12 Mar. 1889; Tate to Auchincloss, 12 Mar. 1889; both Lbk. 28:615, 616 [*TAED* LB028615, LB028616]).

–3057–

From Delos Baker

Cincinnati, O. June 16/87

Dear Sir,

(1)   Shall we ever navigate the air with dirigible craft? (2)   If so will it be by the propulsion of an inclined plane against[a] the ~~sky~~ atmosphere, or by baloon, or by both, or by a combination of both   I would like to print your reply in the Post. Sincerely,

Evening Post, D. R. Baker[1]   R[2]

⟨Inclined plane, apparatus held in air By power and propulsion forward obtained by falling and rising due to the wind— as with birds—[3] E[dison]⟩

L, NjWOE, DF (*TAED* D8705AAG).[a]Obscured overwritten text.

1. Delos R. Baker (1848–1928), a former Methodist minister who took up journalism after he was defrocked for heresy, had edited the *Cincinnati Evening Post* since late 1885. Publisher Edward Wyllis Scripps later wrote that while Baker "was very eccentric, he was very learned, very eloquent, and, in fact, an all-round brilliant character." Scripps hired Baker about 1881 as a writer for the *Penny Post*, shortly after taking control of that paper from his family. He soon changed the name to the *Evening Post*, and by this time the paper claimed to have the largest daily circulation in Ohio. *Williams'* 1886, 114; U.S. Census Bureau 1970 (1880), roll T9_1023, p. 357B, image 0075 (Mill Creek, Hamilton, Ohio); Greenlawn Cemetery Index (Columbus, Ohio), online database accessed through Ancestry.com, 11 Sept. 2013; Stevens 1968, 42, 61, 65–66; Scripps 1966, 53–57; Scripps 1951, 154–58; *ANB,* s.v. "Scripps, E. W."

2. Not identified; presumably a secretary.

3. Edison's marginalia was the basis for a typed reply to Baker on 7 July. The editors have not determined if his letter was published. Lbk. 25:23A (*TAED* LB025023A).

–3058–

Memorandum to Richard Dyer: Gas Lighting Patent

[Harrison, Spring 1887?][1]

Dick—

I want to take out patent on this—[2] The principal point is the use of exceedingly porous infusible materials for the flame to play on— I made this apparatus in 1879 & it worked well—[3] Wessbach[4] or some such name is making great blow about it & is forming large Cos using regular gas[5]—his only claim is very porous[a] infusible matter—[6]

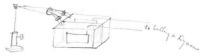

I understand Co formed here of 4 million[7]— make the claim so as to meet him—

What do you think—

Edison

P.S M[olydenite?][8] [C-y------ ---- Th-- -----------tic?][b]

ADS, NjWOE, PS (*TAED* PT032ABA3). [a]Obscured overwritten text. [b]Canceled.

1. The editors have conjectured this date range from several circumstances: Edison's completed patent applications on this subject (see note 2); his uncertainty about Welsbach's name (see note 4); and the formation of Welsbach's company (see note 7).

2. This memorandum was incorporated into the records of Edison's Case 778 for "Gas Incandescents," but it most likely originated in connection with his Case 723, filed on 14 June 1887, for a "Method and Apparatus for the Production of Light." The transcription of a letter from the Patent Office in the file of the latter case summarized Edison's claims for the use of "platinum in a spongy state" and generally "the various earthy oxides as incandescents for gas," a description consistent with this memorandum and with his 1879 experiments (see note 3). The examiner rejected the claims on the basis of earlier U.S. and foreign patents, a decision upheld by the Examiners-in-Chief in February 1888. Edison subsequently abandoned the application in favor of a new one that he filed on 7 June 1888 (Case 778; serial no. 276,385)

after additional experiments by chemist David Trumbull Marshall. (There was also a companion case, identified only by Patent Office serial number 276,386.) In the new application, Edison claimed a "gas incandescent consisting of a decomposed salt of an earthy oxide" and described his lamp design as able to produce incandescence with "much less heat, or smaller flame, and hence will be more economical in the consumption of gas." The Patent Office immediately rejected this new application on the same basis as the prior one. Edison and his attorneys sought to defend it to the patent examiner but abandoned the case after a second rejection in July 1892. Patent Application Files: Production of Light (1887), Gas Incandescent (1888); both PS (*TAED* PT032AAE3, PT032ABA); N-88-01-00, Lab. (*TAED* NB015AAA).

Edison assigned the rights to his 1887 application (Case 723) to John B. Powell in October 1887. Their agreement gave Powell, general manager of the Brush–Swan Electric Co. of New England (with headquarters in New York City), six months to sell the patent. Powell reported in April 1888 that he was negotiating with the Welsbach Co. (probably the New York company; see note 7) and was optimistic that the threat of Edison's competition would induce that firm to purchase the patent. Powell also enclosed with his letter a 23 February 1887 circular of the London Welsbach company containing technical reports on the Welsbach lamp by Conrad Cooke, an electrical engineer, and William Wallace, gas examiner for the city of Glasgow. Edison may have extended Powell's rights to include the June 1888 application (Case 778); Ezra Gilliland conveyed to Powell in August 1888 Edison's intention to start work "very soon" on the "burner matter." Nothing seems to have come of that plan nor of Powell's hope of finding employment with the Edison lighting interests. TAE agreement with Powell, 18 Oct. 1887, Miller (*TAED* HM87AAX); Powell to TAE, 21 Apr. [with enclosure] and 28 Sept. 1888, both DF (*TAED* D8805ACK, D8805ACK1, D8805AGR); Gilliland to Powell, 2 Aug. 1888, LM 22:1 (*TAED* LM022001); regarding Powell, see advertisement for the Brush–Swan Electric Light Co. of New England, *Electrical Review* 8 (6 Mar. 1886): 16; and "Personals. Major John B. Powell," *Electrical World* 10 (10 Sept. 1887): 148; *Trow's* 1888, 31, 1581.

3. Edison likely referred to the form of lamp described in Doc. 1677; regarding related experiments, see Docs. 1669, 1670, and 1672.

4. Carl Auer von Welsbach (1858–1929) was an Austrian chemist who specialized in the study of rare earths, a subject he took up while studying with Robert Bunsen at the University of Heidelberg. Welsbach was the first to find a practical use for rare earth compounds by impregnating them into fabric; the so-called Welsbach mantle produced a bright white glow in a gas flame. Using osmium, he devised the first metal-filament incandescent electric lamp in 1898. Asimov 1982, 575.

5. Welsbach formed the Incandescent Gas Light Co., Ltd., in London in March 1887 (*The Statist* 19 [26 Mar. 1887]: vi). In May, Welsbach Incandescent Gas Light Cos. were formed in New York (see note 7) and Pennsylvania, and by late summer in the South as well ("Home Items," *Light, Heat and Power* 3 [May 1887]: 216; ibid. 3 [Sept. 1887]: 456).

6. Figure label is "to battery or Dynamo." Edison likely read the texts and claims of Welsbach's British Patents 12,286 (1885) and 3,592 (1886) in the 29 April 1887 issue of *Chemical News* (55:192–94).

7. On 17 May 1887, the *New York Times* reported that the Welsbach Incandescent Gaslight Co. had been incorporated with a capital stock of $4,500,000 ("City and Suburban News: New-York," p. 2). Demonstrations for prospective investors were held in May and for the public in November (*Light, Heat and Power* 3 [1887]: 217, 727).

8. During his experiments in 1879 (see note 3), Edison had used a variety of metals and minerals, including earth oxides such as molybdenite, which he described as "fusible with diff[iculty]   gives a splendid light" (N-79-01-21:23, Lab. [*TAED* N016010]). The heavily canceled words that follow in this memorandum are likely other materials he found promising for use as gas incandescents.

–3059–

Draft Caveat: Miscellaneous

[Orange?] June 22 1887[1]

Dyer—

Here is a general sweeping Caveat I want to file in the office for dating purposes   I suppose of course they will reject it as not according to rules, but it will get patent office stamp on it & can lay there.[2]

Improvements in Transformers. I use thin sheet iron punched from sheet in the form of an open ring.

a large number are laid together & insulated from each other by paper & or a coating of Varnish.[3]

A is a bobbin made in one piece from glass hard Rubber or other insulating material   the material employed is one that is difficult to perforate by charges of high potential   Hard Rubber is excellent for this purpose— The bobbins are wound with wire half of them having fine wire & the other half coarse wire the former for the primary & the latter for the secondary— The bobbins are placed in the broken part of the iron ring & shoved over it first a fine wire then a large wire bobbin & so on until the ring is covered with bobbins. They are afterwards so connected together that all the coarse wires Bobbins[a] form a complete winding & the fine ones another complete winding— after[b] this is done all the bobbins are pressed very close to each other & then by a chisel shaped tool Working[b] on the Edges of the thin wire plate ring. ~~they are advance~~ various ones are advanced to different distances thus closing the ring completely the total break being now distributed throughout the entire ring, and are infinitessimal at any one point. The advantage of this form is first that a[b] larger amount of iron can be put in a smaller space than if fine iron wire was used in the usual way   2nd   The [~~ling?~~][c] Lines of force pass through a practically Continuous piece of iron   3rd   The trouble of winding a closed ring is obviated. 4th   The bobbins being placed separately on the core & the wire being on a complete piece of material of high disruptive power to currents of high potential, it is not so liable to be destroyed by lightning. The ordinary converters are very subject to destruction by lightning the primary & secondary acting as a condenser—but with bobbins whose parts are integral The flash must be of enormous potential to burst through the material, while it would find no difficulty were cotton insulation even covered with viscous materials   5   another advantage is that should any bobbin be destroyed by lightning the chisel may be used to bring all the iron rings back to a common opening & the bobbin replaced.—

A method I propose to prevent the destruction of Converters in a system by lightning is as follows[4]

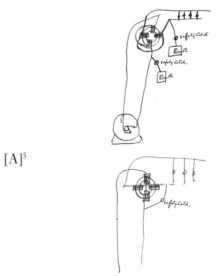

[A][5]

Another Method is to use condensers connected to earth which will be more easily disrupted by lightning than the Coils of the Converter & yet sufficiently nondisruptive to remain intact with 100 percent more Electromotive force than the primary Current of high potential.[6]

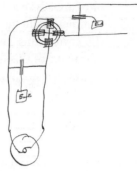

Method of Tanning Leather. This method consists in driving the tanning liquor through the hide by Endmose.[7] The Cell containing the tanning liquor is divided electrically in two parts by the hide   The positive Electrode of Carbon is in[b] the liquor on one side of the hide & the negative Carbon is on the other side, when a battery or other source of Electricity is connected to the two electrodes A[b] current of E is established through the pores of the hide and Endomose is set up the liquor being rapidly Carried from the positive to the negative through the pores, and the liquor in one half begins to fall while the Liquor in the other half commences to rise thus as much tanning liquor is passed through the hide in 4 hours as[b] would naturally pass in a week in the ordinary manner without the assistance of Electricity—

Improvement in Machines for seperating magnetic & diamagnetic substances by altering the trajectory of their fall by a magnet.[8] The improvement consists in Causing the finely divided particles to fall in a thin sheet before the poles of the magnet while in a vacuum so that the air will not interfere to deflect the light particles from falling in a straight line when not magnetic.[9]

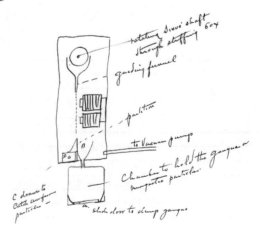

Magnetic Bridge for measuring the magnetic conductability of magnetic substances & for the detection of flaws in steel & iron articles shafts etc.[10]

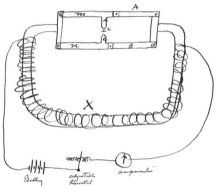

X is the large Electro magnet for magnetising the Bridge. N M are constant & are formed of pure iron   d e is the bridge piece broken in the centre to create a magnetic field   in this is a magnetized needle provided with pointer & scale or mirror & Lamp stand & scale   The needle is held in position by a torsion wire. A is a section of iron to be measured & B is a standard bar   when m n are perfectly equal conductors of magnetic lines & A & B the same, There is no field of force formed in the bridge at C hence the needle is not deflected but should the piece of iron at A have the slightest impurity it will not conduct[b] the magnetic lines so well as the pure bar B hence the balance of fo Lines will be destroyed and the needle deflected in One[b] direction. should the iron be known to be[d] pure but a Cavity was in it, it would for this reason fail to conduct as well. To those acquainted with the Electrical Bridge it will be at once seen that with a number of standard bars Known to be pure & of different sizes The magnetic resistance of any magnetic substance placed at A even the resistance of air spaces etc can be acurately ascertained. This apparatus arranged in a proper manner can be made to rapidly measure the perfection of car wheel axles also the qualities of steel as they depend on the Carbon & structural arrangement.

ADf, NjWOE, PS, Caveat Files (*TAED* PT031AAO). [a]Interlined above. [b]Obscured overwritten text. [c]Canceled. [d]"known to be" interlined above.

1. Mina Edison subsequently signed and dated the first page on 15 November 1887, when she did the same for a large number of Edison's looseleaf technical drawings. Cat. 1152; Unbound Notes and Drawings (1887); both Lab. (*TAED* NM021, NS87).

2. The editors have not found evidence that this caveat was completed or filed.

3. Figure labels are "A," "A," and "A'."

4. Figure labels are "safety Catch," "Earth," "safety Catch," and "Earth."

5. Figure label is "safety Catch."

6. Figure labels are "Earth" and "Earth."

7. Edison probably meant "endosmosis," the passage of a fluid through a membrane from an area of lower concentration to one of higher concentration. *OED*, s.v. "endosmosis."

8. Two days after drafting this caveat, Edison prepared a rough draft of a patent application on this separation process, including two drawings similar to those below (Cat. 1152, Lab. [*TAED* NM021AAC]). He executed the application on the last day of June, and it was filed at the Patent Office on 6 July. The resulting specification (with the two drawings) issued in February 1888 as U.S. Patent 377,518. At the end of July, Edison drafted a related application (Doc. 3074) on a method of preparing powdered pyrite ores for this separation process by heating them.

William K. L. Dickson began a series of experiments on 5 June "To Separate Metallic Gold from Pyrites &c Magnetically." He recorded his trials in a notebook (N-87-06-10, Lab., NjWOE) and later neatly transcribed the records into a different book (E-2610, Lab. [*TAED* NM031]). Dickson started with the material in a sieve, from which he shook it out in a stream past electromagnets. Noting its tendency to scatter, he made a wooden box to shield the stream from drafts. When this proved insufficient, he constructed a chute to narrow the stream and direct the particles in a thin line toward a sharp divider below. By 10 June, he had sketched and tried the essential elements represented here in Edison's drawing and text, except for the use of a vacuum, the genesis of which is uncertain. Dickson worked along these lines through June, adding copper dust to the experimental mix in July. He continued similar experiments for months (including some in a vacuum, in December), transcribing frequent entries into the notebook through the next summer and occasional ones until November 1888 (E-2610:1–4, Lab. [*TAED* NM031001, NM031002, NM031002A, NM031003]).

Apparently buoyed by Dickson's work, Edison had his secretary answer an inquiry on 9 June that he had "recently made great improvements in my ore milling process and am still at work upon it. In the course of a couple of months I will be prepared to put these in practical use." A similar letter went to a different correspondent on the same day Edison drafted this caveat. TAE to Nogales and Sonora Mining and Smelting Co., 9 June 1887; TAE to Thomas Connery, 22 June 1887; Lbk. 24:493, 25:6A (*TAED* LB024493, LB025006A).

9. Figure labels are "rotating Sieve shaft through stuffing box," "guiding funnel," "partition," "B," "to Vacuum pump," "Chamber to

Photograph of Edison's magnetic bridge in August 1887.

hold the gangue or magnetic particles," "C⎯ drawer to Catch auriferous particles," "a⎯slide door to dump gangue."

10. Figure labels (left to right from top) are "M," "d," "A," "C," "N," "e," "B," "X," "Battery," "adjustable Rheostat," and "ampermeter." Edison explained the instrument's design and construction in a paper for the American Association for the Advancement of Science meeting in August 1887. He stated that it had been "devised for the purpose of testing readily the quality of the iron purchased for the construction of dynamos" and was intended as "the exact counterpart of an ordinary Wheatstone bridge," used to measure electrical resistance, "since now whatever is true electrically of the one is true magnetically of the other." Charles Batchelor later placed in his journal a photograph of a magnetic bridge constructed in a similar manner. The photo was probably taken at the AAAS meeting in August. At least one other inventor came forward in August to claim credit for a similar device. Edison 1888a, 93; Cat. 1337:6 (item 465, 13? Sept. 1887), Batchelor (*TAED* MBJ004, image 4); "The Eickemeyer Magnetic Bridge," *Electrical Review* [N.Y.] 10 (27 Aug. 1887): 9.

–3060–

Notebook Entry: Phonograph

[Harrison,] June 22 1887

Cylinders for phonographs[1]

    Take split mould you used for Plaster paris, & mould following substances

    Melt Asphalt in Crucible pour —

    Melt   "   mixed with equal quantity Kaolin.

    Melt Rosin mixed with ¼ weight ɵfor bulk of asphalt.

Melt Rosin mixed with ¼ weight Chalk or Kaolin

"   "   "   ½ "   "   "

Melt Parafine mixed with ½ its bulk Rosin

Melt "   "   ½   " Chalk or Kaolin

"

TAE

X, NjWOE, Lab., Misc. Notebooks, Notebook (1887): 19 (*TAED* NM030019).

1. Batchelor tested a dozen combinations of resin and kaolin (with pine wood tar) within a few days, trying for "a suitable & cheap material to make the phonograph 'sleeve.'" He found enough success with one compound, described as "excellent because it does not stick to tinfoil nor paper," to conduct moulding experiments on 27 June, for which he noted both poor and excellent results. When Batchelor returned to testing kaolin compounds on 30 June, none proved fully satisfactory on a very warm day. Cat.1336:243–47 (items 408, 410, 413; 24, 27, 30 June 1887), Batchelor (*TAED* MBJ003243B, MBJ003245B, MBJ003247B).

–3061–

Alfred Tate to William Perry

NEW YORK, June 27th 1887[a]

My dear Perry:[1]

Re our conversation of today— Ore Milling. Mr Edison's idea is to form under the laws of New Jersey a new corporation to be called the Edison Milling Co'y Capital $2,000,000— one quarter of which, or $500,000 to be paid to the Ore Milling Company[2] which will be absorbed, and the balance to be issued to T.A.E. for new inventions, he to turn back into the Treasury $250,000 for working expenses. Mr Edison has certain experiments which he desires to conduct that will cost a great deal of money. Under agreement the Ore Milling Co'y should furnish this Capital but as you know they have no money— Edison has made some important discoveries and the reorganization will make the Stock valuable.[3]

What day could you come out with me to see T.A.E. If you are at present a grass widower[4] and can leave town for a night you might come out with me any afternoon. I have taken Upton's house[5] for this Summer, and my Wife and I would be very glad to see you. I usually go out at 4.50 but can suit your convenience—then we could see Edison in the evening and I can get back to town with you at any hour you like next morning—thus not encroaching upon your business hours. Yours faithfully

A. O. Tate

L, NjWOE, DF (*TAED* D8748AAL). Letterhead of Thomas A. Edison. Written by John Randolph. ᵃ"NEW YORK," and "188" preprinted.

1. William Sumner Perry (1848–1933), born in Bristol, R.I., was a Wall Street broker who handled stock transactions for Edison and his associates. He had been secretary of the Edison Ore Milling Co. since 1884 and also secretary of the Miocene Mining Co., of which Edison seems to have been a trustee. Perry was a stockholder in the Edison Electric Light Co. of Europe and had served as that company's secretary *pro tempore*. Doc. 2572 n. 3; Perry 1987 [1902], 160; Perry to TAE, 22 May 1886, DF (*TAED* D8635I).

2. The Edison Ore Milling Co. was incorporated in 1879 and capitalized at $350,000. It held broad rights to Edison's inventions for the extraction of precious and base metals from ores, as well as his contracts to work several mining claims in California. It had briefly operated a commercial-scale plant for the separation of iron from black sands but had been largely moribund since the end of 1882, with few assets on its books and debts in excess of $5,500. Edison had been making his ore experiments, mainly for the recovery of gold, at his own expense. See Docs. 1844 n. 5, 2393, and 2591; Report of Edison Ore Milling Co. board of directors, 4 Aug. 1887, CR (*TAED* CG001AAI6).

3. Responding in early June to an inquiry about the company, Edison stated that he had recently improved his process for separating ferrous material from sand and that "developments during the next few months will take the Company out of its present experimental stage and place a value on the stock." Having resigned the presidency and his directorship in the company in late July, he submitted a formal reorganization proposal to the board on terms similar to those outlined in this document. The directors acted favorably in early August and recommended that the company be disbanded if stockholders did not ratify the proposition. The stockholders did so in a special meeting on 13 September, authorizing the directors to recapitalize the company to $2,000,000 (TAE to Thomas Connery, 2 June 1887, Lbk. 24:481 [*TAED* LB024481]; TAE to Edison Ore Milling Co., 23 July 1887, DF [*TAED* D8748AAM]; Edison Ore Milling Co. circular letters to stockholders, n.d. [Aug. 1887] and 5 Aug. 1887, both CR [*TAED* CG001AAM, CG001AAK]). Edison and the company executed a new contract on 14 October to replace their original agreement of January 1880. It provided that Edison should assign ownership in all his patents for the "extraction of metals from Ores, tailings, gravel, or other deposits, whether by the use of Electricity, or of any chemical, mechanical or other process." Edison released the company from all debts owed him under the old contract and agreed to advance up to $25,000 for experiments (including construction of a dedicated laboratory) and to donate 2,500 stock shares (from his total of 14,750, the balance being assigned to the company) for its business expenses. John Tomlinson signed the agreement as president of the reorganized company (TAE agreement with Edison Ore Milling Co., 14 Oct. 1887, Miller [*TAED* HM87AAU]).

4. A man whose wife is away or living apart. *OED*, s.v. "grass-widow," derivatives.

5. Upton left for Europe in mid-July. His house was on Day St. in Orange (opposite Snyder St.), a short distance from Edison's Llewellyn

Park home. TAE letter of introduction, 15 July 1887, Upton (*TAED* MU107); Baldwin 1886, 319.

–3062–

To John Randolph

[Llewellyn Park?] ⟨June 29—1887⟩[1]

Johnny

Starting from next saturday Please send a check for $3.00 Every week to Dot Edison, to Orange or such place as she may direct[2]

T.A.E.

⟨Send for following Catalogues.[3]

~~M. Schuchadt   Mnf. Rare chemicals Berlin~~[4]
~~Powers & Weightman ny   (mfg chemists)~~[5]
~~Chas Pfizer ny chem^a (chemists) 81 M—~~[6]
M Grunow mfr philosophical Apparatus NYork—[7b]
~~Johnson Mathay & Co   Hatton Garden London England~~
~~Patterson Bros—ny~~[8]
~~Peter Frasor & Co—ny—~~[9]

TAE

ALS, NjWOE, DF (*TAED* D8704AAZ). ^aCanceled. ^bFollowed by "over" to indicate page turn.

1. John Randolph wrote this date at the top of the first page and again (with "Received") at the bottom of the second page.

2. Randolph promised in reply to send the checks to Marion each week, but the editors have found in Edison's books records of only three such payments: one each on this date and 11 and 13 July. In late December, at Edison's request, Randolph sent her (at school in Massachusetts) a $30 check and $20 in cash. Cash Book (1 Jan. 1887–30 Mar. 1887): 142, 144; Accts., NjWOE; Randolph to TAE, 29 June 1887; Randolph to Marion Edison, 20 Dec. 1887, Lbk. 25:18, 26:120 (*TAED* LB025018, LB026120).

3. Randolph replied the same day that he had arranged for each of the catalogs listed below to be delivered to Edison's home, except that of J. & W. Grunow, whose address he could not find (Randolph to TAE, 29 June 1887, Lbk. 25:18 [*TAED* LB025018]). Edison reportedly was in Philadelphia on 6 July to buy electrical equipment for his new laboratory ("Whistling Under Water," *Chicago Daily Tribune*, 11 July 1887, 5). He also collected and saved around a dozen catalogs of equipment and supplies (notably rubber goods) during his first years in the Orange lab (NjWOE, Cat. 58594, Ser. 3, Boxes 24–25). When the James W. Queen Co., of Philadelphia, published in an 1887 catalog the names of "a few of the many well-known electricians to whom we have sold electrical apparatus," Edison was at the top of the list, followed by Edward Weston (McQueen "Catalogue and Price-List of Electrical Testing Apparatus," [1887]: 71, in Warner 1993, Vol. 2).

4. Conrad Gideon Theodor Heinrich Schuchardt (1829–1892) trained as a botanist and physician but became a mineral collector as early as 1862. Having secured a position in charge of an alum processing works in the Gorlitz region of Prussia, Schuchardt started a chemical and mineral supply business there in 1865. The firm grew rapidly in size and significance and by this time was advertised as Schuchardt's "celebrated Chemical Manufactory." "Theodore Schuchardt," biographical archive website of the *Mineralogical Record* (mineralogicalrecord .com), accessed 29 Jan. 2014; Advertisement for P. E. Becker & Co., *Nature* 38 (3 May, 1888): i.

5. Powers and Weightman was a leading manufacturer of medicinal chemicals. The Philadelphia business of Thomas H. Powers (1812–1878) and William Weightman (1813–1904) was directly descended from a partnership established in 1818 by John Farr, uncle of Thomas Powers, and Abraham Kunzi, and it took its present name after Farr's death in 1847. England 1922, 31–35.

6. Charles Pfizer (1823–1906), who had trained as an apothecary in his native Württemberg (Germany), emigrated to Brooklyn by 1848 with a cousin, Charles Erhart, a confectioner by trade. Together they formed Charles Pfizer and Co. in 1849 to manufacture chemicals in small quantities. Quickly gaining a reputation for quality, the firm developed a large trade with wholesale and retail druggists, especially in camphor, citric acid, and related products. It greatly expanded its Brooklyn factory and opened a New York City office in 1857. By the mid-1880s it had developed an export trade and employed some 150 people. Obituary, *NYT*, 21 Oct. 1906, 9; Mines 1978, 1–14.

7. J. & W. Grunow made physical instruments, notably microscopes and spectroscopes. It was a partnership of J. and Wilhelm Grunow, brothers who emigrated from Berlin to New Haven, Conn., in 1850 or soon after. Wilhelm seems to have brought with him experience with such instruments, but the business was a new one to his brother. Their partnership was established in New Haven by 1852 and in New York City in 1861. Frey 1872, 83–84; Brachner 1985, 131, 140; Moe 2004, 176.

8. Patterson Bros. was a hardware dealer at 27 Park Row in New York City. Doc. 1197 n. 4.

9. (Peter) Frasse & Co. was an importer and dealer in tools and materials for machinists, jewelers, engravers, and similar trades. It was based at 95 Fulton St. in New York. Letterhead, Frasse & Co. to TAE, 29 July 1878, DF (*TAED* D7811ZAI); *Trow's* 1888, 75.

–3063–

Draft Patent Application: Incandescent Lamp Filaments

[Harrison,] July 1 1887

Dyer—

Patent following—[1]

The obj[ect][a] of this invention is to provide an economical and satisfactory process for seperating fibres from the stalk & surrounding matter in fibrous plants.

The process consists in packing the stalks stems etc after barking or splitting as the case may be in tanks containing a coil of steam pipe— Then allowing a hydrocarbon to percolate upwards through the mass until the whole is covered with the Liquid. The tank being closed, water is passed through the coiled pipes at various temperatures according to the material to be operated upon; for plants containing fibres surrounded by pulpy material like the agaves, The temperature should be about 130 Fahr but for flax etc a temperature of 175 is best, but many very pulpy plants ordinary temperatures will answer. after 1 to 3 weeks the action of the hydrocarbon liquid is such that it displaces the water from the material surrounding the fibres, & destroys all the adhesive properties so that when the material is taken from the tank, running water will carry off all the pulpy matter and leave the fibres perfectly clean They will not be injured in the slightest by this operation, whereas by rotting, hackling[2] & use of acids & alkalis as is now the Case the fibre is very much broken and weakened— Almost any liquid of the nature of a hydrocarbon, which is insoluable in water will answer for this reaction, but I prefer to use petroleum oil being cheaper & as good as[b] any Oxygenated Hydrocarbons insoluable in water[c] Biosulphide Carbon [~~Ether?~~][d] & many others can be used—

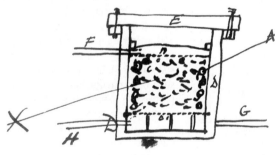

D is the tank holding a ton or more of material C′ & C are plates perforated C′ acting as false bottom, on this rests the material X on top of this is another perforated plate Secured to a frame B & by which is mass of material is pressed down to occupy smaller space; an inlet pipe G. serves to force the hydrocarbon upwards through the material. when[b] the action is finished ~~wa~~ G is shut off & water is forced through H. displacing the Hydrocarbon upwards & [fr?][a] from the material. The Hydrocarbon is drawn off through F & run back into the tank connected to .G. to be used over again. Time must be given the water to displace the oil so that a very small fraction will be [---][d] wasted if care is taken in this respect the oil will perform an almost infinite amount of disintegration thus cheapening the process—

Claim process of seperated fibres from the plant by steep-
ing in a material insoluable in Water—
2nd   use of a hydrocarbon—
3   Heat.
4— displacement.
5   the apparatus—

ADf, NjWOE, Lab., Cat. 1152 (*TAED* NM021AAD). ᵃObscured by ink
blot. ᵇObscured overwritten text. ᶜ"insoluable in water" interlined
above. ᵈCanceled.

1. The editors have not determined if Richard Dyer prepared and
filed a patent application based on this draft. An experiment made by
Arthur Payne in early June is somewhat suggestive of the process described below. Near the beginning of what became a long series of carbonization trials, Payne sealed fibers in a tube containing wood tar and
heated the vessel to 600 degrees Fahrenheit. The tube cracked in the
oven allowing the tar to escape, but on examining the carbons, Payne
found they had "broken split up into elementary fibers, the cementing
material being almost entirely destroyed by the process." This result
did not immediately recommend itself for lamps, but Payne thought
the fibers were "first rate stuff" for making paper. N-87-05-09:23, Lab.
(*TAED* NB002023).

2. That is, to hack, cut, or mangle; more specifically, to use a hackle
to prepare fibers of flax or hemp for spinning by splitting, straightening,
and combing. *OED*, s.v. "hackle" (v. 1, 3).

–3064–

*Draft Patent
Application:
Metalworking*

[Harrison,] July 2 1887

Dyer—
   Patent.[1]
   The object of this invention is to keep metals at a red heat
or other temperature by an electric current, while being acted
upon by hammers, punches dies & rolls & various other machines for working metals.
   The invention consists in the Employment of regulatiable
⟨(Thomson)⟩ᵃ Dynamo Electric Machines giving Currents of
~~great quantity or~~ directly or indirectly by the use of a tension
reducer ⟨(Thomson)⟩ᵃ to heat the metal to the desired temperature which is to be acted upon. ⟨(Thomson (?))⟩ᵃ [~~Clam?~~]ᵇ
also in the use of clamping terminals of metals of high Electrical Conductivity, ⟨(Thomson⟩ᵃ & the circulating of air or
water through the same ⟨(Thomson⟩ᵃ when necessary to keep
the same from reaching too high a temperature. The various
details for applying this invention to specific operations in the
mechanic arts will be the subject of other applications   It

will only be necessary to show a single application to illustrate the invention.

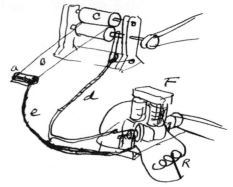

C are rolls for sheet iron several sheets being together. B[c] is the sheet iron A the clamp—d e the conductors from the Dynamo—R the regulating Res to regulate the temperature of .B. As B gets shorter less and less current ⟨E.M.F.⟩[a] will be required which is regulated at R. or by the variation in the speed of the Engine or it may be automatically regulated. all

B after passing towards the rolls until the clamps nearly touch is stopped by the stopping of the rolls one of which is lifted & B pulled back and the operation gone over again until the right thickness is obtained It is obvious from this illustration that bolt[c] rods can be kept at a red heat while being operated upon, that the area around a hole in a Boiler plate can be kept at a welding[c] heat to permit[c] of rivets being welded etc—

Claim the process of keeping Iron & other metals at a red or workable heat by regulatiable Electric Currents in the various Mechanic operations etc. Mention that alternating machines may be used of high Emf & a converter[c] for reducing the tension down to any point— ⟨(Thomson)⟩[a]

E[dison]

ADfS, NjWOE, PS (*TAED* PT032AAF). [a]Marginalia probably written by Richard Dyer. [b]Canceled. [c]Obscured overwritten text.

1. The editors have found evidence of neither a completed application based on this draft nor any related applications. Richard Dyer's marginal insertions of "Thomson" in the text suggest that he believed Edison's ideas were anticipated by Elihu Thomson's patents on electric welding, a technology that Thomson is credited with inventing. See U.S. Patents 347,140; 347,141; and 347,142 (all issued 10 August 1886); and Thomson 1887.

New York, July 6/87.[a]

My Dr Edison

We are organizing a wiring Co for New York City & out-side wiring work—[1] Bergmann Batchelor & I Each go in for 4000$—25% only of which will be called for some time— The Noll Bros[2]—Chinnock—Greenfield[3] & some Others are in— Capital will be 50,000—13,000 of which will remain in the Treasury for the time   It will be a very profitable busi-ness, so I have put you down for 4000$— We have gone on the principle of putting in only people that we think can do some good—except perhaps Hannington[4] who we have put in merely to give him an opportunity—[b]

We will call it Edison Wiring Co.[5]

Bergmann and I will keep the control of it (I mean of course in conjunction with the stock of you Batch & our own)   Yours

E. H. Johnson

⟨The Lord only knows where I am to get the Sheckels[6]— Laboratory is going to be an awful pull on me— Has[c] Coster returned   E[dison]⟩

⟨C[harles] H. C[oster].= When can we take up the Edison 60,000? EHJ⟩[7d]

⟨ EH.J.   will <u>try</u> to do so next week   CHC⟩[e]

ALS, NjWOE, DF (*TAED* D8739AAA). Letterhead of Edison Electric Light Co. [a]*"New York,"* preprinted. [b]Johnson wrote Edison's name in large letters (with what appears to be a flourish) vertically across this paragraph. [c]Obscured overwritten text. [d]Marginalia written by John-son. [e]Marginalia written by Charles Coster.

1. Johnson was planning at this time for construction of the Second District, the new service area in New York City (cf. Docs. 3068 and 3085). He likely was mindful of problems that had arisen from hap-hazard contracting and shoddy work on interior wiring in service areas built by the Edison Construction Dept. in 1884 and 1885 (see Doc. 2424 [headnote]).

2. The Noll Brothers was a partnership of Augustus Noll, Charles A. Noll, and Edwin Greenfield formed in New York on 1 January 1887, with offices at 65 Fifth Ave. The firm began as wiring contractors for Edison central stations, and it recently added a branch at the Edison station in Boston (and installed the wiring at Edison's Llewellyn Park home). It was also a dealer of electric light fixtures. Augustus Noll had worked for the Edison Electric Light Co.'s installation department in 1880, the meter department at Pearl St., and the Edison Machine Works before becoming an independent contractor around 1885. "New York Notes," *Electrical World* 9 (27 Jan. 1887): 57; "New England Notes," ibid., 10 (9 July 1887): 22; Charles Batchelor to Johnson, 28 Dec. 1887, DF (*TAED* D8717ACJ); "Authors and Papers for March," *Engineering Magazine,* 10 (Mar. 1896): 6 [1,202].

3. An electrician and inventor, Edwin Truman Greenfield (1847?–1920) had been general manager of the wiring department at Bergmann & Co. and subsequently did central station wiring work for the Edison Construction Dept. Greenfield had recently been working for the Metropolitan Electric Service Co. and sold Edison a thermostatic device for rooms at Glenmont. Edison declined to have it installed, claiming that he had placed the order only as a favor to Greenfield. Docs. 2125 n. 7 and 2516 n. 6; *TAEB* 6 App. 5.C; "Obituary," *NYT*, 4 Apr. 1920, 22; William Temple to TAE, 5 Dec. 1887, DF (*TAED* D8746AAH).

4. Charles F. Hanington (b. 1842) had been Samuel Insull's clerical assistant, a wiring electrician for Bergmann & Co., and a supervisor of wiring for the Edison Construction Dept. Hanington was working for the Electric Light Co. at this time when Edison summoned him for a one-month assignment; before the year's end he was engaged with the search for fibrous plants in Latin America. Docs. 2171 n. 5, 2450 n. 4, 2464 n. 12; Hanington to TAE, 1 July and 16 Nov. 1887, both DF (*TAED* D8713AAA1, D8731AAQ).

5. Incorporated on 12 July 1887 in New York, the Edison Wiring Co. combined the wire manufacturing operations of the Edison Machine Works with the contracting of the Noll Bros. Capitalized at $50,000, the firm operated out of 19 Dey St. Its founding officers were Johnson (president), Charles Batchelor (vice-president), Sigmund Bergmann (treasurer), Charles Noll (secretary), and Augustus Noll (general superintendent). In November, the firm submitted a bid to wire Edison's new laboratory. It was absorbed by the Edison United Manufacturing Co. in March 1888. Edison Wiring Co. certificate of incorporation, 12 July 1887, NNNCC-AR (*TAED* X119RA); "New York Notes," *Electrical World* 10 (6 Aug. 1887): 70; "New York Notes," ibid. 10 (20 Aug. 1887): 106; Augustus Noll to Batchelor, 23 Nov. 1887; Johnson to TAE, 2 March 1888; both DF (*TAED* D8739AAF, D8837AAE).

6. Shekels were the ancient silver coin of the Hebrews, and the word entered colloquial English as a synonym for coinage or money generally, particularly in large amounts. *OED*, s.v. "shekel."

7. Johnson may have been referring, in round numbers, to the amount owed to Edison in settlement of his claims against the Edison Electric Light Co. See Docs. 2736 (esp. n. 3), 2771, and 2795; Frank Hastings to Coster, 10 Dec. 1887, Miller (*TAED* HM850253).

–3066–

Alfred Tate to
Charles Chinnock

[New York,] July 7th. 1887.

My Dear Mr. Chinnock:—

Mr. Edison has noted your letter of the 2nd instant[1] in reference to a circular which you desire to send to your representatives throughout the United States; he says that he thinks you had better consult Mr. Johnson in regard to the wording of such a circular, and his own opinion is that it should be thoroughly impersonal in its remarks. He means by this that it should state that Mr. Edison has been employed for some time in improving lamps, and in consequence of such improvement certain results have been attained, but that the Manufacturing Co. should simply say that owing to improvements which have been made in connection with lamps they are in a position to offer their customers a superior article, particulars in regard to which you will of course embody in this circular. Mr. Edison's idea is not to bring him personally before your Agents when explaining these recent developments to them.[2] Yours truly,

TL (carbon copy), NjWOE, Lbk. 25:23B (*TAED* LB025023B).

1. Chinnock had been in his former position as the Edison Electric Illuminating Co.'s superintendent at Pearl St. as recently as February 1887 but seems to have taken on a new role as vice president of the Edison United Manufacturing Co. (Association of Edison Illuminating Cos. 1887, 47; Henry Huidekoper to Chinnock, 12 Sept. 1887, DF [*TAED* D8738AAD]). Replying to a 1 July letter from Edison (not found) about new high-resistance lamps, Chinnock wrote: "If we are anxious for 15 lamps per electrical horsepower and require bracing, I am quite sure you will agree with me that our various agents are also in need of the good news. I trust you will not object to our sending a circular letter, embodying all contained in yours, to our representatives throughout the United States." Such a document, he believed, would be decisive in obtaining installation contracts (Chinnock to TAE, 2 July 1887, DF [*TAED* D8738AAA]).

2. The editors have not found a "circular letter" like that envisioned by Chinnock (see note 1). However, the Edison Lamp Co. prepared an undated pamphlet for sales agents, likely about this time. Most of its thirty-one pages promoted the new lamp as "the latest improvements in the manufacture of incandescent lamps." It stated that Edison "has given from the inception of the business, and still continues to give, a large share of his personal attention to the manufacture of his lamps. . . . He is constantly consulted regarding the methods used; and no change is made in the manufacture, except with his consent; thus, the lamps made by THE EDISON LAMP COMPANY OF HARRISON, NEW JERSEY, are manufactured under the direct personal supervision of THOMAS A. EDISON." Edison Lamp Co. brochure, n.d. [1887?], PPC (*TAED* CA041J)

-3067-

Alfred Tate to Samuel Insull

[New York,] July 7th. 1887.

My Dear Insull:—

Mr. Edison has given me a memorandum of the payments which he has to make in connection with the New Laboratory during the ensuing few months.[1]

On July the 30th, he has to meet $5,500. Aug. 30th $5,000. Aug. 31st, $8,500. Sept. 17th $4,800. Sept. 30th $1,000. Oct. 10th $3,000. Nov. 10th $2,609. These dates are approximate so far as absolute certainty is concerned, and depend upon the provisions of the contract with Builders being fulfilled to the letter; but in financing for the payments it will be necessary to consider them correct.[2] Mr. Edison has about $5,000. and can probably get $5,000. more which will meet the first two payments, but after that I believe he is going to look to you to furnish him with money to the extent of the loan which he made you recently, the amount of which I think $15,000.00.[3] Sincerely yours,

TL (carbon copy), NjWOE, Lbk. 25:23E (*TAED* LB025023E).

1. Memorandum not found.

2. The amounts given here total $30,409 and correspond loosely to the schedule of payments to the masons (J. B. Everett) and carpenters (Fairchild & Co.) for the laboratory and library buildings ($36,409 in all). There were in addition four anticipated payments (totaling $14,440) to V. J. Hedden & Sons falling due at various stages of construction of four brick outbuildings at the new laboratory complex (cf. Doc. 3049 n. 2; Henry Holly to Charles Batchelor, 23 June 1887; TAE agreement with Everett, 1 July 1887; C. W. Smith to Charles Batchelor, 1 July 1887; all DF [*TAED* D8755AAE, D8755AAK, D8755AAL]; Hedden to Joseph Taft, 5 Aug. 1887; Hedden to TAE, 10 Aug. 1887; both Vouchers [*TAED* VC87004A, VC87004B]; receipts and certifications for these and other construction-related bills are in Vouchers [*TAED* VC87]). In addition to the building expenses, Edison also ordered about $9,400 in equipment for the new laboratory in August (Samuel Insull to Henry Livor, 16 Aug. 1887, DF [*TAED* D8736ADK]).

3. The editors have not identified the date or amount of this loan to the Edison Machine Works, which owed Edison nearly $150,000 in aggregate loans as of 1 July. Although Samuel Insull was highly pleased by the company's financial state at mid-summer, it had recently been through a straitened period due to the expenses of relocating and expanding in Schenectady. Insull to TAE, 3 Aug. 1887, DF (*TAED* D8736ADC).

−3068−

From Samuel Insull

SCHENECTADY, N.Y., July 13 1887.[a]

My dear Edison,

I have your favor of the 10th. inst. with relation to the manufacture of arc lights.[1]

I did not understand that Carmen[2] controlled the Tesla lamp;[3] I was under the impression that we would be dealing to a certain extent with Tesla,[4] and that I thought, would be most objectionable. We will have the lamps we have got here, tested right away. I agree with you most decidedly, that it is advisable for us to manufacture some kind of an arc lamp, if there is a demand for one. Our present brush holder and switch board department, would certainly form a nucleus for a fine instrument making department. When we get the tube department into the new building, we will have space available so that we can take these two departments out of the main shop and put them into part of the space at present occupied by the tube department. Moreover it would place us in a position to deal with any small stuff which you may get out in the future. I am confident that you will find it to your interest to place your small work with us; we shall certainly do it cheaper than you have heretofore been in the habit of getting it done, and I think the business relation between you and ourselves as regards this class of work, would be far more satisfactory to you than have been those which you have had heretofore with other people.

We have got the thin sheet iron for the converter. We are now trying some experiments on a thin disc armature under instructions from Mr. Batchelor, before we proceed to put through the converter experiment; as soon as these tests are through, we will send you the result. If you have any directions to give us in relation to the converter experiment, please send them along.

Johnson has not yet obtained the permit to lay tubes;[5] he expects it at any moment and we have already started shipping mains and feeders to New York. Yours very truly

Saml Insull

TDS, NjWOE, DF (*TAED* D8736ACX). Typed on letterhead of Edison Machine Works. [a]"SCHENECTADY, N.Y.," preprinted.

1. Not found.

2. William Carman (1849–1926) had been a key member of Edison's office staff at the Menlo Park laboratory, where he served as bookkeeper, paymaster, and sometime secretary. Carman left Edison's employ in 1882 to join the lumber business of his brother Charles in New York City, though he seems to have continued managing Edison's insurance policies. Carman appears to have been connected with the Tesla Electric Light and Manufacturing Co. at this time, but it is unclear whether he held an official position. Before the end of 1887, Edison invited Carman to join the staff of his new laboratory in Orange, but Carman was dealing with his father's estate and declined to move. "Carman, William," Pioneers Bio; Doc. 1652 n. 14; *TAED*, s.v. "Carman, William"; Carman to TAE, 8 Dec. 1887, DF (*TAED* D8713ABN).

3. Nikola Tesla's arc lighting system was described on the front page of the 14 August 1886 issue of *Electrical Review*. Its main features included a dynamo designed to concentrate the developed magnetism on the armature while minimizing Foucault currents, a new method for regulating the dynamo, and a feed mechanism for the lamp carbons that would produce a steady light. The article also reported that the light was being used on the streets of Rahway, N.J. Carman apparently sent

a lamp to Charles Batchelor for tests at the Edison Machine Works in March or April 1887. At Edison's instruction, Batchelor began testing them in March 1888 but the editors have found no other information about that work and nothing further appears to have been done with the Tesla arc light by Edison or his companies. Carman to Batchelor, 2 Apr. 1887 and 29 Mar. 1888; Carman to TAE, 3 Feb. and 16 Mar. 1888; Batchelor to Carman, 27 Mar. 1888; all DF (*TAED* D8704AAO, D8828AAZ, D8828AAG, D8828AAX, D8818AHB).

4. Nikola Tesla (1856–1943) was a Serbian-born American inventor and electrical engineer best known for his contributions to the development of alternating current electrical systems. Tesla acquired his interest in electricity and engineering at the Joaneum Polytechnic School, which he attended from 1875 to 1877; he also studied briefly at Karl-Ferdinand University in Prague in 1880. Needing to support himself, Tesla moved to Budapest to work for a telephone exchange being set up by Ferenc Puskas, the brother of Edison associate Theodore Puskas. The system was not ready to go when he arrived and he instead found work as a draftsman for the Central Telegraph Office of the Hungarian Government. Tesla then spent time in 1882 at Ganz and Co., which had expanded into the manufacture of arc light equipment, before returning to Puskas's telephone exchange in Budapest. He moved to Paris by 1883 at the behest of Theodore Puskas, who got him a job working with the Edison lighting companies, primarily the Société Électrique Edison, under the direction of Charles Batchelor. When Batchelor returned to the United States in 1884 as general manager of the Edison Machine Works, Tesla also came to New York to work for the company at a salary of one hundred dollars a month. Tesla stayed there for about six months, primarily on dynamo design but he also began developing an arc lighting system. W. Bernard Carlson, Tesla's most recent biographer, suggests that Tesla left because the Edison company failed to adopt his arc light. Earlier biographers, relying on the first full-length biography of Tesla, have claimed that he left after Edison reneged on a promise of $50,000 for his generator improvements, telling Tesla, "you don't understand our American humor." In his autobiography, Tesla wrote that the company's manager had offered him this amount and then claimed it was a practical joke. However, he does not refer to Batchelor as the manager and may have meant superintendent W. M. McDougall. In December 1884, two New Jersey businessmen, Benjamin Vail and Robert Lane, established the Tesla Electric Light and Manufacturing Co. The following spring, Tesla applied for patents covering his arc light system; his patent attorney was Edison's former attorney Lemuel Serrell. By 1886, the Tesla system was being used to light streets in Rahway, N.J. Carlson 2013, 34–75; O'Neill 1944, 65; Tesla and Johnston 1982, 72.

5. That is, the underground conductors for the new (Second District) service area in Manhattan.

[New York,] July 14/[8]7

My dear Sir:

Mr. Edison wishes to know if you can send a reporter some evening to his house at Llewellyn Park to write up an article on converters, transformers &c which he says will be very interesting to your readers. Mr. Edison will furnish all the cuts required. Kindly let me know what evening will be convenient for you to do this.[2] Respectfully

A. O. Tate

ALS (letterpress copy), NjWOE, Lbk. 25:39 (*TAED* LB025039).

1. George Worthington edited the *Electrical Review,* a New York weekly, from its inception in 1883, having edited and published a predecessor, *Review of the Telegraph & Telephone.* He was introduced to Edison in 1882 by a mutual acquaintance and shortly afterward accepted the first of several short-term loans from Edison to support his publication. Worthington expressed his thanks with offers of free advertising and editorial space in *Review of the Telegraph & Telephone* for Edison or his companies. William Somerville to TAE, 20 May 1882; Letterhead, Worthington to TAE, 16 Apr. 1883; both DF (*TAED* D8204ZBQ, D8303ZCF); *TAED,* s.v. "Worthington, George."

2. The 17 September *Electrical Review* included an article with excerpts and illustrations from four Edison patents, all issued on 6 September, related to the transmission and reduction of high-voltage current ("Long-Distance Distribution," *Electrical Review* 11 [17 Sept. 1887]: 1–2). The lead article of the 1 October issue, titled "Edison's System of Electrical Distribution," borrowed text and illustrations from four patents awarded in 1882 and 1883. It judged these specifications as the "hard pan" or absolute antecedents of all subsequent conversion systems, including the transformers of Gaulard and Gibbs. While giving Edison credit for that, the article went on to point out that since then, he had made "practically, no change, merely a touch here and there as to mechanical details to fit the gauges of later refinement, and dress the child of 1881 in the fashion of 1887." It also noted that the principles of these converters would be employed in the 1,200-volt distribution system from Edison's new Orange laboratory, which would include converter "sub stations" at his home and several other Llewellyn Park residences (ibid., 11 [1 Oct. 1887]: 1–2; see Doc. 3056). These articles had been preceded by one in the 10 September issue about another patent recently issued to Edison. That specification, arising from an application filed in 1880, pertained broadly to the design and regulation of a multiple-arc distribution system and was only peripherally related to transformers ("Edison's New Patent on Multiple Arc Distribution," ibid., 11 [10 Sept. 1887]: 9).

*Alfred Tate to
A. B. Dick Co.*

[New York,] July 18th. 1887.

Dear Sirs:—

Mr. Edison has succeeded in his experiments with Parraffine paper, and suggests that you send him some of the paper which you use so that he can paraffine one side of it and return it to you to see if it is satisfactory.[1] Yours truly,

TL (carbon copy), NjWOE, Lbk. 25:43A (*TAED* LB025043A).

1. The A. B. Dick Co. was marketing the "Edison mimeograph" by this time. The editors have not found records of Edison's recent experiments on producing paraffin paper (a process he worked on as early as 1875 and as recently as 1885; see Docs. 698 and 2808), but Alfred Tate dispatched samples to the Dick Co. for inspection on 9 August. If they were acceptable, Tate promised to send the machine used to make them (or a model of it). Edison must have later offered to produce the paper himself, but in September Dick declined "to have our paper paraffined so far away from Chicago as Menlo Park, and we are arranging to do it here ourselves. Would be very glad to utilize the machine which you have perfected if it will work satisfactorily." The company did, however, wish to take up Edison's offer to make copying ink and carbon paper. Tate to A. B. Dick Co., 9 Aug. 1887, Lbk. 25:50C (*TAED* LB025050C); A. B. Dick Co. to TAE, 26 Sept. 1887, DF (*TAED* D8702AAY); also cf. Doc. 3085 regarding paper processing at Menlo Park.

To George Gouraud

New York, July 21/1887[a]

Copy[1]

Gouraud,

Your letter recd[2]   Under no circumstances will I have anything to do with Graham Bell[3] with his phonograph pronounced <u>backward</u> graphphone[4b]   I have a much better apparatus and am already building the factory to manufacture & I not only propose to flood England with them at <u>factory prices</u> but I shall come out with a strong letter the moment they attempt to float the Co. there

Edison

L (transcript), NjWOE, DF (*TAED* D8751AAB). Written by Henry Livor; letterhead of Edison Phonoplex System of Telegraphy. [a]"New York," and "188" preprinted. [b]Arrow drawn from end of word to front.

1. According to a faint note written at the bottom of the page, this transcription was made by Henry Livor of the Edison Machine Works sometime in 1887.

2. Gouraud had written on 2 July with several ideas for aligning the graphophone and Edison phonograph interests in Great Britain in a syndicate. He also asked to represent Edison in connection with

any new phonograph patents in Britain. After receiving Edison's reply, Gouraud sent a rather ambiguous congratulatory cable about the new "practical machine" in which he appeared to offer to "drop him [Bell] like hot potato" and to pay for all European patents. Receiving no reply to that message (and evidently others), Gouraud wrote again on 6 August. Gouraud to TAE, 1 and 6 Aug. 1887, both DF (*TAED* D8751AAC, D8704ABR1).

3. Alexander Graham Bell (1847–1922) was a teacher of the deaf who is best known as an inventor of the telephone. Born in Edinburgh, Bell followed in the steps of his father and grandfather, both noted elocutionists, by becoming a professor of vocal physiology and elocution at Boston University in 1873. He had by that time been experimenting with harmonic telegraphy for years, and by June 1875, he developed a telephonic transmitter, for which he received the foundational telephone patents. He and Edison had freely expressed admiration for each other's achievements (*Oxford DNB*, s.v. "Bell, Alexander Graham"; Docs. 1214, 1353, 1354). Bell had been experimenting with the phonautograph as early as 1874; astonished at having failed to hit on the idea of the phonograph, he began working in 1878 on phonograph improvements that he could claim as his own (Bruce 1973, 110–13; Doc. 1260). With money received in 1880 from the French government's Volta Prize, Bell established the Volta Laboratory in Washington, D.C., with Chichester Bell, a cousin, and Sumner Tainter, an instrument maker adept with sound recording and reproducing technologies. There they successfully devised a lateral-cut recorder and a playback machine. After modifying the recorder to cut (rather than simply indent) vertically into a soft blank, Tainter (with Chichester Bell) applied for a broad patent on sound recording and reproduction mechanisms in June 1885 (Martland 2013, 5–6; Bruce 1973, 350–55; Wile 1990a; Carlson and Prezter 2002; U.S. Pat. 341,214).

4. Edison referred to the graphophone, Bell's initial term for a device to reproduce sound from a record created by a phonographic machine. The name had become the market identity, if not yet a trademark, for the Volta Associates' improved versions of machines for recording and reproducing sound on wax records, and its similarity to "phonograph" caused at least one confused inquiry to Edison. Gouraud attempted to trademark the words "phonograph" and "phonogram" for Edison in Great Britain during the fall (Wile 1990a, 212; A. B. Dick Co. to Alfred Tate, 23 July 1887; Gouraud to TAE, 30 Nov. 1887; both DF [*TAED* D8702AAP, D8751AAL]). In 1887, the U.S. and Canadian rights to the Tainter-Bell patents were controlled by the American Graphophone Co., which moved quickly into manufacturing machines for dictation and transcription; the foreign rights were owned by the International Graphophone Co., an investment group organized the same year (Martland 2013, 6 and chap. 2).

Draft Patent
Application: Railway
Signaling

Dyer

Patent[1]   Railway Signals—

I put a large p̶Large permanent magnet on Locomotive with poles of iron flush with Rail in center track   In a closed[b] box between two ties is an Electromagnet. the box is poured full Insulating[c] Compound— The poles project above box & are about 3 or 4 inches from the polar faces of magnet on Locomotive— when Loco is running the passage of the permanent magnet induces a powerful current into the fixed rail magnet & the current passing along the track to any required distance actuates a magnet to control a sempaphore signal   when the train enters it sets up a current which blocks both on[c] emerging the magnet that is fixed has[c] its coils reversed & hence sends reverse current to unblock— an electromagnet could be used on Loco energized by battery or dynamo—and any form of fixed magnet could be used—

I[c] want claim broad principle of actuating signals by induced current without electrical or mechanical contacts & by motion of train[2]

E[dison]

You might glance thro RR Signaling patents to see if new before preparing[3]   E

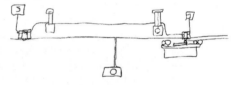

ADfS, NjWOE, PS (*TAED* PT032AAHx). Letterhead of Edison Lamp Co. [a]"HARRISON, N.J.," and "188" preprinted. [b]Interlined above. [c]Obscured overwritten text.

1. According to a docket notation, Edison's draft was received on 26 July at the offices of Dyer & Seely. Edison executed the patent application, which became Case No. 727, on 9 August; it was filed in the Patent Office ten days later. It issued in 1892 as U.S. Patent 470,923 after some modifications to the text, the addition of a small detail drawing (Fig. 2), and new claims to overcome the patent examiner's rejection due to British Patents 2,800 (1871) and 3,802 (1882) and German Patent 4,461. For the original application and correspondence regarding its rejection and alteration, see Pat. App. 470,923.

2. In the original application, Edison sought a general claim for the "method of operating electrical apparatus on railways, by current induced by a magnet on a moving train in a stationary electric circuit." The claims in the issued patent were limited to magnetically actuated signals energized by currents induced by the passage of the magnet on a train over a stationary electromagnet.

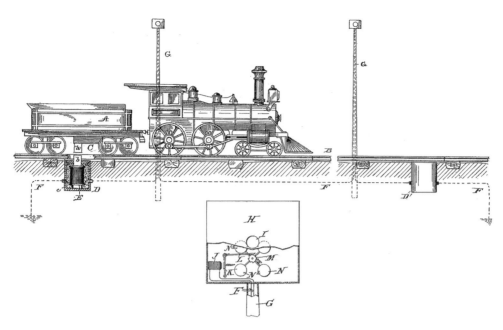

Figures in Edison's railroad signal patent 470,923. E is the electromagnet acted on by the poles (a) of the passing permanent magnet (C). Inset is an electro-mechanical device for moving semaphore signal.

3. In the application as filed (and issued), Edison stated that his invention was an improvement on "the operation of electrically con-trolled or operated signals upon railway-lines such as have heretofore been operated by the closing or opening of circuits by contacts made or broken by the train in its progress; and my object is to do away with the making and breaking of contacts for this purpose" (Pat. App. 470,923). American railroads began in the early 1870s to adapt electric signals to defined blocks of track. Such systems typically operated by sending low-voltage battery current through the rails within a relatively short section or "block" of track to a visual signal mechanism. A train enter-ing a block would cause a short circuit, and the loss of current to the sig-nal would produce a visual indication that the track was occupied. The leading developer and manufacturer of such systems at this time was George Westinghouse's Union Switch & Signal Co. (Solomon 2003, 33–42).

–3073–

From David Homer Bates

NEW YORK, 27 July 1887[a]

Dear Sir:

On March 30th 1886 I wrote you re draft of proposed agreement covering use of phonoplex on B&O lines, to which I have had no reply.[1] I would be glad to have you take this sub-ject up at your earliest convenience in order that we may place the matter in definite formal shape. Yours truly

D. H. Bates Pr[esident] & G[eneral] M[anager]

⟨Tate— See to this[b]

I have been helping Bates on his fire alarm telegh[2] & suggested that we could get a great deal of pleasure by scratching each others back occasionally so if you will follow this up you can get Bates to introduce phonoplex more largely— attend to it promptly pls   Edison⟩

ALS, NjWOE, DF (*TAED* D8752AAK). Letterhead of Baltimore & Ohio Telegraph Co., offices of the president and general manager. [a]"NEW YORK," and "188" preprinted. [b]Followed by dividing mark.

1. Although neither the letter from Bates nor the draft contract has been found, they are likely the items to which Alfred Tate referred in a letter to Edison on 15 March. He told Edison, who was in Florida, that Samuel Insull had given him a letter from Bates and a draft contract for the Baltimore & Ohio Railroad to use the phonoplex on its own lines and the commercial lines of its subsidiary, the B&O Telegraph Co. Tate explained that Bates had sent a memorandum of understanding that was less detailed than the "contract form" for the B&O that Edison had left with him, and that he had met with Bates to reconcile differences between the two documents. Tate put the matter in the hands of attorney John Tomlinson to prepare a final agreement, which he promised to forward to Florida. Tate to TAE, 15 Mar. 1886, Lbk. 21:424 (*TAED* LB021424).

2. Edison had worked extensively on municipal fire alarm telegraph instruments and systems as an inventor and manufacturer early in his career (see, e.g., Docs. 37–38, 78, 226 n. 2, 415 n. 5, 553 n. 5, 601, and 654). After Western Union bought out the Baltimore & Ohio Telegraph Co. in October 1887, Bates became vice president of the Gamewell Fire Alarm Telegraph Co. It was through Edison's "kind intervention" about this time that Gamewell obtained a contract to build a telegraphic fire alarm system in Orange, N.J. (James MacKenzie to TAE, 8 Sept. 1887, DF [*TAED* D8704ACO]; "Sketches of Some of the Members of the Old Time Telegraphers and Historical Association and of the Society of the United States Military Telegraph Corps," *Telegraph Age* 24 [1 Oct. 1906]: 478).

–3074–

Draft Patent Application: Gold Ore Separation

[New York? c. July 27, 1887][1]

To all whom it may concern:

Be it known that I, Thomas A. Edison of Llewelyn Park in the county of Essex and State of New Jersey, have invented a certain new and useful Improvement in Methods of Treating Auriferous Pyrites of which the following is a specification.

The object of my invention is to provide a simple and effective process by which gold may be separated from auriferous pyrites—that is from the common iron pyrites or bi-sulphuret of iron,[2] when this contains gold ores as it very frequently does.

The process which forms my invention consists of several steps, the main feature being the heating of the bi-sulphuret of iron to such a point as to drive out one atom of sulphur and convert the compound to the proto-sulphuret of iron. I have found that while the common pyrites or bi-sulphuret is very slightly magnetic and the oxide ~~found in~~ resulting from[a] the driving off of all the sulphur is but little more so, the proto-sulphuret is highly magnetic, and may be very readily separated from any non-magnetic materials which may be f mixed with it by the process of magnetic separation set forth in my patents Nos[3]   and   that is by altering the trajectory of the magnetic portion of ~~the~~ a[b] falling mass of the mingled materials.

In carrying my invention into effect, I first grind or crush the gold bearing pyrites into a pulverized condition and this pulverized material I heat in [----][c] a suitable manner at[b] a red heat[d] sufficiently to drive from it a single atom of sulphur and convert the compound into the proto-sulphuret ⟨(How do you tell?)⟩[e] [~~This may be determined by showing the amount of sulphur?~~][c]

The heating must be stopped at this point and not continued so as to drive out the remaining atom of sulphur, for the resulting oxide is very little more magnetic than the original bi-sulphuret, whereas the proto-sulphuret is highly magnetic.[4]

The next step in the process is the separation of the magnetic from the non magnetic particles in the finely divided mixture. I prefer to do this in the manner and by the apparatus set forth in my patents above referred to and which are now well understood. That is to say, the mingled mass of material is permitted to fall past the poles of one or more[f] electromagnets, whereby the trajectory of the magnetic portion of the mixture is altered and this is caused to fall on one side of a dividing line partition[g] while the non magnetic portions fall in a straight line upon the other side.

By this means, in the present process, the magnetic proto-sulphuret is separated from the gold and quartz and any other non-magnetic particles which may be mixed with it.

Afterward, the gold may be seperated from the quartz by the well understood processes of free milling.[5]

What I claim is:

First: The herein described[h] process of treating auriferous pyrites, by converting the bi-sulphuret of iron into the proto-sulphuret and then separating this magnetic material from the gold by magnetic attraction. ~~sub~~

<u>Second</u>: The herein-described process of treating auriferous pyrites, consisting in heating the material sufficiently to drive off sulphur enough to convert the bi-sulphuret of iron[i] into the proto-sulphuret, and then separating this magnetic material from the gold by magnetic attraction ~~subs~~

<u>Third</u>: The herein-described process of treating auriferous pyrites, consisting in first pulverizing the material, then heating it to convert the bi-sulphuret of iron into the proto-sulphuret, then separating this magnetic material from the gold and quartz or other non-magnetic[g] substances by magnetic attraction, and finally separating the gold from the quartz.

Df, NjWOE, PS (*TAED* PT032AAG, image 2). Probably written by Henry Seely. [a]"resulting from" interlined above. [b]Oscured overwritten text. [c]Canceled. [d]"at a red heat" interlined above. [e]Marginalia written in unidentified hand. [f]"one or more" interlined above. [g]Interlined above. [h]"herein described" interlined above. [i]"of iron" interlined above.

1. Edison wrote the original draft on 27 July but only one page of that document has been found (PS [*TAED* PT032AAG]), which differs substantially in form (though not substance) from the corresponding portion of this intermediate draft prepared sometime later by his patent attorneys, presumably in their New York office. Their version was retained in Edison's files with an oath that he signed but did not date, and the editors have not determined if a completed application was ever filed at the Patent Office. In any case, Edison did not receive a patent based on it. In May 1888, he applied for and ultimately was granted a patent on a method of extracting gold from a different type of rock (quartz) by first pulverizing the ore, then using a statically charged plate to remove the quartz particles. In September of that year, he also sought patent protection on a process for extracting gold from sulfide ores by using nitric acid to facilitate amalgamation with mercury (U.S. Pats. 476,991 and 474,591).

2. In his original draft, Edison had twice written "Bisulphide," and someone (probably Edison himself) subsequently added "Bi-sulphueret" in parenthesis above those instances.

3. Edison's original magnetic ore separator patent was U.S. Patent 228,329, filed and issued in 1880. By this time, he had obtained two other separator patents. U.S. Patent 248,432 was also filed in 1880 and issued the following year; U.S. Patent 263,131, filed in 1881, issued in 1882.

4. This process is closely related to experiments made by William K. L. Dickson as part of his ongoing summertime work on ore separation (see Doc. 3059 n. 8). On 26 July, Dickson recorded his observations about the "Roasting of Ores" to reduce iron pyrites "to magnetic pyrites, and by a much stronger heat, to a proto sulphide of Iron (FeS) which is not further reducible." He noted that

it appears necessary . . . to roast to a certain point and cease immediately on reaching a decided color for should I go one step further its magnetic property becomes extinct. the change is affected within a few seconds on a dull red pan— for this reason a small hand magnet must be in constant attendance, testing the ore every two or three seconds. [E-2610:12, Lab. (*TAED* NM031012)]

5. Free milling is a process for freeing gold by crushing the ore and amalgamating it with mercury. O'Driscoll 1889, chaps. 4, 6.

–3075–

Draft to Henry Hudson Holly[1]

Llewellyn Park N.J. July 30th, 1887

Dear Sir:

After a careful and thorough inspection of the work on my Labratory, made to day, I find the specifications have been departed from in several particulars, & that the work has been done in a poor and unworkmanlike manner.[2] ~~From which I am satisfied that your inspection of the work has not been proper, and I therefore desire. The work under your inspection. Instead of matters improving~~ [They?][a] ~~have They have become worse & I have therefore concluded to dispense with your services~~—. The walls are bulged & are[b] from ½ an inch to an inch & ½[c] out of plumb in a ~~distance~~ height[d] of six feet & the brick-laying is rough & irregular & projecting in every case from ⅛ to ¼ inch, & in some cases laid up in the walls without any mortar— ~~Your inspection of the~~ Your inspection of the work has not only been inefficient but has even[d] been an injury to me. I desire therefore [~~to pre?~~][a] to have the work done under some other inspections & do not care to further avail myself of your services—[3]

If you [~~the ---- you?~~][e] will furnish me with the detail drawings as per schedule.[f] I will the work done under other inspection from now on. Yours truly

T.A.E.

Df, NjWOE, DF (*TAED* D8755AAT). Written by John Tomlinson. [a]Canceled. [b]"bulged & are" interlined in left margin. [c]Obscured overwritten text. [d]Interlined above. [e]Illegibly interlined above. [f]"as per schedule." interlined.

1. The finished version of this letter (not found) was sent to Holly the same day and acknowledged by him soon afterward. Holly to TAE, 6 Aug. 1887, DF (*TAED* D8755AAU).

2. Edison visited the construction site frequently, often accompanied by an associate and occasionally by his wife. He hired or designated an employee, Jeff Waldron, to attend the site every day, and Waldron recorded his observations in a small notebook. Waldron noticed problems almost immediately, beginning with the composition of concrete in the foundations. Holly condemned the foundations but had a very different understanding from Edison of his role as supervising architect. In the

face of criticism, Holly hired his own inspector to monitor the work but this action failed to stem the troubles. Walls were out of true, built not according to plan and tied incorrectly to the beams, and the mortar was inadequate. Waldron marked Holly as a "schemer" and, amid mounting problems, concluded that "architect & contractors are all in together." He briefed Edison (and Mina) at the new laboratory on 29 July. Edison reportedly "was very angry & said he would have it stopped. He plumbed the wall with one of Fairchild's men Found it away out of plumb. Mad as hops." PN-87-07-16 (entries of 16, 20, 28–29 July 1887), WOL (*TAED* NL005AAA [images 7, 9, 12–13]; Holly to TAE, 20, 25, and 27 July 1887; all DF [*TAED* D8755AAN, D8755AAO, D8755AAP]).

3. Edison swiftly hired Joseph H. Taft (1855?–1911), a young architect with his own office on Broadway in New York, whom Waldron met later the same day at Edison's home. Taft reported to the laboratory the afternoon of the next working day (Monday, 1 August) and conferred there with Edison in the evening. Taft wielded authority over the builders and generally seems to have exercised much firmer supervision than Holly, although the next week Edison was still "very angry about way in which work was done," and problems persisted at least into September (U.S. Census Bureau 1970 [1880], roll T9_910, p. 62D, image 0127 [Cornwall, Orange, N.Y.]; Obituary, *NYT,* 18 July 1911, 9; Taft to John Everett, 1 Sept. 1887; Everett to TAE, 3 Sept. 1887; all DF [*TAED* D8755ABB, D8755ABC, D8755ABE]; PN-87-07-16 [entries of 30 July, 8 Aug., and 5 Sept. 1887], WOL [*TAED* NL005AAA, images 14, 16]). Eight detailed architectural drawings of the four small laboratory buildings, with Taft's stamp, are in NL010, WOL (*TAED* NL010AAA).

–3076–

To John Tomlinson

East Newark N.J. [July 1887?][1]

Tomlinson—

I want agreement between TAE—and Lowell C Briggs[2] & William[a] W Jacques[3] both of Boston,[4] as follows

1st   Exclusive license to make and sell phonograph Dolls and figures,[5] & not phonographs

2n   Royalty to be 10 per cent on the selling price of the Doll or figure whether retailed or wholesaled—in legitimate way—[b]

3rd   Clocks are excepted—[6] ⟨[-------   ---------][c] (?)⟩[c]

4th   They are to have the right to sell in any foreign countries, where there are patents[d] royalties to be the same.

5th   Royalties acruing to Edison for first nine 9[e] months to be not less than $4000. and the contract is thereafter to continue providing that in each year the royalties paid Edison shall not be less than $10,000 per year

Edison to sue the first infringer of toy phonograph apparatus in N York City but[f] not in[b] any other city or state ie[e] make one test[f] case—

Usual clauses about sworn statement access to books, acknowledge validity patent, bonifide Sales etc.[7]

T A Edison

ALS, NjWOE, DF (*TAED* D8750AAZ). [a]"illiam" interlined above. [b]Obscured overwritten text. [c]Marginalia written in unidentified hand and canceled. [d]"where there are patents" interlined below, followed by dividing mark. [e]Circled. [f]Partially obscured by ink blot.

1. Edison's directive to Tomlinson was prompted at least in part by Edward Johnson's 22 June letter (see note 4). Tomlinson had time before 5 August to prepare a "revised phonograph Contract" based on this outline. Tomlinson to TAE, 5 Aug. 1887, Misc. Legal (*TAED* HX87009A).

2. Lowell Chickering Briggs (1853–1917?), a Boston native, became the founding treasurer of the Edison Phonograph Toy Manufacturing Co. near the end of 1887. Briggs birth record, *Massachusetts Town and Vital Records, 1911–1915,* online database accessed through Ancestry.com, 21 Apr. 2014; "Sanity Hearing for Lowell C. Briggs," *NYT,* 6 Apr. 1908, 3; "Auction Notice," *New York Herald,* 13 Jan. 1918, 6; "Miscellaneous Notes," *Electrical World* 10 (26 Nov. 1887): 288.

3. William White Jacques (1855–1932), who became the founding president of the Edison Phonograph Toy Manufacturing Co., was born in Newburyport, Mass., and held a Ph.D. in physics (1879) from Johns Hopkins University. He was a lecturer at the Massachusetts Institute of Technology (1885–91), from which he had also graduated (in 1876), as well as an electrician and experimenter for the American Bell Telephone Co., and the first head of its "Experimental Shop." By this time Jacques had already received patents for cables, insulation compounds, conductors, and couplings (e.g., U.S. Pats. 270,438; 227,168; 240,720; 249,840). Jacques incorporated the Edison phonograph into a talking doll, for which he filed a patent application in October 1887 (U.S. Pat. 383,299). He later did significant research on fuel cell technology and the direct conversion of carbon into electrical energy. "Obituary," *NYT,* 25 June 1932, 13; "Miscellaneous Notes," *Electrical World* 10 (26 Nov. 1887): 288; Hoddeson 1981, 519–21; Brown 1926, 181; Johns Hopkins University 1881, 31; Johns Hopkins University 1914, 100; Fagen 1975, 41; Wile 1987, 7, 29 n. 12; Doc. 2520 (headnote); Suplee 1901, 264.

4. Edward Johnson informed Edison on 22 June that his "Boston friends [who] are at work upon the toy application" had demonstrated "their apparatus" to him. It was, he reported, "in an unfinished Condition—that is why they want another month—but it is very cheap & now they have some good Ideas." He enclosed the two 1878 contracts governing applications of the phonograph to toys and clocks (see note 6), pointing out that they had expired (with the rights presumably reverting to Edison). He queried: "Do you want to continue to deal with them separately or will you pool the whole thing? If the latter I will extend the Call I have given these Gentlemen for another 30 Days— If you do not intend to hand this branch Over to me then I can of course Only refer them to you or act for you in the matter" (Johnson to TAE, 22 June 1887, DF [*TAED* D8750AAA]). Neither of the five-year agreements for toys and clocks had led to any commercial success, although

the toy contract, at least, had been coveted by Uriah Painter, a principal of the Edison Speaking Phonograph Co., in 1879 (see Docs. 1190, 1238, 1448 n. 1, 1618, and 1785 n. 5).

5. Charles Batchelor took up the technical challenges of a small phonograph for dolls between February and April 1888. Wile 1987, 9–10; Cat. 1337, Batchelor (*TAED* MBJ004).

6. In January 1878, Edison signed separate agreements licensing the use of the phonograph in toys and clocks. The two clock licensees were both associated with the Ansonia Clock Co. of Ansonia, Conn. (Doc. 1168 n. 7; TAE agreement with Daniel Somers and Henry Davies, 7 Jan. 1878, DF [*TAED* D7932ZBF13]). They made little progress in the ensuing decade but when, at the end of 1888, another party inquired about making a similar arrangement, Edison replied that he already had a "contract for it with Ansonia Clock Co. & that we have been Expmtg for that end last 6 months" (TAE marginalia on Alfred Tate to TAE, 10 Dec. 1888, DF [*TAED* D8847ADS]).

7. On 1 October, Edison executed a contract on terms similar to those outlined here. It granted to Jacques and Briggs collectively the right to domestic manufacture of phonographic dolls and toys for sale in the U.S. or foreign countries. Applications of the phonograph to anything other than the amusement of children (specifically to devices for business or science) were excluded. The contract also anticipated an extension of manufacturing rights to foreign countries, and a second agreement to that effect was signed on 25 November. Neither document licensed Edison's original phonograph patent or referred to its ownership; both instead authorized the manufacture and sale of the invention described in the specification, and (in the second case) referred to Edison as "the inventor of what is generally known as the 'Phonograph.'" TAE agreements with Jacques and Briggs, 1 Oct. and 25 Nov. 1887, both Misc. Legal (*TAED* HX87009, HX87014).

–3077–

Notebook Entry:
New Laboratory[1]

[Harrison, July 1887?]

Building 25 × 100 one story with 8 stone foundations for galvanometer & acurate Experiments. No iron or iron nails. granet roof— ~~Building~~ closing shutters to make it dark. Instrument cases on one side—brick

Building for stamp mill crusher, furnaces 25 × 150 one story—brick floor—granet roof—brick.

Acid[a] store house dangerous chemicals paints Scrap heap— 25 × 50. 1 story brick—light walls—

⟨Do you want a -------?]⟩[b]
Kind of men wanted—

Watch tool maker
Telegraph instrument makers
Pattern maker on small work
A. Mechanic & one fully understanding Lapidary work, grinding lenses jewel draw plates etc.

Mechanic who understands Tinsmith & Coppersmith work
glass blower fancy for Philosophical & ordinary Lamp ones
Joe Force will answer for fancy man I think.

Engineer for night run—

Johnny Randolph bookeeper & store keeper   one young
man for assistant.

John Ott for head draughtsman Inspector & designer

3 other draughtsmen,[2] one of which must be able draw pictures f[-----][c] Weimer[3] will do

One person familiar with scientific matters to translate
must understand English french german & Italian—

A good operative chemist, organic & inorganic

"   "   good practical mathamatician

A person thoroughly familiar with ~~Lenses~~ Optics & optical
apparatus

Get Bob Spice[4] if possible.

What does Carl Hening[5] do.

photographer— Parrish[6] will do—

Payne—Hamilton, Kenny Dickson and six smart young
fellows as assistants. Couple from College who understand
physical measurement—[d]

Keller Fred Ott,[7] in shop. ~~Ke~~ with 15 good[a] workmen

A general cleaner & sweeper—

good foreman for shop. Wurth[8] might do if not too high
salary

Machinist who understands buffing & polishing

Blacksmith.

[A][9]

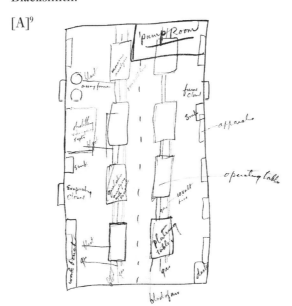

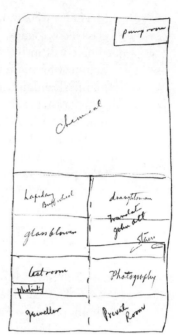

Clamping by deposit in small room in Engine H[ouse] Extension[11]

[C][12]

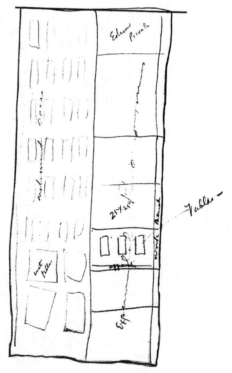

gal & acurate Expmts   Extra bldg 25 × 100 1 story[13]

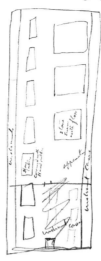

Small bldg   50 × 25—1 story[14]

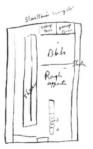

Furnace bldg—25 × 100   1 story—[15]

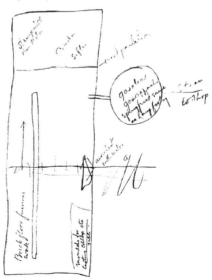

ascertain from Upton what Carbonizing bldg cost[16]

X, NjWOE, WOL, E-4294 (*TAED* NL006031). ^aObscured overwritten text. ^b"Marginalia written in unidentified hand and incompletely erased. ^cCanceled. ^d"who understand physical measurement" interlined above succeeding line.

1. The editors have selected from an undated notebook devoted to Edison's new laboratory in Orange sections containing details about the new buildings and his prospective staff. (The book's remaining portions, including a page [33] preceding the first drawing, consist of lists of materials, equipment, and apparatus to be ordered.) The selected portions contain Edison's early ideas for the layout of the rooms on the upper floors of the main laboratory building and his first designs for the small laboratory buildings. They were most likely written in July 1887, perhaps over a period of days. According to Charles Batchelor's journal, by 4 August the brickwork on the main laboratory building had been "done up to top of windows of first floor" and "Mr. Edison has added other buildings which Mr. Taft (the architect who succeeds Mr. Holly) is now designing." Joseph Taft submitted the drawings for those structures to the contractor, V. J. Hedden & Sons, by 5 August. By that time, the 50 × 25 foot building for chemical storage also included the carpenter shop and had been extended to 100 feet long while the "Building for stamp mill crusher, furnaces" had become the Metallurgical Building. A fourth 25 × 100 foot building, the Chemical Laboratory, had also been added. Cat. 1336:253 (item 422, 4 Aug. 1887), Batchelor (*TAED* MBJ003253A); V. J. Hedden & Sons to Taft, 5 Aug. 1887, Vouchers (*TAED* VC87004A); Architectural Drawings 1887, WOL (NL010AAA).

2. Edison advertised in the *New York Herald* on 7 September (p. 12): "Good Mechanical Draughtsman Wanted: one who has experience in designing special machinery; it will be useless for any but first class man to apply; must come well recommended; hours 8 to 6. State salary required and address M., box 341, Herald office." Responses to this notice can be found in the "Employment" folder for 1887 (*TAED* D8713). At least one of the respondents, John Cook, who had been employed by the Matthiessen and Wiechers Sugar Refining Co. in Jersey City, went to work for Edison. Cook to TAE, 7 Sept. 1887 (with Charles Batchelor maginalia); Cook to Charles Batchelor, 21 Sept. 1887, 6 Feb. 1888; William Roach to TAE, 7 Sept. 1887; L. Duvinage to TAE; 7 Sept. 1887; all DF (*TAED* D8713AAD, D8713AAJ, D8814AAV, D8713AAF, D8713AAG).

3. Not identified.

4. Robert Spice (b. 1845?), an English-born teacher of chemistry and natural philosophy at the Brooklyn High School had given Edison private instruction in chemistry in 1874. Edison hired him again in late 1875 for lessons in acoustical science, and Spice stayed into 1876 as a member of the laboratory staff. U.S. Census Bureau 1982 (1910), roll T624_1036, p. 12A (Manhattan Ward 18, New York, N.Y.); see Docs. 487, 490, and *TAEB* 2:581.

5. Edison likely referred to Carl Hering (1860–1926), who had been involved with the comparative life tests of incandescent lamps in 1884 at the Electrical Exposition in Philadelphia, where he served as assistant electrician. Hering received his mechanical engineering degree from

the University of Pennsylvania in 1880 and then became an instructor in mathematics and drafting and an assistant in mechanical engineering there. Deciding in 1882 to take up electrical engineering, he served for a year as assistant in physics under George Barker before going to study with Erasmus Kittler in Darmstadt. He then became chief electrical engineer for Henry Moehring and Co. in Frankfurt. Hering returned to Philadelphia in 1886 and set up practice as a consulting electrician. "Carl Hering," *Electrochemical and Metallurgical Industry* 4 (1906): 230–31; "Carl Hering," *Electrical Review and Western Electrician* 57 (1910): 819; Cutter 1916, 10:197–202.

6. Possibly Legrand Parish.

7. Frederick P. Ott (1860–1936), John Ott's brother, entered Edison's employ in 1874 as an apprentice at the Ward St. shop in Newark. Ott remained in Newark with Edison's former partner Joseph Murray before rejoining Edison at Bergmann & Co. in 1883. He became part of Edison's lamp factory laboratory staff in 1886 and then went to the Orange laboratory as Edison's mechanical assistant. "Ott, Frederick," Pioneers Bio.; Dyer and Martin 1910, 648.

8. Charles Nicholaus Wurth (1841–1921) was a Swiss machinist who came to the United States in 1869 as a toolmaker for the Singer Sewing Machine Co. in Elizabeth, N.J. He began working for Edison as a machinist the next year. He went back to Switzerland in 1872 but returned to Edison's employ from November 1873 to April 1877. Wurth rejoined Edison at the Orange laboratory in 1888 and stayed there until 1903, when he went to Europe to establish factories for manufacturing phonograph records; he continued working for Edison until 1910. "Wurth, Charles Nicholaus," Pioneers Bio.; Wurth's testimony, *Edison v. Lambert*, p. 255, Lit. (*TAED* QP009255).

9. Figure labels are: (left column from top) "blast," "assay furnace," "distilling & dangerous expts," "blast," "sink," "Evaporating closet," "work bench," "blast," and "gas"; (next column to right) "inorganic analysis," "100 volt line," "qualative analysis with Reagents," "blast," and "gas"; (third column to the right) "pump Room," "10 volt line," "gas," "plating table," "gas," and "blast of air"; (fourth column to the right) "fume closet," "Sink," "apparatus," "operating table" (connected by line to table in third column), and "desk."

10. This drawing likely is an early idea for the arrangement of either the second or third floor of the main laboratory building; another version (possibly earlier), is in N-87-00-00.3:3, Lab. (*TAED* NA003, image 3). Figure labels are: (left column) "chemical," "Lapidary Buff wheel," "glassblower," "test room," "photometer," and "Jeweller"; (right column) "pump room," "draughtsman," "Translator," "John Ott," "Stairs," "Photography," and "Private Room."

11. Edison presumably meant the power house that was later designated Building 6, located next to the far end of the machine shop. It contained three Babcock & Wilcox boilers and two Armington & Sims steam engines (125 and 80 horsepower). The engines drove dynamos that provided electric light and power for the laboratory, Glenmont, and several other houses in Llewellyn Park. Millard 1995, 1:26–28; see also Doc. 3056.

12. Figure labels are: (left column) "instrument cases" and "Inst. tables"; (right column) "Edison Private," "Rooms," "25 × 25," "ap-

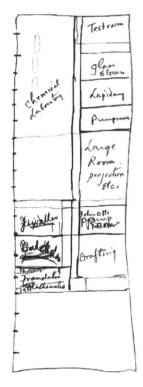

Edison's alternate sketch of the layout of one floor of the main laboratory building.

paratus," and "Experimental"; (narrow column at far right) "work bench"; (in margin connected by line to right column) "Tables—."

13. Figure labels are: (narrow column at far left) "instruments"; (next column to the right) "stone tables   covered with Hard rubber"; (next column to the right) "stone Even with floor," "apparatus," and "instrument case"; (narrow column at far right) "instrument cases."

14. Figure labels are: (at top) "Store House Scrap Etc"; (left column) "shelves"; (right column) "scrap Iron," "scrap Brass," "[Bls?]," "Rough apparatus," and "acids"; (narrow column at far right) "Shelves." In an undated memorandum, probably written later to Charles Batchelor, Edison indicated that the "Woodworking shop is 50 × 25 ½ of one of the small bldgs—so calculate floor space accordingly— we run shafting by an electromotor—" (DF [*TAED* D8755ACJ]).

15. Figure labels are: (top, left to right) "Stamp mill Run Motor," "Crusher," "Sifter," and "wood partition"; (in circle) "gasolene gas apparatus spring [feed?] same as Lamp factry"; (to right of circle) "Steam to shop"; (center, below circle) "air blast with motor"; (bottom, left to right) "Brick floor furnace work," "moulds for casting alloys etc," and "Table."

16. Edison referred to the new carbonizing furnaces installed earlier in the year at the Edison Lamp Co. factory in Harrison ("Specification for the Construction of two Brick Chimneys," 4 Mar. 1887, DF [*TAED* D8734AAF]; "Drawing of Chimney for carbonizing furnaces," 9 Oct. 1886, Oversize Notes and Drawings, Lab. [*TAED* NS7986BAT]). The "Furnace bldg" became the Metallurgical Building. In an undated memorandum on laboratory letterhead, likely written to Francis Upton in December 1887, he stated: "I want a furnace built just like the gas furnace you have at factory send   your bricklayer over with plans to Laboratory as it is to go into the Metallurgical bldg" (DF [*TAED* D8755ACL]).

–3078–

Draft to James Hood Wright

[Harrison? c. August 3, 1887[1]]

Mr Wright—

My laboratory will soon be completed— The dimensions are one building 250 ft long 50 wide & 3 stories   4 other bldgs 25×100 one story high all of brick—

I will have the best equipped & largest Laboratory extant, and the ~~best~~ facilities incomparably superior to any other for rapid & cheap development of an invention, & working it up into Commercial shape with models patterns & special machinery— In fact there is no similar institution in Existence   We do our own castings forgings   Can build anything from a ladys watch to a Locomotive.[2] The machine shop is sufficiently large to employ 50 men & 30 men can be worked in the other parts of the works— Inventions that formerly[a] ~~cost me~~ took months & cost a large sum can now be done ~~in day~~ 2 or 3 days with very small Expense as I shall Carry a stock of almost every Conceivable material of every size and with the latest machinery a man will produce 10 times a much as in a laboratory ~~with~~ which has but little material not of a size, delays of days waiting for Castings and machinery not universal or modern [-----]ᵇ—

You are aware from your long acquaintance with me that I do not fly any financial Kites, or speculate, and that the works which I control are well managed— In the Early days of the shops it was necessary that I should largely manage them 1st because the art had to be created 2nd because, I could get no men who were competent in such a new business,—but as soon as it were possible, I put other persons in charge. ~~I also think I know my place wh I am not competent~~. I am perfectly well aware of the fact that my place is in the Laboratory, but I think you will admit that I know how a shop should be managed ~~also~~ [--]ᵇ & also how to select men to manage them—

~~Have~~With this prelude I will Come to business—

My Ambition is to ~~start~~ build up a great Industrial works in the Orange Valley starting in a small way & gradually working up— The Laboratory supplying the perfected invention ~~will~~ models pattern & fitting up necessary special machinery in the factory for each invention

~~My p~~[---]ᵇ My Plan Contemplates to making of only that class of inventions which require but small investments for each and of a highly profitable nature & also of that character that the articles are only Sold to [~~traders?~~] Jobbers Dealersᵃ Etc— No cumbersome inventions like the Electric Light.

Such a works in time would beᶜ running ~~of~~ on 30 or 40 special things of so diversified nature that the average profit would scarcely ever be varied by competition Etc—

My plan is to form a Company with a large authorized Capital but all held in the treasury—to have a board directors Etc— Having formed the Co I bring before it say ~~25~~ or ~~36~~ small inventions the board decides which ones they think will pay Say 4 are selected, Estimates are made for bldg grounds & necessary machinery to start mfg these 4 inventions. Say 35 000 The Costs of Experimenting were say, 65 000ᵈ The subscribers to the stock take ~~35~~40 000 at par and give me ~~35~~40 000 of stock—

Thus of the profits ½ would go to the money & ½ to the inventor—but to insureᵃ the money. The money stock to have a 5 pct preferential div ieᵉ if the Co earn ~~20 pct~~ only 5 pct the inventor gets nothing if it Earn 180 percent the inventor gets 5

& the money 5. if the earnings are 20 pct the money of course gets 10 & Inventor 10—

Now this would look as if the Co would have to earn a large profit to yield the investors 10 pct but if you remember that the usual royalty paid[a] inventors is 15 pct on the selling price, ~~this is what I ask is low~~ & this is always added & Called part of[f] Cost & the dividends & profits are made over & above this you see that, what is paid me is not ac[coun]t[ed] as a[g] reduction in dividend but may properly be charged to Cost of production

Again a great number of these small inventions require scarcely any investment to mfr— $1000 invested in some special tool will turn out articles which would pay to the Co. several thousand dollars a year profit—

The money would earn after paying my part 100 to 200 pct—

When an invention by reason of Competition ~~or~~Etc fails to pay a satisfactory profit we can drop it & substitute something Else. = ~~I would not bind myself to give them Every invention but that I might devise but~~

~~The~~   ~~All inventions~~ — All inventions which[a] the Co can manufacture devised[a] in Laboratory to go to them but any process etc not capable of being mfd to belong to E & no Expenses for Experimenting to be charged Co — All[a] other ~~inventions~~ Experiments to be paid by Co.— Total Experimental Expense per year limited to $25 000. without the directors decide to spend more

Inventions relating to EL   WU Telegh—Telephone— phonograph—Ore milling Excepted—[3]

I honestly believe I can build up work in 15 or 20 years that will employ[a] 10 to 15 000 persons & yield 500 pct to the poineer stockholders—

I propose starting with $30 @ 35 000 [-------][b] & 3 or 4 small inventions. Now Mr W do you think ~~you can~~ this practicable   if so can you help me along with it[4]

E[dison]

ADfS, NjWOE, Lab., N-87-11-15:5 (*TAED* NA011005). Text written over two unrelated and incomplete sketches. [a]Obscured overwritten text. [b]Canceled. [c]"in time would be" interlined above. [d]"The Costs . . . 65000" interlined above. [e]Circled. [f]"part of" interlined above. [g]"act as a" interlined above.

1. This draft was written in a laboratory notebook, probably at the lamp factory laboratory in Harrison. The conjecture of the date is based on Wright's later acknowledgment of letters from Edison dated 4 and 12 August. Edison likely wrote them out himself, bypassing Alfred

Tate and the usual process of making copies, and neither letter has been found. Wright was away from New York City for an extended period but promised to bring Edison's communications before Charles Coster at Drexel, Morgan & Co. at the start of September. The editors suppose that Edison's 4 August letter was based on this draft and that, having no reply, he followed up five days later. Wright to TAE, 17 Aug. 1887, DF (*TAED* D8703AAD); see also Doc. 3088.

2. A variation on this sentence found its way into an editorial paraphrase of a newspaper interview with Edison in late October or early November. According to the article, Edison went on to explain (in his own words) that his new laboratory complex would not be "a factory in any sense, for nothing will actually be produced here but models. It is to be a factory hatcher, so to speak. We shall devise things to be manufactured elsewhere." "A Talk with Edison," *Mechanical Engineer* 14 (5 Nov. 1887): 105.

3. Commercial rights to these inventions were covered by separate contracts.

4. Before hearing back from Wright, Edison floated a similar proposal to William Lloyd Garrison, Jr. (see Doc. 3080). Early in 1888, he also sought funding for both the laboratory and the proposed industrial works from Henry Villard, for whom he provided both a list of prospective inventions and draft contracts. TAE to Villard, 19 Jan. 1888; Alfred Tate to Samuel Insull, 18 Feb. 1888; both DF (*TAED* D8805AAI, D8818AEJ); TAE draft agreements with Villard, 1887 and 1888, both Misc. Legal (*TAED* HX87017, HX88036); Israel 1998, 268–69, 334.

–3079–

Charles Batchelor
Journal Entry

[Glenmont?] Aug. 12th 1887.

432[1]   Tellurides of Gold.

Prof. Barker and I dined at Edison's to night & slept there.[2] During the evening Edison asked Barker how they found out that Tellurides of Gold existed,[3] as he said he believed there was no such a thing— He said he had powdered the so called Tellurides very fine so that it would go though 150 mesh[4] (roughly), and then washed it well and panned it on a slab[a] of glass and scparatcd gold that was so finc it was microscopic. This he separated and the residue he found no gold in at all.

AD, NjWOE, Batchelor, Cat 1336:67 (*TAED* MBJ003267B). [a]Obscured overwritten text.

1. Charles Batchelor consecutively numbered each entry in this journal.

2. George Barker was probably en route to New York City, where he read Edison's paper on the pyro-magnetic dynamo (Edison 1888b) to the American Association for the Advancement of Science on 15 August. "Scientists and Prayers," *New York World*, 16 Aug. 1887.

3. Tellurides of gold (native alloys of gold with tellurium) appeared in multiple editions of the standard chemical dictionaries by Henry

Watts to which Edison frequently referred. One early edition identified them under the heading "sylvanite" (also known as yellow tellurium or aurotellurite), found in Transylvania (Watts 1872, s.v. "Sylvanite"). The 1875 edition contained a separate entry for "Gold, Telluride of" (Watts 1875), and a subsequent revision listed the ores under the main entry for tellurium, at that point linking them with the Calaveras mountains of California (Watts 1879b, s.v. "Tellurium"). A more recent edition identified one of these compounds as calaverite and associated it with the analytic work of Frederick Genth (Watts 1881, 1899), a mineralogist and chemist at the University of Pennsylvania and a colleague of George Barker, who had facilitated at least one small scientific interchange between him and Edison in 1878 (*ANB*, s.v. "Genth, Frederick Augustus"; see Doc. 1437). Edison had also gathered limited information about tellurium and its association with gold in the course of his search for platinum in 1879 (see Doc. 1734 n. 1; A. Coan to TAE, 4 June 1879; Decker and Jewett to TAE, 8 July 1879; E. A. Chase to TAE, 19 July 1879; all DF [*TAED* D7928ZAK, D7928ZEW, D7928ZHP]).

Edison sent Genth an inquiry (not found) on 6 August about obtaining samples of rare minerals. In answer, Genth offered the possibility of selling portions of his own collection in the near future, though he made no promise of selling to Edison. His reply focused on the availability of tellurium itself rather than precious metals often found with it, and he advised Edison to apply to a Leipzig dealer or to government authorities in Austria (Genth to TAE, 9 Aug. 1887, DF [*TAED* D8747AAD]). Edison apparently sent broader inquiries to at least three other potential sources in August and September. One was Arthur Foote, a Philadelphia mineral dealer, to whom he gave a list (not found) in August of desired minerals, including large quantities of pyrite and chalcopyrite, probably for ore milling experiments. Edison made inquiries to another Philadelphia dealer in September, who acknowledged his interest in "procuring information about the ores you want for some of the rarer metals" and minerals, notably tellurium but also bismuth, tungsten, cerite, and cobalt. Edison also wrote to Paul Dyer, a former canvasser for his central station construction department, who was then working in Colorado, and whom he hoped to engage to collect ores from the region. Dyer regretted his inability "to mail samples of minerals as you desire" but suggested the local availability of telluride ores which "have never been worked for Tellurium but it exists with other minerals more precious" (Foote to TAE, 23 Aug. 1887; George L. English & Co. to TAE, 10 Sept. 1887; Dyer to TAE, 19 Sept. and 21 Oct. 1887; all DF [*TAED* D8756AAW, D8747AAH, D8713AAI, D8704ADU]).

4. Screens and sieves are standardized by the number of meshes per inch. A 150 mesh screen has that many meshes to the linear inch, or 22,500 holes per square inch. See, e.g., Great Exhibition 1851, 2:632 (Section III, Class 22, no. 332 for Nicklin & Sneath).

Orange, Aug. 13th, 1887.

Copy.ᵃ

Friend Garrison:[1]

I suppose you know that I am building a large Laboratory on the Valley Road, consisting of one building 50 × 250, 3 stories, and 4 other buildings 25 × 100, one story. It will be equipped with every modern appliance for cheap and rapid experimenting, and I expect to turn out a vast number of useful inventions and appliances in industry. I have already several good things, and propose putting up a factory to manufacture, making only such articles that can be sold through jobbers, and yield large profit.

I write to you to ask if you think Boston parties would put money in such a factory? The plan is that 51 per cent. of the stock be paid me for the inventions, the 49 to be sold at par. This would seem to be asking too much on my part, but when one considers that some of the small inventions will yield a good many thousand dollars profit, and the cost of machinery to produce will in some cases only cost $500. you will see that it is equitable. Of course the Company should have the privilege of rejecting (ie) not manufacturing any article presented. They could make only those which required small investment—in other words the Laboratory turns out the invention in a commercial shape with patterns, models, etc, ready to manufacture, and the money part of the Company can either decide to make it or not. I should want to start small at first, say for buildings and land $20,000; $15,000. for machinery to start on 2 or 3 things I already have. In time I think it would grow into a great industrial work with thousands of men. You know Mr. G. that I have already 1300 men under me. The works at Schenectady, of which I own 75 per cent, employ 900 men, and turn out $1,500,000. per year, and earn 15 per cent. on the cost of labor, material and general expense. I am proud of those works, and I believe them to be the finest in the country, and any expert will say they are <u>well</u> managed. I do not manage shops myself as I am incompetent for that class of work, but I do know how to select the right kind of men to do it for me. I also own 59 per cent. of the Lamp Factory, which industry with all its appliances I built up. This I believe is considered by all to be a well managed factory, and pays 12 per cent. and has surplus for extending; and pays all experimenting expenses thus freeing Light Company from that part. I expended $21,000. on new Lamp, and necessary appliances to make it. I also built up Bergmann & Co., and own ⅓ of capital

$750,000. They did pay 20 per cent. on $300,000. but now pay 8% on $750,000, so you see I have created the industry, and run the works successfully. Owing to the rapid spread of our business in last 6 months, our works have over a million dollars worth of work on books, and it has taken all my money to help carry the stock, etc., so my intention of starting the new works myself is defeated. My ambition is to build up gradually and surely a great Industrial Works in the Orange Valley. With my Laboratory and skilled men as the creator of highly profitable specialties, but not big cumbersome things like a system of Electric Lighting. Now Mr. G. what do you think, would people invest in it? Yours,

Edison.

L (copy), NjWOE, DF (*TAED* D8704ABW). [a]Followed by dividing mark.

    1. Cf. Doc. 3078.

−3081−

From William Lloyd Garrison, Jr.

Wianno, Mass.[1a] Aug. 19, 1887.

Dear Mr. Edison.

Pardon my delay in answering your kind letter of the 13th[2]   I have been unwell this past week & have come down to Cape Cod to rest a little—

I have read with great interest your plan to establish the electrical works in the Orange valley & should be proud to be of service to you in the line suggested. The amount of capital asked is modest, too modest in my opinion. It seems to me wiser to start with enough, so that dividends & not assessments may follow.— At present writing, owing to the peculiar state of the money markets, it is not easy to interest capital in new enterprises.[3] My business is with investors & I know the stringent and cautious feeling prevalent. Many apprehend a severe commercial revulsion & the general attitude of capital is that of caution & self protection. A short time will either make a change for better or worse.— With ordinary financial confidence I should consider the task an easy one[a] to place the capital stock of your company on the terms proposed,—provided I see some adequate remuneration for myself.— The Brockton Co. has not been remunerative to my stockholders, although I think it promising under the new organization.[4] I have labored & worried to make it a success nearly four years, without one cent of recompense in any shape, & have locked up, besides, $13 000.—in the stock.— The moment I make the stock worth more than par I shall unlock considerable capital available for further use, excluding my own.— Until I do that my friends whom I induced to invest may justly decline to repeat the operation.— Not that I have used up my constituency, but disappointment & experience make[b] me more circumspect & less enthusiastic in urging new enterprises on friends, especially ~~when~~ since[a] "the war has[a] failed to come up to the proclamation."[5]

But after all is said & done, I believe in you & your star, although I have had ocasion to question the quality & the conscience of some of your lieutenants. And I should like further time to consider your scheme & to talk with you about it.— Sincerely & cordially,

W. L. Garrison

I shall be in Boston on Monday & any letters may be addressed there.—[c]

ALS, NjWOE, DF (*TAED* D8704ABY). [a]Interlined above. [b]Obscured overwritten text. [c]Postscript written in left margin.

1. Garrison kept a summer residence at Wianno, a section of Osterville, on Cape Cod. His late father, the abolitionist, had summered at Osterville at least several times in the late 1870s. Mitchell 1883, 658; Sargent (*New England*) 1916, 569; Merrill and Ruchames 1971–1981, 6:411.

2. Doc. 3080.

3. According to the *Financial Review*, U.S. stock markets suffered a "prolonged attack of indigestion" in the latter half of 1887. Money markets "were greatly unsettled" in early summer by the bursting of a speculative "coffee bubble" in New York and a Chicago "wheat bubble." Those events produced a temporary sharp tightening of credit and a broader "fear of tight money" that was abated only by the federal government's bond purchases in the fall. "Retrospect of 1887," *Financial Review* 1888, 1, 4.

4. Operations of the Edison Illuminating Co. of Brockton had been reorganized earlier in the year by John Vail, general superintendent of the Edison Electric Light Co. of New York. According to the *Electrical World*, Vail effected "a consolidation of the arc and incandescent lighting, avoiding any conflicting interests, and largely reducing the cost of running the stations." Vail also modernized and enlarged the plant (including the addition of a second floor) and added Sprague motors and other equipment to increase its capacity. "New England Notes," *Electrical World* 9 (21 May 1887): 246.

5. The editors have not identified a specific source for Garrison's quote, possibly a reference by his abolitionist father to Lincoln's Emancipation Proclamation.

New York, Aug. 24th, 87.[a]

Dear Sir:—

Referring to the difficulties which are experienced in the blackening of the lamp globes, it has occurred to me to suggest that the globes of the lamps might be made somewhat larger than they now are, with good results as to less blackening.

If we have a globe of a certain size with the carbon inserted therein, ~~they~~ there[b] are exposed to the ~~decomposition~~ deposition[b] of carbon a certain number of square inches of the interior surface of the globe; consequently, within a certain length of time, the smut on the globe will reach a certain thickness which cuts down the candle power of the lamp to a specified amount. Now then, if we put the same carbon under the same conditions in a larger globe, there is more surface to be covered by the smut, and it consequentially will not reduce the candle power so rapidly.

This is simply offered as a suggestion. Yours truly

J. H. Vail. Gen'l Supt.

⟨EHJ would kill us if we did this— He is all time trying get them smaller   Ive made them small as I dare already— What you say is correct   E[dison]⟩[1]

TLS, NjWOE, Vail (*TAED* MEo10). Letterhead of Edison Electric Light Co. executive offices, J. H. Vail, Gen'l Supt. [a]"*New York*," preprinted. [b]Interlined above by hand.

1. Edward Johnson wrote to Vail a few days later regarding a memo from Edison (not found) reportedly showing that "in a fall of 3 candles but $\frac{1}{2}$ c was Due to Blackening & $2\frac{1}{2}$ C to increase of resistance." He questioned whether the "infinitesimal" decrease in blackening from depositing the carbon over a larger glass area would "Compensate for the great Annoyance of large globes not to speak of their unsightliness." Johnson to Vail, 29 Aug. 1887, Vail (*TAED* MEo11).

Harrison, N.J., Aug 24/87[a]

Dyer & Seely—

New patent— System of Electrical distribution for Light heat & power[b]

The object of this invention is to economically & effectually distribute Electric Energy at a practically constant potential over extended areas without the use of large and expensive [Copper?][c] Conductors & the employment of regulating apparatus consuming[d] electrical energy— The system being Specially valuable [----][c] in Towns & Cities where there is a minimum amount of lighting over a maximum area & where the ordinary 3 wire system of patentee is too expensive—[1]

The invention consists in dividing the ~~Tow~~ City or town in two or more areas each worked with the Ordinary[d] 3 wire system with feeders arranged in the ordinary manner—but in this system all the feeders which I call sub feeders[e] centering to a main feeder.— The areas being small the ~~fee~~ sub feeders are very short from the connection with the mains to the main feeder and are calculated for only 2 per cent drop of potential   They[d] have no lamp connected to them— The main feeder being[d] long is calculated for a drop of 10 or 15 per cent.

The 2nd area is arranged in a similar manner and the two areas are connected together in series, the two areas together forming ~~with~~ as a whole a 5 wire system   AA A′A′ are the mains of the two areas. BBB° is a sub feeder BBB another Connected to the one side of[f] the main feeder C′— B′B′B′ & B²B²B² are two sub feeders connected to the mains & other side of the[g] main feeder C′ the portion of the main feeder C² serves to connect the two areas together   balancing wires d e g extend from the sub feeders back to the station— Thus each area is worked on the 3 wire system but the 2 areas together form a 5 wire system—

4 Dynamos are employed Each seperately regulatable, and have a surplus capacity for increased ~~p~~Electrical pressure that if any section of either area is thrown greatly out of balance the [--][c] pressure can be kept constant at all points notwithstanding a great drop in the ½ size balance wire d e & g

[S--------][c] in my patent _____ lately issued I shew what would seem at the 1st glance to be the same system but the distinction in this case is the use of a greater[h] number of sub feeders ~~shorter than a~~ & thus shorten them up so that even small wires will only give a drop of 2 per cent in pressure & thereby do away with the necessity of feeder regulators & attendance at the point where the main feeders is connected to the sub feeders[2]   ~~Another~~ [def--- ~~is the carrying~~?][c] X X are pressure wires leading back to the central station.

fig 1 merely Illustrates the simple principle—but fig 2 & 3[i] shews the plans I [~~prefer to~~?][c] use in practice.

~~Now~~ The mains are not shewn in this diagram—only the connected network of sub-feeders, the extremities c. d. e. f g. h in one area & c′ d e′ f′ g′ h′ in the other area connect with[d] the mains   The main feeder & balance wires connect to the subfeeders network at A. &. B. This network of subfeeders does away with individual feeders which owing to inequality of consumption of Electric[d] Energy in different parts of the mains causes great variations in pressure and substitutes one Single network of subfeeders ~~which~~ through which the pressure can be Equalized & if there is a great tendency to a[h] drop at one point the The pressure is equalized through the Subfeeder network & not through the mains as in the regular system ~~again~~ [~~control of the?~~----------][c] fig 3 & 4 illustrate only subfeeder network Etc Etc— The main e

Claim is for a connected subfeeder network connected to the station by a main feeder[b]

2 or more[j] areas in Series arranged with network mains— network Sub-feeders[d]—& main feeders—Etc

T A Edison

ADfS, NjWOE, PS (*TAED* PT032AAI1). Letterhead of Edison Lamp Co. [a]"Harrison, N.J.," preprinted. [b]Followed by dividing mark. [c]Canceled. [d]Obscured overwritten text. [e]"which I call sub feeders" interlined above. [f]"one side of" interlined above. [g]"other side of the" interlined above. [h]Interlined above. [i]"& 3" interlined above. [j]"or more" interlined above.

1. Edison's draft was the basis for an application that he executed on 13 September. The application was filed ten days later and issued in March 1888 as U.S. Patent 380,101. The design was part of his ongoing effort to overcome the economic disadvantages of low-voltage electrical transmission and distribution (see Doc. 3002 [headnote]). Edison did not indicate in the draft that he would achieve higher economy by using a relatively high voltage from the dynamos, but the completed application and specification stated that intention explicitly. To make the system operable for standard lamps (at around 110 volts) without using converters or transformers, Edison divided the electrical load into two three-wire distribution subcircuits and placed these in series with each other in the larger circuit from the dynamos. Because of the series connection between them, each subcircuit would experience only part of the full dynamo voltage in proportion to its share of the total resistance (see drawing p. 795). Pat. App. 380,101.

Edison called this design a five-wire system because each of the two subcircuits included conductors connected with the dynamos (**H**) and the compensating line (**10**), as in the three-wire plan; they also shared another compensating line (**13**) for balancing the subcircuits against each other. In addition, "pressure wires" (**G**) connected with voltage indicators at the central station, one instrument for each of the four divisions in the system; see Doc. 3089 regarding regulation.

2. Edison's U.S. Patent 365,978 was for a similar conception of distribution sub-systems (preferably of the three-wire type) connected in series, although he did not use the term "five-wire" in it. Edison filed that application on 29 November 1886, amid a burst of work on high-voltage distribution (see Doc. 3002 [headnote]), and it issued on 5 July 1887. The specification described the junctions between the dynamo circuit and intermediate circuits occurring in specialized substations, each with a set of indicators, regulating devices, and their attendants.

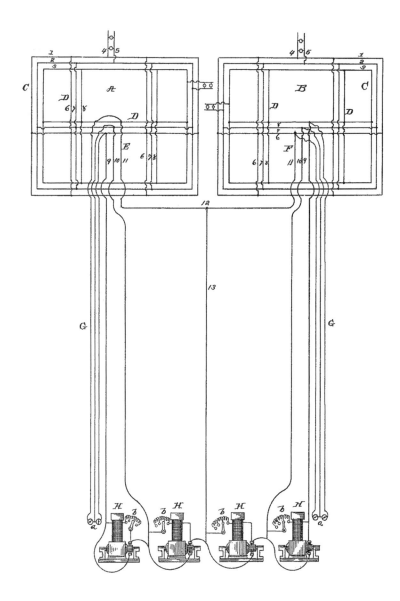

(An even earlier Edison patent, filed in November 1882, contained the idea of linking two essentially discrete distribution systems through a compensating line back to the generators; see U.S. Pat. 274,290, esp. fig. 4.) In the present five-wire patent, Edison stipulated that because the "sections into which the entire area to be lighted is divided being small, the sub-feeders will be short." The sub-feeders were arranged into their own crisscrossing network, and their brief length enabled him to design them for small losses without using a "great mass of copper" (U.S. Pat. 380,101). The related subject of feeder equalizers had recently been brought to Edison's attention by John Vail in the context of the three-wire system. Pointing out that they were expensive

(\$7,000 to \$12,000 for a large urban station) as well as "very cumbersome and unsightly," Vail asked if Edison could devise an alternative to the regulator for future stations (Vail to TAE, 18 Aug. 1887, DF [*TAED* D8732ABD]).

–3084–

Notebook Entry: Phonograph and Electric Lamp

[Harrison,] Aug 28 1887

Phonograph

Try soap cylinder & after indentation harden with an acid[a]

Try gum Camphor as a base—[a]

Mix Kaolin with Vasalene[a]

Mix Kaolin with glue water—[a]

Kaolin with boiled Linseed—

Soap cylinder coated thin glue—[a]

Kaolin made plastic with water & put around cylinder—[a]

Venice[b] turpentine as a softener—[a]

Cannola Balsam as a softener—[a]

Vasaline with Parafine to soften—[1]

Cut cylinder thus with extra tool—then make a Knife Edge recorder

This shape so needle will record hiss easy but ~~low~~ great amplitude, less costly   Experimenting will develope proper shape for perfect articulation

See how thin wax beeswax can be squirted into sheet say 002 thick & use on cylinder like tin foil—might add Rosin to harden it if at 1st it dont work

How thin can sheets of glue be made by squirting & pouring on glass—greased if necessary—[a]

See if Vulcanized fibre can be drawn through Saphire dies—[a]

Have Ott draw some Agave—[Amgou?][2c] Bamboo—Bast—& Palm & have them Carbonzed   see how look under Micro

X, NjWOE, Lab., PN-87-08-28 (*TAED* NP029A). [a]Followed by dividing mark. [b]Obscured overwritten text. [c]Illegible.

1. Edison drew a line across this section of the page.

2. Edison may be referring to amadou, a spongy material made from a tree fungus known today as *Fomes fomentarius*. Hard amadou was commonly used as tinder but softer amadou, also made from other large polypores, had an appearance like felt or pliable leather and was used as a fiber for a variety of purposes including clothing, chest protectors, hat linings, and household objects. It was also used as absorbent material by doctors and dentists. Dodge 1897, 167.

–3085–

To John Kruesi

Orange N.J. Aug 29 87

Dear Kruzi—

There is nothing the matter with your present Compound. it is splendid— The whole trouble is <u>Air Bubbles</u>=[a] The hotter it is poured the greater the amount of air bubbles   at 212 it can be put on rods & there is no bubbles— I have man experimenting and testing all the time— until I[b] get at the proper method of pouring & getting rid of the air bubbles it will be waste of time to Experiment with the other Asphalts—[1] Resin oil distills off Easily—it may answer but parafine or[c] other similar substance must be put in to prevent Brittleness— <u>One thing is Certain</u>[a] & that is Everything must be poured in <u>layers</u>[a] not only the boxes but the tubes— The tube itself should have a thin Coating the rope should also have a Coating—the rods also[d]   the whole of rods & rope when ready for the tube should have another Coat & then placed in tube & filled— This <u>will do the business</u>—

Regarding the paper for Dynamos we are trying a set of elaborate Experiments—which will be finished next week— so far we find paper should soak 10 days then air dried in closed place & then re dipped for a moment only & redried— Its like glass & spark does not go through= the moment we are ready & know what to do[2]—I will arrange to make you[c] a big supply in the brick shop at Menlo Park—[3] Yours

Edison

ALS, NjWOE, Kruesi (*TAED* MK005). ᵃMultiply underlined. ᵇObscured overwritten text. ᶜInterlined above. ᵈFollowed by "over" to indicate page turn.

1. Edison's letter was part of continuing correspondence about insulation for underground wires, which the Edison Electric Illuminating Co. was preparing to install in the new (Second District) service area in New York City. In a note addressed to Edison at New York's Hotel Normandie on 10 August (and transcribed by Charles Batchelor), Kruesi urged him to choose quickly from among the numerous asphalt samples already sent the one that he wished to use in production. (Most of the samples came from outside Trinidad, whose large supply was under monopoly control.) Kruesi asked Edison on 24 August to experiment with a compound made of "Cuban (Prime) Asphalt" and resin oil. In a reply drafted in the margins, Edison remarked that he intended to delve into questions of insulation as soon as he could "get into new Laboratory," about 15 October. Kruesi to TAE, 10 Aug. 1887; Samuel Insull to John Randolph, 19 July 1887; John Vail to Kruesi, 8 Sept. 1887; all DF (*TAED* D8733AAL, D8736ACY, D8733AAO); Kruesi to TAE (with TAE marginalia), 24 Aug. 1887, Kruesi (*TAED* MK004); "Asphalt," *The Cottage Hearth* 11 (July 1885): 227.

Insulation for underground lines was a long-running concern of Edison's lighting work. In this most recent iteration of research, Kruesi had been experimenting with asphalt compounds since at least June, apparently with an eye to obtaining uniform solidity. He discovered that when the liquid was applied in layers, each being allowed to cool in turn, no seams were formed. Toward the end of June, Samuel Insull forwarded samples prepared this way and asked whether Edison could "detect any trace of layers in them." In early August, Kruesi sent Edison ten pounds of an insulating compound made from Trinidad asphalt and boiled linseed oil mixed at a ratio of eight to one. He also forwarded ten pounds each of the asphalt and oil (Insull to TAE, 23 June 1887; Kruesi to TAE, 4 Aug. 1887; both DF [*TAED* D8737ABA, D8736ADD]). As Edison noted above, air bubbles were a recurring problem. Kruesi's standard compound performed well on bare copper rods but failed when applied over metal wrapped in cotton. Edison concluded that when the insulation was applied to a surface that was not clean or perfectly dry "or on surfaces covered with material which the heat of the Compound turns into gas we shall have Microscopic Vent holes invisible to the naked Eye" (an analysis similar to his observation in 1879 of destructive gas pockets in metal lamp filaments; see Doc. 1796). He instructed that cotton-wrapped copper rods should be "put in a iron trough containing Compound in liquid state but not hot enough to Carbonize the Rope say 250," in order to drive moisture from the fiber. On 24 August, he noted that tests of compounds at the lamp factory laboratory showed that "if you can pour compound at 300 Fahr it will be better than at higher temperature as when very hot it gives off too much gas" (TAE to Kruesi, 14 Aug. 1887; TAE marginalia on Kruesi to TAE, 24 Aug. 1887; both Kruesi [*TAED* MK003, MK004]).

2. Charles Batchelor, who had been testing coated paper at the Edison Machine Works, noted on 5 August that it "deteriorated to such an extent that it is no good at all. ¹⁄₁₆" spark goes easily through it anywhere and armatures (Municipal) made with it burn out as fast as made," and he directed Henry Walter to investigate the cause (Cat. 1336:255 [item

425, 5 Aug. 1887], Batchelor [*TAED* MBJ003255]). It is not apparent when Edison became aware of the problem. In an undated letter to Kruesi (probably about that time) he remarked, "You are in a pretty Scrape— The Linseed oil you are using in tubes & on armature paper is very acid   if you take your armature paper & put on tongue it will taste sour." He requested a gallon of oil "from every dealer & I will test." He instructed Kruesi to add five percent paraffin to the compound to compensate for its acidity until he could devise a permanent solution. "This is very serious," he noted, and it revealed "the necessity of having some one up there to test your materials=under every condition" (TAE to Kruesi, n.d. [Aug. 1887?] Kruesi [*TAED* MK006]). On 26 August, Batchelor had a dust-free room set up at Schenectady where the paper could be coated. He noted better electrical results when the paper was dipped twice in linseed oil. Dipping three times was better still though this made the paper brittle, a difficulty apparently diminished by repeatedly heating it (Cat. 1336:271 [item 443, 26 Aug. 1887], Batchelor [*TAED* MBJ003271B]).

3. Edison meant the machine shop at his former laboratory in Menlo Park. Although some problems persisted with the oiled paper, Legrand Parish was fixing up the shop building for manufacturing by early September. Batchelor noted that Parish could make treated paper there for use in the mimeograph, condensers, and "anything else that comes along." Cat. 1336:277 (item 452, 5 Sept. 1887); Walter to Batchelor, 6 Sept. 1887; both Batchelor (*TAED* MBJ003277A, MB231).

–3086–

To Edwin Houston[1]

[New York,] Aug 30th [188]7

Dear Sir:

In reference to your letter of the 12th. instant,[2] I think it is quite possible that between the dates you mention I may be able to give you a paper on my Pyro-Magnetic Motor.[3]

It would be impracticable for me to comply with your request to deliver a lecture in person, as my time for the next few months will be fully occupied in connection with my new laboratory   Yours truly

L (letterpress copy), NjWOE, Lbk. 25:71 (*TAED* LB025071). Written by John Randolph.

1. Electrical engineer Edwin James Houston (1847–1914) chaired the department of Natural Philosophy and Physical Geography at Central High School in Philadelphia. He served on the board of managers of the Franklin Institute from 1874 to 1897 and was an editor of its journal at this time; he became the first president of its electrical section in 1891. A co-inventor (with his colleague Elihu Thomson) of arc lights and dynamos, Houston helped form the Thomson-Houston Electric Co. in 1879 but left the business in 1882. He and Thomson had been involved in several public disputes with Edison in the 1870s, including an acrimonious one over credit for a carbon microphone; more cordial

relations had since been restored with Houston, at least, who brought the phenomenon of the tripolar "Edison Effect" lamp to the attention of the American Institute of Electrical Engineers in 1884. Doc. 2684 n. 2; *ANB*, s.v. "Houston, Edwin James"; Franklin Institute 1914, 13; Franklin Institute 1892, 1; Doc. 1884 n. 3.

2. Edison seems to have been mistaken about the date. Houston wrote on 20 August to ask if Edison would deliver a lecture on the pyromagnetic motor to the Franklin Institute sometime between October 1887 and March 1888. Houston to TAE, 20 Aug. 1887, DF (*TAED* D8704ABZ).

3. Edison's paper on the pyro-magnetic dynamo (Edison 1888b), read by George Barker to a section of the American Association for the Advancement of Science on 15 August, had received some notice in the daily press ("Scientists and Prayers," *New York World,* 16 Aug. 1887; "The News This Morning," *New York Tribune,* 16 Aug. 1887, 4). Houston and Elihu Thomson reportedly had devised a "thermo-magnetic motor" on a similar principle in 1879 ("Abstracts from the Secretary's Report," *Journal of the Franklin Institute* 125 [3rd ser.; Feb. 1888]: 150–51).

Houston answered Edison by encouraging him to pick a date for the lecture. William Wahl, secretary of the Institute and chair of its lecture committee, separately urged Edison to do the same, noting that the full lecture program had to be fixed by 10 September. Edison answered each man in similar noncommittal terms on 12 September, telling Wahl that the writing would "depend upon my new Laboratory being finished during the next month or two. Until this latter work is perfected my time will be wholly occupied with the same." The editors have found no evidence that Edison prepared a presentation for the Franklin Institute. Houston to TAE, 2 Sept. 1887; Wahl to TAE, 3 Sept. 1887; both DF (*TAED* D8704ACK, D8704ACL); TAE to Houston, 12 Sept. 1887; TAE to Wahl, 12 Sept. 1887; both Lbk. 25:89D, 89G (*TAED* LB025089D, LB025089G).

–3087–

Samuel Insull to Alfred Tate

SCHENECTADY, N.Y., Sept. 1 1887.[a]

Dear Sir,

I have your favor of the 30th. inst. and if I am not mistaken, your letter is the first intimation I have received of the exact amount which is to be paid to Mr. Holly.[1] I am under the impression that I either wrote or [askyou?][b] asked you if Mr. Holly could not be made to take a draft or short note. I don't think that we should be in any particular hurry to pay a man who fails so signally in doing the square thing by Mr. Edison.

In regard to office rent, I gave Randolph a Cheque for it last Saturday.[2] I understand that Mr. Tomlinson informed the Manhattan Bank[3] people, that the reason the rent was not paid, was that Mr. Edison had not sent a Cheque for his share. Before making any complaint about the matter, I propose to ask Mr. Tomlinson if he made such a statement; if so, I shall certainly not hesitate to inform Mr. Edison. I remember about a year ago, I paid not only our share of the rent, but also Mr. Tomlinson's.

I saw Mr. Edison last Sunday week and had a considerable talk with him about his finances.[4] He of course wants a great deal more money than he at first anticipated, but this is simply a repetition of what has occurred so frequently before. The trouble is, that Mr. Edison does not seem to have anyone with him who urges him to curtail his expenses on his new laboratory. Exactly how I am going to carry out his wishes and give him what he requires, I don't know. Heretofore, when I have had to provide money, I have always had something to say about how much should be spent. My position now is somewhat different and I propose to wait and see how the experiment turns out. The insurance notice which you enclose,[5] was left by me in the office by mistake; it is a matter which I have attended to. Yours very truly

Samuel Insull

TLS, NjWOE, DF (*TAED* D8719ABD). Letterhead of Edison Machine Works. [a]"SCHENECTADY, N.Y." preprinted. [b]Canceled.

1. Tate asked in his 30 August letter if Insull had paid $945.42 due architect Henry Hudson Holly (Lbk. 25:72A [*TAED* LB025072A]). Holly had requested this amount from Edison on 6 August as his fee based on 2.5 percent of the laboratory's contract price, and Edison instructed Tate to "get this from Insull— if possible." A check for $800 was sent to Holly on 16 August, though in early September the architect recalculated his fee based on 3.5 percent of the contract, or $1,323.59, leaving a balance of $523.59. Holly was paid in full in October but his correspondence with Edison about the laboratory construction, including an investigation of the large number of bricks used, continued sporadically through 1888 (Holly to TAE, 6 [with TAE marginalia] and 17 Aug. 1887; both DF (*TAED* D8755AAU, D8755AAW); Vouchers [Laboratory, 1887] nos. 346 and 370 [enclosing Holly to TAE, 9 Sept 1887]; [*TAED* VC87006Z1, VC87009]).

2. Tate pointed out in his 30 August letter (see note 1) that $250 was due for Edison's portion of the quarterly rent on the Wall St. office shared with John Tomlinson and Dyer & Seely. He explained that he routinely advanced "certain small amounts" from the Edison Lamp Co. to pay Edison's bills but preferred to have Insull's explicit permission in this case. Insull, for his part, kept close tabs on Edison's balances from Schenectady and apparently arranged to transfer money with some frequency to meet the most pressing needs. He gave John Randolph the $250 check on Saturday, 27 August (one of at least two transactions that month involving the rent; see Vouchers [Laboratory] no. 237 [1887] for Tomlinson and no. 252 [1887] for Edison Machine Works. Soon afterward, Insull upbraided both Tate and Randolph for paying ordinary

bills from Edison's checking account instead of using the funds to meet notes coming due, as he had intended (Tate to Insull, 30 Aug. 1887, Lbk. 25:72A [*TAED* LB025072A]; Insull to Tate, 2 Sept. 1887; Insull to Randolph, 3 Sept. 1887; both DF [*TAED* D8719ABE, D8719ABF]).

3. The Manhattan Co., incorporated in 1799, had long owned the 40 Wall St. property that was now its headquarters. The bank shared the present building at 40–42 Wall St. (erected in 1883 between Nassau and William Sts. and known as the Merchants' and Manhattan Building) with the Merchants' National Bank. King 1893, 706.

4. In his 30 August letter (see note 1), Tate asked if Insull had, during a recent visit to New York, talked with Edison about "financing in connection with his Laboratory? It looks to me as though he was going to require a good deal more money than was at first anticipated."

5. Not found.

–3088–

Charles Batchelor
Journal Entry

[New York, September 5, 1887?[1]]

454[2]   Manufactories

Edison's idea now for the future is to get up processes for manufacture & start factories— He has corresponded with Drexel Morgan & Co[3] about it and he thinks they will like it and take it up. It is:— Company formed that can be drawn on for $25,000,000 in 25 years but only such money called up as is actually necessary for the present inventions— the Co to have the right to take (or reject) all his inventions except ore milling, Phonoplex.

He proposes to buy a 30 acre tract of land on the small RR that runs to New York from Orange[4] and gradually cover it up with new manufacturing industries   He spoke of two that are now ready to go right into & on which from what he says and from his applications for patents I should say he has been working on a long time— 1st   Drawing fine wire such as is now imported by annealing in a vacuum or hydrogen gas between each drawing operation.[5] 2nd   Depositing metals in a vacuum by heating them to vaporisation, ('or the arc') on articles placed in the chamber to be deposited on.[6] This would take the place of electroplating and any metal or any alloy of a metal could be equally well deposited—

Immediately the new laboratory is finished these will be commenced in earnest.

AD, NjWOE, Batchelor, Cat. 1336:277 (*TAED* MBJ003277B).

1. Batchelor did not put a date on this entry but wrote it immediately below and above others dated 5 September.

2. Batchelor consecutively numbered each entry in this journal.

3. See Doc. 3078.

4. Beginning in November 1888, Edison acquired a series of parcels in the vicinity of Silver Lake, roughly between Bloomfield and Belleville, N.J., several miles east of Orange. He reportedly owned forty-seven acres by the end of the following year (by which time the eighteenth-century dam impounding the lake had been destroyed in a storm), and eventually built a number of major plants there, including the Edison Storage Battery Co.'s factory and the Silver Lake Chemical Works (Hill 2007, chap. 9; see, e.g., TAE agreement with Lydia Ropes, 1 Nov. 1888, Miller [*TAED* HM89ABJ] and similar property deeds in Miller 1889 [*TAED* HM89]). The rail line referred to was likely the Watchung Branch Railroad, a subsidiary of the New York and Greenwood Lake Railroad Co. It later became known as the Orange Branch, a spur of the Erie Railroad ("Pulling Up the Tracks," *NYT,* 27 Mar. 1887, 8; U.S. Interstate Commerce Commission 1889, 72).

5. Edison signed a patent application for this process on 17 October. It covered a continuous process of passing wire, "before it reaches a die or draw-plate, through a closed chamber filled with a non-oxidizing gas, such as hydrogen," where it was heated (preferably by electricity). In response to objections by the Patent Office, Edison made several amendments to the application, among them to specify that the wire would also be cooled in the presence of the gas so as to prevent oxidation before it reached the draw plates. Edison stated that this process, an extension of those he had already patented (U.S. Pats. 436,968 and 436,969), was "especially adapted to the economical production of exceedingly fine wire" because it prevented the loss of metal due to oxidation and also allowed the wire to be annealed without the use of pickling tubs. The application resulted in U.S. Patent 563,462 in July 1896.

6. Edison had completed a patent application on a vacuum deposition process in January 1884 but, after several rounds of rejections, it was still pending at the Patent Office (see Doc. 2587 n. 2). He drafted a caveat (Doc. 3101) in October 1887 for making phonogram copies by a similar process.

–3089–

*Draft Patent
Application: Electric
Light and Power*

[Harrison,] Sept 5/87

Dyer—
    Patent—[1]

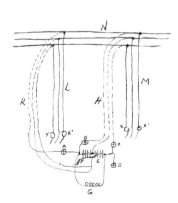

Sep bat[tery] on each one in shunt[2]

Improvements in methods of indicating Electric pressure in Central Electric Light stations=

The invention consists in the employment of a standard battery giving the same electromotive force as that required at the ends of the feeders combined with several sets of pressure or return wires multiple arcd across such battery through galvanometers A B C D[3]   The battery is so connected that it gives a contrary electromotive force to that due to the Dynamos [--][a] so that when the pressure at the pressure wires is 100 volts positive the standard battery will give 100 volts N[egative] & hence no Current will pass through the galvanometers A B C D & they will stand at Zero indicating that the pressure at the end of the feeders K M is correct, but if the load on the mains N decreases[b] on one of the feeders & not in the other—and rises to say 101 volts, then there will be a current due to a potential difference of one volt & this will deflect the indicator   it may then be brought back to zero by means of the feeder regulator X X′   a Resistance[b] G is placed around the battery to keep it in good action as by this method the acuracy of the results do not depend on the indicating apparatus which [----][a] need not be delicate and a deflection of several inches may be had due to a diference of one volt wh   The maximum diference that is possibly obtainable in current is obtained by this method, hence friction & other defects which render ordinary indicating apparatus so liable to [--][a] incorrect reading owing principally to the [-----][a] Necessity[b] of great delicacy is obviated— As all indicating apparatus is originally standardized by the battery it follows that a properly constructed battery is the best possible device for securing Constant pressures in a Station over[b] a long period of time— The battery I prefer to use is the ordinary gravity battery[4] with moderately[c] pure zinc & sulphate of Copper

Dyer make Broad claims for this method—[5]

E[dison]

ADfS, NjWOE, Lab., Cat. 1152 (*TAED* NM021AAE). [a]Canceled. [b]Obscured overwritten text. [c]Added in right margin.

1. Edison signed the completed application on 14 September. It was filed nine days later (with the application arising from Doc. 3083) and

issued in March 1888 as Edison's U.S. Patent 380,102. Edison's stated aim was "to accomplish the indication of pressures for a system of electrical distribution by means which will be accurate and not liable to get out of order." He stipulated that although the drawings showed a three-wire distribution arrangement, the invention could be applied "to any multiple-arc system."

Correct indication of voltage on the lines was only one step toward actually regulating the line conditions, a dynamic process that Edison had learned was hard enough on a three-wire system (see Doc. 2305 [headnote]). The five-wire plan, essentially composed of two geographically distinct three-wire systems connected electrically end-to-end in series, promised to be more challenging. In late October, Edison drafted a patent application for a networked set of switches to keep the load in balance. He pointed out in his draft that it was "essential in any extension of the 3 wire system to a 4 5 or more wire system that the balance should be very even" because differences of pressure would in some cases not be "regulatable by varying the resistance of the feeders and as it is a great saving to use very small neutral wires in the feeders it is necessary that the balance should be such at all times that only a comparatively small current shall ever pass through them." That application covered switches located on poles along the lines that could transfer the electrical load of houses (or groups of them) from one branch to another. Every switch could be controlled individually from the central station and, by virtue of a unison device in each, could also be set in conjunction with all the others. The unison mechanism at the heart of the patent was an adaptation of one Edison developed in 1871 to enable a central telegraph transmitter to synchronize the simultaneous operation of multiple stock printers (see Doc. 158 [headnote]). Evidently anticipating that a networked system of switches could have wider uses (as the final patent made clear), Edison instructed his attorney to draw up a broad specification or, as he put it, "Claim the Earth." He signed the finished application on 5 November but it was not filed until 9 December; it issued in June 1888 as Edison's U.S. Patent 385,173 (see drawing p. 806). There is at least one other related drawing, dated 18 October. Cat. 1152; Unbound Notes and Drawings (1887); both Lab. (*TAED* NM021AAG, NS87AAT); see also Doc. 3104 n. 2.

2. Edison's first sketch (above) appears to show a battery connected to each galvanometer between a pair of pressure wires, an arrangement that was illustrated and described in the patent as an alternative. The second sketch (below) seems to have been the basis for the first drawing in the patent, where that arrangement was identified as Edison's preferred form. In it, several differentially wound galvanometers were depicted, "two for each set of pressure-wires, and having each two separate windings or coils. One coil of a galvanometer is in circuit with two pressure-wires, . . . while the other coils of the two or more galvanometers are in circuit with a standard battery . . . the current of which serves to oppose at the galvanometers the current flowing on the pressure-wires, bringing the needles to zero" at normal voltage. U.S. Pat. 380,102.

3. The labeled galvanometers are at the bottom of the first drawing, below feeder regulators "X" and "X'." Feeders "K" and "M" are at the left and right sides, respectively, connected to the main "N" at top.

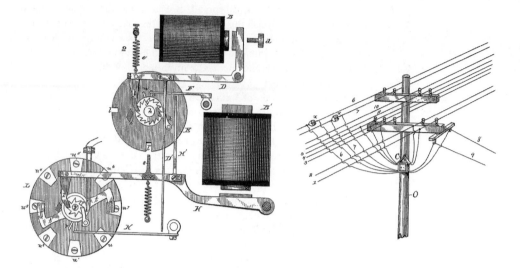

Edison's unison switch mechanism for remotely moving electrical loads from one part of a distribution network to another.

4. That is, any form of battery (such as Fuller or Callaud cells) in which the electrolytic fluids are separated within a single vessel by their different specific gravities. *KNMD*, s.v. "Gravity Battery"; Reid 1886, 868–69.

5. The patent issued with four claims essentially the same as those in the application. Pat. App. 380,102.

–3090–

Notebook Entry: New Laboratory

[Harrison, Summer 1887?][1]

List of schemes to work at in new Laboratory—

Heating Metals by Current for Mechanics Operations—[2]

Extraction fibre—

Ore Seperator—

Industrial process for mfg electro chemically by Electrolysis & dynamo machines—[a]

Aluminum—

Tanning—[3]

Magnothermic motor & Battery[4]

Magnetic bridge—[5]

Lipmann Mercury Meter[6]

Ozone for bleaching—Nitric acid—Ammonia etc by Houzean apparatus   big silent discharge machines run by power[7]— Alternating etc

Converters—

Application in telegraphy telephony Electric Light & instruments of precision of effect of heat in altering[b] magnetism of Nickel Cobalt & iron.

ditto that magnetism diminishes the <u>resistance</u> of iron & the
magnetic metals   See if it effects iron salts with unpolariz-
able Electrodes[a]
Motograph to be worked up in all possible applications
Artificial silk[8]
Moulding Compounds for use in the arts—[9a]
infusible flexible insulation.[a]
Ampliphone—[10]

X, NjWOE, Lab., N-87-00-00.3 (*TAED* NA003004). [a]Followed by dividing mark. [b]Obscured overwritten text.

1. This undated entry immediately follows a drawing that appears to
be an early arrangement for either the second or third floor of the new
Orange laboratory (see Doc. 3077 for what is likely an alternate version
of the sketch). That drawing, in turn, comes after two pages of Edison's
notes on the itinerary of a planned trip to Philadelphia, Baltimore, and
Washington. He did visit Philadelphia in early July to buy equipment
for the new laboratory; the editors have not determined if he continued
his journey southward from there ("Whistling Under Water," *Chicago
Daily Tribune*, 11 July 1887, 5). Doc. 3091 follows this entry.

2. See Doc. 3064.

3. See Doc. 3059.

4. That is, Edison's pyromagnetic motor and generator.

5. See Doc. 3059.

6. This instrument, more commonly known as a capillary electrometer, is especially useful for detecting small electrical charges of short
duration. It was conceived by Gabriel Lippmann while studying in
Gustav Kirchhoff's laboratory at the University of Heidelberg, where
he received his Ph.D. in 1873. The capillary electrometer operates on
the principle that a change in the electric potential between mercury and
dilute sulfuric acid produces a change in the surface tension between
them, causing the mercury to move and thus act as a detector of very
small electric charges. The device became especially useful for physiological studies; Augustus Désiré Waller used it to produce the first human electocardiogram in 1887. Wilhelm Ostwald began making extensive use of the electrometer in his laboratory at the University of Leipzig
in that same year. Stock 2004; Burch and DePasquale 1990, 97–108.

7. See Doc. 3038.

8. Edison's Fort Myers notebooks included ideas for methods of producing artificial silk (see Docs. 2917 and 2935).

9. Edison may have been referring to his efforts to produce artificial
mother of pearl (see Doc. 3020).

10. Edison may have been referring to his efforts to develop a hearing
aid (see Doc. 2906 n. 1) or to what he called (in Doc. 2935) an "Iron wire
Non-E Telephone."

Watts Dic. 3rd Sup. Pt 1[2]—p 708— Resistance[a] ~~d~~increase or diminish by heat of batteries   Cooling Daniel cell to—10 Cent[igrade],[a] Res increased 25 times. ~~hence~~ Carbon[a] cell at 15°C 20 times & Emf diminished $\frac{1}{10}$— hence try both in sealed glass[b] tubes with pressure valve to blow at 3050—& heat & get internal Res. also a table to be published of the increase of conductivity of a Daniel Bunsen & other batteries by rise of temperature also change in Emf—on commercial cells & on pure Zinc & chemical cells— starting 3 or 4 degs from Freezing point.[c]

An investigation as ~~the~~ to the change of resistance in batteries due to increase in size of positive & neg Electrodes so as to fully show which Electrode diminishes internal Resistance the greatest by increase of size—[3c]

Lot of syphon glass tubes to replace porous cup & by this means prevent difusion

[A][4]

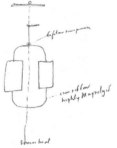

Closed magnetic ckt $\frac{1}{1000}$ pure iron or $\frac{1}{5000}$ if possible short magnetized needle on bifilar suspension—in front of spot where heat beam is focussed—   the heat arising diminishes magnetic permeability at this spot & makes free poles—

[B][5]

about 1000 ohms fine insulated iron wire surrounded with copper wire & magnetized   Magnetization ~~increases~~decreases conductivity of iron—increased hard steel[d]   Make a table of rise per ampere spires[6] etc—to get value of this phenomenon for practical application—

Watts 3rd Sup pt 1 p 719—states res diminishes at square Current[7] but gives no data for starting point of value or amount of Change with given current.

Collect[a] as[a] Literature of Effect light on Selenium & see if it has ever been arranged in Vacuo so as to get Constancy of Condition if not make[a] a Vacuum Cell & experiment.[c]

Read up see if glass has Ever been pressed hot through dies if not ascertain if it can be pressed out through long flat die to make window glass same as lead pipe is made—

Acurate & scientifically Conducted Experiments on the exact melting points of metallic filiments in vacuo arranged like a Carbon fil & in Volt ampere apparatus to read continuous as heat rises— also McLeod guage to measure gas coming out due to heating—[8c]

Prepare about 1 @ 2 lbs of every Soluable Carbzonable material for dipping—its probable that Every material will on Carbztn give a diferent surface   No 6 160 cp hp   No 17 178 cp hp—[9] prepare Asphaltene way up as far as Can go without Carbonzation   Try and disolve that black residual Coke from Oil Cos retort in parafine or other high boiling point coal Tar product—

~~make some Carbon monosulphide   also graphitic acid d[itt]o oxides~~

Take common asphalt. Melt[a] and stir for several days until it becomes impossible to stir it & not burn it more than possible. a hand stirer should be rigged up

Then pour & scrape it out to Cool— take a portion say Exactly one lb   powder finely 100 mesh then act on by following menstrua to disolve out the various compounds—

Alcohol,—then Ether—then mixture Alcohol & Ether, then Ammonia, ~~Then Water~~, Then Gasolene, then Chloroform—then several of the Essential oils, then [-][c] ~~Naptha Bisulph~~ Water, Strong hot Alkali—strong HCl—Then Naptha, Bisulphide[a] Carbon Benzol Spirits turpentine.

Use a small portion of the powder (not the lb) in test tube for a preliminary trial as I want to reserve the Major part of the lb for the remaining & best Solvents. object a final Bitumen that <u>neither softens or melts</u>

Take another lb of Asphaltene, dissolve the whole in Benzol & allow it to settle perfectly then distil Benzol off— ~~dis~~ this will eliminate the Ash & inorganic materials as well as solid Carbon—

Then act on residue with Every solvent which does not appear to go for the general bulk of the residue.

Try boiling Vasalene, parafine—

Make solution of amber[a] in any solvent that can be distilled off—& is best general solvent.

Then after distillation treat with the various inferior solvents to eliminate More fusible resins. Same with Zanzibar Copal[c]

Try Caouchanoc[10] in Benzine Chloroform or best solvent, dip

Varnishes— Powdered Copal gradually[b] added to sulphuric Ether in in flask with cork shaken at times for 24 hours. 5 pts Copal 2 pts Ether—used for repairing the glazing of enamels—[c]

In 17 use small trough and mix as much ignited lampblack as it will hold in suspension— ditto Electrotypers Lampblack   ditto ignited charcoal—[c]

form fils by using single fibre Ramie or Silk, Shellac & successive dips in 17. ~~A~~ also others which only soften but not melt. also ignited charcoal or Lampblack held to saturation in suspension— also Oxides Zircon—Magnesia—Alumina—Calcium—[a]

Mix Linseed with 17.

X, NjWOE, Lab. N-87-00-00.3:6–17 (*TAED* NA003006). [a]Obscured overwritten text. [b]Interlined above. [c]Followed by dividing mark. [d]"increased hard steel" interlined in right margin and circled. [e]Canceled.

1. This set of notes immediately follows Doc. 3090. In these notes, unlike those he would make in November and December 1887, Edison did not assign experimental projects to members of his laboratory staff. This fact suggests that Edison made this entry during the summer, while his plans for staffing the lab were incomplete (cf. Doc. 3077).

2. Watts 1879.

3. Figure label is "Syphon."

4. Figure labels are "mirror," "bifilar suspension," "iron ribbon   highly magnetized," and "source heat."

5. Figure label is "to battery."

6. That is, ampere turns.

7. That is, the square of the magnetizing current in a helix of wire surrounding a piece of iron or steel. Watts wrote about the effect of magnetization on the electrical resistance of iron and steel, but Edison paraphrased him to write about its inverse effect on conductivity.

8. Edison had tried to detect and remove gas from metallic filaments in 1879 but had not attempted to quantify its pressure or volume. See Docs. 1665–1666, 1669–1670, and esp. 1675–1676.

9. Edison meant candlepower per horsepower. The editors have not identified "No 6" or "No 17"; those descriptions likely referred to particular chemical treatments for filaments, possibly including those listed among the experiments assigned to Hugh De Coursey Hamilton in Doc. 3041.

10. Edison probably meant "caoutchouc."

–3092–

From Samuel Insull

SCHENECTADY, N.Y., Sept. 22 1887.[a]

My dear Edison,

Gouraud informs me that you agreed with him generally, that he should handle the phonograph in England, but that you left it to me to decide the method in which he should handle it. I did not commit myself one way or the other, as it occurred to me that the "illustrious Colonel" has construed your silence, or your simply remarking that he must deal with me, as meaning that you were satisfied that he should have it if I were. I have got to meet the gentleman on Monday. I have been talking broad platitudes to him all day and have put him off until Monday. Immediately you get this letter, please write me to 40 Wall St., where I shall be Monday morning, saying whether you agree to give him anything or not.[1]

With relation to the Gower-Bell telephone matter, after the talk I have had with him and from my knowledge of the affairs of the Company,[2] I shall advise you to join him in suing the people who have "gobbled" up this concern. It looks as if we have a good fighting case, and the only way that we will ever get a single, solitary cent out of Gower-Bell telephone shares, is by fighting Winslow, Lanier & Co's.[3] friends in London. Incidentally, I found out that Col. Gouraud owes you quite a block of this Gower-Bell Telephone Co's. stock, in addition to the shares belonging to the Edison Telephone Co. of Europe,[4] and which are now in London. I will see you either Monday or Tuesday with relation to this matter. I am compelled to go to Chicago to-night on underground work. Yours very truly

Samuel Insull   G[5]

P.S.— Do not on any account fail to send me word to 40 Wall St., so that I will get it Monday morning, exactly what you had to say to Gouraud with relation to the phonograph. I cannot conceive it to be possible that you stated at all, what Gouraud would wish me to believe.[b]

TL, NjWOE, DF (*TAED* D8704ACX). Letterhead of Edison Machine Works. [a]"SCHENECTADY, N.Y." preprinted. [b]Entirety of postscript typed in upper case.

1. The editors have found neither a reply from Edison nor other evidence of meetings around this time between Gouraud and Insull. Insull had started his business career in London as private secretary to Gouraud, a veteran of the American Civil War who was breveted Lieutenant Colonel six month's after the war's end (Docs. 159 n. 6, 1947 n. 2; Heitman 1903, 466). Gouraud was still in the vicinity several weeks later, and Charles Batchelor noted on 10 October that he "Met Gouraud this afternoon and came in from Orange with him— He informs me that he has got from Edison the phonograph for all countries outside of America and he is going right at it to systematically cover the world with agents." The contract that Edison signed four days later, though originally written to license Gouraud to sell phonographs throughout the world (except the U.S., Canada, China, and Japan), was altered to pertain only to the United Kingdom and Ireland. Gouraud did not sign that document, and Edison's signature was not witnessed. Edison's late-October agreement with the Edison Phonograph Co., however, limited that company to the U.S. and Canada and referred to a contract with Gouraud for "foreign countries" (Cat. 1337:17 [item 486, 10 Oct. 1887], Batchelor [*TAED* MBJ004017]; TAE agreement with Gouraud, 14 Oct. 1887; TAE agreement with Edison Phonograph Co., 28 Oct. 1887; both Miller [*TAED* HX87010, HX87013]). When Richard Dyer reported at the end of the month that Edison's European telephone patents covering the phonograph appeared to be in force in Belgium, France, Germany, Italy, Russia, and Norway (he lacked information about Spain), Edison appended a directive for Alfred Tate to send the list to Gouraud in London. In December, Gouraud submitted Edison's British Patent 17,175 (1887) for "Phonographs and Phonograms," taking care to do so only after several of Edison's American applications had been filed at the U. S. Patent Office ("New Patents—1887," *Teleg. J. and Elect. Rev.* 21 [23 Dec. 1887]: 643; Dyer to TAE, 31 Oct. 1887; Gouraud to TAE, 30 Nov. 1887; both DF [*TAED* D8749AAO, D8751AAI]).

2. Gouraud had helped (in 1881) to organize the Edison Gower-Bell Telephone Co. of Europe, Ltd., to control telephone patents (including Edison's) in continental Europe, excluding France, Turkey, and Greece (Doc. 2079 n. 2). Gouraud resigned from its board in 1884 but remained involved in ongoing disputes over the company's management and prospective liquidation. Insull had recently suggested that Edison negotiate a buyout of his interest in the company rather than align with Gouraud in a fight (Docs. 2079 esp. n. 3 and 2127; Charles Fitzgerald to TAE, 10 Jan. 1885; Insull to TAE, 22 July 1887; both DF [*TAED* D8547D, D8754AAN]).

3. A New York investment bank since 1849, Winslow, Lanier & Co. had close ties to Drexel, Morgan & Co. and to the Edison lighting companies (and was one of the early Edison light customers). Edward Dean Adams, a partner, was a director of the Edison Electric Light Co. and the Edison Illuminating Co. of New York (Cassis 2006, 60; Hausman, Herter, and Wilkins 2008, 340 n. 46; Docs. 2189 n. 3, 2278 n. 9, 2356). In January 1887, Insull inquired whether the firm's London "friends" (to whom Edison, Charles Batchelor, and Edward Johnson had previously sold shares in the Oriental Telephone Co.) were behind an anonymous offer for Edison's shares in the Edison Gower-Bell Telephone Co. of Europe. Adams concluded that the Edison Gower-Bell shares were unsaleable on the open market and that the offer must have come from an inside party (Insull to Alfred Tate, 12 Jan. 1887; Adams to Insull [with enclosure] 1 June 1887; all DF [*TAED* D8719AAA3, D8754AAK, D8754AAL]).

4. The Edison Telephone Co. of Europe, Ltd., incorporated in New York in 1879, controlled Edison's telephone patents in Europe, excluding the United Kingdom, France, Turkey, and Greece. Many of its patent rights were assigned in 1879 to the Edison Gower-Bell Telephone Co. of Europe, Ltd. Doc. 1731 n. 9; TAE agreement with Edison Telephone Co. of Europe, Ltd., et al., 10 Nov. 1881, DF (*TAED* D8148ZCV).

5. Probably William Gilmore.

–3093–

Notebook Entry:
Phonograph

Filim coating for phonogh cylinders

Parafine Base— shellac in alcohol   Syrupy Linseed oil d[itt]o thinned with Benzol   do Turpentine—do Rhigolene shellac in Benzole   do Ether   do Al & Ether—Collodion com[mercial]—Collodion flexible—Collodion diluted with Ether—Copal in Turpentine—do benzol—Rubber in Benzol—do gasolene—do Rhigolene—Rubber & thick Linseed in Benzol—guttapercha in Chloroform—do Bisulphide—do gasolene—do Rhigolene—gum Balata in every solvent— believe this is going to be best thing

Soap Base— glue thin—Russian issinglass—Am[er]ic[a]n —gelatine—Blood & Egg albumen—Tragacanth—glue & molasses

Bases— Spermacetti—white wax   Sterine oleac acid— Beeswax—Tallow—Lard   Camphor   Napthaline—ozokerit   Benzoic acid. Caoca butter german in Lump—Japan Wax   Carnauba Wax   BayWax

for cylinder   Sol soda & Ox Zinc[a]   Combine so you can roll it in Sheets—[1] See if solvent for Silk   Dry some collodion & redisolve in something not as volitile— disolved Solid

Linseed[a] by Chlorate & Mange O$_3$ ditto manganes peroxide
alone—

Recording point

Receiver—

Try magnesium & aluminum for reproducers— also hollow
wire also $^{8}/_{1000}$ Solid bamboo— glass fibre filiment carbon??
very light elastic & flexible—

X, NjWOE, Lab., N-87-09-24 (*TAED* NP030A). [a]Obscured overwritten text.

1. In addition to the acoustical and mechanical advantages of a pliable recording surface (see, e.g., Doc. 3101), Edison probably also had in mind the commercial advantage of material plastic enough to be rolled into sheets. As he was quoted by the *New York Post* about a month later (in an interview republished a number of times elsewhere), he intended that recordings on such sheets could be surrogates for postal correspondence:

> The merchant or clerk who wishes to send a letter has only to set the machine in motion and talk in his natural voice. . . . When he has finished, the sheet, or "phonogram," as I call it, is ready for putting into a little box made on purpose for the mails. We are making the sheets in three sizes—one for letters of from 800 to 1,000 words, another size for 2,000 words, another size for 4,000 words. I expect that arrangements may be made with the post-office authorities enabling the phonogram boxes to be sent at the same rate as a letter. ["Edison's Perfect Phonograph," *Chicago Daily Tribune*, 25 Oct. 1887, 12; cf. Doc. 3042 n. 1]

–3094–

From Marion Edison

Bradford.[1] [September 1887?][2]

My darling Papa.

I realy ment to write you before but I have been so busy ever since I got here that I have not had a minute to myself. I am very much pleased with my school and I would be very ungrateful if I did not study as hard as I could. I practice two hours a day and take a music-lesson twice a week and if I cant play at the end of this year then I will give up. It is simply lovly here every body is so kind and we all get along so nicely, the climate is simply lovly, it has already begun to get cold and if it keeps on I realy think we will have snow. I hope the labratory is nearly finished and that you are happy in its completion. Mina was very kind to me when she was here and I realy think papa that I love here truly and that every day she takes more and more the place of my own mother, you ought to be very happy in having such a noble woman to love you and I hope it will ever continue so. Did you darling papa ever here such good[a] advice from a daughter before, one would think that I had gone over the trials[b] of life and were try to make you profits by my experience. I have just written to Mina and she will probably get it tomorrow. I hope the boys are well and that something will be done about Willie, it is realy a shame how is growing up and I dont think it is doing him any good to go out west, perhaps you will find the truth of these words to late but I hope not. Has Mina decided where she is going to send the little boys?[3] Dear, darling papa I hope that I shall be such a nice woman and so accomplished that you will not altogether think that I am underserving of all you have done for me   I study all day now and not only for knowledge but to show you that I love you. Give my love to Mina and the boys and keep lots of love and kiss from   Your daughter

George.[4]

ALS, NjWOE, FR (*TAED* FB002AAB). [a]Interlined above. [b]"a" in "tri-als" interlined above.

1. The Bradford Academy, thirty miles by rail north of Boston, occupied twenty-five partially wooded acres along the Merrimac River. Founded in 1803, the school had been a single-sex institution since the mid-1830s, offering to high-school aged young women a broad curriculum that included theology, Latin and modern languages, literature, mathematics, art, and music (with special instruction in piano). It also offered classes in several sciences and counted astronomer Charles Young among its lecturers. Its principal for twenty years from 1855 was Abby H. Johnson, who subsequently ran the Boston school that Mina Miller attended. Edison paid the school $93.13 in early December for the balance of Marion's board, use of a piano, and music lessons. Pond 1930, 54–68, 112–29, 142, chap. 12; Hill 1903; Bradford circular and course of study for 1885–1886, Kingsbury 1885, Voucher (Laboratory) no. 557 (1887), NjWOE.

2. Based on her acknowledgment of not having written to her father, this letter appears to be the first one that Marion sent him from Bradford Academy. Her fresh impressions of the school and the fact that the weather had "already begun to get cold" strongly suggest that she composed it early in her first term there, likely from about mid-September to early October.

3. Thomas, Jr., and William Leslie Edison attended the Dearborn-Morgan School as recently as June 1887 and probably continued there until enrolling at St. Paul's School in Concord, N.H., in 1889 or 1890. Dearborn-Morgan receipt to TAE, 23 June 1887, DF (*TAED* D8714AAQ).

4. See Doc. 3035 n. 13.

PROSPECTUS AND PLAN OF ORGANIZATION

Draft Prospectus:
Edison Industrial Co.

of the

EDISON INDUSTRIAL COMPANY.

In the course of his many and varied experiments, Mr Edison has discovered various processes, and made many inventions, by the use of which a great number of articles of commerce could be manufactured at a much less cost than by the means now employed. In other cases at the same cost, the quality and efficiency of the articles could be greatly improved. He has also invented a number of new devices and articles of commerce, for which there could be very large sales if proper facilities were possessed for manufacturing and introducing them. Some of these inventions have been patented, and are so far developed as to be capable of immediate introduction. Others have been perfected, but have not been patented, and still others require further investigation and experiment.

Now that his electric light, which has largely occupied his time for several years past, no longer requires his active attention, he proposes devoting himself to such industries as offer the most promising field for invention and experiment, and with this view has recently erected and equipped, at a great cost, a laboratory, containing every facility for scientific research and experiment.

Instead of, as heretofore, merchandizing his inventions through different channels, Mr Edison is desirous of securing one organized avenue for the manufacture and introduction of all of his inventions. To accomplish this, he proposes forming a corporation, to be called the EDISON INDUSTRIAL COMPANY, which shall have sufficient capital at its disposal to erect and equip factories for the manufacture of all of his inventions, the manufacture and sale of which the Company may deem it profitable to undertake. They are not to deal in Patents or patent rights, but are to be restricted to manufacturing such new articles as Mr Edison has invented, or may invent, or such old articles as may have been improved, or their cost of manufacture cheapened.

Annexed to this prospectus, as an exhibit, is a proposed form of contract between Mr Edison and the Company, which explicitly defines the nature and character of the inventions to be assigned to the Company, the conditions upon which they are to be assigned, and the extent of the Company's ownership of the same.[2]

The Company is to be organized under the laws of New

Jersey, with an authorized capital of   Dollars. It shall begin business with an actual capital of   Dollars, paid in cash.

Upon an invention being submitted to the Company, if it shall decide to undertake its manufacture and sale, bonds shall be issued ~~at par~~ for such sum as shall be required to erect and equip a factory for manufacturing such invention, and starting the business. These bonds shall run for twenty years, and shall bear interest at five per cent, and shall be offered for sale to the bond-holders of the Company proportionately to their holdings.

There shall also be issued to Mr Edison as a consideration for the assignment of the inventions, stock to double the amount of the issue of the bonds, one half of which he in turn will assign to the bond-holders in proportion to their holdings of the bonds.

When a new factory is started to manufacture another invention, a second issue of bonds shall be made sufficient to meet the actual cost of erection and the equipment of this factory and the expenses of starting the business. Double the amount of this second issue shall be issued to Mr Edison in stock, one half of which, as before, shall be returned to the bond-holders, in proportion to their holdings of the second issue of bonds; and this plan shall be followed in the case of every new factory until the entire capital stock has been issued. The bonds shall be secured by a single deed of trust, covering the entire property of the Company, real and personal, without preference of one issue over the other. A reserve fund shall be established out of the earnings of the Company for the redemption of the bonds at maturity. With the exception of the earnings so reserved, the entire profits of the Company shall be distributed each year in the form of dividends. If at any time any one manufacture should be found unprofitable, the factory shall be altered to carry on some other industry.

The contract between the Company and Mr Edison shall continue for a period of four years. If, however, after two years from the date thereof the Company shall have failed to pay the interest on its outstanding bonds and five percent on such stock as may have been issued, either party shall have the right, upon giving to the other sixty days notice of his or its desire and intention so to do, to terminate and cancel the same.

TDf, NjWOE, DF (*TAED* D8704AGT).

1. This draft prospectus describes the type of company for which Edison sought financing from James Hood Wright and William Lloyd Garrison, Jr., in August (see Docs. 3078 and 3080). He probably wrote

it in connection with Wright's promise to bring the matter to Charles Coster in early September.

2. The draft agreement with the proposed Edison Industrial Co. was written on terms generally similar to those given in this prospectus. It would have required Edison, during its four-year lifetime, to explain to the board of directors all inventions suitable for manufacture, specifically exempting those covered by other agreements: "the generation, regulation or application of electricity to light, heat or power, to electro-heating or deposition, to the telegraph, the telephone, the ocean cable, the milling of ores, the phonograph or duplicating processes." The company would have exclusive rights to inventions submitted to it but would relinquish control if it did not accept them within three months. The Edison Industrial Co. would have the right to patent inventions it chose to manufacture and would assume all costs of obtaining and defending patents (including those already taken out by Edison), and it would finance the construction of a dedicated factory for each invention by the sale of bonds. The firm's factories would be under Edison's management, though its sales and "general business" would be controlled by a board of directors. TAE draft agreement with the Edison Industrial Co., n.d. [1887], DF (*TAED* D8704AGU).

8

October–December 1887

As his family receded into the background, Edison spent most of the fall focused on getting his new phonograph in shape for manufacturing and readying his new laboratory for experimental work. He also continued his efforts to meet the challenge coming from the alternating current systems offered by the Westinghouse and Thomson-Houston electrical companies. And he renewed his search for a better natural plant fiber from which to make filaments.

Edison had at least one of two experimental models of the new wax-cylinder phonograph at hand by the end of September. His experiments with the new machine led to additional modifications over the course of the next two months. The changes included an improved feed and return mechanism for repeating sections of a recording during playback, and an apparatus to soften the wax cylinder before recording on it so as to reduce scratching noises. He also decided to make his phonogram cylinders entirely out of wax rather than using a hard base covered by a softer wax layer. In late November and early December, Edison executed patent applications to cover these modifications as well as the phonograph itself.

Edison's experiments with the phonograph involved efforts to make recordings of music which could then be copied. He executed a caveat in late October in which he described experiments for making copies of recordings, and in late November he completed a patent application that incorporated a vacuum deposition process he had developed in 1884.[1] As he told a reporter from the *New York World*,

> phonographic opera will cost nothing, because the phonogram can be passed through the phonograph, if necessary,

a thousand times in succession, and once the machine is bought there is no other cost, beyond the trifle for phonograms. For books the phonogram will come in the shape of a long roll wound upon a roller. To make the first phonographic copy of a book some good reader must of course read it out to the instrument; once that is done, duplication to any number of thousand or million copies is a simple mechanical work, easy and cheap.[2]

The phonograph factory at Bloomfield, New Jersey, was being readied by Ezra Gilliland in early November. In an interview with the *New York Post*, Edison expressed optimism that it would soon be turning out about twenty-five phonographs a day, with the first 500 machines ready before the end of January.[3] However, in mid-December Gilliland found that "the present form Machine that we are at work upon in Bloomfield, would not compare favorably in any respect with the Graphophone."[4]

Besides overseeing the phonograph factory, Gilliland was appointed the exclusive sales agent of the Edison Phonograph Company, which Edison had organized at the end of September to manufacture and sell phonographs in the United States and Canada. Edison's decision to form a new company arose from a desire to circumvent the old Edison Speaking Phonograph Company, whose management wanted the commercial rights to his new machine. Believing that the original phonograph patent under which that company operated was no longer valid, Edison offered only to convert stock in the old company to that of the new entity. Some (but not all) of the principals in the old firm rejected the offer.

Edison seems to have settled into the still-uncompleted laboratory complex by mid-November, when he demonstrated the new phonograph to members of the National Academy of Sciences and to a reporter from the *New York Post*. The main building and two of the smaller lab buildings had been completed by early October, and by this time it is likely that the two remaining smaller buildings were also completed. However, work continued on the interior, which had to be finished and equipped with machinery and apparatus.[5] While awaiting these final touches, Edison spent more than a week sketching out experimental plans in one of his standard-size laboratory notebooks. He then went through the book to assign tasks to individual staff members, in some cases writing more careful instructions in notebooks that he then gave to his assistants. He also began to hire new laboratory assistants, among them several who had worked at his lamp factory laboratory, including John Ott, David and John Marshall, and William Dickson, who was placed in charge of the metallurgical laboratory. New members of the staff included Erwin von Wilmowsky, who took charge of the chemical laboratory, and Arthur Kennelly, whom Edison placed in charge of the electrical laboratory known as the Galvanometer Room. Other notable new staff members included physicist Franz Schulze-Berge and chemist Jonas Aylsworth. Edison's hopes for the new research facility quickly became apparent in an ambitious list of projects he drew up just after the turn of the new year.[6]

1. See Docs. 3101 and 3119.
2. "The Wizard Edison Talks," *New York World,* 6 Nov. 1887, 28.
3. "The Phonograph at Work," *New York Evening Post,* 18 Nov. 1887, 8.
4. Gilliland to TAE, 16 Nov. 1887, DF (*TAED* D8750AAU).
5. Cat. 1337:15, 31 (items 483, 515; 7 Oct. and 25 Nov. 1887), Batchelor (*TAED* MBJ004, images 8 and 16).
6. N-88–01-03.2, Lab. (*TAED* NA021AAF).

–3096–

From Samuel Insull

SCHENECTADY, N.Y., Oct. 6 1887.[a]

My dear Edison,

Since writing to you early this morning,[1] I have a letter from Tate,[2] which is almost like a thunder clap, and in which he says you require $8,000.00 on Thursday or Friday; at the same time he asks me to arrange to pay the Lamp Co. about $2000.00, you having overdrawn your account $8000.00 with them, and they being short of money. Now I have no objection to struggling to finance for everybody, but I have a very strong objection to so short a notice being given of your requirements; you promised to let me know a week ahead what you would require. I have $65,000.00 to meet within the next 15 working days, that is at the rate of about $3200.00 a day. To suddenly have $10,000.00 put on the top of this, without even 24 hours notice, is a little more than I can stand without protesting. I have sent Tate to-day, a note of the Illuminating Co. of Detroit, which has about two months and three weeks to run, amounting to $10,116.96; I imagine the Lamp Co. can discount this in their Bank.[3] You had better put your signature on the note underneath ours, and then with the Lamp Co's. signature, it will make extremely good paper. Tate can then pay you $8000.00, keep about $2000 for himself, and if you still want enough money to meet the interest on your mortgage, I will send you a Cheque for this. Since July 1st. I have paid out for you the sum of $11,647.72. This is made up partly in small amounts for rent, office pay roll, and small bills at 40 Wall St.; partly in large amounts of cash to meet notes which you yourself have given out, and one amount of a note for $2,087.47, which I gave to Eimer & Amend.[4] I am, as you well know, perfectly willing to run all round the country to hunt up money for you, and I can by hook or by crook, supply your wants, but I cannot do it unless you advise me ahead of your requirements. The note I have sent Tate to discount to-day, I had reckoned on getting discounted some time next week in order to take care of some of my notes for the week following. You will see that the $15,000.00, which you stated[b] was a debt of honor on my part, so to speak, has not only been paid back, but 33–1/3% more than this.[5] Yours very truly

Samuel Insull

I have also sent check to Randolph today for $287 to pay Jennie Stilwells School bill[c]

TLS, NjWOE, DF (*TAED* D8704ADG). Letterhead of Edison Machine Works. [a]"SCHENECTADY, N.Y.," preprinted. [b]"d" added by hand. [c]Postscript written by Insull.

1. Letter not found.

2. Insull referred to a letter from Alfred Tate dated 5 October (not found) about the financial matters described below, to which he made a direct reply on 6 October. DF (*TAED* D8719ABI).

3. Insull directed his instructions about this transaction to Tate (see note 2), who was acting as treasurer of the Edison Lamp Co. He referred in that letter to a "very stiff note on this subject" that he had sent to Edison.

4. Established in 1851, Eimer & Amend was a New York City importer, wholesaler, and manufacturer of drugs, chemicals, chemical apparatus, and laboratory materials. It had supplied Edison and his associates on occasion since at least 1874 and provisioned some needs at his laboratory in Florida in early 1887, but in October, Eimer & Amend found itself "overlooked entirely" as a supplier for the new Orange laboratory. Bernard G. Amend (1821–1911), a cofounder of the firm, was a native of Darmstadt, where as a chemistry student he had a teacher in common with Justus von Liebig, for whom he later worked as an assistant. Eimer & Amend was acquired by Fischer Scientific in 1940. Advertisement, *Medical Directory* 1887, 6; Eimer & Amend to Charles Batchelor, 27 May 1874, Cat. 1143:81, Scraps. (*TAED* SB143081); TAE to Eimer & Amend, 12 Jan. and 8 Apr. 1887, both Lbk. 23:186, 24:328 (*TAED* LB023186, LB024328); Eimer & Amend to TAE, 1 Oct. 1887, DF (*TAED* D8756ACU); "Obituary," *Journal of Industrial and Engineering Chemistry* 3 (11 May 1911): 351–52; "Eimer & Amend Will Open Store In East 34th St.," *N.Y. Herald Tribune,* 11 June 1941, 36.

5. Tate mentioned a loan of this size in Doc. 3067.

To John Vail

Friend Vail—

I think Orange is a go[a] providing East Orange[1] don't make us put too much underground— Please send me a blue print shewing mains & feeders   with this I can ascertain just what they will permit us to do—

Regarding regulation of feeders Howell has been figuring on it & says that the regulation on 5 wire system is ok   says that Feeder Reg[ulator]s are only on outside wires and the Center one thus

that its just as good a system as the 3 wire, to regulate— You better go ahead & get up Estimate for station on [-][b] 3,000 &[a] also on 5,000 Light basis ~~leaving out~~ all poles[c]   After getting this we can afterwards add the costs of portions underground— Trees are plentiful here hence calculate on very high & 1st class straight poles— a 12 inch brick wall and inexpensive foundation is all thats needed here—its hard gravel deposit.

In Re Greenwich—I have marked mains & feeders for 5 wire system.[c] The two small places have seperate systems, are connected to station by a single feeder each[2]   A seperate Engine & Jackshaft with 4 small dynamos is to be used & the two feeders provided with Feeder Regs[c] & other apparatus just as if it was another system— It can be thrown on the regular system after the loads go off & in case of accident you could throw it on ~~with the~~ the main system if you provide Extra Capacity— The reason I do not put these two outlaying places on main system is that they might cause too much regulation on Main system and make an unnecessary waste in feeder Regulation possibly. The load will permit of permanently connecting therefore arrange so that if its found in practice to work OK The Extra Little Dynamos may act [----][b] another capacity. I merely send the map back to your man so I can get a rough idea of cost of Copper & if found too much will study it further & may Change the system—

[~~Las?~~][b]   Your

Edison

ALS, NjWOE, Vail (*TAED* ME012). [a]Obscured overwritten text. [b]Canceled. [c]Multiply underlined.

1. The township of East Orange, N.J., was separated politically from Orange in 1863, but it (like parts of nearby West Orange) was often referred to simply as "Orange." *Ency. NJ*, s.v. "East Orange"; *Am. Cycl.*, s.v. "Orange [N.J.]."

2. Having recently installed an isolated station for electric light and power at his country home and fifty-acre estate in Greenwich, Conn., overlooking Long Island Sound, Edward Johnson commissioned a canvass of the town that he hoped to give to Edison in early October for purposes of planning a central station. Johnson specifically requested figures for the "two pieces added   One on the North East & One on the South West" ("Electricity in the House Beautiful," *Electrical World* 10 [16 July 1887]: 27–28; Johnson to Vail, 3 Oct. [1887], DF [*TAED* D8732ABL]). The Greenwich plant became a focal point for a general critique of the Edison electrical distribution system (see Doc. 3118 n. 1).

–3098–

Patent Application: Phonograph

[New York, October 14, 1887][1]

⟨—734—⟩[2a]

To all whom it may concern:

Be it known that I, Thomas A. Edison, of Llewellyn Park, in the County of Essex and State of New Jersey, have invented a certain new and useful Improvement in Phonographs,[3] (Case No. 734), of which the following is a specification:[4]

The object I have in view is to produce a better blank for receiving the recording indentations of the phonograph. This I accomplish by the employment of a surface of wax or wax-like material, such as waxes, gums, lacs, paraffine or other wax-like hydrocarbon. The material is coated upon a suitable backing or support of tougher material such as paper, and if required the surface is made smooth and true in any suitable manner. The recording and reproducing surface and its backing form a blank which is an integral article of manufacture and use and which is designed to be removed from one phonograph to another. This blank may be given any form which may be required by the construction of the phonograph. It may be in the form of a disc, a sheet, an endless belt, a cylinder, a roller, or a belt or strip, and the indentations may be in straight, spiral or zig-zag lines, according to the organization of the particular phonograph.[5]

In the accompanying drawing forming a part hereof,—Figure 1, is an enlarged sectional view of a part of the blank:

Fig. 1.[b]

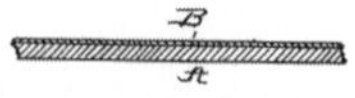

Fig. 2.[b]

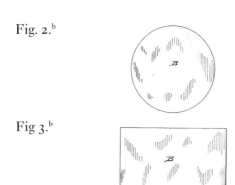

Fig 3.[b]

Figures 2, and 3, represent ~~blank~~ the blank as a disk and sheet respectively:

Fig. 4.[b]

Figure 4, represents the blank as a cylinder:

Fig. 5.[b]

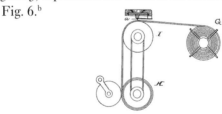

Figure 5, represents the blank as an endless belt; and

Fig. 6.[b]

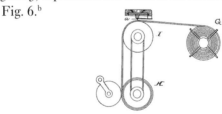

Figure 6, represents the blank as a strip.

A is a suitable backing or support of tough material such as paper which is given a sufficient thickness to produce the desired stiffness. Upon this backing is the coating B of wax or wax like material as before stated. Together, the backing and coating form a phonograph blank. In use the indenting point a of the recording instrument acts directly upon the wax or wax-like surface, and the reproducing point follows these indentations. The phonograph blank A B is mounted in use removably upon the moving blank carrier of the phonograph designed to receive it, so that it can be removed from the machine after the record is made and can be placed upon another similar machine for reproduction or again upon the same machine for reproduction.

The disk of figure 2 will require a plate blank carrier for its reception; while the sheet of figure 3 may be mounted upon a plate or wrapped on a cylinder. The hollow cylinder of figure 4 will be slipped over a supporting cylinder C of metal which is the blank carrier of the phonograph. The endless belt of figure 5 will pass around rollers D, E, F, forming the blank carrier. The strip of figure 6 will be wound from a reel G on to a reel H passing over an intermediate roller I where the surface is acted on by the indenting or reproducing point.

The wax or wax-like surface receives the indentations bodily without requiring an underlying groove and the reproduced sounds have the minimum amount of scraping and other foreign noises.

What I claim is:—

First: A phonograph blank having a surface of wax or wax-like material, substantially as set forth.

Second: A phonograph blank having a surface of wax or wax-like material and a backing of tough material such as paper, substantially as set forth.

Third: In a phonograph, the combination with a blank carrier and a recording instrument, of a blank having a surface of wax or wax-like material, substantially as set forth.

Fourth: In a phonograph, the combination with a blank carrier and a recording instrument, of a blank adapted to be removably supported by such blank carrier, said blank being composed of a surface of wax or wax-like material and a backing of tougher material, such as paper, substantially as set forth.

This specification signed and witnessed this   day of   1887.

Inventor.[c] Thomas A. Edison

By his Attorneys   Dyer & Seely

Witnesses   E C Rowland[6]   William Pelzer[7]

TD, NjWOE, PS (*TAED* PT032AAK). Drawings grouped together with signatures on a separate page. [a]Marginalia written by hand, possibly by Richard Dyer. [b]Figure designations written by hand, probably by Edward Rowland. [c]Signatories' names and roles written by hand.

1. The Patent Office file wrapper indicates that Edison signed the application and inventor's oath on this date. The application was prepared at the offices of Dyer & Seely.

2. This is Edison's patent application case number.

3. In December, Edison's attorneys asked to amend the application

in the Patent Office by substituting the phrase "phonogram blank" for the word "phonographs" in the title. They also sought to insert "phonogram" in place of each occurrence of "blank" in the text (Dyer & Seely to Commissioner of Patents, 6 Dec. 1887, Pat. App. [*TAED* PT032AAK, image 7]). The word "phonogram" was not new and had been used by Edison in 1878 to designate the piece of material on which a recording was impressed (see Doc. 1341; *OED*, s.v. "phonogram"; Doc. 3093 n. 1; "News of the Week," *Spectator* 60 [22 Oct. 1887]: 3).

4. This application was filed on 21 October. The Patent Office advised that the claims would likely be rejected on the basis of having been anticipated by Edison's British Patent 1,644 of 1878, and it declined to consider the case further until Edison executed a new oath regarding the relevance of his foreign patents. There ensued a protracted process of legal wrangling that also ensnared a contemporary Edison application for the regulation of phonograph drive mechanisms (Case 373). The examiner insisted on a particular form of declaration and finally, in December 1891, he invited Edison to appeal to the patent commissioner. The editors have not determined whether Edison did so but in any case, the application did not result in a patent (nor did Case 373). Edison did, however, file applications in November and early December pertaining to the construction of phonogram blanks of a hard underlying base matrix and a softer outer layer for indenting. See Docs. 3098 and 3101; U.S. Pats. 430,570 and 382,462.

5. This inclusive language echoes that in two phonograph caveats from 1878, in which Edison enumerated and illustrated a variety of physical arrangements for moving the indenting point over the recording surface. Docs. 1227 and 1341.

6. Edward C. Rowland (1863–1926) was a former Edison laboratory employee now working as a draftsman for Dyer & Seely. Doc. 2673 n. 5.

7. William Pelzer (1870?–1955), a New York City native, began working as an office boy at one of the Edison lighting companies at age fourteen. He studied law at night at Cooper Union and at this time was working in some capacity for Dyer & Seely, with whom he later practiced. Pelzer subsequently served as vice president of the National Phonograph Co. (from 1907), secretary of the General Film Co. (from 1910), and president of the Motion Picture Patents Co., all firms connected with Edison. Obituary, *NYT,* 25 Apr. 1955, 23.

–3099–

To George Churchill[1]

[New York,] October 18, [188]7.

My Dear Sir:—

I have received your note of the 15th instant.[2]

The Convertors which you speak of are not in use anywhere, as yet, but as soon as I get into my new Laboratory, which will be about the 10th of November, I am going to build several of them and operate them in connection with an electric railroad at Orange.[3] If at that time you are in America, I should be very glad to have you come over and see these machines in operation, and also inspect my Laboratory.

Believing it may interest you to see our Central Station at Pearl St., New York, I take pleasure in enclosing a note to the Sup't.,[4] who will give you any information you may desire. Yours truly,

(Signed)[a]   Thos. A Edison

TL (carbon copy), NjWOE, Lbk. 25:92E (*TAED* LB025092E). Signed for Edison by Alfred Tate. [a]Written by Alfred Tate.

1. George Charles Spencer-Churchill (1844–1892) was, since 1883, the eighth Duke of Marlborough. He arrived in New York on 28 August and spent a week or more at Newport, R.I., where his presence caused something of a scandal. At the time of his death, he was chairman of the New York Telephone Co. and the Brush Electric Co. "A Real Duke in Town," *NYT*, 29 Aug. 1887, 8; "Food for the Gossips," ibid., 10 Sept. 1887, 5; "The Duke Makes a Mem.," ibid., 11 Sept. 1887, 2; "The Duke of Marlborough," ibid., 10 Nov. 1892, 1; "Sudden Death of the Duke of Marlborough," *Birmingham Daily Post,* 10 Nov. 1892, 2.

2. In reply to a request from an intermediary in New York, Edison had invited Churchill to visit his laboratory at the lamp factory in Harrison. After meeting Edison there on 14 October, Churchill wrote that he was "much interested with your mention of the new form of Electrical converter," which he had seen only in a "complicated contrivance" by a "London firm of Italians" (possibly that of Sebastian Ferranti). He asked if it would be "possible to see these machines in your Electric Light works," where he also hoped to learn something about winding dynamos and insulating wire. Years later, Alfred Tate recalled that Churchill's visit was the first of several with Edison, and that the Duke was well informed on technical matters. According to Tate, Edison subsequently called him "the most accomplished scientist I've ever known." W. E. Connor to TAE, 1 and 15 Sept. and 7 Oct. 1887; Churchill to TAE, 15 Oct. 1887; all DF (*TAED* D8730A, D8730B, D8730C, D8704ADQ); Tate to Connor, 14 Sept. 1887; TAE to Connor, 13 Oct. 1887; both Lbk. 25:91B, 94B (*TAED* LB025091B, LB025094B); Tate 1938, 134–35.

3. Edison may have had in mind the Crosstown Railroad, which by April 1887 was running Daft electrical cars (from overhead wires) through Orange, N.J. Several years later, he did design and test at the Orange laboratory a "Converter System for Electric Railways." High-voltage current was supplied along the length of the line but reduced to very low pressure before reaching the street-level conductors from which the motors would draw their power. "The Daft Electric Railway, Orange, N.J.," *Electrical World* 9 (23 Apr. 1887): 191; Israel 1998, 333; U.S. Pats. 468,949 and 509,518.

4. The editors have neither found Edison's note nor determined who succeeded Charles Chinnock as superintendent at the Pearl St. station earlier in the year. Jonathan Vail was general superintendent for the Edison Electric Light Co.

–3100–

From William Jacques

Dear Mr. Edison:

I have received your letter of Oct. 19.[1] & thank you very much. You may be sure that we shall not use your name in any objectionable way=nor in any way without consulting you.[2]

We have sold ⅙ interest for $25 000[3] & this, together with what we have, will enable us to carry on the business for some time though the prospects are a large and immediate market.

We feel that our proper policy is to assure and maintain in every way that the phonograph is your invention and that we are working under your patent[4] &, to this end, we desire to organize our business into a stock company to be called the Edison Phonograph Toy Manufacturing Co.[5] whose business will be the manufacture and sale of phonograph dolls. This company, as you see, will be owned & managed by Mr. Briggs and myself. I have written you thus fully that you may understand all of our plans and I would be very glad if you would kindly telegraph us your approval of the above mentioned name for our corporation.

We desire to begin manufacturing as soon as possible and, as we have now plenty of cash and an abundant credit with our Boston bank and [~~and?~~][b] the doll is sufficiently perfect for use, we are only awaiting the closure of these formalities. Yours very truly

W. W. Jacques

ALS, NjWOE, DF (*TAED* D8750AAH). Letterhead of 95 Milk St. [a]"*Boston, Mass.,*" and "*188*" preprinted. [b]Canceled.

1. Not found.

2. Jacques had recently written with optimism about the prospects for "manufacturing several hundred samples" of a talking doll. He sought to confirm that Edison had "consented to stand god father and allow us to do you the honor of calling it the Edison Phonographic Doll or some similar name." Jacques to TAE, 17 Oct. 1887, DF (*TAED* D8750AAG).

3. The editors have not identified the toy phonograph backers, who seem to have been telephone investors in Boston. Wile 1993, 181, 190 n. 29.

4. The validity of Edison's basic phonograph patent in the U.S. was in doubt, a fact tacitly recognized in his contracts with Jacques and Lowell Briggs. See Doc. 3076 n. 7.

5. This company was incorporated on 27 October in Portland, Maine, with a capital stock of $600,000, to make dolls and other toys with phonograph attachments. Wile 1987, 9, 30 n. 17; "Miscellaneous Notes," *Electrical World* 10 (26 Nov. 1887): 288.

Caveat.

The object of this invention is to reproduce or multiply Copies of phonographic Records Called phonograms

The following is a record of Experiments tried, being tried and untried.

After the cylinder or plate ~~is~~ covered with[a] wax or similar material is indented by the action of the voice, The Cylinder is placed in a vacuous chamber Containing two electrodes of Gold or other unoxidizable metal and an arc being formed by bringing the points together & seperating the same the electric Current Causes the metal of the points to be vaporized & this is deposited in a perfect manner over the whole of the Exposed surface of the cylinder[2]   The latter if necessary is rotated while being deposited on by Clockworth in the Vacuum— Thus the non conducting surface is perfectly Coated with a Conducting surface more perfect than by any other means. it is then taken out of the vacuum chamber placed in a mould & backed up by either Type metal[3] or plaster of paris or it may be placed in a plating Bath & the deposit may[a] be increased to a ~~Quarter~~ ¼ of an inch ~~by~~ with Copper nickel etc in the usual manner—When so prepared the mould is cut in [---][b] 3 parts[c] by an exceedingly thin saw it is jointed with hinges so that when closed it will come acurately together, the space left by the saw being replaced by a strip of metal,—the loss of of 2 or 3 indentations ~~does not~~ is not noticable to the Ear— after this is done, The cylinders may be reproduced by Type metal, plaster paris; Sealing wax under a slight pressure, while soft; Gums resins; Oxycholoride Zinc; Chloride & oxide magnesia, [-][b] Thick polished tin foil placed in the cylinder against[a] the inner face & then poured with a semi solid like putty Viscous Tar or pieces of[d] Rubber, Gutta percha etc & the whole subjected to pressure which forces the foil into the indentation over the whole surface thus reproducing the record— on removing the semisolids, plaster Paris may be poured in to back up the foil— The cylinder opened the record taken out and the process gone over again to produce the Second cylinder—

If the cylinder reproduced has a plaster of paris face; The cutting quality of the surface which tends to produced scratching & wearing away of the point may be modified & reduced by dipping in a thin solution of a gum or wax in a volitile solvent, like parafine in Benzol[a] Etc, or Gelatine—fish glue Tragacanth, in water; gutta percha—in Chloroform or flexible collodion

It is even possible to take a cylinder of Very hard wax and by raising its temperature to a certain point it is recorded upon when it cools it hardens very materially & this ~~r~~cylinder may be used to reproduce on the surface of a similar cylinder covered with a hard wax softened by heat, The indenting cylinder being kept cool— If the two cylinders are placed on rotating shafts & the two surfaces be brought acurately together with considerable pressure, ~~the~~a single rotation will perfectly transfer the record of one to the other. This second cylinder may itself be used to reproduce the record audably but it is preferably used to indent a great number of cylinders thus reproducing the records almost indefinitely.

The recorded cylinder may have[a] its surface covered with plumbago so as to cause it to become an electrical conductor and this placed in a plating bath & covered as thickly as required by Copper or other plateable metal, ~~The a~~ but in this case it is essential to add one feature to ensure perfect results which feature is new & that is the graphite always used by electrotypers must be ground very much finer than now used otherwise the particles of graphite[e] are themselves in many cases as larger if not larger than many of the indentations due[a] to speaking especially in the hissing sounds & overtones of stringed instruments, but the graphite may be ground so fine that this defect will not arise & almost[f] perfect surface is obtained but it is never so perfect & acurate as that due to electro vacuous deposits.[4]

On hard wax cylinders recorded upon while softened by a rise of temperature, Tin foil very thin & polished is then placed over the entire surface—over this again is poured plaster paris. After[a] setting The wax is melted or disolved out by its proper solvent The Mould is then split in 2 or more sections & Molten waxes gums ~~Etc~~ Bitumens Soap—~~Ha~~ Setting Compounds Etc Can be run in—

The cylinder or plate[g] having a wax surface is made of several[a] pieces of thin paper ~~like~~ straw paper for instance wound up on a mandril to the proper size the end being fastened by glue ~~or~~ Tragacanth, Gum or adhesive[a] ~~surf~~ substances not effected by the hot wax or any solvent thereof.

To insure a clean smooth inner surface to each phonogram so it will register acurately[a] & pass on the holder in the phonograph properly[5] I take a sheet of smooth material for instance hard Callendered card board—of such as size that will just wrap around the mandril & the two Ends butt together, over this the body of the phonogram is wound of[a] the coarse quality[a] paper.

Thus when the paper phonogram is pulled off the mandril it has a smooth inner surface;— This inner sheet may have the surface which goes over the mandril coated with a smooth surface of Varnish Asphalt for instance or shellac after the Ends of the cylinder so formed is Cut off by a Cutter (the mandril having 2 grooves in it at each end of the cylinder the whole is pulled off and another cylinder may be formed—

Another method of forming Cylinders is to have a mould in which plaster paris, Lime & waterglass, Oxide of Zinc & Chloride Zinc; Oxide Magnesia, & Chloride Magnesia, or other setting material may be poured this forming the hollow cylindrical shell ready for putting in[h] the thin wax surface for recording   This mould may be greatly strengthened and powders such as Earthy Oxides mixed with tar, or asphalt may be forced into the mould when softened by heat & cylinders produced by[a] this method. Where the Cylinders are to be very light, Light porous bodies like Lampblack Charcoal Cork may be mixed with powdered Asphalt, Pitch or other materials which will soften by a gentle heat, & thus the whole be forced into the mould the Asphalts etc hardening immediately as the Temperature drops binds the whole together. The cylinders may Even be formed of a hard wax or asphalt itself—& afterwards surfaced with the softer material for ~~which~~ recording—

Another method of forming plates or shells is the use of the Leadpipe press[6] whereby compounds softened by heat such as asphalts mixed with various inert substances like brick dust Chalk, Lime etc may be squirted out with great rapidity & be remarkably smooth & acurate. Paper or wood or straw board[i] pulp mixed with gums, Bitumens in a soluable or viscous condition, glue; may be softened by heat making a plastic mass the Temperature not rising high enough to char the ~~paper~~ pulp; can be squirted through the Lead pipe press in any required form the surface afterwards being waxed to receive the record.

Boric acid may be squirted into cylinders or plates when the proper heat is applied the whole being waxed to prevent the action of the atmosphere. Even glass itself may be squirted in this manner a special glass having an excess of alkali & Lead oxide rendering it extremely fusible[j]

It is essential to transmit phonogram[a] cylinders by mail, if these are made of fraible substances they are very liable to be

Crushed, indented, or otherwise injured in transit. Cylinders of a slightly increased size may be used made of Lime & asphalt moulded hot will stand a great pressure & thus preserve the enclosed phonogram from injury.

Metallic phonogram shells say of Zinc, Type metal or brass tubing may be used in many cases.

~~Cylinder~~ shells may be turned out of wood ~~Hard rubber should~~ Sawdust mixed with binding substances can be moulded or pressed.

TA Edison

ADfS, NjWOE, PS (*TAED* PT031AAP). ªObscured overwritten text. ᵇCanceled. ᶜ"3 parts" interlined above. ᵈ"or piece of" interlined above. ᵉ"of graphite" interlined above. ᶠInterlined above. ᵍ"or plate" interlined above. ʰ"putting in" interlined above. ⁱ"or wood or straw board" interlined above. ʲFollowed by dividing mark.

1. Henry Seely marked this draft "Recd" at the offices of Dyer & Seely on 3 November. The editors have found no evidence that a corresponding completed caveat was filed at the Patent Office, but major elements of it were included in several patent applications in subsequent months.

2. Edison had filed a patent application in January 1884 covering a process for creating a thin coating by the electro-deposition of vaporized metal in a vacuum. That application was still pending and soon became entangled with other applications that he filed for the use of that process in reproducing phonograph recordings. See Docs. 2587 n. 2 and 3119.

3. "Type metal" refers to a broad class of alloys that would melt readily for casting as type or decorative objects. These usually consisted mainly of lead and antimony, with smaller amounts of copper or other metals often mixed in. Krupp and Wildberger 1889, 286–87.

4. In the patent resulting from a January 1888 application on the metal-deposition process, Edison noted that plumbago or similar substances required in conventional electroplating "do not bring out the fine vibrations and produce rough reproductions. . . . The vacuous deposit, however, adheres uniformly to the wax surface and reproduces the record with great perfection." U.S. Pat. 484,582.

5. Edison meant that the recording cylinder should slide smoothly and center easily on the mandril or "holder" of the machine. He executed one patent application dealing generally with this subject at the end of January 1888 and another in late February. U.S. Pats. 382,417 and 382,418.

6. As its name suggests, the lead pipe press was a machine used to form pipes, usually of considerable length. A hydraulic-powered piston forced molten metal from a cylinder into the space between a solid mandril and the wall of an enclosing larger tube. *KNMD*, s.v. "Lead pipe."

To Edward Johnson

EH.J.

If I come to NYork to meet Painter[2] & Hubbard[3] I lose a whole day which I cannot afford— My proposition is very plain and I will reiterate it—

Capital of new Co is 1,200,000, of this I give to the holders of the stock of the old Co[4] including what is in treasury $400,000   I take $800,000[b]   They paying one dollar per share to me which I in turn pay into the treasury of the new Co—[5]

2nd— I make a contract with the new Co to take for my ro~~l~~yalty 15 percent on the Cost price of the machine which will be about $2.15 instead of 10.00 or 20 pct on the selling price, as[c] old agreement called for—

3rd   I agree to put up a factory & supply all the phonographs & appliances for 20 pct on cost of Labor material & general expenses without adding any salary in for myself—

4   To give all improvements for 5 years—

5th   Turn in the new contract with Jacques & Co for the phonograph Dolls just made & which they guarantee 10,000 yearly as royalty—

They[c] sold last week ⅙ of this Contract for 25,000. so they have plenty of Cash & have a real practical Toy—

Now I consider that I am doing the fair thing by the old stockholders. They get ⅓ & more of the whole thing for the 10 000 originally paid me ~~by t~~ for the Option to purchase— Even Mr Cheever who controls a considerable quantity of the old stock thinks it is fair & just & accepts the proposition.[6] I have started in to make this a success Scientifically & Commercially   [-][d]   Mr Hubbard or Painters insinuation that I intend to do anything dishonest is gratuitous[7]

I have a strong impression that Our patents are not worth a cent so I am going on the basis that we are going to hold the field only by making a better & cheaper machine. If Mr Hubbard or our Friend Painter ~~think~~ want to go into the field I think they could do so and Copy my machine & I have no patent that would stop them. The phono & graphone is unfortunately killed by that d——d infernal ~~B~~Second British patent.[8] Yours

Edison

The above proposition based on actual biz is better than If The old contract was alive[9] but of course in a stock speculation view it is not but this is not going to be a stock speculating Co Like the graphophone If I can prevent it   E

ALS, PHi, UHP (*TAED* X154A6BL). Letterhead of the Laboratory of Thomas A. Edison. ª"*Orange, N.J.*," preprinted. ᵇ"I take $800,000" interlined above. ᶜObscured overwritten text. ᵈCanceled.

1. This letter was marked received on 24 October.

2. A business promoter, Washington lobbyist, and former journalist, Uriah Hunt Painter (1837–1900) had a prominent role in forming and running the Edison Speaking Phonograph Co., of which he was the largest stockholder. Painter had recently expressed his opposition to what he called Edison's breach of faith and "annihilation" of that company by vesting recent and future phonograph improvements in a new entity. Painter became president of the Edison Speaking Phonograph Co. in 1888. Docs. 672 n. 2, 2746 n. 1; Edison Speaking Phonograph Co. list of stockholders, 30 Apr. 1887; Johnson to TAE, 14 Oct. 1887; Edison Speaking Phonograph Co. minutes, 25 July 1888; all UHP (*TAED* X154A6AG, X154A6AS, X154A7DK); Wile 1991, 8–16; Israel 1998, 282–83.

3. Boston patent attorney Gardiner Greene Hubbard (1822–1897) was one of the original investors in the Edison Speaking Phonograph Co. and its current president. Hubbard, who also organized the telephone business of Alexander Graham Bell (his son-in-law), had been trying since 1885 to combine the business interests of the Edison phonograph and the Bell-Tainter graphophone. See Docs. 1190, 2442 n. 3; Wile 1991, 8–16; Johnson to Painter, 27 July 1885; Hubbard to Johnson, 13 Oct. 1887; Hubbard to TAE, 28 Oct. 1887; all UHP (*TAED* X154A4EF, X154A6BI, X154A6BR).

4. The Edison Speaking Phonograph Co., incorporated in 1878 in Connecticut, held the American manufacturing and marketing rights to Edison's phonograph, exclusive of its use in toys, dolls, and clocks. After about two years of licensing exhibitions and making limited sales, it became essentially dormant. Edison was vice president. Docs. 1305 n. 2, 1657 nn. 2–3; Wile 1976; Wile 1974, ix–xi; Hubbard to TAE, 28 Oct. 1887, UHP (*TAED* X154A6BR).

5. The Edison Phonograph Co. was organized on 30 September 1887 to manufacture and sell phonographs in the United States and Canada. It was registered in New Jersey on 8 October and held its first stockholder meeting the same day in Harrison, where Edison, Ezra Gilliland, John Tomlinson, and Alfred Tate were chosen as directors. At the next meeting, on 17 October, Edison was elected president, and formal discussions of purchasing his phonograph patents began. A contract to that effect was made eleven days later, subject to any existing rights such as those of the Edison Speaking Phonograph Co., although these were "believed to have expired and become void." The contract granted the company full ownership not only of the patents themselves, whose validity was in some doubt, but of "the inventions therein severally described." In return, Edison was to receive 11,960 shares of its stock and a 20 percent royalty on the cost of each phonograph sold. Edison Phonograph Co. memorandum of association, 30 Sept. 1887; Edison Phonograph Co. minutes, 17 and 28 Oct. 1887; all FFmEFW (*TAED* X104A001, X104A004 [images 1–3, 4–8, 14–16, 19]); TAE agreement with Edison Phonograph Co., 28 Oct. 1887; Gilliland agreement with Edison Phonograph Co., 28 Oct. 1887; both Misc. Legal (*TAED* HX87013, HX87013A).

6. After receiving Edison's proposal about the new company from John Tomlinson, Charles Cheever commented that the terms proposed for stockholders of the old phonograph company "are not as liberal as I had been led to expect"; as of 18 October, he was seeking details about the earlier contracts and still questioning the proposition, which would be "made to the old stockholders individually and not as a company." By 21 October, according to Painter, Cheever was content to "make his own deal" for shares in the new company. Probably about this time, and with some of Cheever's questions in mind, Edison sought further information about the old company from Tomlinson. Cheever to Hubbard, 18 Oct. 1887; Hubbard to Johnson, 19 Oct. 1887; Painter to Hubbard, 21 Oct. 1887; all UHP (*TAED* X154A6AW, X154A6AV, X154A6BD, X154A6BE); TAE to Tomlinson, n.d. [c. 19 Oct. 1887], DF (*TAED* D8848AFL).

7. Hubbard had recently written to Johnson that Edison's maneuvers with regard to the new phonograph company "would not be an honorable way of treating" shareholders of the older company. Johnson, who had been the conduit for numerous exchanges with Hubbard and Painter, let Edison know his objection to "being made a receptacle for the complaints of others." Edison replied that Johnson had "better refer them to me on Phonograph matters—I have made a perfectly fair offer to them. . . . [Hubbard] better not harp too much on other peoples honesty." Around the same time, Edison asked Tomlinson to review a draft letter to Hubbard containing many elements of his proposition for the participation of stockholders from the old company. Painter to Johnson, 6 and 9 Oct. 1887; Johnson to Painter, 7 Oct. 1887; Hubbard to Johnson, 13 Oct. 1887; Johnson to TAE, 14 Oct. 1887; TAE to Johnson, 17 Oct. 1887; all UHP (*TAED* X154A6AM, X154A6AR, X154A6AP, X154A6BI, X154A6AS, X154A6BB); TAE to Tomlinson, n.d. [c. 19 Oct. 1887], DF (*TAED* D8848AFL).

8. Edison's "Second British patent" on the phonograph was 1,644 (1878) for "Recording and Reproducing Sounds." The legal implications of this specification (and possibly that of an earlier patent [2,909 of 1877] from which the phonograph had subsequently been excised) were muddy (Holmes 2006, s.v. "Patents"; Andrews 1986, xii, xx n. 3; Wile 1990b, xiii–xiv). Under one interpretation of U.S. law, the lifetime of his existing American phonograph patents was tied to that of foreign specifications issued before them, so that they would expire with the British one instead of running for a full seventeen years (see Doc. 3120). In 1885, Lemuel Serrell had prompted Edison to pay the seven-year renewal tax on his British Patent 1,644 (1878) in order to keep it in force. Serrell's notice arrived while Edison was in Florida, and the editors have found no response to it. The specification seems to have expired, with George Gouraud referring in July 1887 to "the fact that there is no longer any British Phonograph patent." If such was the case it was not the only lapse. Edison had already declined to pay annuities on phonograph patents in France, Italy, and Belgium; several Swedish patents (including one for the phonograph) had lapsed in 1883, precipitating fear then about the consequence for Edison's U.S. patents (Serrell to TAE, 13 Mar. 1885 and 12 May 1884; Gouraud to TAE, 2 July 1887; all DF [*TAED* D85441I, D8468ZAX, D8751AAA]; Doc. 2626 n. 4).

9. That contract (Doc. 1190) was to "continue during the existence of the [American] patent" for which Edison had applied in December 1877 (U.S. Pat. 200,521).

Orange, N.J., Oct 25 *1887*[a]

Dyer—
    Patent—[1]

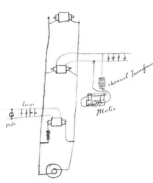

It is not practical to work motors on Alternating circuits— Currents of the same direction are essential.[2]

This invention consists in a method of causing alternating Currents to be all in one direction in various circuits which are to Contain motors. I accomplish this by using a large number of cells containing in each cell a large number of very thin lead places[3] close together immersed in Sulphuric acid & water— Enough cells are Employed so the Counter Emf ~~is just~~ is only slightly less than the prime Emf— Hence one alternation is almost balanced by the battery & only a weak Current passes while the opposite alternation is added to the battery which doubles the Emf—The motor being wound to work with this Emf   The Lamps may be in a second circuit or the same circuit ~~as~~   The Volts will be 2.5 and no depreciation or weakening of solution takes place—the internal resistance is remarkably low—lower than it is possible with a storage battery & the converting power per pound of lead is enormous— [~~Th--~~ large?][b] ~~can be used with strength~~
    Claim the Earth[4]

Edison

ADfS, NjWOE, PS (*TAED* PT032AAL). Letterhead of the Laboratory of Thomas Edison; drawing made on separate sheet. [a]"*Orange, N.J.*," and "*188*" preprinted. [b]Canceled.

1. Figure labels are "Motor," "Lamps," "Motor," and "Chemical Transformer."

2. The application of alternating current to a motor wound on the general plan of a direct current dynamo, because of the reversal of magnetic fields, would tend to produce oscillating mechanical motion rather than continuous revolution. When the Chicago Electric Club debated the merits of AC currents, Elmer Sperry, then an aspiring inventor-entrepreneur, contributed a paper in May 1888 on the vexing AC motor problem and the economic incentives for solving it. "There is probably no field of electrical engineering in which there is a greater activity at the present time than that of the alternating current motor," Sperry began, pointing out that nearly all this research was carried out in secret. Raising the question whether the possible results could merit the "expenditure of any considerable amount of time or money upon the problem," he pointed out that a "large part" of total generating capacity went unused during the day, but the development of a sizeable market for electric power would create economies of scale for producers and consumers. Reviewing the state of the art, Sperry noted that motors adapted to AC either could not start by themselves or were so carefully synchronized to the current's frequency that they could accommodate only a fixed load at a predetermined speed (Sperry 1888; Hughes 1971, 43–45; Martin and Wetzler 1888 [chap. 25] provide another contemporary overview of AC motor research). Nikola Tesla had by this time largely surmounted the fundamental engineering problems by designing a polyphase motor in which the magnetic field itself rotated. Tesla filed what his biographer calls a "comprehensive patent application" in the U.S. for his motor on 12 October 1887; it was the first of several filed in close succession before the end of the year, and the whole lot issued on 1 May 1888, just ahead of the publication of Sperry's paper. The polyphase motor could not be run on existing AC systems, however, and in response to some pressure from his financial backers, Tesla had also been developing a single-phase AC motor that could do so, though he delayed having the patents drawn up (Carlson 2013, chaps. 2 and 4, quoted p. 95).

3. Edison probably intended "plates."

4. Edison did not receive a patent based on this draft. The editors have found no indication that he signed a finished application, although it could be one of the two missing from the sequence of patent cases completed in November (see App. 2.A).

–3104–

Memorandum to Richard Dyer: Electric Lighting Distribution System

Orange, N.J., Oct 29 *1887*[a]

Dick—

Please fix up another Case With the shifting ratchet wheel device & apply it to transformers[1] The object being to disconnect & connect transformers as the load diminishes or increases all from the station[2] Thus[3]

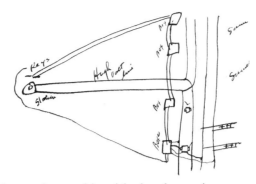

You know the trouble with the alternating system is that while the sale of light is usually all in 4 hours The loss in the transformers which is 8 @ 12 per cent[4] according to construction goes on during the whole 24 hours & the losses in some cases in the transformers equal the sales so you see it is a great stride in the direction of economy to connect[b] & disconnect transformers as the load varies— get a strong claim on this   howl on the great economy etc as[b] I am certain they will run a foul of this before long— Bill to light Co as its for protection purposes

Edison

PS— You might show seperate circuits running to each transformer from station with magnet to control switch also a hand switch at each transformer to disconnect—

also a time clock controlling a switch to disconnect at any prearranged time— If this cant be put in one patent make 2.[5] E

ADS, NjWOE, Lab., Cat. 1152 (*TAED* NM021AAH). Letterhead of the Laboratory of Thomas A. Edison. [a]"*Orange, N.J.*," and "*188*" preprinted. [b]Obscured overwritten text.

1. Edison referred to the ratchet wheel unison mechanism in a patent application that Dyer was preparing from a draft that Edison made on 20 October. That application (discussed in Doc. 3089 n. 1) covered a network of switches throughout a distribution system, each one separately controllable through a common circuit, to transfer electrical load from one part of the system to another. Cat. 1152, Lab. (*TAED* NM021AAG).

2. Edison signed the resulting patent application on 4 December but it was not filed at the Patent Office until 27 December; it issued in May 1888. It referred specifically (by serial number and date) to the application filed on 9 December for the switching apparatus, mentioned in note 1.

The specification included six drawings, none adapted directly from Edison's sketch, and four claims. Edison explained that his intention was to minimize electrical losses in a high-voltage AC distribution system by ensuring that only so many converters will be in circuit at any time as are required to supply the lamps or translating devices actually in use. I do this by providing the converters, or certain of them, with switches whereby their primary and secondary circuits may be opened or closed, as desired, and thus any desired number of the converters may be removed from or maintained in connection with the system, according to the amount of current required to be used at any time. [U.S. Pat. 382,415]

3. Figure labels are "Keys," "Station," "High Volt line," four iterations of "Box," and two iterations of an illegible word (possibly "Same").

4. The specification referred to a transformer loss of "seven to twelve per cent" but conformed generally to Edison's suggestions. It noted that "In some cases the current used in the converters—which of course is a dead loss to those operating the plant—will be equal in amount to that sold to the consumers. Evidently this results in a great diminution of the profits of the business." U.S. Pat. 382,415.

Electrical engineers were just then trying to articulate rules for the optimal design of transformers. One attempt was an August 1887 article by Rankin Kennedy presenting formulae for calculating the proper relationship among the voltage and frequency of the current and the lengths of the primary and secondary wires. A crucial variable was the coefficient of induction, or the induction in volts per foot of wire. Where this value was too low in practice, Kennedy noted, "a considerable current passes in the primary [coil] when no secondary current is flowing"; that current could, in theory, be reduced nearly to zero. Kennedy 1887, 301–2.

5. Each of Edison's suggested variations was represented in the patent.

FIRST WAX-CYLINDER PHONOGRAPH   Doc. 3105

By 21 October, Edison had two machines embodying the basic design of his first wax-recording phonograph, which he planned to put on the market by the end of January.[1] At least one may have been finished as early as the end of September. In mid-November, only one of the machines was still at the laboratory, where it was being "taken apart and put together . . . by the machinists who are making tools for the wholesale manufacture."[2] The other was likely at the offices of Edison's patent attorneys, Dyer and Seely, as a model for making drawings for the patent application that Edison executed on 22 November and filed four days later (U.S. Pat. 386,974). Edison described the principal features of the new phonograph in his application:

my object is to improve the recording and reproducing qualities of the instrument, to provide a proper motor for

running the phonogram cylinder at a uniform speed, to
provide a suitable phonogram cylinder and phonogram,
and to arrange and construct the several parts of the
machine so that its manipulation will be simple, readily
understood and convenient.[3]

In interviews with reporters he elaborated on these aspects of
the instrument.

Edison asserted that he had "more difficulty in getting a
motor to suit me than any other part of the apparatus." After
experimenting with clockwork and spring motors, he finally
devised an electric motor "which runs at a perfectly regular
rate of speed, is noiseless, and start or stops at the touch of
a spring."[4] It was to be operated by a four-cell battery that
would only need to be replenished monthly at a cost of less
than a dollar if the motor ran four to five hours a day. Because
the electric motor automatically turned the cylinder for recording and playback, the new phonograph was easier to operate than the old hand-cranked tinfoil phonographs, even
though it had a more complicated mechanical design.

The sound quality of the wax-recording phonograph was
also vastly superior to that of the tinfoil machines. Unlike the
earlier models, in this one Edison used a separate diaphragm
and needle to record and play back sound. As a result, he
claimed that it had a "tone more distinct, clearer, more characteristic of the voice of the writer than any telephone you or
I ever heard. . . .The present apparatus will satisfy any one
who is half satisfied with the telephone."[5] A reporter for the
New York Evening Post agreed with Edison's claim after hearing him read "a list of geographical names, many of which the
reporter had never heard before, but which were perfectly
distinct." He found only "one word out of every six or eight
was not perfectly distinct, largely owing to the noise" of workmen in the laboratory. He also reported the experience of an
"editor of scientific paper, who listened while the phonograph
read to him one page of 'Nicholas Nickelby' resulted in his
getting 80 per cent. of the words the first time."[6]

As useful as the new machine might be for typesetting and
business correspondence, Edison found its musical qualities
more impressive:

> I have got the playing of an orchestra so perfectly that each
> instrument can be heard distinct from the rest; you can
> even tell the difference between two pianos of different
> makes; you can tell the voice of one singer from another;

you can get a reproduction of an operatic scene in which the orchestra, the choruses and the soloists will be as distinct and as satisfactory as opera in this sort of miniature can ever be made.[7]

Edison also made an experimental triple recording of himself whistling "Yankee Doodle" over a recording of him singing "Hail Columbia," recorded over the list of geographical names. The *Post* reporter described it as "a most curious combination in which each part was perfectly distinct."[8]

Edison planned on having three sizes of phonograms for use with the phonograph:

> The phonograms will be sold in the shape of small cylinders one and a quarter inch in diameter and from one inch to four inches in length. The one-inch phonograms will contain 200 words, or what is considered quite sufficient for an ordinary business letter; they will cost 15 cents a dozen. The full-size phonograms of four inches in length will contain from 800 to 1,000 words, according to the rate of speed of the speaker, and will cost about 36 cents a dozen.[9]

He also hoped to make arrangements with the U.S. Post Office to have phonogram boxes sent at the same rate as letters.

1. "A Wonderful Workshop," *New York Evening Post*, 21 Oct. 1887, 1.
2. "The Wizard Edison Talks," *New York World*, 6 Nov. 1887, 28.
3. This text was deleted before the patent issued. Pat. App. 386,974.
4. "A Wonderful Workshop," *New York Evening Post*, 21 Oct. 1887, 1; "The Wizard Edison Talks," *New York World*, 6 Nov. 1887, 28.
5. "The Wizard Edison Talks," *New York World*, 6 Nov. 1887, 28.
6. "The Phonograph at Work," *New York Evening Post*, 18 Nov. 1887, 8.
7. "The Wizard Edison Talks," *New York World*, 6 Nov. 1887, 28.
8. "The Phonograph at Work," *New York Evening Post*, 18 Nov. 1887, 8. Edison also told the *Post* reporter that he planned to test a variety of "sound condensers or funnels" being made by his assistants, which he thought would "be necessary for recording the music of an orchestra or the voices of a number of speakers." To a *New York World* reporter, he claimed to have recorded an orchestra so that "each instrument can be perfectly distinguished." "The Wizard Edison Talks," *New York World*, 6 Nov. 1887, 28.
9. "The Phonograph at Work," *New York Evening Post*, 18 Nov. 1887, 8. In another interview, Edison described the three sizes of phonograms as "one for letters of from 800 to 1,000 words, another size for 2,000 words, another size for 4,000 words." "A Wonderful Workshop," *New York Evening Post*, 21 Oct. 1887, 1.

Experimental Model:
Phonograph[1]

M (28 cm × 25. cm × 30.5 cm), NjWOE, Cat. 2196.

1. See headnote above.

2. See headnote above for the dating of this design. This machine has an 1888 modification of the spectacle unit that held the recording and repeating diaphragms. It is unclear whether the metal bar, on which the end of the spectacle unit rested, is original or modified. An illustration from the 31 December issue of *Scientific American* shows the bar with units of measure etched on it.

From Lewis Miller

Dear Sir:

We are now using 50 Volt 32 C.P. Edison incandescent lamps on our Brush circuit, three in place of one arc lamp. Can you send us a lamp of which we can use <u>five</u> in place of one arc light? If so send us 100. If not send us 50 of the 55 Volt 32 C.P. Yours Truly

Lewis Miller   Supt   Means[1]

⟨Upton— Please attend to this= I dont think they want the arcs   You could easily work up a big Lamp trade in going round the country where parties have arc machines only & substitute municipals all over their shop—& constant loads   You know there must be 1000 isolated arc plants in the US[2b]

It occurs to me that the Lamp Co does not do enough outside biz— It seems to me that if I was running the Lamp Co to <u>make a record</u> that I wouldnt depend on the Light Co or any one Else. I would put a man or two on the road with a list of all our plants and work up the <u>increase</u> of all our plants made[c] possible by the new Lamp. it is simply folly to depend on our Local agents also there is no need to wait because we havent a supply   The changes will take time, and your men cant visit the whole inside of 6 or 8 months— I would also put[c] a man on specially to do just what Mr Miller wants ie[d] substitute 20 32 & 50 cp municipals for arc isolated.= Insull is working up everything—he has buillt up a trade of 150 000 a yr in wire & has travellers— I want to talk with you about this matter— I think <u>Card</u>[3] would be a good man—for arc biz— I think we could use Steringer also— E[dison]⟩

L, DSI-AC (*TAED* X098A025). Letterhead of Aultman, Miller & Co. [a]"AKRON, O.," preprinted; "NOV 10 1887" stamped. [b]Followed by "over" as page turn. [c]Obscured overwritten text. [d]Circled.

1. Walter K. Means (1863–1931) was a private secretary and purchasing agent for Aultman, Miller & Co. until 1903. U.S. Census Bureau 1882? (1900), roll T623_1323, p. 15A, image 1241323 (Akron, Summit, Ohio); Letterhead, Means to TAE, 4 Apr. 1893, DF (*TAED* D9308AAI); "Personal," *Iron Trade Review* 20 (14 May 1903): 37; Death record for Walter K. Means, *Ohio Deaths, 1908–1932, 1938–2007*, online database accessed through Ancestry.com 24 Feb. 2014.

2. A contemporary directory listed fourteen pages of isolated arc lighting plants in the U.S. and Canada. *American Electrical Directory* 1886, 159–72.

3. Benjamin F. Card (1838?–1921) became an agent for the Edison Isolated Co. in New York City and Long Island about 1882, having previously corresponded with Edison about his own electric meter and electric railway inventions. (Edison took a partial ownership interest in a Card meter patent in 1881.) Edison complained in 1882 that he "talks

a great deal too much" and should be given a special assignment far from New York. In January 1888, Card was dispatched to replace the lights in Miller's shop with Edison's municipal lamps, and he seems to have received a broader mandate like that outlined in this document. By mid-1888, he was identified as an agent of the Edison Lamp Co. and had recently reported to Francis Upton "that if this municipal business is persisted in it will bring every station we have got to a paying dividend basis, and effectually keep out the other fellow." Death record for Benjamin F. Card, *New York, New York, Deaths Index, 1862–1948*, online database accessed through Ancestry.com, 24 Feb. 2014; Card to TAE, 4 Nov. 1878 and 23 May 1883; Upton to TAE, 14 May 1888; all DF (*TAED* D7802ZZJJ, D8325ZAH, D8833ABG); TAE agreement with Card and Edison Electric Light Co., 7 Feb. 1887, Miller (*TAED* HM87AAF); TAE to Sherburne Eaton, 2 Aug. 1882, Lbk. 7:836 (*TAED* LB007836); Lewis Miller to TAE, 23 Jan. 1888; CEF (*TAED* X018C6A3); "Eighth Meeting of the Edison Illuminating Companies," *Electrical World* 12 (18 Aug. 1888): 82.

–3107–

*Charles Batchelor
Journal Entry*

[Orange?][1] Nov 11th 1887

510[2]   Phonograph.

Edison brought the model phonograph to the laboratory and exhibited it to a number of members of the National Academy of Sciences who paid him a visit—[3]

AD, NjWOE, Batchelor, Cat. 1337:27 (*TAED* MBJ004, image 14).

1. Edison's new laboratory complex was far from finished, as evidenced by Batchelor's lists of construction projects around this time, which included covering the galvanometer building, putting in the second floor of Edison's library, and installing boilers and heavy machinery. Even so, it was probably Orange to which Batchelor referred in this entry, partly because he was spending most of his time there. He had also noted two days earlier that the new phonograph factory had "not received the model from the Laboratory at East Newark yet but expect it every day." Presumably it was from there that Edison brought the model for the event described below. Cat. 1337:25, 31 (items 506, 508, 515; 5, 10, 25 Nov. 1887), Batchelor (*TAED* MBJ004; images 13, 16).

2. Charles Batchelor consecutively numbered each entry in this journal.

3. The National Academy of Sciences was concluding its four-day meeting at Columbia College in New York City on this date. According to one newspaper report, "attendance was smaller than on any of the previous days, as several of the members had accepted an invitation from Professor Edison to visit." George Barker referred to this demonstration a year later when he invited Edison to show the improved phonograph to Academy members at their 1888 meeting in New Haven (National Academy of Sciences 1888, 5; "Scientific Papers Read," *New York Tribune*, 12 Nov. 1888, 7; Barker to TAE, 31 Oct. 1888, DF [*TAED* D8847ADA]). Founded in 1863, the NAS had exhibited Edi-

son's electromotograph in 1874, as well as his phonograph and carbon telephone in 1878. Edison had also invited Academy members to his Pearl St. station in November 1882 (True 1913, 1–6; see Docs. 500, 504, 1291, 1293, 2367).

–3108–

From Charles
Chinnock

NEW YORK, November 12, 1887.[a]

Mr. Edison:—

I have delayed answering yours of the 7th inst.[1] in order to forward you enclosed statement of October business.[2] You will notice by it that we are earning at least $7,000 per month (41 plants closed, 10,400 lamp capacity as against, in 1886, 28 plants of 4,600 lamp capacity) and that in itself should be a sufficient answer to predictions made by any one. Of course you are well aware of the wild and awful prophecies uttered at my expense while at Pearl Street. Many of my friends holding their breath waiting for a grand and terrible collapse.

I have one man traveling in Maine and New Hampshire, another in Connecticut and a third one will start Monday morning through New York. A synopsis of what they report I will forward you within a day or two. I am in communication with a Mr. Shain[3] of Chicago, employed by the United States Company, and I am quite certain that he is an A1 man in every respect. It has been reported to me that Mr. Warren[4] owes all his success to this man's endeavors. Of course he is high price but if he is the right man that will not deter my action. Mr. Griffith[5] called on me and I used my best endeavors to have him cancel his engagement with the Springfield Gas Company[6] but he held off, stating he could not come with us before January. I am in hopes of securing one or two good men from other Companies, as I believe that men experienced, even at double the salary, are better than green men for immediate results. While we are doing double the business ever accomplished before at a large profit, taking our October statement as a criterion, I am still of the opinion that four times that amount should be the rate and the limit to the number of plants that can be sold is governed by the amount of money expended in canvassing for the orders.

It has been reported to me by several outsiders that the Westinghouse Company are sweeping the West in the Central Station business on the Cash Basis.[7] I am satisfied that a concentrated effort should be made to check this wholesale business on the part of Westinghouse, if reports are true. How would it do to take the "bull by the horns" and give intending purchasers the choice between a fac simile of the West-inghouse converter system and our own and trust to the intel-ligence of our agents to prove to them the superiority of the Edison apparatus. Messrs. Bergmann & Livor have appointed a Committee to consult with myself and try and devise some comprehensive scheme that will cause part of this business to revert to this Company. I certainly would like to be in posses-sion of your ideas concerning this branch of the business.

Mr. Stern[8] has certainly done very well in Pittsburgh and in consideration of this success we have concluded to increase his territory by giving him all of Pennsylvania West of the Al-leghenys, and if he will remain true I am quite sure he will accomplish good work in the field allotted him. His specifi-cations are certainly very complete and comprehensive and I have written him to forward me a model according to his ideas and will use this as a sample to forward to our agents.

After having one or two interviews with parties operating small central stations on the cash basis, I am convinced that one of the best ways to boom that particular branch of the business would be to make existing small stations thoroughly successful. For instance, a Mr. Marshall,[9] having charge and being the largest stockholder in a small central station in La-conia, N.H., has just left this office and in that interview he states that he has unlimited power, that gas is selling for $4.50 per M[10] and that he could place 1,000 lights (at the present time he has but 300 in operation) and yet he offers to sell the entire plant to this Company for 25% of what he paid for it, and in addition to that will furnish all power we require, up to 150 h.p., free of charge for three years. It is as much as I can do to refrain from buying this entire plant myself and taking a man of ordinary intelligence, send him on there and prove conclusively that, with a little judgement, at least 20% could be made on the investment. If this is a sample of what other small central stations have to put up with, I do not wonder that Westinghouse, or anybody else, are closing orders by the wholesale. Truly[b]

C. E. Chinnock

P.S.   I wanged away at the [Red lines?][c] but they went too [que----][c]   C[hinnock][11d]

TLS, NjWOE, DF (*TAED* D8738AAL). Letterhead of Edison United Manufacturing Co. [a]"NEW YORK," and "18" preprinted. [b]Written by Chinnock. [c]Illegible. [d]Postscript written by Chinnock.

1. Not found.
2. Two versions of the statement, nearly identical, found their way

into Edison's files. Edison United Mfg. Co. statement, Oct. 1887, both DF (*TAED* D8738AAM, D8738AAK1).

3. Charles D. Shain (1855–1934), an 1879 graduate of Philadelphia's Central High School, worked as a civil and construction engineer before entering the electrical field in 1883. Shain joined the United States Electric Lighting Co. that year when it opened a Chicago office. Except for a stint as sales agent for the electric light department of George Westinghouse's Union Switch and Signal Co., he remained with the United States Co. until it was purchased by the Westinghouse Electric Lighting Co. in 1889. He then joined the United Edison Manufacturing Co., ultimately becoming assistant general manager. Shain left the orbit of Edison interests (though not electric lighting) when General Electric was formed in 1892. "Pioneer Hustlers," *Electricity* 8 (20 Feb. 1895): 75; Obituary, *NYT,* 14 Apr. 1934, 15.

4. Charles C. Warren (d. 1905) began his business career in Michigan as a merchant. From 1881 to 1890, he managed the western affairs of the United States Electric Lighting Co. in Chicago, where he hired Charles Shain. Warren later took on similar responsibilities for the Edison United Manufacturing Co. He subsequently purchased and operated an electrical manufacturing business in Sandusky, Ohio. "Mr. Charles C. Warren," *Electrical Review* 46 (1 Apr. 1905): 554; "Death of C. C. Warren," *Western Electrician* 36 (1 Apr. 1905): 247.

5. Not identified.

6. Chinnock referred to the former Springfield Gas Machine Co., incorporated since 1870 as the Gilbert and Barker Manufacturing Co. by businessman Charles N. Gilbert (b. 1843) and inventor John Francis Barker (b. 1839). Located in Springfield, Mass., the firm manufactured and installed the so-called Springfield gas machine for the on-site production of illuminating gas from gasoline or naptha. (Both fuels were byproducts of kerosene refining, which led to the company's acquisition by a Standard Oil subsidiary in 1884.) Marketed as the Springfield gas machines, they were not the only isolated illuminated gas plants available for producing carbureted gasoline. They were the most popular, however, and they were widely adopted in the 1870s and 1880s by factories and affluent country homes outside the service of city mains supplying coal gas. The Edison Machine Works and the Edison Lamp Co. were reportedly among the firm's customers, and Edison contracted in August or September for one of its large machines to generate 1,200 cubic feet of gas per day at the new Orange laboratory. The basic Springfield machine consisted of a fuel reservoir buried at some distance from the building to be lighted; a weight-driven fan blew air through the tank to create a flammable fuel-air mixture that was forced through pipes into the building. Linebaugh 2011, xvii–xxii, chaps. 2–3; Gilbert and Barker Mfg. to TAE, 13 Aug. and 1 Oct. 1887; Gilbert and Barker Mfg. to Charles Batchelor, 13 Sept. 1887; all DF (*TAED* D8756AAS, D8756ACV, D8756ABW); Vouchers (Laboratory) no. 767 (1887) for Gilbert and Barker Mfg.

7. The Westinghouse Electric Co. was incorporated in 1886 in Pennsylvania to consolidate into a dedicated firm the growing electrical manufacturing, sales, and installation enterprises of George Westinghouse. The company sold its equipment to local central station operators for cash; unlike the Edison Electric Light Co., it did not take stock

or other interests in the illuminating companies. Passer 1972 [1953], 136, 140; "The Westinghouse Electric Co.," *Elec. and Elec. Eng.* 5 (Mar. 1886): 120.

8. William A. Stern (1860–1914), an Edison associate from Menlo Park days, was connected with a family printing business in Philadelphia in 1883. He had at that time at least two patents pending or issued for electric lighting of railroad cars, in which Edison at some point acquired a half-interest. Having tried unsuccessfully to interest Westinghouse and the Pennsylvania Railroad, he continued to correspond with Edison about the matter for several years. By 1884, Stern was with the Pittsburgh Electric Co., an agent for the Edison Isolated Co. (among others). He remained in Pittsburgh, where Stern & Silverman became agents for Edison lighting, at least through 1888, but within a few years he was back in Philadelphia, where he became deeply involved in electric traction. Obituary, *Philadelphia Record*, 5 Sept. 1914, SC14; *TAEB* 6 App. 5.C nn. 2–3; *TAEB* 7 App. 4.C. esp. n. 6.

9. Possibly William C. Marshall (b. 1843), the operator of prosperous woollen yarn mills in Laconia, N.H. Marshall became involved with a local electric power company in 1895; his son, Lyman C. Marshall, was also later in the electric power business. *Biographical Review* 1897, 21:377–78.

10. That is, per thousand cubic feet. Cf. Doc. 1897.

11. To "whang" (variously "wang") is to strike heavily or to throw, but the editors have not determined the meaning of Chinnock's partially illegible postscript. Partridge 1984, s.v. "whang."

–3109–

Notebook Entry:
Miscellaneous[1]

[Orange,] Nov 15 87

New Laboratory—

In rotary thermo use alloy of Arsenic & Cobalt, Misprickel[2] Cobalt glance—[3] Cobalt & Sulphur[a]

See if Cobaltoso[b]-Cobaltic oxide[4] which is steel gray crystals under some conditions if is a good conductor & test its other properties for thermo—for telephone for[b] storage, etc—its insol[uble] in Nitric—HCl & Aqua Reg[ia]   ⟨John Ott—⟩[c]

Spegilheisen[5]—for rotating Thermo   ⟨English[6]⟩[d]

Alternate a powerful Current through heated wrot & cast iron   see changes— ⟨Batch⟩[e]

Photograph Scintillations   ⟨Dixon[7]⟩[f]

For carbon forms: put wires in & press & carbonize & afterwards disolve wires out. ⟨John Ott⟩[8g]

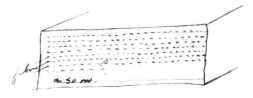

Possibility of forcing[b] out wrot iron brot to plastic condition into pipe by Lead press lined with fire brick & with fire
brick plunger— ⟨Batch⟩[h]

Make Curve of Lamps for fall candle power & life at 80 & 16
CP—at several points of lower vacuum than we usually use—
to do this perhaps a definite Kind of residual gas must be used
again ⟨Entered in Marshalls book[9]⟩ the carrying & center wire
current may be too strong at 80 to entirely vitiate result while
at 16 it may be entirely absent. There must be a point at normal where there can be no infinitesimal arcs & hence CP will
go up & life be great. Bad Vac Lamps go up in CP & give good
life— ⟨Entered Marshall book   Expmts to determine⟩[d]

Measure the resistance of pressed buttons under constant
pressure & electrode surface of every Known oxide or compound & their behavior for future use   ⟨Shultz-Berge[10]⟩[d]

Experiment on the puffy gas hood to get a sensitive but
hardy hood— hold it on pressed (Lead press) fibres of aluminum etc.[a]

[A][11]

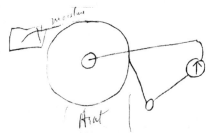

Motograph for generating Current   Heat altering Capilliar-
ity gives Current. Work up & study why turning ~~MEMG~~ gives
Current   ~~Stop turning~~   ⟨Shultz Berge—or Kennelly[12]⟩

Make the Torpedo Lamp for submersion for Murdock[13]
⟨Force⟩[h]

Photograph through metals & other thin plates Opaque to
Light— ⟨Dixon⟩[h]

Get a locking material for magnetite, that requires only a
trace & is a flux as well, Lime water—[---][i] Water glass, Lime—
Lime in Sugar—brick clay—etc. ⟨Chemist [Building?]⟩[14b]
Ent[ered]⟩[h]

Meter   pages old expmt[15]   ⟨John Ott⟩[16]

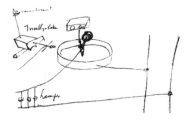

Meter ⟨John Ott⟩[17]

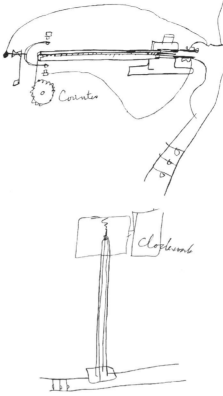

Use a reverser & Wire[b] in Calorometer, get heat with Constant Current & then heat when reversed— I think heat will be as great with alternating & ¼ of for insulation on break wheel as with constant Current ie[j] Orienting Molecules require Energy which is lost in heat. ⟨Schultz-Berge⟩[h]

also get one of those wire testing machines same as Gilliland has at Boston & Alternate Currents through Copper soft & hard & take breaking strain after different times. ⟨Batchelor⟩[h]

Deposit Silver on Muslin Lace silk etc in Vacuum— ⟨~~Marshall~~ Select another man for these Expmts⟩[d]

Study the Carbon from the action of Dehydrizing agents on organic material ⟨Chemist Marshall[18]⟩[d]

Deposit Chromium on glass in Vacuo— ⟨Marshall another man⟩[d]

Carbo preliminarze at 300 in asphalt 3 days then bring to 600 and take out dip & run reg— ⟨Young Man in Kinneys Room[19]⟩[d]

Measure resistance ~~of~~ between 2 plat wires in porcelain tube white hot wires say ¼ inch apart also one plat 1 carbon— probably with latter get a current. ⟨Shultz-Berge⟩[h]

Try Chlorine in in Lamp with paste clamps for low Vac residual—use ChlChromium—also use Cyanide Silver but only at normal so as not to decomp the Cyanogen—put on U tube with Hg graduated in Milimeters— ⟨Entered in Marshalls book⟩[h]

on new phonograph   Have small magnet work the ~~ind~~ Recorder at the other End of the telegh line   have a Vibrator & break it up into dots & dashes record this on cylinder ~~at~~ & speed it to send at[k] 3 or 400 words per minute then put on reproducer & let cylinder run slow to Copy out Morse— Use Motograph & also[l] Talk over wire & record—[a]

See if plumbagoed wax phonogram or rather Lamp blacked will when reproducer passes over vary pressure & retransmit   ⟨Hamilton⟩[h]

~~Try Experiment of sinking well &or use some well a~~[m]

Connect wire from a high to low place on Earth get dif potential, also wells—[a]

Connect tin roof Laboratory with wire to Earth get potential— ⟨Shultz B⟩[h]

Try heating one side of static Cylinder & then take off Electrification[b] by a metal which reduces temperature back to start, a great variety of materials can be used— I believe Electrification by rubbing is due to heat & is a Thermo pile action ie[i] a Thermo pile of tremendous EMF— This can be got into a Current form by transformers & Condensers reducing potential to workable quantity E   ⟨John Ott⟩[f]

[C][20]

Try metals connected to non conductors   also The P&N non conductors which are rubbed together with collectors to take off E, Either by rubbing contact[b] or at a distance— ⟨Shultz Berge⟩[f]

In a meter use a magnet which will have no stick like Iron reduced by H mixed with shellac—parafin etc. ⟨Ott⟩[f]

Try an induction coil on telephone with core of Parafine & Iron Reduced by H   put as much Iron[b] in as it will stand & still be an insulator. Make quick changes of reg iron wire core & this to get Volume of Sound   if this works well perhaps it will work in receiver— ⟨Ott⟩[e]

X, NjWOE, Lab., N-87-11-15 (*TAED* NA011039). Document multiply dated; miscellaneous sketches and calculations on left-hand pages not reproduced. [a]Followed by dividing mark. [b]Obscured overwritten text. [c]"In rotary . . . .Aqua Reg[ia]" enclosed by line at left and followed by dividing mark; marginalia written above these entries. [d]Paragraph enclosed by line at left and followed by dividing mark; marginalia written in left margin. [e]Paragraph enclosed by line at left; marginalia written in left margin. [f]Marginalia written in left margin. [g]Drawing followed by dividing mark. [h]Followed by dividing mark; marginalia written in left margin. [i]Canceled. [j]Circled. [k]"& speed it to send at" interlined above. [l]Interlined above. [m]Followed by "over" to indicate text continues following three pages of unrelated sketches.

1. This entry marks the start of a series of Edison's notes for experiments to be undertaken at the new laboratory that Edison made in this notebook during the last two weeks of November (the other entries are Docs. 3111, 3112, 3113, 3116, and 3117). Edison assigned individual assistants to carry out these tasks by writing their names next to the entries; these names appear to have been added later and the editors have treated them as marginalia. In some cases, Edison also indicated that he had copied entries into separate notebooks for his experimenters. David Trumbull Marshall later recalled that on his first day at the laboratory Edison, "handed me a note book of the size and style of the hundreds of note books he and his assistants have used and I presume, still use. In this were orders to 'make' or 'do' enough different things to keep me busy for a long time" (Marshall 1931, 58–59). Marshall's book, along with those for John Ott and Erwin von Wilmowsky, are still extant (N-87-12-10.1, N-87-12-10.2, N-87-12-08.1; all Lab. [*TAED* NA015AAA, NA016AAA, NB007001]). What may be additional undated experimental plans are in N-87-00-00.3:6–17, Lab. (*TAED* NA003006).

2. Edison meant "mispickel," a term (now archaic) for the mineral arsenopyrite. *OED*, s.v. "mispickel."

3. That is, a form of ore whose lustre indicates its metallic character. *OED*, s.v. "glance" n. 2.

4. According to Watts 1872 (Supp., 477), "Any oxide of cobalt ignited for about a quarter of an hour in a half-covered crucible over an ordinary gas-flame, and then quickly cooled, is converted into cobaltoso-cobaltic oxide, CO_3O_4."

5. Edison meant "spiegeleisen," a lustrous manganese-iron compound used in making steel. *OED*, s.v. "spiegeleisen."

6. John C. English (b. 1855) was at this time a New York broker in chemicals and chemical apparatus. In reply to an inquiry about English earlier in November, Edison wrote that he "is my broker and buys all my chemicals for Laboratory & various shops with which I am connected— I have known him a number of years." He entered Edison's employ in 1888 as purchasing agent for the Orange laboratory and was appointed the following June as manager of the Edison Phonograph Works, for which he had also acted as purchasing agent. English left Edison's employ in 1890 to join the New York Phonograph Co.; after six months, he began manufacturing sound recording machines based on his own patents. He later conducted experiments for the Victor Talking Machine Co. English's testimony, *American Graphophone Co. v. Na-*

tional Phonograph Company, pp. 457–59, 462, Lit. (*TAED* QP003457); TAE marginalia on Gustav Gennert to TAE, 2 Nov. 1887, DF (*TAED* D8704AEI); Charles Batchelor to English, 21 June 1889, Lbk. 30:477 (*TAED* LB030477).

7. William Kennedy Laurie Dickson (1860–1935) was now Edison's primary experimenter on ore milling technology. He started working for Edison in 1883 at the Testing Room in the Edison Machine Works and by late 1885 was experimenter at Edison's New York laboratory. He moved with Edison to the laboratory at the lamp factory in Harrison in 1886 and then accompanied him to the new laboratory in Orange. There he worked not only as an experimenter but also as Edison's photographer. Dickson was especially important in the work on ore milling and motion pictures and is often credited as co-inventor of the latter technology. Doc. 2452 n. 2.

8. Figure labels are "fibres" and "& so on."

9. John Trumbull Marshall (1860–1909) went to work in the Edison Lamp factory at Menlo Park after graduating from Rutgers College in 1881. He continued working at the lamp factory after it moved to Harrison in 1882 until his death. Marshall had charge of photometric work at the factory and also conducted experimental work. His brothers William, Bryun, and David also worked for Edison. Docs. 2129 n. 3 and 2697 n. 1; Marshall 1931, 51–53; "Obituary," *General Electric Review* 13 (1910): 96.

10. Franz Schulze-Berge (1856–1894) studied mathematics and physics in the universities of Heidelberg and Strasburg, and then under Hermann von Helmholtz and Gustav Kirchoff in Berlin, where he received his Ph.D. in 1880. He remained in von Helmholtz's laboratory, but in May 1887 he applied for a job in Edison's laboratory. He was working by 13 December at Orange, where he conducted many of the experiments on recording materials and had charge of the library. He left Edison's employ in June 1891 to set up a private laboratory in Brooklyn. Obituary, *American Institute of Electrical Engineers Transactions* 11 (1894): 873–74; John Randolph's testimony, *National Phonograph Co. v. American Graphophone Company,* p. 103, Lit. (*TAED* QP006102); Schulze-Berge to TAE, 26 May 1887, enclosing letter of recommendation from von Helmholtz (15 Jan. 1886); both DF (*TAED* D8713AAA, D8613A); Thomas Maguire to Alfred Tate, 17 Dec. 1891, Lbk. 54:630 (*TAED* LB054630).

11. Figure labels are "moisture" and "Heat."

12. Arthur Edwin Kennelly (1861–1939) joined Edison's staff in November 1887 as head of the Galvanometer Room, the new electrical laboratory. Kennelly was born in Bombay, India, where his father was harbor master, but he was sent to England in 1864 after the death of his mother. He attended school until age fourteen, when he became an office boy and assistant secretary in the London office of the Society of Telegraph Engineers. The following year he became a telegraph clerk for the Eastern Telegraph Co. and later rose through the ranks, achieving the position of senior chief electrician by 1887, by which time he had published several articles in the *Electrician.* In September 1887, he inquired of Ezra Gilliland about employment in the American electrical industry, which led to the offer of employment from Edison in early November. He was Edison's principal electrical assistant until early 1894,

when he left to join Edwin Houston in Philadelphia to form the firm of Houston and Kennelly, consulting electrical engineers. Kennelly was appointed professor of electrical engineering at Harvard in 1902 and he remained there until 1930. Between 1913 and 1925, he was also professor of electrical communication at the Massachusetts Institute of Technology, where he directed electrical engineering research. Kennelly is best known for his important contributions to alternating-current circuit theory and the discovery, with Oliver Heaviside, of the ionized layer of the upper atmosphere known as the Kennelly-Heaviside layer. *DAB*, s.v. "Kennelly, Arthur Edwin"; Bush 1940; Kennelly to Gilliland, 16 Sept., 11 Oct., and 5 Nov. 1887, all DF (*TAED* D8713AAH, D8713AAS, D8713AAY, D8713AAZ); "Houston and Kennelly," Feb. 1894, Cat. 1346, Batchelor (*TAEM* 227:760).

13. Joseph Ballard Murdock (1851–1931), a lieutenant at the U.S. Navy's Torpedo Station in Newport, R.I., had served on the examining committee of the lamp tests made in connection with the 1884 Electrical Exhibition in Philadelphia. His 4 December letter requesting this lamp has not been found, but Edison stated in reply that "About the first of the New Year my Laboratory here will be in running order and I will then be able to give you what you want." Murdock wrote again on 13 January giving details for the "diving lamp." *DAB*, s.v. "Murdock, Joseph Ballard"; "Proposed Code for Duration Test of Incandescent Lamps to be Made by the Franklin Institute" enclosed with John Burkitt Webb to Philadelphia Electrical Exhibition, 27 Sept. 1884, Weston (*TAED* X250AAB); TAE to Murdock, 12 Dec. 1887, Lbk. 26:93 (*TAED* LB026093); Murdock to TAE, 13 Jan. 1888, DF (*TAED* D8828AAD).

14. The notebook into which Edison entered experiments for Erwin von Wilmowsky (see note 1) was labeled "Order Book for C. Wilmowsky's Dpt." According to Israel 1998 (pp. 272–73), Wilmowsky took charge of the chemical laboratory when Edison opened his new lab complex.

15. It is unclear to what earlier meter experiments Edison referred; John Ott had been working on various forms of electric meters since 1881.

16. Figure labels are, clockwise from top left, "Siemens Escapement," "Trevellyn Koks," and "Lamps."

17. Figure labels are "Counter" and "clockwork."

18. David Trumbull Marshall (1860–1909) was a graduate of Rutgers College. He joined Edison at the lamp factory laboratory in early 1887 after leaving his job at the chemical testing laboratory of the New York, Erie, and Western Railroad in East Buffalo, N.Y. He moved with Edison at the end of the year to the new Orange laboratory, where he conducted many of the experiments on insulation and squirted filaments. He later made magnetic surveys for Edison of mining properties in New Jersey. He left the laboratory in 1890. Marshall 1931, 57–59, 71–72, 97.

19. Unidentified.

20. Figure labels, from top right, are "Transfer," "Leyden," "Heat," and "Collectors."

*Memorandum to
Richard Dyer:
Phonograph*

Dyer—

Be sure & incorporate the following in foreign & US patent—[1] it works beautifully diminishing the roughness of the
surface & giving better articulation[2]

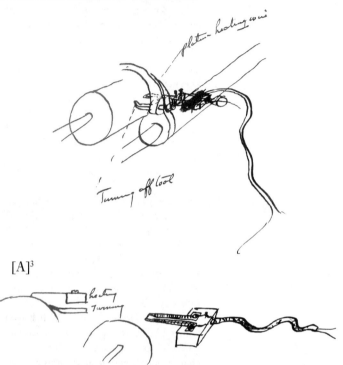

[A][3]

both the turning tool & heating polisher are adjustable. I
find that the wire is Kept sufficiently hot by using a[a] shunt
around [----- of ----][b] two of the 4 cells of the battery &
the speed of the motor is not diminished by reason of it   It
should be just hot enough to melt the outer[a] surface of the
wax, it obliterates the tool marks & gives a burnished like surface & is a great improvement. It can be cut out of sheet or a
short length of .008 wire used—

You better mention that an Aluminium or Silver wire of ⅛
dia reduced to 1/32 at the wax may be used & heated continuously by a minature alcohol Lamp travelling with the traveller may be used the heat passing to the point & wax by conduction.[4]

AD, NjWOE, Lab., Cat. 1152 (*TAED* NM021AAI). [a]Obscured over-written text. [b]Canceled.

1. This memorandum became the basis of a patent application that Edison executed on 22 November. The application moved swiftly through the Patent Office and issued in May 1888 with five claims. In it, Edison explained that his purpose was "to smooth out the tool-marks produced in turning the surface to a true cylinder, and to reduce greatly the scratching noise heard in the recorder, and consequently made a part of the record." The specification also included four figures, the third of which was based closely on Edison's third sketch. U.S. Pat. 382,414.

2. Figure labels are "platina heating wire" and "Turning off tool."

3. Figure labels are "heating" and "Turning."

4. Figure labels are "silver" and "swivel."

-3111-

Notebook Entry:
Miscellaneous[1]

[Orange,] Nov 18 87

Make a telephone transmitter with lampblack button like the recorder of phono   ⟨Ott⟩[2a]

If induction coil with Iron reduced by H wks—then make a transformer with core this way & also with Higher[b] melting point   Stuff for armature of Dynamo for High speeds— ⟨Ott⟩[c]

Try some of that non magnetic Manganeese iron (they have recently discovered) in connection with Thermo Experiments   ⟨English or Read[3] [Composition?][d] & make⟩[e]

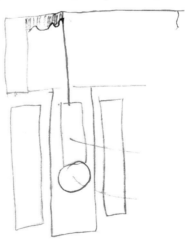

Iron-parafine Core in water bal by bulb— Lamp in meter gives light & blue print paper to photo [position?][d] scale— use clock work   work paper Suction magnet or pivot parafine core[4]   ⟨Ott⟩[f]

Try a telephone diaphragm of celluloid coated with Shellac[b] dusted over with iron reduced by H, also double dia with Iron powder in between.   ⟨Ott⟩

It may be that to electrify a piece of rubber by heat it would be essential to radiate it from Sealing wax or touch cold rubber with Hot seating wax, etc[b] using in same manner the two non-Conductors which when rubbed together give opposite E or two wheels one of Rubber other of shellac rotated between a disk of hot water with Collectors Leyden j[a]rs & transformer to reduce it to same deg of ampere Capacity—

There is something very strange that just as you go to worse & worse conductors you get higher volts on Thermo thus. Bis Tellurium Selenium Amorph phos—if you keep going on you go to perfect insulators ie[g] better and better transformers of heat into E, but the conditions of getting a Current grows beautifully less but if the right conditions ie[f] induction & transformers intervene you should get over this bug—

There may be such a thing as polarization at the ends of magnets, hence coat over surface with various things   finely divided iron—Coat one pole also diferent from the other. ⟨Shultz Berge⟩[f]

Will a galvanometer deflecting 10 degs say on one ampere give more or less in a Vacuum— ⟨Kennelly⟩[c]

Investigate the black matter on one wire[b] on inside part of E Lamp   also analyse the deposit on globe— ⟨New Chemist   Ent⟩[e]

Method of using transformers & motors by alternation on transformer but one direction on main   ⟨Have Kennelly figure this⟩[5]

Effect of alternat[or?][6d]

On Rubbing hard rubber it is electrified[b] and then on passing your hand over it it isnt discharged shewing that the electrification goes to some depth ie[f] a diaelectric stress—hence use metal plates with very very thin solutions or varnis[h] of the electrifying material[f]

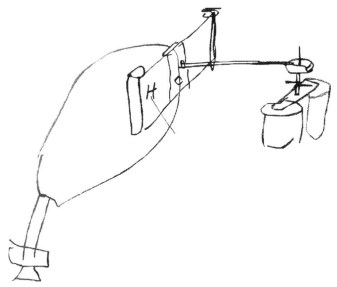

H hard rubber with foil each side connected to telephone vibrate great rapidity by crank & belt or motor to produce heat & cold[b] & thus Elec[t]rification   waves hear in telephone ⟨Ott⟩

Perhaps Nickel wire gains & loses its magnetism quicker than iron—try it in phono & telephone Induction Coil— ⟨Shultz-B⟩[f]

In Thermo Electrification by heat expmt perhaps we dont want DiaElectric[b] like Hard Rubber that hold their electrification so long—they would be like a perm mag   we want something that will be electrified & deelectrifd almost instantly[f]

X, NjWOE, Lab., N-87-11-15 (*TAED* NA011069). Document multiply dated. Miscellaneous calculations not reproduced. [a]Drawing followed by dividing mark. [b]Obscured overwritten text. [c]Paragraph fol-

lowed by dividing mark and enclosed by line in left margin. ᵈIllegible.
ᵉParagraph enclosed by line at left; marginalia written in left margin.
ᶠFollowed by dividing mark. ᵍCircled.

1. See Doc. 3109 n. 1.
2. Figure label is "button."
3. Unidentified.
4. Line drawn from preceding text to center of drawing.
5. Figure labels from top to bottom are "Continuous Current," "Alternator," and "Motor."
6. Figure label is "Continuous."

–3112–

Notebook Entry:
Miscellaneous[1]

[Orange,] Nov 19 87

I have noticed that in vacuo it takes a <u>very</u> high temperature
to decompose cyanogen, also that in low Vac lamp the CP goes
up— now as there is certainly arcs going on in the filiment
even at normal these arcs must have sufficiently high temp to
decompose cyanogen & yet the main body has too low a temp
to do so— hence by using Cyanide of Silver in side tube &
getting high Vac & working lamp up in reg way ready to seal
the Silver Salt may be heated & enough Cyanogen driven off
to bring the Vac down to the point required using a U shaped
guageᵃ or a spark if the latter dont decompose too much Cyanogen or produce other bad results— Thus if lamp set up at
normal most arcs be prevented by low vac & those do occur
will be stopped by deposit Carbon from decomp of Cy

If we rotate a bismuth Cylinder with a fixed Antimony rubber we get a constant current as shewn by my expermnt with
rubbing these together & listening to telephone[2]

Now if friction E is a Thermo electric action then

Using two metals A which is shellaced very thin or other varnish like collodion & B a sheet to give great surface & Microfarad capacity [~~the?~~]ᵇ it should act as a thermo pile condenser
& be constantly Charged & give a current. ⟨Ott⟩ᶜ

according to Faraday[3] hot Dryᵈ Oxalate of Calcium stirred
by plat or glass becomes positively electrified in a powerful
degree so that it may be lifted out of the dish

Erdman[4] states (Sullivan archives of E x. 480) that if any thermo pile be rubbed at its juncture it gives a current that it cannot be <u>due to heat</u> as it commences instantly & ceases instantly the friction commences & stops— This is <u>very</u> important if true. another expmt in Watt article on E confirms this

May it not be probable that more Energy can be transformed when a thermo juncture is rising & falling in temperature than when it is kept at constant temperature higher than the other junction= this is so in pyroelectricity of Crystals. no electrification while temp constant but only shews in on rising & falling temp—[c]

a cold wire (same metal) touching for an instant a hot wire gives momentary Current.[c]

<u>good?</u>[e] ⟨Ott⟩[5]

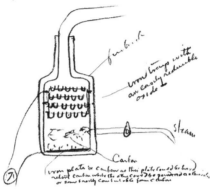

The flame conducts inside[c]

Metuecci[6a] says combution Charcoal in pure O gives no E but with H_2O charcoal plate charged with Neg E & Carbonic Oxide or anhydride to Pos E   In above the steam is decomped by hot carbon which becomes Neg & CO_2 rising is reduced by the oxide to CO & the flame g[–][h] is good condr—

Iron pyrites are good conductors   hence in ore milling use heavy amperes & pack 3 or 400 lbs in a Earthenware tube with end electro[de?] put on pressure & send current to roast—[f]

Try roating reg way but in Lump—think twill pulverize Easier— ⟨Batch—This way make [an?]⟩[h]⟩[c]

Study the repulsion of a flame by a magnet. ⟨Edison⟩[g]

Monosulphide of Carbon is a maroon colored powder stable but decomp at 200 C into its elements. This might be good as the sulphur would take out the mercury & not attack the plat wires ⟨Chemist Marshall   Entered⟩

Why shouldnt Iodide or Carbonates of Aluminum heated to intense heat in a Vac be disassociated & give the metal— There being no oxygen present. ⟨Marshall⟩[g]

See 3rd Sulpliment Watts page 1746[7]   Colophthalin= a Hydrocarbon $C_{10}H_6O_2$ Resembles Alumina—its infusible— non volitile & not Decomp by Nitro or Chlorates at 1000 Centgrade— ⟨Chemist Marshall   Entered⟩

I believe all chemical Elements which give more than one line in the Spectrum are like the Hydrocarbon series   for instance we have the several parafines— Why should there not be several Hydrogens—if three lines of H then there are 3 molecules of H each diferently arranged   I think for instance that, take an oxide of a metal that gave widely spread lines say 4 lines that by ~~dis~~ means of solvents it the oxide could be seperated pretty[a] fairly into 4 qualities of oxides—& ~~these~~ Each would give one particular line brighter than the other. Shaded lines are simply oxides with slight Changes only—

Carbonize Elder pith—Cork & other light matter to get a finely divided easily powdered charcoal for Lamp[a] filiment by squirting— ⟨Batchelor⟩[g]

See if pure Anthracene will mix & harden parafine with Carnauba—or leave out Carnauba ⟨Hamilton⟩[g]

Try the Experiment of putting Crude Anthracene in High Vac & volatilzing   See if commercial & examine residue ⟨Edison⟩[g]

Distill hardest parafine until the highest boiling point parafine is left— ⟨Chem Bldg⟩[f]

Phenylnapthal—Carbazol $C_{16}H_{11}N$ fuses at 330 C & has a higher boiling point than Sulphur— ⟨Chem—Marshall Entered⟩[c]

Crude Anthracene contains all the higher melting point HCarbons— by subliming in a vac probably or[a] boiling in open air I cld probably get a residue that would be good for filiments— ⟨Chemist   Ent⟩[g]

In Lamp Expmts with dif deg vac to see effect on Drop CP—make 20 Lamps ~~2[—]~~[b] ~~cp hp~~ 320 cp hp—of each deg pressure & see up at normal   use Cyanogen as a residual ⟨Entered—Marshall⟩[f]

Test to determine if anything will retard lines of force— ⟨Ott—& Kennelly or Shultz–Berge⟩[8c]

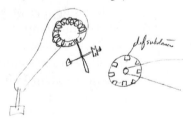

Experiment on the static charging of rotated matter in concentrated magnetic fields—[9c]

As platinum readily amalgamates when Hg is allowed to run into a Vacuum I think it probable that clean sheet iron in vacuumo can be beautifully Covered with tin[a] by deposit ie[i] Vacuous with arc & allowing hot tin to run in—[c]

investigate the substance produced by heating the filiment when the globe has a mixture of Napthaline & parafine in it— it seems to have tremendous high melting point & the most dazzling fluorescence— ⟨probably by crinkling of film only Edison⟩[f]

also make a still thus[10]

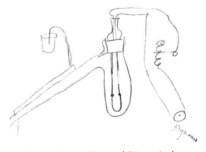

Try diferent Hydrocarbons Etc= ⟨Chemist⟩[c]

Try if Roasted auriferous pyrites mixed with carbon & salt water & left to itself in warm place isnt entirely decomposed— ⟨Chemist—⟩[c]

Monkey with strong Current under various conditions through Sulphides to see if there could be such a thing as a transformation   ⟨Batch & Edison⟩[c]

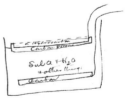

The Hydrocarbon is the Combustable[11]   ⟨Edison⟩[c]

Galleïn or gallin carbon[i]zes without change— ⟨Chem— Marshall   Ent⟩

Plaster paris mixed with H_2O Containing some Patassic

Sulphite—in Molecular proportions when mixed & stirred it hardens before it can be poured—

—Cream Tartar also quickens—[c]

Try the Carbon Rod[a] in a very narrow porous cup with tellurium rubbed on it. the H Combines to form tellurilled[12a] H & on contact with air decomp & Metallic[a] Tellurium falls back to be used over again. or this ⟨Shultz-B⟩[13f]

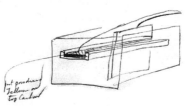

in Watts Dic 1875 2nd Suplement p 638[14] speaks of decomp of BiSulphide Carbon by <u>heated platinum Spiral</u>—with decomposition into Carbon & Sulphur   also page 657[j]

Watt 2nd sup p. 722[15] Hydracrylamic acid—Sol water on heating to 170C turns brown sublimes slowly in feathery needles at higher temperature it decomp completely leaving a very <u>difficulty Combustable charcoal</u>   ⟨Chem—Marshall Ent⟩

X, NjWOE, Lab., N-87-11-15 (*TAED* NA011091). Document multiply dated. [a]Obscured overwritten text. [b]Canceled. [c]Followed by dividing mark. [d]Interlined above. [e]Question mark written backwards. [f]Paragraph enclosed by line in left margin. [g]Paragraph enclosed by line in left margin and followed by dividing mark. [h]Illegible. [i]Circled. [j]"page 657" overwritten on a horizontal line and followed by a dividing mark.

1. See Doc. 3109 n. 1.

2. Figure labels are "Bis" and "Ant."

3. Here and elsewhere in this document Edison refers to statements in the article on electricity in Watts 1882 (2:374–481). The reference to Faraday appears on p. 406.

4. Edison mistakenly indicated the reference in Watts 1882 (2:410) as Erdman, but Watts refers to the work of Paul Ermann. Edison also confused the citation which reads "Sullivan, *Archives d'Electricité*, x. 480" and "Ermann (ibid. v. 477)." It is not clear what other experiment mentioned in the electricity article Edison had in mind at the end of this paragraph.

5. Figure labels, from top, are "fire brick," "iron trays with an easily reducible oxide—," "steam," "Carbon," and "iron plate or carbon as this plate could be hard retort carbon while the other could be powdered anthracite or some easily combustible form carbon."

6. Edison meant Carlo Matteucci (1811–1868), Italian physicist and physiologist. The reference is to Watts 1882, 2:482. *Complete DSB*, s.v. "Matteucci, Carlo."

7. Watts 1881.

8. Figure label is "dif substances." Edison drew this experimental

apparatus with a more detailed description in one of his Fort Myers notebooks (Doc. 2935).

9. Edison overwrote the sketch above with this paragraph. The editors have rotated the image 180 degrees, which appears to be the orientation in which Edison made the drawing. The smaller device on the left is likely an electromagnet in a cylinder, presumably to produce the concentrated magnetic field, with an induction coil above, presumably to produce the static charge. The larger figures on the right appear to be a partial exploded view of the cylinder containing the magnet.

10. Figure label is "dynamo."

11. In the image above, figure labels are "a Hydrocarbon," "Carbon porous," "Sul A & H_2O & other things," and "Carbon."

12. That is, hydrogen telluride.

13. Figure label is "put powdered Tellurium on top Carbon—."

14. Watts 1875.

15. Watts 1875.

-3113-

Notebook Entry: Miscellaneous[1]

[Orange,] Nov 20/87

Idea for a steam Engine— Use two boilers—one containing ~~lik~~ say sulphate of Copper In[a] powder— ~~Surround thi~~ heat this with fire to dehydrize & make steam—or to make feed water to 2nd boiler— use it in Engine & then exhaust into the dehydrized sulphate in other words procure & also condense you water from a substance like Sulphate Copper utilizing the heat of all the reactions.[b]

Fill large boiler with several tons of Cocoanut charcoal, allow it to absorb a[c] dry gass & then Close & heat to obtain pressure to work Engine. Thus we can Condense Atmosphere air to a ~~solid~~ liquid almost.[b]

Make a vacuous glass box

boil it in hot Linseed oil while exhausting use McLeod Gauge & run from 7 am till 6 pm— highest obtainable Vacuum— This chamber is to be used in expmts on retarding Lines of force or stress etc ⟨Force⟩[b]

With spectrum tube end on and side heating tube— get Spectrum by gently heating say Iodide Lead— Record Spectrum. Then heat little more get & record spectrum— then so on until all the Iodine is driven off   my idea is that Iodine is a mixture of <u>Iodine Molecules</u> & spectrum will change as higher heats are used to decompose the compound—PbI may not be best ⟨Edison⟩

It may be necessary to put this tube in a metallic heated bath & use a Thermo to measure the different disassociations points— If this works with Iodine, Chlorine, Bromine, Try it with Hydrogen use Palladium & get spectrum at several Exhaustions & temperatures. ~~also use~~ ⟨Edison⟩ [-][d]

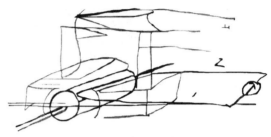

Use Hard rubber cylinder rotated in powerful field   Resting on this near top is a rubbing wire with 2 leading wires— my idea is the rubber will be charged &[c] will discharge & give a Current through wire 1 & 2   ⟨John Ott⟩

X, NjWOE, Lab., N-87-11-15 (*TAED* NA011121). [a]Obscured overwritten text. [b]Followed by dividing mark. [c]Interlined above. [d]Illegible; followed by dividing mark.

1. See Doc. 3109 n. 1.

–3114–

Draft to Edward Johnson

[Orange, November 21, 1887[1]]

You probably got your impression from the fact that at first I proposed sub stations in places like orange with converters.[2] afterwards I told you I could do better, that it was better not to use them if we cld wk direct, which we can now do[3]— I guarantee the converter will work perfectly & beyond your expectations & it will be usefull for bringing in power at a distance in many cases

E[dison]

ADfS, NjWOE, DF (*TAED* D8732ABS).

1. The date is taken from the docket on Johnson's 17 November memorandum. Johnson to TAE, 17 Nov. 1887 (*TAED* D8732ABR).

2. Edison drafted this reply on the back of Johnson's 17 November memorandum (see note 1), in which Johnson complained that the Machine Works was ready "to make and deliver the 'Converter'" but that "Edison fails to notify the Light Co that such a system is available &

further on its applying to him always recommends something else." A slightly revised typed response was sent to Johnson on 26 November. TAE to Johnson, 26 Nov. 1887, DF (*TAED* D8717AAN).

3. That is, with the five-wire system.

–3115–

Edison Laboratory   Orange NJ   Nov 21 1887

Phonograph—
  Repeating device[1]

Repeating device[2]

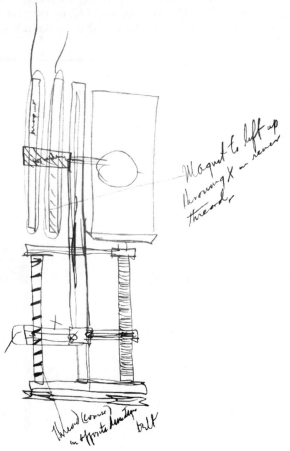

good repeating device—[3] raising lever throws it into gear with Coarse reverse screw which sets it back—arm on this is disconnectable   a thread or magnet can reach to distance to manipulate for instance on printers Case—[4]

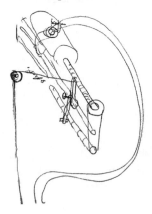

TAE

X, NjWOE, Lab., N-87-11-21 (*TAED* NA012AAA). Document multiply signed and dated.

1. Figure labels are "or Copper cylinder over Iron" and "dash pot."

2. "Magnet," "[for recording?]," "Magnet to lift up throwing X on rever[ser] thread," "X," "Thread (coarse) in opposite direction," and "belt."

3. The following day Richard Dyer received from Edison a drawing similar to the device shown here (Cat. 1152, Lab. [*TAED* NM021AAI1]). It likely accompanied Edison's draft for the patent application that he executed a week later for a "Feed and Return Mechanism for Phonographs." For unknown reasons the application was not filed until 5 January 1888, but it issued only five months later as U.S. Patent 382,416. According to the specification, Edison sought "to produce a simple and efficient mechanism for setting back automatically on the depression of a key or treadle the reproducer of a phonograph for reproducing [i.e., playing back] the whole or a portion of the record." According to a late October interview, Edison had then been experimenting to enable "printers to set type directly from the dictation of the phonograph" by using a foot pedal to find and replay select portions of a recording ("A Wonderful Workshop. The Resources of Edison's Laboratory," *N. Y. Evening Post*, 21 Oct. 1887, 1).

4. Figure label is "3 feet." Edison's reference is not clear, but most likely he meant a printer's chase. That is the iron frame used in letterpress printing to hold a page of type, which is locked into the chase by adjustable wedges called quoins (*The Printer* 1884, 24–27, 36–37). However, this is not the kind of adjusting mechanism Edison describes in this note or the patent:

> For lifting the reproducer from the phonogram and causing the arm **K** to engage the screw-shaft **I**, I connect the finger *d* of the eye of the spectacles carrying the reproducer with a cord, *k*, passing up over a wheel, *l*, and down to a lever, *m*, which may be a hand-key or foot-treadle. This cord preferably has an elastic section, *k'*, which will yield to permit the retracting movement of the reproducer.

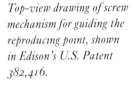

Top-view drawing of screw mechanism for guiding the reproducing point, shown in Edison's U.S. Patent 382,416.

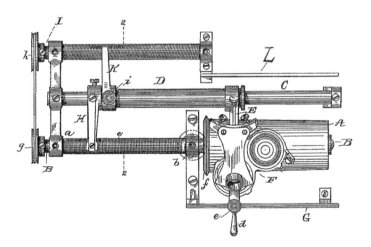

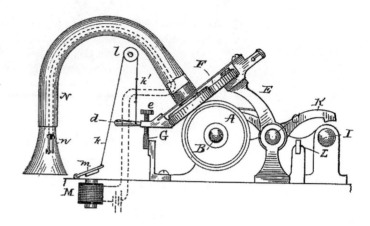

End-view drawing from Edison's patent, showing the mechanism for tracking the reproducer point in the cylinder's indented grooves.

An electro-magnet, **M**, acting on an armature on the lever *m* and controlled by a circuit-controller, *n*, on the listening-tube N, may be used for lifting the rocking holding-arm to set it back. [U.S. Pat. 382,416]

–3116–

Notebook Entry: Miscellaneous[1]

[Orange, c. 22–24 November 1887]

Time Expmt— Put in Large bottle, a sheet of every kind of metal seperated from each other & then fill with Bisulphide Carbon[a]   ditto another bottle[b]— Terchloride of Carbon & other Chl of Carbon—[c] Set these 2 bottles away sealed for several years to see if Diamonds formed by decomp & crystallization[a]   ⟨~~Chemist~~ Edison⟩[d]

Get some Iodide Carbon also—

Watts says in some expmts on spectra of Carbon that with spark tubes no deposit of Carbon takes place with diminished pressures but at pressure atmosphere spark is white & Carbon set free— I think that Carbon fils could be deposited upon a dull red if in a Hydrocarbon <u>Vapor</u> under 5 or 6 atmosphere pressure—[2a]

In Electrolytic experiments for the Comcl production of various chemicals Note that any insulating Liquid can be decomposed by <u>great surface</u> small distance & high rolls

Schonbein[3] says that all peroxides formed electrolytically are formed by ozone—especially Silver— Marshall to work this up— ⟨Chemist   Ent⟩[a]

Welsbach hood[4] replaced by falling Zirconium powder perhaps under proper conditions The powder (very light) on an eletrified base at bottom flame may be drawn up through flame & fall back   Certainly with a chimney a whirl operated by the heat could be arranged to do Mechanical work sufficient to pass continuously the oxide— Work this scheme up   E

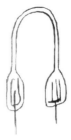

Bohemian glass[5] that will stand yellow heat with capilliary bore filled with dif gas & ~~unde~~ & transformers with sharp currents— also bore filled with Fluffy Earthy Oxides.[a]

~~Pyridine~~ Try some Nitrogen Hydro[carbon?][e] in vac   The heating of filiment probably produce a good residual & deposit. ⟨Marshall   Edison[f] to name these Compounds⟩[g]

Telimeter[6]

To get temperature at distance turn R until Gal balance Read of Temp in terms of units of change of Resistance—[a]

X, NjWOE, Lab., N871115:141 (*TAED* NA011141). Miscellaneous sketches on facing page not reproduced. [a]Followed by dividing mark. [b]"another bottle" interlined above. [c]"& other Chl of Carbon" interlined above. [d]Marginalia written in left margin, followed by dividing line. [e]Illegible. [f]Obscured overwritten text. [g]"Marshall" written in left margin with line to "Edison . . . Compounds" written after dividing mark.

1. See Doc. 3109 n. 1.
2. Edison probably referred to Watts 1881, 1810.
3. For Christian Friedrich Schönbein's work on ozone, see Rubin 2001.
4. That is, the Welsbach mantle. See Doc. 3058.
5. Bohemian glass was extensively used for chemical apparatus

because it was generally unaffected by chemical reagents. Spon et al. 1879–82, 3:1067.

6. Figure labels are "Ice house" and "R." Edison meant telemeter, a system for conveying measurements, including temperature, over a distance. Originally invented by Robert Hewitt, the system had been improved by Edison's former employee Charles Clarke, who, together with several Edison associates, incorporated the Telemeter Co. of New York in January 1886. The incorporators included Erastus Wiman as president, Francis Upton as vice-president, William Brewster as general manager, and Clarke as superintendent. Upton read a paper on the system at the Manchester meeting of the British Association for the Advancement of Science in March 1887. Doc. 2585 n. 15; Edgecomb 1884; "The New Apparatus of the Telemeter Company of New York," *Electrical Review* 8 (3 Apr. 1886): 3; Upton 1887; "British Association for the Advancement of Science," *Industries and Iron* 3 (1887): 277.

–3117–

Notebook Entry: Miscellaneous[1]

[Orange,] Nov 25 87

Phonogh

Hard rubber cylinder polished   Record by a rubber point increased area & strength of electrification on surface represents increased amplitudes— use a delicate reprodcg dia[phragm] & work by dif friction as the Electrification attraction don't seem to work well on former Expmts—[a]

For duplicating process for phonograms to make conducting surface dissolve phosphorous [~~water?~~][b] very small amount in BiSulC or ether dip cylinders in then at moment solvent has evaporated dip in Solution Nitrate Silver— ⟨Chemist Ent[ered]⟩[a]

In compounded preliminary use No 17 well dried and instead No 6[2] try—the Aniline in H_2O—Dextrine, caramel, Hard burnt Sugar—Grape sugar—Milk sugar—Sacharine—Albumen—blood & egg—Catechin— ⟨~~Hickman & Force~~ Hickman & F⟩[a]

Mix with 17 some photographers plumbago—also after put on dust over & when dry burnish with brush—[a]

After Treatment with 17 & final Carbzn put in gas tube with hydrocarbon & deposit on[a]

Instead 17—disolve the Hard Cubian in hot Linseed & then let it dry by air—6 days before sending to final— Edison[c]

Silver carbamide when heated gives off Ammonia— ⟨Chemist & Marshall   Entered—also Marshall⟩[a]

Iodide Carbon— Tetrachloride carbon necessary to make it— Iodide Carbon decomposes slowly at <u>ordinary tempera-</u><u>ture</u> but more rapidly at 200 fahr when exposed to air into CO & CO_2[d]

Monosulphide Carbon a powder decomp by heat at 206 cent. Its made by exposing DiSulphide to strong sunlight— also by putting lot iron wire in sealed tube with DiSulphide for 6 weeks   See Watts—3rd Suplement Pt 1^3 P 406—[a]

Try Bromide of Silicon—is in crystals which sublime at 240 C   This is good for <u>depositing</u> on filiments   See Watts 2nd Sup^4 P 1086   ⟨Chem—Marshall⟩[e]

Watts 2nd Sup p 1215   Bromide of Zirconium—$ZrBr_4$— Volitile at heat of gas burner good for depositing on fili-ments—

Iodide Silicon Tetrac[h]l[o]ride—melt at 120 C boils 200 C Sesquioxdide—Melts 250 C

Amonia-chloride boron $3NH_32BCl_3$ This may decom-pose— ⟨Chem Marshall Ent Make⟩[f]

In 17 mix Chloride Platinum—also Chl Iridium—also soak in Chl Iridium first then dry—then 17 it   Idea to fill gaps by a good condr to make good contact in a crack—

Idea for photoghg red or heat end of spectrum   Coat a plate with material which at ordinary temperature decom-pose but do not below zero— Now absorbtion of heat rays will raise temperature on certain points up to combining point & thus photogh lines of spectra etc   ⟨Dixon⟩[g]

Carbolate of Ammonia— for heating in Vac expmts— Car-bolates of Baryta Lime & Earthy Salts having no water should decomp & gives various HdyroCbns—[a]

Picramic acid for a No 17— Chlorine passed thro Aniline Oil gives a Tar good for a 17^a Cyanilinc— ⟨Chem Marshall⟩[h]

Curious— on heating Cyanide of silver it gives off Cy-anogen & leaves paracyanide Silver which is a grey porous very refractory mass which when struck with hammer ac-quires perfectly <u>metallic</u> appearance like Bismuth   Try for Thermo—Telephone—etc— ⟨Mar[i] Chemist⟩[j]

This paracyanide of silver combines with Hg to form an extremely hard Crystalline Amalgam ⟨Chemist⟩[j]

Azulmic acid is an interesting Nitrogen compound of Car-bon   Work at it for a base for fils[a]

Comenic acid—by manipulation gives a charcoal like glance Coal— Ammonia disolves it   See Gmelin Organic Vol V P 384 also Comenatic Ammonia[5]   ⟨Chem—Marshall⟩[h]

I think for 17s The destructive distillation of various Organic Compound just short of Carbonization will give me a great number for instance Citric acid by heating at a certain point turns black & a <u>little further</u>[k] gives an extremely lustrous Charcoal— ⟨Chemist <u>Ent</u>⟩[j]

by stopping short of Carbonization Solvents can be generally found for all residues & thus a large number of 17s can be obtained— ⟨Chemist⟩[h]

See if it is possible to Carbonze Hair in asphalt Linseed oil—Chloride Zinc, etc   not only hair but put a piece of every kind animal matter, Silk horn hair—Albumen glue gelitin aniline crystals etc. use slowly rising heat in prelim chamber— ⟨Boy in Kinneys Room⟩[h]

Mucic acid on carbonization leaves a charcoal with an almost <u>metallic lustre</u>, Gmelin V p 503— ⟨Chem Marshall⟩[j]

Saccharic acid—Des Distillation[a]

<u>After preliminarizing</u> & H_2O driven off so there is scarcely any O left it might be acted on by Chlorine at the same temperature as it was prelimzd at so as to take the H out without forming Hydrocarbons & thus prevent porosity—good?? Pent Chl phos[i]   Edison[g]

Anyhydrous Phophoric acid is especially valuable for dehydrizing filiments as it doesnt part with its O Like SO_4 & carbonize— This is <u>good point</u>   boil fils in phos acid— ⟨Chem Marshall making chem carbon⟩[h]

All Nitrogenized organic Compounds when treated with excess aqu[e]ous fixed alkali in manner adapted to their peculiar nature give off all their Nitrogen as NH. 2 atoms H_2O Take their place hence if this is so hair horn etc boiled with an alkali should after treatment be capable of Carbonization without swelling or puffing— ⟨Boy in Kinneys R⟩[h]

Carb Lime decomp at red heat in air will stand melting if in a closed chamber where it is under pressure hence in any vacuum expmts its probable that it wld M̶edecomp at 6 or 700 fahr & other <u>salts in proportion</u>

See Watts Dic 2nd Supplement P 265   Pentacarbon Bisulphide a <u>solid & black</u>   ⟨Chem Marshall⟩[a]

Have a continous transformer drawn out & made on basis of a 10 ampere machine with double surface velocity & thin plates inclosed in iron box to put on poles—box forming part field mag   1000 Volts pressure— big diameter armature short length, work up to saturation of cores— put it in Centers if possible—Agate if possible—perfectly balanced—

also large ones with Manhole boxes end feeders running

wires down pole in tube to transformer in manhole— use 3
wire or double transformer. in this case put motor armature
horizontal so as to run in oil on agate Centers— ⟨Batchelor⟩[j]

Artfcl fils— ⟨Edison⟩

Carbon bases— First to be used to form a frame to prevent
early fusible carbonizing substances from running such as asphalt The amount of Carbon being small in proportion to the
fusible Carbnzble Material & Second where the greater per
centage of the whole is carbon the fusible material being used
only to lock the particles together [—][b]

Carbon to be very finely divided like Electrotypers or rather
like photographers plumbago—

1=	Carbonized cork
+2	" Pith
3	" cocoanut shell
4	Carbonized cotton.
5	" bamboo filaments ground up
6	Lampblack Eddys
7	Telephone Lampblack—
8	Carbonized residue from Standard oil retort
9	Carbonized Asphalt.
10	" Sugar
11	" Albumen
12	" Tragacanth

13 Carbon from action of SO_4 on various organic material—Sugar—Dextrin, paper, Cotton, Milk sugar—
anilines, oils—

14 Carbon from action Chloride Zinc on above organic
materials.

15 From pine wood charcoal.

16 Tartaric, Citric Oxalic & other organic acids carbonized also from Tannin, pyrogallic—

17 ~~Mucic acid—Carminic acid—Hydro~~

Material[i] which is to form principle part of the fil   1st series—

1 Hard Cubian asphalt, in liquid carbon mixed & solvent
evaporated,—preliminized & redipped in Reg 17

2 Hard Cubian with Carbon & 20 pct Napthaline to grease
mould & worked hot—prelimzd & dipd 17

3 Hard C. & Coal tar

4 " Pitch of coal tar

5 " Wood pitch.

I will only mention now materials to be got together as I can then work out proportion to suit results on Squirter—

All the Analines—

Gum Dextrin

   "   Arabic

   "   Shellac

all the other gums

glue albumen— Tragcanth, Syrian asphalt, Anthracene Crude— Petroleum Tar— Petroleum Pitch Linseed oil, Castor oil— olive oil Cotton Seed glycerine— ~~Mucic acid Comenic ac~~

Mucic acid,— Comenic acid, Hydro[xyclamic?]^{m} acid[6]— anthracene crude—Colophthalin see Watts 2nd Sup page 1746—[7] infusible non-volitile Hydrocarbon—

Monosulphide Carbon—decomposed at 200 C^{a}

A great many Hydrocarbons pyrene etc are carbozd by Sul-Acid—get this form of Carbon— ⟨Chemist Marshall—⟩^{8n}

X, NjWOE, Lab., N-87-11-15:151 (*TAED* NA011151). ^{a}Followed by dividing mark. ^{b}Canceled. ^{c}"In compounded . . . sending to final" all enclosed by line in left margin; marginalia "Edison" written on line. ^{d}"Chemist . . . also Marshall" interlined at top of page with lines drawn to this and the preceding paragraph. ^{e}"Iodide Carbon . . . p 1086" enclosed by line in left margin, "Chem-Marshall" written on line; followed by dividing mark. ^{f}"Bromide of Zirconium . . . This may decompose—" enclosed by line in left margin, marginalia written on line. ^{g}Marginalia written in left margin. ^{h}Enclosed by line in left margin, marginalia written on line; followed by dividing mark. ^{i}Obscured overwritten text. ^{j}Enclosed by line in left margin, marginalia written on line. ^{k}Multiply underlined. ^{l}"Pent Chl phos" written in left margin. ^{m}Illegible. ^{n}Enclosed by line at left with line drawn to marginalia written below dividing mark.

1. See Doc. 3109 n. 1.

2. The editors have not determined what these numbers represent, but see Doc. 3091 n. 9.

3. Watts 1881.

4. Watts 1875.

5. Volume 11 of Gmelin 1848–71 is subtitled "Organic Chemistry, Vol. 5 Organic Compounds Containing Ten and Twelve Atoms of Carbon."

6. Edison likely meant hydroxymaleic acid. Watts 1881, 1256.

7. Watts 1881, 1746.

8. This is likely the last entry made by Edison on 25 November. The following page begins "Notes Continued," suggesting that he picked up where he had left off. These additional notes are related to materials for filaments and filament coatings based on his readings in Gmelin 1848–71, Volumes 2 (Organic Compounds Containing Two and Four Atoms of Carbon), 4 (Containing Eight and Ten Atoms of Carbon), 6 (Containing Fourteen Atoms of Carbon), and 10 (Containing Twenty-Four

to Thirty-four Atoms of Carbon). He also made additional notes from his reading of Miller 1878. He assigned these filament experiments to either David Marshall or the unknown chemist. He also assigned John Ott to

make some 2One inch ~~stap~~ stabs of Carbon with holes thus

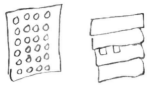

Scratch no on each hole which is ½ dia by ½ deep weigh out equal quantities of everything Carbonzable & then Carbonze slowly & test the charcoal for quantity shine & combustibility.

In connection with these experiments he further indicated, "In depositing Carbon on filiments in closed iron tube— 1st Coat, then take out & weigh—then recharge tube cool & reweigh & so on to ascertain if heavy deposits can be got on." N-87-11-15:181–97, Lab. (*TAED* NA011151, images 91–99).

–3118–

To John Vail

[Orange,] Nov 28— '87

Friend Vail—

Johnson tells me you are coming out here to see me about new scheme about Central station—[1] If you do so I shall want you to come at 930 in the morning and stay until 10 or 11 at night—it is utterly useless to come out and spend an hour— also I want you figuring man also— <u>all</u>[a] The data as to costs of at least <u>20 central stations</u> with maps originally made and maps just as the mains are bring the Engineering plans; tables of sizes & Costs of everything, The <u>actual</u> costs of poles; setting, stringing wire of a dozen stations; I want to be well provided with every question that may arise—bring <u>Steringer</u> if po[s-s]ible—[b] Also Westinghouse Costs sheet ie[c] of det[ails?]—[b] also[d] How he works his system in <u>practice</u>. Now when can you arrange to come out. The sooner the better— Yours

Edison

ALS, NjWOE, Vail (*TAED* ME013). [a]Multiply underlined. [b]Paper torn. [c]Circled. [d]Interlined above.

1. The reported "new scheme" and the urgent tone of Edison's response were informed by competition from the Westinghouse interests. In particular, William Jenks, having received complaints from his agents about the popularity of the rival system, addressed a 12 November memo to Vail nominally on the subject of the proposed Edison plant in Greenwich, Conn., but in reality critiquing Edison's municipal and

five-wire systems more generally. "Every agent and all our Municipal stations show the same anxiety," he told Vail, quoting one agent who had asked if he was "to sit still and be called 'old fashioned,' 'fossils,' &c., and let the other fellows get a lot of the very best paying business?" Jenks offered this assessment:

> Evidently there is an enormous pressure everywhere for a system to cover distances. You desired me some time ago, to consider superintending the construction of a "Five-Wire" system in Greenwich. Before the time comes for action I wish to carefully review the subject with Mr. Edison. From him I received my first conceptions of Three-Wire distribution and Municipal control, and it is possible I may get some of my doubts on this five wire plan cleared up. But I have talked carefully with the best inspectors, central station managers and specialists we have today, and I find them all of one mind in deploring its complications.
>
> I do not hesitate to say I shrink from carrying out your plans as proposed.

Rather than putting up a five-wire system in view of the public and "under the merciless criticism of our competitors," Jenks proposed making the initial installation a private one at "some favored locality" such as the Edison Machine Works. Jenks to Vail, 12 Nov. 1887, DF (*TAED* D8732ABP).

Vail, Jenks, and Luther Stieringer spent "all day" in Orange with Edison on 30 November. Stieringer returned the next day with W. D. McQueston (who would oversee construction of the Edison central station in nearby Paterson, N.J., in 1888) to discuss the five-wire system. Edison (according to Charles Batchelor) "explained his method of throwing over a house on any side from station." Cat. 1337:31 (item 518, 1 Dec. 1887), Batchelor (*TAED* MBJ004 [images 16–17]); "Special Correspondence. New York Notes," *Electrical World* 11 (30 June 1888): 334.

–3119–

Patent Application: Phonograph

[New York, November 29, 1887][1]

⟨743⟩[2a]

To all whom it may concern:

Be it known that I, Thomas A. Edison, of Llewellyn Park, in the Country of Essex and State of New Jersey, have invented a certain new and useful Process for Duplicating Phonograms, (Case No. 743), of which the following is a specification.[3]

The object I have in view is to produce a practical process for the duplication of phonographic records, so that the new art of phonographic publication can be established.

Generally I propose to construct a suitable matrix preferably in metal and by its use mould duplicate phonograms with the phonographic records thereon such phonograms or the surface thereof being preferably constructed of a material too hard for the satisfactory indentation thereof by the phonograph recorder, but the duplicate phonograms may be made of a softer material.

For the construction of the matrix I preferably employ the process of vacuous deposit described in my application No. 118942, filed January 28, 1884.[4] The original phonogram is preferably constructed with a surface of wax or a similar material. This is placed in a suitable phonograph and the phonographic record produced thereon. The phonogram so impressed with the phonographic record is placed in a high vacuum in which an electrode arc continuous or discontinuous is produced between electrodes of metal, or in which metal vapor is otherwise produced. The electric arc produces a vapor of the metal of which the electrodes are composed, which vapor, or a metallic vapor otherwise produced within or supplied to said chamber, is deposited on the indented surface of the phonogram forming a layer of metal thereon which follows accurately all the indentations of the record however minute, owing to the highly comminuted condition of the metal deposited. The phonogram while the deposit is taking place in the vacuum chamber is revolved slowly by a suitable power connection, and this is especially necessary when the form of the phonogram is cylindrical which it preferably is. The vacuous deposit is continued until the layer of metal is sufficiently thick, when the covered phonogram is removed from the vacuum chamber and is further covered by a more rapid process to give strength and body to the covering. A further covering of metal may be produced by electro-plating a metal upon the vacuous deposit in the usual manner of electro-plating, or the vacuous deposit may be backed up by casting upon it type metal or other metal or alloy having a lower fusing point than the vacuous deposit, or this may be done after electro-plating upon the vacuous deposit, or the vacuous deposit may be backed up by a cement, or gum, or by plaster of Paris, but a metal backing is preferred.

The material of the original phonogram is then dissolved off of the metal covering leaving in the case of cylindrical phonograms a hollow metal cylinder or one internally faced with metal carrying the phonographic record in relief upon its inner surface. This metal cylinder is then split longitudinally by a very thin saw into a number of parts, say for illustration three parts, which are suitably mounted upon levers, so that a mould is formed which can be closed to receive the material to be moulded and opened to permit of its being taken out.

The duplicate phonograms are produced by means of this

mould by pouring therein and preferably around a suitable core placed in the mould, suitable substances such as wax, or wax-like material, resin, or plaster of Paris, the material being preferably too hard to be satisfactorily indented by the phonograph, or the duplicate phonograms may be made by taking sheets of smooth material like waxed paper or tin foil and pressing them upon the surface of the mould by a plunger or otherwise, the sheets being afterwards backed up by a wax, resin or cement. The latter way of making the duplicate phonograms is especially applicable to flat-surface phonograms although it may be used for phonograms with cylindrical surfaces.

Instead of employing the vacuous deposit for first covering the record of the original phonogram, I may employ the process of electro-plating for this purpose. A specially prepared plumbago of exceedingly great fineness might be employed to cover the wax like surface as a basis for the electro-plating, or gold-leaf, or silver-salts reduced by chemical reagents to the metallic state might be used for the same purpose. But the plumbago and gold-leaf do not bring out the fine vibrations and produce rough reproductions while the silver-salts do not run well on the wax-like surface.

The vacuous deposit however adheres uniformly to the wax-like surface and reproduces the record with great perfection, and hence I prefer to employ it in the production of the matrix.

The invention is illustrated for convenience in connection with a cylindrical phonogram.[5]

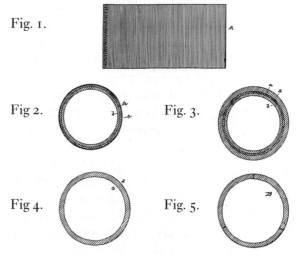

Fig. 1.

Fig 2. Fig. 3.

Fig 4. Fig. 5.

Fig. 6.

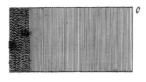

Fig. 7.

In the accompanying drawing forming a part hereof,—

Figure 1, is an elevation of an original phonogram:

Figure 2, a cross section of the original phonogram with a thin vacuous deposit thereon:

Figure 3, a view similar to figure 2 with a further backing:

Figure 4, a view the same as figure 3, with the original phonogram dissolved out:

Figure 5, a sectional view of the divided mould or matrix:

Figure 6, an elevation of a duplicate phonogram produced by the mould; and

Figure 7, a cross section of such duplicate phonogram.

A is the original phonogram having a relatively soft wax or wax-like surface a and the backing of harder material b. The phonographic record is produced upon the surface of a. The metallic vacuous deposit is shown at c and the further backing preferably of metal is shown at d.

B is the divided mould produced as has been stated and having the phonographic record in relief. C is the duplicate phonogram, produced by the mould and having a surface e indented with the phonographic record and preferably of harder material than could be practically or satisfactorily indented directly by the phonograph.

I do not claim the specific invention of duplicating phonographic records or constructing matrices therefor wherein the phonograph record is first covered by a vacuous deposit, that specific subject-matter being reserved for a separate application for patent which I propose to file.[6]

What I claim, is:

First: The process of duplicating phonograms carrying a phonographic record, consisting first in indenting the original record upon a phonogram, second constructing a matrix or mould of such original record, and third producing duplicate phonograms from such matrix, substantially as set forth.

Second: The process of duplicating phonograms carrying a phonographic record, consisting first in indenting the original record upon a phonogram having a wax-like surface, second constructing a matrix or mould of such original record, and third producing duplicate phonograms from such matrix, substantially as set forth.

Third: The process of forming a matrix or mould for the duplication of phonographic records, consisting first in indenting the original record upon a phonogram, second covering the recording surface of such phonogram with a deposit of hard material and then removing the original phonogram, substantially as set forth.

Fourth: The process of forming a matrix or mould for the duplication of phonographic records, consisting first in indenting the original record upon a phonogram, second covering the recording surface of such phonogram with a deposit of metal and then removing the original phonogram, substantially as set forth.

TD, NjWOE, PS (*TAED* PT032AAM). Oath omitted. ªMarginalia written in unidentified hand.

1. Edison signed the completed application on this date, according to the inventor's oath filed at the Patent Office. It was prepared at the offices of Dyer & Seely.

2. This is the case number assigned by Edison's attorneys.

3. This application became bogged down at the Patent Office amid a labyrinthine examination process involving a related application (Case 744, PS [*TAED* PT032AAN]) that Edison signed and filed simultaneously with this one. It also became entangled with an application, pending since 1884, covering the same process more broadly (Case 615; see note 4). Case 743 was divided shortly after filing and a new application (Case 751, executed on 17 January 1888), was submitted for examination. All these applications languished as the Patent Office raised an interlocking series of substantive and procedural objections. The new one (Case 751) did issue in 1892 as U.S. Patent 484,582. According to a reference in that specification, Edison intended the original application to secure broad claims on the process, but he abandoned the remaining portion of Case 743. Case 744, which described the same reproducing process but claimed only the duplicate phonogram itself, was also abandoned.

Extant correspondence about these applications (kept in overlapping case files) shows the Patent Office unable or unwilling to understand the new process. The examiners objected that Edison had explained neither how the vaporized metal could be deposited uniformly nor why its heat did not destroy the wax original, and they requested a practical demonstration. Edison's 1884 application had included a sample phonogram coated in this way, to which his attorneys belatedly referred (Patent Office to TAE, 14 Mar. 1888; Dyer and Seely to Patent Office, 8 Mar. 1890; both Case 743 file [images 11–14]). But the examiners contended on the basis of the specimen that metal coating, so thin it could be scraped off with a thumbnail, would not survive the melting of the wax

(D. G. Purman to TAE, 2 May 1890, Pat. App. 484,582). Edison responded by swearing that he had successfully employed the deposition process "many times." He brushed aside the argument that the coating was too slight, noting that its "sole design . . . is to furnish an electrically conducting surface on which a further deposit can be made by electrolysis." To the objection that the original wax record would be damaged, he answered that "Although the temperature of the vapor is very high . . . the process is so slow that the heat energy available at any point of the surface of the cylinder at any time is not sufficient to impair" the surface (TAE affidavit, 11 June 1890, Case 743 file [images 23–24]). Unpersuaded, the examiners argued that the cylinder's ability to withstand the heated metal vapor appeared to violate the laws of nature. Edison's attorneys countered that the process was similar to that occurring in a Geissler vacuum tube, where a layer of vaporized platinum could be deposited on glass having a lower melting point than the metal. The examiners finally allowed the divided application to go forward in August 1890, but Edison's failure to pay the required fee further delayed its issue (N. A. Seely to TAE, 3 and 25 July 1890; Dyer and Seely to the Commissioner of Patents, 7 Aug. 1890; TAE to Commissioner of Patents, 16 Mar. 1892; W. E. Simonds to TAE, 5 Apr. 1892; all Pat. App. 484,582).

4. Edison had filed Case 615 shortly after he and assistant Edward Acheson sketched several experimental arrangements for depositing an exceedingly thin layer of vaporized metal in a vacuum. The idea was not entirely original, which partly accounts for the application's ten-year odyssey in the Patent Office before it issued as U.S. Patent 526,147, but the process later became very useful in several industries, including the manufacture of semiconductor circuits. Doc. 2587 n. 2.

5. All seven figures were traced onto a separate sheet on 29 November. The paper (signed by witnesses William Pelzer and draftsman Edward Rowland) was marked "—743—744—751— same drawing" and placed in the file of Case 744 (PS [*TAED* PT032AAN]).

6. This process was the subject of an application that Edison filed on 8 March 1888 (Case 765) that became his U.S. Patent 382,419 in May 1888.

–3120–

To Uriah Painter

Orange N.J. Dec 5 [188]7

U.H.P—

My whole action in the phonograph matter is based on the assumption that the phonograph patents are void.[1] They cannot be set right by a decision of the supreme court like our electric lamp patents[2] because They were filed after the granting of the foreign patent while the lamp patent was filed before   The statue is clear, and the foreign patents have expired—[3] After ascertaining this I determined to go into it as a business enterprise with the risks and struggles of competition. I did not want to see the old phono people left out so I offered them a ⅓ interest in the business— My feelings were outraged when in a letter to EH.J. Hubbard[4] insinuated dishonesty and further by you flatly refusing the proposition, as I under all the circumstances thought I was doing a generous action, ~~and~~   After these events I submitted the whole thing to disinterested persons, who also considered [-------][a] that I had acted very generously   After your refusal I could conceive of no other method you would adopt than the legal one, hence my referring you to Tomlinson—[5] If you can show me that you or any other person can be stopped from selling phonographs by [~~the?~~][a] the[b] patents you better do so immediately & I will make a new proposition. It was never my intention to make yourself or EH.J pay the assessment asked but only the outsiders. on my original list in addition to the ⅓ to the old stockholders are Batchelor, Painter Johnson & Reiff[6] down for a large proportion of the remainder= With what I have left and a royalty down to about $3. per machine, the furnishing all the money to experiment etc[c] It will relatively Speaking be myself who will come out of the little end of the horn[7]   Now Painter you know perfectly well that you can see me any time out here, why didnt you come out and learn the whole story before refusing as you did— I never have time to go to NY havent been there but 7 times in two & half years   Yours

Edison

ALS, PHi, UHP (*TAED* X154A6BW). [a]Canceled. [b]Interlined above. [c]Obscured overwritten text.

1. Edison's letter was the latest in a chain of increasingly agitated correspondence between him and Painter, the immediate cause of which seems to have been an exchange between Painter and Edward Johnson in late November, following Edison's expression of exasperation to Edward Johnson. On 2 December, Edison had written that "You must know that the Phonograph can no longer be held by patents in this country." Referring to his proposition to stockholders of the Edison Speaking Phonograph Co., he chastised Painter for a "point blank refusal to accept what under all the circumstances is a gift," and (for the second time) referred him to attorney John Tomlinson, who would "explain The whole situation patents   you will see that you have treated me outrageously." In a terse reply on 4 December, Painter refused to "concede that the Phonograph patents in this country are void." TAE to Johnson, 22 Nov. 1887; Painter to Johnson (with TAE and Johnson marginalia), 27 Nov. 1887; Painter to TAE, 4 Dec. 1887; all DF (*TAED* D8750AAK, D8750AAL, D8750AAO); Painter to TAE, 30 Nov. 1887; TAE to Painter, 2 Dec. 1887; both UHP (*TAED* X154A6BU, X154A6BV).

2. Edison's exclusive claims on the high-resistance lamp filament were hampered by varying interpretations of how the expiration of for-

eign patents affected U.S. patents under section 4887 of the Revised Statutes of the United States, passed in 1870. These general questions were also the subject of complex litigation involving the Bate Refrigerator Co., assignee of John J. Bate's U.S. Patent 197,314 (for preserving meat during transportation and storage), that had produced contradictory rulings in federal courts. The Edison Electric Light Co. watched the Bate cases with interest and, at some point, contributed money and legal counsel to his cause. When a U.S. Supreme Court justice dissolved an injunction against Bate in August 1887, Charles Batchelor pasted a press account into his journal and, under the heading "Edison Patents (Light)," he commented: "This we expected would be in our favour." A related case, *Bate Refrigerating Co. v. Hammond*, was later heard by the U.S. Supreme Court. The court ruled in Bate's favor in January 1889, after which the decision's possible implications for Edison's phonograph patents came up for discussion almost immediately. Cf. App. 1.C.19; Pope 1889, 74–75; Williams 1889, 645–46; Cat. 1336:267, (item 438, 14 Aug. 1887), Batchelor (*TAED* MBJ003267C); "The Edison People Happy," *New York Tribune,* 23 Jan. 1889, 2; Cat. 1085:77, Scraps. (*TAED* SM085077); "The Supreme Court Decision," *Electrical Review* 13 (2 Feb. 1889): 10; Edison Electric Light Co. executive committee minutes, 13 Dec. 1888, p. 4; Dyer & Seely to Alfred Tate, 25 Jan. 1889; both DF (*TAED* D8830ACV, D8954AAJ).

3. Section 4887 stated:

No person shall be debarred from receiving a patent for his invention or discovery, nor shall any patent be declared invalid, by reason of its having been first patented or caused to be patented in a foreign country, unless the same has been introduced into public use in the United States for more than two years prior to the application. But every patent granted for an invention which has been previously patented in a foreign country shall be so limited as to expire at the same time with the foreign patent, or, if there be more than one, at the same time with the one having the shortest term; and in no case shall it be in force more than seventeen years." [Williams 1889, 649]

4. Gardiner Hubbard, for his part, had declined to accept Edison's offer on his own behalf or to recommend it to the Edison Speaking Phonograph Co. Since then, he had resigned as company president, but Uriah Painter continued (as recently as 20 November) to negotiate with Edison. Painter hoped to organize a new company that would acquire rights to phonographic toys and clocks and defend Edison's patents against infringement by both Western Electric (which was making graphophones) and "a new Company that I understand from Mr. Gilliland that he has organized to manufacture them [phonographs] in New Jersey." Hubbard to TAE, 27 (draft) and 28 Oct. 1887, both UHP (*TAED* X154A6BP, X154A6BR); Painter to Johnson, 20 Nov. 1887, DF (*TAED* D8750AAJ).

5. Painter had objected that he had "no desire to see Mr. Tomlinson or to make his acquaintance any further as it has neither been pleasant nor profitable to me what I do know of him." Painter to TAE, 4 Dec. 1887, DF (*TAED* D8750AAO).

6. Railroad financier and telegraph entrepreneur Josiah Custer Reiff (1838–1911), a longtime business associate of Edison, became an investor in the Edison Speaking Phonograph Co. (and also a director) in November 1878. He was listed among the original stockholders of the new Edison Phonograph Co. as of 17 October 1887. See Docs. 141 n. 7, 1574, and 1583; Edison Phonograph Co. stock ledger, p. 8, CR (*TAED* CK002).

7. "Coming out at the little end of the horn" is a slang expression for producing small results after great effort and boasting, or for a failure more generally. The phrase reportedly originated as a symbolic reversal of the cornucopia, but Edison may also have have been alluding to the shape of the phonograph's reproducing horn. Barrère 1889, s.v. "Little end of the horn"; "Sir Oracle [Notes and Queries]," *The Era: An Illustrated Monthly* 8 (Oct. 1901): 654.

–3121–

From Francis Upton

Harrison, N.J., Dec. 9, 87.[a]

Dear Sir:—

In the Testing Room to-day there has been several cases of lamps arcing. In looking over the Pump Room records I find a number of lamps arcing, more than should. Mr. Saxelby[1] feels that the new clamp working only with one current in one way does not have all the gas taken out of it, and that when the current is reversed afterwards that it will make an arc. The arcing of lamps in practice is one of the most serious faults that can occur. A case was brought to my notice where a lamp was placed over a printing press in the "N.Y. Evening Post" office,[2] and which arced. The proprietor of the journal ordered the whole plant taken out, as he said it would spoil his presses if the glass got into them. This was remedied by putting plain glass shades over all the lights, making a very clumsy fixture. There has been other instances in the N.Y. Station, when starting new installations, ~~and~~ where[b] one or more lamps have arced as soon as the light was turned on. We would like your opinion as to the cause of the arcing, and the remedy which we should apply. Yours truly,

Francis R. Upton

⟨Upton= The Cause of arcing is due to crystals of sulphate of Copper in clamp near hot part of Carbon or on Carbon itself   The cause of arcing on new Lamps will be one of the clamps not yet hot.[c]

There are two ways of stopping this   1st. Reversing lamp on pump—or heating clamp by a flame after it is [de?][d] perfectly dry— Reversing lamp for an instant would probably be better=as at high C.P The clamp is brought to red or white heat almost instantly— Baking the lamps for ½ hour at 800 or even 750 ought to [for?]ᵈ Carbonize the gelatin in the paste

~~You~~ [----]ᵈ Force tells me he touches the clamp with a flame but I consider this a dangerous practice. on the <u>whole I advise that</u> when you reach <u>such an exhaustion</u> as to work the ~~pump~~ Lamp up high to heat the clamp that you alternate one instant on 1 clamp the other on the next & in this way work out air on both simultaneously then work blue off on one clamp only³ a very simple plain switch could be made or you could work a wholeᵉ row and work high then reverse the whole row & work high this would dispense with many switches⁴ Edison⟩

TLS, NjWOE, Upton (*TAED* MU119). Letterhead of the Edison Lamp Co. ᵃ"*Harrison, N.J.*," preprinted. ᵇInterlined above by hand. ᶜFollowed by "over" and index marker to indicate page turn. ᵈCanceled. ᵉObscured overwritten text.

1. Frederick Saxelby (b. 1885), a native of Shrewsbury, England, had been a real estate clerk in Rahway, N.J., before starting work at the Edison lamp factory, likely in 1881. Saxelby had charge of the Exhausting Dept. by 1890 and remained connected with the Harrison factory until shortly after its takeover by General Electric in 1892. He was elected an associate of the American Institute of Electrical Engineers in 1888. Doc. 2329 n. 1; U.S. Census Bureau 1970 (1880), roll T9_801, p. 544C, image 0491 (Rahway, Union, N.J.); ibid. 1982 (1910), roll T624_883, p. 14A (East Orange Ward 4, Essex, N.J.); Saxelby birth record, Shrewsbury, Shropshire, England, Vol. 6a:617, *FreeBMD Birth Index, 1837–1915*, online database accessed through Ancestry.com, 27 May 2014; Whipple *Directory* 1890, 511; *Teleg. J. and Elec. Rev.* 23 (13 July 1888): 30; "Notes," ibid., 23 (10 Aug. 1888); "New York Notes," *Electrical Engineer* 14 (30 Nov. 1892): 538.

2. The Evening Post Publishing Co., in which financier Henry Villard had a controlling interest, published the *Evening Post* six days a week. Its editorial offices (occupied by E. L. Godkin and Horace White) were at 210 Broadway, at the corner of Fulton St., in New York. Rowell 1887, 463; Doc. 2439 n. 1.

3. The "blue" is the phenomenon described in Doc. 2966 n. 4.

4. Upton pinned Edison's response to a cover letter he sent to William Holzer with instructions to "take his [Edison's] remarks regarding reversing the current, starting with 'On the whole I advise,' as orders for running the Pump Room." Upton to Holzer, 13 Dec. 1887, Upton (*TAED* MU117).

–3122–

To Arthur Payne

New York, Dec 12th 1887ᵃ

Arthur Payne¹

Send freight ten leaf longest largest fibre select leaves without flaws then proceed southward Panama for Bamboo and other hard fibre.² If Bamboo found equal large we have here arrange ship several thousand joints ten to twelve inch long Telegraph from City Mexico address and money required will telegraph funds[3]

Edison.

⟨Charge Lamp Coy⟩[b]

L (telegram), NjWOE, Vouchers (Laboratory), no. 151 (1888) for Western Union (*TAED* VC88026A). Letterhead of Edison Phonoplex System of Telegraphy; written by John Randolph. [a]"New York," pre-printed. [b]Marginalia likely written by Randolph.

1. This telegram was addressed to Payne at "Aqua Calienties   Mexico," meaning Aguascalientes, the capital of the central Mexican state of the same name. *WGD*, s.v. "Aguascalientes."

2. Edison sent Payne to Mexico to look for plant fibers suitable for lamp filaments. On or before 3 September, Payne arrived in Aguascalientes, where he discovered the agave lechugilla plant, which had hard, dense leaves from which fibers could be extracted. Payne scoured the area around the town for a radius of 150 miles to find the best samples, and he sent leaves and fibers to Edison at the lamp factory. "Fibres not dense enough," Edison wired on 30 November, after receiving the first shipment; "[D]ont understand your failure to send ten or fifteen pounds clear fibrers Each kind by express." Edison urged Payne to go further south and look for bamboo and other fibers. "Am very sorry you are dissatisfied with my work," Payne responded on 7 December. He explained that the lechugilla grew in a vast, uninhabited wilderness where he had endured difficult conditions and slept in the open to reach the plant, once going without food or water for forty-eight hours. Even after bringing the leaves to a populated area, Payne had trouble finding kettles large enough for boiling them to draw out the fibers. Payne to TAE, 3 Sept., 20 Oct., 10 Nov. and 7 Dec. 1887; all DF (*TAED* D8704ACM, D8704ADT, D8731AAL, D8731AAO); TAE to Payne, 30 Nov. 1887, Vouchers (*TAED* VC87015A).

Payne had relocated to Eagle Pass, Tex., in early December and seems not to have received Edison's instructions. On 13 December, Edison wired him again at Aguascalientes and San Antonio, Tex., urging him to head to Panama by way of Mexico City or to the Magdalena River in Colombia. Payne may not have received those messages either, but he had written that he would be in San Antonio by 23 December to receive telegrams, and Edison wired him there on the 24th to return home. On that same day, Alfred Tate reported to Charles Hanington in Brazil that the assai palm fibers received from there were "good" and that Hanington should send ten thousand right away and a hundred thousand to follow. Payne was again working in Edison's laboratory by March 1888. Payne to TAE, 7 Dec. 1887, DF (*TAED* D8731AAO); TAE to Payne, both 13 Dec. 1887, Lbk. 26:102–3 (*TAED* LB026102, LB026103); TAE to Payne, 24 Dec. 1887; Tate to Hanington, 24 Dec. 1887, Vouchers (Laboratory), no. 151 (1888) for Western Union (*TAED* VC88026C, VC88026B1); N-88-03-06 [5 Mar. 1888], Lab. (*TAED* NB026AAK).

In addition to seeking new fibers, Payne also had his eye out for sources of gum and rubber, but these, he reported, grew farther south.

He visited a number of Mexican mines and forwarded samples of gold and silver. He also described in detail how the mine operators extracted the metals from ore tracings and noted their "willingness to adopt a new cheap method," if Edison could devise one. In addition, Payne reported on large deposits of asphalt in California and in Tancanesque, Mexico. Payne to TAE, 3 Sept. and 20 Oct. 1887, both DF (*TAED* D8704ACM, D8704ADT).

3. Payne had reported in November that he had sufficient funds to go to Coahuila, from where he hoped to ship as many lechugilla fibers as Edison might desire, but that he would need more money to supply fibers from Aguascalientes. On 7 December, he indicated that he planned to travel south, bypassing Mexico City, but Edison did not receive that letter for at least six days. At some time during this period, Edison asked Francis Upton to see about wiring money to Payne at Mexico City. Payne to TAE, 10 Nov. and 7 Dec. 1887, both DF (*TAED* D8731AAL, D8731AAO); TAE to Upton, n.d. [Dec. 1887?], Upton (*TAED* MU121).

–3123–

From Ezra Gilliland

NEW YORK, Dec. 14, 1887.[a]

My Dear Edison:—

I have tried to get out to your house some night to get your approval of the Agency contract. I find, however, it is necessary for me to work every night in order to keep up my office work.

As it now stands the cost of the apparatus to the selling Agents will be about $30.00 and the selling price has been fixed at $60.00. It is equivalent to 50% discount from the selling price. I have drawn up a contract on this plan, and have submitted it to several good men who propose to become agents for us and they are all satisfied with it. According to the plan we first adopted, as set forth in my contract, the selling Agents profit is determined by a sliding scale dependent upon the cost, and it is therefore never twice[b] alike.[1] It places us at a disadvantage with the selling agents, giving them all the advantage of any saving that we may make by improved facilities, buying material cheap, etc. The contract provides for giving this benefit to the public by a reduction of the selling price, but of course we would not want to reduce the selling price $.50 or $1.00 at a time.

I suggest that we adopt the plan of fixing the selling price, and agree to give to the Agents, 50% discount upon the apparatus and 25% upon supplies. As the sale of the supplies are guaranteed to the selling Agent, and as ~~they~~ supplies[c] are a running expense against the use of the Phonograph, he should not receive so much profit as upon the sale of the apparatus.

There is a clause in the contract which provides for not only changing the list or selling price, but provides for changing the discount, consequently, if this plan does not work well, it can be changed. It will be satisfactory to the selling agents and I think it would be to our advantage to adopt this plan. I enclose a copy.[2]

You will remember that I told you sometime ago that I had certain notions as to how the Phonograph should be made. I have been working along on it nights and at odd times, and am making one of them. If it works all right, I will show it to you; if it does not I will drown it.[3]

Everything seems to be going all right at the Factory.[4] There is no use making apologies and explanations to you, as no one knows better than you do, the delays and disappointments in the manufacture of a new article and the starting of a new Factory. Do not forget that in this case we are doing both. We are doing the best we can. Yours truly,

E. T. Gilliland

TLS, NjWOE, DF (*TAED* D8750AAT). Letterhead of Edison Phonograph Co., Ezra Gilliland, general agent. [a]"NEW YORK," preprinted. [b]Interlined above. [c]Interlined above by hand.

1. Under an agreement of 28 October, Gilliland was appointed the exclusive sales agent of the Edison Phonograph Co. The contract's second and third articles provided for the division of the company's revenues among Gilliland and his sub-agents, and for adjustments to the selling price of phonographs. Miller (*TAED* HM89ABV).

2. Not found.

3. Edison evidently expressed to Frank Toppan his concern or displeasure at Gilliland's efforts to improve the phonograph while simultaneously supervising the start of its manufacture. Gilliland responded directly to Edison, reassuring him that he had in "no way neglected or delayed the work in the Factory." He explained that while Edison's prototype "would not compare favorably in any respect with the Graphophone," his own was the equal or better of the rival machine. "Toppan says you particularly dwelt upon the point, that no man could invent and do business at the same time," Gilliland continued, before pointing to his past success doing both for American Bell Telephone. He closed the typed letter with a short handwritten paragraph assuring Edison that if he had "said or done anything to displease you it has been due to my over anxiety to do the most and best I can for you." Gilliland to TAE, 16 Dec. 1887, DF (*TAED* D8750AAU).

4. Edison had designated Gilliland to set up a shop to manufacture instruments for the Edison Phonograph Co. By early November, Albert Keller had also been engaged and was supervising twelve to fifteen men setting up tools in a rented building on the corner of Liberty and Broad Sts. in Bloomfield, a few miles from Orange. Cat. 1337:19, 25 (items 490, 506; 15 Oct. and 5 Nov. 1887), Batchelor (*TAED* MBJ004; images 10, 13); Koenigsberg 1990, xxvi.

New York, December 14th, *1887*[a]

Gentlemen:—

I take pleasure in announcing that the Standardizing Bureau, one of the most valuable adjuncts of a growing industry involving constant development, has been revived.[2]

Its members include Mr. Edison and Mr. Batchelor of the laboratory, such of the leading officers of this company as by their experience are familiar with the practical workings of the system, and one representative of each of the licensed shops.

The detail of the work has been committed to Mr. W. J. Jenks,[3] who has been appointed Director of the Bureau, and will act as my representative in receiving statements of any cases of dissatisfaction regarding operation of Edison apparatus, or the results of any faulty methods of construction employed by contractors or others having to do with the Edison system, with the object of investigating and correcting causes of difficulty.

The work thus proposed will necessarily be one of time. As a first step, I desire that Mr. Jenks procure from the records of your various departments, such material as he may consider useful in promoting the objects of the Bureau. You are also requested to forward to him all information as to imperfections of apparatus or methods either in the form of statements of agents, or letters of criticism or complaint (or copies thereof) which may come to your notice from any source.

It is also desirable that each of the shops appoint some practical delegate whose duty it shall be to attend the meetings of the Bureau whenever called upon to the end that each manufacturing branch may be represented at every session. Yours truly

Edward H. Johnson   President.

Please address all communications on this subject to W. J. Jenks, Director of the Standardizing Bureau, Room 53, 16 Broad Street.

TL, NjWOE, DF (*TAED* D8732ABY). Letterhead of Edison Electric Light Co. [a]"*New York*," and "*188*" preprinted.

1. This letter was addressed "To the Officers of the Edison Electric Light Company, The Edison United Manufacturing Company, The Edison Lamp Company, The Edison Machine Works and Bergmann & Company." This copy was forwarded to Edison by William Jenks five days later. Jenks to TAE, 19 Dec. 1887, DF (*TAED* D8732ABX).

2. The antecedent Standardizing Committee, consisting of the heads of all departments in the Edison lighting business, existed prior to the end of 1883. Created in an effort to improve the lighting system, it was charged with testing new devices and responding to complaints. In Jan-

uary 1887, Johnson put forward a plan (as described by Charles Batchelor) "for a special testing and standardizing shop to be in the first district to get lots of current etc & be where the officers of the Light Co can easily get at it." Batchelor was initially suggested to head this branch, but William Jenks was named instead because "no one of the older and wiser heads can now devote the necessary attention and time." Informing Edison of his appointment, Jenks promised that the group would "carefully weigh the practical opinions of the experimenter, the manufacturer and the practitioner . . . and avoid the friction which would otherwise be inevitable in making any necessary changes of standards." The Bureau's first meeting took place at Edison's Orange laboratory on 25 January 1888. Doc. 2564 n. 2; Cat. 1336:167 (item 298, 27 Jan. 1887), Batchelor (*TAED* MBJ003167C); Jenks to TAE, 19 Dec. 1887; Jenks to Batchelor, 21 Jan. 1888, both DF (*TAED* D8732ABX, D8830AAA).

3. Presently the manager of the Edison Electric Light Co.'s Municipal Dept., William Johnson Jenks (1852?–1918) possessed a wide range of experience in the telephone and electric lighting businesses. According to his testimony in a later patent lawsuit, Jenks started as a telegrapher in 1872 and became involved in the telephone industry when he built an exchange in Brockton, Mass., in 1878. Between 1880 and 1882, while still involved with the Brockton Telephone Co., he was also connected with the Brush system of electric lighting. He quit the telephone company in 1883 to manage the Edison Electric Illuminating Co. of Brockton and subsequently joined W. J. Paine, formerly of the Edison Co. for Isolated Lighting, to start the Brockton Wiring Co. (subsequently the New England Wiring and Construction Co.), which he headed until 1887. Jenks started the first Edison municipal lighting system, in Portland, Me., in October 1885. He also worked outside of the Edison orbit for parts of 1885 and 1886, managing the Incandescent Dept. of the American Electrical Manufacturing Co., before being named to the Edison Municipal Dept. in July 1886. Jenks served as director of the Standardizing Bureau until 1889, when he focused his attention on patent litigation for the Edison General Electric Co. Doc. 2963 n. 3; Jenks testimony, pp. 26–28, *Edison Electric Light Company v. F. P. Little* 1896; National Electric Light Association 1918, 357–58.

–3125–

To Alfred Southwick[1]

[New York,] Decr 19th [188]7

Dear Sir:

I am in receipt of y[our][a] letter 5th inst. in further reference to E[lec]tricity[a] as an agent to supplant the gallows, and have carefully considered your remarks.[2]

Your points are well taken and though I would join heartily in an effort to totally abolish capital punishment, I at the same time realize that while the system is recognized by the State, it is the duty of the latter to adopt the most humane method available for the purpose of disposing of criminals under sentence of death.

The best appliance in this connection is, to my mind, the one which will perform its work in the shortest space of time, and inflict the least amount of suffering upon its victim. This, I believe, can be [ac]complished[a] by the use of Electricity, and the most suitable apparatus for the purpose is that class of dynamo Electric machinery which employs intermittent currents.

The most effective of these are known as "Alternating Machines," manufactured principally in this country by Mr Geo. Westinghouse, Pittsburg, and the cost of Engine, Alternating dynamo and appliances sufficient for the work above referred to would hardly exceed $2500.; the cost of maintenance would be a mere trifle owing to the infrequent use made of the apparatus.

The passage of the current from these machines through the human body, even by the slightest contacts, produces instantaneous death, practical evidence of which has been supplied during the past six months in the city of New Orleans, where two men have been killed and others injured by this quality of current. The details of this circumstance I cannot myself furnish, but doubtless the New Orleans authorities would provide you with accurate data, if you consider it would assist you in arriving at a solution of the problem you have commenced to work out.[3] Yours very truly

Thos A Edison

ALS (letterpress copy), NjWOE, Lbk. 26:116 (*TAED* LB026116). Written by Alfred Tate. [a]Paper torn.

1. Alfred Porter Southwick (1826–1898), a Buffalo dentist and former steamboat mechanic, first became interested in electricity as a possible means of alleviating pain during dental procedures. A supporter of capital punishment, he hoped to make the process more humane in an effort to counter opposition to it. After learning about an accidental electrocution at the local Brush lighting plant in 1881, Southwick started experimenting with electricity on dogs and other animals toward that goal. In 1885, New York governor David Hill appointed him to a three-man commission charged with recommending an alternative to hanging for execution. *ANB*, s.v. "Southwick, Alfred Porter"; Brandon 1999, 14; Essig 2003, 91–94; Moran 2002, 70–75.

2. Southwick's letter was his second to Edison on this subject. He first wrote on 8 November on behalf of the commission on capital punishment, appealing to Edison as "a scientist and especially an electrician" for technical information pertinent to electrocution as a possible means of execution, including recommendations on strength of current and type of machine to use, and its probable cost. Edison's reply to that letter has not been found, but he seems to have declined, on moral grounds, to provide the information requested. When Southwick wrote

again on 5 December, he urged Edison to consider that the question to be addressed was not the moral justification for capital punishment but only whether science could discover "some more humane method than the rope," which he called "a relic of barbarism." He argued that the legislature would revise the law "if you and a few others will but assist the commission. . . . Civilization, science and humanity demand a change." Edison's response reportedly persuaded at least one skeptical commissioner, and portions of it were quoted directly in the commission's January 1888 report to the state legislature, which soon passed a bill to substitute electrocution for hanging. Southwick to TAE, 8 Nov. 1887 and 5 Dec. 1887, both DF (*TAED* D8704AEP, D8704AFJ); Gerry, Southwick, and Hale 1888, 80; Moran 2002, 74–85, Essig 2003, 90–99, 116–24; "Death by Electricity," *NYT,* 5 June 1888, 2.

3. Edison likely heard of these incidents from Edward Johnson, who was apprised of them by William Mottram, general superintendent of the Edison company in New Orleans. A Brush-affiliated company there was operating what Mottram designated as "with one exception the largest and most important station now supplying incandescent lights on the Westinghouse Alternating system," and he cheerfully described a series of accidents it had suffered. "The alternating current has killed two men and has injured several others" in New Orleans, Mottram reported; one of those killed was a telephone lineman whose employer was now seeking an injunction against the Brush company putting more wires near its own. Mottram to Johnson, 20 Nov. 1887, PPC (*TAED* CA019B5).

–3126–

Draft Patent Application: Electric Power

Dyer patent.[2]

Orange, N.J., Dec 20, 1887[1a]

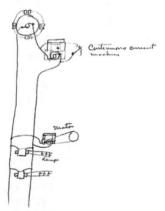

The object of this invention is to work motors on alternate current circuits. The invention consists of sending alternating Currents into a circuit containing transformers & motors in multiple arc, & including in the primary circuit a continuous current generating[b] machine of the same Voltage as the alternating so that the direction of the[c] Currents on the primary are never reversed direction through the wire but instead of sending a wave of 1000 volts in one direction and above the Zero line of no current & a waves of 1000 volts below the Zero line, the continuous current causes the ~~rise from the Zero line~~ wave to rise 2000 volts above the Zero line & the opposite wave to fall to the zero line. thus no reversal of the Iron or Current takes place in the motors & as the waves follow with such extreme rapidity the motor runs as if with a Continuous Current, but the Effect on the transformers is still to cause alternating Currents to be sent into the secondary lamp circuits.[3]

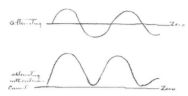

Storage batters may be charged—etc—[4]

Edison

ADfS, NjWOE, PS (*TAED* PT032AAO1). Letterhead of the Laboratory of Thomas A. Edison. [a]"*Orange, N.J.*," and "*188*" preprinted. [b]Interlined above. [c]"direction of the" interlined above.

1. Henry Seely indicated that Dyer received the draft the following day.

2. Figure labels are "Continuous current machine," "Motor," and "Lamps." This application was Edison's Patent Case 748 and was assigned serial number 259,899 when it was filed on 5 January. Edison later abandoned the application. Case 748 file wrapper, PS (*TAED* PT032AAO).

3. Figure labels on the upper drawing are "alternating" and "Zero"; on the bottom drawing they are "alternating with continuous Current" and "Zero."

4. The charged storage batteries would then be used to drive the motors with direct current.

–3127–

Memorandum: Laboratory Materials[1]

[Orange? Fall 1887]

Nice specimen is wanted of each kind for cabinet, but of small size—[a]

Considerable quantities are wanted for experimental purposes—

please give [sp?][b] prices for the Cabinet specimen & price per lb for the common varities in quantities and probable quantity in ore— I have dropped the formula as unnecessary— This List is from Dana—[2] I am particularly desirous

of optaining ~~mineral~~ fossil Resins—Lignites Bitumens etc—
In many cases I shall want 50 lbs of a mineral=

T A Edison

Mark the 2 prices opposite name of mineral

ADS, NjWOE, Lab., N-88-06-01.1 (*TAED* NA024AAA). ªFollowed by dividing mark. ᵇCanceled.

1. Immediately following this note is an undated forty-eight page list of minerals, resins, and hydrocarbon compounds totaling many hundreds in all. In a different notebook, Edison created a separate twenty-page list (also undated) of other materials to be stocked, with instructions to get prices for each. Many of those items came from the natural world, including various animal skins, hairs, and teeth, but the list also included manufactured articles such as glass, inks, paper, and fabrics, among many others. Both lists were likely made sometime during the fall as Edison contemplated completion of his new laboratory. N-88-06-01.2, Lab. (*TAED* NA025AAA).

2. Edison likely referred to the *System of Mineralogy*, the definitive work of distinguished American geologist and mineralogist James Dwight Dana (1813–1895). Originally published in 1837, it had been revised and reissued repeatedly; a separate volume of the appendices to the fifth edition by Dana's son, Edward Dana, and George Brush appeared in 1882. *ANB*, s.v. "Dana, James Dwight"; Brush, Dana, and Dana 1882.

Appendix 1

Edison's Autobiographical Notes

From 1907 to 1909 Edison wrote a series of autobiographical notes to assist Thomas C. Martin and Frank L. Dyer in their preparation of his authorized biography.[1] Edison produced Document D, including notes on queries posed by Martin, probably about October 1907.[2] Those recollections were followed by those in books A and G, made in September and October of 1908. This material was incorporated into the initial chapters of the biography, which were complete by February 1909; Martin then requested additional personal reminiscences from Edison in order to flesh out the remaining chapters.[3] William Meadowcroft, who was coordinating the project, acknowledged in May 1909 that the continuing lack of Edison's additional material was a "very serious affair," and the next month Edison produced the notes in books E and F.[4] Some of these formed the basis for oral interviews with Martin, the typed transcripts of which became documents B and C; together, these four documents served as the basis for anecdotes related in later chapters of the published biography.

Five of the documents contain sections related to events in the period of Volume Eight; those sections are published here.[5] Edison sometimes referred in the same paragraph to events in periods covered by more than a single volume; these paragraphs will be reprinted as appropriate. Each document has been designated by a letter and each paragraph has been sequentially numbered. A few individual items that were inadvertently omitted from previous volumes are presented here. Items that are either solely by the interlocutor or completely indistinct as to time have not been transcribed.

1. Dyer & Martin 1910. The designations A through F were assigned to these documents in Volume One, which also contains a general editorial discussion of them. See *TAEB* 1 App. 1; document G was discovered later.

2. An Edison notebook entry from this time reads "Martins book take Lab note bk 4, 1 Extra . . . answer Martins immediate notes." PN-07-09-15, Lab. (*TAED* NP077).

3. Martin to TAE, 23 Feb. 1909, Meadowcroft (*TAED* MM001BAP).

4. Meadowcroft to Martin, 24 May 1909, Meadowcroft (*TAED* MM001BAQ).

5. The autobiographical documents designated A and E do not refer to the period of this volume. The sections from A published in Volumes One and Four were drawn from a typed version of Edison's notes prepared by William Meadowcroft. However, a copy of Edison's original manuscript, in a notebook labeled "Book No. 1   September 1, 1908   Mr. Edison's notes re. Biography," was published in Part IV of the microfilm edition. Meadowcroft (*TAED* MM002).

B. FIRST BATCH

The following is a transcription of a typescript that Edison revised. At the top of the first page is a handwritten note: "First Batch   Notes dictated by Mr Edison to T. C. Martin June, 1909.— Pencil indicates Mr. Edison's revision." Four of its eighty-one paragraphs likely pertain to the period covered by this volume. Section twenty-two refers to a celebratory dinner that probably occurred in late 1887 or early 1888. Section seventy-five refers specifically to the Schenectady period of the Edison Machine Works, but Edison may have remembered a similar arrangement with the Ansonia Brass & Copper Co. reached in 1885 (Doc. 2515), when the Machine Works was still in New York City. Section seventy-six probably refers to events in 1882 and should have been published in Volume Six.

A Dinner With Childs.

[22] George Washington Childs was very anxious I should go down to dine with him. I seldom went to dinners. He insisted that I should go, that a special car would leave New York. It was for me to meet Mr. Joseph Chamberlain. We had the private car of Mr. Roberts, president of the Pennsylvania Railroad. We had one of those celebrated dinners that only Mr. Childs could give, and I heard speeches from Charles Francis Adams and different people. When I came back to the depot, Mr. Roberts was there and insisted on carrying my satchel for me. I never could understand that.

LABOR TROUBLES.

[64] After our works at Goerck street got too small, we had labor troubles also. It seems I had rather a socialistic strain in me and I raised the pay of the workmen 25 cents an hour above the prevailing rate of wages, whereupon Hoe & Co. our near neighbors, complained at our doing this. I said I thought it was all right. But the men having got a little more wages thought they would try coercion and get a little more, as we were considered soft marks. Whereupon they struck at a time that was critical. However, we were short of money for pay rolls and we concluded it might not be [----]ᵃ so badᵇ after all, as it would give us a couple of weeks to catch up. So when the men went out they appointed a committee to meet us. But for two weeks they could not find us so they became somewhat more anxious than we were. Finally they said they would like to go back. We said all right and back they went. It was quite a novelty to the men not to be able to find us when they wanted, and they didn't relish it at all. What with these troubles and the lack of room we decided to find a factory somewhere else, and decided to try the locomotive works up at Schenectady. It seems that the people interested there had had a falling out among themselves, and one of the directors had started opposition works, but before he had completed all the buildings and put machinery in some compromise was made, and the works were for sale. We bought them very reasonably and moved everything there. These works were owned by me and my assistants until sold to the General Electric Company. At one time we employed several thousand men, and since then the works have been greatly expanded.

BUYING COPPER FOR SCHENECTADY.

[75] In operating the Schenectady Works, Mr. Insull and I had a terrible burden. We had enormous orders and little money, and great difficulty to meet our pay rolls, and buy supplies. At one time we had so many orders on hand we wanted $200,000 of copper and didn't have a cent to buy it. We went down to the Ansonia Brass & Copper Co. and told Mr. Cowles just how we stood. He said: "I will see what I can do. Will you let my bookkeeper look at your books?" We said: "Come right up and look them over." He sent his man up and found we had the orders and were all right, although we didn't have the money. He said: "I will let you have the copper" and for years he trusted us for all the copper we wanted even if we didn't have the money to pay for it.

[76] I remember one time when we were short of money, Insull said the only possible way would be to sell a draft on London and get the money from Drexel, Morgan, but, of course, you couldn't do that unless you had the right to make the draft on somebody. "Well" I said "Would it be all right if we told Drexel Morgan that those drafts were being sold in reality for nobody?" He said we would be making fools of ourselves. I said I would go down and tell Morgan we wanted to draw on London and wanted to get the money, and we would gain time and could cable over to meet the drafts. We told him what we were driving at and he let us have the money.

TD (transcript), NjWOE, Meadowcroft (*TAED* MM003). ªCanceled. ᵇ"so bad" interlined above in pencil.

C. SECOND BATCH

The following is a transcription of a typescript that includes Edison's revisions. At the top of the first page is a handwritten note: "Second Batch   Mr Edison's notes dictated Mr Martin June 1909   Pencil indicates revision by Mr Edison." Two of its twenty-four sections pertain to the period of this volume.

TRAIN WIRELESS EXPERIMENTS.

[7] I got up a wireless telegraph system for trains and introduced it on the Lehigh Railroad. That came from experiments tried previously at Menlo Park. I got up a megaphone there by means of which with ear tubes and very large funnels I could talk over distances of 2-¹/₂ miles. Then I started to telegraph by induction, using kites, but we did not succeed in getting over 2-¹/₂ miles with induction. Afterwards I introduced this induction system with telephone on the Lehigh Valley road for trains in motion. It was employed there for about a year on a construction train. The first experiments were done on Staten Island by King, of whom I have already spoken.

THE COST OF A COMMA.

[19] All of my telephone and electric light patents were lost or ended because when they codified the American patent laws in Congress a clerk had misplaced a comma. As the law read, American patents were made to expire with the ~~term~~ expirationª of the foreign patents. That had nothing to do with the case, and didn't mean anything, but it killed a lot of pat-

ents that had also been taken out abroad. Had the comma been put in right it would not have gone that way.

TD (typescript), NjWOE, Meadowcroft (*TAED* MM004). ªInterlined above in pencil.

D. BOOK NO. 2

This undated notebook, labeled "Book No. 2," contains a mix of narrative passages, questions, and notes in Edison's hand. The first two pages are a memo by Meadowcroft, dated 9 January 1920, recounting the preparation and use made of this material between 1907 and 1910. The next sixty-six pages alternately present narrative passages and brief references to various anecdotes. The next nine-page section is labeled "Martin's Questions." The remaining twenty-one pages contain only notes. Six sections may pertain to the period of this volume.

[347] Works moved schenectady—
[353] Strike at Goerck disappeared Com couldnt find— Hoe complained
[365] Ansonia letting us have ~~money~~ copper after seeing our books— terrible time with payrolls— selling drafts on London & Cabling money to pay them—
[375] Phonoplex—
[378] Sims Edison Torpedo.
[385] Lost patents by foreign because Himalayian [Anderson?] jackass misplaced a comma when pat law was codified—

AD (photocopy), NjWOE, Meadowcroft (*TAED* MM005).

F. NOTES (JUNE 1909)

This notebook includes sixteen pages in an unlabeled section in Edison's hand relating to the Dyer and Martin biography. These pages are preceded by a memo to Edison from William Meadowcroft dated "June 28/09" stating that these notes had been copied. There is a typed version of the notes in the William H. Meadowcroft Collection at the Edison National Historic Site. Four of its twenty-four items pertain specifically to the period covered by this volume; another (section twenty-two) is more general but also included here.

[14] I also perfected the a system of telegraphy between stations & trains in motion whereby messages could be sent from the moving train to the Central offices ~~This was ada~~ & was the forerunner of wireless telegphy This system was used for a number of years on the Lehigh Valley RR on their construction trains. The Electric wave passed from a piece of metal on top of the car across the air to the telegh wires & then proceeded to the dispatchers office.

[15] In my first Experiments with this system—I tried it on the staten Island RR & employed our Opr named King to do the Experimenting ~~when he got corned~~ he reported results Every day & recd instructions by mail but for some reason He could send messages all right when the Trains went in one direction but could not make it go in the Contrary direction ~~I sent every~~ I made suggestions of Every Kind to get around this phenomenon finally I teleghed King if he had any suggestions & I received a reply that the only way he could suggest to get around the difficulty was to put the Island on pivots ~~& turn~~ so it could be turned around— I finally found the trouble & its practical introduction on the Lehigh Valley was the result.

[17] Meadocroft you can find the press reports, a message was sent from a spcl train to England & recd answer while train going

[18] This system was sold to a very wealthy man & he never would sell any rights or even answer letters. he subsequently became a spiritualist ~~when the first system of~~ [--------------]ᵃ which probably Explains it.

[22] In any trade ~~we~~ any of my boys made with Bergmann he always got the best of us.

AD (photocopy), NjWOE, Meadowcroft (*TAED* MM007). ᵃEntire line of text canceled.

G. "MR. EDISON'S NOTES" (OCTOBER 1908)

These reminiscences come from a notebook labeled "Book No. 2, Mr. Edison's notes re. Biography October 1908." The sections from G published in Volumes Two through Five were drawn from a typed version prepared by William Meadowcroft. However, a copy of Edison's original manuscript was located and published in Part IV of the microfilm edition, and it is the basis for the transcription below. Only two paragraphs pertains to the period covered in this volume.

[35] Experiments were also made with Kites & induction coils to endeavor to transmit morse signals to a distance & receive them on a telephone, ~~afterwards~~ one mile was the greatest distance we could send signals without a wire. afterwards I applied this idea to moving trains to permit sending messages from a moving train to the terminal station on the road. This system was installed on the Lehigh Valley road and was used for a number of years on Construction Trains. (meadowcroft can get newspaper account of the test from a passenger train.)

[36] I also applied it to ships, for sending messages at sea. After, Hertz' paper, Marconi applied the knowledge thus conveyed to wireless teleghy with marvelous results. The Marconi Company purchased my patent.

AD (photocopy), NjWOE, Meadowcroft (*TAED* MM008).

Appendix 2

Edison's Patent Applications, 1885–1887

A. EDISON'S U.S. APPLICATIONS

Edison maintained a prolific rate of patent activity over the three years covered by Volume Eight. One measure of this activity is the number of United States patent applications to which his attorneys assigned a Case Number before filing at the Patent Office. Not every application resulted in a patent.

The following list of Edison's U.S. patent cases from this period is a companion to similar appendices in previous Edison Papers volumes. It is arranged by case number with such information about unsuccessful applications as the editors have been able to learn.[1] Patents issued jointly to Edison and Ezra Gilliland do not have case numbers in this sequence and are listed by execution date. Not included here is a telephone application that Edison executed on 17 February 1886 as a division of Case 141, originally executed on 9 July 1877. The new application issued on 3 May 1892 as U.S. Patent 474,232; the companion patent issued the same day as U.S. Patent 474,231.

Case	*Exec. Date*	*Appl. Date*	*Issue Date*	*Pat. No.*	*Title*
644	01/02/85	10/14/85	04/27/86	340,709	Telephone-Circuit (with Ezra Gilliland)
645	01/14/85	10/14/85	10/14/90	438,305	Fuse-Block
646	Missing	10/14/85	02/14/88	378,044	Telephone-Transmitter
647	01/09/85	10/14/85	08/24/86	348,114	Electrode for Telephone-Transmitters
A	01/12/85	04/07/85	11/22/92	486,634	System of Railway Signaling (with Ezra Gilliland)
B	03/27/85	04/07/85	10/05/86	350,234	System of Railway Signaling (with Ezra Gilliland)
648	03/27/85	05/08/85	12/29/85	333,289	Telegraphy
649	04/27/85	05/08/85	12/29/85	333,290	Duplex Telegraphy
650	04/30/85	05/16/85	12/29/85	333,291	Way-Station Quadruplex Telegraphy
651	05/06/85	Missing			
652	05/14/85	05/23/85	12/29/91	465,971	Means for Transmitting Signals Electrically
653	10/07/85	10/23/85	02/25/90	422,072	Telegraphy
654	10/07/85	10/23/85	09/30/90	437,422	Telegraphy
655	11/12/85	11/24/85	02/25/90	422,073	Telegraphy
656	11/24/85	02/19/86	02/25/90	422,074	Telegraphy
657	11/30/85	02/19/86	09/02/90	435,689	Telegraphy
658	Missing				
659	Missing				
660	Missing				
661	12/22/85	02/19/86	10/14/90	438,306	Telephone (with Ezra Gilliland)
662	12/28/85	01/13/86	10/05/86	350,235	Railway-Telegraphy (with Ezra Gilliland)
663	01/28/86	02/19/86	07/09/89	406,567	Telephone
664	Drawing traced on 01/29/86		Abandoned		Insulated Three-Wire Conductor
665	01/29/86	02/16/86	Abandoned (1892/1893)		Railway Signaling Apparatus
	05/11/86	05/15/86	09/20/87	370,132	Telegraphy
	07/07/86	07/10/86	Abandoned (1891?)		Telegraphy

666	07/15/86	07/17/86	09/17/89	411,018	Manufacture of Incandescent Electric Lamps
667	Missing				
668	Draft by TAE (07/01/86)				
669	07/15/86	07/17/86	10/14/90	438,307	Manufacture of Incandescent Electric Lamps
670	Drawing traced (07/12/86)		Abandoned		Straightening Filaments in Vacuo
671	07/20/86	07/29/86	09/17/89	411,019	Manufacture of Incandescent Electric Lamps
672	Missing				
673	07/15/86	07/19/86	03/24/91	448,779	Telegraph
674	07/15/86	07/19/86	Abandoned (1892?)		Telegraph
675	Drawing traced (08/03/86)		Abandoned		Manufacture of Incandescent Electric Lamps
676	Drawing traced (07/30/86)		Abandoned		Manufacture of Incandescent Electric Lamps
677	Missing				
678	08/06/86	08/11/86	07/02/89	406,130	Manufacture of Incandescent Electric Lamps
679	09/30/86	10/05/86	11/02/86	351,856	Incandescent Electric Lamp
680	10/26/86	10/27/86	06/16/91	454,262	Incandescent-Lamp Filament
681	10/26/86	10/27/86	01/05/92	466,400	Cut-Out for Incandescent Lamps (with John Ott)
682	Missing				
683	10/26/86	10/27/86	10/11/92	484,184	Manufacture of Carbon Filaments
684	Missing				
685	Missing				
686	11/02/86	11/06/86	01/31/93	490,954	Manufacture of Carbon Filaments for Electric Lamps
687	Missing				
688[2]	Possible draft by TAE (10/8/86)				Manufacture of Carbon Filaments
689	Possible draft by TAE (10/8/86)				Manufacture of Carbon Filaments
690	Possible draft by TAE (10/8/86)				Manufacture of Carbon Filaments
691	11/09/86	12/06/86	10/14/90	438,308	System of Electrical Distribution
692	11/09/86	12/06/86	08/14/94	524,378	System of Electrical Distribution
693[3]	Missing				

No.	Executed	Filed	Issued / Status	Patent No.	Title
694	11/22/86	12/06/86	09/06/87	369,439	System of Electrical Distribution
695	Missing				
696	Missing				
697	Missing				
698	11/24/86	11/29/86	06/19/88	384,830	Railway Signaling (with Ezra Gilliland)
699	11/26/86 [Drawing traced (11/03/86)]	12/06/86	09/17/89	411,020	Manufacture of Carbon Filaments
700	11/29/86		Abandoned		System of Electrical Distribution
701	Missing				
702	11/26/86	12/06/86	03/27/88	379,944	Commutator for Dynamo-Electric Machines
703[+]	Missing				
704	12/06/86		Allowed (12/18/97; abandoned?)		System of Electrical Distribution
705	12/06/86	12/15/86	11/08/92	485,615	Manufacture of Carbon Filaments
706	12/06/86	12/15/86	11/08/92	485,616	Manufacture of Carbon Filaments
707	12/06/86	12/15/86	08/28/94	525,007	Manufacture of Carbon Filaments
708	12/10/86	12/16/86	09/06/87	369,441	System of Electrical Distribution
709	12/16/86	12/22/86	09/06/87	369,442	System of Electrical Distribution
710	12/16/86	12/27/86	09/06/87	369,443	System of Electrical Distribution
711	Missing				
712	12/20/86	12/27/86	10/11/92	484,185	Manufacture of Carbon Filaments
713	12/20/86	12/27/86	02/12/95	534,207	Manufacture of Carbon Filaments
714	12/21/86	12/28/86	11/22/87	373,584	Dynamo-Electric Machine
715	Missing				
716	Missing				
717	02/07/87	02/16/87	Abandoned (09/30/95)		Material for Ornamental Purposes
718	02/07/87	03/08/87	02/16/92	468,949	Converter System for Electric Railways
719	05/24/87	06/04/87	Abandoned (1891)		Manufacture of Wrought Iron
720	05/24/87	06/13/87	03/27/88	380,100	Pyromagnetic Motor

721	05/24/87	06/13/87	476,983	Pyromagnetic Generator
722	06/01/87	06/04/87	476,530	Incandescent Electric Lamp
723	06/14/87	Abandoned		
724	Missing			
725	06/30/87	07/06/87	377,518	Magnetic Separator
726	Missing			
727	08/09/87	08/19/87	470,923	Railway-Signaling
728	Missing			
729	08/26/87	08/29/87	545,405	System of Electrical Distribution
730	09/13/87	09/23/87	380,101	System of Electrical Distribution
731	09/14/87	09/23/87	380,102	System of Electrical Distribution
732	09/26/87	09/30/87	470,924	Electric Conductor
733	10/13/87	10/19/87	Abandoned (1891/1892)	Phonographs
734	10/14/87	10/21/87	Abandoned (1891)	Phonographs
735	10/17/87	10/21/87	563,462	Method of and Apparatus for Drawing Wire
736	11/09/87	11/11/87	506,215	Method of Making Plate Glass
737	11/05/87	12/09/87	385,173	System of Electrical Distribution
738	Missing			
739	Missing			
740	11/22/87	11/26/87	430,570	Phonogram-Blank
741	11/22/87	11/26/87	386,974	Phonograph
742	11/22/87	11/26/87	382,414	Burnishing Attachment for Phonographs
743	11/29/87	01/05/88	(Divided 01/88; abandoned)	Duplicating Phonograms
744	11/29/87	01/05/88	Abandoned (1892)	Phonograms
745	11/29/87	01/05/88	382,416	Feed and Return Mechanism for Phonographs
746	12/04/87	12/27/87	382,415	System of Electrical Distribution
747	12/05/87	01/05/88	382,462	Phonogram-Blank

1. The information in this chart is compiled mainly from the list of Edison's patents on the Thomas Edison Papers website, a collection of patent application files and drawings, a set of Abstracts of Edison's Abandoned Applications from 1876 to 1885 (all PS [*TAED* PT2A, PT023, PT032, PT004]), and occasional references in patent specifications to pending applications. The full text and drawings of Edison's U.S. patents are on the Edison Papers website at http://edison.rutgers .edu/patents.htm, where they can be searched by execution date, patent number, or subject area.

2. See Doc. 2994 n. 2 regarding possible Edison drafts of Cases 688–90.

3. This application may have been the one designed by Patent Office serial number 219,358; it was filed 19 November 1886 and subsequently referenced in the *Zipernowski v. Edison* interference, pp. 11–12 (*TAED* W100DMA).

4. Referenced in Case 710.

B. U.S. PATENTS BY EDISON EMPLOYEES AND ASSOCIATES, EXECUTED JANUARY 1885 TO DECEMBER 1887[1]

This list identifies U.S. patents obtained by individuals with whom Edison or one of his enterprises had a significant relationship in this period. Some of these patents are clearly related to inventive work done in connection with Edison's own, while others are not; the editors have not discriminated on this basis. The names are given alphabetically; under each name, the patents appear by execution date (where known). The list is an extension of similar ones in *TAEB* 6 App. 5.C and *TAEB* 7 App. 4.C. The U.S. Patent and Trademark Office maintains full-text images of all issued patents, searchable by patent number, at: http://patft.uspto.gov/; images may also be obtained through Google Patents at http://www.google .com/patents.

Patentee	Patent No.	Title	Executed	Filed	Issued	Assigned to
Andrews, William S., and Thomas Spencer	318,157	System of Electric Lighting		17 Jan. 1885	19 May 1885	
Batchelor, Charles	338,383	Dynamo Electric Machine	13 Nov. 1885	16 Nov. 1885	23 Mar. 1886	
Batchelor, Charles	341,990	Dynamo Electric Machine	14 Jan. 1886	20 Jan. 1886	18 May 1886	
Batchelor, C., and Henry E. Walter	339,839	Dynamo Electric Machine	23 June 1885	6 July 1885	13 Apr. 1886	
Bergmann, Sigmund	344,938	Telephone	6 Mar. 1886	9 Mar. 1886	6 July 1886	
Dyer, Richard N.	332,649	Printing Telegraph	8 Sept. 1885	10 Sept. 1885	15 Dec. 1885	Commercial Telegram Co., New York
Dyer, Richard N.	335,275	Printing Telegraph	14 Oct. 1885	15 Oct. 1885	2 Feb. 1886	Commercial Telegram Co., New York
Dyer, Richard N.	348,155	Night-Lamp for Electric Lighting Systems	25 Nov. 1885	5 Dec. 1885	24 Aug. 1886	John W. Howell, Charles S. Van Nuis, and Dyer and Seely
Gilliland, Ezra	324,678	Electrical Contact-Point	13 May 1885	15 May 1885	18 Aug. 1885	
Gilliland, Ezra	327,080	Junction Between Cabled and Uncabled Conductors	10 July 1885	13 July 1885	29 Sept. 1885	
Gilliland, Ezra	334,014	Automatic Circuit-Changer	29 Sept. 1885	31 Oct. 1885	12 Jan. 1886	American Bell Telephone Co., Boston
Gilliland, Ezra	335,693	Electric Call-Generator	28 Oct. 1885	31 Oct. 1885	9 Feb. 1886	American Bell Telephone Co., Boston
Gilliland, Ezra	336,562	Electric Signaling Apparatus	28 Oct. 1885	31 Oct. 1885	23 Feb. 1886	American Bell Telephone Co., Boston
Gilliland, Ezra	336,563	Looping-In Switch	5 Oct. 1885	31 Oct. 1885	23 Feb. 1886	American Bell Telephone Co., Boston
Gilliland, Ezra	343,449	Telephone	25 Aug. 1885	14 Sept. 1885	8 June 1886	American Bell Telephone Co., Boston
Gilliland, Ezra	356,197	Telephone-Receiver	21 July 1886	2 Aug. 1886	18 Jan. 1887	
Greenfield, Edwin T.	355,446	Safety-Circuit for Electric Lights	28 June 1886	3 July 1886	4 Jan. 1887	Himself and Sigmund Bergmann

Inventor	Patent	Title	Date 1	Date 2	Date 3	Assignee
Haid, Alfred	321,493	Electric-Lamp Holder	21 Feb. 1885	24 Feb. 1885	7 July 1885	Excelsior Electric Apparatus Co, New York
Hammer, William J.	363,332	Device for Attaching and Detaching Electric Lamps	18 Nov. 1886	24 Nov. 1886	17 May 1887	Himself and Francis R. Upton
Hammer, William J.	363,333	Indicator for Electric-Lighting Systems	10 Dec. 1886	16 Dec. 1886	17 May 1887	Himself and Francis R. Upton
Hammer, William J.	363,334	Electrical Switch	28 Dec. 1886	30 Dec. 1886	17 May 1887	Himself and Francis R. Upton
Holzer, William	356,199	Incandescent Electric Lamp	22 Mar. 1886	24 Mar. 1886	18 Jan. 1887	
Johnson, Edward H.	335,285	Device for Transmitting Power from Electric Motors	3 Feb. 1885	7 Apr. 1885	2 Feb. 1886	
Johnson, Edward H.	360,223	Electrical Apparatus for Heating and Cooling Buildings	6 Oct. 1886	8 Oct. 1886	29 Mar. 1887	
Kenny, Patrick	339,558	Stock-Quotation Telegraph	8 Oct. 1885	15 Oct. 1885	6 Apr. 1886	
Kruesi, John, and John Langton, Jr.	334,708	Underground Electrical Cable	10 June 1885	23 June 1885	19 Jan. 1886	
Kruesi, John, and John Langton, Jr.	334,709	Machine for Making Electrical Cables	10 June 1885	23 June 1885	19 Jan. 1886	
Sprague, Frank J.[2]	315,183	Electro-Dynamic Motor	8 Jan. 1885	19 Jan. 1885	7 Apr. 1885	Sprague Electric Railway and Motor Co.
Sprague, Frank J.	321,147	Electro-Dynamic Motor	24 Feb. 1885	27 Feb. 1885	30 June 1885	Sprague Electric Railway and Motor Co.
Sprague, Frank J.	321,148	Electro-Dynamic Motor	16 Feb. 1885	3 Mar. 1885	30 June 1885	Sprague Electric Railway and Motor Co.
Sprague, Frank J.	321,149	Electric Railway System	19 Mar. 1885	21 Mar. 1885	30 June 1885	Sprague Electric Railway and Motor Co.
Sprague, Frank J.	321,150	Electro-Dynamic Motor	15 May 1885	20 May 1885	30 June 1885	Sprague Electric Railway and Motor Co.
Sprague, Frank J.	323,460	Electro-Dynamic Motor	16 Feb. 1885	12 Mar. 1885	4 Aug. 1885	
Sprague, Frank J.	324,891	Electro-Dynamic Motor	28 Feb. 1885	12 Mar. 1885	25 Aug. 1885	Sprague Electric Railway and Motor Co.
Sprague, Frank J.	324,892	Electric Railway-Motor	23 May 1885	25 May 1885	25 Aug. 1885	Sprague Electric Railway and Motor Co.

Name	Patent No.	Title				Notes
Sprague, Frank J.	328,821	Electric-Railway System	22 Aug. 1885	27 Aug. 1885	20 Oct. 1885	Edison Electric Light Co., New York
Sprague, Frank J.	335,045	System of Electrical Distribution	2 Sept. 1885	19 Sept. 1885	26 Jan. 1886	
Sprague, Frank J.	335,781	Electro-Dynamic Motor	23 Sept. 1885	26 Sept. 1885	9 Feb. 1886	
Sprague, Frank J.	337,793	Electro-Dynamic Motor	2 May 1885	6 July 1885	9 Mar. 1886	
Sprague, Frank J.	337,794	Electro-Dynamic Motor	2 May 1885	6 July 1885	9 Mar. 1886	
Sprague, Frank J.	338,619	Electric Railway	4 May 1885	6 May 1885	23 Mar. 1886	
Sprague, Frank J.	340,684	Electric Railway	7 Nov. 1885	24 Nov. 1885	27 Apr. 1886	
Sprague, Frank J.	340,685	Electric Railway	7 Nov. 1885	24 Nov. 1885	27 Apr. 1886	
Sprague, Frank J.	353,829	Electrical Propulsion of Vehicles	1 June 1886	12 June 1886	7 Dec. 1886	
Stern, William A.	333,975	Municipal Alarm Service	2 Jan. 1885	5 Jan. 1885	5 Jan. 1886	Half to Eugene Ingold
Stern, William A.	358,949	Apparatus for Detecting Leaks in Gas-Pipes	3 July 1886	7 July 1886	8 Mar. 1887	Half to Isidore Coblens
Stieringer, Luther	341,778	Incandescing Electric Lamp	4 Jan. 1886	13 Jan. 1886	11 May 1886	Edison Lamp Co., Harrison, N.J.
Vail, Jonathan H.	331,924	Combined Engine and Dynamo-Electric Machines	25 June 1885	11 July 1885	3 Dec. 1885	
Vail, Jonathan H.	357,050	Lightning-Protector for Electrical Conductors	29 May 1886	2 June 1886	1 Feb. 1887	
Walter, Henry E.	351,544	Dynamo-Electric Machine	29 Apr. 1886	1 May 1886	26 Oct. 1886	Half to Charles Batchelor
Walter, Henry E., and Charles Batchelor	360,258	Dynamo-Electric Machine	1 Nov. 1886	6 Nov. 1886	29 Mar. 1887	
Walter, Henry E., and Charles Batchelor	360,259	Dynamo-Electric Machine	1 Nov. 1886	6 Nov. 1886	29 Mar. 1887	
Wirt, Charles	345,755	Electrical Indicator	21 Sept. 1885	2 Oct. 1885	20 July 1886	
Wirt, Charles	345,756	Electric Battery	21 Sept. 1885	2 Oct. 1885	20 July 1886	

1. Compiled from yearly supplements of the U.S. Patent Office's *Index of Patents Relating to Electricity* (Washington, D.C.: GPO).

2. Frank Sprague is a special case in the context of this list. He was no longer employed by Edison or any of the Edison companies, but the Sprague Electric Railway and Motor Co. (to which he was devoting all his time and talent) enjoyed close personal and financial ties with Edison enterprises (notably through Edward Johnson). It carried out part of its manufacturing business under license from the Edison Electric Light Co., and the Edison Machine Works built and tested a number of its motors.

Bibliography

AAAS Programme. 1885. *Programme of the Thirty-Fourth Meeting of the American Association for the Advancement of Science.* Ann Arbor, Mich.: AAAS.

Abbott, Benjamin Vaughan, comp. 1886. *The Patent Laws of All Nations.* Vol. 2. Washington, D.C.: Charles R. Brodix.

Abernethy, J[ohn]. P[atterson]. 1887. *The Modern Service of Commercial and Railway Telegraphy, in Theory and Practice.* 6th ed. Cleveland: [J. P. Abernethy?].

Acts and Resolutions Adopted by the Legislature of Florida, at its First Session, under the Constitution of A.D. 1885. 1887. Tallahassee: N. M. Bowen.

Acts of the Parliament of the Dominion of Canada. 1880. Sess. 2, Chap. 66. "The Great North-Western Telegraph Company of Canada Act." Ottawa: Brown Chamberlain.

Adams, Stephen B., and Orville R. Butler. 1999. *Manufacturing the Future: A History of Western Electric.* Cambridge: Cambridge University Press.

Albion, Michele Wehrwein. 2008. *The Florida Life of Thomas Edison.* Gainesville: University of Florida Press.

Alglave, Émile and J. Boulard. 1884. *The Electric Light: Its History, Production, and Applications.* Translated by T. O'Conor Sloane. New York: D. Appleton and Co.

Almqvist, Ebbe. 2003. *History of Industrial Gases.* New York: Kluwer Academic/Plenum Publishers.

Alphandéry, Marie-Fernande. 1962. *Dictionnaire des inventeurs français.* Paris: Éditions Seghers.

Alumni Record of Wesleyan University. 1883. Middletown, Conn.: Wesleyan University.

American Electrical Directory. (Printed annually.) Fort Wayne, Ind.: Star Iron Tower Co.

American Missionary Association. 1894. *The Forty-Eighth Annual Report of the American Missionary Association.* New York: American Missionary Association.

Andrews, Frank. 1986. *The Edison Phonograph: The British Connection.* Rugby, U.K.: City of London Phonograph and Gramophone Society.

Annual Register: A Review of Public Events at Home and Abroad, for the Year 1884. 1885. London: Rivingtons.

Anuario del comercio, de la industria, etc. de Venezuela. 1885. Caracas: Rojas Hermanos.

Appel, Toby A. 1987. *The Cuvier-Geoffroy Debate: French Biology in the Decades before Darwin.* New York: Oxford University Press.

Appleton's Dictionary of New York and Its Vicinity. 1889. New York: D. Appleton & Co.

Appleton's General Guide to the United States and Canada. 1882. New York: D. Appleton & Co.

Appleton's Handbook of Winter Resorts. 1884. New York: D. Appleton & Co.

Armstrong, George Elliot. 1896. *Torpedoes and Torpedo-vessels.* London and New York: George Bell & Sons.

Asimov, Isaac. 1982. *Asimov's Biographical Encyclopedia of Science and Technology: The Lives and Achievements of 1510 Great Scientists from Ancient Times to the Present Chronologically Arranged.* Garden City, N.Y.: Doubleday.

Association of Edison Illuminating Companies. 1887. *Minutes of Stated Meetings: With an Appendix, Containing Tables of Load Diagrams, Statistics of Economy, and "The Edison Standard Gauge."* New York: Burgoyne's "Quick" Print.

Atchison, Topeka and Santa Fe Railway Co. 1898. *A Colorado Summer.* Chicago: Atchison, Topeka and Santa Fe.

Atkinson, E[dmund]., trans. and ed. 1883. *Elementary Treatise on Physics, Experimental and Applied.* From *Ganot's Éléments de Physique.* 11th ed. New York: William Wood and Co.

Atkinson, Philip. 1889. *The Elements of Electric Lighting: Including Electric Generation, Measurement, Storage and Distribution.* New York: D. Van Nostrand & Co.

———. 1896. *Elements of Static Electricity.* 2nd ed. New York: W. J. Johnston Co.

Ayrton, W[illiam]. E., and John Perry. 1883. "Electro-Motors and Their Government." *Journal of the Society of Telegraph-Engineers and Electricians* 7 (10 May): 301–72.

Babe, Robert E. 1990. *Telecommunications in Canada: Technology, Industry, and Government.* Toronto: University of Toronto Press.

Bacon, Edwin Munroe. 1886. *Bacon's Dictionary of Boston.* Boston: Houghton, Mifflin and Co.

Baker, Edward C. 1976. *Sir William Preece, F.R.S.: Victorian Engineer Extraordinary.* London: Hutchinson & Co.

Baldwin, Isaac P. (Printed annually.) *Baldwin's Orange Directory: Containing the Names and Business Addresses of All Residents of Orange, East Orange, West Orange, South Orange, Bloomfield and Montclair.* Orange, N.J.: Isaac P. Baldwin.

Baldwin, Neil. 1995. *Edison: Inventing the Century.* New York: Hyperion.

Banker's Almanac and Register. 1881. New York: Bankers Magazine.

Barbour, George M. 1884. *Florida for Tourists, Invalids, and Settlers:*

Containing Practical Information Regarding Climate, Soil, and Produc-tions . . . Routes of Travel, etc., etc. New York: D. Appleton and Co.

Barbour, Lucius Barnes, Lorraine Cook White, and Marie Schlum-brecht Crossley, comps. 2000. *The Barbour Collection of Connecticut Town Vital Records: Middletown—Part I, A–J, 1651–1854.* Baltimore: Genealogical Publishing Co.

Barnard, Frederick A. P. 1870 [1869]. *Machinery and Processes of the Industrial Arts, and Apparatus of the Exact Sciences.* In *Reports of the United States Commissioners to the Paris Universal Exposition, 1867.* Vol. 3. Washington, D.C.: GPO.

Barrère, Albert. 1889. *Dictionary of Slang, Jargon and Cant: Embracing English, American, and Anglo-Indian Slang, Pidgin English, Tinker's Jargon and Other Irregular Phraseology.* London: Ballantyne Press.

Barrett, Richard and Joseph Gross. 1994. *The Illustrated Encyclopedia of Railroad Lighting.* Rochester, N.Y.: Railroad Research Publications.

Barrett, W. F. 1873. "On Certain Remarkable Molecular Changes Oc-curring in Iron Wire at a Low Red Heat." *Philosophical Magazine* 4th ser., 46:472–78.

Barsantee, Harry. 1926. "The Telephone in Wisconsin." *Wisconsin Magazine of History* 10 (Dec.): 150–63.

Bates, Cyrus H. 1908. "Address on the Life and Character of Hon. Ed-ward W. Kinsley." In *History of the Forty-fifth Regiment, Massachu-setts Volunteer Militia,* edited by Albert W. Mann, 444–51. Boston: W. Spooner.

Bates, David Homer. 1907. *Lincoln in the Telegraph Office: Recollections of the United States Military Telegraph Corps During the Civil War.* New York: The Century Co.

Baughman, James P. 1972. *The Mallorys of Mystic: Six Generations in American Maritime Enterprise.* Middletown, Conn.: Wesleyan Uni-versity Press.

Bayles, Richard M. 1887. *History of Richmond County (Staten Island), New York: From Its Discovery to the Present Time.* New York: L. E. Preston.

Bealey, R. 1884. "A Remedy for Sore Heads on Chicks." *The Poultry Keeper* 1 (Nov.): 123.

Beauchamp, Christopher. 2010. "Who Invented the Telephone? Law-yers, Patents, and the Judgments of History." *Technology and Culture* 51 (4): 854–78.

Bedell, Frederick. 1896. "The Evolution of the Transformer." *Proceed-ings of the Electrical Society of Cornell University* 3:33–43.

Beeching, Wilfred A. 1974. *Century of the Typewriter.* New York: St. Martin's Press.

Beers, J. H. & Co., comp. 1906. *Commemorative Biographical Record of the County of Lambton, Ontario.* Toronto: Hill Binding Co.

Bell, Alexander Graham. 1881. "Upon the Production of Sound by Ra-diant Energy." *Electrician* 7 (21 May): 9–13.

Bell, Herbert C. 1891. *History of Northumberland County Pennsylvania.* Chicago: Brown, Runk & Co.

Bender, David A., and Arnold E. Bender. 2001. *Benders' Dictionary of*

Nutrition and Food Technology. 7th ed. Abington, U.K.: Woodhead Publishing, Ltd.

Berger, Daniel. 1910. *History of the Church of the United Brethren in Christ.* Dayton, Ohio: Otterbein Press.

Berly, J. A. (Printed annually.) *J. A. Berly's British, American and Continental Electrical Directory and Advertiser.* London: Wm. Dawson & Sons.

Billington, David P., and David P. Billington, Jr. 2006. *Power, Speed, and Form: Engineers and the Making of the Twentieth Century.* Princeton, N.J.: Princeton University Press.

Biographical Review: This Volume Contains Biographical Sketches of Leading Citizens of Cumberland County, Maine. 1896. Vol. 14. Boston: Biographical Review Publishing Co.

Biographical Review: Containing Life Sketches of Leading Citizens of Stafford and Belknap Counties, New Hampshire. 1897. Vol. 21. Boston: Biographical Review Publishing Co.

Bishop, Farnham. 1918. *The Story of the Submarine.* New York: Century Co.

Black, Robert M. 1983. *The History of Electric Wires and Cables.* London: Peter Peregrinus, Ltd., in association with the Science Museum.

Board, Prudy Taylor. 2006. *Remembering Lee County: Where Winter Spends the Summer.* Charleston, S.C.: The History Press.

Board of Railroad Commissioners [Massachusetts]. 1892. *Annual Report.* Boston: Wright & Potter Printing.

Boase, Frederic. 1892–1921. *Modern English Biography, Containing Many Thousand Concise Memoirs of Persons Who Have Died Since the Year 1850, with an Index of the Most Interesting Matter.* Truro, U.K.: Netherton and Worth.

Bohm, Arnd. 2004. "Goethe and the Romantics." In *The Literature of German Romanticism,* edited by Dennis F. Mahoney, 35–60. Woodbridge, U.K.: Camden House.

Bolton, Sarah Knowles. 1885. *How Success is Won.* Boston: D. Lothrop & Co.

The Book of Discipline of the United Methodist Church. 2008. Nashville, Tenn.: United Methodist Publishing House.

Boston Directory Containing the City Record, A Directory of the Citizens, and Business Directory. (Printed annually.) Boston: Sampson, Murdock, & Co.

Brachner, A. 1985. "German Nineteenth-Century Scientific Instrument Makers." In *Nineteenth-Century Scientific Instruments and Their Makers: Papers Presented at the Fourth Scientific Instrument Symposium, Amsterdam 23–26 October 1984,* edited by P. R. de Clercq, 117–57. Amsterdam: Rodopi.

Braden, Susan R. 2002. *The Architecture of Leisure: The Florida Resort Hotels of Henry Flagler and Henry Plant.* Gainesville: University Press of Florida.

Bradford, R[oyal]. B[ird]. 1882. *Notes on Movable Torpedoes.* Newport, R.I.: U.S. Torpedo Station.

Bramwell, Byrom. 1889. *The Treatment of Pleurisy and Empyema*. Edinburgh: Young J. Pentland.

Brandon, Craig. 1999. *The Electric Chair: An Unnatural American History*. Jefferson, N.C.: McFarland & Co.

Brannt, William T. 1900. *India Rubber, Gutta-Percha, and Balata*. Philadelphia: H. C. Baird & Co.

Bright, Arthur A., Jr. 1972 [1949]. *The Electric Lamp Industry: Technological Change and Economic Development from 1800 to 1947*. New York: Arno Press.

Brin, Arthur, and Leon Quentin Brin. 1884. "The Industrial Production of Oxygen and Nitrogen." *Scientific American* 51 (18 Oct.): 243–44.

Brock, William H. 2008. *William Crookes (1832–1919) and the Commercialization of Science*. Aldershot, U.K.: Ashgate.

Brookes, Martin. 2004. *Extreme Measures: The Dark Visions and Bright Ideas of Francis Galton*. London: Bloomsbury.

Brown, Canter, Jr. 1989. "The International Ocean Telegraph." *Florida Historical Quarterly* 68 (No. 2, Oct.): 135–59.

Brown, John Howard, ed. 1900–1903. *Lamb's Biographical Dictionary of the United States*. Boston: James H. Lamb Co.

Brown, W. Norman, comp. 1926. *Johns Hopkins Half-Century Directory*. Baltimore: Johns Hopkins University.

Browne, Jefferson Beale. 1912. *Key West: The Old and the New*. St. Augustine, Fla.: The Record Co.

Browning, John. 1883. *How to Work with the Spectroscope*. 2nd ed. London: John Browning.

Bruce, Robert V. 1973. *Bell: Alexander Graham Bell and the Conquest of Solitude*. Boston: Little, Brown & Co.

Brush, George J., Edward S. Dana, and James Dwight Dana. 1882. *Appendixes to the Fifth Edition of Dana's Mineralogy*. New York: J. Wiley & Sons.

Buchanan, W. M. 1846. *A Technological Dictionary: Explaining the Terms of the Arts, Sciences, Literature, Professions, and Trades*. London: Printed for W. Tegg.

Bud, Robert, and Deborah Jean Warner, eds. 1998. *Instruments of Science: An Historical Encyclopedia*. New York: Science Museum, London, and National Museum of American History, Smithsonian Institution, in association with Garland Publishing

Burch, George E., and Nicholas P. DePasquale. 1990. *A History of Electrocardiography*. San Francisco: Norman Pub.

Burchfield, Joe D. 1975. *Lord Kelvin and the Age of the Earth*. New York: Science History Publications.

Bush, Vannevar. 1943. "Biographical Memoir of Arthur Edwin Kennelly, 1861–1939." *National Academy of Sciences Biographical Memoir* 22:83–119.

Butler, Frank. 1901. *The Story of Paper-Making: An Account of Paper-Making from Its Earliest Known Record Down to the Present Time*. Chicago: J. W. Butler Paper Co.

Buttrick, John. 1952. "The Inside Contract System." *Journal of Economic History* 12 (no. 3): 205–21.

Buxton, Andrew. 2004. *Cash Carriers in Shops*. Princes Risborough, U.K.: Shire.

Callahan, Edward William, ed. 1901. *List of officers of the Navy of the United States and of the Marine Corps, from 1775 to 1900*. New York: L. R. Hamersly & Co.

Campin, Francis. 1883. *Details of Machinery, Comprising Instructions for the Execution of Various Works in Iron in the Fitting-Shop, Foundry, & Boiler-Yard*. London: Crosby Lockwood and Co.

Canadian Club. 1885. *Constitution and By-Laws of the Canadian Club with a List of its Officers and Members*. New York: A. D. Smith Press.

Cantor, Geoffrey. 1991. *Michael Faraday: Sandemanian and Scientist: A Study of Science and Religion in the Nineteenth Century*. London: Macmillan.

Carlson, W. Bernard. 1991a. *Innovation as a Social Process: Elihu Thomson and the Rise of General Electric*. Cambridge: Cambridge University Press.

———. 1991b. "Building Thomas Edison's Laboratory at West Orange, New Jersey: A Case Study in Using Craft Knowledge for Technical Invention, 1886–1888." *History of Technology* 13: 150–67.

———. 2013. *Tesla: Inventor of the Electrical Age*. Princeton: Princeton University Press.

Carlson, W. Bernard, and William S. Pretzer. 2002. "Thinking and Doing at Menlo Park: Edison's Development of the Telephone, 1876–1878." In *Working at Inventing*, edited by William S. Prezter, 84–99. Baltimore: Johns Hopkins Press.

Caroli, Betty Boyd. 1987. *First Ladies: From Martha Washington to Laura Bush*. New York: Oxford University Press.

Caron's Directory of the City of Louisville. (Printed annually.) Louisville, Ky.: C. K. Caron.

Carosso, Vincent P., and Rose C. Carosso. 1987. *The Morgans: Private International Bankers, 1854–1913*. Cambridge, Mass.: Harvard University Press. [The name of Rose C. Carosso was mistakenly omitted from citations of this work in *TAEB* 5 and 7].

Carpenter, R. C. 1909. "The High-Pressure Fire-Service Pumps of Manhattan Borough, City of New York." *American Society of Mechanical Engineers Transactions* 31 (Sept.): 839–64.

Cassis, Youssef. 2006. *Capitals of Capital: A History of International Financial Centres 1780–2005*. Cambridge: Cambridge University Press.

Chamberlain, Joshua L[awrence]., ed. 1900. *Universities and Their Sons: History, Influence and Characteristics of American Universities*. Vol. 5. Boston: R. Herndon Co.

———, ed. 1901. *University of Pennsylvania: Its History, Influence, Equipment and Characteristics*. Vol. 1. Boston: R. Herndon Co.

Chamberlin, John Edgar. 1969 [1930]. *The Boston Transcript: A History of its First Hundred Years*. Freeport, N.Y.: Books for Libraries Press.

Chandler, Alfred D. 1990. *Scale and Scope: The Dynamics of Industrial Capitalism*. Cambridge, Mass: Harvard University Press.

Chautauqua Historical and Descriptive: A Guide to the Principal Points of Interest on Lake Chautauqua. 1884. Fairbanks, Palmer & Co.

Chernow, Ron. 1997. *Titan: The Life of John D. Rockefeller, Sr.* New York: Random House.

Chokki, Toshiaki. 1988. "Modernization of Technology and Labor in Pre-War Japanese Electrical Machinery Enterprises." *Japanese Yearbook on Business History* 4:26–49.

Clayton, Lawrence A. 1985. *Grace: W. R. Grace & Co., the Formative Years, 1850–1930.* Ottawa, Ill.: Jameson Books.

Cleveland, Rose Elizabeth. 1885. *George Eliot's Poetry, and Other Studies.* New York: Funk & Wagnalls.

Coe, George H. 1919. "How a Great Export Trading House Handles Machinery Sales," *Compressed Air Magazine* 24 (Oct.): 9363–64.

Coe, Lewis. 2003 [1993]. *The Telegraph: A History of Morse's Invention and Its Predecessors in the United States.* Jefferson, N.C.: McFarland & Co., Inc.

Cokayne, George Edward. 1887. *Complete Peerage of England, Scotland, Ireland, Great Britain and the United Kingdom, Extant, Extinct, or Dormant.* 5 vols. London: G. Bell and Sons.

Cole, Arthur Stanley. 1908. *The Scott Family of Shrewsbury, N.J.: Being the Descendants of William Scott and Abigail Tilton Warner.* Red Bank, N.J.: Register Press.

Coleman, William. 1964. *Georges Cuvier, Zoologist: A Study in the History of Evolution Theory.* Cambridge, Mass.: Harvard University Press.

Coles, Roswell S. 1964. "Tantallon, A Staten Island 'Castle.'" *The Staten Island Historian,* old series, 25 (no. 1, Jan.–Mar.): 1–5.

Collins, Wilkie. 1879. *A Rogue's Life: From His Birth to His Marriage.* New York: D. Appleton & Co.

Columbia College. 1888. *Catalogue of the Officers and Graduates of Columbia College (Originally King's College) in the City of New York, 1754–1888.* New York: Columbia College.

Coman, Martha, and Hugh Weir. 1925. "The Most Difficult Husband in America." *Collier's* 76 (18 July): 11.

Connelley, William Elsey. 1919. *A Standard History of Kansas and Kansans.* Chicago: Lewis Publishing Co.

Conot, Robert. 1979. *A Streak of Luck: The Life and Legend of Thomas Alva Edison.* New York: Seaview Books.

Cooke, James Francis. 1910. *Standard History of Music.* Philadelphia: Theodore Presser Co.

Coopersmith, Jennifer. 2010. *Energy, the Subtle Concept: The Discovery of Feynman's Blocks from Leibniz to Einstein.* Oxford: Oxford University Press.

Cowan, Diane F. 1999. "Method for Assessing Relative Abundance, Size Distribution, and Growth of Recently Settled and Early Juvenile Lobsters (Homarus Americanus) in the Lower Intertidal Zone." *Journal of Crustacean Biology* 19:738–51.

Cranston, Maurice. 1991. *The Noble Savage: Jean-Jacques Rousseau, 1754–1762.* Chicago: University of Chicago Press.

Croffut, W[illiam]. A[ugustus]. 1886. *The Vanderbilts and the Story of Their Fortune.* New York: Belford, Clarke & Co.

Crompton, Rookes Evelyn Bell. 1922. "Reminiscences and Experiences . . ." *Proceedings of the Institution of Electrical Engineers* 60 (Apr.): 392–95.

———. 1928. *Reminiscences.* London: Constable & Co., Ltd.

Crookes, William. 1881. *"Radiant Matter": A Resume of the Principal Lectures and Papers of Prof. William Crookes, on the "Fourth State of Matter."* Philadelphia: James W. Queen & Co.

Cross, Charles R. 1886. "Experiments on the Melting Platinum Standard of Light." *Proceedings of the American Academy of Arts and Sciences* 22 (1885): 220–26.

Crossby, P[eter]. A[lfred]., ed. 1881. *Lovell's Gazetteer of British North America.* Montreal: John Lovell & Son.

Crowell, W. F. 1918. "Pioneer Struggle of the Telephone." *State Service* 2 (5): 14–25.

Crowley, Julia M. 1890. *Echoes from Niagara: Historical, Political, Personal.* Buffalo, N.Y.: Charles Wells Moulton.

Csendes, Peter. 1998. *Historical Dictionary of Vienna.* Lanham, Md: Scarecrow Press.

Cudahy, Brian J. 1990. *Over and Back: The History of Ferryboats in New York Harbor.* New York: Fordham University Press.

Current, Richard N. 1954. *The Typewriter and the Men Who Made It.* Urbana: University of Illinois Press.

Curti, Merle, and Vernon Carstensen. 1949. *The University of Wisconsin: A History.* 2 vols. Madison: University of Wisconsin Press.

Cutter, William Richard. 1916. *American Biography: A New Cyclopedia.* New York: Published under the direction of the American Historical Society.

Dalzell, Frederick. 2010. *Engineering Invention: Frank J. Sprague and the U.S. Electrical Industry.* Cambridge, Mass.: MIT Press.

Dana, James Dwight. 1837. *A System of Mineralogy: Including an Extended Treatise on Crystallography. . . .* New Haven: Durrie & Peck and Herrick & Noyes.

Daniel Hovey Association. 1913. *The Hovey Book: Describing the English Ancestry and American Descendants of Daniel Hovey of Ipswich, Massachusetts.* Haverhill, Mass: Press of Lewis R. Hovey.

Daniels, Bettie Marie, and Virginia McConnell. 1964. *The Springs of Manitou.* Denver: Sage Books.

Daniels, Rudolph L. 2000. *Trains Across the Continent: North American Railroad History.* Bloomington, Ind.: Indiana University Press.

Dante Alighieri, and Thomas William Parsons. 1843. *The First Ten Cantos of the Inferno of Dante Alighieri.* Boston: W. D. Ticknor [private printing].

David, Paul A. 1991. "The Hero and the Herd in Technological History: Reflections on Thomas Edison and the Battle of the Systems." In *Favorites of Fortune: Technology, Growth, and Economic Development since the Industrial Revolution,* edited by Patrice Higonnet, David S. Landes, and Henry Rosovsky, 72–119. Cambridge, Mass.: Harvard University Press.

De Secada, C. Alexander G. 1985. "Arms, Guano, and Shipping: The W. R. Grace Interests in Peru, 1865–1885." *Business History Review* 59 (4): 597–621.

Decisions of the Commissioner of Patents for the Year 1883. 1884. Washington, D.C.: GPO.

Decisions of the Commissioner of Patents and of United States Courts in Patent Cases. 1887. Washington, D.C.: GPO.

Dennett, Andrea Stulman. 1997. *Weird and Wonderful: The Dime Museum in America.* New York: New York University Press.

Denny, Mark. 2013. *Lights On! The Science of Power Generation.* Baltimore: Johns Hopkins Press.

Denzel, Markus A. 2010. *Handbook of World Exchange Rates, 1590–1914.* Farnham, Surrey: Ashgate.

Deschanel, A. Privat. 1873. *Elementary Treatise on Natural Philosophy.* Pt. 4. *Sound and Light.* Translated and edited by J. D. Everett. New York: D. Appleton and Co.

Dodge, Charles Richards. 1897. *A Descriptive Catalogue of Useful Fiber Plants of the World, Including the Structural and Economic Classifications of Fibers.* Washington, D.C.: GPO.

Dodson, Pat, ed. 1972. "Cruise of the Minnehaha." *Florida Historical Quarterly* 50 (Apr.): 385–413.

Dorflinger, Don, and Marietta Dorflinger. 1999. *Orange: A Postcard Guide to Its Past.* Charleston, S.C.: Arcadia Publishing.

Doukas, Dimitra. 2003. *Worked Over: The Corporate Sabotage of an American Community.* Ithaca, N.Y.: Cornell University Press.

Downey, Gregory John. 2002. *Telegraph Messenger Boys: Labor, Technology, and Geography, 1850–1950.* New York: Routledge.

Downs, Jim. 2006. "The Other Side of Freedom: Destitution, Disease, and Dependency among Freedwomen and Their Children during and after the Civil War." In *Battle Scars: Gender and Sexuality in the American Civil War,* edited by Catherine Clinton and Nina Silber, 78–103. Oxford: Oxford University Press.

Dredge, James. 1882–1885. *Electric Illumination.* 2 vols. London: Engineering.

Du Moncel, Theodose. 1883. *Electric Lighting.* Translated by Robert Routledge. London: G. Routledge.

Dunbaugh, Edwin L. 1992. *Night Boat to New England, 1815–1900.* New York: Greenwood Press.

Duncan, Andrew. 1819. *The Edinburgh New Dispensatory: Containing I. The Elements of Pharmaceutical Chemistry. II. The Materia Medica. . . . III. The Pharmaceutical Preparations. . . .* Edinburgh: Bell and Bradfute.

Dunn, Jacob Piatt. 1919. *Indiana and Indianans: A History of Aboriginal and Territorial Indiana and the Century of Statehood.* 5 vols. Chicago: American Historical Society.

Duyckinck, Evert A. 1856. "Biographical Memoir." In *Wit and Wisdom of the Rev. Sydney Smith,* by Sydney Smith, 9–104. New York: Redfield.

Dyer, Frank Lewis, and Thomas Commerford Martin. 1910. *Edison: His Life and Inventions.* 2 vols. New York: Harper & Bros.

Dyer, Richard [N.]. 1897. "The Evolution of Electric Lighting Fixtures." *American Gas Light Journal* 67 (13 Sept.): 410–12.

Edgecomb, D. W. 1884. "Telemetry." *Electrician and Electrical Engineer* 3 (December): 265–66.

Edison Electric Light Company vs. F. P. Little Electrical Construction and Supply Company. 1896. Equity Case File 6204, U.S. Circuit Court for the Northern District of New York. New York: C. G. Burgoyne.

Edison, Thomas A. 1886. "The Air Telegraph: System of Telegraphing to Trains and Ships." *North American Review* 142 (Mar.): 285–91. In Pub. Works (*TAED* PA014).

———. 1888a. "On a Magnetic Bridge or Balance for Measuring Magnetic Conductivity." *Proceedings of the American Association for the Advancement of Science* 36 (Mar.): 92–94. In Pub. Works (*TAED* PA015).

———. 1888b. "On a Pyromagnetic Dynamo: A Machine for Producing Electricity Directly from Fuel." *Proceedings of the American Association for the Advancement of Science* 36 (Mar.): 94–98. Alternate version in Cat. 1038:62, Scraps. (*TAED* SM038062).

———, and Dagobert D. Runes. 1948. *The Diary and Sundry Observations of Thomas Alva Edison.* New York: Philosophical Library.

"Edisonia": A Brief History of the Edison Electric Lighting System. 1904. New York: Association of Edison Illuminating Companies.

Emerson, Edgar C. 1898. *Our County and Its People. A Descriptive Work on Jefferson County, New York.* [Boston]: Boston History Co.

Encyclopædia of Chemistry: Theoretical, Practical, and Analytical, as Applied to the Arts and Manufacturers. 1877. Philadelphia: J. B. Lippincott & Co.

Enders, Kathleen L. 2006. *Akron's "Better Half": Women's Clubs and the Humanization of the City, 1825–1925.* Akron, Ohio: University of Akron Press.

England, Joseph W., ed. 1922. *The First Century of the Philadelphia College of Pharmacy, 1821–1921.* [Philadelphia]: Philadelphia College of Pharmacy and Science.

Englund, Steven. 2004. *Napoleon: A Political Life.* New York: Scribner.

Essig, Mark. 2003. *Edison & the Electric Chair: A Story of Light and Death.* New York: Walker Publishing Co.

Fagen, M. D., ed. 1975. *A History of Engineering and Science in the Bell System: The Early Years (1875–1925).* [New York]: Bell Telephone Laboratories, Inc.

Fahie, J[ohn]. J[oseph]. 1900. *A History of Wireless Telegraphy.* 2nd ed. New York: Dodd, Mead, and Co.

———. 1971 [1901]. *History of Broadcasting: Radio to Television.* New York: Arno Press.

Faraday, Michael. 1965 [1855]. *Experimental Researches.* Vol. 3. New York: Dover Publications.

Farwell, Byron. 1985. *Eminent Victorian Soldiers: Seekers of Glory.* New York: W. W. Norton.

Feaster, Patrick. 2007. "Speech Acoustics and the Keyboard Telephone: Rethinking Edison's Discovery of the Phonograph Principle." *Association for Recorded Sound Collections Journal* 38:10–43.

Fennell, C. A. M., ed. 1892. *The Stanford Dictionary of Anglicised Words and Phrases.* Cambridge: University Press.

Fenton, E. Dyne. 1873. "Miss Sprauleverer Poe." In *Eve's Daughters*, 415–41. London: Tinsley Brothers.

Finance and Industry: The New York Stock Exchange. 1886. New York: Historical Publishing Co.

Financial Review: Commerce, Banking, Investments. (Printed annually.) New York: William B. Dana & Co.

Fiske, John. 1940. *The Letters of John Fiske.* Edited by Ethel F. Fiske. New York: Macmillan Co.

Fleming, J[ohn]. A[mbrose]. 1886. *Short Lectures to Electrical Artisans: Being a Course of Experimental Lectures Delivered to a Practical Audience.* London: E. & F. N. Spon.

———. 1892. *The Alternate Current Transformer in Theory and Practice.* 3rd ed.; 2 vols. New York: D. Van Nostrand.

———. 1911. "Telegraphy." *Encyclopaedia Britannica* (11th ed.) 26:510–41. New York: Encyclopaedia Britannica Co.

Florida State Census. 1971 [1885]. *Schedules of the Florida State Census of 1885.* RG 29. Washington, D.C.: National Archives, National Archives Microfilm Publication M845.

Fownes, George, and Henry Watts. 1883. *A Manual of Chemistry.* 13th ed.; vol. 1. London: J. & A. Churchill.

Fouché, Rayvon. 2003. *Black Inventors in the Age of Segregation: Granville T. Woods, Lewis H. Latimer & Shelby J. Davidson.* Baltimore: Johns Hopkins Press.

Franklin Institute. 1885a. *On the Efficiency and Duration of Incandescent Lamps.* Philadelphia: Franklin Institute. Published as a supplement to the September issue of the *Journal of the Franklin Institute*, 120:4–127.

———. 1885b. *Competitive Tests of Dynamo-Electric Machines.* Philadelphia: Franklin Institute. Published as a supplement to the November issue of the *Journal of the Franklin Institute*, 120:5–59.

———. 1885c. *General Report of the Chairman of the Committee on Exhibitions.* Philadelphia: Franklin Institute.

———. 1889. "Report of the Committee on Science and the Arts. The Mimeograph Duplicating System and Apparatus. [No. 1410]." *Journal of the Franklin Institute* 127 (May): 381–83.

———. 1892. *Proceedings of the Electrical Section of the Franklin Institute.* Vol. 1. Philadelphia: Press of Edward Stern.

———. 1914. *Year Book.* Philadelphia: Franklin Institute.

Fresenius, C. Remigius, and Samuel W. Johnson. 1876. *Manual of Qualitative Chemical Analysis.* New York: J. Wiley & Son.

Frey, Heinrich. 1872. *The Microscope and Microscopical Technology.* Translated by George R. Cutter. New York: William Wood & Co.

Friedel, Robert, and Paul Israel. 2010. *Edison's Electric Light: The Art of Invention.* Baltimore: Johns Hopkins Press.

Fritz, Florence. 1949. *Bamboo and Sailing Ships: The Story of Thomas A. Edison and Fort Myers, Florida.* Fort Myers, Fla.

Galiana, Thomas de, and Michel Rival. 1996. *Dictionnaire des inventeurs et inventions.* Paris: Larousse.

Garnet, Robert W. 1985. *The Telephone Enterprise: The Evolution of the*

Bell System's Horizontal Structure, 1876–1909. Baltimore: Johns Hopkins Press.

Gay, Peter. 1986. *The Tender Passion: The Bourgeois Experience, Victoria to Freud.* Oxford: Oxford University Press.

Gellerman, Robert F. 1985. *Gellerman's International Reed Organ Atlas.* Vestal, N.Y: Vestal Press.

Gerry, Elbridge T., Alfred P. Southwick, and Matthew Hale. 1888. *Report of the Commission to Investigate and Report the Most Human and Practical Method of Carrying into Effect the Sentence of Death in Capital Cases.* Albany, N.Y.: Argus Co.

Gibbon, Edward. 1831. *The History of the Decline and Fall of the Roman Empire.* 5th American ed. New York: J. & J. Harper.

Gibbs, James W. 1954. *The Dueber-Hampden Story.* Exeter, N.H.: Adams Brown.

Gibson, Jane Mork. 1980. "The International Electrical Exhibition of 1884: A Landmark for the Electrical Engineer." *IEEE Transactions on Education* E-23 (Aug.): 169–76.

———. 1984. "The International Electrical Exhibition of 1884 and the National Conference of Electricians: A Study in Early Electrical History." M.A. thesis, University of Pennsylvania.

Gillham, Nicholas W. 2001. *A Life of Sir Francis Galton: From African Exploration to the Birth of Eugenics.* New York: Oxford University Press.

Gmelin, Leopold. 1848–71. *Hand-Book of Chemistry.* Translated by Henry Watts. London: Printed for the Cavendish Society.

Goethe, Johann Wolfgang von. 1855. *Wilhelm Meister's Apprenticeship.* Translated by R. Dillon Boylan. London: H. G. Bohn.

Gold, Kenneth M., and Lori R. Weintrob. 2011. *Discovering Staten Island: A 350th Anniversary Commemorative History.* Charleston, S.C.: The History Press.

Gooday, Graeme J. N. 2004. *The Morals of Measurement: Accuracy, Iron, and Trust in Late Victorian Electrical Practice.* Cambridge: Cambridge University Press.

Goodwin, William Watson. 1900. *Memoir of George Martin Lane.* Cambridge, Mass.: J. Wilson and Son.

Gordon, Robert B. 1996. *American Iron, 1607–1900.* Baltimore: Johns Hopkins Press.

Gore, G. 1868. "On the Relation of Mechanical Strain of Iron to Magnetic Electric Induction." *Philosophical Magazine* 36:446–47.

———. 1869. "On the Development of Electric Currents by Magnetism and Heat." *Proceedings of the Royal Society* 17:265–67.

———. 1873. "On the Molecular Movements and Magnetic Changes in Iron at Different Temperatures." *Philosophical Magazine* 46:170–77.

Gould, Joseph E. 1961. *The Chautauqua Movement: An Episode in the Continuing American Revolution.* Albany: State University of New York.

Graham, John. 1961. "Lavater's Physiognomy: A Checklist." *The Papers of the Bibliographic Society of America* 55:297–308.

Graham, Thomas. 2003. *Flagler's St. Augustine Hotels.* Sarasota, Fla.: Pineapple Press.

Gray, John. 1890. *Electrical Influence Machines.* London: Whittaker and Co.

Great Exhibition of the Works of Industry of All Nations. 1851. *Official Descriptive and Illustrated Catalogue.* 3 vols. London: Spicer Bros.

Greene, Welcome Arnold. 1886. *The Providence Plantations for Two Hundred and Fifty Years.* Providence, R.I.: J. A. & R. A. Reid.

Grismer, Karl H. 1949. *The Story of Fort Myers: The History of the Land of the Caloosahatchee and Southwest Florida.* St. Petersburg, Fla.: St. Petersburg Printing Co.

Groft, Tammis Kane. 1984. *Cast with Style: Nineteenth Century Cast-iron Stoves from the Albany Area.* Albany, N.Y.: Albany Institute of History and Art.

Guarini, Emile. 1903. "The Development of Marconi's System of Wireless Telegraphy." *Scientific American Supplement* 55 (25 Aug.): 22,833.

Guillemin, Amédée. 1891. *Electricity and Magnetism.* Translated, revised, and edited by Silvanus P. Thompson. London: Macmillan and Co.

Hackett, Horatio B., and Alvah Hovey. 1882. *A Commentary on the Acts of the Apostles.* Rev. ed. Philadelphia: American Baptist Publication Society.

Hadley, F. B., and B. A. Beach. 1913. "Controlling Chicken Pox, Sorehead, or Contagious Epithelioma by Vaccination." *Proceedings of the American Veterinary Medical Association* 50:704–12.

Hall, A[lexander]. Wilford. 1880 [1877] (20th rev. ed.). *The Problem of Human Life: Embracing the "Evolution of Sound" and "Evolution Evolved," with a Review of the Six Great Modern Scientists.* New York: Hall and Co.

Hall, Edwin H. 1931. *Biographical Memoir of John Trowbridge, 1843–1923.* Washington, D.C.: National Academy of Sciences.

Hall, Henry, ed. 1895. *America's Successful Men of Affairs: An Encyclopedia of Contemporaneous Biography.* New York: New York Tribune.

Hallerberg, Paul John Frederick. 1939. "Charles D. Deshler, Versatile Jerseyman. . . ." M.A. thesis, Rutgers University.

Hammer, W[illiam]. J. 1889. "Edison and His Exhibits at the Paris Exposition. I." *Electrical World* 14 (31 Aug. 1889): 151–53.

Hammond, John Winthrop. 1941. *Men and Volts: The Story of General Electric.* Philadelphia: J. B. Lippincott Co.

Handy, Moses P. 1893. *Official Directory of the World's Columbian Exposition.* Chicago: W. B. Conkey Co.

Harman, P[eter]. M[ichael]. 1982. *Energy, Force, and Matter: The Conceptual Development of Nineteenth-Century Physics.* Cambridge: Cambridge University Press.

Hatch, Alden. 1956. *Remington Arms in American History.* New York: Rinehart & Co.

Hausman, William, Peter Hertner, and Mira Wilkins. 2008. *Global Electrification: Multinational Enterprise and International Finance in* the History of Light and Power, *1878–2007*. Cambridge: Cambridge University Press.

Hawks, Ellison. 1974 [1927]. *Pioneers of Wireless.* New York: Arno Press.

Hayes, P. S. 1872. "Electrical Instruments for Medical Use." *Chicago Medical Journal* 29 (no. 11, November): 665–79.

Hays, Will[iam]. S[hakespeare]. 1895. *Poems and Songs.* Louisville, Ky.: Charles T. Dearing.

Hazell's Annual Cyclopaedia. (Printed annually.) London: Hazell, Watson, and Viney.

Hazeltine, Gilbert W. 1887. *The Early History of the Town of Ellicott, Chautauqua County, N.Y.* Jamestown, N.Y.: Journal Printing Co.

[Head, Joseph H., ed.]. 1884. *Favorite Poems Selected from English and American Authors.* New York: Thomas Y. Crowell & Co.

Heaviside, Oliver. 1894a. "The Energy of the Electric Current. VIb." *Electrical Papers,* 1:297–303. New York: Macmillan. Originally published as "Current Energy.—VI," *Electrician* 11 (11 Aug. 1883): 294–96.

———. 1894b. "Electromagnetic Induction and Its Propagation. I: Rough Sketch of Maxwell's Theory." *Electrical Papers,* 1:429–34. New York: Macmillan. Originally published in *Electrician* 14 (3 Jan. 1885): 148–50.

———. 1894c. "Electromagnetic Induction and Its Propagation. II: On the Transmission of Energy Through Wires by the Electric Current." *Electrical Papers,* 1:434–41. New York: Macmillan. Originally published in *Electrician* 14 (10 Jan. 1885): 178–80.

Hedges, Killingworth. 1879. *Useful Information on Practical Electric Lighting.* London: E. F. N. Spon.

———. 1892. *Continental Electric Light Central Stations; with Notes on the Methods in Actual Practice for Distributing Electricity in Towns.* London and New York: E. & F. N. Spon.

Hedley, James. 1905. "Canadian Celebrities, No. 59—Harvey P. Dwight." *The Canadian Magazine* 24 (Feb.): 312–14.

Heilbron, J. L. 2005. *Oxford Guide to the History of Physics and Astronomy.* Oxford: Oxford University Press.

Heitman, Francis B. 1903. *Historical Register and Dictionary of the United States Army, from Its Organization, September 29, 1789, to March 2, 1903.* Washington, D.C.: GPO.

Helmholtz, Hermann L. F. 1875. *On the Sensations of Tone as a Physiological Basis for the Theory of Music.* Translated by Alexander J. Ellis from the 3rd German edition. London: Longman, Green, and Co.

Hendrick, Ellwood. 1925. *Lewis Miller: A Biographical Essay.* New York: G. P. Putnam's Sons.

Henshall, James A. 1884. *Camping and Cruising in Florida.* Cincinnati: R. Clarke & Co.

Hentschel, Klaus. 2002. *Mapping the Spectrum: Techniques of Visual Representation in Research and Teaching.* Oxford: Oxford University Press.

Hering, Carl. 1886. "Practical Deductions from the Franklin Institute Tests of Dynamos." *Electrician and Electrical Engineer* 5 (Nov.): 423–27.

————. 1893. *Electricity at the Paris Exposition of 1889.* New York: W. J. Johnston Co., Ltd.

Herring, Phillip F. 1895. *Djuna: The Life and Work of Djuna Barnes.* New York: Viking.

Herron, Kristin S[tacey]. 1998. *The House at Glenmont: Edison National Historic Site, West Orange, New Jersey.* 2 vols. West Orange, N.J.: National Park Service.

Herschel, John. 1833. "Observations of Beila's Comet." *Memoirs of the Royal Astronomical Society* 6:99–110.

Hill, George J. 2007. *Edison's Environment: Invention and Pollution in the Career of Thomas Edison.* Morristown, N.J.: New Jersey Heritage Press.

Hill, Mabel. 1903. "Bradford Academy: A Jubilee Sketch." *The New England Magazine* n.s. 28 (May): 345–68.

Hill, Richard A. 1994. "You've Come a Long Way, Dude: A History." *American Speech* 69 (Autumn): 321–27.

Hill, R[oger]. B. 1953. "Early Work on Dial Telephone Systems." *Bell Laboratories Record* 31 (no. 1, Jan.): 22–29.

Hobson, Anthony. 1991. *Lanterns That Lit Our World: How to Identify, Date, and Restore Old Railroad, Marine, Fire, Carriage, Farm, and Other Lanterns.* New York: Golden Hill Press, Inc.

Hochfelder, David. 2012. *The Telegraph in America, 1832–1920.* Baltimore: Johns Hopkins Press.

Hoddeson, Lillian. 1981. "The Emergence of Basic Research in the Bell Telephone System, 1875–1915." *Technology and Culture* 22 (3): 512–44.

Hoke, Donald R. 1990. *Ingenious Yankees: The Rise of the American System of Manufactures in the Private Sector.* New York: Columbia University Press.

Holbrook, A. Stephen. (Printed annually.) *Holbrook's Newark City and Business Directory.* Newark, N.J.: A. Stephen Holbrook. [Cited in *TAEB* 4 and 7 as *Holbrook's Newark City Directory.*]

Holden, Luther L., ed. 1883. *White & Franconia Mountains.* Boston: Boston, Concord, Montreal, and White Mountains Railroad.

Holley, Donald. 2000. *The Second Great Emancipation: The Mechanical Cotton Picker, Black Migration, and How They Shaped the Modern South.* Fayetteville, Ark: University of Arkansas.

Holmes, Oliver Wendell. 1867. *The Guardian Angel.* Boston: Ticknor and Fields.

Holmes, Thom. 2006. *The Routledge Guide to Music Technology.* New York: Routledge.

Homer. *Iliad.* 1796. Translated by Alexander Pope, with additional notes in a new edition by Gilbert Wakefield. London: Baldwin.

Hong, Sungook. 2001. *Wireless: From Marconi's Black-Box to the Audion.* Cambridge, Mass.: MIT Press.

Hooper, S. K., William Abraham Bell, and Stanley Wood. 1890. *The Story of Manitou.* Denver: Denver & Rio Grande Railroad.

Hospitalier, É[douard]. 1883. *The Modern Applications of Electricity.* 2nd ed.; 2 vols. Translated and edited by Julius Maier. London: Kegan Paul, Trench & Co.

Hounshell, David A. 1984. *From the American System to Mass Produc-*

tion, 1800–1932: The Development of Manufacturing Technology in the United States. Baltimore: Johns Hopkins Press.

Houston, Edwin J. 1884. "Notes on Phenomena in Incandescent Lamps." *Transactions of the American Institute of Electrical Engineers* 1:1–8.

———, and Arthur E. Kennelly. 1902. *Electric Arc Lighting.* New York: Electrical World & Engineer.

Houstoun, Matilda Charlotte. 1872. *First in the Field.* London: Hurst and Blackett.

Hovey, William A. *Mind-Reading and Beyond.* Boston: Lee and Shepard.

Howell, George R., and John H. Munsell, et al. 1886. *History of the County of Schenectady, N.Y., from 1662 to 1886.* New York: W. W. Munsell & Co.

Howell, John W., and Henry Schroeder. 1927. *History of the Incandescent Lamp.* Schenectady, N.Y.: The Maqua Co.

Hufbauer, Karl. 1991. *Exploring the Sun: Solar Science since Galileo.* Baltimore: Johns Hopkins Press.

Hughes, Ivor, and David Ellis Evans. 2011. *Before We Went Wireless: David Edward Hughes, FRS: His Life, Inventions, and Discoveries (1829–1900).* Bennington, Vt.: Images from the Past.

Hughes, Thomas Parke. 1962. "British Electrical Industry Lag: 1882–1888." *Technology and Culture* 3 (1): 27–44.

———. 1971. *Elmer Sperry: Inventor and Engineer.* Baltimore: Johns Hopkins Press.

———. 1983. *Networks of Power: Electrification in Western Society, 1880–1930.* Baltimore: Johns Hopkins Press.

Hughes. W. S. 1887. "Modern Aggressive Torpedoes." *Scribner's Magazine* 1 (Apr.): 427–37.

Hunt, Bruce J. 1991. *The Maxwellians.* Ithaca, N.Y.: Cornell University Press.

———. 2010. *Pursuing Power and Light: Technology and Physics from James Watt to Albert Einstein.* Baltimore: Johns Hopkins Press.

Hunter, John. 1870. "Note on the Absorption of Mixed Vapours by Charcoal." *Journal of the Chemical Society* 23 (Mar.): 73–74.

———. 1872. "On the Effects of Temperature on the Absorption of Gases by Charcoal." *Journal of the Chemical Society* 25 (Aug.): 649–51.

Hurst, John F. 1887. "Mars Hill and the Oldest Athens." *The Chautauquan* 7 (Jan.): 226–28.

Hyde, Charles K. 1998. *Copper for America: The United States Copper Industry from Colonial Times to the 1990s.* Tucson: University of Arizona Press.

Indiana. 1883. *Thirty-second Annual Report of the Indiana State Board of Agriculture, Including the Proceedings of the Annual Meeting, 1883.* Indianapolis: Wm. B. Buford.

———. 1885. *Thirty-fourth Annual Report of the Indiana State Board of Agriculture, Including the Proceedings of the Annual Meeting, 1885.* Indianapolis: Wm. B. Buford.

———. 1886. *Thirty-fifth Annual Report of the Indiana State Board of*

Agriculture, Including the Proceedings of the Annual Meeting, 1886. Indianapolis: Wm. B. Buford.

Indianapolis Board of Trade. 1990 [1857]. *A. C. Howard's Directory for the City of Indianapolis: Containing a Correct List of Citizen's Names, Their Residence and Place of Business.* 1857. Indianapolis: Genealogical Society of Marion County.

Insull, Samuel. 1992. *The Memoirs of Samuel Insull: An Autobiography.* Edited by Larry Plachno. Polo, Ill.: Transportation Trails.

International Textbook Co. 1901. *Telegraphy.* Vol. 2 of *International Library of Technology.* Scranton, Pa.: International Textbook Co.

Israel, Paul. 1998. *Edison: A Life of Invention.* New York: John Wiley & Sons.

James, Marquis. 1993. *Merchant Adventurer: The Story of W. R. Grace.* Wilmington, Del.: SR Books.

Jeffrey, Thomas E. 2008. *From Phonographs to U-Boats: Edison and His "Insomnia Squad" in Peace and War, 1911–1919.* Bethesda, Md.: LexisNexis.

Jehl, Francis. 1937–41. *Menlo Park Reminiscences.* 3 vols. Dearborn, Mich.: Edison Institute.

Jenks, W[illiam]. J. 1887. "The Development and Future of the Edison Municipal System: A Paper Read before the Association of Edison Illuminating Companies" [10 Aug.]. Republished in *Edison System of Electric Railways,* 171–90. 1891. New York: Edison General Electric Co.

Jensen, William B. 2009. "The Origin of the Brin Process for the Manufacture of Oxygen." *Journal of Chemical Education* 86 (11): 1266–67.

"John Heyl Vincent Papers: A Guide to the Collection." [n.d.]. Dallas, Tex.: Southern Methodist University (http://www.lib.utexas.edu/taro/smu/00143/smu-00143.html [accessed 16 May 2013]).

Johns Hopkins University. (Printed annually.) *Annual Report of the Johns Hopkins University.* Baltimore: Johns Hopkins University.

———. 1914. *Graduates and Fellows of the Johns Hopkins University, 1876–1913.* Baltimore: Johns Hopkins Press.

Johnson, Rossiter, and John Howard Brown. 1904. *The Twentieth Century Biographical Dictionary of Notable Americans.* Boston: Biographical Society.

Johnson's Steam Vessels of the Atlantic Coast. (Printed annually.) New York: Eads Johnson Publishing Co.

Jones, Francis Arthur. 1907. *Thomas Alva Edison: Sixty Years of an Inventor's Life.* New York: Thomas Y. Crowell & Co.

Jones, Payson. 1940. *A Power History of the Consolidated Edison System, 1878–1900.* New York: Consolidated Edison Co. of New York.

Jonnes, Jill. 2003. *Empires of Light: Edison, Tesla, Westinghouse, and the Race to Electrify the World.* New York: Random House.

Jordan, John W. 1978 [1911]. *Colonial and Revolutionary Families of Pennsylvania: Genealogical and Personal Memoirs.* 3 vols. Baltimore: Genealogical Publishing Co.

Josephson, Matthew. 1959. *Edison: A Biography.* New York: McGraw-Hill.

Juergens, George. 1966. *Joseph Pulitzer and the New York World*. Princeton, N.J.: Princeton University Press.

Jussen, Edmund. 1886. "Electric Lighting in Vienna." In *Reports from the Consuls of the United States*, 477–80. Washington, D.C.: GPO.

Kales, David. 2007. *The Boston Harbor Islands: A History of Urban Wilderness*. Charleston, S.C.: History Press.

Karl Baedeker. 1882. *Norway and Sweden: Handbook for Travellers*. Leipsic: Karl Baedeker.

Kennedy, Rankin. 1887. "Electrical Distribution by Alternating Currents and Transformers." [3 parts]. *Electrician and Electrical Engineer* (June, July, Aug.): 187–89, 257–62, 299–302.

Kenney, Kimberly A. 2003. *Canton: A Journey Through Time*. Portsmouth, N.H.: Arcadia Publishing.

Kenny, D. J. 1879. *Cincinnati Illustrated: A Pictorial Guide to Cincinnati and the Suburbs*. Cincinnati: Robert Clarke & Co.

King, Moses, ed. 1893. *King's Handbook of New York City: An Outline History and Description of the American Metropolis*. Boston: Moses King.

Kingsbury, J. D. 1885. *A Historical Sketch of Harriette Briggs Stoddard: Bradford Academy*. Lawrence, Mass.: American Printing House.

Kleber, John E., ed. 2001. *The Encyclopedia of Louisville*. Louisville: University Press of Kentucky.

Klein, Maury. 2008. *The Power Makers: Steam, Electricity, and the Men Who Invented Modern America*. New York: Bloomsbury Press.

Kobbé, Gustav. 1891. *New York and Its Environs*. N.Y.: Harper & Bros.

Kobrak, Christopher. 2007. *Banking on Global Markets: Deutsche Bank and the United States, 1870 to the Present*. Cambridge: Cambridge University Press.

Koenigsberg, Allen. 1990. *The Patent History of the Phonograph, 1877–1912: A Source Book*. . . . Brooklyn, N.Y.: APM Press.

Kohlrausch, Friedrich Wilhelm Georg. 1883. *An Introduction to Physical Measurements: With Appendices on Absolute Electrical Measurement, Etc.* London: Churchill.

Kolbe, Hermann, and T. S. Humpidge. 1884. *A Short Text-Book of Inorganic Chemistry*. New York: J. Wiley & Sons.

Krass, Peter. 2007. *Ignorance, Confidence, and Filthy Rich Friends: The Business Adventures of Mark Twain, Chronic Speculator and Entrepreneur*. New York: John Wiley & Sons.

Krauth, Charles P. 1878. *A Vocabulary of the Philosophical Sciences*. New York: Sheldon & Co.

Krupp, A., and Andreas Wildberger. 1889. *The Metallic Alloys. A Practical Guide for the Manufacture of All Kinds of Alloys, Amalgams, and Solders, Used by Metal Workers*. . . . Translated and edited by William T. Brannt. Philadelphia: Henry Carey Baird & Co.

La Rochefoucauld, François. 2001. *Maxims*. Edited and translated by Stuart D. Warner and Stéphane Douard. South Bend, Ind.: St. Augustine's Press.

Lamar, William B., reporter. 1896. *Cases Argued and Adjudicated in the Supreme Court of Florida, at the January Term, A.D. 1896*. Tallahassee, Fla.: Tallahasseean Book and Job Office.

Landau, Sarah Bradford, and Carl W. Condit. 1996. *Rise of the New York Skyscraper, 1865–1913.* New Haven: Yale University Press.

Lane, Samuel A. 1892. *Fifty Years and Over of Akron and Summit County.* Akron, Ohio: Beacon Job Dept.

Lanthier, Pierre. 1988. "Les Constructions Électriques en France: Financement et Stratégies de Six Groupes Industriels Internationaux, 1880–1940." Ph.D. diss. [3 vols.], University of Paris X (Nanterre).

Lathrop, Henry Burrowes. 1904. *The University of Wisconsin: A Study of Higher Education by the State.* Madison: Louisiana Purchase Exposition.

Lee, Alfred E. 1892. *History of the City of Columbus, Capital of Ohio.* New York: Munsell & Co.

Lee, Alfred McClung. 1973. *The Daily Newspaper in America: The Evolution of a Social Instrument.* New York: Octagon Books.

Leng, Charles W., and William T. Davis. 1930. *Staten Island and Its People: A History, 1609–1929.* New York: Lewis Historical Publishing Co.

Lenormant, Amélie. 1867. *Madame Récamier and Her Friends.* Translated by Isaphene M. Luyster. Boston: Knight and Millet.

Lerer, Seth. 2003. "Hello, Dude: Philology, Performance, and Technology in Mark Twain's 'Connecticut Yankee.'" *American Literary History* 15 (Autumn): 471–503.

Leupp, Francis E. 1918. *George Westinghouse: His Life and Achievements.* Boston: Little, Brown, and Co.

Lilienberg, Nils. 1915. "The Properties of Swedish Wrought Iron." *Iron Age* 95 (8 Apr.): 788–89.

Lindley, David. 2004. *Degrees Kelvin: A Tale of Genius, Invention and Tragedy.* London: Aurum Press.

Lindquist, Charles N. 2004. *Adrian: The City That Worked: A History of Adrian, Michigan, 1825–2000.* Adrian, Mich: Lenawee County Historical Society.

Linebaugh, Donald W. *The Springfield Gas Machine: Illuminating Industry and Leisure, 1860s–1920s.* Knoxville: University of Tennessee Press.

Lockwood, Thomas Dixon. 1883. *Electricity, Magnetism, and Electric Telegraphy: A Practical Guide and Hand-book of General Information for Electrical Students, Operators, and Inspectors.* New York: D. Van Nostrand.

Lockyer, Norman. 1873. *The Spectroscope and Its Applications.* London: Macmillan.

Lomax, Michael E. 2003. *Black Baseball Entrepreneurs, 1860–1901: Operating by Any Means Necessary.* Syracuse, N.Y.: Syracuse University Press.

Longfellow, Henry W. 1848. *Hyperion: A Romance.* London: H. D. Clarke.

Loomis, Elias. 1870. *A Treatise on Astronomy.* New York: Harper & Brothers.

Lyman, Theodore. 1925. "John Trowbridge (1843–1923)." *Proceedings of the American Academy of Arts and Sciences* 60:651–54.

MacKelvey, Blake. 1995. *Snow in the Cities: A History of America's Urban Response.* Rochester, N.Y.: University of Rochester Press.

MacLaren, Malcolm. 1943. *The Rise of the Electrical Industry During the Twentieth Century.* Princeton, N.J.: Princeton University Press.

MacLeod, Christine, and Alessandro Nuvolari. 2010. "Patents and Industrialization: An Historical Overview of the British Case, 1624–1907." *Report to the Strategic Advisory Board for Intellectual Property Policy (SABIP).* London: Intellectual Property Office.

McLeod, H. 1874. "Apparatus for Measurement of Low Pressure of Gas." *Proceedings of the Physical Society* 1:30–34.

Maguire, Edward. 1884. *The Attack and Defence of Coast-Fortifications.* New York: D. Van Nostrand.

Maguire, Frank Z. 1886. "The Graphophone." *Harper's Weekly* 30 (17 July): 458–59.

Mahon, Basil. 2009. *Oliver Heaviside: Maverick Mastermind of Electricity.* IET History of Technology Series 36. London: Institution of Engineering and Technology.

Maier, Julius. 1886. *Arc and Glow Lamps: A Practical Handbook on Electric Lighting.* London: Whittaker & Co.

Maltbie, Milo R[oy]. 1911. *Franchises of Electrical Corporations in Greater New York: A Report Submitted to the Public Service Commission for the First District.* New York: Public Service Commission for the First District.

Mansfield, E. S. 1901. "The Edison System in Boston—Its Development and Present Status." *Electrical World and Engineer* 37 (18 May): 797–822.

Manufacturing and Mercantile Resources of Indianapolis, Indiana: A Review of its Manufacturing, Mercantile & General Business Interests, Advantageous Location, &c. 1883. N.p.

Maril, Nadja. 1989. *American Lighting, 1840–1940.* West Chester, Pa: Schiffer Publishing.

Marshall, David Trumbull. 1931. *Recollections of Edison.* Boston: Christopher Publishing House.

Martland, Peter. 2013. *Recording History: the British Record Industry, 1888–1931.* Lanham, Md.: Scarecrow Press.

Martin, S. Walter. 2010. *Florida's Flagler.* Athens: University of Georgia.

Martin, Thomas C. 1922. *Forty Years of Central Station Service, 1882–1922.* New York: New York Edison Co.

———, and Joseph Wetzler. 1888. *The Electric Motor and Its Applications.* New York: W. J. Johnston.

Mason, E. A. 1991. "From Pig Bladders and Cracked Jars to Polysulfones: an Historical Perspective on Membrane Transport." *Journal of Membrane Science* 60:125–45.

Mason, Edward T., ed. 1884. *Personal Traits of British Authors.* New York: Charles Scribner's Sons.

Maver, William, Jr. 1892. *American Telegraphy: Systems, Apparatus, Operation.* New York: J. H. Bunnell & Co.

———. 1904. *Maver's Wireless Telegraphy: Theory and Practice.* New York: Maver Publishing Co.

———, and Minor M. Davis. 1890. *The Quadruplex*. New York: W. J. Johnston Co.

Maxwell, James Clerk, Elizabeth Garber, Stephen G. Brush, and C. W. F. Everitt. 1986. *Maxwell on Molecules and Gases*. Cambridge, Mass.: MIT Press.

McCagg, William O. 1992. "Jewish Wealth in Vienna, 1670–1918." In *Jews in the Hungarian Economy, 1760–1945: Studies Dedicated to Moshe Carmilly-Weinberger on His Eightieth Birthday*, edited by Moshe Carmilly-Weinberger and Michael K. Silber, 53–91. Jerusalem: Magnes Press.

McClure, J[ames]. B[aird]. 1889. *Edison and His Inventions*. Chicago: Rhodes and McClure Publishing Co.

McGaha, Ruth K. 1990. "The Journals of S. Elizabeth Dusenbury, 1852–1857: Portrait of a Teacher's Development." Ph.D. diss., Iowa State Univ.

McGucken, William. 1969. *Nineteenth-century Spectroscopy: Development of the Understanding of Spectra, 1802–1897*. Baltimore: Johns Hopkins Press.

McIver, Stuart B. 1994. *Dreamers, Schemers, and Scalawags*. Sarasota, Fla.: Pineapple Press.

McPartland, Donald Scott. 2006. "Almost Edison: How William Sawyer and Others Lost the Race to Electrification." Ph.D. diss., City University of New York.

Medical Directory of the City of New York. 1887. New York: Medical Society of the County of New York.

Meikle, Jeffrey L. 1995. *American Plastic: A Cultural History*. New Brunswick, N.J.: Rutgers University Press.

Mellor, Joseph William. 1822. *A Comprehensive Treatise on Inorganic and Theoretical Chemistry*. London: Longmans, Green and Co.

Mendenhall, Thomas Corwin. 1886. "On the Electrical Resistance of Soft Carbon under Pressure." *American Journal of Science*, 3rd ser., 32:218–23.

Mercantile Publishing Co. 1889. *Leading Business Men of Lewiston, Augusta and Vicinity*. Boston: Mercantile Publishing Co.

Meredith, Roy. 1974. *Mr. Lincoln's Camera Man: Mathew B. Brady*. New York: Dover Publications.

Merrill, Walter McIntosh, and Louis Ruchames, eds. 1971–1981. *The Letters of William Lloyd Garrison*. 6 vols. Cambridge, Mass.: Harvard University Press.

Millard, Andre. 1990. *Edison and the Business of Innovation*. Baltimore: Johns Hopkins Press.

———, Duncan Hay, and Mary Grassick. 1995. *Historic Furnishings Report: Edison Laboratory*. 2 vols. Harpers Ferry, Va.: National Park Service.

Miller, Raymond C. 1957. *Kilowatts at Work: A History of the Detroit Edison Company*. Detroit: Wayne State University Press.

Miller, William Allen. 1855. *Elements of Chemistry: Theoretical and Practical*. London: J. W. Parker and Son.

———. 1878. *Elements of Chemistry: Theoretical and Practical. Part II.*

Inorganic Chemistry. Revised by Charles E. Groves. London: Longmans, Green, Reader, and Dyer.

Mines, Samuel. 1978. *Pfizer . . . An Informal History.* New York: Pfizer Inc.

Minutes and Reports of the Boards of Commissioners of Electrical Subways in the City of Brooklyn, 1885–1889, and 1892–1896. 1896. New York: Martin B. Brown Press.

Mitchell, F. 1883. "Cape Cod." *Century Magazine* 26 (Sept.): 657–58.

Mitchell, James J. 1891. *Detroit in History and Commerce: A Careful Compilation of the History, Mercantile and Manufacturing Interest of Detroit. . . .* Detroit: Rogers & Thorpe.

Moe, Harald. 2004. *The Story of the Microscope.* Translated by David Stoner. [London]: Rhodos.

Molinari, Ettore, and Thomas H. Pope. 1920. *Treatise on General and Industrial Inorganic Chemistry.* Philadelphia: Blakiston.

Moran, Richard. 2002. *Executioner's Current: Thomas Edison, George Westinghouse, and the Invention of the Electric Chair.* New York: Alfred A. Knopf.

Morrison, Theodore. 1974. *Chautauqua: A Center for Education, Religion, and the Arts in America.* Chicago: University of Chicago Press.

Mortimer, John. 2005. *Zerah Colburn, The Spirit of Darkness.* Suffolk, U.K.: Arima Publishing.

Mossman, Susan. 1997. "Perspectives on the History and Technology of Plastics." In *Early Plastics: Perspectives, 1850–1950,* edited by Susan Mossman, 15–71. London: Leicester University Press.

Mott, Frank Luther. 1941. *American Journalism: A History of Newspapers in the United States through 250 Years, 1690 to 1940.* New York: Macmillan Co.

Mott, Henry Augustus. 1885. *The Fallacy of the Present Theory of Sound.* New York: J. Wiley & Sons.

Moyer, Albert E. 1983. *American Physics in Transition: A History of Conceptual Change in the Late Nineteenth Century.* Los Angeles: Tomash Publishers.

Muir, Ross L., and Carl J. White. 1964. *Over the Long Term: The Story of J. & W. Seligman & Co.* New York: J. & W. Seligman.

Murray, John. 1872. *A Handbook for Travellers in Greece.* 4th ed. London: John Murray.

Nahin, Paul J. 2002. *Oliver Heaviside: The Life, Work, and Times of an Electrical Genius of the Victorian Age.* Baltimore: Johns Hopkins Press.

Nason, Elias. 1878. *A Gazetteer of the State of Massachusetts.* Boston: B. B. Russell.

Nason, Henry B., ed. 1887. *Biographical Record of the Officers and Graduates of the Rensselaer Polytechnic Institute, 1824–1886.* Troy, N.Y.: William H. Young.

National Academy of Sciences (U.S.). 1888. *Report of the National Academy of Sciences.* Washington, D.C.: GPO.

National Civic Federation. 1907. *Municipal and Private Operation of Public Utilities: Report to the National Civic Federation Commission on Public Ownership and Operation.* New York: National Civic Federation.

National Electric Light Association. (Printed Annually.) *Convention.* N.p.

National Publishing Co. 1883. *Commerce, Manufactures and Resources of Boston, Mass.: A Historical, Statistical & Descriptive Review.* [Boston]: National Publishing Co., Ltd.

Nelson, Richard R. 1993. *National Innovation Systems: A Comparative Analysis.* New York: Oxford University Press.

Nerney, Mary Childs. 1934. *Thomas A. Edison: A Modern Olympian.* New York: H. Smith and R. Haas.

"New Brighton in the Guilded [sic] Age: Social Life at 'Tantallon.'" *Staten Island Historian,* n.s.,1 (1983): 15.

New Jersey Dept. of State. 1914. *Corporations of New Jersey: List of Certificates to December 31, 1911.* Trenton: MacCrellish & Quigley.

New York Edison Co. 1913. *Thirty Years of New York, 1882–1912: Being a History of Electrical Development in Manhattan and the Bronx.* New York: New York Edison Co.

New York State Senate. 1885. *Documents of the Senate of the State of New York.* Vol. 2 no. 8. Albany: Weed, Parsons and Co.

New York's Great Industries. Exchange and Commercial Review, Embracing also Historical and Descriptive sketch of the City. . . . 1885. New York: Historical Publishing Co.

Newell, William Wells. 1883. *Games and Songs of American Children.* New York: Harper & Brothers.

Niaudet, Alfred. 1884. *Elementary Treatise on Electric Batteries.* Translated by L. M. Fishback. New York: John Wiley and Sons.

Nishikawa, Makoto. 2002. "Sasaki Takayuki to Kōbushō." In *Kōbushō to Sono Jidai,* edited by Jun Suzuki, 229–60. Tokyo: Yamakawa Shuppan.

Noad, Henry M. and W[illiam]. H. Preece. 1879. *The Student's Text-Book of Electricity. A New Edition, Carefully Revised.* London: Crosby Lockwood and Co.

Nörgaard, Victor A. 1916. "Division of Animal Husbandry." *Hawaiian Forester and Agriculturalist* 13 (May): 152–55.

Normand, J. A. 1885. "A Study of Torpedo-Boats." In *Papers on Naval Operations During the Year Ending July, 1885,* 147–63. Translated by W. L. Rogers. Washington, D.C.: Navy Dept. Office of Naval Intelligence.

North, John. 2008. *Cosmos: An Illustrated History of Astronomy and Cosmology.* Chicago: University of Chicago Press.

Norton, C[harles]. B. 1883. *Official Catalogue: Foreign Exhibition.* Boston: George Coolidge.

Norton, Charles Ledyard. 1892. *A Handbook of Florida.* New York: Longmans, Green, & Co.

Nothnagel, Hermann, and Michael Joseph Rossbach. 1884. *A Treatise on Materia Medica. (Including Therapeutics and Toxicology).* Translated from the 4th edition by H. N. Heineman and H. W. Berg. New York: Bermingham & Co.

Nursey, Perry F. 1885. "Modern Bronze Alloys for Engineering Purposes." In *Society of Engineers (London) Transactions for 1884,* 127–66. London: E. & F. N. Spon.

Nye, Bill. 1884. "The True Story of Damon and Pythias." In *Bill Nye and Boomerang; or, the Tale of a Meek-Eyed Mule*, 13–17. Chicago: Belford, Clarke & Co.

Obenzinger, Hilton. 2005. "Going to Tom's Hell in *Huckleberry Finn*." In *A Companion to Mark Twain*, ed. by Peter Messent and Louis J. Budd, 401–15. Malden, Mass.: Blackwell Publishing.

Oberhuber, Konrad. 2001. "Raphael's Vision of Women." In *Raphael: Grace and Beauty*, edited by Patrizia Nitti, Marc Restellini, et. al., 39–55. Milan: Skira.

Odagiri, Hiroyuki, and Akira Goto. 1996. *Technology and Industrial Development in Japan: Building Capabilities by Learning, Innovation, and Public Policy*. Oxford: Clarendon Press.

O'Driscoll, Florence. 1889. *Notes on the Treatment of Gold Ores*. London: Offices of "Engineering."

Oeser, Marion Edison. [1956]. "The Wizard of Menlo Park." CEFC (*TAED* Xo18A5Z). A variant typescript, dated 1956 and titled "Early Recollections of Mrs. Marion Edison Oser," is in Box 16, Edison Biographical Collection, NjWOE. Incorrectly spelled "Oser" in *TAEB* 7 citations and bibliography.

Okamura, Sōgo. 1994. *History of Electron Tubes*. Tokyo: Ohmsha.

Olegario, Rowena. 2006. *A Culture of Credit: Embedding Trust and Transparency in American Business*. Cambridge, Mass.: Harvard University Press.

Oliver, Thomas. 1902. *Dangerous Trades: The Historical, Social, and Legal Aspects of Industrial Occupations as Affecting Health, by a Number of Experts*. London: J. Murray.

O'Neill, John J. 1944. *Prodigal Genius: The Life of Nikola Tesla*. New York: I. Washburn.

Orcutt, Samuel. 1886. *A History of the Old Town of Stratford and the City of Bridgeport, Connecticut*. Part 2. New Haven: Fairfield County Historical Society.

Ord-Hume, Arthur W. J. G. 1986. *Harmonium: The History of the Reed Organ and Its Makers*. London: David & Charles.

———. 2005. *Perpetual Motion: the History of an Obsession*. Kempton, Ill.: Adventures Unlimited Press.

Parker, Leonard F. 1893. *Higher Education in Iowa*. Washington, D.C.: GPO.

Parker, Richard Green. 1862. A School Compendium of Natural and Experimental Philosophy. New York: Collins & Brothers.

Partridge, Eric. 1984. *A Dictionary of Slang and Unconventional English*. 8th ed. Edited by Paul Beale. New York: Macmillan.

Passenger Lists of Vessels Arriving at New York, 1820–97. 1962. Washington, D.C.: National Archives Microfilm Publication M237. Accessed through Ancestry.com.

Passer, Harold C. 1972 [1953]. *The Electrical Manufacturers, 1875–1900: A Study in Competition, Entrepreneurship, Technical Change, and Economic Growth*. Cambridge, Mass.: Harvard University Press.

Payen, Anselme, and B. H. Paul. 1878. *Industrial Chemistry: A Manual for Use in Technical Colleges or Schools and for Manufacturers &c. Based upon a Translation (partly by Dr. T. D. Barry) of Stohmann and*

Engler's German edition of Payen's 'Précis de chimie Industrielle.' London: Longmans, Green, and Co.

Pennsylvania Railroad Co. 1881. *Summer Excursion Routes: Season of 1881*. Philadelphia: Pennsylvania Railroad.

The People's Telephone Company et al., Appellants, vs. the American Bell Telephone Company et al., Appellees; the Overland Telephone Company et al., Appellants, vs. the American Bell Telephone Company et al., Appellees. Appellants' Supplemental Brief. 1886. Chicago: Chicago Legal News Co.

Perry, Calbraith B. 1987 [1902]. *Charles D'Wolf of Guadaloupe, His Ancestors and Descendants: Being a Complete Genealogy of the "Rhode Island D'Wolfs"*. . . . Salem, Mass.: Higginson Genealogical Books.

Petroski, Henry. 2008. *The Toothpick: Technology and Culture*. New York: Vintage Books.

Pettengill, George W. Jr. 1952. *The Story of the Florida Railroads, 1834–1903*. Boston: The Railway & Locomotive Historical Society, Inc.

Philadelphia. 1887. *Ordinances of the City of Philadelphia, From January 1 to December 31, 1886, and Opinions of the City Solicitor*. Philadelphia: Dunlap & Clark, Printers.

Phillips, Morris. 1891. *Abroad and at Home: Practical Hints for Tourists*. New York: Brentano's.

Pierce, Bessie Louise. 1957. *A History of Chicago*. Vol. 3, *The Rise of a Modern City, 1871–1893*. Chicago: University of Chicago Press.

Pierce, Frederick Clifton. 1901. *Field Genealogy*. Chicago: Hammond Press.

Pierson, David Lawrence. 1922. *History of the Oranges to 1921*. New York: Lewis Historical Publishing Co.

Pitkin, Thomas M. 1973. *The Captain Departs: Ulysses S. Grant's Last Campaign*. Carbondale: Southern Illinois Univesity Press.

Plunz, Richard. 1990. *A History of Housing in New York City: Dwelling Type and Social Change in the American Metropolis*. New York: Columbia University Press.

Polk, R. L. & Co. (Printed annually.) *R. L. Polk & Co.'s Indianapolis City Directory*. Indianapolis: R. L. Polk & Co.

Pond, Jean Sarah. 1930. *Bradford: A New England Academy*. Bradford, Mass.: Bradford Academy Alumnae Association.

Poor's Manual of the Railroads of the United States. (Printed annually.) New York: H. V. & H. W. Poor.

Pope, Franklin Leonard. 1889. *Evolution of the Electric Incandescent Lamp*. Elizabeth, N.J.: Henry Cook.

Porter, Charles T. 1908. *Engineering Reminiscences Contributed to "Power" and "American Machinist."* New York: John Wiley & Sons.

Poulos, Terrence. 2004. *Extreme War: The Biggest, Best, and Worst in Warfare*. New York: Citadel Press.

Pratt, William A. 1879. *The Yachtman and Coaster's Book of Reference*. Hartford, Conn.: Case, Lockwood & Brainard Co.

Preece, W[illiam]. H[enry]. 1878. "The Telephone and Its Application to Military and Naval Purposes." *Journal of the Royal United Services Institute for Defence* 12 (94): 209–16.

———. 1884. "The Watt and Horse-Power. *Electrician* 13 (4 Oct.): 473.

————. 1885. "On a Peculiar Behavior of Glow–Lamps When Raised to High Incandescence." *Proceedings of the Royal Society of London* 38 (Mar.): 219–30.

————. 1886. "Long-Distance Telephony." *Journal of the Society of Telegraph-Engineers and Electricians* 15 (13 May): 274–303.

Preis, S. 2007. "History of Ozone Synthesis and Use for Water Treatment." In *Ozone Science and Technology,* edited by Rein Munter, in *Encyclopedia of Life Support Systems (EOLSS),* developed under the Auspices of the UNESCO. Eolss Publishers, Oxford [http://www.eolss.net].

Prescott, George B. 1877. *Electricity and the Electric Telegraph.* New York: D. Appleton and Co.

————. 1878. *The Speaking Telephone, Talking Phonograph and Other Novelties.* New York: D. Appleton & Co.

————. 1879. *The Speaking Telephone, Electric Light, and Other Recent Electrical Inventions.* New York: D. Appleton & Co.

————. 1884. *Dynamo Electricity: Its Generation, Application, Transmission, Storage and Measurement.* New York: D. Appleton & Co.

————. 1892. *Electricity and the Electric Telegraph.* 8th ed. New York: D. Appleton and Co.

————. 1972 [1884]. *Bell's Electric Speaking Telephone: Its Invention, Construction, Application, Modification and History.* New York: Arno Press.

Printer, The: A Description of His Business and All that Pertains to It. 1884. London: Houlston and Sons.

Pritchard, O. G. 1890. *The Manufacture of Electric Light Carbons.* London: "The Electrician" Printing and Publishing Co.

Prominent Men of Staten Island. 1893. New York: A. Y. Hubbell.

Prout, Henry G. 1921. *A Life of George Westinghouse.* New York: American Society of Mechanical Engineers.

Putnam, J. Pickering. 1886. *The Open Fireplace in All Ages.* Boston: Ticknor.

Quain, Richard. 1883. *A Dictionary of Medicine, Including General Pathology, General Therapeutics, Hygiene, and the Diseases Peculiar to Women and Children.* New York: D. Appleton and Co.

Quinn, Terry. 2012. *From Artefacts to Atoms: The BIPM and the Search for the Ultimate Measurement Standards.* Oxford: Oxford University Press.

Rammelkamp, Julian S. 1967. *Pulitzer's Post-Dispatch, 1878–1883.* Princeton, N.J.: Princeton University Press.

Rapalje, Stewart, and Robert L. Lawrence. 1888. *Dictionary of American and English Law with Definitions of the Technical Terms of the Canon and Civil Laws. . . .* Jersey City, N.J.: F. D. Linn & Co.

Rasmussen, R. Kent. 2007. *Critical Companion to Mark Twain: A Literary Reference to His Life and Work.* New York: Facts on File, Inc.

Rathbun, Richard. 1886. "Notes on Lobster Culture." *Bulletin of the United States Fish Commission* 6 (8 Feb.): 17–32.

Récamier, Jeanne Françoise Julie Adélaïde Bernard. 1867. *Memoirs and Correspondence of Madame Récamier.* Translated and edited by Isaphene M. Luyster and Amélie Lenormant. Boston: Roberts Bros.

Registrations of Deaths, 1869–1936. Toronto: Archives of Ontario. MS 935; 576 reels. Accessed through Ancestry.com.

Reid, James D. 1886. *The Telegraph in America.* 2nd. ed. New York: John Polhemus.

Rens, Jean-Guy. 2001. *The Invisible Empire: A History of the Telecommunications Industry in Canada, 1846–1956.* Translated by Käthe Roth. Montreal and Kingston: McGill-Queen's University Press.

Rensselaer Polytechnic Institute. 1875. *Catalogue of Officers and Students of the Rensselaer Polytechnic Institute, 1824–1874.* Troy, N.Y.: Wm. H. Young.

Report of the Chief of Engineers, U.S. Army. (Printed annually.) Washington, D.C.: GPO.

Reports of Deaths of American Citizens Abroad, 1835–1974. Provo, Utah: Ancestry.com Operations, Inc. Online database accessed through Ancestry.com.

Richardson, Darcy G. 2004. *Others: Third-Party Politics from the Nation's Founding to the Rise and Fall of the Greenback-Labor Party.* New York: iUniverse, Inc.

Rieser, Andrew C. 2003. *The Chautauqua Moment: Protestants, Progressives, and the Culture of Modern Liberalism.* New York: Columbia University Press.

Roberts, Oliver Ayer. 1901. *History of the Military Company of the Massachusetts, Now Called the Ancient and Honorable Artillery Company of Massachusetts, 1637–1888.* Boston: Alfred Mudge & Son.

Roger, Jacques. 1997. *Buffon: A Life in Natural History.* Ithaca, N.Y.: Cornell University Press.

Roscoe, Henry E. 1873. *Spectrum Analysis: Six Lectures, Delivered in 1868, Before the Society of Apothecaries of London.* London: Macmillan and Co.

Rose Polytechic Institute. 1909. *Rose Polytechnic Institute Memorial Volume, Embracing a History of the Institute, a Sketch of the Founder, Together with a Biographical Dictionary and Other Matters of Interest.* Terra Haute, Ind.: [Montefort, typographers].

Rosebrugh, A[bner]. M. 1886. "Telegraphing to and from Railway Trains." *Proceedings of the Canadian Institute* 4 (3 Apr.): 177–80.

Rosenberg, Chaim M. 2007. *Goods for Sale: Products and Advertising in the Massachusetts Industrial Age.* Amherst, Mass: University of Massachusetts Press.

Rosenblum, Martin Jay, and Associates. 2000. *Edison Winter Estate Historic Structures Report.* 2 vols. Philadelphia: Martin Jay Rosenblum, R. A. and Associates.

Ross, Peter, and William S. Pelletreau. 1903–1905. *A History of Long Island: From its Earliest Settlement to the Present Time.* 3 vols. New York: Lewis Publishing Co.

Rousseau, Jean-Jacques. 1968. *La Nouvelle Heloise.* Translated by Judith H. McDowell. University Park, Pa.: Pennsylvania State University Press.

Rowell, George Presbury, comp. (Printed annually.) *Rowell's American Newspaper Directory.* New York: Geo. P. Rowell & Co.

Rubin, Mordecai B. 2001. "The History of Ozone. The Schönbein Period, 1839–1868." *Bulletin of the History of Chemistry* 26:40–56.

———. 2002. "The History of Ozone. II. 1869–1899 (1)." *Bulletin of the History of Chemistry* 27:81–106.

Rudd, C[harles]. H. 1887. "Telegraphing to Moving Trains." *Western Electrician* 1 (3 Dec.): 269–70.

Ruse, Michael. 1979. *The Darwinian Revolution.* Chicago: University of Chicago Press.

Sargent, Porter. (Printed annually.) *A Handbook of New England.* Boston: Porter Sargent.

———. (Printed annually.) *A Handbook of Private Schools for American Boys and Girls: An Annual Survey.* Boston: Porter Sargent.

Scharf, J. Thomas. 1883. *History of Saint Louis City and County, From the Earliest Periods to the Present Day.* Vol. I. Philadelphia: Louis H. Everts and Co.

Schellen, Heinrich. 1872. *Spectrum Analysis in its Application to Terrestrial Substances, and the Physical Constitution of the Heavenly Bodies.* Translated from 2nd German edition by Jane and Caroline Lassell; edited by William Huggins. London: Longmans, Green.

———. 1884. *Magneto-Electric and Dynamo-Electric Machines: Their Construction and Practical Application to Electric Lighting and the Transmission of Power.* Translated by Nathaniel S. Keith and Percy Neymann. New York: D. Van Nostrand.

Schermerhorn, Nicholas I. [1928?]. "Early History of the Electrical Industry in Schenectady." Typescript in Box 1, Folder 11, Rice Papers, NScIS. Date inferred from "RECEIVED" stamp.

Schilling, Nicolaus Heinrich. 1886. *The Present Condition of Electric Lighting. A Report Made at Munich 26th September, 1885.* Boston: Cupples, Upham and Co.

Scribner, Charles Ezra. 1919. "History of the Engineering Department." *Western Electric News* 8 (Nov.): 10–12.

Scripps, E[dward]. W[yllis]. 1951. *Damned Old Crank: A Self-Portrait of E. W. Scripps Drawn from His Unpublished Writings.* Edited by Charles R. McCabe. New York: Harper & Bros.

———. 1966. *I Protest: Selected Disquisitions of E. W. Scripps.* Edited by Oliver Knight. Madison: University of Wisconsin Press.

Segawa, Hideo, ed. 1998 [1933]. *Kōgaku Hakushi Fujioka Ichisuke den.* Tokyo: Yumani Shobo.

Seiler, Carl. 1884. "A New Battery Solution." *The Medical and Surgical Reporter* 50 (5 Jan.): 8.

Sellon, R. Percy. 1888. "Continuous-Current Transformation." *Electrician* 21 (1 June): 107–8.

Sen, S. N. 1990. *Acoustics, Waves and Oscillations.* New York: Wiley.

Sewall, Charles Henry. 1903. *Wireless Telegraphy; Its Origins, Development, Inventions, and Apparatus.* New York: D. Van Nostrand Co.

Shurtleff, Nathaniel B. 1871. *A Topographical and Historical Description of Boston.* Boston: Boston City Council.

Siemens, Georg. 1957. *History of the House of Siemens.* Translated by A. F. Rodger. Freiburg: Karl Alber.

Siemens, Wilhelm. 1885. "On Improvements in Glow Lamps. *Teleg. J. and Elec. Rev.* 17 (19 Dec.): 514–16.

Silliman, Benjamin. 1871. *Principles of Physics, or Natural Philosophy.* New York: Ivison, Blakeman, Taylor & Co.

Simon, Josep, and Pedro Llovera. 2009. "Between Teaching and Research: Adolphe Ganot and the Definition of Electrostatics (1851–1881)." *Journal of Electrostatics* 67 (May): 536–41.

Sipes, William B. 1875. *The Pennsylvania Railroad: Its Origin, Construction, Condition, and Connections.* Philadelphia: Pennsylvania Railroad.

Slack, Charles. 2005. *Hetty: The Genius and Madness of America's First Female Tycoon.* New York: Harper Perennial.

Sloane, T. O'Conor. 1887. "Standard Daniell's Battery." *Sci. Am.* 56 (11 June): 370.

———. 1892. *The Standard Electrical Dictionary: A Popular Dictionary of Words and Terms Used in the Practice of Electrical Engineering.* New York: George D. Hurst.

Smith, Chauncey. 1890. "A Century of Patent Law." *The Quarterly Journal of Economics* 5 (Oct.): 44–69.

Smith, Crosbie. 1998. *The Science of Energy: A Cultural History of Energy Physics in Victorian Britain.* Chicago: University of Chicago.

———, and M. Norton Wise. 1989. *Energy and Empire: A Biographical Study of Lord Kelvin.* Cambridge: Cambridge University Press.

Smith, Joseph P., ed. 1898. *History of the Republican Party in Ohio.* Chicago: Lewis Publishing Co.

Smith, Percy H. 1883. *Glossary of Terms and Phrases.* London: Kegan Paul, Trench & Co.

Smoot, Tom. 2004. *The Edisons of Fort Myers: Discoveries of the Heart.* Sarasota, Fla.: Pineapple Press.

Smythe, Ted Curtis. 2003. *The Gilded Age Press: 1865–1900.* Westport, Conn: Praeger.

Snell, James P., comp. 1881. *History of Hunterdon and Somerset Counties, New Jersey.* Philadelphia: Everts & Peck.

Solomon, Brian. *Railroad Signaling.* St. Paul, Minn.: MBI Publishing.

Sperry, Elmer A. 1888. "Motors and Alternating Currents." *Western Electrician* 2 (26 May): 264 65.

Spies, Stacy E. 2000. *Metuchen.* Charleston, S.C.: Arcadia.

Spon, Edward, Francis N. Spon, George Guillaume Andre, and Charles G. Warnford Lock. 1879–82. *Spons' Encyclopædia of the Industrial Arts, Manufactures, and Commercial Products.* London: E. & F. N. Spon.

Sprague, John T. 1884. *Electricity: Its Theory, Sources, and Applications,* 2nd ed. London: E. & F. N. Spon.

Stanton, Curtis Harvey. 1967. "A History of the Steamer Manatee." Eaton Collection, Manatee County (Fla.) Public Library System.

Stemmler, Joan K. 1993. "The Physiognomical Portraits of Johann Caspar Lavater." *The Art Bulletin* 75 (Mar.): 151–68.

Stevens, Alfred Edgar. 1885. "My Traveling Companion." *The Current* 4 (15 Aug.): 105–6.

Stevens, George Edward. 1968. "A History of the Cincinnati Post." Ph.D. diss., Univ. of Minnesota.

Stewart, Randall. 1935. "Hawthorne in England: The Patriotic Motive in the Note-Books." *The New England Quarterly* 8 (Mar.): 3–13.

Stieringer, Luther. 1901. "A Brief History of Luminous Fountain Development." *Electrical World and Engineer* 38 (14 Sept.): 417–23.

Stillé, Alfred. 1885. *Cholera: Its Origin, History, Causation, Symptoms, Lesions, Prevention, and Treatment.* Philadelphia: Lea Brothers & Co.

Stimson, A. L. 1881. *History of the Express Business: Including the Origin of the Railway System in America.* New York: Parker & Godwin.

Stock, John T. 2004. "Gabriel Lippmann and the Capillary Electrometer." *Bulletin of the History of Chemistry* 29:16–20.

Strouse, Jean. 1999. *Morgan: American Financier.* New York: Random House.

Strutt, John William. 1945. *The Theory of Sound.* Vol. 1. Edited by Robert Bruce Lindsay. New York: Dover Publications.

Stuart, William H. 1909. "Business-Getting Methods of the Edison Electric Light & Power Company, of Amsterdam, N.Y." *Electrical World* 53 (1 Apr.): 807.

Sullivan, Joseph Patrick. 1995. "From Municipal Ownership to Regulation: Municipal Utility Reform in New York City, 1880–1907." Ph.D. diss., Rutgers University.

Suplee, Henry Harrison, ed. 1901. *The Engineering Index.* Vol. 3. New York: Engineering Magazine.

Supreme Court [N.Y.] Appellate Div. 1897. *W. Preston Hix, Plaintiff and Respondent, against Edison Electric Light Co. Case on Appeal.* New York: C. G. Burgoyne.

Swanberg, W. A. 1967. *Pulitzer.* New York: Scribner.

Swander, John I. 1885. "The Future of Substantialism—No. 1." *The Microcosm* 5 (Oct.): 11–14.

————, and Alexander Wilford Hall. 1887. *A Text-Book on Sound: The Substantial Theory of Acoustics Adapted to the Use of Schools, Colleges, etc.* New York: Hall & Co.

Swann, Leonard Alexander. 1980 [1965]. *John Roach, Maritime Entrepreneur.* New York: Arno Press.

Sweetser, M. F., ed. 1882. *The White Mountains: A Handbook for Travellers.* Boston: James R. Osgood and Co.

————, and Moses King, eds.. 1888. *King's Handbook of Boston Harbor.* 3rd ed. Boston: Moses King Corp.

Takahashi, Yuzo. 1990. "William Edward Ayrton at the Imperial College of Engineering in Tokyo—The First Professor of Electrical Engineering in the World." *IEEE Transactions on Education* 33 (no. 2, May): 198–205.

Taltavall, John B. 1893. *Telegraphers of Today.* New York: John B. Taltavall.

Tate, Alfred O. 1938. *Edison's Open Door: The Life Story of Thomas A. Edison, a Great Individualist.* New York: E. P. Dutton.

"Telegraph Street." 1871. *Temple Bar* 22 (July): 501–18.

Tesla, Nikola, and Ben Johnston. 1982. *My Inventions: The Autobiography of Nikola Tesla.* New York: Barnes & Noble.

Thompson, Seymour D. 1891. *The Law of Electricity: A Treatise on the*

Rules of Law Relating to Telegraphs, Telephones, Electric Lights, Electric Railways, and Other Electric Appliances. St. Louis: Central Law Journal.

Thompson, Silvanus P. 1886. *Dynamo-Electric Machinery: A Manual for Students of Electrotechnics.* 2nd ed. London: E. & F. N. Spon.

———. 1888a. *The Development of the Mercurial Air-Pump.* London: E. &. F. N. Spon.

———. 1888b. "The Influence Machine, from 1788 to 1888." *Journal of the Society of Telegraph-Engineers and Electricians* 17:569–635.

———. 1904. *Dynamo-Electric Machinery: A Manual for Students of Electrotechnics.* 7th ed., 2 vols. London: E. & F. N. Spon.

———. 1910. *The Life of William Thomson, Baron Kelvin of Largs.* London: Macmillan.

Thomson, Elihu. 1887. "The New Art of Electric Welding." *Proceedings of the Society of Arts* 25:35–40.

Thomson, J. J. 1886. "Report on Electrical Theories." In *Report of the Fifty-fifth Meeting of the British Association for the Advancement of Science; Held at Aberdeen in September 1885,* 97–155. London: British Association.

Tidy, Charles Meymott. 1887. *Handbook of Modern Chemistry.* 2nd ed. London: Smith, Elder & Co.

Tolles, Bryant F., Jr. 1898. *The Grand Hotels of the White Mountains: A Vanishing Architectural Legacy.* Boston: David R. Godine.

Toole, Kenneth Ross. 1950. "The Anaconda Copper Mining Company: A Price War and a Copper Corner." *Pacific Northwest Quarterly* 41 (Oct.): 312–29.

Tosiello, Rosario J. 1979 [1971]. *The Birth and Early Years of the Bell Telephone System, 1876–1880.* New York: Arno Press.

Trowbridge, John. 1884. *The New Physics: A Manual of Experimental Study for High Schools and Preparatory Schools for College.* New York: D. Appleton and Co.

———. 1885a. "What Is Electricity?" [Vice presidential sectional address.] *Proceedings of the American Association for the Advancement of Science* 33:81–95.

———. 1885b. "A Standard of Light." *Proceedings of the American Academy of Arts and Sciences* 20:494–99.

———. 1896. *What Is Electricity?* New York: D. Appleton and Co.

Trow's New York City Directory. (Printed annually.) New York: Trow City Directory Co.

True, Frederick William. 1913. *A History of the First Half-Century of the National Academy of Sciences, 1863–1913.* Washington, D.C.: National Academy of Sciences.

Tucker, D. G[ordon]. 1974. "Phonophore and Phonoplex: F.D.M. Telegraph Systems Used on Railways in the Late 19th Century." *Proceedings of the IEE* 121 (Dec.): 1603–8.

———. 1978. "François Van Rysselberghe: Pioneer of Long-Distance Telephony." *Technology and Culture* 19 (4): 650–74.

Tupper, Alfredo Puelma. 1887. *La industria azucarera en Chile y establecimiento de una nueva fábrica nacional de azúcar de betarraga en Santa Fé.* Santiago: Gutenberg.

Turner, Gerard. 1983. *Nineteenth-Century Scientific Instruments.* Berkeley: University of California Press.

Turner, Gregg. 2003. *A Short History of Florida's Railroads.* Charleston, S.C.: Arcadia Publishing.

Tyndall, John. 1863. *Heat Considered as a Mode of Motion.* London: Longman, Green, Longman, Roberts & Green.

——. 1867. *Sound: A Course of Eight Lectures Delivered at the Royal Institution of Great Britain.* New York: D. Appleton and Co.

——. 1977 [1895]. *Heat: A Mode of Motion.* Oceanside, N.Y.: Dabor Science Publications.

Ullery, Jacob G. 1894. *Men of Vermont: An Illustrated Biographical History of Vermonters and Sons of Vermont.* Brattleboro, Vt.: Transcript Publishing Co.

Upson, Irving Strong, comp. 1885. *Catalogue of the Officers and Alumni of Rutgers College in New Brunswick, N.J., 1770 to 1885.* Trenton, N.J.: John L. Murphy.

Upton, Francis. 1887. "The Telemeter System." *Scientific American Supplement* 24 (19 Nov.): 9900–1.

Urquhart, John W. 1881. *Electro-Typing.* London: Crosby, Lockwood and Co.

U.S. Army. 1886. *Report of the Chief of Engineers.* Washington, D.C.: GPO.

U.S. Census Bureau. 1963? (1850). *Population Schedules of the Seventh Census of the United States, 1850.* National Archives Microfilm Publication Microcopy M432. Washington, D.C.: National Archives. Accessed through Ancestry.com.

——. 1967? (1860). *Population Schedules of the Eighth Census of the United States, 1860.* National Archives Microfilm Publication Microcopy M653. Washington, D.C.: National Archives. Accessed through Ancestry.com.

——. 1965 (1870). *Population Schedules of the Ninth Census of the United States, 1870.* National Archives Microfilm Publication Microcopy M593. Washington, D.C.: National Archives. Accessed through Ancestry.com.

——. 1970 (1880). *Population Schedules of the Tenth Census of the United States, 1880.* National Archives Microfilm Publication Microcopy T9. Washington, D.C.: National Archives. Accessed through Ancestry.com.

——. 1982? (1900). *Population Schedules of the Twelfth Census of the United States, 1900.* National Archives Microfilm Publication Microcopy T623. Washington, D.C.: National Archives. Accessed through Ancestry.com.

——. 1982 (1910). *Population Schedules of the Thirteenth Census of the United States, 1910.* National Archives Microfilm Publication Microcopy T624. Washington, D.C.: National Archives. Accessed through Ancestry.com.

——. 1992 (1920). *Population Schedules of the Fourteenth Census of the United States, 1920.* National Archives Microfilm Publication Microcopy T625. Washington, D.C.: National Archives. Accessed through Ancestry.com.

————. 2002 (1930). *Population Schedules of the Fifteenth Census of the United States, 1930.* National Archives Microfilm Publication Microcopy T626. Washington, D.C.: National Archives. Accessed through Ancestry.com.

U.S. Department of Interior. (Printed annually.) *Annual Reports. Report of the Commissioner of Education.* Washington, D.C.: GPO.

U.S. Department of State. 1887. "Report of Secretary of State Relative to the Relations of Certain Telegraph and Cable Companies." 3 Mar. 1887, Senate Executive Document 122, 49th Cong., 2nd sess.

————. n.d. *Passport Applications, 1795–1905.* National Archives Microfilm Publication M1372. Washington, D.C.: National Archives. Accessed through Ancestry.com.

U.S. Electrical Commission. 1886. *Report of the Electrical Conference at Philadelphia, in September, 1884.* Washington, D.C.: GPO.

U.S. House of Representatives. 1887. "Reports of the Commission Appointed Under an Act of Congress Approved July 7, 1884, to 'Ascertain and Report Upon the Best Modes of Securing More Intimate International and Commercial Relations Between the United States and the Several Countries of Central and South America.'" 49th Congress, 1st session. Ex. Doc. No. 50. Washington, D.C.: GPO.

U.S. Interstate Commerce Commission. (Printed annually.) *Annual Report on the Statistics of Railways in the United States for the Year Ending. . . .* Washington, D.C.: GPO.

U.S. Navy. 1884. *Annual Report of the Secretary of the Navy.* Washington, D.C.: GPO.

U.S. Official Postal Guide. 1886. Chicago: Callaghan & Co.

U.S. Patent Office. (Printed annually.) *Annual Report of the Commissioner of Patents.* Washington, D.C.: GPO.

Usselman, Steven W. 2002. *Regulating Railroad Innovation: Business, Technology, and Politics in America, 1840–1920.* Cambridge: Cambridge University Press.

Van Allen, Elizabeth J. 1999. *James Whitcomb Riley: A Life.* Bloomington: University of Indiana Press.

Van Lunteren, F. H. 1988. "Gravitation and Nineteenth-Century Physical Worldviews." In *Newton's Scientific and Philosophical Legacy,* edited by P. B. Scheurer and G. Debrock, 161–73. Dordrecht: Kluwer Academic Publishers.

Varnum, John P. 1885. *Jacksonville, Florida: A Descriptive and Statistical Report.* Issued by the Jacksonville Board of Trade. Jacksonville: Times-Union Book and Job Printing House.

Verity, John B. 1889. "Underground Conduits and Electrical Conductors." *Journal of the Institution of Electrical Engineers* 18 (11 Apr.): 338–84.

Vincent, George E. 1893. "The Evolution of a Summer Town." *The Chautauquan* 7 (n.s.; Mar.): 697–703.

————, ed. 1899. *Theodore W. Miller, Rough Rider: His Diary as a Soldier Together with the Story of His Life.* Akron, Ohio: privately printed.

Vincent, John Heyl. 1886. *The Chautauqua Movement.* Boston: Chautauqua Press.

Vincent, Leon H. 1925. *John Heyl Vincent: A Biographical Sketch*. New York: MacMillan Co.

Vroom, Garret D. W., comp. 1886. *Reports of Cases Argued and Determined in the Supreme Court and, at Law, in the Court of Errors and Appeals, of the State of New Jersey*. Vol. 18. Trenton, N.J.: W. S. Sharp Printing Co.

Wada, Masanori. 2007. "Engineering Education and the Spirit of Samurai at the Imperial College of Engineering in Tokyo, 1871–1886." M.S thesis, Virginia Polytechnic Institute.

Waltenhofen, A. v. 1886. "Bericht über die Accumulatoren von Farbaky und Schenek in Schemnitz." *Centralblatt für Elektrotechnik* 8:600–605.

Ward, Henry W. 1911. *Western–Leander Clark College, 1856–1911*. Dayton, Ohio: Otterbein Press.

Warner, Deborah Jean, ed. 1993. *The Queen Catalogues*. 2 vols. San Francisco: Norman Publishing.

Washington and Lee University. 1888. *Catalogue of the Officers and Alumni of Washington and Lee University, Lexington, Virginia, 1749–1888*. Baltimore: John Murphy & Co.

Watts, Henry A. 1872. *A Dictionary of Chemistry and the Allied Branches of Other Sciences: Vols. 1–5 and Supplement*. New York: D. Van Nostrand.

———. 1875. *A Dictionary of Chemistry and the Allied Branches of Other Sciences: Second Supplement*. London: Longmans, Green, and Co.

———. 1879a. *A Dictionary of Chemistry and the Allied Branches of Other Sciences: Third Supplement, Part 1*. London: Longmans, Green, and Co.

———. 1879b. *A Dictionary of Chemistry and the Allied Branches of Other Sciences*. New ed. London: Longmans, Green, and Co.

———. 1881. *A Dictionary of Chemistry and the Allied Branches of Other Science: Third Supplement, Part 2*. London: Longmans, Green, and Co.

———. 1882. *A Dictionary of Chemistry and the Allied Branches of Other Science*. New Edition. London: Longmans, Green, and Co.

Weaver, James Riley. 1884. "Electric Street Lighting in Vienna." In *Reports from the Consuls of the United States*, 20–21. Washington, D.C.: GPO.

Weeks, Joseph D. 1886. *Report on the Statistics of Wages in Manufacturing Industries*. . . . Washington, D.C.: GPO.

Weise, Arthur James. 1886. *The City of Troy and its Vicinity*. Troy, N.Y.: E. Green.

Welch, Walter L., and Leah Brodbeck Stenzel Burt. 1994. *From Tinfoil to Stereo: The Acoustic Years of the Recording Industry, 1877–1929*. Gainesville: Univ. Press of Florida.

Wells College. 1894. *General Catalogue of the Officers and Students with Historical Sketches of the Founder*. . . . [Aurora, N.Y.?]: Wells College.

Whelpley, Samuel. 1808. *A Compend of History, From the Earliest Times*. 1st rev. ed. Philadelphia: Kimber & Conrad.

Whipple, Fred H. 1888. *Municipal Lighting*. Detroit: [Free Press Print].

———, comp. (Printed annually.) *Whipple's Electric, Gas and Street Railway Financial Reference Directory*. Detroit: Fred H. Whipple Co.

Whitehead, John, ed. 1901. *The Passaic Valley, New Jersey, in Three Centuries.* New York: New York Genealogical Co.

Whittaker, James T. 1889. "Erysipelas," *Cincinnati Lancet-Clinic: A Weekly Journal of Medicine and Surgery* 22 (n.s.): 621–26.

Whittemore, Henry, comp. 1889. *Free Masonry in North America from the Colonial Period to the Beginning of the Present Century. . . .* New York: Artotype Printing and Publishing Co.

———. 1896. *The Founders and Builders of the Oranges. . . .* Newark, N.J.: L. J. Hardham.

Wiechmann, Ferdinand Gerhard. 1906. *Notes on Electrochemistry.* New York: McGraw Publishing Co.

Wile, Raymond R. 1974. "Introduction," *Proceedings of the 1890 Convention of Local Phonograph Companies.* Nashville: Country Music Foundation.

———. 1976. "The Rise and Fall of the Edison Speaking Phonograph Company, 1877–1880." *Association for Recorded Sound Collections Journal* 7 (3): 4–31.

———. 1987. "Jack Fell Down and Broke His Crown: The Fate of the Edison Phonograph Toy Manufacturing Company." *Association for Recorded Sound Collections Journal* 19 (2–3): 5–36.

———. 1990a. "The Development of Sound Recording at the Volta Laboratory." *Association for Recorded Sound Collections Journal* 21 (2): 208–25.

———. 1990b. "The Seventeen-Year Itch: The Phonograph and the Patent System." In *The Patent History of the Phonograph, 1877–1912: A Source Book,* xii–xvi, edited and compiled by Allen Koenigsberg. Brooklyn, N.Y.: APM Press.

———. 1991. "Edison and Growing Hostilities." *Association for Recorded Sound Collections Journal* 22 (1): 8–34.

———. 1993. "The Launching of the Gramophone in America 1890–1896." *Association for Recorded Sound Collections Journal* 24 (2): 176–92.

Wilkins, Mira. 1970. *The Emergence of Multinational Enterprise: American Business Abroad from the Colonial Era to 1914.* Cambridge, Mass.: Harvard University Press.

———. 1989. *The History of Foreign Investment in the United States to 1914.* Cambridge, Mass: Harvard University Press.

Williams, Albert, Jr., ed. 1885. *Mineral Resources of the United States (1883–1884).* Washington, D.C.: GPO.

Williams' Cincinnati Directory. (Printed annually.) Cincinnati: Williams & Co.

Williams, Stephen K., comp. 1889. *U.S. Supreme Court Reports. Cases Argued and Decided in the Supreme Court of the United States.* Book 32. Rochester, N.Y.: Lawyers' Co-Operative Publishing Co.

Wills, Ian. 2007. "Edison, Science, and Artefacts." *Integrated History and Philosophy of Science.* 1:1–11.

———. 2009. "Edison and Science: A Curious Result." *Studies in History and Philosophy of Science.* 40:157–66.

Wilson, John F. 1988. *Ferranti and the British Electrical Industry, 1864–1930.* Manchester: Manchester University Press.

Wilson's New York City Co-Partnership Directory. 1879. Vol. 27 (Mar.). New York: Trow City Directory Co.

Wiman, Erastus. [1885]. *The Canadian Club, Its Purpose and Policy, As Set Forth in the Speech of Erastus Wiman, President, Dominion Day Dinner, July 1, 1885.* [New York: Canadian Club.]

————. 1893. "Jay Gould Made 10 Millions." In *Chances of Success: Episodes and Observations in the Life of a Busy Man.* New York: American News Co.

Withey, Henry F., and Elsie Rathburn Withey. 1970. *Biographical Dictionary of American Architects (Deceased).* Los Angeles: Hennessey & Ingalls, Inc.

Woman's Board of Missions. (Printed annually.) *Annual Report.* Boston, Mass: Frank Wood, Printer.

Wood, George B., and Franklin Bache. 1885. *The Dispensatory of the United States of America.* 15th ed., revised by H. C. Wood, Joseph P. Remington, and Samuel P. Sadtler. Philadelphia: J. B. Lippincott & Co.

Woodard, Colin. 2004. *The Lobster Coast: Rebels, Rusticators, and the Struggle for a Forgotten Frontier.* New York: Viking.

Woodbury, David O. 1949. *A Measure for Greatness: A Short Biography of Edward Weston.* New York: McGraw Hill.

Woolley, Edward Mott. 1910. "Why This Store Has Lived 70 Years." *System: The Magazine of Business.* March: 243–45.

Yates, Austin A. 1902. *Schenectady County, New York: Its History to the Close of the Nineteenth Century.* [New York?]: New York History Co.

Yates, W. Ross. 1992. *Lehigh University: A History of Education in Engineering, Business, and the Human Condition.* Bethlehem, Pa.: Lehigh University Press.

Yavetz, Ido. 1995. *From Obscurity to Enigma: The Work of Oliver Heaviside, 1872–1889.* Basel, Switzerland: Birkhäuser Verlag.

Yocum, Barbara A. 1998. *The House at Glenmont : Edison National Historic Site, West Orange, New Jersey.* 2 vols. Lowell, Mass.: National Park Service.

Zacharias, J. 1887. "Central Electric Light Stations at Berlin." *Journal of the Society of Telegraph-Engineers and of Electricians.* 16 (24 Feb.): 161–62.

Credits

Courtesy of the AT&T Archives & History Center: Doc. 2798. Reproduced with permission of the Edison-Ford Winter Estates: Docs. 2860, 2899, and 2903. Courtesy of the Charles Edison Fund: Doc. 2904. Reproduced with permission of the Henry Ford Museum and Greenfield Village: Doc. 3034. From the collection of the late David Heitz: Doc. 2781. Reproduced with permission of the Historical Society of Pennsylvania (Uriah Hunt Painter Papers): Docs. 3102 and 3120. From the collection of Charles Hummel: Doc. 2868. Courtesy of the National Archives and Records Administration: Docs. 2780, 2786, and 3055. From the collections of the Port Huron (Mich.) Museum: Doc. 2787. Reproduced with permission of David E. E. Sloane: Doc. 2852. Courtesy of the Smithsonian American Art Museum: frontispiece (Longworth Powers bust of Thomas A. Edison; Museum purchase in memory of Ralph Cross Johnson). Courtesy of the Smithsonian Institution's National Museum of American History (William J. Hammer Collection): Doc. 3106. Reproduced with permission of Thomas Whitney: Doc. 2789.

Reprinted from "Electrical Pioneers," *Electricity: The Popular Journal of Electricity* 3 (27 July 1892), 16: illustration on p. 663. Reproduced from Michael Faraday's *Experimental Researches*, Vol. 3, para. 3084, fig. 1: illustration on p. 479. Reprinted from Arnold Lewis and Keith Morgan, *American Victorian Architecture: A Survey of the 70's and 80's in Contemporary Photographs* (Dover, 1975), 109: photograph on p. 315. Reproduced from United States patents: 333,289 (illustration p. 94); 437,422 (illustrations pp. 252–53); 466,400 (illustration

p. 594); 380,101 (illustration p. 795); 385,173 (illustrations p. 806).

Courtesy of the Thomas Edison National Historical Park (designations are to *TAED* notebook or volume number): illustrations on pp. 251 (N311011, image 7); 369 (N314023); 381 (NSUN08, image 108); 382 (N314123, image 61); 430 (N315009, images 6–7); 438 (N317019, image 12); 477 (N319127), 478 [top] (N319127, image 62); 478 [bottom] (N319127, image 66); also photographs on pp. 679 (MBJ004, item 466, image 4), (MBJ004, item 463, image 3); 753 (MBJ004, item 465, image 4).

Index

Boldface page numbers signify primary references or identifications; italic numbers, illustrations. Page numbers refer to headnote or document text unless the reference appears only in a footnote.

Anderson, John, 334 n. 4

Andrew, John, 265 n. 5

Andrews, William S., **579**; and Edison United, 658; and generators, 617 n. 4; and New York second district, 579; patents, 617 n. 4, 626 n. 20, 910; and TAE's memo on AC, 643–45; and transformers, 626 n. 20

Anglo-Austrian Brush Electric Light Co., 146 n. 9

Anson, Thomas Francis, **634**, 635

Ansonia Brass & Copper Co., 52 n. 8, 227 n. 2, 308 n. 5, 310 n. 3, 899, 901

Ansonia Clock Co., 778 n. 6

Armeda, Nicholas (Nick), **694**

Armington, Pardon, 334 n. 3

Armington & Sims, 334 n. 3, 630 n. 4

Armington & Sims Engine Co., 186 n. 15

Arnold, Constable & Co., 314 n. 1, 315 n. 5

Artificial materials: mother of pearl, 665–66, 807 n. 9; silk, 135, 394, 472–73, 807. *See also* Electric lighting incandescent lamps: filament materials (artificial)

Associated Press, 164

Association of Edison Illuminating Companies, 422 n. 18, 580 n. 3, 660 n. 5, 742 n. 1

Astronomy: atmospheric magnetism, 533 n. 1; comets, 537; cosmology, 509–10; instruments, 459, 538; Kennelly-Heaviside layer, 854 n. 12; planetary magnetism, 533 n. 1, 539 n. 2; solar eclipse expedition, 484 n. 1; solar system, 307, 481–82, 485–88, 493–94, 497, 505, 511 n. 1, 532, 537–39, 551–53; telescopes, 302, 459; theory of gravity, 363, 482–84, 487–88, 491–92, *495*, 529–33, 551–53

Atchison, Topeka, & Santa Fe Railroad, 578 nn. 1–2

Atkinson, Edmund, 207 n. 4

Atlanta, 342, 426 n. 1

Atlantic Monthly, 135 n. 11, 174 n. 10

Atlantic & Pacific Telegraph Co., 357 nn. 10–11

Atomic theory, 487, 500–504

Auchincloss, Henry B., **744**

Auchincloss Bros., 744 n. 3

Aultman, Miller & Co., 163, **180**, 246 n. 1, 297 n. 2, 694 n. 4, 844

Aultman (C.) & Co., 182 n. 17, 246 n. 1, 303 n. 5, 582 n. 3

Australasian Electric Light Power and Storage Co., Ltd., 317 n. 3, 517 n. 16

Automatic Signal Telegraph Co., 230 n. 6

Automatic Telegraph Co., 328 n. 1

Ayers, Brown, 108 n. 5

Aylsworth, Jonas, 821

Ayrton, William, **648**

Babcock & Wilcox, 208 n. 2, 783

Baggs, Anna W., **240**, 241

Baggs, Frederick, 245 n. 29

Bailey, Joshua, 160 n. 11

Baker, Delos, **745**

Baker, George, 127 n. 11

Baker & Co., **85**

Balmoral Hotel, 216 n. 2

Baltimore & Ohio Telegraph Co.: Phelps system demonstrated for, 34; phonoplex agreement with, 771; phonoplex demonstrated for, 263; phonoplex problems, 351 n. 3, 516 n. 15; phonoplex used by, 92, 266–67, 308, 354, 419 n. 3, 578 nn. 1–2

Bancroft, George W., **124**

Barker, George Frederick, **218**, 782 n. 5; advises on spectroscopy, 564; member of Committee on Standard of Light, 65 n. 1; and NAS phonograph demonstration, 845 nn. 2–3; reads TAE's AAAS paper, 800 n. 3; and Sprague Electric, 616 n. 3; visits TAE, 787

Barker, John F., 848 n. 6

Barr, William, **743**, 744

Barrett, William, **501**

Bartholdi, Frédéric, 628 n. 3

Batchelor, Charles, 8, 309, 798 n. 1; and arc lighting, 765 n. 3; attends railway telegraph demonstration, 320; attends TAE's wedding, 339; and Boston block plant, 421 n. 14; comments on lamp life, 558 n. 1; comments on TAE's health, 666, 668, 672–73, 675–76, 686, 698; and converters, 603, 638–39, 646–47, 659, 874–75; Edison Electric stock, 8, 77–78; and Edison United, 553–54, 658; and Edison Wiring, 761; and electric-

powered torpedo, 6 n. 3, 30–31, 153, 305, 333, 541 n. 2; and electric railway, 351, 417; electric railway stock, 355 n. 3; electric telegraph stock, 103 n. 2; and electromotograph, 133 n. 4; and European negotiations, 140 n. 1; in Fort Myers, 668, 678; and French companies, 159 n. 8, 160 n. 12; as general manager of Edison Machine Works, 142, 151–53, 247, 295, 310 n. 1, 512–13, 525, 541, 549, 616, 659, 661–62, 676, 714 n. 1, 766 n. 4; and generators, 149, 151–53, 351 n. 3, 429 n. 3, 450 n. 4, 617, 659, 663–64, 765, 798 nn. 2–3; interest in Walter patent, 645 n. 15, 912; iron experiments, 849–50; journal entries, 320, 333, 339, 541, 549, 553, 558 nn. 1&4, 579, 616–17, 661, 666, 672–73, 675–76, 684–86, 698, 703, 715, 718, 727, 741 n. 1, 742 n. 1, 744 n. 5, 753 n. 10, 782 n. 1, 787, 802, 812 n. 1, 845, 877 n. 1, 884 n. 2; and lamp filaments, 396 n. 16, 862; and municipal street lighting, 296 n. 1, 715, 742 n. 1; and New York second district, 579; and ore milling, 861, 863; Oriental Telephone shares, 813 n. 3; patents, 155 n. 13, 910, 912; and phonograph, 669, 714, 718, 732 nn. 2–3, 740–41, 754 n. 1, 778 n. 5, 884; physiognomic traits, 236, 238–39; and pyromagnetism, 678, 684–85; and Railway Telegraph, 45 n. 2, 320; in Schenectady, 714 n. 1; and snow-clearing machine, 327 n. 15; and Standardizing Bureau, 891; and TAE's claims against Edison Electric, 50, 74–79; TAE's instructions to, 149; and Taylor & Co., 403 n. 6, 664 n. 4; telephone stock, 203 n. 3; and wax paper, 125 n. 5; and (West) Orange lab, 659, 669, 685, 703, 727, 736, 782 n. 1, 845 n. 1, 862; wire tests, 851;

—letters: from TAE, 151–53, 646–47, 663–64, 685–86

Batchelor, Rosanna, 77, 78–79, 238–39

Bate, John J., 884 n. 2

Bate Refrigerator Co., 884 n. 2

Bates, David H., **354**, 771–72

Bates, Edwin G., 377 n. 2

Batteries, 302; Bergmann & Haid, 138; Bunsen, 74 n. 3, 138; Callaud, 338, 806 n. 4; carbon, 138, 277; Daniell, 337; De la Rue's, 324; De Luc dry pile, 396, 402; for electrical distribution system, 895; fuel cells, 677, 777 n. 3; Fuller, 253, 384, 806 n. 4; gravity, 339 n. 7, 804; Grove, 139 n. 3; Leclanche, 139 n. 4; Meadowcroft & Guyon, 74 n. 3; for miner's lamps, 73, 82–83; phonograph, 856; for phonoplex, 253, 261 n. 1, 268–71; for pressure indicators, 804; Trouvé's, 73 n. 1; for wireless telegraph, 431; zinc for, 8, 139 nn. 3–4, 302, 327 n. 8, 384, 396 n. 7, 397 n. 33, 446, 804

—storage, 363; for electric lighting, 146 n. 9; Farbaky & Schenk, 146 n. 9; McDougall's, 661 n. 10; for miner's lamps, 82–83; plates, 67

Beall, John E., 245 n. 32

Beck, Frances M., **240**, 241

Beck, Joseph W., **240**, 241

Beggs, John I., **658**

Bell, Alexander Graham, 69 n. 6, **768**, 835 n. 3; graphophone, 603, 611 n. 1, 714 n. 1, 768; patents, 7 n. 4, 87 n. 2, 138 n. 2, 200 n. 4, 357 n. 16, 358 n. 16, 611 n. 1, 769 n. 3; photophone, 478 n. 25; telephone inventions, 7 n. 4, 87 n. 2, 138 n. 2, 200 n. 4, 358 n. 16, 361, 769 n. 3; Volta Laboratory, 603, 611 n. 1, 769 n. 3

Bell, Chichester, 611 n. 1, 714 n. 1, 769 n. 3

Bell, William, 54 n. 16

Bell (William) & Co., **49**, 50

Benedict, Henry H., 692 n. 10

Bennett, James Gordon, 52 n. 9

Bentley, Henry, 48

Bergmann, Sigmund, **58**, 350 n. 14, 902; attends TAE's wedding, 339; batteries, 138; and consolidation of shops, 515 n. 8; and Edison United, 553–54, 847; and Edison Wiring, 761; finances, 789–90; loan to Edison Machine Works, 676 n. 4; patents, 138 n. 2, 200 n. 4, 209–10, 211 n. 7, 601 n. 1, 910; and phonograph, 611 n. 1; and Railway Telegraph, 45 n. 2; and Sprague Electric, 616; and

telephone, 138, 200 n. 4, 209–10, 230, 601 n. 1

Bergmann & Co., **553**, 783 n. 7, 789; brass work, 400; carbon buttons for, 42 n. 6; and Edison United, 526, 553–54; finances, 512; Insull's role, 658; location, 2; magneto call bells, 58; makes Graphophones, 611 n. 1; manufactures phonoplex, 274; orders, 512, 673; shipments to Japan, 156 n. 1; ships municipal lamps, 673; and Sprague Electric, 616 n. 3; and Standardizing Bureau, 891; TAE's lab, 3; telephone manufacture, 231 n. 8, 369, 419; typewriters, 60; wiring department, 762 nn. 3–4. *See also* Laboratories: New York

Berlin, 158, 195, 548 n. 1, 555, 566–67

Berliner, Emil, 724 n. 12, 727 n. 8

Bethlehem, 217 n. 1, 218, 221

Bethlehem (N.H.), 217 n. 1, 218, 221

Biedermann, Ernst, 121 n. 4, 140 n. 1, **144**, 145

Biedermann (M.L.) & Co., 146 n. 7

Bijou Theater, 421 n. 14

Bingham, George, 690 n. 5

Black, George, 290 n. 5

Blackledge, Frank H., 242 n. 11

Blackwood's Magazine, 25 n. 15

Bláthy, Otto, 621, 645 n. 16

Bliss, George, 20, 308, **309**

Bolton, Francis, 215 n. 12

Bolton, Sarah, 185 n. 10

Bonaparte, Napoleon, 184 n. 7, 205

Book of Discipline, 342

Borgia, Lucrezia, **190**

Borneo, 85

Bossart, Paul W., **577**, 578

Boston, 197, 199, 256; Boston Museum, 192 n. 3; central station, 418, 626 n. 17, 659, 761 n. 2; Electrical Exhibition, 6 n. 7; Gilliland's home, 165–66, 179; Gilliland's office, 181 n. 4, 182–83, 186 n. 20; M. Force in, 40; map, 166; Mina in, 5, 151 n. 6, 163–65, 229, 344 n. 8, 815 n. 1; phonoplex experiments on line to, 92, 267, 273, 277, 289; TAE in, 4–6, 6–7, 27 n. 6, 34, 91, 138, 139 n. 1, 142, 144, 149, 151–53, 162, 164–65, 165, 172 n. 1, 182–83, 203 n. 1, 208 nn. 1–2, 230 n. 2, 233

Boston, Revere Beach and Lynn Railroad, 181 n. 7

Boston, Winthrop and Shore Railroad, 181 n. 7

Boston Daily Globe, 165

Boston Electrical Exhibition (1885), 6 n. 7, 168 n. 17

Boston Evening Transcript, 186 n. 19

Boston Herald, 187, 196

Boston & Maine Railroad, 199 n. 6

Boston Traveller, 202

Boston University, 769 n. 3

Bowen, Herbert, 615 n. 2

Boyesen, Hjalmar Hjorth, *Goethe and Schiller: Their Lives and Works*, 177

Boylan, R. Dillon, 184 n. 5

Bradford Academy, **814**

Bradford (Mrs.) School, 330 n. 1

Bradstreet & Sons, **48**

Brady, Juliet Handy, 304 n. 1

Brady, Matthew, 304

Brewster, William F., 872 n. 6

Bridgeport Forge Co., **663**

Briggs, Lowell C., **776**, 777, 829

Brin, Arthur, 440 n. 1

Brin, Leon Quentin, 440 n. 1

British Association for the Advancement of Science, 109 n. 17, 872 n. 6

British Post Office Telegraph, 65 n. 1, 264

Brockton Telephone Co., 892 n. 3

Brockton Wiring Co., 892 n. 3

Broderick Copygraph Co., 692 n. 12

Brontë, Charlotte, **171**

Brooke, Charles Morgan, **16**

Brooklyn Bridge, 14 n. 18, 55 n. 23

Brooklyn Electric Subway Commission, 141 n. 2

Brooklyn High School, 17 n. 3, 782 n. 4

Brown, Danforth, 242 n. 10

Brown, Edwin, 178 n. 12

Bruch, Charles, 339 n. 2

Brummel, George Bryan (Beau), **183**

Brush, Charles, 437 n. 2

Brush, George, 896 n. 2

Brush Electric Co., 25 n. 6, 828 n. 1, 844, 892 n. 3, 893 n. 1, 894 n. 3

Brush-Swan Electric Light Co. of New England, 746 n. 2

Bryn, Knud, 549 n. 4

Buffon, Georges-Louis, **194**

Bunnell, Jesse H., 304 n. 7

Bunnell (J.H.) & Co., **302**

Bunsen, Robert Wilhelm, 139 n. 3, **380**, 565 n. 2, 747 n. 4

Burbank, A. P., 738

Burg Theatre, 146 n. 9

Burke, John, **744**

Burke (E.&J.), 744 n. 4

Butler, Julius W., 124 n. 1, 125 n. 4

Butler, Oliver, 124 n. 1

Butler & Hunt, 124 n. 1

Butler (J.W.) Paper Co., **124**

Byllesby, Henry, 622, 635 n. 3

Callaud, Armand, 339 n. 7

Canada: Custom House, 300; electric lighting in, 234 n. 5; mining, 50, 85; phonograph in, 820; phonoplex in, 91–93, 150, 212–13, 234, 247, 253, 263–64, 277–78, 288–89, 295, 300–301, 356 n. 5, 369, 417, 516 n. 15; railway telegraphy in, 128

Canadian Club, **213**

Canadian Institute (Toronto), 93 n. 4

Canadian Pacific Railroad, 578 n. 2

Cantor, Geoffrey, 485 n. 6

Card, Benjamin F., **844**

Carley, Francis D. (Frank), **580**

Carlson, W. Bernard, 766 n. 4

Carlyle, Thomas, **176**, 184 n. 5

Carman, Charles, 765 n. 2

Carman, William, **764**

Carré, Ferdinand Philippe, **598**

Carus, Lucretius, **132**

Cases (patent): list of TAE's, 904–8; No. 132, 255 n. 2; No. 187, 220 n. 8; No. 373, 827 n. 4; No. 615, 882 n. 3, 883 n. 4; No. 663, 318 n. 2; No. 667, 572 n. 2; No. 668, 560, 570–71; No. 670, 396 n. 16; No. 672, 572 n. 2; No. 674, 278 n. 3; No. 678, 586 n. 2; No. 680, 595–97; No. 683, 613 n. 2; No. 685, 597–98; No. 686, 597–98; No. 687, 597–98; No. 688, 612; No. 689, 612; No. 690, 612; No. 693, 625 n. 16; No. 698, 613 n. 2; No. 699, 625 n. 16; No. 703, 625 n. 16; No. 704, 625 n. 16; No. 706, 650–52; No. 717, 666 n. 1; No. 719, 716–17; No. 720, 724–26; No. 721, 719–23; No. 723, 746 n. 2; No. 727, 770 n. 1; No. 734, 824–26; No. 743, 878–82; No. 744, 882 n. 3, 883 n. 5; No. 748, 895 n. 2; No. 751, 882 n. 3, 883 n. 5; No. 765, 883 n. 6; No. 778, 746 n. 2

Cash carrier system, 363, 371–74, 376–77

Caveats, 686 n. 3, 737, 748–51, 803 n. 6, 819, 827 n. 5, 830–33

Cedar Key (Fla.), 21–22, 63

Central Edison Light Co. of Cincinnati, 422 n. 16, 619 n. 2

Central High School (Philadelphia), 799 n. 1, 848 n. 3

Central Railroad, 178 n. 5

Chadbourne, Edwin R., 55 n. 27, **355, 738**

Chamberlain, Joseph, 898

Chance, Norman H., 243 n. 20

Chandler, Albert B., **354**

Chardonnet, Comte Hilaire de, 137 n. 2

Charles Edison Fund, 695 n. 9

Chautauqua House Hotel, 694 n. 2

Chautauqua Institution, **189**, 244 n. 21, 342; founding of, 148, 163–64, 185 n. 11, 191 n. 2, 246 n. 1; invitation to speak at, 185 n. 11; Miller family at, 148, 217 n. 1, 236–37, 527, 559 n. 1, 580–81; TAE at, 148, 164–65, 217–18, 221 n. 1, 236–37; TAE's family at, 527, 559 n. 1

Cheever, Charles A., **48**, 49, 834

Cheever, John H., 53 n. 13

Chemistry: arc light carbons, 335–36, 428, 431; artificial silk, 135, 472–73; asphalt compounds, 809–10; batteries, 82–83, 384, 808; book ink, 367; chemical recording, 88 n. 1, 89 n. 2; commercial production of chemicals, 870; energy research, 345, 374, 383; fuel cells, 861; gold ore separation, 772–74, 863; incandescent lamps, 365–67, 378–81, 402, 414–15, 429, 561–65, 569–72, 574–76, 579, 584–89, 597–600, 699–702, 809–10, 852, 860–61, 872, 886; insulation, 797; light and, 698; and luminosity, 70, 73; nitrogen compounds, 135; oxygen production, 135–36, 435–36, 439; ozone production, 402, 698, 806; photophone, 472; production of lampblack, 367–69, 428; record materials, 718 n. 2, 753–54, 796–97, 813–14; research plans, 849–52, 860–64, 870–71; spectroscopic analysis, 390–91, 498, 537, 564–65, 862, 865–66, 873; storage batteries, 82–83; theory, 487, 500–

504, 862; thermoelectricity, 849; in vacuum process, 526, 560–65, 569–76, 579, 588–89, 599–600; varnishes, 810; water distillation, 136; wire insulation, 797, 807, 855 n. 18; XYZ experiments, 118–19, 281–87, 345, 378, 385–88, 400, 466

Chess, Amelia J., **580**

Chess, Carley, and Co., 582 nn. 5&7

Chess, William Edward, 582 n. 4

Chess and Wymond, 582 n. 4

Chester & Co., 377 n. 9

Chicago, 215 n. 12; baseball team, 198 n. 4; Columbian Exposition, 64 n. 4; Edison Isolated agent, 309 n. 4; electric light convention in, 20; Insull in, 358 n. 13, 369 n. 1, 370 n. 8, 419 n. 1, 811; Marion in, 24 n. 5; phonoplex experiments in, 527; press reports, 124, 230; railway telegraph exhibition in, 420 n. 9; railway telegraph experiments, 353–54, 511; TAE in, 4, 20, 24 n. 5, 40–41, 45 n. 1, 47; Tate in, 689 n. 1; underground conductors in, 675 n. 4, 811; U.S. Electric Lighting agents in, 846. *See also* Western Edison Light Co.

Chicago, Milwaukee & St. Paul Railway, 356 n. 6, **417**, 600

Chicago Daily Tribune, 231 n. 9

Chicago Electric Club, 838 n. 2

Childs, George W., 898

Chinnock, Charles Edward, 11; attends TAE's wedding, 339; and Edison Wiring, 761; as general manager of Edison United, 658, 762–63, 846–47; and New York second district, 579; as superintendent of Pearl Street, 11, 828 n. 4

Chistiana Telephone Co., 549 n. 4

Cholera, 176

Churchill, George Charles Spencer, **827**, 828

Cincinnati, 20–21, 47, 216

Cincinnati, Indianapolis, St. Louis, and Chicago Railroad, 421 n. 15

Cincinnati Evening Post, 745 n. 1

Civil War, 22, 32 n. 3, 63 n. 3, 243 n. 16, 265 n. 5, 304 n. 7, 356 n. 8, 515 n. 9, 523 n. 6, 812 n. 1

Clarendon Hotel, 235 n. 1

Clark, Henry A., **355**, 422 n. 17

Clark, Josiah Latimer, 95 n. 17

D'Israeli, Isaac, *Curiosities of Literature,* 182

Dixon, Theron, **218**

Dolbear, Amos, 34

Dominion Telegraph Co., 214 n. 6

Douty, William, **226**

Draper, Henry, 484 n. 1

Drexel, Harjes & Co., 160 n. 11, **647**

Drexel, Morgan & Co., **295**, 555, 647 n. 1, 813 n. 3; agreement with, 546 n. 8; and Edison Industrial, 786 n. 1, 802; European lighting interests, 159 n. 4; interest in Electric Tube, 298 n. 11; loan to TAE, 550 n. 4, 676 n. 4, 900; partners, 55 n. 24, 352 n. 5, 355 n. 3, 630 n. 4, 735, 786 n. 1

Dueber, John, 619 n. 1

Dueber Watch Case Co., 619 n. 1

Duncan, Andrew, 347 n. 4

Dwight, Harvey P., **212**–13, 234, 253 n. 1, 290 n. 4, 300–301, 417

Dyer, Paul, 787 n. 3

Dyer, Philip, **142**

Dyer, Richard N.: and armature patent, 450 n. 4; *Edison: His Life and Inventions,* 897; and electrical system patents, 737, 838–39; and European telephone patents, 812 n. 1; and foreign patents, 230, 544; and gas lamp patent, 746; and incandescent lamp patents, 218, 561, 563 n. 2, 595–98, 757–59, 838–39; and metal working patent, 759–60; and municipal lamp patent, 742; offices, 516 n. 14; and ore separation patent, 737; partnership with Seely, 40 n. 3, 103 n. 2, 516 n. 14, 650 n. 2; patents, 910; and phonograph patents, 730–31, 856, 869 n. 3; and phonoplex patents, 96–98, 103, 232, 255 nn. 1&2, 340–41; and pressure indicator patent, 737, 803–4; and pyromagnetic generator patent, 721–23; and pyromagnetic motor patent, 726; and railway signaling patent, 737, 770; and railway telegraph patents, 33, 38–39, 317–18; railway telegraph stock, 103; and way-station telegraph patents, 100–102, 232; and wireless telegraph patent, 121–22; writes to TAE in Fort Myers, 360 n. 2. *See also* Dyer & Seely

Dyer & Seely: and electrical system patents, 626 n. 20, 792–94; and incandescent lamp patents, 586–87, 597 n. 1, 598 n. 3; offices, 516 n. 14, 801 n. 2; partnership, 40 n. 3, 103 n. 2, 650 n. 2; patent assignment to, 910; and phonograph patents, 824–26, 833 n. 1, 840, 882 nn. 1–3; and railway signaling patent, 770 n. 1; and wireless telegraph patent, 123 n. 7

Earle, Ferdinand, 230 n. 4

Eastern Telegraph Co., 854 n. 12

Eaton, Sherbourne Blake, **9**; and Edison Construction, 9–10, 75, 226; as general manager of Edison Ore Milling, 54 n. 17; as president of Edison Electric, 9–10, 75, 86 n. 8

Ecclesine & Tomlinson, 7 n. 1

Eckert, Thomas, 213 n. 3, 357 n. 11

École Centrale, 146 n. 9

Eden Musée, **198**

Edinburgh New Dispensary, 347 n. 4

Edinburgh Review, 174 n. 14

Edinburgh University, 347 n. 4

Edison, Ellen J., 345, **696**

Edison, Marion (Dot), 4, **170**, 592; at Chautauqua, 148, 527, 581; education, 5, 248, 308 n. 9, 330, 541 n. 4, 604, 814; entries in notebooks, 307, 363; health, 667, 674 n. 3, 676 n. 3; in Menlo Park, 170–72; money from TAE, 737, 756; nicknamed George, 694, 815; piano playing, 180; and railway telegraph demonstration, 320 n. 3; relations with Mina, 190, 668, 682, 693–94, 697, 814–15; reminiscences, 149; travels with TAE, 20–21, 148–49, 162–63, 177, 179, 183, 188, 190, 197, 217 n. 1, 306, 330 n. 1, 348 n. 1, 667–68; witnesses TAE's proposal to Mina, 221 n. 2

—letters: from, 693–94, 814–15; to, 235

Edison, Mary Stilwell, 175 n. 19, 242 n. 8; account of wedding, 185 n. 10; children, 55 n. 26, 174 n. 11, 345 n. 2; death, 4, 20, 163, 198 n. 1, 306, 330; father, 595 n. 3; Florida vacation, 26 n. 2, 57 n. 6; mother, 540 n. 3, 594 n. 1; sister, 541 n. 4

Edison, Mina Miller, **180**, 302, 736; birth, 163; in Boston, 5, 151 n. 6,

163–65, 229, 344 n. 8, 815 n. 1; in Chautauqua, 217 n. 1, 527; comments on TAE's health, 675 n. 3; conducts experiment with TAE, 363, 377; courtship, 4–6, 21, 148–49, 163–67, 169, 172, 182, 188–90, 197–98, 218, 221 n. 2, 279; education, 5, 163–64, 815 n. 1; expresses anxiety, 527, 604, 668, 694, 696–97; in Fort Myers, 306–7, 362–64, 368, 371, 377, 396, 403, 406, 437 n. 1, 496 n. 2, 514 n. 2, 525, 667–68; honeymoon, 306, 342–43, 525; interview, 165; marriage proposal, 221 n. 2, 246; money from TAE, 737; opinion of TAE, 165; other suitors, 164, 244 n. 21, 279 n. 1; physiognomic traits, 236–40; pregnancy, 668, 682–83, 735, 737–38; relations with Marion, 190, 668, 682, 693–94, 697, 814–15; reminiscences, 149, 314 n. 2; TAE's opinion of, 696–97; travels in Florida, 306, 342–43, 525, 667–68, 693; visits Exposition Universelle, 604; visits family, 527, 592; wedding, 305–6, 339, 342–43; witnesses notebook entries, 307, 362, 368, 371, 377, 396, 403, 406, 526, 560, 703 n. 4, 752 n. 1; writes notebook entries, 362, 364, 404–6, 437 n. 1, 526, 560, 669, 703 n. 4, 713 nn. 12–13

—letters: from father, 695–97; to Insull, 592; from Jane, 279, 580–81, 682–83; to mother, 342–43; from TAE, 337; from TAE, Jr., 345; from William, 559

Edison, Samuel (father), **15**, 261, 527, 736

Edison, Thomas Alva: attends theater, 190–91, 738; book purchases, 177; chemistry instruction, 17 n. 3, 782 n. 4; cigar smoking, 25 n. 15, 150, 169–70, 201; claims against Brockton Edison, 28 n. 2; Construction Dept. debts, 28 n. 2; courtship of Mina, 4–6, 21, 148–49, 163–67, 169, 172, 182, 188–90, 197–98, 218, 221 n. 2, 279; deafness, 191, 346 n. 1; description of, 130; diary, 148, 162–94, 196–98, 201–2, 205–7; Dutch ancestry, 261 n. 2; and electrical engineering education, 223; fame, 306; health, 24 n. 5,

Index